# Selected Titles in This Series

28 **Liviu I. Nicolaescu,** Notes on Seiberg-Witten theory, 2000
27 **J. C. McConnell and J. C. Robson,** Noncommutative Noetherian rings, 2000
26 **Rolf Berndt,** An introduction to symplectic geometry, 2000
25 **Thomas Friedrich,** Dirac operators in Riemannian geometry, 2000
24 **Helmut Koch,** Number theory: Algebraic numbers and functions, 2000
23 **Alberto Candel and Lawrence Conlon,** Foliations I, 2000
22 **Günter R. Krause and Thomas H. Lenagan,** Growth of algebras and Gelfand-Kirillov dimension, 2000
21 **John B. Conway,** A course in operator theory, 2000
20 **Robert E. Gompf and András I. Stipsicz,** 4-manifolds and Kirby calculus, 1999
19 **Lawrence C. Evans,** Partial differential equations, 1998
18 **Winfried Just and Martin Weese,** Discovering modern set theory. II: Set-theoretic tools for every mathematician, 1997
17 **Henryk Iwaniec,** Topics in classical automorphic forms, 1997
16 **Richard V. Kadison and John R. Ringrose,** Fundamentals of the theory of operator algebras. Volume II: Advanced theory, 1997
15 **Richard V. Kadison and John R. Ringrose,** Fundamentals of the theory of operator algebras. Volume I: Elementary theory, 1997
14 **Elliott H. Lieb and Michael Loss,** Analysis, 1997
13 **Paul C. Shields,** The ergodic theory of discrete sample paths, 1996
12 **N. V. Krylov,** Lectures on elliptic and parabolic equations in Hölder spaces, 1996
11 **Jacques Dixmier,** Enveloping algebras, 1996 Printing
10 **Barry Simon,** Representations of finite and compact groups, 1996
9 **Dino Lorenzini,** An invitation to arithmetic geometry, 1996
8 **Winfried Just and Martin Weese,** Discovering modern set theory. I: The basics, 1996
7 **Gerald J. Janusz,** Algebraic number fields, second edition, 1996
6 **Jens Carsten Jantzen,** Lectures on quantum groups, 1996
5 **Rick Miranda,** Algebraic curves and Riemann surfaces, 1995
4 **Russell A. Gordon,** The integrals of Lebesgue, Denjoy, Perron, and Henstock, 1994
3 **William W. Adams and Philippe Loustaunau,** An introduction to Gröbner bases, 1994
2 **Jack Graver, Brigitte Servatius, and Herman Servatius,** Combinatorial rigidity, 1993
1 **Ethan Akin,** The general topology of dynamical systems, 1993

# Notes on Seiberg-Witten Theory

# Notes on Seiberg-Witten Theory

Liviu I. Nicolaescu

Graduate Studies in Mathematics

Volume 28

American Mathematical Society
Providence, Rhode Island

2000 *Mathematics Subject Classification.* Primary 57R57, 57R19, 57R15, 58D27, 14J80, 53C55, 58J05, 58J52.

ABSTRACT. This text is intended to be an introduction to gauge theory and its applications in geometry and topology. The goal is to present in great detail, and with many examples, a basic collection of principles, techniques and applications needed to conduct independent research in gauge theory. In particular, we present complete and self-contained computations of the Seiberg-Witten invariants of most simply connected algebraic surfaces and we discuss at great length a new approach to cutting and pasting of Seiberg-Witten invariants. Familiarity with basic algebraic topology and differential geometry is assumed.

---

**Library of Congress Cataloging-in-Publication Data**

Nicolaescu, Liviu I.
Notes on Seiberg-Witten theory / Liviu I. Nicolaescu.
p. cm. — (Graduate studies in mathematics ; v. 28)
Includes bibliographical references and index.
ISBN 0-8218-2145-8
1. Seiberg-Witten invariants. 2. Global analysis. 3. Four-manifolds (Topology) 4. Mathematical physics. I. Title. II. Series.
QA614.N53 2000
514′.2—dc21 00-044761

---



♾ The paper used in this book is acid-free and falls within the guidelines
established to ensure permanence and durability.
Visit the AMS home page at URL: http://www.ams.org/

10 9 8 7 6 5 4 3 2 1 05 04 03 02 01 00

*To my parents, with love and gratitude*

# Contents

# Introduction

> *My task which I am trying to achieve is by the power of the written word, to make you hear, to make you feel - it is, before all, to make you* see. *That - and no more, and it is everything.*
>
> Joseph Conrad

Almost two decades ago, a young mathematician by the name of Simon Donaldson took the mathematical world by surprise when he discovered some "pathological" phenomena concerning *smooth* 4-manifolds. These pathologies were caused by certain behaviours of instantons, solutions of the Yang-Mills equations arising in the physical theory of gauge fields.

Shortly after, he convinced all the skeptics that these phenomena represented only the tip of the iceberg. He showed that the moduli spaces of instantons often carry nontrivial and surprising information about the background manifold. Very rapidly, many myths were shattered.

A flurry of work soon followed, devoted to extracting more and more information out of these moduli spaces. This is a highly nontrivial job, requiring ideas from many branches of mathematics. Gauge theory was born and it is here to stay.

In the fall of 1994, the physicists N. Seiberg and E. Witten introduced to the world a new set of equations which according to physical theories had to contain the same topological information as the Yang-Mills equations.

From an analytical point of view these new equations, now known as the Seiberg-Witten equations, are easier to deal with than the Yang-Mills equations. In a matter of months many of the results obtained by studying instantons were re-proved much faster using the new theory. (To be perfectly honest, the old theory made these new proofs possible since it created the right mindset to think about the new equations.) The new theory goes one step further, since it captures in a more visible fashion the interaction geometry-topology.

The goal of these notes is to help the potential reader share some of the excitement afforded by this new world of gauge theory and eventually become a player him/herself.

There are many difficulties to overcome. To set up the theory one needs a substantial volume of information. More importantly, all this volume of information is processed in a nontraditional way which may make the first steps in this new world a bit hesitant. Moreover, the large and fast-growing literature on gauge theory, relying on a nonnegligible amount of "folklore"[1], may look discouraging to a beginner.

To address these issues within a reasonable space we chose to present a few, indispensable, key techniques and as many relevant examples as possible. That is why these notes are far from exhaustive and many notable contributions were left out. We believe we have provided enough background and intuition for the interested reader to be able to continue the Seiberg-Witten journey on his/her own.

It is always difficult to resolve the conflict clarity vs. rigor and even much more so when presenting an eclectic subject such as gauge theory. The compromises one has to make are always biased and thus may not satisfy all tastes and backgrounds. We could not escape this bias, but whenever a proof would have sent us far astray we tried to present all the main concepts and ideas in as clear a light as possible and make up for the missing details by providing generous references. Many technical results were left to the reader as exercises but we made sure that all the main ingredients can be found in these notes.

Here is a description of the content. The first chapter contains preliminary material. It is clearly incomplete and cannot serve as a substitute for a more thorough background study. We have included it to present in the nontraditional light of gauge theory many classical objects which may already be familiar to the reader.

The study of the Seiberg-Witten equations begins in earnest in Chapter 2. In the first section we introduce the main characters: the monopoles, i.e. the solutions of the Seiberg-Witten equations and the group of gauge transformations, an infinite dimensional Abelian group acting on the set of monopoles. The Seiberg-Witten moduli space and its structure are described in Section 2.2 while the Seiberg-Witten invariants are presented in Section 2.3. We have painstakingly included all the details concerning orientations

[1] That is, basic facts and examples every expert knows and thus are only briefly or not at all explained in a formal setting. They are usually transmitted through personal interactions.

because this is one of the most confusing aspects of the theory. We conclude this chapter with two topological applications: the proof by P. Kronheimer and T. Mrowka of the Thom conjecture for $\mathbb{CP}^2$ and the new proof based on monopoles of Donaldson's first theorem, which started this new field of gauge theory.

In Chapter 3 we concentrate on a special, yet very rich, class of smooth 4-manifolds, namely the algebraic surfaces. It was observed from the very beginning by E. Witten that the monopoles on algebraic surfaces can be given an explicit algebraic-geometric description, thus opening the possibility of carrying out many concrete computations. The first section of this chapter is a brief and informal survey of the geometry and topology of complex surfaces together with a large list of examples. In Section 3.2 we study in great detail the Seiberg-Witten equations on Kähler surfaces and, in particular, we prove Witten's result stating the equivalence between the Seiberg-Witten moduli spaces and certain moduli spaces of divisors. The third section is devoted entirely to applications. We first prove the nontriviality of the Seiberg-Witten invariants of a Kähler surface and establish the invariance under diffeomorphisms of the canonical class of an algebraic surface of general type. We next concentrate on simply connected elliptic surfaces. We compute all their Seiberg-Witten invariants following an idea of O. Biquard based on the factorization method of E. Witten. This computation allows us to provide the complete *smooth* classification of simply connected elliptic surfaces. In §3.3.3, we use the computation of the Seiberg-Witten invariants of $K3$-surfaces to show that the smooth $h$-cobordism theorem fails in four dimensions. We conclude this section and the chapter with a discussion of the Seiberg-Witten invariants of symplectic 4-manifolds and we prove Taubes' theorem on the nontriviality of these invariants in the symplectic world.

The fourth and last chapter is by far the most technically demanding one. We present in great detail the cut-and-paste technique for computing Seiberg-Witten invariants. This is a very useful yet difficult technique but the existing written accounts of this method can be unbalanced as regards their details. In this chapter we propose a new approach to this technique which in our view has several conceptual advantages and can be easily adapted to other problems as well. Since the volume of technicalities can often obscure the main ideas we chose to work in a special yet sufficiently general case when the moduli spaces of monopoles on the separating 3-manifold are, roughly speaking, Bott nondegenerate.

Section 4.1 contains preliminary material mostly about elliptic equations on manifolds with cylindrical ends. Most objects on closed manifolds have cylindrical counterparts which often encode very subtle features. We discovered that a consistent use of cylindrical notions is not only æsthetically desirable, but also technically very useful. The cylindrical context highlights and coherently organizes many important and not so obvious aspects of the whole gluing problem. An important result in this section is the Cappell-Lee-Miller gluing theorem. We adapt the asymptotic language of [**110**], which is extremely convenient in gluing problems. This section ends with the long subsection §4.1.6 containing many useful and revealing examples. These are frequently used in gauge theory and we could not find any satisfactory reference for them.

In Section 4.2 we study the finite energy monopoles on cylindrical manifolds. The results are very similar to the ones in Yang-Mills equations and that is why this section was greatly inspired by [**96, 133**].

Section 4.3 is devoted to the local study of the moduli spaces of finite energy monopoles. The local structure is formally very similar to that in Yang-Mills theory with a notable exception, the computation of the virtual dimensions, which is part of the folklore. We present in detail this computation since it is often relevant. Moreover, we describe some new exact sequences relating the various intervening deformation complexes to objects covered by the Cappell-Lee-Miller gluing theorem. These exact sequences represent a departure from the mainstream point of view and play a key role in our local gluing theorem.

Section 4.4 is devoted to the study of global properties of the moduli spaces of finite energy monopoles: generic smoothness, compactness (or lack thereof) and orientability. The orientability is no longer an elementary issue in the noncompact case and we chose to present a proof of this fact only in some simpler situations we need for applications.

Section 4.5 contains the main results of this chapter dealing with the process of reconstructing the space of monopoles on a 4-manifold decomposed into several parts by a hypersurface. This manifold decomposition can be analytically simulated by a neck stretching process. During this process, the Seiberg-Witten equations are deformed and their solutions converge to a singular limit. The key issue to be resolved is whether this process can be reversed: given a singular limit can we produce monopoles converging to this singular limit?

In his dissertation [**99**], T. Mrowka proved a very general gluing theorem which provides a satisfactory answer to the above question in the related context of Yang-Mills equations. In §4.5.2, we prove a local gluing theorem, very similar in spirit to Mrowka's theorem but in an entirely new context. The main advantage of the new approach is that all the spectral estimates needed in the proof follow immediately from the Cappell-Lee-Miller gluing theorem. Moreover, the Mayer-Vietoris type local model is just a reformulation of the Cappell-Lee-Miller theorem. The asymptotic language of [**110**] has allowed us to provide intuitive, natural and explicit descriptions of the various morphisms entering into the definition of this Mayer-Vietoris model.

The local gluing theorem we prove produces monopoles converging to a singular limit at a certain rate. If all monopoles degenerated to the singular limit set at this rate then we could conclude that the entire moduli space on a manifold with a sufficiently long neck can be reconstructed from the local gluing constructions. This issue of the surjectivity of the gluing construction is conspicuously missing in the literature and it is quite nontrivial in non-generic situations. We deal with it in §4.5.3 by relying on Łojasewicz's inequality in real algebraic geometry.

In §4.5.4 we prove two global gluing theorems, one in a generic situation and the other one in a special, obstructed setting.

Section 4.6 contains some simple topological applications of the gluing technique. We prove the connected sum theorem and the blow-up formula. Moreover, we present a new and very short proof of a vanishing theorem of Fintushel and Stern.

These notes were written with a graduate student in mind but there are many new points of view to make it interesting for experts as well (especially our new approach to the gluing theorem). The minimal background needed to go through these notes is a knowledge of basic differential geometry, algebraic topology and some familiarity with fundamental facts concerning elliptic partial differential equations. The list of contents for Chapter 1 can serve as background studying guide.

* * *

**Personal note.** I have spent an exciting time of my life thinking and writing these notes and I have been supported along the way by many people.

The book grew out of a year long seminar at McMaster University and a year long graduate course I taught at the University of Notre Dame. I want to thank the participants at the seminar and the course for their patience, interest, and most of all, for their many useful questions and comments.

These notes would perhaps not have seen the light of day were it not for Frank Connolly's enthusiasm and curiosity about the subject of gauge theory which have positively affected me, personally and professionally. I want to thank him for the countless hours of discussions, questions and comments which helped me crystallize many of the ideas in the book.

For the past five years, I have been inspired by Arthur Greenspoon's passion for culture in general, and mathematics in particular. His interest in these notes kept my enthusiasm high. I am greatly indebted to him for reading these notes, suggesting improvements and correcting my often liberal use of English language and punctuation.

While working on these notes I benefited from the conversations with Andrew Sommese, Stefan Stolz and Larry Taylor, who patiently answered my sometimes clumsily formulated questions and helped clear the fog.

My wife has graciously accepted my long periods of quiet meditation or constant babbling about gauge theory. She has been a constant source of support in this endeavor. I want to thank my entire family for being there for me.

*Notre Dame, Indiana 1999*

Chapter 1

# Preliminaries

*The last thing one knows in constructing a work is what to put first.*

Blaise Pascal, *Pensées*

The first chapter contains a fast and unavoidable biased survey of some basic facts needed in understanding Seiberg-Witten theory. The choices in this minimal review reflect the author's background and taste and may not answer everyone's needs. We hope the generous list of references will more than make up for the various omissions.

This introductory chapter has only one goal, namely to familiarize the reader with the basic terms and points of view in the Seiberg-Witten world and cannot serve as a substitute for a solid background.

## 1.1. Bundles, connections and characteristic classes

**1.1.1. Vector bundles and connections.** Smooth vector bundles formalize the notion of "smooth family of vector spaces". For example, given a smooth manifold $M$ and a vector space $F$ we can think of the Cartesian product

$$\underline{F} = \underline{F}_M := F \times M$$

as a smooth family $(F_x)_{x\in M}$ of vector spaces. This trivial example is not surprisingly called the trivial vector bundle with fiber $F$ and base $M$.

We can obtain more interesting examples by gluing these simple ones using *gluing data.* These consist of

A. an open cover $(U_\alpha)$ of a smooth manifold $M$,

B. a *gluing cocycle* i.e. a collection of smooth maps

$$g_{\beta\alpha} : U_{\alpha\beta} \to \mathrm{Aut}\,(F)$$

(where $U_{\alpha\beta} = U_\alpha \cap U_\beta$), such that

$$g_{\alpha\alpha}(x) \equiv \mathbf{1}_F,\;\; g_{\gamma\alpha} \equiv g_{\gamma\beta}(x)\cdot g_{\beta\alpha}(x)\; \forall x \in U_{\alpha\beta\gamma} := U_\alpha \cap U_\beta \cap U_\gamma \neq \emptyset.$$

The open cover $U_\alpha$ is also known as a *trivializing cover.* We will also say it is the *support* of the $g_{\beta\alpha}$.

The map $g_{\beta\alpha}$ describes the "transition from $\underline{F}_\alpha := \underline{F}_{U_\alpha}$ to $\underline{F}_\beta$" in the sense that for every $x \in U_{\alpha\beta}$ the element $(v, x) \in \underline{F}_\alpha$ is identified with the element $(g_{\beta\alpha}(x)v, x) \in \underline{F}_\beta$. Pasting together the trivial bundles $\underline{F}_\alpha$ following the instructions given by the gluing cocycle we obtain a smooth manifold $E$ (called the *total space*), a smooth map $\pi : E \to M$ (called *the canonical projection*) and diffeomorphisms

$$\psi_\alpha : \pi^{-1}(U_\alpha) \to \underline{F}_\alpha$$

(called local *trivializations*) such that for all $x \in U_{\alpha\beta}$, $v \in V$

$$\psi_\beta \circ \psi_\alpha^{-1}(v, x) = (g_{\beta\alpha}(x)v, x).$$

$E \xrightarrow{\pi} M$ as above is called a *vector bundle* over $M$. The rank of $E$ is by definition the dimension of the standard fiber $F$ (over its field of scalars). Rank-one bundles are also known as *line bundles.*

**Example 1.1.1.** Consider the projective space $\mathbb{CP}^n$ defined as the set of one-dimensional complex subspaces of $\mathbb{C}^{n+1}$. There is a natural projection

$$\pi : \mathbb{C}^{n+1} \setminus \{0\} \to \mathbb{CP}^n$$

where $\pi(x) :=$ *the one-dimensional subspaces spanned by* $x$. The fibers

$$\pi^{-1}(p),\;\; p \in \mathbb{CP}^n,$$

are vector subspaces of $\mathbb{C}^{n+1}$. The family $\pi^{-1}(p)$ is indeed a smooth family of vector spaces in the sense described above. It is called the *tautological (or universal) line bundle* over the projective space and is denoted by $\mathbb{U}_n$.

**Exercise 1.1.1.** Describe a gluing cocycle for $\mathbb{U}_n$.

Suppose that $X \xrightarrow{f} Y$ is a smooth map and $E \to Y$ is a smooth vector bundle given by a gluing cocycle $g_{\beta\alpha}$ supported by an open cover $(U_\alpha)$ of $Y$. Then $f$ induces a vector bundle on $X$ called the *pullback of* $E$ *by* $f$ and denoted by $f^*E$. It is given by the open cover $(V_\alpha = f^{-1}(U_\alpha))$ and gluing cocycle $h_{\beta\alpha} = g_{\beta\alpha} \circ f$.

The following exercise describes a very general procedure of constructing smooth vector bundles.

**Exercise 1.1.2.** Consider a smooth map $P$ from a compact, connected, smooth manifold $X$ to the space $\mathrm{End}\,(V)$ of endomorphisms of a vector space $V$ such that $P^2(x) = P(x)\ \forall x \in X$, i.e. $P(x)$ is a smooth family of projectors of $V$.

(a) Show that $\dim\ker P(x)$ is independent of $x \in X$. Denote by $k$ this common dimension.

(b) Show that the assignment $x \mapsto \ker P(x)$ defines a rank-$k$ smooth vector bundle over $X$.

(c) Provide a projector description of the tautological line bundle over $\mathbb{CP}^n$.

(d) Show that any map $X \to V^* \setminus \{0\}$ defines in a canonical way a vector bundle over $X$ of rank $\dim V - 1$.

**Remark 1.1.2.** Denote by $G_k(V)$ the Grassmannian of $k$-dimensional subspaces of an $n$-dimensional vector space $V$. Assume $V$ is equipped with an inner product. For each $k$-dimensional subspace $U \subset V$ denote by $P_U$ the orthogonal projection onto $U^\perp$. The smooth family

$$G_k(V) \ni U \mapsto P_U$$

defines according to the previous exercise a rank-$k$ vector bundle over $G_k(V)$ called the *universal vector bundle* and denoted by $\mathbb{U}_{k,n}$. When $k = 1$ this is precisely the tautological line bundle over $\mathbb{RP}^{n-1}$ or $\mathbb{CP}^{n-1}$.

**Exercise 1.1.3.** Suppose that $x \mapsto P(X)$ is a smooth family of projectors of a vector space $V$ parameterized by a connected smooth manifold $X$. Set $k = \dim\ker P(x)$ and $n = \dim V$ and denote by $f$ the map

$$f : X \to G_k(V), \quad x \mapsto \ker P(x) \in G_k(V).$$

Show that $f$ is smooth and that the pullback of $\mathbb{U}_{k,n}$ by $f$ coincides with the vector bundle defined by the family of projections $P(x)$.

A smooth map $s$ from a smooth manifold $X$ to a vector space $F$ is a smooth selection of an element $s(x)$ in each fiber $F \times x$ of $\underline{F}$. In other words, it is a smooth map $s : X \to \underline{F}_X$ such that $\pi \circ s = \mathbf{1}_X$ where $\pi : \underline{F}_X \to X$ is the natural projection. Replacing $\underline{F}_X$ with any smooth vector bundle $E \xrightarrow{\pi} X$ we get the notion of smooth section of $E$. The space of smooth sections of $E$ will be denoted by $\Gamma(E)$ or $C^\infty(E)$. In terms of gluing cocycles we can describe a section as a collection of smooth maps

$$s_\alpha : U_\alpha \to F$$

such that

$$s_\beta(x) = g_{\beta\alpha}(x)s_\alpha(x), \quad \forall x \in U_\alpha \cap U_\beta.$$

The functorial operations in linear algebra have a vector bundle counterpart. Suppose $E_i \stackrel{\pi_i}{\to} X$ $(i = 1, 2)$ are two vector bundles over $X$ with standard fibers $F_i$, $i = 1, 2$, given by gluing cocycles $G_{\beta\alpha}(i)$ along the same support. For example the direct sum $F_1 \oplus F_2$ corresponds to the *direct (Whitney) sum* $E_1 \oplus E_2$ given by the gluing cocycle $G_{\beta\alpha}(1) \oplus G_{\beta\alpha}(2)$.

The dual vector bundle $E_1^*$ is defined by the gluing cocycle $(G^*_{\beta\alpha}(1))^{-1}$ where "$*$" denotes the conjugate transpose.

We can form tensor products, symmetric, exterior products of vector bundles, etc. In particular, the bundle $E_1^* \otimes E_2$ will be denoted by $\mathrm{Hom}\,(E_1, E_2)$. Its sections are *bundle morphisms*, i.e. smooth maps $T : E_1 \to E_2$ mapping the fiber $E_1(x)$ of $E_1$ linearly to the fiber $E_2(x)$ of $E_2$. When $E_1 = E_2 = E$ we use the notation $\mathrm{End}\,(E)$. If the induced morphisms $T(x)$ are all isomorphisms then $T$ is called a bundle isomorphism. A bundle automorphism of a vector bundle $E$ is also called a *gauge transformation*. The group of bundle automorphisms of $E$ is denoted by $\mathfrak{G}(E)$ and is known as the *gauge group of* $E$.

**Exercise 1.1.4.** Suppose $L \to X$ is a smooth complex line bundle over $X$. Show that

$$\mathfrak{G}(L) \cong C^\infty(M, \mathbb{C}^*).$$

The line bundle $\Lambda^{\mathrm{rank}(E_1)}E_1$ is called the *determinant line bundle of* $E_1$ and is denoted by $\det E_1$.

If $E \to X$ is an $\mathbb{R}$-vector bundle then a *metric* on $E$ is a section $h$ of $\mathrm{Symm}^2(E^*)$ such that $h(x)$ is positive definite for every $x \in X$. If $E$ is complex one defines similarly Hermitian metrics on $E$. A *Hermitian bundle* is a vector bundle equipped with a Hermitian metric.

The next exercise will show how to use sections to prove that any complex line bundle over a compact manifold is the pullback of the universal line bundle over a complex projective space.

**Exercise 1.1.5.** Suppose $M$ is a smooth compact manifold and $E \to M$ is a complex line bundle. A subspace $V \subset C^\infty(E)$ is said to be *ample* if for any $x \in M$ there exists $u \in V$ such that $u(x) \neq 0$.

(a) Show that there exist finite-dimensional ample subspaces $V \subset C^\infty(E)$.

(b) Let $V$ be a finite-dimensional ample subspace of $C^\infty(E)$. For each $x \in M$ set

$$V_x = \{v \in V;\ v(x) = 0\}.$$

Equip $V$ with a Hermitian metric and denote by $P(x) : V \to V$ the orthogonal projection onto $V_x$. Show that $\dim \ker P_x = 1$ and the family of projections $\{P(x);\ x \in M\}$ is smooth. As in Exercise 1.1.2(b) we obtain a complex line bundle $E_V \to V$.

(c) Show that the line bundle $E$ is isomorphic to $E_V$. In particular, this shows that $E$ is the pullback of a universal line bundle over a projective space.

(d) Suppose that $f, g : M \to \mathbb{CP}^n$ are two (smoothly) homotopic maps. Denote by $E_f$ (resp. $E_g$) the pullbacks of the universal line bundle $\mathbb{U}_n$ via $f$ (resp. $g$). Show that $E_f \cong E_g$.

**Remark 1.1.3.** For every smooth manifold $M$ denote by $\mathrm{Pic}^\infty(M)$ the space of isomorphism classes of smooth complex line bundles over $M$ and by $[M, \mathbb{CP}^n]_\infty$ the set of (smooth) homotopy classes of smooth maps $M \to \mathbb{C}^n$. This is an inductive family

$$[M, \mathbb{CP}^1]_\infty \hookrightarrow [M, \mathbb{CP}^2]_\infty \hookrightarrow \cdots$$

and we denote by $[M, \mathbb{CP}^\infty]_\infty$ its inductive limit. The above exercise shows that if $M$ is compact we have a bijection

$$\mathrm{Pic}^\infty(M) \cong [M, \mathbb{CP}^\infty]_\infty.$$

The tensor product of line bundles induces a structure of Abelian group on $\mathrm{Pic}^\infty(M)$. Since the inductive limit $\mathbb{CP}^\infty$ of the $\mathbb{CP}^n$'s is a $K(\mathbb{Z}, 2)$-space we can conclude that we have an isomorphism of groups

$$c_1^{top} : \mathrm{Pic}^\infty(M) \to H^2(M, \mathbb{Z}).$$

For any $L \in \mathrm{Pic}^\infty(M)$ the element $c_1^{top}(L)$ is called the *topological first Chern class of L*.

One is often led to study families of vector spaces satisfying additional properties such as vector spaces in which vectors have lengths and pairs of vectors have definite angles (as in Euclidean geometry). According to Felix Klein's philosophy, this is the same as looking at the *symmetry group*, i.e. the subgroup of linear maps which preserve these additional features. In the above case this is precisely the orthogonal group. If we want to deal with families of such spaces then we must impose restrictions on the gluing maps: they must be valued in the given symmetry group. Here is one way to formalize this discussion. Suppose we are given the following data.

• A Lie group $G$ and a representation

$$\rho : G \to \mathrm{End}\,(F).$$

• A smooth manifold $X$ and open cover $U_\alpha$.

• A *G-valued gluing cocycle*, i.e. a collection of smooth maps

$$g_{\beta\alpha} : U_{\alpha\beta} \to G$$

such that $g_{\alpha\alpha}(x) = \mathbf{1} \in G \; \forall x \in U_\alpha$ and

$$g_{\gamma\alpha}(x) = g_{\gamma\beta}(x) \cdot g_{\beta\alpha}(x) \;\; \forall x \in U_{\alpha\beta\gamma}.$$

Then the collection

$$\rho(g_{\beta\alpha}) : U_{\alpha\beta} \to \mathrm{End}\,(F)$$

defines a gluing cocycle for a vector bundle $E$ with standard fiber $F$ and symmetry group $G$. The vector bundle $E$ is said to have a $G$-structure.

**Remark 1.1.4.** Differential geometers usually phrase the above construction in terms of *principal G-bundles*. Given a gluing $G$-cocycle as above we can obtain a smooth manifold $P$ as follows. Glue the product $G \times U_\alpha$ to $G \times U_\beta$ along $U_{\alpha\beta}$ using the following prescription: for each $x \in U_{\alpha\beta}$ the element $(g, x)$ in $G \times U_\alpha$ is identified with the element $(g_{\beta\alpha}(x) \cdot g, x)$ in $G \times U_\beta$. We obtain a smooth manifold $P$ and a smooth map $\pi : P \to X$ whose fibers $\pi^{-1}(x)$ are diffeomorphic to the Lie group $G$. This is called the principal $G$-bundle determined by the gluing $G$-cocycle $g_{\beta\alpha}$. The above vector bundle $E$ is said to be *induced from* $P$ via the representation $\rho$ and we write this as $P \times_\rho F$. For more details we refer to vol. 1 of [**64**].

**Exercise 1.1.6.** Show that the above manifold $P$ comes with a natural free, right $G$-action and the space of orbits can be naturally identified with $X$.

**Exercise 1.1.7.** Regard $S^{2n+1}$ as a real hypersurface in $\mathbb{C}^{n+1}$ given by the equation $|z_0|^2 + |z_1|^2 + \cdots |z_n|^2 = 1$. The group

$$S^1 = \{e^{\mathbf{i}t} \; ; \; t \in \mathbb{R}\} \subset \mathbb{C}^*$$

acts on $S^{2n+1}$ by scalar multiplication. The quotient of this action is obviously $\mathbb{CP}^n$.

(a) Show that $S^{2n+1} \to \mathbb{CP}^n$ is a principal $S^1$-bundle. (It is known as the *Hopf bundle.*

(b) Show that the line bundle associated to it via the tautological representation $S^1 \to \mathrm{Aut}\,(\mathbb{C}^1)$ is precisely the universal line bundle $\mathbb{U}_n$ over $\mathbb{CP}^n$.

**Exercise 1.1.8.** Show that any metric on a rank-$n$ real vector bundle naturally defines an $O(n)$-structure.

To exist as a subject, *differential* geometry requires a way to *differentiate* the objects under investigation. This is where connections come in. A *connection* (or *covariant derivative*) $\nabla$ on a vector bundle $E \stackrel{\pi}{\to} M$ is a map which associates to every section $s \in \Gamma(E)$ and any vector field $X$ on $M$ a new section $\nabla_X s$ such for every $f \in C^\infty(M)$

$$\nabla_X(fs) = df(X)s + f\nabla_X s.$$

$\nabla_X s$ is the derivative of $s$ in the direction $X$. One usually forgets the vector field $X$ in the above definition and thinks of $\nabla$ as a map

$$\nabla_\bullet : \Gamma(E) \to \Gamma(T^*M \otimes E)$$

satisfying Leibniz' rule

$$\nabla_\bullet(fs) = df(\bullet) \otimes s + f\nabla_\bullet s.$$

Note the following fact.

**Proposition 1.1.5.** *There exists at least one connection $\nabla^0$ on $E$. Moreover, any other connection can be obtained from $\nabla^0$ by the addition of an* $\mathrm{End}\,(E)$*-valued 1-form $A \in \Omega^1(\mathrm{End}\,(E))$ where by definition, for any vector bundle $F \to M$ we set*

$$\Omega^k(F) := \Gamma(\Lambda^k T^*M \otimes F).$$

*In particular, the space $\mathfrak{A}(E)$ of connections on $E$ is an affine space modeled by $\Omega^1(\mathrm{End}\,(E))$.*

The trivial bundle $\underline{F}$ admits a natural connection $\Theta$ called the *trivial connection.* To describe it recall that sections of $\underline{F}$ can be regarded as smooth functions $s : M \to F$. Define

$$\Theta s = ds \in \Omega^1(M) \otimes F.$$

Any other connection $\nabla$ on $\underline{F}$ will differ from $\Theta$ by a 1-form $A$ with coefficients endomorphisms of $F$, i.e.

$$\nabla = \Theta + A, \quad A \in \Omega^1(M) \otimes \mathrm{End}\,(F).$$

Now if $E$ is obtained by gluing the trivial bundles $\underline{F}_\alpha := \underline{F}_{U_\alpha}$ using the cocycle $g_{\beta\alpha}$ then any connection on $E$ is obtained by gluing connections $\nabla^\alpha$ on $\underline{F}_\alpha$. More precisely, if $\nabla^\alpha = \Theta + A_\alpha$ then on the overlaps $U_{\alpha\beta}$ the 1-forms $A_\alpha$ and $A_\beta$ satisfy the compatibility rules

$$A_\beta = -dg_{\beta\alpha}g_{\beta\alpha}^{-1} + g_{\beta\alpha}A_\alpha g_{\beta\alpha}^{-1} = g_{\alpha\beta}^{-1}dg_{\alpha\beta} + g_{\alpha\beta}^{-1}A_\alpha g_{\alpha\beta}. \tag{1.1.1}$$

**Exercise 1.1.9.** Prove (1.1.1).

**Exercise 1.1.10.** Consider a smooth family $P : x \mapsto P_x$ of projectors of the vector space $F$ parameterized by the connected smooth manifold $X$. Show that $(\mathbf{id} - P)\Theta$ defines a connection on the subbundle $\ker P \subset \underline{F}_X$.

Imitating the above local description of a connection we can define a notion of connection compatible with a $G$-structure. Thus, let us suppose the vector bundle $E \to M$ has a $G$-structure defined by the gluing cocycle

$$g_{\beta\alpha} : U_{\beta\alpha} \to G$$

and the representation $\rho : G \to \operatorname{Aut}(F)$. Denote by $\mathfrak{g}$ the Lie algebra of $G$. The gluing cocycle defines a principal $G$-bundle $P \to M$. A connection on $P$ is a collection of 1-forms

$$A_\alpha \in \Omega^1(U_\alpha) \otimes \mathfrak{g}$$

satisfying (1.1.1), where $g_{\beta\alpha} A_\alpha g_{\beta\alpha}^{-1}$ denotes the adjoint action of $g_{\beta\alpha}(x)$ on $\mathfrak{g}$ while $-dg_{\beta\alpha} g_{\beta\alpha}^{-1} = g_{\alpha\beta}^{-1} dg_{\alpha\beta}$ is the pullback via $g_{\alpha\beta}$ of the Maurer-Cartan form on $G$. (This is the $\mathfrak{g}$-valued, left invariant 1-form on $G$ whose value at $\mathbf{1}$ is the tautological map $T_{\mathbf{1}}G \to \mathfrak{g}$.)

Given a connection on the principal bundle we can obtain a genuine connection (i.e. covariant derivative) on $E = P \times_\rho F$ given by the $\operatorname{End}(F)$-valued 1-forms $\rho_*(A_\alpha)$, where

$$\rho_* : T_{\mathbf{1}}G \to \operatorname{End}(F)$$

denotes the differential of $\rho$ at $\mathbf{1} \in G$.

A *gauge transformation* of a bundle $E$ with a $G$-structure is a collection of smooth maps $T_\alpha : U_\alpha \to G$ subject to the gluing conditions

$$T_\beta = g_{\beta\alpha} T_\alpha g_{\beta\alpha}^{-1}.$$

(From a more invariant point of view, a gauge transformation is a special section of the bundle of endomorphisms of $E$.) The set of such gauge transformations forms a group which will be denoted by $\mathfrak{G}_G(E)$.

To a bundle $E$ with a $G$-structure one can naturally associate a vector bundle $\operatorname{Ad}(E)$ defined by the same gluing $G$-cocycle as $E$ but, instead of $\rho$, one uses the adjoint representation $Ad : G \to \operatorname{End}(\mathfrak{g})$.

**Proposition 1.1.6.** *The space $\mathfrak{A}_G(E)$ of $G$-compatible connections on a vector bundle $E$ with a $G$-structure is an affine space modeled by $\Omega^1(\operatorname{Ad}(E))$. Moreover, the group of gauge transformations $\mathfrak{G}_G(E)$ acts on $\mathfrak{A}_G(E)$ by conjugation*

$$\mathfrak{G}_G(E) \times \mathfrak{A}_G(E) \ni (\gamma, \nabla^A) \mapsto \gamma \nabla^A \gamma^{-1} \in \mathfrak{A}_G(E).$$

For more details about principal bundles and connections from a gauge theoretic point of view we refer to the very elegant presentation in [**116**].

If $E$ is a complex vector bundle of complex rank $r$ equipped with a Hermitian metric $\langle \bullet, \bullet \rangle$ then it is equipped with a natural $U(r)$-structure. A

*Hermitian connection* $\nabla$ on $E$ is by definition a connection compatible with this $U(r)$-structure or, equivalently,

$$L_X\langle s_1, s_2\rangle = \langle \nabla_X s_1, s_2\rangle + \langle s_1, \nabla_X s_2\rangle, \ \ \forall X \in \mathrm{Vect}\,(M), \ s_1, s_2 \in C^\infty(E).$$

There is a natural (left) action of $\mathfrak{G}_G(E)$ on $\mathfrak{A}_G(E)$ given by

$$T \cdot \nabla := T\nabla T^{-1}.$$

The covariant method of differentiation has a feature not encountered in traditional calculus in $\mathbb{R}^n$. More precisely, the classical result "partial derivatives commute"

$$\frac{\partial^2 f}{\partial x \partial y} = \frac{\partial^2 f}{\partial y \partial x}$$

no longer holds in this more general context because of deep geometric reasons. One is led to quantify the extent of this noncommutativity and this is usually encoded by the *curvature* of a connection.

Suppose $\nabla$ is a connection on a vector bundle $E \to M$. For any vector fields $X$, $Y$ on $M$ and any section $u \in \Gamma(E)$ define

$$F(X,Y)u = F_\nabla(X,Y)u := [\nabla_X, \nabla_Y]u - \nabla_{[X,Y]}u$$

$$= (\nabla_X\nabla_Y - \nabla_Y\nabla_X)u - \nabla_{[X,Y]}u \in \Gamma(E).$$

Note that for all $f \in C^\infty(M)$

$$F(fX,Y)u = F(X,fY)u = F(X,Y)(fu) = fF(X,Y)u.$$

Thus the map

$$u \mapsto F(X,Y)u$$

is an endomorphism of $E$ for all $X, Y$. We denote it by $F(X,Y)$. Note that the map

$$TM \otimes TM \to \mathrm{End}\,(E), \ \ X \otimes Y \mapsto F(X,Y)$$

is a skew-symmetric bundle morphism. Thus we can regard the object $F(\cdot,\cdot)$ as a section of $\Omega^2(\mathrm{End}\,(E)) = \Lambda^2 T^*M \otimes \mathrm{End}\,(E)$. $F(\bullet,\bullet)$ is called the curvature of $\nabla$. When $F_\nabla = 0$ we say $F$ is *flat.*

**Exercise 1.1.11.** Suppose $E$ is a vector bundle equipped with a $G$-structure and $\nabla$ is a $G$-compatible connection. Show that $F_\nabla \in \Omega^2(\mathrm{Ad}\,(E))$. In particular, if $E$ is a Hermitian vector bundle and $\nabla$ is Hermitian then the curvature of $\nabla$ is a 2-form with coefficients in the bundle of skew-Hermitian endomorphisms of $E$.

**Exercise 1.1.12.** (a) Consider the trivial bundle $\underline{F}_M$. Then the trivial connection $\Theta$ is flat.

(b) If $A \in \Omega^1(\mathrm{End}\,(\underline{F}))$ then the curvature of $\Theta + A$ is

$$F_A = dA + A \wedge A.$$

Above, $A$ is thought of as a matrix of with entries smooth 1-forms $\omega_{ij}$. Then $dA$ is the matrix with entries the 2-forms $d\omega_{ij}$ and $A \wedge A$ is a matrix whose $(i,j)$-entry is the 2-form

$$\sum_k \omega_{ik} \wedge \omega_{kj}.$$

If $E$ is given by a gluing cocycle $g_{\beta\alpha}$ and $\nabla$ is given by the collection of 1-forms $A_\alpha \in \Omega^1(\mathrm{End}\,(\underline{F}_\alpha))$ then the above exercise shows that $F$ is locally described by the collection of 2-forms $dA_\alpha + A_\alpha \wedge A_\alpha$.

**Example 1.1.7.** Suppose $L \to M$ is a complex line bundle given by a gluing cocycle $z_{\beta\alpha} : U_{\alpha\beta} \to \mathbb{C}^*$. Then a connection on $L$ is defined by a collection of complex valued 1-forms $\omega_\alpha$ satisfying

$$\omega_\beta = \frac{dz_{\alpha\beta}}{z_{\alpha\beta}} + \omega_\alpha.$$

The curvature is given by the collection of 2-forms $d\omega_\alpha$.

If $L$ has a $U(1)$-structure (i.e. is equipped with a Hermitian metric) then the gluing maps belong to $S^1$:

$$z_{\beta\alpha} : U_{\alpha\beta} \to S^1.$$

The connection is *Hermitian* (i.e. compatible with the metric) if $\omega_\alpha \in \Omega^1(U_\alpha) \otimes \underline{u}(1) \cong \mathbf{i}\mathbb{R}$. Thus we can write

$$\omega_\alpha = \mathbf{i}\theta_\alpha, \;\; \theta_\alpha \in \Omega^1(U_\alpha).$$

They are related by

$$\theta_\beta - \theta_\alpha = -\mathbf{i}\frac{dz_{\alpha\beta}}{z_{\alpha\beta}} = z^*_{\alpha\beta}(d\theta)$$

where $d\theta$ denotes the angular form on $S^1$.

**Exercise 1.1.13.** Consider a Hermitian line bundle $L \to M$ and denote by $P \to M$ the corresponding principal $S^1$-bundle. For each $p \in P$ denote by $i_p$ the injection

$$S^1 \ni e^{\mathbf{i}t} \mapsto p \cdot e^{\mathbf{i}t} \in P.$$

Suppose $\nabla$ is a Hermitian connection as in the above example. Show that $\nabla$ naturally defines a 1-form $\omega \in \Omega^1(P)$ such that

$$i_p^*\omega = d\theta, \;\; \forall p \in P.$$

$\omega$ is called the *global angular form* determined by $\nabla$. Conversely, show that any angular form uniquely determines a Hermitian connection on $L$.

**Example 1.1.8.** Consider the unit sphere $S^2 \subset \mathbb{R}^3$ with its canonical orientation as the boundary of the unit ball in $\mathbb{R}^3$. Define the open cover $\{U_\alpha, U_\beta\}$ by

$$U_\alpha = S^2 \setminus \{\text{south pole}\}$$

and

$$U_\beta = S^2 \setminus \{\text{north pole}\}.$$

The we can identify in an orientation preserving fashion

$$U_{\alpha\beta} \cong \mathbb{C}^*.$$

Denote by $z$ the complex coordinate on $\mathbb{C}^*$. For each $n \in \mathbb{Z}$ denote by $L_n$ the complex line bundle defined by the gluing cocycle

$$z_{\beta\alpha} : \mathbb{C}^* \cong U_{\alpha\beta} \to \mathbb{C}^*, \quad z \mapsto z^n.$$

Suppose $\nabla$ is a connection on $L$ defined locally by $\omega_\alpha$, $\omega_\beta$ where

$$\omega_\beta = -n\frac{dz}{z} + \omega_\alpha.$$

Denote by $F$ its curvature. It is a complex valued 2-form on $S^2$ and thus it can be integrated over the 2-sphere. Denote by $D_\pm$ the upper/lower hemisphere. $D_+$ is identified in an orientation preserving fashion with the unit disk $\{|z| \leq 1\} \subset \mathbb{C}$. We have

$$\int_{S^2} F = \int_{D_+} d\omega_\alpha + \int_{D_-} d\omega_\beta = \int_{\partial D_+} (\omega_\alpha - \omega_\beta)$$

$$= n\int_{\partial D_+} \frac{dz}{z} = 2\pi\mathbf{i}n.$$

We arrive at several amazing conclusions.

• The integral of $F_\nabla$ is independent of $\nabla$ !!!

• The integral of $F_\nabla$ is an *integer* multiple of $2\pi\mathbf{i}$ !!!

• The line bundle $L_n$ with $n \neq 0$ cannot admit flat connections so that the noncommutativity of partial derivatives is present for *any* covariant method of differentiation !!!

• The line bundle $L_n$ with $n \neq 0$ is not isomorphic to the trivial line bundle $\underline{\mathbb{C}}$ which admits a flat connection !!!

**Exercise 1.1.14.** Prove that the line bundle $L_1$ in the above example is isomorphic to the universal line bundle over $\mathbb{CP}^1 \cong S^2$.

The above conclusions do not represent an isolated occurrence. They are manifestations of a more general construction called *Chern-Weil theory.* Below we describe a few particular cases of this construction.

**1.1.2. Chern-Weil theory.** Consider a complex vector bundle $E \to M$ and $\nabla$ an arbitrary connection on it. Set $n = \operatorname{rank}(E)$. The curvature $F(\nabla)$ can be viewed either as a 2-form on $M$ whose coefficients are endomorphisms of $E$ or as a $n \times n$ matrix with entries complex valued 2-forms on $M$. The multiplication of even-dimensional forms is commutative so we can speak of determinants of such matrices. Then

$$c(E,\nabla) := \det\left(\mathbf{1}_E + \frac{\mathbf{i}}{2\pi}F(\nabla)\right)$$

is a nonhomogeneous element,

$$c(E,\nabla) \in \Omega^{even}(M) \otimes \mathbb{C} := \bigoplus_{k \geq 0} \Omega^{2k}(M) \otimes \mathbb{C}.$$

The component of degree $2k$ is denoted by $c_k(E,\nabla)$ and is called the *k-th Chern form* of $E$ corresponding to the connection $\nabla$. Note that

$$c_1(E,\nabla) = \frac{\mathbf{i}}{2\pi}\operatorname{tr}(F(\nabla)) \in \Omega^2(M) \otimes \mathbb{C},$$

$$c_n(E,\nabla) = \left(\frac{\mathbf{i}}{2\pi}\right)^n \det(F(\nabla)) \in \Omega^{2n}(M) \otimes \mathbb{C}.$$

**Example 1.1.9.** Consider again the line bundle $L_n \to S^2$. The computations in Example 1.1.8 show that

$$\int_{S^2} c_1(L_n,\nabla) = -n$$

for *any connection* on $L_n$.

The above nice accident is a special case of the following theorem.

**Theorem 1.1.10. (Chern-Weil)** *(a) The Chern forms $c_k(E,\nabla)$ are closed for any $k$ and any connection $\nabla$ on $E$.*

*(b) For any connections $\nabla^0$, $\nabla^1$ on $E$ and any $k \in \mathbb{Z}_+$ there exists a $(2k-1)$-form $T(\nabla^1,\nabla^0)$ on $M$ such that*

$$c_k(E,\nabla^1) - c_k(E,\nabla^0) = dT(\nabla^1,\nabla^0).$$

For a proof of this theorem we refer to [**105**]. Part (a) of this theorem shows that $c_k(E,\nabla)$ defines a cohomology class in $H^{2k}(M,\mathbb{R})$ which by part (b) is *independent* of $\nabla$. We denote this class by $c_k(E)$ and we call it the $k$-th *Chern characteristic class* of $E$. The element

$$c(E) = 1 + c_1(E) + c_2(E) + \cdots$$

is called the *total Chern class* of $E$. Note that if $E$ is trivial then all classes $c_k(E)$ vanish. We can turn this statement around and conclude that if one of the classes $c_k(E)$ is not trivial then $E$ is certainly not trivial. Thus these classes provide a measure of nontriviality of a complex vector bundle.

**Remark 1.1.11.** The computations in Example 1.1.8 show that

$$\int_{S^2} c_1(L_n) = -n$$

so that in particular $L_n$ is nontrivial and $c_1(L_n) \in H^2(S^2, \mathbb{Z})$. One can show that for any smooth manifold $M$ and any complex vector bundle $E \to M$ the characteristic class $c_k(E)$ belongs to the image of $H^{2k}(M, \mathbb{Z})$ inside $H^{2k}(M, \mathbb{R})$. If we denote by $\imath$ the natural morphism $H^2(M, \mathbb{Z}) \to H^2(M, \mathbb{R})$ then one can show that

$$c_1(L) = \imath(c_1^{top}(L))$$

where the topological first Chern class was defined in Remark 1.1.3.

Define the *Chern polynomial* of $E$ by

$$c_t(E) = \sum_{k \geq 0} c_k(E) t^k \in H^*(M, \mathbb{R})[t].$$

**Exercise 1.1.15.** Show that

$$c_t(E_1 \oplus E_2) = c_t(E_1) \cdot c_t(E_2)$$

where for simplicity we denoted by "$\cdot$" the $\wedge$-multiplication in $\Omega^{even}(M)$. Show that if $E = \bigoplus_{i=1}^n L_i$, where $L_i$ are complex line bundles then

$$c_k(E) = \sigma_k(y_1, \cdots, y_n) := \sum_{1 \leq i_1 < \cdots i_k \leq n} y_{i_1} \cdots y_{i_k}$$

where $y_i := c_1(L_i)$.

**Exercise 1.1.16.** Consider a complex line bundle $L$ over a compact, closed, oriented Riemann surface $\Sigma$.

(a) Show that the quantity

$$\deg L := \int_{\Sigma} c_1(L)$$

is an integer.

**Hint:** Use the fact that the restriction of $L$ over the complement of a small disk in $\Sigma$ is trivial.

(b) Suppose $u$ is a section of $L$ with only nondegenerate zeros, i.e. for any $x \in u^{-1}(0)$ the adjunction map

$$\mathfrak{a}_x : T_x\Sigma \to L_x, \ \ T_x\Sigma \ni \zeta \mapsto (\nabla_\zeta u)\,|_x \in L_x$$

($\nabla$ some connection on $L$) is invertible. For each $x \in u^{-1}(0)$ set

$$\deg(x) := \text{sign}\,\det \mathfrak{a}_x.$$

Show that

$$\deg L := \sum_{x \in u^{-1}(0)} \deg(x).$$

**Hint:** Use the fact that $L$ is trivial outside $\bigcup_{x \in u^{-1}(0)} D_x$, where $D_x$ denotes a very small disk centered at $x$.

Define the *Chern character* of a vector bundle to be the cohomology class

$$\mathbf{ch}\,(E) := \text{tr}\,\exp(\frac{\mathbf{i}}{2\pi}F(\nabla)) = \sum_{k \geq 0} \frac{1}{k!}\text{tr}\left(\frac{\mathbf{i}}{2\pi}F(\nabla)\right)^k.$$

Again this is a closed form whose cohomology class is independent of $\nabla$.

**Exercise 1.1.17.** (a) Show that if $L \to M$ is a complex line bundle then

$$\mathbf{ch}\,(L) = \exp(c_1(L)).$$

(b) Show that

$$\mathbf{ch}(E_1 \oplus E_2) = \mathbf{ch}\,(E_1) + \mathbf{ch}\,(E_2)$$

and

$$\mathbf{ch}\,(E_1 \otimes E_2) = \mathbf{ch}\,(E_1) \cdot \mathbf{ch}\,(E_2).$$

The construction of the Chern character has a multiplicative counterpart. Suppose that $f(T)$ is a formal power series

$$f(T) := \sum_{n \geq o} a_n T^n \in \mathbb{C}[[T]]$$

such that $a_0 = 1$. If $E \to M$ is a complex vector bundle then $f(E) \in H^*(M)$ is the cohomology class represented by

$$f(E, \nabla) := \det\Bigl(\sum_{n \geq 0} a_n \bigl(\frac{\mathbf{i}}{2\pi}F(\nabla)\bigr)^n\Bigr).$$

A situation frequently encountered in geometry is that of

$$\mathbf{td}\,(T) := \frac{T}{1 - \exp(-T)} = 1 + \frac{1}{2}T + \sum_{k=1}^{\infty}(-1)^{k-1}\frac{B_k}{(2k)!}T^{2k}$$

where the coefficients $B_k$ are known as the *Bernoulli numbers*. Here are a few values of these numbers

$$B_1 = \frac{1}{6},\quad B_2 = \frac{1}{30},\quad B_3 = \frac{1}{42},\quad \cdots$$

The cohomology class obtained in this manner is called the *Todd genus* of $E$ and is denoted by $\mathbf{td}\,(E)$.

Both **ch** and **td** decompose into homogeneous parts

$$\mathbf{ch}(E) = \sum_{i\geq 0} \mathbf{ch}_i(E), \quad \mathbf{td}\,(E) = \sum_{i\geq 0} \mathbf{td}_i(E)$$

expressible in terms of the Chern classes $c_i$. For example

$$\mathbf{ch}_0(E) = \mathrm{rank}\,(E), \tag{1.1.2}$$
$$\mathbf{ch}_1(E) = c_1(E), \quad \mathbf{ch}_2(E) = \tfrac{1}{2}(c_1(E)^2 - 2c_2(E))$$

$$\mathbf{td}_0(E) = 1, \quad \mathbf{td}_1(E) = \frac{1}{2}c_1(E), \quad \mathbf{td}_2(E) = \frac{1}{12}\Big(c_1(E)^2 + 2c_2(E)\Big). \tag{1.1.3}$$

So far we have considered only complex vector bundles. There is a real theory as well. Consider a real vector bundle $E \to M$ and $\nabla$ an arbitrary connection on it. We define the *total Pontryagin form* associated to $E(\nabla)$ by

$$p(E, \nabla) = \det(\mathbf{1} - \frac{1}{2\pi}F(\nabla)).$$

Again one can prove that this is a closed form whose cohomology class is independent of $\nabla$. This time a new phenomenon arises.

**Lemma 1.1.12.** *The components of $p(E, \nabla)$ of degree $4k + 2$ are exact.*

**Exercise 1.1.18.** Prove the above lemma.

The cohomology class $p(E)$ decomposes as

$$p(E) = 1 + p_1(E) + p_2(E) + \cdots + p_k(E) + \cdots$$

where

$$p_k(E) \in H^{4k}(M, \mathbb{R}).$$

The cohomology classes are called the *Pontryagin classes* of the real vector bundle $E$. For example, $p_1(E)$ can be represented by the form

$$p_1(E, \nabla) = -\frac{1}{8\pi^2}\mathrm{tr}\,(F(\nabla) \wedge F(\nabla)).$$

**Exercise 1.1.19.** Suppose $E \to M$ is a real vector bundle and denote by $E^c$ its complexification $E \otimes \mathbb{C}$. Show that

$$c_{2k+1}(E^c) = 0 \text{ and } c_{2k}(E^c) = (-1)^k p_k(E).$$

## 1.2. Basic facts about elliptic equations

Before we begin talking about elliptic equations we must first define the notion of partial differential operator (p.d.o. for brevity) on a smooth manifold and explain the basic operations one can perform on such objects. We refer again to [**105**] for more details.

Consider a smooth, oriented Riemannian manifold $(M, g)$ and $E, F \to M$ complex Hermitian vector bundles over $M$. We will denote the Hermitian metrics on $E$ (resp. $F$) by $\langle \cdot, \cdot \rangle_E$ (resp. $\langle \cdot, \cdot \rangle_F$).

Denote by $\mathbf{Op}\,(E, F)$ the space of $\mathbb{C}$-linear operators

$$T : C^\infty(E) \to C^\infty(F).$$

Denote by $C^\infty(M)$ the space of complex valued smooth functions on $M$. The spaces $C^\infty(E)$ and $C^\infty(F)$ have natural structures of $C^\infty(M)$-modules and we will be interested in a subspace of **Op** consisting of operators interacting in a nice way with these module structures.

For each $f \in C^\infty(M)$ and each $T \in \mathbf{Op}\,(E, F)$ define $ad(f)(T) \in \mathbf{Op}(E, F)$ by

$$(ad(f)T)u = [T, f]u := T(fu) - f(Tu), \ \ \forall u \in C^\infty(E).$$

Note that the maps $T \mapsto [T, f]$ and $f \mapsto [T, f]$ behave like derivations, i.e. they satisfy the Leibniz rule

$$[ST, f] = [S, f]T + S[T, f] \ \text{ and } \ [T, fg] = [T, f]g + f[T, g] \tag{1.2.1}$$

for all $f, g, T, S$ for which the above operations make sense.

Now define inductively an increasing sequence of subspaces

$$\mathbf{PDO}^{(0)}(E, F) \subset \mathbf{PDO}^{(1)}(E, F) \subset \cdots \subset \mathbf{PDO}^{(k)}(E, F) \subset \cdots$$

following the prescriptions

$$\mathbf{PDO}^{(0)}(E, F) := \mathrm{Hom}\,(E, F)$$

and

$$\mathbf{PDO}^{(k+1)}(E, F)$$

$$:= \Big\{ T \in \mathbf{Op}\,(E, F);\ [T, f] \in \mathbf{PDO}^{(k)}(E, F),\ \forall f \in C^\infty(M) \Big\}.$$

The elements of $\mathbf{PDO}^{(k)}(E, F)$ will be called *partial differential operators of order* $\leq k$.

**Example 1.2.1.** (a) Let $E = F = \underline{\mathbb{C}}$ and let $X$ be a smooth vector field on $M$. Then the Lie derivative $L_X : C^\infty(M) \to C^\infty(M)$, $u \mapsto L_X u$, is a p.d.o. of order at most 1. Indeed, for any $u, f \in C^\infty(M)$ we have

$$[L_X, f]u = L_X(fu) - f(L_X u) = (L_X f)u$$

so that $[L_X, f]$ is the endomorphism $(L_X f)\bullet$.

(b) Let $E = F = \Lambda^* T^* M$. Then the exterior derivative

$$d : \Omega^*(M) \to \Omega^*(M)$$

is a p.d.o. of order at most 1. Indeed, for any $f \in C^\infty(M)$ and any $\omega \in \Omega^*(M)$ we have

$$[d, f]\omega = d(f\omega) - f(d\omega) = df \wedge \omega.$$

Thus $[d, f]$ is the endomorphism $df \wedge \bullet$ of $\Lambda^* T^* M$.

(c) Consider the Laplacian $\Delta = -\partial_x^2$ on $C^\infty(\mathbb{R})$. Then $\Delta$ is a p.d.o. of order at most 2. Indeed, for any $f \in C^\infty(\mathbb{R})$ we deduce from the Leibniz rule (1.2.1)

$$[\partial_x^2, f]\bullet = 2[\partial_x, f]\partial_x \bullet + (\partial_x^2 f) \bullet .$$

$(\partial_x^2 f)$ is the zeroth order operator defined as multiplication by $\partial_x^2 f$. The computation in part (a) shows that $[\partial_x, f]$ is the operator of multiplication by $\partial_x f$. Hence the commutator $[\partial_x^2, f]$ is the Lie derivative along the vector field $\frac{df}{dx}\partial_x$ which by part (a) is a first order p.d.o.

Suppose $L \in \mathbf{PDO}^{(k)}(E, F)$ and choose $f_1, \cdots, f_k \in C^\infty(M)$. Then

$$A_L(f_1, \cdots f_k) := [[L, f_1], \cdots, f_k] \in \mathrm{Hom}\,(E, F).$$

One can prove the following.

- $A_L(f_1, \cdots, f_k)$ is symmetric in its arguments.
- If $df_i(x_0) = dg_i(x_0)$ for all $i = 1, \cdots, k$ then

$$A_L(f_1, \cdots, f_k)\mid_{x_0} = A_L(g_1, \cdots, g_k)\mid_{x_0} .$$

Thus $A_L(f_1, \cdots, f_k)\mid_{x_0}$ depends only on the quantities $\xi_i := df_i(x_0)$ and the symmetry property shows that it is completely determined by

$$\sigma_L(\xi) := \frac{1}{k!} A_L(\xi, \cdots, \xi).$$

The quantity $\sigma_L(\cdot)$ is called the *symbol* of $L$. It is a bundle morphism

$$\sigma_L(\cdot) : \pi_k^* E \to \pi_k^* F$$

where $\pi_k : S^k T^* M \to M$ denotes the canonical projection of the $k$-th symmetric power of $T^*M$. A p.d.o. $L \in \mathbf{PDO}^{(k)}$ is said to have order $k$ if its symbol is not trivial. The set of $k$-th order operators will be denoted by $\mathbf{PDO}^k$.

**Proposition 1.2.2.** *If $L_1 \in \mathbf{PDO}^{(k_1)}(E_1, E_2)$ and $L_2 \in \mathbf{PDO}^{(k_2)}(E_2, E_3)$ then*

$$L_2 \circ L_1 \in \mathbf{PDO}^{(k_1+k_2)}(E_1, E_3)$$

*and*

$$\sigma_{L_2 \circ L_1}(\xi) = \sigma_{L_2}(\xi) \circ \sigma_{L_1}(\xi), \ \ \forall x \in M, \ \forall \xi \in T_x^* M \setminus \{0\}.$$

**Example 1.2.3.** Suppose $\nabla : C^\infty(E) \to C^\infty(T^*M \otimes E)$ is a linear connection. Then setting $\xi = df$ we deduce

$$A_\nabla(\xi)u = [\nabla, f]u = \xi \otimes u, \ \ \forall u \in C^\infty(E).$$

Thus $\sigma_L(\xi) = \xi \otimes \bullet$. Similarly, for the exterior derivative

$$d : \Omega^*(M) \to \Omega^*(M)$$

the symbol is given by

$$\sigma_d(\xi) = \xi \wedge \bullet.$$

If $\Delta := -\sum_{i=1}^N \partial_i^{\,2} : C^\infty(\mathbb{R}^N) \to C^\infty(\mathbb{R}^N)$ is the (geometers') Laplacian on $\mathbb{R}^N$ then

$$\sigma_\Delta(\xi) = -|\xi|^2 \bullet = -\Big(\sum_i |\xi_i|^2\Big) \bullet .$$

**Definition 1.2.4.** A *generalized Laplacian* on a vector bundle $E$ over a Riemannian manifold $(M, g)$ is a second order operator $L : C^\infty(E) \to C^\infty(E)$ such that

$$\sigma_L(\xi) = -|\xi|_g^2 \mathbf{1}_E.$$

**Definition 1.2.5.** If $L \in \mathbf{PDO}(E, F)$ is a p.d.o. acting between two Hermitian vector bundles then a *formal adjoint* is a p.d.o. $L^* : C^\infty(F) \to C^\infty(E)$ such that

$$\int_M (Lu, v)_F dv_g = \int_M (u, L^* v)_E dv_g$$

for all compactly supported sections $u \in C^\infty(E)$ and $v \in C^\infty(F)$.

For a proof of the following result and examples we refer to [**105**].

**Proposition 1.2.6.** *Every $k$-th order operator $L$ admits a unique formal adjoint $L^*$ which is a $k$-th order operator whose symbol is given by*

$$\sigma_{L^*}(\xi) = (-1)^k \sigma_L(\xi)^*.$$

A p.d.o. $L$ is called *formally selfadjoint* if $L = L^*$.

**Example 1.2.7.** (a) Suppose $E \to F$ is a Hermitian vector bundle over a Riemannian manifold $(M, g)$ and $\nabla$ is a Hermitian connection on $E$. Then for every vector field $X$ on $M$ the covariant derivative $\nabla_X$ is a first order p.d.o. $C^\infty(E) \to C^\infty(E)$ with formal adjoint

$$\nabla_X^* = -\nabla_X - \mathbf{div}_g(X)$$

where $\mathbf{div}_g(X)$ is the scalar defined by

$$L_X dv_g = \mathbf{div}_g(X) \cdot dv_g.$$

(b) If $E$, $\nabla$ are as above then $\nabla^*\nabla : C^\infty(E) \to C^\infty(E)$ is a generalized Laplacian called the *covariant Laplacian* determined by the connection $\nabla$.

(c) The formal adjoint of the exterior derivative

$$d : \Omega^k(M) \to \Omega^{k+1}(M)$$

is the operator

$$d^* = (-1)^{\nu_{n,k}} * d* : \Omega^{k+1}(M) \to \Omega^k(M)$$

where $n = \dim M$, $\nu_{n,k} = nk + n + 1$ and $*$ is the Hodge $*$-operator.

(d) The operator $(d + d^*)^2 = dd^* + d^*d : \Omega^*(M) \to \Omega^*(M)$ is a generalized Laplacian called the *Hodge Laplacian.*

The covariant Laplacian in the above example is in some sense the basic example of generalized Laplacian. More precisely, we have the following result. We refer to [**12**] for a different proof.

**Proposition 1.2.8.** *Suppose $L : C^\infty(E) \to C^\infty(E)$ is a formally selfadjoint generalized Laplacian. Then there exists a* Hermitian *connection on $E$ and a* symmetric *endomorphism $\mathcal{R} : E \to E$ such that*

$$L = \nabla^*\nabla + \mathcal{R}.$$

*We will refer to such a presentation of a generalized Laplacian as a* Weitzenböck presentation. *The endomorphism $\mathcal{R}$ is called the* Weitzenböck remainder *of $L$.*

**Proof** Choose an arbitrary Hermitian connection $\nabla$ on $E$. Then $L_0 = \nabla^*\nabla$ is a generalized Laplacian so that $L - L_0$ is a formally selfadjoint first order operator which can be represented as

$$L - L_0 = A \circ \nabla + B$$

where

$$A : C^\infty(T^*M \otimes E) \to C^\infty(E)$$

is a bundle morphism and $B$ is an endomorphism of $E$. We will regard $A$ as an End $(E)$-valued 1-form on $M$. Hence

$$L = \nabla^*\nabla + A \circ \nabla + B. \tag{1.2.2}$$

The connection $\nabla$ induces a connection on End$(E)$ which we continue to denote by $\nabla$:

$$\nabla : C^\infty(\mathrm{End}\,(E)) \to \Omega^1(\mathrm{End}\,(E)).$$

We define the *divergence* of $A$ by

$$\mathbf{div}_g(A) := -\nabla^* A.$$

If $(e_i)$ is a local *synchronous* frame at $x_0$ and if $A = \sum_i A_i e^i$ then, at $x_0$, we have

$$\mathbf{div}_g(A) = \sum_i \nabla_i A_i.$$

Since $L - L_0$ is formally selfadjoint we deduce

$$A_i^* = -A_i, \quad \mathbf{div}_g(A) = B^* - B.$$

We seek a Hermitian connection $\tilde{\nabla} = \nabla + C$ , $C \in \Omega^1(\mathrm{End}\,(E))$ and a symmetric endomorphism $\mathcal{R}$ of $E$ such that

$$\tilde{\nabla}^*\tilde{\nabla} + \mathcal{R} = \nabla^*\nabla + A \circ \nabla + B.$$

To determine the terms $C$ and $\mathcal{R}$ we work locally, using a *synchronous* local frame $(e_i)$ at $x_0$. Then

$$\tilde{\nabla} = \sum_i e^i \otimes (\nabla_i + C_i), \quad C_i^* = -C_i, \quad \forall i.$$

Then, as in [**105**], Example 9.1.26, we deduce that, at $x_0$,

$$\tilde{\nabla}^*\tilde{\nabla} = -\sum_i (\nabla_i + C_i)(\nabla_i + C_i)$$

$(\langle C_i \rangle^2 := C_i C_i^* = -C_i^2)$

$$= -\sum_i \nabla_i^2 - \sum_i \nabla_i C_i - 2\sum_i C_i \nabla_i + \sum_i \langle C_i \rangle^2$$

$(\langle C \rangle^2 = \sum_i \langle C_i \rangle^2)$

$$= \nabla^*\nabla - 2C \circ \nabla - \mathbf{div}_g(C) + \langle C \rangle^2 = \nabla^*\nabla + A \circ \nabla + B - \mathcal{R}.$$

We deduce immediately that

$$C = -\frac{1}{2}A, \quad \mathcal{R} = B - \frac{1}{2}\mathbf{div}_g(A) - \langle C \rangle^2 = \frac{1}{2}(B + B^*) - \frac{1}{4}\langle A \rangle^2. \tag{1.2.3}$$

The proposition is proved. ■

Besides their nice algebraic properties, the generalized Laplacians enjoy many nice analytic features. They all derive from the *ellipticity* of these operators.

**Definition 1.2.9.** Let $E, F \to M$ be two smooth vector bundles over the smooth manifold $M$. A p.d.o. $T \in \mathbf{PDO}^k(E, F)$ is said to be *elliptic* if for any $x \in M$ and any $\xi \in T_x^*M \setminus \{0\}$ the linear map $\sigma_T(\xi) : E_x \to F_x$ is an isomorphism.

Clearly, the generalized Laplacians are elliptic second order operators. The operator $d + d^*$ of Example 1.2.7 (d) is elliptic because $(d + d^*)^2$ is a generalized Laplacian. This feature is so frequently encountered in geometry that it was given a name.

**Definition 1.2.10.** A *Dirac operator* is a first order operator $D : C^\infty(E) \to C^\infty(F)$ such that $D^2$ is a generalized Laplacian.

Frequently, the Dirac operators are obtained from an operator $D \in \mathbf{PDO}^1(E, F)$ such that both $D^*D$ and $DD^*$ are generalized Laplacians. Then

$$\tilde{D} := \begin{bmatrix} 0 & D^* \\ D & 0 \end{bmatrix} : C^\infty(E \oplus F) \to C^\infty(E \oplus F)$$

is a Dirac operator.

To discuss the basic analytic properties of elliptic operators we need to introduce a suitable analytical framework. For geometric applications the Sobolev and Hölder spaces provide such a framework.

To define these spaces we need two things: an oriented Riemannian manifold $(M, g)$ and a $\mathbb{K}$-vector bundle $\pi : E \to M$ endowed with a metric $h = \langle \cdot, \cdot \rangle$ and a connection $\nabla = \nabla^E$ compatible with $h$. The metric $g = (\cdot, \cdot)$ defines two important objects:

(i) the Levi-Civita connection $\nabla^g$ and

(ii) a volume form $dv_g = *1$. In particular, $dv_g$ defines a Borel measure on $M$. We denote by $L^p(M, \mathbb{K})$ the space of $\mathbb{K}$-valued $p$-integrable functions on $(M, dv_g)$ (modulo the equivalence relation of equality almost everywhere).

**Definition 1.2.11.** Let $p \in [1, \infty]$. An $L^p$-section of $E$ is a Lebesgue measurable map $\psi : M \to E$ (i.e. $\psi^{-1}(U)$ is Lebesgue measurable for any open subset $U \subset E$) such that:

(i) $\pi \circ \psi(x) = x$ for almost all $x \in M$ except possibly a negligible set.

(ii) The function $x \mapsto |\psi(x)|_h$ belongs to $L^p(M, \mathbb{R})$.

The space of $L^p$-sections of $E$ (modulo equality almost everywhere) is denoted by $L^p(E)$. The space $L^p_{loc}(E)$ consists of measurable sections $u$ of $E$ such that $u\varphi \in L^p(E)$ for any smooth, compactly supported function on $M$.

**Proposition 1.2.12.** *$L^p(E)$ is a Banach space with respect to the norm*

$$\|\psi\|_{p,E} = \begin{cases} \left(\int_M |\psi(x)|^p dv_g(x)\right)^{1/p} & if\ p < \infty \\ \operatorname{ess\,sup}_x |\psi(x)| & if\ p = \infty \end{cases}.$$

For each $m = 1, 2, \cdots$ define $\nabla^m$ as the composition

$$\nabla^m : C^\infty(E) \xrightarrow{\nabla^E} C^\infty(T^*M \otimes E) \xrightarrow{\nabla^{T^*M\otimes E}} \cdots \xrightarrow{\nabla} C^\infty(T^*M^{\otimes m} \otimes E)$$

where we used the symbol $\nabla$ to generically denote the connections in the tensor products $T^*M^{\otimes j} \otimes E$ induced by $\nabla^g$ and $\nabla^E$.

The metrics $g$ and $h$ induce metrics in each of the tensor bundles $T^*M^{\otimes m} \otimes E$, and in particular, we can define the spaces $L^p(T^*M^{\otimes m} \otimes E)$.

**Definition 1.2.13.** (a) Let $u \in L^1_{loc}(E)$ and $v \in L^1_{loc}(T^*M^{\otimes m} \otimes E)$. We say that $\nabla^m u = v$ weakly if

$$\int_M \langle v, \phi\rangle dv_g = \int \langle u, (\nabla^m)^*\phi\rangle dv_g, \quad \forall u \in C_0^\infty(T^*M^{\otimes m} \otimes E).$$

(b) Define $L^{m,p}(E)$ as the space of sections $u \in L^p(E)$ such that $\forall j = 1, \ldots, m$ there exist $v_j \in L^p(T^*M^{\otimes j} \otimes E)$ such that $\nabla^j u = v_j$ weakly. We define the Sobolev norm $\|\cdot\|_{m,p}$ by

$$\|u\|_{m,p} = \|u\|_{m,p,E} = \sum_{j=1}^{p} \|\nabla^j u\|_p.$$

**Proposition 1.2.14.** *$(L^{k,p}(E), \|\cdot\|_{k,p,E})$ is a Banach space which is reflexive if $1 < p < \infty$.*

**Exercise 1.2.1. (Kato's Inequalities)** Suppose $E \to M$ is a Hermitian vector bundle over an oriented Riemannian manifold $(M, g)$. Fix a Hermitian connection $\nabla$ on $E$.

(a) Show that for every $u \in L^{1,2}_{loc}(E)$ the function

$$M \ni x \mapsto |u(x)|$$

is in $L^{1,2}_{loc}(M)$ and moreover

$$|d|u(x)|| \le |(\nabla u)(x)|$$

almost everywhere on $M$.

(b) Set $\Delta_E := \nabla^*\nabla$ and denote by $\Delta_M$ the Laplacian on $M$. Show that for all $u \in L^{2,2}_{loc}(E)$ we have

$$\Delta_M(|u|^2) = 2\mathfrak{Re}\langle \Delta_E u, u\rangle - 2|\nabla u|^2$$

so that

$$\Delta_M|u|^2 \le 2\mathfrak{Re}\langle \Delta_E u, u\rangle$$

almost everywhere on $M$.

The Hölder spaces can be defined on manifolds as well. If $(M, g)$ is a Riemannian manifold then $g$ canonically defines a metric space structure on $M$ and, in particular, we can talk about the oscillation of a function $u : M \to \mathbb{K}$. On the other hand, defining the oscillation of a section of some bundle over $M$ requires a little more work.

Let $(E, h, \nabla)$ as before. We assume the *injectivity radius* $\rho_M$ *of* $M$ *is positive.* Set $\rho_0 = \min\{1, \rho_M\}$. If $x, y \in M$ are two points such that $\mathrm{dist}_g(x, y) \leq \rho_0$ then they can be joined by a unique minimal geodesic $\gamma_{x,y}$ starting at $x$ and ending at $y$. We denote by $T_{x,y} : E_y \to E_x$ the $\nabla^E$-parallel transport along $\gamma_{x,y}$. For each $\xi \in E_x$ and $\eta \in E_y$ we set by definition

$$|\xi - \eta| = |\xi - T_{x,y}\eta|_x = |\eta - T_{y,x}\xi|_y.$$

If $u : M \to E$ is a section of $E$ and $S \subset M$ has diameter $< \rho_0$ we define

$$\mathrm{osc}\,(u\ ;\ S) = \sup\{|u(x) - u(y)|\ ;\ x,\ y \in S\}.$$

Finally set

$$[u]_{\alpha,E} = \sup\{r^{-\alpha}\mathrm{osc}\,(u\ ;\ B_r(x))\ ;\ 0 < r < \rho_0,\ x \in M\}.$$

For any $k \geq 0$ define the *Hölder norm*

$$\|u\|_{k,\alpha,E} = \sum_{j=0}^{k} \|\nabla^j u\|_{\infty,E} + [\nabla^m u]_{\alpha,T^*M^{\otimes m}\otimes E}$$

and set

$$C^{k,\alpha}(E) = \{u \in C^k(E)\ ;\ \|u\|_{k,\alpha} < \infty\}.$$

**Theorem 1.2.15.** *Let $(M, g)$ be a* **compact**, *$N$-dimensional, oriented Riemannian manifold and $E$ a vector bundle over $M$ equipped with a metric $h$ and compatible connection $\nabla$. Then the following are true.*

*(a) The Sobolev space $L^{m,p}(E)$ and the Hölder spaces $C^{k,\alpha}(E)$ do not depend on the metrics $g$, $h$ and on the connection $\nabla$. More precisely, if $g_1$ is a different metric on $M$ and $\nabla^1$ is another connection on $E$ compatible with some metric $h_1$ then*

$$L^{m,p}(E, g, h, \nabla) = L^{m,p}(E, g_1, h_1, \nabla^1) \ \ \textit{as sets of sections}$$

*and the identity map between these two spaces is a Banach space isomorphism. A similar statement is true for the Hölder spaces.*

*(b) If $1 \leq p < \infty$ then $C^\infty(E)$ is dense in $L^{k,p}(E)$.*

*(c) (Sobolev) If $(k_i, p_i) \in \mathbb{Z}_+ \times [1, \infty)$ $(i = 0, 1)$ are such that*

$$k_0 \geq k_1 \ \textit{and}\ \sigma_N(k_0, p_0) = k_0 - N/p_0 \geq k_1 - N/p_1 = \sigma_N(k_1, p_1)$$

*then $L^{k_0,p_0}(E)$ embeds continuously in $L^{k_1,p_1}(E)$. If moreover*

$$k_0 > k_1 \ \textit{and}\ k_0 - N/p_0 > k_1 - N/p_1$$

*then the embedding* $L^{k_0,p_0}(E) \hookrightarrow L^{k_1,p_1}(E)$ *is compact, i.e. any bounded sequence of* $L^{k_0,p_0}(E)$ *admits a subsequence convergent in the* $L^{k_1,p_1}$*-norm.*

*(d) (Morrey) If* $(m,p) \in \mathbb{Z}_+ \times [1,\infty)$ *and* $(k,\alpha) \in \mathbb{Z}_+ \times (0,1)$ *and*

$$m - N/p \geq k + \alpha$$

*then* $L^{m,p}(E)$ *embeds continuously in* $C^{k,\alpha}(E)$*. If moreover*

$$m - N/p > k + \alpha$$

*then the embedding is also compact.*

The proofs of all the above results can be found in [**105**].

Suppose now that $L$ : $\mathbf{PDO}^k(E,F)$ is a $k$-th order elliptic operator over an oriented Riemannian manifold $(M,g)$. Let $v \in L^p_{loc}(F)$. A *weak* $L^p$*-solution* of the equation

$$Lu = v$$

is a section $u \in L^p_{loc}(E)$ such that for any smooth, compactly supported section $\varphi$ of $F$ the following holds

$$\int_M \langle v, \varphi\rangle_F dv_g = \int_M \langle u, L^*\varphi\rangle_E dv_g.$$

The following result describes the fundamental property of elliptic operators. For simplicity we state it only in the special case when $M$ *is compact.* We refer to [**105**] and the references therein for proofs of more general statements.

**Theorem 1.2.16.** *Suppose* $M$ *is a compact, oriented Riemannian manifold without boundary.*

*(a) Let* $p \in (1,\infty)$ *and* $m \in \mathbb{Z}_+$*. Then there exists a constant*

$$C = C(L,m,p,g,E,F) > 0$$

*such that if* $u$ *is a weak* $L^p$*-solution* $u$ *of* $Lu = v$, $v \in L^{m,p}(F)$ *then*

$$u \in L^{m+k,p}(E)$$

*and*

$$\|u\|_{m+k,p;E} \leq C(\|u\|_{p;E} + \|v\|_{m,p;F}).$$

*(b) Let* $\alpha \in (0,1)$ *and* $m \in \mathbb{Z}_+$*. Then there exists a constant*

$$C = C(L,m,\alpha,g,E,F) > 0$$

*such that if* $u$ *is a weak* $L^p$*-solution* $u$ *of* $Lu = v$, $v \in C^{m,\alpha}(F)$ *then*

$$u \in C^{m+k,\alpha}(E)$$

*and*

$$\|u\|_{m+k,\alpha;E} \leq C(\|u\|_{0,\alpha;E} + \|v\|_{m,\alpha;F}).$$

The above result has a famous corollary.

**Corollary 1.2.17. (Weyl's Lemma)** *Let $L$ be as above. If*

$$Lu \in C^\infty(F)$$

*weakly then $u \in C^\infty(E)$.*

From the a priori inequalities in the above theorem one can deduce the following important result.

**Theorem 1.2.18.** *Suppose $M$ is a compact, oriented Riemannian manifold, $E_0$, $E_1$ are Hermitian vector bundles over $M$ and*

$$L : C^\infty(E_0) \to C^\infty(E_1)$$

*is a $k$-th order elliptic operator. We define the* analytical realization of $L$ *as the* unbounded *linear operator*

$$L_a : L^2(E_0) \to L^2(E_1)$$

*with domain* $\mathrm{Dom}\,(L_a) := L^{k,2}(E_0)$ *and acting according to*

$$L^{k,2}(E_0) \ni u \mapsto Lu \in L^2(E_1).$$

*Then the following hold.*

*(i) $L_a$ is a closed operator, i.e. its graph is a closed subspace of $L^2(E_0) \times L^2(E_1)$.*

*(ii) The functional adjoint of $L_a$ (i.e. the adjoint as a closed linear operator acting between Hilbert spaces) coincides with the analytical realization of the formal adjoint $L^*$, i.e.*

$$(L_a)^* = (L^*)_a.$$

*(iii) The ranges of both $L_a$ and $(L^*)_a$ are closed subspaces in $L^2(E_1)$, respectively $L^2(E_0)$. Moreover* $\ker L_a \in C^\infty(E_0)$, $\ker L_a^* \in C^\infty(E_1)$ *and*

$$\mathrm{Range}\,(L_a) = (\ker L_a^*)^\perp, \;\; \mathrm{Range}\,(L_a^*) = (\ker L_a)^\perp.$$

*(iv) The kernels of both $L_a$ and $(L^*)_a$ are finite dimensional.*

*(v) Denote by $P : L^2(E_0) \to L^2(E_0)$ the orthogonal projection onto* $\ker L_a$. *Then for every $1 < p < \infty$ and every $m \in \mathbb{Z}_+$ there exists a constant $C = C(m,p,L > 0)$ such that*

$$\|u - Pu\|_{k+m,p} \leq C\|Lu\|_{m,p}, \;\; \forall u \in L^{k+m,p}(E_0).$$

The properties (iii) and (iv) in the above theorem are succinctly referred to as the *Fredholm property* of elliptic operators on compact manifolds. The quantity

$$\dim_{\mathbf{F}} \ker L_a - \dim_{\mathbf{F}} \ker L_a^*$$

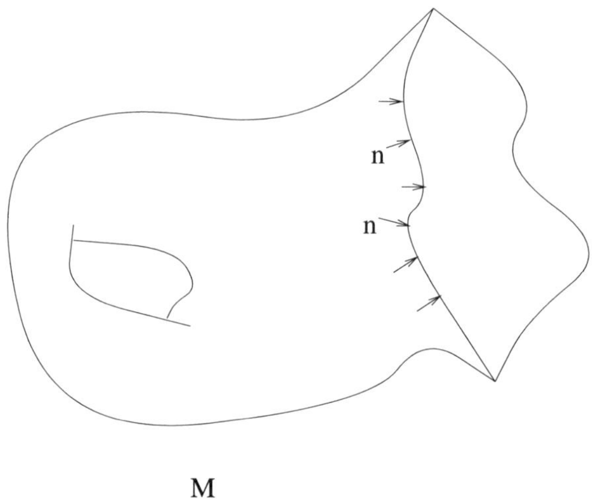


**Figure 1.1.** *Riemannian manifold with boundary*

($\mathbf{F}$= $\mathbb{R}$, $\mathbb{C}$) is called the $\mathbf{F}$-*Fredholm index* of $L$ and is denoted by $\mathrm{ind}_{\mathbf{F}}L$.

The Fredholm index of an elliptic operator $L$ is remarkably stable under deformations. For example, one can show (see [**105**]) that it depends only on the symbol of $L$.

We conclude this section with an exercise which describes the Green formulæ for various p.d.o.'s. These are more sophisticated versions of the usual integration-by-parts trick.

**Exercise 1.2.2.** Consider a compact Riemannian manifold $(M, g)$ with boundary $\partial M$. Denote by $\vec{n}$ the unit outer normal along $\partial M$ (see Figure 1.1). Let $E, F \to M$ be Hermitian vector bundles over $M$ and suppose $L \in \mathbf{PDO}^k(E, F)$. Set $g_0 = g\,|_{\partial M}$, $E_0 = E\,|_{\partial M}$ and $F_0 = F\,|_{\partial M}$. The Green formula states that there exists a sesquilinear map

$$B_L : C^\infty(E) \times C^\infty(F) \to C^\infty(\partial M)$$

such that

$$\int_M \langle Lu, v\rangle dv(g) = \int_{\partial M} B_L(u, v) dv(g_0) + \int_M \langle u, L^* v\rangle dv(g).$$

Prove the following.

(a) If $L$ is a zeroth order operator (i.e. $L$ is a bundle morphism) then $B_L = 0$.

(b) If $L_1 \in \mathbf{PDO}(F, G)$ and $L_2 \in \mathbf{PDO}(E, F)$ then

$$B_{L_1 L_2}(u, v) = B_{L_1}(L_2 u, v) + B_{L_2}(u, L_1^* v).$$

(c)

$$B_{L^*}(v, u) = -\overline{B_L(u, v)}.$$

(d) Suppose $\nabla$ is a Hermitian connection on $E$ and $X \in \mathrm{Vect}(M)$. Set $L = \nabla_X : C^\infty(E) \to C^\infty(E)$. Then

$$B_L(u, v) = \langle u, v \rangle g(X, \vec{n}).$$

(e) Let $L = \nabla : C^\infty(E) \to C^\infty(T^*M \otimes E)$. Then

$$B_L(u, v) = \langle u, i_{\vec{n}} v \rangle_E$$

where $i_{\vec{n}}$ denotes the contraction by $\vec{n}$.

(f) Denote by $\vec{\nu}$ the section of $T^*M\mid_{\partial M}$ $g$-dual to $\vec{n}$. Suppose $L$ is a first order p.d.o. and set $J := \sigma_L(\vec{\nu})$. Then

$$B_L(u, v) = \langle Ju, v \rangle_F.$$

(g) Using (a) – (f) show that for all $u \in C^\infty(E)$, $v \in C^\infty(F)$ and any $x_0 \in \partial M$ the quantity $B_L(u, v)(x_0)$ depends only on the jets of $u, v$ at $x_0$ of order at most $k - 1$.

## 1.3. Clifford algebras and Dirac operators

**1.3.1. Clifford algebras and their representations.** Suppose $E \to M$ is a smooth, Hermitian vector bundle over a Riemannian manifold $(M, g)$ and $D : C^\infty(E) \to C^\infty(E)$ is a Dirac operator, i.e. $D^2$ is a generalized Laplacian. Denote by $\mathbf{c}$ the symbol of $D$. It has the remarkable property that

$$\mathbf{c}(\xi)^2 = -|\xi|_g^2 \mathbf{1}_{E_x}, \quad \forall x \in M, \ \forall \xi \in T_x^*M.$$

If we set $V_x := T_x^*M$ then the above identity implies that for every $x \in M$ we have a linear map

$$\mathbf{c} : V_x \to \mathrm{End}\,(E_x) \tag{1.3.1}$$

with the property

$$\{\mathbf{c}(u), \mathbf{c}(v)\} = -2g(u, v)\mathbf{1}, \quad \forall u, v \in V \tag{1.3.2}$$

where $\{A, B\}$ denotes the anticommutator $AB + BA$ of two elements $A, B$ in an associative algebra.

Now, given a Euclidean vector space $(V, g)$, we denote by $\mathsf{Cl}(V) := \mathsf{Cl}(V, g)$ the associative $\mathbb{R}$-algebra with $\mathbf{1}$ generated by $V$ subject to the relations (1.3.2). It is not difficult to prove the existence and uniqueness of such an algebra. It will be called the *Clifford algebra* associated to the Euclidean space $(V, g)$. We see that the map in (1.3.1) extends to a representation

$$\mathbf{c} : \mathsf{Cl}(V) \to \mathrm{End}\,(V)$$

of the Clifford algebra called the *Clifford multiplication*. The maps in (1.3.1) can be assembled together to form a bundle morphism

$$\mathbf{c} : T^*M \to \mathrm{End}\,(E)$$

such that

$$\{\mathbf{c}(\alpha), \mathbf{c}(\beta)\} = -2g(\alpha, \beta)\mathbf{1}_E, \ \ \forall \alpha, \beta \in \Omega^1(M).$$

A map $\mathbf{c}$ as above will be called a *Clifford structure* on the bundle $E$. Thus the existence of a Dirac operator implies the existence of a Clifford structure. Conversely, if $\nabla$ is any connection on a bundle $E$ equipped with a Clifford structure $\mathbf{c}$ then the composition

$$C^\infty(E) \stackrel{\nabla}{\to} C^\infty(T^*M \otimes E) \stackrel{\mathbf{c}}{\to} C^\infty(E)$$

is a Dirac operator. Thus the existence of a Dirac operator is equivalent to an algebraic-topological problem, that of the existence of Clifford structures. We will be interested in a structure finer than Clifford.

**Definition 1.3.1.** Suppose $(M, g)$ is a Riemannian manifold. A *Dirac structure* on $M$ is a quadruple $(E, \mathbf{c}, \nabla^E, \nabla^M)$ where $E$ is a Hermitian vector bundle, $\mathbf{c} : T^*M \to \mathrm{End}\,(E)$ is a *selfadjoint* Clifford structure, i.e.

$$\mathbf{c}(\alpha)^* = -\mathbf{c}(\alpha), \ \ \forall \alpha \in \Omega^1(M), \tag{1.3.3}$$

$\nabla^M$ is a connection on $T^*M$ compatible with the Riemannian metric and $\nabla^E$ is a Hermitian connection on $E$ compatible with the Clifford multiplication, i.e.

$$\nabla^E_X(\,\mathbf{c}(\alpha)u) = \mathbf{c}(\nabla^M_X \alpha)u + \mathbf{c}(\alpha)\nabla^E_X u, \tag{1.3.4}$$

$\forall u \in C^\infty(E),\ \alpha \in \Omega^1(M),\ X \in \mathrm{Vect}\,(M)$.

When $\nabla^M$ is the Levi-Civita connection we say that $(E, \mathbf{c}, \nabla^E)$ is a *geometric Dirac structure* on $M$. The Dirac operator associated to a geometric Dirac structure will be called a *geometric Dirac operator*.

**Proposition 1.3.2. (Weitzenböck formula for geometric Dirac operators)** *If $D$ is a geometric Dirac operator associated to the geometric Dirac structure $(E, \mathbf{c}, \nabla^E)$ then $D = D^*$ and*

$$D^2 = (\nabla^E)^*\nabla^E + \mathbf{c}(F(\nabla^E)).$$

*The last term should be understood as follows. The curvature $F(\nabla^E)$ is an $\mathrm{End}\,(E)$-valued 2-form. Locally it is a $C^\infty(M)$-linear combination of terms of the form $\omega \otimes T$, $\omega \in \Omega^2(M)$ and $T \in \mathrm{End}\,(E)$. Then $\mathbf{c}(\omega \otimes T)$ is the endomorphism $\mathbf{c}(\omega) \cdot T$.*

**Exercise 1.3.1.** Prove the above proposition.

To describe the Dirac structures on a given manifold $M$ we need a better understanding of the representation theory of the Clifford algebra associated to a Euclidean space $(V, g)$. If $\dim V = n$ and $\{e_1, \cdots, e_n\}$ is an orthonormal basis of $V$ then the monomials

$$e_{i_1} \cdots e_{i_k}, \;\; 1 \leq i_1 < \cdots < i_k \leq n, \;\; e_{\emptyset} := \mathbf{1}$$

form a basis of $\mathsf{Cl}(V)$. Thus $\dim \mathsf{Cl}(V) = 2^{\dim V}$. Since the only invariant of a Euclidean space is its dimension we will often use the notation $\mathsf{Cl}_n$ to denote the Clifford algebra associated to an $n$-dimensional Euclidean space.

Note first there is a natural representation

$$T : \mathsf{Cl}\,(V) \to \operatorname{End} \Lambda^* V$$

induced by the correspondence

$$V \ni v \mapsto T_v := e(v) - i_{v^*}$$

where $e(v)$ denotes the (left) exterior multiplication by $v$ while $i_{v^*}$ denotes the contraction by the co-vector $v^* \in V^*$, the metric dual of $v$. The Cartan identity

$$\{e(v), i_{v^*}\} = |v|^2$$

shows that the above correspondence does indeed extend to a representation of the Clifford algebra. The *symbol map*

$$\sigma : \mathsf{Cl}(V) \to \Lambda^* V$$

is defined by

$$\sigma(\omega) := T_\omega \cdot \mathbf{1}, \;\; \forall \omega \in \Lambda^* V.$$

For example, if $\{e_1, \cdots, e_n\}$ is an orthonormal basis of $V$ then

$$\sigma(e_{i_1} \cdots e_{i_k}) = e_{i_1} \wedge \cdots \wedge e_{i_k}.$$

We see that the symbol map is a bijection. Its inverse is called the *quantization map* and is denoted by $\mathfrak{q}$. Set

$$\mathsf{Cl}^{\pm}(V) := \mathfrak{q}(\Lambda^{even/odd} V).$$

The splitting $\mathsf{Cl}(V) = \mathsf{Cl}^+(V) \oplus \mathsf{Cl}^-(V)$ makes $\mathsf{Cl}(V)$ a *superalgebra*, i.e.

$$\mathsf{Cl}^{\pm}(V) \cdot \mathsf{Cl}^{\pm}(V) \subset \mathsf{Cl}^+(V), \;\; \mathsf{Cl}^{\pm}(V) \cdot \mathsf{Cl}^{\mp}(V) = \mathsf{Cl}^-(V).$$

Given $x \in \mathsf{Cl}(V)$ we denote by $x_{\pm}$ its even (odd) components, $x = x_+ + x_-$.

To understand the complex representations of $\mathsf{Cl}(V)$ we need to distinguish two cases.

**A.** $\dim V$ *is even*, $\dim V = 2n$.

**FUNDAMENTAL FACT** *There exist a $\mathbb{Z}_2$-graded complex vector space $\mathbb{S}(V) = \mathbb{S}_{2n} = \mathbb{S}_{2n}^+ \oplus \mathbb{S}_{2n}^-$ and a $\mathbb{C}$-linear isomorphism*

$$c : \mathsf{Cl}(V) \otimes \mathbb{C} \to \operatorname{End}(\mathbb{S}_{2n})$$

*with the following properties.*

*(a)* $\dim_{\mathbf{C}} \mathbb{S}_{2n}^+ = \dim_{\mathbf{C}} \mathbb{S}_{2n}^- = 2^{n-1}$.

*(b)*

$$c(\mathsf{Cl}^+(V) \otimes \mathbb{C}) = \operatorname{End}(\mathbb{S}_{2n}^+) \oplus \operatorname{End}(\mathbb{S}_{2n}^-).$$

$$c(\mathsf{Cl}^-(V) \otimes \mathbb{C}) = \operatorname{Hom}(\mathbb{S}_{2n}^+, \mathbb{S}_{2n}^-) \oplus \operatorname{Hom}(\mathbb{S}_{2n}^-, \mathbb{S}_{2n}^+).$$

*The above pair $(\mathbb{S}_{2n}, c)$ is unique up to isomorphism and is called the* complex spinor representation *of* $\mathsf{Cl}(V)$.

**Sketch of proof** We will produce an explicit realization of the pair $(\mathbb{S}_{2n}, c)$ using an additional structure on $V$.

Fix a complex structure on $V$ compatible with the metric. This is a linear map $J : V \to V$ such that

$$J^2 = -1, \quad J^* = -J.$$

Then $V \otimes \mathbb{C}$ splits as

$$V \otimes \mathbb{C} = V^{1,0} \oplus V^{0,1}$$

where $V^{1,0} = \ker(\mathbf{i} - J)$ and $V^{0,1} = \ker(\mathbf{i} + J)$. Set

$$\mathbb{S}(V) := \Lambda^{*,0} V = \Lambda^* V^{1,0}.$$

Note that the Euclidean metric on $V$ induces Hermitian metrics on $\Lambda^{p,q} V$ and thus a Hermitian metric on $\mathbb{S}(V)$.

A morphism $\mathsf{Cl}(V) \to \operatorname{End}(\mathbb{S}(V))$ is uniquely defined by its action on $V^{1,0}$ and $V^{0,1}$. For $v \in V^{1,0}$ define $c(v) := \sqrt{2} e(v)$, i.e.

$$c(v)(u_1 \wedge \cdots \wedge u_k) = \sqrt{2} v \wedge u_1 \wedge \cdots \wedge u_k.$$

Any $\bar{v} \in V^{0,1}$ can be identified (via the metric $g$) as a complex linear functional on $V^{1,0}$. Define $c(\bar{v}) = -\sqrt{2} i(\bar{v})$, i.e.

$$c(\bar{v})(u_1 \wedge \cdots \wedge u_k) = \sqrt{2} \sum_{j=1}^{k} (-1)^j g_c(u_k, \bar{v}) u_1 \wedge \cdots \wedge \hat{u}_j \wedge \cdots \wedge u_k.$$

Above, $g_c(\cdot, \cdot)$ denotes the extension by *complex bilinearity* of $g$ to $V \otimes \mathbb{C}$. Now choose an orthonormal basis $(e_1, f_1; \cdots ; e_n, f_n)$ of $V$ such that $f_j = J e_j$, $\forall j$ and then set

$$\varepsilon_j := \frac{1}{\sqrt{2}}(e_j - \mathbf{i} f_j), \quad \bar{\varepsilon}_j = \frac{1}{\sqrt{2}}(e_j + \mathbf{i} f_j).$$

Then $(\varepsilon_j)$ is a unitary basis of $V^{1,0}$, $(\bar{\varepsilon}_j)$ is a unitary basis of $V^{0,1}$ and $\langle \varepsilon_i, \bar{\varepsilon}_j \rangle = \delta_{ij}$. We deduce

$$c(e_j) = e(\varepsilon_j) - i(\bar{\varepsilon}_j), \;\; c(f_k) = \mathbf{i}(e(\varepsilon_k) + i(\bar{\varepsilon}_k)).$$

One can now check that $c$ induces a map with all the required properties. In this case

$$\mathbb{S}^+(V) = \Lambda^{even,0}V, \;\; \mathbb{S}^-(V) = \Lambda^{odd,0}V. \;\blacksquare$$

**Example 1.3.3.** Suppose that $V$ is the four-dimensional Euclidean space $\mathbb{R}^4$ with coordinates $(x^1, y^1, x^2, y^2)$. Set $e_i = \frac{\partial}{\partial x_i}$ and $f_j = \frac{\partial}{\partial y_j}$ and define $J$ by $Je_i = f_i$. Set $z^j = x^j + \mathbf{i}y^j$. We identify $V^{1,0}$ with $(V^*)^{0,1}$ so that

$$\varepsilon_i = \frac{1}{\sqrt{2}}d\bar{z}^i, \;\; \bar{\varepsilon}_i = \frac{1}{\sqrt{2}}dz^i.$$

Then

$$\mathbb{S}_4 \cong \Lambda^{0,*}V^* \cong \mathbb{C} \oplus \Lambda^{0,1}V^* \oplus \Lambda^{0,2}V^*$$

and

$$\mathbb{S}_4^+ \cong \mathbb{C} \oplus \Lambda^{0,2}V^*, \;\; \mathbb{S}_4^- \cong \Lambda^{0,1}V^*.$$

Define $\omega = dx^1 \wedge dy^1 + dx^2 \wedge dy^2$ and orient $V^*$ using $\omega \wedge \omega$. Denote by $*$ the Hodge $*$ operator on $V^*$ defined by the metric $g$ and the above orientation. Note that

$$*(\Lambda^2 V^*) \subset \Lambda^2 V^*$$

and $*^2 = 1$ on $\Lambda^2 V^*$. Thus we have a splitting

$$\Lambda^2 V^* = \Lambda^2_+ V^* \oplus \Lambda^2_- V^*$$

where $\Lambda^2_\pm = \ker(1 \mp *)$. The above choice of basis defines a nice orthonormal basis of $\Lambda^2_+$, $\{\eta_0, \eta_1, \eta_2\}$ where

$$\eta_0 = \frac{1}{\sqrt{2}}\omega = \frac{\mathbf{i}}{2\sqrt{2}}(dz^1 \wedge d\bar{z}^1 + dz^2 \wedge d\bar{z}^2),$$

$$\eta_1 = \frac{1}{\sqrt{2}}(dx^1 \wedge dx^2 - dy^1 \wedge dy^2) = \frac{1}{2\sqrt{2}}(d\bar{z}^1 \wedge d\bar{z}^2 + dz^1 \wedge dz^2),$$

$$\eta_2 = \frac{1}{\sqrt{2}}(dx^1 \wedge dy^2 + dy^1 \wedge dx^2) = -\frac{\mathbf{i}}{2\sqrt{2}}(d\bar{z}^1 \wedge d\bar{z}^2 - dz^1 \wedge dz^2).$$

We see that $\eta_0$, $dz^1 \wedge dz^2$ and $d\bar{z}^1 \wedge d\bar{z}^2$ form a complex basis of $\Lambda^2_+ V^* \otimes \mathbb{C}$.

The metric isomorphism $V \cong V^*$ defines an action of $\Lambda^* V^* \otimes \mathbb{C}$ on $\mathbb{S}(V) = \Lambda^{0,*}(V)$ generated by

$$c(dz^j) = -\sqrt{2}i(dz^j), \;\; c(d\bar{z}^j) = \sqrt{2}e(d\bar{z}^j)$$

where $i(dz^j)(d\bar{z}^k) = 2\delta_{jk}$. Since $d\bar{z}^1$ and $d\bar{z}^2$ are orthogonal we deduce

$$c(d\bar{z}^1 \wedge d\bar{z}^2) = c(d\bar{z}^1)c(d\bar{z}^2) = 2e(d\bar{z}^1)e(d\bar{z}^2)$$

and

$$c(dz^1 \wedge dz^2) = 2i(dz^1)i(dz^2).$$

To determine the action of $\eta_0$ we use the *real* description

$$c(\eta_0) = \frac{1}{\sqrt{2}}\Big\{c(dx^1)c(dy^1) + c(dx^2)c(dy^2)\Big\}$$

$$= \frac{\mathbf{i}}{4\sqrt{2}}\Bigg\{ \Big(c(d\bar{z}^1) + c(dz^1)\Big)\Big(c(d\bar{z}^1) - c(dz^1)\Big)$$

$$+ \Big(c(d\bar{z}^2 + c(dz^2)\Big)\Big(c(d\bar{z}^2) - c(dz^2)\Big) \Bigg\}$$

$$= \frac{\mathbf{i}}{2\sqrt{2}}\Bigg\{\Big(e(d\bar{z}^1) - i(dz^1)\Big)\Big(e(d\bar{z}^1) + i(dz^1)\Big)$$

$$+\Big(e(d\bar{z}^2) - i(dz^2)\Big)\Big(e(d\bar{z}^2) + i(dz^2)\Big)\Bigg\}.$$

Now it is not difficult to see that $c(\eta_i)d\bar{z}^j = 0$, $\forall i = 0, 1, 2$ so that

$$c(\Lambda^2_+ V^*) \subset \operatorname{End}(\mathbb{S}^+(V)).$$

With respect to the *unitary* basis $1$, $\varepsilon_1 \wedge \varepsilon_2 = \frac{1}{2} d\bar{z}^1 \wedge d\bar{z}^2$ of $\mathbb{S}^+(V)$ we have the following matrix descriptions:

$$c(\omega) = \sqrt{2}c(\eta_0) = 2\begin{bmatrix} -\mathbf{i} & 0 \\ 0 & \mathbf{i} \end{bmatrix},$$

$$c(\varepsilon_1 \wedge \varepsilon_2) = \frac{1}{2}c(d\bar{z}^1 \wedge d\bar{z}^2) = e(d\bar{z}^1)e(d\bar{z}^2) = 2\begin{bmatrix} 0 & 0 \\ 1 & 0 \end{bmatrix},$$

$$c(\bar{\varepsilon}_1 \wedge \bar{\varepsilon}_2) = \frac{1}{2}c(dz^1 \wedge dz^2) = -c(d\bar{z}^1 \wedge d\bar{z}^2)^* = 2\begin{bmatrix} 0 & -1 \\ 0 & 0 \end{bmatrix}.$$

Note that for any *real* form $\varphi \in \Lambda^2_+ V^*$ the Clifford multiplication $c(\varphi)$ is a traceless, skew-symmetric endomorphism of $\mathbb{S}^+(V)$.

There is a quadratic map $q : \mathbb{S}^+_4 \to \operatorname{End}(\mathbb{S}^+_4)$ defined by

$$q(\psi) := \bar{\psi} \otimes \psi - \frac{1}{2}|\psi|^2\mathbf{id}$$

i.e. $q(\psi) : \phi \mapsto \langle \phi, \psi\rangle\psi - \frac{1}{2}|\psi|^2\phi$. (The Hermitian metric is *complex linear in the first argument and complex antilinear in the second.*) Using the *bra-ket* notation of quantum mechanics in which we think of the spinors in $\mathbb{S}_4^+$ as bra-vectors then

$$q(\langle\psi|) = |\psi\rangle\langle\psi| - \frac{1}{2}\langle\psi|\psi\rangle.$$

We can decompose $\psi \in \mathbb{S}^+(V)$ as

$$\psi = \alpha \oplus \beta, \quad \alpha \in \Lambda^{0,0}V^*, \ \beta \in \Lambda^{0,2}V^*.$$

With respect to the basis $\{1, \varepsilon_1 \wedge \varepsilon_2\}$ of $\mathbb{S}^+(V)$ the endomorphism $q(\psi)$ has the matrix description

$$q(\psi) = \left[\begin{array}{cc} \frac{1}{2}(|\alpha|^2 - |\beta|^2) & \alpha\bar{\beta} \\ \bar{\alpha}\beta & \frac{1}{2}(|\beta|^2 - |\alpha|^2) \end{array}\right].$$

We see that $q(\psi)$ is traceless and symmetric. We will often identify $q(\psi)$ with a 2-form via the Clifford multiplication $c : \Lambda^2V^* \otimes \mathbb{C} \to \operatorname{End}(\mathbb{S}_4)$. More precisely

$$q(\psi) \sim \frac{\mathbf{i}}{4}(|\alpha|^2 - |\beta|^2)\omega + \frac{1}{2}(\bar{\alpha}\beta - \alpha\bar{\beta}) \in \mathbf{i}\Lambda^2_+V^* \subset \Lambda^2V^* \otimes \mathbb{C}. \tag{1.3.5}$$

**Exercise 1.3.2.** Using the notation in the previous example show that

$$q(\psi) = \frac{1}{4}\sum_{i=0}^{2}\langle\psi, \mathbf{c}(\eta_k)\psi\rangle\mathbf{c}(\eta_k).$$

**Exercise 1.3.3.** Using the notation in Example 1.3.3 show that for every $\omega \in \Lambda^2V^*$ we have

$$\mathbf{c}(\omega) = \mathbf{c}(*\omega)$$

*as endomorphisms of* $\mathbb{S}^+(V)$.

Since $\mathsf{Cl}_{2n} \otimes \mathbb{C}$ is isomorphic with an algebra of matrices $\operatorname{End}(\mathbb{S}_{2n})$ we can invoke Wedderburn's theorem ([**122**]) to conclude that any complex $\mathsf{Cl}_{2n}$-module $V$ has the form $\mathbb{S}_{2n} \otimes V$.

The odd dimensional situation follows from the even one using the following fact.

**Lemma 1.3.4.** *Let $V$ be a Euclidean space and $u \in V$ such that $|u| = 1$. Set $V_0 = \langle u\rangle^\perp$. Then there is an isomorphism of algebras*

$$\phi : \mathsf{Cl}(V_0) \to \mathsf{Cl}^+(V)$$

*defined by*

$$\phi : \ x \mapsto x_+ + ux_-.$$

**Exercise 1.3.4.** Prove the above lemma.

We deduce from the above result and the **Fundamental Fact** that

$$\mathsf{Cl}_{2n-1} \otimes \mathbb{C} \cong \mathrm{End}\,(\mathbb{S}_{2n}^{+}) \oplus \mathrm{End}\,(\mathbb{S}_{2n}^{-}).$$

Thus the complex representation theory of $\mathsf{Cl}_{2n-1}$ is generated by two, non-isomorphic, irreducible modules.

**1.3.2. The *Spin* and *Spin*$^c$ groups.** To produce a Dirac bundle on an $n$-dimensional Riemannian manifold we need several things.

(a) A bundle of Clifford algebras, i.e. a bundle $\mathcal{C} \to M$ of associative algebras and an injection $\imath : T^*M \hookrightarrow \mathcal{C}$ such that for all $x \in M$ and all $u, v \in T_x^*M$

$$\{\imath(u), \imath(v)\} = -2g(u,v)\mathbf{1}$$

and the induced map $\imath_x : \mathsf{Cl}(T_x^*M) \to \mathcal{C}_x$ is an isomorphism.

(b) A bundle of complex Clifford modules, i.e. a complex vector bundle $\mathcal{E} \to M$ and a morphism $c : \mathcal{C} \to \mathrm{End}\,(\mathcal{E})$.

We want all these bundles to be associated to a common principal $G$-bundle. $G$ is a Lie group with several special properties.

Denote by $(V, g)$ the standard fiber of $T^*M$ and denote by $\mathrm{Aut}_V$ the subgroup of algebra automorphisms $\varphi$ of $\mathsf{Cl}(V)$ such that

$$\varphi(V) \subset V (\subset \mathsf{Cl}(V)\,).$$

First we require that there exists a Lie group morphism

$$\rho : G \to \mathrm{Aut}_V.$$

With such a $\rho$ fixed we notice that it tautologically induces a representation

$$\rho : G \to \mathrm{Aut}\,(V).$$

Denote by $E$ the standard fiber of $\mathcal{E}$. We require there exists a representation $\mu : G \to \mathrm{End}\,(E)$ such that the diagram below is commutative for all $g \in G$ and all $v \in V$.

$$\begin{array}{ccc} E & \xrightarrow{c(v)} & E \\ {\scriptstyle g}\downarrow & & \downarrow{\scriptstyle g} \\ E & \xrightarrow{c(g\cdot v)} & E \end{array} \tag{1.3.6}$$

This commutativity can be given an invariant theoretic interpretation as follows. View the Clifford multiplication $c : V \to \mathrm{End}\,(E)$ as an element $c \in V^* \otimes E^* \otimes E$. The group $G$ acts on this tensor product and the above commutativity simply means that $c$ is invariant under this action.

To produce a Dirac bundle all we now need is a principal $G$-bundle $P \to M$ such that the associated bundle $P \times_\rho V$ is isomorphic to $T^*M$. (This may not be always feasible due to possible topological obstructions.)

Any connection $\nabla$ on $P$ induces by association metric connections $\nabla^M$ on $T^*M$ and $\nabla^E$ on the bundle of Clifford modules $\mathcal{E} = P \times_\mu E$. (In practice one often requires a little more, namely that $\nabla^M$ is precisely the Levi-Civita connection on $T^*M$. This leads to significant simplifications in many instances.) With respect to these connections the Clifford multiplication is covariant constant, i.e.

$$\nabla^E(c(\alpha)u) = c(\nabla^M\alpha) + c(\alpha)\nabla^E u, \quad \forall\alpha \in \Omega^1(M),\ u \in C^\infty(E).$$

This follows from the following elementary invariant theoretic result.

**Lemma 1.3.5.** *Let $G$ be a Lie group and $\rho : G \to \mathrm{Aut}\,(E)$ a linear representation of $G$. Assume there exists $e_0 \in E$ such that $\rho(g)e_0 = e_0$, $\forall g \in G$. Consider an arbitrary principal $G$-bundle $P \to X$ and an arbitrary connection $\nabla$ on $P$. Then $e_0$ canonically determines a section $u_0$ on $P \times_\rho E$ which is covariant constant with respect to the induced connection $\nabla^E = \rho_*(\nabla)$, i.e.*

$$\nabla^E u_0 = 0.$$

**Exercise 1.3.5.** Prove the above lemma.

Apparently the chances that a Lie group $G$ with the above properties exists are very slim. The very pleasant surprise is that all these things (and even more) happen in most of the geometrically interesting situations.

**Example 1.3.6.** Let $(V, g)$ be an oriented Euclidean space. Using the universality property of Clifford algebras we deduce that each $g \in SO(V)$ induces an automorphism of $\mathsf{Cl}(V)$ preserving $V \hookrightarrow \mathsf{Cl}(V)$. Moreover, it defines an orthogonal representation on the canonical Clifford module

$$c : \mathsf{Cl}(V) \to \mathrm{End}\,(\Lambda^*V)$$

such that

$$c(g \cdot v)(\omega) = g \cdot (c(v)(g^{-1} \cdot \omega)) \quad \forall g \in SO(V), v \in V, \omega \in \Lambda^*V,$$

i.e. $SO(V)$ satisfies the equivariance property (1.3.6).

If $(M, g)$ is an oriented Riemannian manifold we can now build our bundle of Clifford modules starting from the principal $SO$ bundle of its oriented orthonormal coframes. As connections we can now pick the Levi-Civita connection and its associates. The corresponding Dirac operator is the Hodge-deRham operator.

The *Spin* and *Spin*$^c$ groups provide two fundamental examples of groups with the properties listed above. Here are their descriptions.

Let $(V, g)$ be a Euclidean vector space. Define

$$Spin(V) := \{x \in \mathrm{Cl}^+ \ ; \ x = v_1 \cdots v_{2k}, \ \ v_i \in V, \ |v_i| = 1\}.$$

Equip it with the induced topology as a closed subset of $\mathrm{Cl}^+$. First note there exists a group morphism

$$\rho : Spin(V) \to SO(V), \ \ x \mapsto \rho_x \in SO(V)$$

$\rho_x(v) = xvx^{-1}$. We must first verify that $\rho$ is correctly defined, i.e. $\rho_x$ is an orthogonal map of determinant 1. To see this note that for every $u \in V$, $|u| = 1$ the map

$$R_u : V \to V, \ \ v \mapsto -uvu^{-1}$$

satisfies $R_u(V) \subset V$ and more precisely, $R_u$ is the orthogonal reflection in the orthogonal complement of $\langle u \rangle := \mathrm{span}\,(u)$. We see that for every $x = v_1 \cdots v_{2k} \in Spin(V)$ we have

$$\rho_x = R_{v_1} \circ \cdots \circ R_{v_{2k}}$$

is the product of an *even* number of orthogonal reflections so that $\rho_x \in SO(V)$. Since any $T \in SO(V)$ can be written as the product of an even number of reflections we conclude that the map $\rho$ is actually onto. We leave it to the reader to prove the following elementary fact.

**Exercise 1.3.6.** Show that $\ker \rho = \{\pm 1\}$.

This implies that $\rho$ is a covering map. If $\dim V \geq 3$ one can show that $Spin(V)$ is simply connected (because the unit sphere in $V$ is so) and thus

$$\rho : Spin(V) \to SO(V)$$

is the universal cover of $SO(V)$. In particular, this shows that $\pi_1(SO(V)) = \mathbb{Z}_2$. By pullback one obtains a smooth structure on $Spin(V)$. Hence $Spin(V)$ is a compact, simply connected Lie group. Its Lie algebra is isomorphic to the Lie algebra $\underline{so}(V)$ of $SO(V)$. We want to present a more useful description of the Lie algebra of $Spin(V)$. To do this we need to assume the following not so obvious fact.

**Exercise 1.3.7.** Show that $Spin(V)$ with the smooth structure induced from $SO(V)$ is a smooth submanifold of $\mathrm{Cl}(V)$.

The Lie algebra of $Spin(V)$ can be identified with a closed subspace of $\mathrm{Cl}^+(V)$. More precisely,

$$\underline{spin}(V) = \rho_*^{-1}(\underline{so}(V))$$

where $\rho_*$ denotes the differential at $\mathbf{1} \in Spin(V)$ of the covering map $Spin(V) \to SO(V)$. A basis of $\underline{spin}(V)$ can be obtained from a basis of

$\underline{so}(V)$. Choose an orthonormal basis $e_1, \cdots, e_n$ of $V$. For each $i < j$ we have a path $\gamma_{ij} : (-\varepsilon, \varepsilon) \to Spin(V)$ given by

$$\gamma_{ij}(t) = -(e_i \cos\frac{t}{2} + e_j \sin\frac{t}{2})(e_i \cos\frac{t}{2} - e_j \sin\frac{t}{2}) = \cos t + (\sin t)e_ie_j.$$

The orthogonal transformation $\rho_{\gamma_{ij}(t)} \in SO(V)$ acts trivially on the orthogonal complement of $V_{ij} = \mathrm{span}\,(e_i, e_j)$, while on $V_{ij}$, oriented by $e_i \wedge e_j$, it acts as the counterclockwise rotation of angle $2t$.

Now denote by $\mathbf{J}_{ij}$ the element of $\underline{so}(V)$ given by

$$e_i \mapsto e_j \mapsto -e_i, \quad e_k \mapsto e_k, \quad k \neq i, j.$$

The family $(\mathbf{J}_{ij})_{i<j}$ is a basis of $\underline{so}(V)$. We deduce

$$\frac{d}{dt}|_{t=0}\, \rho_{\gamma_{ij}(t)} = 2\mathbf{J}_{ij}.$$

Hence

$$\rho_*^{-1}(J_{ij}) = \frac{1}{2}\frac{d}{dt}|_{t=0}\, \gamma_{ij}(t) = \frac{1}{2}e_ie_j.$$

In particular if $A \in \underline{so}(V)$ has the matrix description

$$Ae_j = \sum_i a^i_j e_i, \quad a^i_j = -a^j_i = a_{ij} = a^{ij}$$

then (*notice the crucial negative sign!!!*)

$$A = -\sum_{i<j} a^{ij}\mathbf{J}_{ij}$$

and

$$\rho_*^{-1}(A) = -\frac{1}{2}\sum_{i<j} a^{ij}e_ie_j = -\frac{1}{4}\sum_{i,j} a^{ij}e_ie_j. \tag{1.3.7}$$

**Example 1.3.7.** $Spin(3) \cong SU(2)$.

To see this, regard $SU(2)$ as the group of unit quaternions (so that, in particular, $SU(2)$ is diffeomorphic to $S^3$). There is a map

$$SU(2) \to SO(3), \quad q \mapsto T_q,$$

where $T_q$ acts on $\mathbb{R}^3 \cong \mathfrak{Im}\,\mathbb{H}$ by

$$x \mapsto T_qx = qxq^{-1}.$$

It is not difficult to see that $q \mapsto T_q$ is a double cover.

**Example 1.3.8.** $Spin(4) \cong SU(2) \times SU(2)$. Again identify $SU(2)$ with the group of unit quaternions and define

$$T : SU(2) \times SU(2) \to SO(4), \quad (p, q) \mapsto T_{p,q}$$

where $T_{p,q}$ acts on $\mathbb{R}^4 \cong \mathbb{H}$ by

$$T_{p,q}x = pxq^{-1}.$$

Again one checks $(p,q) \to T_{p,q}$ is a double cover.

There is a natural embedding $Spin(3) \hookrightarrow Spin(4)$ which can be described as the diagonal embedding

$$SU(2) \hookrightarrow SU(2) \times SU(2), \quad q \mapsto (q,q).$$

This embedding is compatible with the natural embedding $SO(3) \hookrightarrow SO(4)$ in the sense that the diagram below is commutative.

(↻)

$$\begin{array}{ccc} Spin(3) & \hookrightarrow & Spin(4) \\ \downarrow & & \downarrow \\ SO(3) & \hookrightarrow & SO(4) \end{array}$$

Suppose now that $(V,g)$ is a $2n$-dimensional Euclidean space. Fix a compatible complex structure $J$. This defines an isomorphism of $\mathbb{Z}_2$-graded algebras

$$\rho : \mathrm{Cl}_{2n} \otimes \mathbb{C} \to \mathrm{End}\,(\mathbb{S}_{2n}^+ \oplus \mathbb{S}_{2n}^-).$$

Since $Spin(2n) \subset \mathrm{Cl}_{2n}^+$ we obtain two complex representations

$$\rho_\pm : Spin(2n) \to \mathrm{Aut}\,(\mathbb{S}_{2n}^\pm).$$

These are irreducible and not isomorphic (as $Spin(2n)$-representations). These are called the even/odd complex spinor representations of $Spin(2n)$. The complex $Spin(2n)$-module $\mathbb{S}_{2n}^+ \oplus \mathbb{S}_{2n}^-$ is denoted by $\mathbb{S}_{2n}$.

When $(V,g)$ is a Euclidean space of odd dimension $2n+1$ then

$$Spin(2n+1) \subset \mathrm{Cl}_{2n+1}^+ \cong \mathrm{Cl}_{2n}.$$

Thus $Spin(2n+1)$ acts naturally on $\mathbb{S}_{2n}$. This action

$$\rho : Spin(2n+1) \to \mathrm{Aut}\,(\mathbb{S}_{2n})$$

is called the fundamental spinor representation and the corresponding $Spin(2n+1)$-module will be denoted by $\mathbb{S}_{2n+1}$.

**Exercise 1.3.8.** Using the isomorphism

$$\mathrm{Cl}_{2n}^+ \cong \mathrm{Cl}_{2n-1}$$

constructed in the previous subsection show that the induced representations of $Spin(2n-1)$ on $\mathbb{S}_{2n}^\pm$ are both isomorphic to $\mathbb{S}_{2n-1}$.

**Example 1.3.9.** Using the isomorphism $Spin(4) \cong SU(2) \times SU(2)$ we can describe the complex spinor representations as follows.

$$\rho_{\pm} : SU(2) \to SU(2) \to \mathrm{Aut}\,(\mathbb{C}^2),$$

$$\rho_{+}(p,q) : \quad \mathbb{C}^2 \cong \mathbb{H} \ni v \mapsto p \cdot v \in \mathbb{H} \cong \mathbb{C}^2$$

(where $\mathbb{H}$ is equipped with the complex structure induced by the right multiplication by $\mathbf{i} \in \mathbb{H}$)

$$\rho_{-}(p,q) : \quad \mathbb{C}^2 \cong \mathbb{H} \ni v \mapsto v \cdot q^{-1} \in \mathbb{H} \cong \mathbb{C}^2$$

(where $\mathbb{H}$ is equipped with the complex structure induced by the left multiplication by $\mathbf{i} \in \mathbb{H}$).

Define the group $Spin^c(V)$ by

$$Spin^c(V) = (Spin(V) \times S^1)/\mathbb{Z}_2$$

where $\mathbb{Z}_2$ denotes the subgroup $\{(1,1),(-1,-1)\}$ of $Spin(V) \times S^1$. Assume for simplicity $\dim V = 2n$. We obtain two representations

$$\rho_{\pm}^{c} : Spin^c(V) \to \mathrm{Aut}\,(\mathbb{S}_{2n}^{\pm})$$

by

$$\rho_{\pm}^{c}(gz) = z\rho_{\pm}(g)$$

where $\rho_{\pm}$ denote the complex spinor representations of $\mathrm{Spin}(V)$.

**Exercise 1.3.9.** Show that $Spin^c(3) \cong U(2)$.

Exactly as in the case of the *spin*-groups we have a commutative diagram

$$(\circlearrowright^c) \qquad \begin{array}{ccc} Spin^c(3) & \hookrightarrow & Spin^c(4) \\ \downarrow & & \downarrow \\ SO(3) & \hookrightarrow & SO(4) \end{array}$$

There is an intimate relationship between the group $Spin^c(V)$ and almost complex structures on $V$. Suppose $J$ is an almost complex structure compatible with the metric $g$ and denote by $U(V,J)$ the group of unitary automorphisms, i.e. orthogonal transformations of $V$ which commute with $J$.

**Proposition 1.3.10.** *There exists a canonical group morphism* $\xi = \xi_J : U(V,J) \to Spin^c(V)$ *such that the diagram below is commutative.*

$$\begin{array}{ccc} U(V,J) & \xrightarrow{\ \xi\ } & Spin^c(V) \\ & \searrow^{\imath} & \downarrow \\ & & SO(V) \end{array}$$

*The vertical arrow is the composition* $Spin^c(V) \to Spin(V) \to SO(V)$.

**Idea of Proof** Let $\omega \in U(V)$ and consider a path $\gamma : [0,1] \to U(V)$ connecting 1 to $\omega$. Viewed as a path in $SO(V)$, $\gamma$ admits a *unique* lift $\tilde{\gamma} : [0,1] \to Spin(V)$, $\tilde{\gamma}(0) = 1$. Using the double cover

$$S^1 \to S^1, \quad z \mapsto z^2$$

we can produce a unique lift $\delta(t)$ of the path $\det \gamma(t) \in S^1$. Now define $\xi(\omega) := (\tilde{\gamma}(1), \delta(1))$. We let the reader verify that $\xi$ is a well defined morphism $U(V) \to Spin^c(V)$. ■

Next, we need to explain how to use these groups to produce Dirac structures on a manifold. This requires a topological interlude, to discuss the notion of *spin* and *spin*$^c$ structures.

**1.3.3.** ***Spin* and *spin*$^c$ structures.** Consider an oriented $n$-dimensional Riemannian manifold $(M, g)$. The tangent bundle $TM$ can be described via a gluing cocycle $g_{\alpha\beta} : U_{\alpha\beta} \to SO(n)$ supported by a *good cover*, that is, an open cover $(U_\alpha)$ of $M$ where all the multiple intersections $U_{\alpha\beta\cdots\gamma}$ can be assumed to be contractible (or even better, geodesically convex). A *spin structure* is a collection of lifts

$$\tilde{g}_{\alpha\beta} : U_{\alpha\beta} \to Spin(n)$$

of $g_{\alpha\beta}$ satisfying the cocycle condition

$$\tilde{g}_{\alpha\beta}\tilde{g}_{\beta\gamma}\tilde{g}_{\gamma\alpha} \equiv \mathbf{1}.$$

A manifold admitting spin structures is called *spinnable*. *Spin* structures may or may not exist. Let's see what can go wrong. Clearly, each map $g_{\alpha\beta} : U_{\alpha\beta} \to SO(n)$ admits at least one lift (in fact precisely two of them)

$$\tilde{g}_{\alpha\beta} : U_{\alpha\beta} \to Spin(n).$$

Since $g_{\alpha\beta}$ satisfies the cocycle condition we deduce

$$w_{\alpha\beta\gamma} := \tilde{g}_{\alpha\beta}\tilde{g}_{\beta\gamma}\tilde{g}_{\gamma\alpha} \in \mathbb{Z}_2 = \ker(Spin(n) \to SO(n)).$$

The collection $w_{\alpha\beta\gamma}$ satisfies the cocycle condition

$$w_{\beta\gamma\delta} - w_{\alpha\gamma\delta} + w_{\alpha\beta\delta} - w_{\alpha\beta\gamma} \equiv 0 \in \mathbb{Z}_2$$

for all $\alpha, \beta, \gamma, \delta$ such that $U_{\alpha\beta\gamma\delta} \neq \emptyset$. In other words, the collection $w_{\dots}$ is a $\mathbb{Z}_2$-valued Čech 2-cocycle. By choosing different lifts $\tilde{g}_{\alpha\beta}$ we only change $w_{\dots}$ *within its Čech cohomology class.* Hence, this cohomology class is a topological invariant of the smooth manifold $M$. It is called the *second Stiefel-Whitney* class and will be denoted by $w_2(M)$. It lives in $H^2(M, \mathbb{Z}_2)$. The above discussion shows that if $w_2(M) \neq 0$ then $M$ does not admit spin structures. The converse is also true. More precisely, we have the following result.

**Proposition 1.3.11.** *The oriented manifold $M$ is spinnable if and only if $w_2(M) = 0$. If this is the case there is a bijection between the set of isomorphism classes of spin structures and $H^1(X, \mathbb{Z}_2)$.*

**Remark 1.3.12.** The definition of isomorphism of *spin*-structures is rather subtle (see [**92**]). More precisely, two spin structures defined by the cocycles $\tilde{g}_{\bullet\bullet}$ and $\tilde{h}_{\bullet\bullet}$ are isomorphic if there exists a collection $\varepsilon_\alpha \in \mathbb{Z}_2 \subset Spin(n)$ such that the diagram below is commutative for all $x \in U_{\alpha\beta}$

$$\begin{array}{ccc} Spin(n) & \xrightarrow{\varepsilon_\alpha} & Spin(n) \\ \Big\downarrow{\scriptstyle \tilde{g}_{\beta\alpha}(x)} & & \Big\downarrow{\scriptstyle \tilde{h}_{\beta\alpha}(x)} \\ Spin(n) & \xrightarrow{\varepsilon_\beta} & Spin(n) \end{array}$$

The group $H^1(M, \mathbb{Z}_2)$ acts on $Spin(M)$ as follows. Take an element $\varepsilon \in H^1(M, \mathbb{Z}_2)$ represented by a Čech cocycle, i.e. a collection of *continuous maps* $\varepsilon_{\alpha\beta|} : U_{\alpha\beta} \to \mathbb{Z}_2 \subset Spin(n)$ satisfying the cocycle condition

$$\varepsilon_{\alpha\beta} \cdot \varepsilon_{\beta\gamma} \cdot \varepsilon_{\gamma\alpha} = 1.$$

Then the collection $\varepsilon_{\bullet\bullet} \cdot \tilde{g}_{\bullet\bullet}$ is a $Spin(n)$ gluing cocycle defining a *spin* structure we denote by $\varepsilon \cdot \sigma$. It is easy to check that the isomorphism class of $\varepsilon \cdot \sigma$ is independent of the various choice, i.e Čech representatives for $\varepsilon$ and $\sigma$. Clearly the correspondence

$$H^1(M, \mathbb{Z}_2) \times Spin(M) \ni (\varepsilon, \sigma) \mapsto \varepsilon \cdot \sigma \in Spin(M)$$

defines a left action of $H^1(M, \mathbb{Z}_2)$ on $Spin(M)$. This action is transitive and free.

**Exercise 1.3.10.** Prove the above proposition and the statement in the above remark.

**Exercise 1.3.11.** Describe the only two *spin* structures on $S^1$.

There is a very efficient topological machinery which can be used to decide whether $w_2(M) = 0$. We refer to [**93**] for details. We only want to mention a few examples.

**Example 1.3.13.** A compact, simply connected 4-manifold admits spin structures if and only if its intersection form is even. A compact, simply connected manifold $M$ of dimension $\geq 5$ admits *spin* structures if and only if every compact oriented surface $S$ embedded in $M$ has trivial normal bundle.

Let $(M^n, g)$ be an oriented, $n$-dimensional Riemannian manifold. As above, we can regard the tangent bundle as associated to the principal bundle $P_{SO(M)}$ of oriented orthonormal frames. Assume $P_{SO(M)}$ is defined by a good open cover $\mathfrak{U} = (U_\alpha)$ and transition maps

$$g_{\alpha\beta} : U_{\alpha\beta} \to SO(n).$$

The manifold $M$ is said to possess a *spin*$^c$ structure if there exist smooth maps $\tilde{g}_{\alpha\beta} : U_{\alpha\beta} \to Spin^c(n)$, satisfying the cocycle condition such that

$$\rho^c(\tilde{g}_{\alpha\beta}) = g_{\alpha\beta}.$$

As for spin structures, there are obstructions to *spin*$^c$ structures as well which clearly are less restrictive. Let us try to understand what can go wrong. We stick to the assumption that all the overlaps $U_{\alpha\beta\cdots\gamma}$ are contractible.

Since $Spin^c(n) = (Spin(n) \times S^1)/\mathbb{Z}_2$, lifting the $SO(n)$-structure $(g_{\alpha\beta})$ reduces to finding smooth maps

$$h_{\alpha\beta} : U_{\alpha\beta} \to Spin(n)$$

and

$$z_{\alpha\beta} : U_{\alpha\beta} \to S^1$$

such that

$$\rho(h_{\alpha\beta}) = g_{\alpha\beta}$$

and

$$(\epsilon_{\alpha\beta\gamma}, \zeta_{\alpha\beta\gamma}) \stackrel{def}{=} (h_{\alpha\beta}h_{\beta\gamma}h_{\gamma\alpha}\,,\, z_{\alpha\beta}z_{\beta\gamma}z_{\gamma\alpha}) \in \{(-1,-1),(1,1)\}. \tag{1.3.8}$$

If we set $\lambda_{\alpha\beta} = z_{\alpha\beta}^2 : U_{\alpha\beta} \to S^1$ we deduce from (1.3.8) that the collection $(\lambda_{\alpha\beta})$ should satisfy the cocycle condition. In particular, it defines a principal $S^1$-bundle over $M$ or, equivalently, a complex line bundle $\mathcal{L}$. This line bundle should be considered as part of the data defining a *spin*$^c$ structure. The collection $(\epsilon_{\alpha\beta\gamma})$ is an old acquaintance: it is a Čech 2-cocycle representing the second Stiefel-Whitney class.

We can represent the cocycle $\lambda_{\alpha\beta}$ as

$$\lambda_{\alpha\beta} = \exp(\mathbf{i}\theta_{\alpha\beta}), \;\; \theta_{\alpha\beta} : U_{\alpha\beta} \to \mathbb{R}.$$

The collection

$$n_{\alpha\beta\gamma} = \frac{1}{2\pi}(\theta_{\alpha\beta} + \theta_{\beta\gamma} + \theta_{\gamma\alpha})$$

defines a 2-cocycle of the constant sheaf $\mathbb{Z}$ which represents the topological first Chern class of $\mathcal{L}$. The condition (1.3.8) shows that

$$n_{\alpha\beta\gamma} = \epsilon_{\alpha\beta\gamma} \pmod 2.$$

To summarize, we see that the existence of a *spin*$^c$ structure implies the existence of a complex line bundle $\mathcal{L}$ such that

$$c_1^{top}(\mathcal{L}) = w_2(M) \pmod 2.$$

It is not difficult to prove that the above condition is also sufficient. In fact one can be more precise.

Denote by $Spin^c(M)$ the collection of isomorphism classes of *spin*$^c$ structures on the manifold $M$. Any $\sigma \in Spin^c(M)$ is defined by a lift $(h_{\alpha\beta}, z_{\alpha\beta})$ as above. We denote by $\det(\sigma)$ the complex line bundle defined by the gluing data $(z_{\alpha\beta})$. We have seen that

$$c_1^{top}(\det(\sigma)) \equiv w_2(M) \pmod 2.$$

Denote by $\mathcal{L}_M \subset H^2(M, \mathbb{Z})$ the "affine" subspace consisting of those cohomology classes satisfying the above congruence modulo 2. Such elements are called *characteristic* (not to be confused with the characteristic classes of Chern and Pontryagin). We thus have a map

$$Spin^c(M) \to \mathcal{L}_M, \quad \sigma \mapsto c_1^{top}(\det(\sigma)).$$

**Proposition 1.3.14.** *The above map is a surjection.*

**Exercise 1.3.12.** Show that if $H^2(M, \mathbb{Z})$ has no 2-torsion (e.g. $M$ is simply connected) then the above map $Spin^c(M) \to \mathcal{L}_M$ is one-to-one.

**Exercise 1.3.13.** Complete the proof of the above proposition.

The smooth Picard group $\mathrm{Pic}^\infty(M)$ acts on $Spin^c(M)$ by

$$Spin^c(M) \times \mathrm{Pic}^\infty(M) \ni (\sigma, L) \mapsto \sigma \otimes L.$$

More precisely, if $\sigma \in Spin^c(M)$ is given by the cocycle

$$\sigma = [h_{\alpha\beta}, z_{\alpha\beta}] : U_{\alpha\beta} \to Spin\,(n) \times S^1/\sim$$

and $L$ is given by the $S^1$ cocycle

$$\zeta_{\alpha\beta} : U_{\alpha\beta} \to S^1$$

then $\sigma \otimes L$ is given by the cocycle

$$[h_{\alpha\beta}, z_{\alpha\beta}\zeta_{\alpha\beta}].$$

Note that

$$\det(\sigma \otimes L) = \det(\sigma) \otimes L^2$$

so that

$$c_1^{top}(\sigma \otimes L) = c_1^{top}(\sigma) + 2c_1^{top}(L).$$

**Proposition 1.3.15.** *The above action of* $\mathrm{Pic}^\infty(M)$ *on* $Spin^c(M)$ *is free and transitive.*

**Proof** Consider two $spin^c$ structures $\sigma^1$ and $\sigma^2$ defined by the good cover $(U_\alpha)$ and the gluing cocycle

$$[h_{\alpha\beta}^{(i)}, z_{\alpha\beta}^{(i)}], \ \ i = 1, 2.$$

Since $\rho^c(h_{\alpha\beta}^{(1)}) = \rho^c(h_{\alpha\beta}^{(2)}) = g_{\alpha\beta}$ we can assume (possibly modifying the maps $h_{\alpha\beta}^{(2)}$ by a sign) that

$$h_{\alpha\beta}^{(1)} = h_{\alpha\beta}^{(2)}.$$

This implies that

$$\zeta_{\alpha\beta} = z_{\alpha\beta}^{(2)}/z_{\alpha\beta}^{(1)}$$

is an $S^1$-cocycle defining a complex line bundle $L$. Obviously $\sigma^2 = \sigma^1 \otimes L$. This shows the action of $\mathrm{Pic}^\infty(M)$ is transitive. We leave the reader verify this action is indeed free. The proposition is proved. ■

The group of orientation preserving diffeomorphisms of $M$ acts in a natural manner on $Spin^c(M)$ by pullback.

Given two $spin^c$ structures $\sigma_1$ and $\sigma_2$ we can define their "difference" $\sigma_2/\sigma_1$ as the unique line bundle $L$ such that $\sigma_2 = \sigma_1 \otimes L$. This shows that the collection of $spin^c$ structures is (noncanonically) isomorphic with $H^2(X, \mathbb{Z}) \cong \mathrm{Pic}^\infty$. It is a sort of affine space modeled on $H^2(X, \mathbb{Z})$ in the sense that the "difference" between two $spin^c$ structures is an element in $H^2(X, \mathbb{Z})$ but there is no distinguished origin of this space. A structure as above is usually called an $H^2(M, \mathbb{Z})$-**torsor**.

We will list below (without proofs) some examples of $spin^c$ manifolds.

**Example 1.3.16.** (a) Any *spin* manifold admits a $spin^c$ structure.

(b) Any almost complex manifold has a natural $spin^c$ structure.

(c) (Hirzebruch-Hopf, [**55**]; see also [**98**]) Any oriented manifold of dimension $\leq 4$ admits a $spin^c$ structure.

Let us analyze the first two examples above. If $M$ is a spin manifold then the lift

$$\tilde{g}_{\alpha\beta} : U_{\alpha\beta} \to Spin(n)$$

of the $SO$-structure to a spin structure canonically defines a $spin^c$ structure via the trivial morphism

$$Spin(n) \to Spin^c(n) \times_{\mathbb{Z}_2} S^1, \quad g \mapsto (g, 1) \text{ mod the } \mathbb{Z}_2\text{-action}.$$

We see that in this case the associated complex line bundle is the trivial bundle. This is called the **canonical** $spin^c$ structure of a *spin* manifold. Thus on a *spin* manifold the torsor of $spin^c$-structures does in fact possess a "canonical origin" so in this case there is a canonical identification

$$Spin^c(M) \cong \text{Pic}^\infty \cong H^2(M, \mathbb{Z}).$$

To any complex line bundle $L$ defined by the $S^1$-cocycle $(z_{\alpha\beta})$ we can associate the $spin^c$ structure defined by the gluing data

$$\{(\tilde{g}_{\alpha\beta}, z_{\alpha\beta})\}.$$

Clearly, the line bundle associated to this structure is $L^2 = L^{\otimes 2}$. In particular, this shows that a *spin* structure on a manifold $M$ canonically determines a square root $\det(\sigma)^{1/2}$ of $\det(\sigma)$, for any $\sigma \in Spin^c(M)$ .

**Exercise 1.3.14.** Show that any two spin structures on a manifold $M$ such that $H^2(M, \mathbb{Z})$ has no 2-torsion are isomorphic *as $spin^c$ structures.*

**Exercise 1.3.15.** Suppose $N$ is a closed, oriented, Riemannian 3-manifold. Denote by $Fr_N$ the bundle of oriented, orthogonal frames of $TN$. $Fr_N \to N$ is a principla $SO(3)$-bundle. Denote by $\mathcal{S}_N$ the set of cohomology classes $c \in H^2(Fr_N, \mathbb{Z})$ such that their restriction to any fiber coincides with the generator of $H^2(SO(3), \mathbb{Z}) \cong \mathbb{Z}_2$. Prove that there exists a natural bijection

$$Spin^c(N) \to \mathcal{S}_N.$$

The commutative diagram ($\circlearrowleft^c$) shows that given a $spin^c$-structure $\sigma$ on a closed, oriented 3-manifold $N$ canonically induces a $spin^c$ structure $\hat{\sigma}$ on $\mathbb{R} \times N$. We will often use the notations

$$\hat{\sigma} := \mathbb{R} \times \sigma, \quad \sigma := \hat{\sigma}\,|_N \,.$$

Conversely, the $SO(4)$-structure on $T(\mathbb{R} \times N)$ naturally reduces to a $SO(3)$-structure (split the longitudinal tangent vector $\partial_t$), and invoking the diagram ($\circlearrowleft^c$) again we deduce that any $spin^c$ structure $\hat{\sigma}$ on $\mathbb{R}$ induces a $spin^c$ structure on $N$ or, more precisely, the map

$$Spin^c(N) \to Spin^c(\mathbb{R} \times N), \quad \sigma \mapsto \mathbb{R} \times \sigma$$

is an isomorphism.

In the conclusion of this subsection we would like to explain in some detail why an almost complex manifold (necessarily of even dimension $n = 2k$) admits a canonical *spin*$^c$ structure. Recall that the natural morphism $U(k) \rightarrow SO(2k)$ factors through a morphism

$$\xi : U(k) \rightarrow Spin^c(2k).$$

The $U(k)$-structure of $TM$, defined by the gluing data

$$h_{\alpha\beta} : U_{\alpha\beta} \rightarrow U(k)$$

induces a *spin*$^c$ structure defined by the gluing data $\xi(h_{\alpha\beta})$. Its associated line bundle is given by the $S^1$-cocycle

$$\det_{\mathbf{C}}(h_{\alpha\beta}) : U_{\alpha\beta} \rightarrow S^1$$

and it is precisely the determinant line bundle

$$\det_{\mathbf{C}} T^{1,0}M = \Lambda^{k,0}TM.$$

The dual of this line bundle, $\det_{\mathbf{C}}(T^*M)^{1,0} = \Lambda^{k,0}T^*M$ plays a special role in algebraic geometry. It usually denoted by $K_M$ and it is called the *canonical line bundle.* Thus the line bundle associated to this *spin*$^c$ structure is $K_M^{-1} \stackrel{def}{=} K_M^*$.

**Exercise 1.3.16.** Show that an almost complex manifold $M$ admits a spin structure if and only if the canonical line bundle $K_M$ admits a square root, i.e. there exists a complex line bundle $L$ such that $L^{\otimes 2} \cong K_M$. (Traditionally such a line bundle is denoted by $K_M^{-1/2}$, although the square root may not be unique.)

**1.3.4. Dirac operators associated to *spin* and *spin*$^c$ structures.** Suppose $(M, g)$ is an oriented Riemannian manifold of dimension $n$ equipped with a *spin* structure. To describe it we assume the tangent bundle $TM$ is defined by the open cover $(U_\alpha)$ and transition maps

$$g_{\alpha\beta} : U_{\alpha\beta} \rightarrow SO(n).$$

These define a principal $SO(n)$-bundle $P_{SO(n)} \rightarrow M$. The spin structure is given by the lifts

$$\tilde{g}_{\alpha\beta} : U_{\alpha\beta} \rightarrow Spin(n)$$

which define a principal $Spin(n)$-bundle $P_{Spin(n)} \rightarrow M$. Using the representation

$$\tau : Spin(n) \rightarrow \mathrm{Aut}\,(\mathbb{S}_n)$$

we can construct the associated vector bundle $P_{Spin(n)} \times_\tau \mathbb{S}_n$ with structure group $Spin(n)$ and fiber $\mathbb{S}_n$ given by the gluing cocycle

$$\tau(\tilde{g}_{\alpha\beta}) : U_{\alpha\beta} \rightarrow \mathrm{Aut}\,(\mathbb{S}_n).$$

It is called the *bundle of complex spinors associated to the given spin structure* and will be denoted by $\mathbb{S}_0 = \mathbb{S}_0(M)$.

**Exercise 1.3.17.** As indicated in the Exercise 1.3.11, there are two *spin* structures on $S^1$, $\bullet$ and $\circ$. Denote by $\mathbb{S}_\bullet$ and $\mathbb{S}_\circ$ the associated bundles of complex spinors. These are complex line bundles over $S^1$ and as such they must be isomorphic. What bit of information do the spin structures add to these bundles which will allow us to distinguish them?

**Exercise 1.3.18.** The bundle $\mathbb{S}_0$ has a natural selfadjoint Clifford structure $\mathbf{c} : T^*M \to \mathrm{End}\,(\mathbb{S}_M)$.

The Levi-Civita connection $\nabla^M$ on $T^*M$ is induced by a connection on $P_{SO(n)}$. This is given by a collection of $\underline{so}(n)$-valued 1-forms $\omega_\alpha \in \Omega^1(U_\alpha)\otimes \underline{so}(n)$ satisfying the transition rules (1.1.1). Using the double covering map $\rho : Spin(n) \to SO(n)$ we obtain a $Spin(n)$-connection given by the collection

$$\tilde{\omega}_\alpha = \rho_*^{-1}(\omega_\alpha) \in \Omega^1(U_\alpha) \otimes \underline{spin}(n).$$

Then the collection of $\mathrm{End}\,(\mathbb{S}_0)$-valued 1-forms $\tau_*(\tilde{\omega}_\alpha)$ defines a connection $\tilde{\nabla}^M$ on $\mathbb{S}_M$, compatible with the $Spin(n)$-structure. The proof of the following result is left to the reader as an exercise.

**Proposition 1.3.17.** *$(\mathbb{S}_0, \mathbf{c}, \tilde{\nabla}^M)$ is a geometric Dirac bundle.*

The geometric Dirac operator associated to the above Dirac structure is called the *spin Dirac operator* associated to the given *spin*-structure on $M$. We will denote it by $\mathfrak{D}_M$.

It is useful to have a local description of this Dirac operator. Suppose $(e_i)$ is a local, oriented, orthonormal frame of $TM$ over $U_\alpha$ and denote by $(e^i)$ the dual coframe. Then the Levi-Civita connection on $TM$ is given by

$$\nabla e_j = \sum_i \omega_{ij} e_i, \quad \omega_{ij} \in \Omega^1(U_\alpha), \quad \omega_{ij} = -\omega_{ji}$$

and on $T^*M$ by

$$\nabla e^j = \sum_i \omega_{ij} e^i = \sum_{k,i} e^k \otimes \omega_{kij} e^i.$$

Using (1.3.7) we obtain

$$\tilde{\nabla}^M = d - \frac{1}{4}\sum_{i,j} \omega_{ij}\mathbf{c}(e^i)\mathbf{c}(e^j) = d - \frac{1}{4}\sum_{i,j,k} e^k \otimes \omega_{kij}\mathbf{c}(e^i)\mathbf{c}(e^j).$$

We deduce

$$\mathfrak{D}_M = \sum_k \mathbf{c}(e^k)\partial_{e_k} - \frac{1}{4}\sum_{i,j,k} \omega_{kij}\mathbf{c}(e^k)\mathbf{c}(e^i)\mathbf{c}(e^j). \tag{1.3.9}$$

The curvature of the connection $\tilde{\nabla}^M$ can be obtained as follows. The Riemannian curvature $R$ of $M$ (or equivalently, the curvature of the Levi-Civita connection on $TM$) is given by the collection of $\underline{so}(n)$-valued 2-forms

$$R_\alpha = d\omega_\alpha + \frac{1}{2}\omega_\alpha \wedge \omega_\alpha = \sum_{k<\ell} e^k \wedge e^\ell R_{k\ell}$$

where $R_{k\ell} : U_\alpha \to \underline{so}(n)$ is given by

$$R_{k\ell} e_j = R^i_{jkl} e_i = R_{ijk\ell} e_i.$$

Then the curvature of the connection $(\tilde{\omega}_\alpha)$ on $P_{Spin(n)}$ is given by

$$\tilde{R} = \rho_*^{-1}(R) = \sum_{k<\ell} e^k \wedge e^\ell \rho_*^{-1}(R_{k\ell}) \overset{(1.3.7)}{=} -\frac{1}{4}\sum_{k<\ell} e^k \wedge e^\ell \sum_{i<j} R^i_{jk\ell} e_i \cdot e_j.$$

The curvature of $\tilde{\nabla}^M$ is then

$$F(\tilde{\nabla}^M) = -\frac{1}{4}\sum_{k<\ell} e^k \wedge e^\ell \sum_{i<j} R^i_{jk\ell} \mathbf{c}(e_i) \cdot \mathbf{c}(e_j).$$

Using Proposition 1.3.2 and the above expression one can prove the following important result.

**Theorem 1.3.18. (Lichnerowicz)** *$\mathfrak{D}_M$ is a formally selfadjoint operator and*

$$\mathfrak{D}_M^2 = (\tilde{\nabla}^M)^* \tilde{\nabla}^M + \frac{s}{4} \tag{1.3.10}$$

*where $s$ denotes the scalar curvature of the Riemannian manifold $M$.*

**Remark 1.3.19.** Suppose $\nabla^M$ is a metric connection on $T^*M$, *not necessarily the Levi-Civita connection.* Choosing an orthonormal coframe $(e^i)$ as above we can represent

$$\nabla^M e^j = \sum_{k,i} \Omega_{kij} e^k \otimes e^i.$$

Using again the isomorphism $\tau$ we obtain a connection $\hat{\nabla}^M = \tau^* \nabla$ on $\mathbb{S}_0$, locally described by

$$\hat{\nabla}^M = d - \frac{1}{4} \sum_{i,j,k} e^k \otimes \Omega_{kij} \mathbf{c}(e^i)\mathbf{c}(e^j).$$

It satisfies the following compatibility relation:

$$\hat{\nabla}^M_X \mathbf{c}(\alpha) = \mathbf{c}(\nabla^M_X \alpha), \quad \forall X \in \operatorname{Vect}(M), \alpha \in \Omega^1(M).$$

Then $(\mathbb{S}_0, \mathbf{c}, \nabla^M, \hat{\nabla}^M)$ is a Dirac structure called the Dirac structure induced by the connection $\nabla^M$. As explained in Sec. 1.3.1, this Dirac structure determines a Dirac operator we will call the Dirac operator induced by the connection $\nabla^M$.

**Exercise 1.3.19.** Suppose $(M, g)$ is a Riemannian *spin*-manifold and $\nabla^M$ is a metric connection. The trace of its torsion is the 1-form $\operatorname{tr}(T)$ locally defined by

$$\operatorname{tr}(T)(e_i) = \sum_k g(e_k T(e_k, e_i))$$

where $(e_i)$ is a local orthonormal frame. Show that the induced Dirac operator is formally selfadjoint if and only if the torsion of $\nabla^M$ is traceless, $\operatorname{tr}(T) \equiv 0$.

The above construction can be generalized as follows. Given a Hermitian vector bundle $E \to M$ and a Hermitian connection $\nabla^E$ we can define a geometric Dirac structure

$$(\mathbb{S}_M \otimes E, \mathbf{c}_E, \nabla)$$

on $M$ where

$$\mathbf{c}_E : \Omega^*(M) \xrightarrow{\mathbf{c}} \operatorname{End}(\mathbb{S}_M) \overset{\otimes \mathbf{1}_E}{\hookrightarrow} \operatorname{End}(\mathbb{S}_M \otimes E)$$

and $\nabla$ is the connection on $\mathbb{S}_M \otimes E$ induced by the connection $\tilde{\nabla}^M$ on $\mathbb{S}_M$ and the connection $\nabla^E$ on $E$. We denote by $\mathfrak{D}_{M,E}$ the associated geometric Dirac operator. We say that $\mathfrak{D}_{M,E}$ is obtained from $\mathfrak{D}_M$ by twisting with the pair $(E, \nabla^E)$.

**Exercise 1.3.20.** Prove that the above triple $(\mathbb{S}_M \otimes E, \mathbf{c}_E, \nabla)$ is indeed a geometric Dirac structure on $M$.

The curvature of $\nabla$ is

$$F(\nabla) = F(\tilde{\nabla}^M) \otimes \mathbf{1}_E + \mathbf{1}_{\mathbb{S}_M} \otimes F(\nabla^E).$$

From the Weitzenböck formula we deduce

$$\mathfrak{D}^2_{M,E} = \nabla^*\nabla + \frac{s}{4} + \mathbf{c}(F(\nabla^E)). \tag{1.3.11}$$

The endomorphism $\mathcal{R} = \frac{s}{4} + \mathbf{c}(F(\nabla^E))$ is the Weitzenböck remainder of the generalized Laplacian $\mathfrak{D}^2_{M,E}$.

At this point we want to discuss some features of the above formula when $\dim M$ is even. In this case $\mathbb{S}_M$ is $\mathbb{Z}_2$-graded

$$\mathbb{S}_M = \mathbb{S}^+_M \oplus \mathbb{S}^-_M$$

and in particular we obtain a splitting

$$\mathbb{S}_M \otimes E = \mathbb{S}^+_M \otimes E \oplus \mathbb{S}^-_M \otimes E.$$

With respect to the above grading the operator $\mathfrak{D}_{M,E}$ has the block decomposition

$$\mathfrak{D}_{M,E} = \begin{bmatrix} 0 & \not{\mathfrak{D}}^*_{M,E} \\ \not{\mathfrak{D}}_{M,E} & 0 \end{bmatrix}$$

where $\mathfrak{D}\!\!\!/_{M,E} : C^\infty(\mathbb{S}_M^+ \otimes E) \to C^\infty(\mathbb{S}_M^- \otimes E)$. Then

$$\mathfrak{D}_{M,E}^2 = \begin{bmatrix} \mathfrak{D}\!\!\!/_{M,E}^*\mathfrak{D}\!\!\!/_{M,E} & 0 \\ 0 & \mathfrak{D}\!\!\!/_{M,E}\mathfrak{D}\!\!\!/_{M,E}^* \end{bmatrix}.$$

We conclude that the Weitzenböck remainder $\mathcal{R}$ of $\mathfrak{D}_{M,E}^2$ has the block decomposition

$$\mathcal{R} = \begin{bmatrix} \mathcal{R}_+ & 0 \\ 0 & \mathcal{R}_- \end{bmatrix}$$

When $\dim M = 4$ we can be more specific. Using the computation in the Example 1.3.3 we deduce

$$\mathfrak{D}\!\!\!/_{M,E}^*\mathfrak{D}\!\!\!/_{M,E} = \nabla^*\nabla + \frac{s}{4} + \mathbf{c}\big(F^+(\nabla^E)\big), \tag{1.3.12}$$

$$\mathfrak{D}\!\!\!/_{M,E}\mathfrak{D}\!\!\!/_{M,E}^* = \nabla^*\nabla + \frac{s}{4} + \mathbf{c}\big(F^-(\nabla^E)\big) \tag{1.3.13}$$

where $F^\pm(\nabla^E)$ denotes the self/antiself-dual part of the curvature of $\nabla^E$.

Assume now that $(M, g)$ is an oriented, $n$-dimensional Riemannian manifold equipped with a *spin*$^c$ structure $\sigma \in Spin^c(M)$. Denote by $(g_{\alpha\beta})$ a collection of gluing data defining the $SO$ structure $P_{SO(M)}$ on $M$ with respect to some good open cover $(U_\alpha)$. Moreover, we assume $\sigma$ is defined by the data

$$h_{\alpha\beta} : U_{\alpha\beta} \to Spin^c(n).$$

Denote by $\rho^c$ the fundamental complex spinorial representation

$$\rho^c : Spin^c(n) \to \operatorname{Aut}(\mathbb{S}_n).$$

We obtain a complex bundle

$$\mathbb{S}_\sigma(M) = P_{Spin^c} \times_{\rho^c} \mathbb{S}_n$$

which has a natural Clifford structure. This is called the *bundle of complex spinors* associated to $\sigma$. We want to point out that if $M$ is equipped with a *spin* structure then

$$\mathbb{S}_\sigma \cong \mathbb{S}_0 \otimes \det(\sigma)^{1/2}.$$

We will construct a family of geometric Dirac operators on $\mathbb{S}_\sigma(M)$.

Consider for warm-up the special case when $TM$ is trivial. Then we can assume $g_{\alpha\beta} \equiv \mathbf{1}$ and

$$h_{\alpha\beta} = (\mathbf{1}, z_{\alpha\beta}) : U_{\alpha\beta} \to Spin(n) \times S^1 \to Spin^c(n).$$

The $S^1$-cocycle $(z^2_{\alpha\beta})$ defines the line bundle $\det(\sigma)$. In this case something more happens. The collection $(z_{\alpha\beta})$ is also an $S^1$-cocycle defining the complex Hermitian line bundle $\hat{L} = \det(\sigma)^{1/2}$. Now observe that

$$\mathbb{S}_{M,\sigma} = \mathbb{S}_M \otimes \det(\sigma)^{1/2}.$$

We can now twist the Dirac operator $\mathfrak{D}_M$ with a pair $(\det(\sigma)^{1/2}, A)$, where $A$ is a Hermitian connection on $\det(\sigma)^{1/2}$ and obtain a Dirac operator on $\mathbb{S}_{M,\sigma}$. Notice that if the collection

$$\{\omega_\alpha \in \underline{u}(1) \otimes \Omega^1(U_\alpha)\}$$

defines a connection on $\det(\sigma)$, i.e.

$$\omega_\beta = \frac{dz^2_{\alpha\beta}}{z^2_{\alpha\beta}} + \omega_\alpha \text{ over } U_{\alpha\beta}$$

then the collection

$$\hat{\omega}_\alpha = \frac{1}{2}\omega_\alpha$$

defines a Hermitian connection on $\hat{L} = \det(\sigma)^{1/2}$. Moreover if $F$ denotes the curvature of $(\omega_\alpha)$ then the curvature of $(\hat{\omega}_\alpha)$ is given by

$$\hat{F} = \frac{1}{2}F. \tag{1.3.14}$$

Hence any connection on $\det(\sigma)$ defines in a unique way connection on $\mathbb{S}_\sigma(M)$.

Assume now that $TM$ is not necessarily trivial. We can however cover $M$ by open sets $(U_\alpha)$ such that each $TU_\alpha$ is trivial. If we pick from the start a connection on $\det(\sigma)$ this induces a Clifford connection on each $\mathbb{S}_{U_\alpha,\sigma}$. These can be glued back to a Clifford connection on $\mathbb{S}_{M,\sigma}$ using partitions of unity. We let the reader check the connection obtained in this way is independent of the various choices.

**Exercise 1.3.21.** Suppose $(M, g)$ is an oriented Riemannian manifold equipped with a *spin*$^c$ structure $\sigma$ and $A$ is a Hermitian connection on $\det(\sigma)$. Denote by $\hat{\nabla}^A$ the connection on $\mathbb{S}_\sigma$ induced by $A$. Given a smooth map $\gamma : M \to U(1) \subset \mathbb{C}^*$ we can construct a new connection $\gamma\hat{\nabla}^A\gamma^{-1}$. Show that this connection is induced by the connection $A - 2(d\gamma)\gamma^{-1}$ on $\det(\sigma)$. In particular, the assignment

$$(\gamma, A) \mapsto A - 2(d\gamma)\gamma^{-1}$$

defines a smooth left action of the gauge group $\mathfrak{G}_{U(1)}(\det(\sigma))$ on the space of Hermitian connection on $\det(\sigma)$.

Let $A$ be a connection on $\det(\sigma)$. Denote by $\nabla^A$ the Clifford connection it induces on $\mathbb{S}_{M,\sigma}$ and by $\mathfrak{D}_A := \mathfrak{D}_{M,A}$ the geometric Dirac operator associated to the geometric Dirac structure $(\mathbb{S}_{M,\sigma}, \mathbf{c}, \nabla^A)$. The Weitzenböck remainder of $\mathfrak{D}^2_{M,A}$ is a *local object* so in order to determine its form we can work on $U_\alpha$ where $\mathbb{S}_{U_\alpha,\sigma} = \mathbb{S}_{U_\alpha} \otimes \det(\sigma)\,|^{1/2}_{U_\alpha}$. Using the equalities (1.3.11) and (1.3.14) we deduce

$$\mathfrak{D}^2_{\sigma,A} = (\nabla^A)^*\nabla^A + \frac{1}{4}s + \frac{1}{2}\mathbf{c}(F_A) \tag{1.3.15}$$

where $F_A$ denotes the curvature of the connection $A$ on $\det(\sigma)$. If $M$ is four-dimensional then we have a splitting

$$\mathbb{S}_{M,\sigma} = \mathbb{S}^+_{M,\sigma} \oplus \mathbb{S}^-_{M,\sigma}$$

and

$$\mathfrak{D}^*_A\mathfrak{D}_A = (\nabla^A)^*\nabla^A + \frac{s}{4} + \frac{1}{2}\mathbf{c}(F^+_A). \tag{1.3.16}$$

**Exercise 1.3.22.** Suppose $M$ is a Riemannian manifold equipped with a *spin*$^c$ structure $\sigma$ and $A$ is a Hermitian connection on $\det(\sigma)$. Show that for any imaginary 1-form $\mathbf{i}a \in \mathbf{i}\Omega^1(M)$ we have

$$\mathfrak{D}_{A+\mathbf{i}a} = \mathfrak{D}_A + \frac{1}{2}\mathbf{c}(\mathbf{i}a).$$

The space $Spin^c(M)$ of *spin*$^c$ structures on $M$ is equipped with a natural involution $\sigma \mapsto \bar{\sigma}$. It can be described as follows. Suppose $\sigma$ is a *spin*$^c$ structure given by a cocycle $(h_{\alpha\beta}, z_{\alpha\beta})$. Then $\bar{\sigma}$ is the *spin*$^c$-structure defined by the cocycle $(h_{\alpha\beta}, \bar{z}_{\alpha\beta})$. We let the reader verify that the isomorphism class of $\bar{\sigma}$ depends only on the isomorphism class of $\sigma$. This involution enjoys several nice features.

**Exercise 1.3.23.** (a) For every $\sigma \in Spin^c(M)$ there exists a natural isomorphism of complex line bundles.

$$\det(\bar{\sigma}) \cong \overline{\det(\sigma)}$$

(b)* If $\dim M = 4$ then there exist natural isomorphisms of complex vector bundles

$$\vartheta : \bar{\mathbb{S}}^+_\sigma \to \mathbb{S}^+_{\bar{\sigma}},\ \ \vartheta^\sharp : \bar{\mathbb{S}}^-_\sigma \to \mathbb{S}^-_{\bar{\sigma}} \tag{1.3.17}$$

such that for every 1-form $\alpha$ on $M$ we have the equality

$$\vartheta^\sharp(\overline{\mathbf{c}_\sigma(\alpha)\psi}) = \mathbf{c}_{\bar{\sigma}}(\alpha)\vartheta(\bar{\psi})$$

where $\mathbf{c}_\sigma$ denotes the Clifford multiplication on the bundle $\mathbb{S}_\sigma$. Moreover, for every $\psi \in C^\infty(\mathbb{S}^+_\sigma)$ we have the equality

$$q(\vartheta(\bar{\psi})) = -q(\psi) \tag{1.3.18}$$

where $q(\psi)$ denotes the endomorphism $\phi \mapsto \langle \phi, \psi \rangle \psi - \frac{1}{2}|\psi|^2 \phi$. (The Hermitian metric is assumed to be complex linear in the *first* variable.)

(c) Show that for every Hermitian connection $A$ on $\det(\sigma)$ and for every $\psi \in C^\infty(\mathbb{S}_\sigma^+)$ we have the identity

$$\vartheta^\sharp(\overline{\mathfrak{D}_A \psi}) = \mathfrak{D}_{A^*} \vartheta(\bar{\psi}) \tag{1.3.19}$$

where $A^*$ denotes the connection induced by $A$ on $\det \bar{\sigma} \cong (\det \sigma)^*$.

**Hint for (b).** If $\rho_\pm : Spin(4) \to SO(\mathbb{S}_4^+)$ denotes the even/odd spinor representation then there exists a complex linear isomorphism $C_\pm : \mathbb{S}_4^\pm \to \overline{\mathbb{S}}_4^\pm$ such that $C_\pm \circ \rho_\pm = \bar{\rho}_\pm$. More precisely, if we identify $Spin(4)$ with $SU(2) \times SU(2)$ and $SU(2)$ with the group of unit quaternions then $\mathbb{S}_4^+$ is the space of quaternions $\mathbb{H}$ equipped with the complex structure given by $R_{\mathbf{i}}$, the right multiplication by $\mathbf{i}$. For $(q_+, q_-) \in Spin(4)$ the map $\rho(q_+, q_-) \in SO(\mathbb{H})$ is described by $L_{q_+}$, the left multiplication by $q_+$. The morphism $C_\pm$ is then given by $R_{\mathbf{j}}$, the right multiplication by $\mathbf{j}$. The description of $C_-$ is obtained from the above by making the changes

$$\text{left} \leftrightarrow \text{right} \quad \text{and} \quad \rho_+(q_+, q_-) = L_{q_+} \leftrightarrow \rho_-(q_+, q_-) = R_{q_-^{-1}}.$$

Suppose now that $M$ is a closed, compact, oriented 4-manifold equipped with a *spin*$^c$ structure $\sigma$. Upon choosing a connection $A$ on the associated line bundle $\det \sigma$ we obtain a Dirac operator

$$\mathfrak{D}_A : C^\infty(\mathbb{S}_\sigma^+) \to C^\infty(\mathbb{S}_\sigma^-).$$

This is an elliptic operator which has a finite index

$$\operatorname{ind}_{\mathbb{C}}(\mathfrak{D}_A) = \dim_{\mathbb{C}} \ker \mathfrak{D}_A - \dim_{\mathbb{C}} \ker \mathfrak{D}_A^*.$$

According to the celebrated Atiyah-Singer index theorem this index can be expressed in purely topological terms. More precisely, we have the following equality:

$$\operatorname{ind}_{\mathbb{C}} \mathfrak{D}_A = \frac{1}{8}\left( \int_M c_1(\det \sigma) \wedge c_1(\det \sigma) - \tau(M) \right) \tag{1.3.20}$$

where $\tau(M)$ denotes the signature of the manifold $M$.

## 1.4. Complex differential geometry

We present here a very brief survey of some basic differential geometric facts about complex manifolds in general, and complex surfaces in particular. We will return to this subject later on, in Section 3.1. This is an immense research area and our selection certainly does not do it justice. For more details and examples we refer to [**9, 10, 49, 54, 63**] and the sources therein.

**1.4.1. Elementary complex differential geometry.** An *almost complex structure* on a manifold $X$ is an endomorphism $J$ of the tangent bundle $TX$ such that $J^2 = -1$. Note in particular that such a structure can exist only on *orientable* even-dimensional manifolds. By duality we get a similar endomorphism of the cotangent bundle $T^*X$ which we continue to denote by $J$.

The operator $J$ extends by complex linearity to an endomorphism of the complexified tangent $TX \otimes \mathbb{C}$. It defines two eigenbundles corresponding to the eigenvalues $\pm\mathbf{i}$ and thus it produces a splitting of complex bundles

$$TX \otimes \mathbb{C} = (TX)^{1,0} \oplus (TX)^{1,0}$$

where the $(1,0)$ superscript indicates the $\mathbf{i}$-eigenbundle while the $(0,1)$-superscript indicates the $-\mathbf{i}$-eigenbundle. Note that $(TX, J)$ is isomorphic to $(TX)^{1,0}$ as *complex* vector bundles. Denote by $P^{1,0}$ (resp. $P^{1,0}$) the projection onto $(TX)^{1,0}$ (resp. $(TX)^{0,1}$) corresponding to the above splitting.

For any vector field $X$ on $M$ define $X_c := P^{1,0}X = \frac{1}{2}(X - \mathbf{i}JX)$ and $\bar{X}_c := P^{0,1}X = \frac{1}{2}(X + \mathbf{i}JX)$. By duality, these induce projectors of $T^*X \otimes \mathbb{C}$ and thus we get a similar splitting

$$T^*X \otimes \mathbb{C} = (T^*X)^{1,0} \oplus (T^*X)^{0,1} \tag{1.4.1}$$

which leads to a decomposition

$$\Lambda^k T^*X \otimes \mathbb{C} = \bigoplus_{p+q=k} \Lambda^{p,q} T^*X \tag{1.4.2}$$

where

$$\Lambda^{p,q} T^*X \cong \Lambda^p (T^*X)^{1,0} \otimes \Lambda^q (T^*X)^{0,1}.$$

The sections of $\Lambda^{p,q}T^*X$ are called *(p,q)-forms* on $X$. For example, if $\alpha \in \Omega^1(M) \otimes \mathbb{C}$ then $\alpha$ extends to a $C^\infty(M, \mathbb{C})$-linear map

$$\mathrm{Vect}\,(M) \otimes \mathbb{C} \to C^\infty(M, \mathbb{C})$$

and

$$\alpha = \alpha^{1,0} + \alpha^{0,1}$$

where $\alpha^{1,0}(X) := \alpha(P^{1,0}X)$ and $\alpha^{0,1}(Y) := \alpha(P^{0,1}Y)$.

**Example 1.4.1.** Consider the manifold $\mathbb{C}^n$ with coordinates $z_j = x_j + \mathbf{i}y_j$, $j = 1, \cdots, n$. It is equipped with a natural almost complex structure defined by

$$J : \frac{\partial}{\partial x_j} \mapsto \frac{\partial}{\partial y_j} \mapsto -\frac{\partial}{\partial x_j}.$$

The complex bundle $(T\mathbb{C}^n)^{1,0}$ (resp $(T^*\mathbb{C}^n)^{0,1}$) admits a global trivialization defined by

$$\frac{\partial}{\partial z_j} := \frac{1}{2}(\frac{\partial}{\partial x_j} - \mathbf{i}\frac{\partial}{\partial y_j})$$

and respectively

$$dz_j := dx_j + \mathbf{i}dy_j.$$

Similarly $(T\mathbb{C}^n)^{0,1}$ (resp. $(T^*\mathbb{C}^n)^{0,1}$) is globally trivialized by

$$\frac{\partial}{\partial \bar{z}_j} := \frac{1}{2}(\frac{\partial}{\partial x_j} + \mathbf{i}\frac{\partial}{\partial y_j})$$

and respectively

$$d\bar{z}_j = (dx_j - \mathbf{i}dy_j).$$

A $(p,q)$-form on $\mathbb{C}^n$ has the form

$$\alpha = \sum_{I,J} \alpha_{IJ} dz^I \otimes \wedge d\bar{z}^J$$

where the summation is carried over all ordered multi-indices

$$I: \;\; 1 \leq i_1 < \cdots < i_p \leq n, \;\; J: \;\; 1 \leq j_1 < \cdots < j_q \leq n$$

and $\alpha_{IJ}$ is a complex valued function on $\mathbb{C}^n$.

The exterior derivative extends by complex linearity to an operator

$$d : C^\infty\big(\Lambda^k T^* X \otimes \mathbb{C}\big) \to C^\infty\big(\Lambda^{k+1} T^* X \otimes \mathbb{C}\big).$$

It is not difficult to check that

$$d(\Lambda^{p,q}) \subset \Lambda^{p+2,q-1} \oplus \Lambda^{p+1,q} \oplus \Lambda^{p,q+1} \oplus \Lambda^{p-1,q+2}.$$

Accordingly, we get a decomposition of $d$

$$d = d^{2,-1} + d^{1,0} + d^{0,1} + d^{-1,2}.$$

Traditionally one uses the notation

$$\partial := d^{1,0}, \;\; \bar{\partial} := d^{0,1}.$$

The almost complex structure is said to be *integrable* if $d^{2,-1} = 0$ and $d^{-1,2} = 0$.

**Proposition 1.4.2.** *Consider an almost complex manifold $(M, J)$.The following conditions are equivalent.*

*(a) The almost complex structure is integrable.*

*(b) $d^{2,-1}\alpha = d^{-1,2}\alpha = 0$ for all $\alpha \in \Omega^1(M) \otimes \mathbb{C}$.*

*(c) $\bar{\partial}^2 f = 0 = \partial^2 f$, $\forall f \in C^\infty(M)$.*

*(d) The Nijenhuis tensor $N \in \Omega^2(TM)$ defined by*

$$N(X,Y) = \frac{1}{4}([JX,JY] - [X,Y] - J[X,JY] - J[JX,Y]),$$

$\forall X,Y \in \mathrm{Vect}\,(M)$, *is identically zero.*

**Proof** Clearly (a) $\Rightarrow$ (b). Using a partition of unity it is not difficult to prove the converse, (b) $\Rightarrow$ (a).

Clearly (b) $\Rightarrow$ (c). Using partitions of unity we can replace the condition "$\forall \alpha \in \Omega^1(M)$" in (b) by the condition "$\forall \alpha = fdg,\ f,g \in C^\infty(M)$". This weaker, equivalent version of (b) is clearly implied by (c). To establish the remaining equivalences we need to establish several identities of independent interest.

Let $f \in C^\infty(M)$. Then

$$\partial^2 f(X_c, Y_c) = d\partial f(X_c, Y_c) = X_c \partial f(Y_c) - Y_c \partial f(X_c) - \partial f([X_c, Y_c])$$

$$= X_c df(Y_c) - Y_c df(X_c) - df([X_c, Y_c]_c).$$

We compute each of the terms separately.

$$X_c df(Y_c) = \frac{1}{4}\Big\{ X df(Y) - JX df(JY) - \mathbf{i}(X df(JY) + JX df(Y)) \Big\}.$$

$$Y_c df(X_c) = \frac{1}{4}\Big\{ Y df(X) - JY df(JY) - \mathbf{i}(Y df(JX) + JY df(X)) \Big\}.$$

$$df([X_c, Y_c]_c) = \frac{1}{2} df([X_c, Y_c] - \mathbf{i}J[X_c, Y_c])$$

$$= \frac{1}{8} df([X - \mathbf{i}JX, Y - \mathbf{i}JY]) - \frac{\mathbf{i}}{8} df(J[X - \mathbf{i}JX, Y - \mathbf{i}JY])$$

$$= \frac{1}{8} df([X,Y] - [JX,JY] - \mathbf{i}[JX,Y] - \mathbf{i}[X,JY])$$

$$-\frac{\mathbf{i}}{8} df(J[X,Y] - J[JX,JY] - \mathbf{i}J[JX,Y] - \mathbf{i}J[X,JY])$$

$$= \frac{1}{8} df([X,Y] - [JX,JY] - J[JX,Y] - J[X,JY])$$

$$-\frac{\mathbf{i}}{8} df(J[X,Y] - J[JX,JY] + [JX,Y] + [X,JY]).$$

At this point we use the equality $d^2 f = 0$ which implies

$$U df(V) - V df(V) = df([U,V]), \ \ \forall U, V \in \mathrm{Vect}\,(M).$$

We deduce

$$X_c df(Y_c) - Y_c df(Y_c) = \frac{1}{4}\Big\{ df([X,Y]) - df([JX,JY]) \Big\}$$

$$-\frac{\mathbf{i}}{4}\Big\{df([X,JY]+df([JX,Y])\Big\}.$$

Putting together all of the above we deduce

$$\partial^2 f(X,Y)=\partial^2 f(X_c,Y_c)=\frac{1}{8}df([X,Y]-[JX,JY]+J[JX,Y]+J[X,JY])$$

$$+\frac{\mathbf{i}}{8}df(J[X,Y]-J[JX,JY]+J[JX,Y]+J[X,JY])$$

$$=-df(\overline{N(X,Y)_c})=-\bar{\partial}f(N(X,Y)).$$

Similarly

$$\bar{\partial}^2 f(X,Y)=-\partial f(N).$$

It is now clear that (c) $\Longleftrightarrow$ (d). ■

It is very easy to show that if $M$ is a complex manifold (i.e. admits local coordinates $U\to\mathbb{C}^n$ with holomorphic transition maps) then the induced almost complex structure is integrable. The converse is also true but it is highly nontrivial. It is known as the *Newlander-Nirenberg theorem.*

Suppose now that $M$ is an *almost Hermitian* manifold, i.e. $TM$ is equipped with a Riemannian metric $g$ and a compatible almost complex structure $J$, i.e. $J^*=-J$. Extend $J$ to an almost complex structure $J^\flat$ on $T^*X$ via the metric duality so that

$$(J^\flat\alpha)(X)=-\alpha(JX).$$

We obtain an eigenbundle decomposition

$$T^*X\otimes\mathbb{C}\cong\ker(\mathbf{i}-J^\flat)\oplus\ker(\mathbf{i}+J^\flat)\cong(T^*X)^{1,0}\oplus(T^*X)^{0,1}$$

which coincides with the splitting in (1.4.1). Now define $\omega\in\Omega^2(M)$ by

$$\omega(X,Y)=g(JX,Y),\ \ \forall X,Y\in\mathrm{Vect}\,(M).$$

Note that $\omega\in\Omega^{1,1}(M)$. We can now define a Hermitian metric on the complex bundle $(TX,J)$ by

$$h(X,Y)=g(X,Y)-\mathbf{i}\omega(X,Y).$$

It is often very useful to have local descriptions of the various notions. Pick a local orthonormal frame of $TM$

$$\{e_1,f_1;\cdots;e_n,f_n\},\ \ f_k=Je_k.$$

Then $\varepsilon_j=\frac{1}{\sqrt{2}}(e_j-\mathbf{i}f_j)$ form a local, complex, *unitary* frame of $T^{1,0}$ while $\bar{\varepsilon}_j=\frac{1}{\sqrt{2}}(e_i+\mathbf{i}f_j)$ form a local, complex, unitary frame of $T^{0,1}$. If we denote by $(e^j,f^j)$ the dual basis of $(e_j,f_j)$ then

$$\varepsilon^j=\frac{1}{\sqrt{2}}(e^j+\mathbf{i}f^j)$$

is a local unitary frame of $(T^*X)^{1,0}$ and

$$\bar{\varepsilon}^k := \frac{1}{\sqrt{2}}(e^k - \mathbf{i}f^k)$$

is a local unitary frame of $(T^*X)^{0,1}$. Then

$$\omega = \mathbf{i}\sum_j \varepsilon^j \wedge \bar{\varepsilon}^j.$$

If $D$ denotes the Levi-Civita connection then we have the following identity (see [**64**, IX, §4, vol.2]):

$$(D_X\omega)(Y,Z) = g((D_XJ)Y,Z)$$

$$= -\frac{1}{2}d\omega(X,JY,JZ) + \frac{1}{2}d\omega(X,Y,Z) + 2g(N(Y,Z),JX). \tag{1.4.3}$$

**Exercise 1.4.1.** Prove the identity (1.4.3).

Suppose now that $d\omega = 0$. The identity (1.4.3) simplifies dramatically to

$$(D_X\omega)(Y,Z) = g((D_XJ)Y,Z) = 2g(N(Y,Z),JX). \tag{1.4.4}$$

**Definition 1.4.3.** An almost Hermitian manifold $(M,g,J)$ is said to be *almost Kähler* if the form $\omega$ is closed. An almost Kähler manifold $(M,g,J)$ is said to be *Kähler* if the almost complex structure $J$ is integrable.

**Exercise 1.4.2.** Suppose $(M^{2n},\omega)$ is a symplectic manifold , i.e $\omega$ is a closed 2-form amd $\omega^n$ is a volume form on $M$. Show that there exist almost Káhler structures $(g,J)$ on $M$ such that

$$\omega(X,Y) = g(JX,Y), \ \ \forall X,Y \in \operatorname{Vect}(M).$$

In this case both $g$ are said to be *adapted* to $\omega$. Moreover, show that when $n = 2$ the symplectic form $\omega$ is self-dual with respect to any adapted metric.

Using the metric duality we can regard any tensor $B \in \Omega^2(TM)$ as a $T^*M$-valued 2-form

$$\langle B(X,Y),Z\rangle := g(B(X,Y),Z), \ \ \forall X,Y,Z \in \operatorname{Vect}(M)$$

where $\langle\cdot,\cdot\rangle$ denotes the duality between $T^*M$ and $TM$. Now define the *Bianchi projector*

$$\mathfrak{b}B(X,Y,Z) = \langle B(X,Y),Z\rangle + \langle B(Y,Z)X\rangle + \langle B(Y,Z),X\rangle.$$

Then $\mathfrak{b}N$ is a 3-form. If $d\omega = 0$ then using the elementary identity

$$N(JY,JZ) = -N(Y,Z)$$

we deduce

$$D_X\omega(Y,Z) = -2g(N(JY,JZ),JX)$$

so that $\forall X,Y,Z \in \mathrm{Vect}\,(M)$

$$\mathfrak{b}N(JX,JY,JZ) = -\frac{1}{2}(\mathfrak{b}D\omega)(X,Y,Z) = -\frac{1}{2}d\omega(X,Y,Z) = 0 \tag{1.4.5}$$

where at the second step we have use the following identity (see Exercise 1.4.4 for a more general situation)

$$d\eta(X,Y,Z) = \mathfrak{b}(D\eta)(X,Y,Z), \ \ \forall \eta \in \Omega^2(M), \ X,Y,Z \in \mathrm{Vect}\,(M).$$

Consider now an almost Hermitian manifold $(M,g,J)$. A connection $\nabla$ on $TX$ is said to be Hermitian if $\nabla g = 0$ and $\nabla J = 0$.

If $\nabla$ is such a connection then its *torsion* is the $TM$-valued 2-form $T \in \Omega^2(TM)$ defined by

$$T(X,Y) = \nabla_X Y - \nabla_Y X - [X,Y], \ \ \forall X,Y \in \mathrm{Vect}\,(M).$$

**Proposition 1.4.4.** *Suppose $\nabla$ is a Hermitian connection on an almost Hermitian manifold $(M,g,J)$ and denote by $T$ its torsion. Then $\forall X,Y \in$* Vect $(M)$

*(a)*

$$\begin{aligned} 4N(X,Y) &= T(X,Y) + JT(JX,Y) + JT(X,JY) - T(JX,JY) \\ &= N(X,Y) + JN(JX,Y) + JN(X,JY) - N(JX,JY). \end{aligned} \tag{1.4.6}$$

*(b) If $(M,g,J)$ is almost Kähler then there exists a* unique Hermitian *connection $\nabla$ on $TM$ such that*

$$T_\nabla = N.$$

**Proof** (a) We prove only the first equality in (1.4.6). It all begins with the identity

$$[X,Y] = \nabla_X Y - \nabla_Y X - T(X,Y).$$

Then

$$\begin{aligned} [JX,JY] &= \nabla_{JX}(JY) - \nabla_{JY}(JX) - T(JX,JY) \\ &= J(\nabla_{JX}Y - \nabla_{JY}X) - T(JX,JY). \\ J[X,JY] &= J(\nabla_X(JY) - \nabla_{JY}X - T(X,JY)) \\ &= -\nabla_X Y - J\nabla_{JY}X - JT(X,JY). \\ J[JX,Y] &= J(\nabla_{JX}Y - \nabla_Y(JX) - T(JX,Y)) \end{aligned}$$

$$= J\nabla_{JX}Y + \nabla_Y X - JT(JX, Y).$$

We deduce

$$4N(X,Y) = T(X,Y) + JT(JX,Y) + JT(X,JY) - T(JX,JY).$$

(b) We first need to prove an auxiliary result.

**Lemma 1.4.5.** *For any $TM$-valued 2-form $T$ there exists a unique connection on $TM$ compatible with the metric whose torsion is precisely $T$.*

**Proof of the lemma** Denote by $D$ the Levi-Civita connection on $M$. Then any other metric connection has the form

$$\nabla = D + A, \quad A \in \Omega^1(\mathrm{End}_-(TM))$$

where $\mathrm{End}_-(TM)$ denotes the bundle of skew-symmetric endomorphisms of $TM$. Since $D$ has no torsion we deduce that the torsion of $\nabla$ is

$$T_\nabla(X,Y) = A_X Y - A_Y X, \quad \forall X, Y \in \mathrm{Vect}\,(M)$$

where $A_X$ denotes the contraction of $A$ with $X$. We can regard $A$ as a $T^*M$-valued 2-form using the identification

$$\langle A(X,Y), Z\rangle := g(A_Z X, Y).$$

Thus we deduce

$$\langle T(X,Y), Z\rangle = \langle A(Y,Z), X\rangle + \langle A(Z,X), Y\rangle. \tag{1.4.7}$$

A cyclic summation leads to the identity

$$\mathfrak{b}T = 2\mathfrak{b}A.$$

We can now rewrite (1.4.7) as follows:

$$\langle T(X,Y), Z\rangle = (\mathfrak{b}A)(X,Y,Z) - \langle A(X,Y), Z\rangle$$

$$= \frac{1}{2}\mathfrak{b}T(X,Y,Z) - \langle A(X,Y), Z\rangle.$$

Hence

$$A = -T + \frac{1}{2}\mathfrak{b}T.$$

The lemma is proved. ■

According to Lemma 1.4.5 there exists a unique metric connection $\nabla$ on $TM$ such that $T = N$. It is explicitly defined by

$$\nabla = D - N + \frac{1}{2}\mathfrak{b}N.$$

We have to show that when $(M, g, J)$ is almost Kähler this connection is also Hermitian, i.e.

$$\nabla J = 0.$$

Note first that in this case, according to (1.4.5), we have

$$\nabla = D - N,$$

that is,

$$g(\nabla_X Y, Z) = g(D_X Y, Z) - g(N(Y, Z), X), \quad \forall X, Y, Z \in \mathrm{Vect}\,(M).$$

We have to show that

$$g(D_X JY, Z) - g(N(JY, Z), X) = -g(D_X Y, JZ) + g(N(Y, JZ), X)$$

or equivalently

$$g(D_X JY, Z) + g(D_X Y, JZ) = g(N(JY, Z), X) + g(N(Y, JZ), X). \tag{1.4.8}$$

Note that

$$N(JY, Z) = N(Y, JZ) = -JN(Y, Z)$$

and

$$g(D_X JY, Z) + g(D_X Y, JZ) = g((D_X J)Y, Z)$$

so that (1.4.8) is equivalent to

$$g((D_X J)Y, Z) = 2g(N(Y, Z), JX)$$

which is precisely (1.4.4). The proposition is proved. ■

**Remark 1.4.6.** (a) If $J$ is integrable (so that $M$ is Kähler) then $N = 0$ so that the connection constructed in the above proposition is precisely the Levi-Civita connection.

(b) One can show (see [**64**]) that on any almost complex manifold there exist many connections compatible with the almost complex structure and torsion $N$. We refer to the survey [**46**] for additional facts on Hermitian connections.

**Definition 1.4.7.** The *Chern connection* of an almost Kähler manifold $(M, g, J)$ is the unique Hermitian connection with torsion $N$.

**Exercise 1.4.3.** Suppose that $(M, g, J)$ is an almost Kähler manifold and $D$ is the Levi-Civita connection of $g$. Show that the Chern connection $\nabla$ associated to the almost Kähler structure can be described as

$$\nabla_X = D_X - \frac{1}{2} J(D_X J), \quad \forall X \in \mathrm{Vect}\,(M).$$

**Exercise 1.4.4.** Suppose $(M, g)$ is a Riemannian manifold and $\nabla$ is a connection on $TM$ compatible with the metric $g$ with torsion $T$. Define $\mathrm{tr}\,(T) \in \Omega^1(M)$ by

$$\mathrm{tr}\,(T)(X) = \sum_i g(e_i, T(e_i, X)), \;\; \forall X \in \mathrm{Vect}\,(X)$$

where $e_i$ denotes a local orthonormal frame on $M$. Show that for any $\eta \in \Omega^p(M)$ we have

$$d\eta(X_0, \cdots, X_p) = \sum_{j=0}^{p} (-1)^j (\nabla_{X_j}\eta)(X_0, \cdots, \hat{X}_j, \cdots, X_p)$$

$$+\sum_{j<k} (-1)^{j+k}\eta(T(X_j, X_k), X_0, \dots, \hat{X}_j, \dots, \hat{X}_k, \cdots, X_p), \tag{1.4.9}$$

$$d^*\eta(X_1, \dots, X_{p-1}) = -\sum_{i=1}^{\dim M} (\nabla_{e_i}\eta)(e_i, X_1, \cdots, X_{p-1})$$

$$+\eta((\mathrm{tr}\, T)^\flat, X_1, \dots, X_{p-1}) \tag{1.4.10}$$

$$-\sum_{j=1}^{p-1} (-1)^j \langle\, g(X_j, T)\,,\, \eta(\bullet, \bullet, X_1, \cdots, \hat{X}_j, \cdots, X_{p-1})\,\rangle$$

where $(e_i)$ is a local orthonormal frame, $\mathrm{tr}\,(T)^\flat$ denotes the vector field dual to $\mathrm{tr}\,(T)$, $g(X_j, T)$ denotes the 2-form $(X, Y) \mapsto g(X_j, T(X, Y))$ and the pairing $\langle \bullet, \bullet \rangle$ refers to the inner product of two forms. (Observe that the above identities extend by complex linearity to complex valued forms and vectors.)

**Exercise 1.4.5.** Suppose $(M, g, J)$ is an almost Kähler manifold and $\nabla$ is the associated Chern connection.

(a) Show that $\mathrm{tr}\,(N) = 0$.

(b) Show that if $X, Y \in C^\infty(T^{0,1}M)$ then $N(X, Y) \in C^\infty(T^{1,0}M)$.

(c) Denote by $g_c$ the extension of $g$ by complex bilinearity to $TM \otimes \mathbb{C}$. Show that for every $X \in C^\infty(T^{0,1}M)$ the 2-form $\omega_X$ defined by

$$\omega_X(Y, Z) = g_c(X, N(Y, Z))$$

has type $(0, 2)$, i.e.

$$\omega_X(JY, Z) = \omega_X(Y, JZ) = -\mathbf{i}\omega_X(Y, Z), \;\; \forall Y, Z \in \mathrm{Vect}\,(M).$$

(d) Show that for any $\eta \in \Omega^{0,p}(M)$ and any $Z_0, \cdots, Z_p \in C^\infty(T^{0,1}M)$ we have the identities

$$\bar{\partial}\eta(Z_0, \ldots, Z_p) = \sum_{j=1}^{p}(-1)^j(\nabla_{Z_j}\eta)(Z_0, \cdots, \hat{Z}_j, \cdots, Z_p), \tag{1.4.11}$$

$$\bar{\partial}^*\eta(Z_1, \cdots, Z_{p-1}) = -\sum_{i=1}^{\dim M}(\nabla_{e_i}\eta)(e_i, Z_1, \cdots, Z_{p-1}) \tag{1.4.12}$$

where $e_i$ denotes a local, orthonormal frame of $TM$. (For a generalization of these identities we refer to [**46**].)

**Hint:** Use that fact that for any $Z_0, \cdots, Z_p \in C^\infty(T^{0,1}M)$ and $\eta \in \Omega^{0,p}(M)$ we have

$$(\bar{\partial}\eta)(Z_0, \cdots, Z_p) = d\eta(Z_0, \cdots, Z_p)$$

and

$$(\bar{\partial}^*\eta)(Z_1, \cdots, Z_{p-1}) = d^*\eta(Z_1, \cdots, Z_{p-1}).$$

In the remainder of this section we will assume $(M, g, J)$ is an almost Kähler manifold. Denote by $\omega$ the associated symplectic form

$$\omega(X, Y) = g(JX, Y), \ \ \forall X, Y \in \operatorname{Vect}(M).$$

Set $2n = \dim M$. We orient $M$ using the nowhere vanishing $2n$-form $\omega^n$. Note that

$$dv_g = \frac{1}{n!}\omega^n.$$

Using the metric $g$ and the above orientation we obtain a Hodge operator

$$* : \Omega^p(M) \to \Omega^{2n-p}(M)$$

which we extend by complex *anti-linearity* to an operator

$$* : \Omega^p(M) \otimes \mathbb{C} \to \Omega^{2n-p}(M).$$

**Exercise 1.4.6.** Let $\varphi \in \Omega^{p,q}(M)$. Prove that

$$*\varphi \in \Omega^{n-p,n-q}(M)$$

and

$$\varphi \wedge *\varphi = |\varphi|^2 dv_g$$

where $|\bullet|$ denotes the Hermitian metric induced by $(g, J)$ on $\Lambda^{p,q}T^*M$.

Exterior multiplication by $\omega$ defines a bundle morphism

$$L : \Omega^{p,q}(M) \to \Omega^{p+1,q+1}(M).$$

Its adjoint, $L^* = \Lambda : \Omega^{p+1,q+1}(M) \to \Omega^{p,q}(M)$, is called the *contraction* by the symplectic form.

**Exercise 1.4.7.** Suppose $(M, g, J)$ is an almost Kähler manifold and $(e_i, f_i)$ is a local orthonormal frame such that $f_i = Jf_i$ for all $i$. Its dual coframe will be denoted by $(e^i, f^i)$ and, as usual, set

$$\varepsilon_i = 2^{-1/2}(e_i - \mathbf{i}f_i),\ \ \bar{\varepsilon}_i = 2^{-1/2}(e_i + \mathbf{i}f_i),$$

$$\varepsilon^i = 2^{-1/2}(e^i + \mathbf{i}f^i),\ \ \bar{\varepsilon}^i = 2^{-1/2}(e^i - \mathbf{i}f^i).$$

For $k = 1, 2, \cdots, n$ we denote by $\imath_k$ and $\bar{\imath}_k$ the (locally defined) odd derivations of $\Omega^{*,*}(M)$ uniquely determined by

$$\imath_k\varepsilon^i = \delta^i_k = \bar{\imath}_k\bar{\varepsilon}^i,\ \imath_k\bar{\varepsilon}^i = \bar{\imath}_k\varepsilon^i = 0$$

where $\delta$ denotes the Kronecker symbol. Show that locally

$$\Lambda = -\mathbf{i}\sum_k \bar{\imath}_k\imath_k.$$

Denote by $\Pi^{p,q}$ the natural projection $\Omega^*(M) \otimes \mathbb{C} \to \Omega^{p,q}(M)$ and set

$$\Pi = \sum_{p,q} \mathbf{i}^{p-q}\Pi^{p,q} : \Omega^*(M) \otimes \mathbb{C} \to \Omega^*(M) \otimes \mathbb{C},$$

$$H = \sum_{p,q}(n - p - q)\Pi^{p,q}.$$

Observe that $\Pi$ is bijective and $\Pi^* = \Pi^{-1}$. Now define $d_c, d_c^* : \Omega^*(M) \otimes \mathbb{C} \to \Omega^*(M) \otimes \mathbb{C}$ by $d_c = \Pi^{-1}d\Pi$ and $d_c^* = \Pi^{-1}d\Pi$.

**Example 1.4.8.** Consider the space $\mathbb{C}^n$ (with coordinates $z^1, \cdots, z^n$) equipped with the canonical Kähler structure

$$\omega_0 = \frac{\mathbf{i}}{2}\sum_i dz^i \wedge d\bar{z}^i.$$

Set $\varepsilon^i = \frac{1}{\sqrt{2}}dz^i$ and $\bar{\varepsilon}^i = \frac{1}{\sqrt{2}}d\bar{z}^i$. For every pair of ordered multi-indices

$$I = (i_1 < \cdots < i_k),\ \ J = (j_1 < \cdots < j_m)$$

we set

$$\varepsilon^I = \varepsilon^{i_1} \wedge \cdots \wedge \varepsilon^{i_k},\ \ \bar{\varepsilon}^J = \varepsilon^{j_1} \wedge \cdots \wedge \varepsilon^{j_m}.$$

Denote by $I^c$ the ordered multi-index complementary to $I$, i.e. as unordered sets, we have the equality $I^c = \{1, \cdots, n\}\backslash I$. Also denote by $\sigma_I$ the signature of the permutation obtained by writing the multi-indices $I$ and $I^c$ one after the other.

We can rewrite

$$\omega_0 = \mathbf{i}\sum_i \varepsilon^i \wedge \bar{\varepsilon}^i$$

so that

$$\frac{1}{n!}\omega_0^n = \mathbf{i}^{n^2}\varepsilon^1 \wedge \cdots \wedge \varepsilon^n \wedge \bar{\varepsilon}^1 \wedge \cdots \wedge \bar{\varepsilon}^n.$$

Observe that

$$*\varepsilon^i = \frac{1}{\sqrt{2}}(*dx^i - \mathbf{i} * dy^i)$$

$$= \frac{1}{\sqrt{2}}(dx^1 \wedge dy^1) \wedge \cdots \wedge \widehat{(dx^i \wedge dy^i)} \wedge \cdots \wedge (dx^n \wedge dy^n) \wedge (dy^i + \mathbf{i}dx^i)$$

(hat $\longleftrightarrow$ missing term)

$$= \mathbf{i}^{(n-1)^2+1}(-1)^{n-i}\varepsilon^1 \wedge \cdots \wedge \widehat{\varepsilon^i} \wedge \cdots \wedge \varepsilon^n \wedge \bar{\varepsilon}^1 \wedge \cdots \wedge \bar{\varepsilon}^n$$

Using the above exercise we deduce

$$*\omega_0 = \frac{1}{(n-1)!}\omega_0^{n-1}$$

and, more generally,

$$*(\varepsilon^I \wedge \bar{\varepsilon}^J) = (-1)^{|J|(n-|I|)}\mathbf{i}^{n^2}\sigma_I\sigma_J\varepsilon^{I^c} \wedge \bar{\varepsilon}^{J^c}.$$

The operators we have introduced above satisfy a series of important identities. For a proof of the following proposition we refer to [**146**].

**Proposition 1.4.9.** *Suppose $(M, g, J)$ is an almost Kähler manifold. Then*

$$\Pi^2 = *^2 = \sum_{p,q}(-1)^{p+q}\Pi^{p,q},$$

$$\Lambda = *^{-1}L*, \quad d^* = - * d*,$$

$$[L, \Lambda] = H,$$

$$[L, d] = [\Lambda, d^*] = [L, d_c] = [\Lambda, d_c^*] = 0,$$

$$[L, d^*] = d_c, \quad [\Lambda, d] = -d_c^*, \quad [L, d_c^*] = -d, \quad [\Lambda, d_c] = d^*.$$

When $M$ is Kähler the above list of identities can be considerably enriched. For a proof of the following important identities we refer to [**49**].

**Proposition 1.4.10.** *Suppose $(M, g, J)$ is a* Kähler *manifold. Then*

$$\partial^* = -*\partial*, \quad \bar{\partial}^* = -*\bar{\partial}*, \quad d^* = \partial^* + \bar{\partial}^*,$$

$$[L, \partial] = [L, \bar{\partial}] = [\Lambda, \partial^*] = [\Lambda, \bar{\partial}^*] = 0,$$

$$[L, \partial^*] = \mathbf{i}\bar{\partial}, \quad [L, \bar{\partial}^*] = -\mathbf{i}\partial,$$

$$[\Lambda, \partial] = \mathbf{i}\bar{\partial}^*, \quad [\Lambda, \bar{\partial}] = -\mathbf{i}\partial^*,$$

$$\partial\bar{\partial}^* = -\bar{\partial}^*\partial = -\mathbf{i}\partial^* L\partial^* = -\mathbf{i}\partial\Lambda\partial$$

$$\bar{\partial}\partial^* = -\partial^*\bar{\partial} = -\mathbf{i}\partial^* L\partial^* = \mathbf{i}\bar{\partial}\Lambda\partial.$$

*If we set $\Delta_d = dd^* + d^*d$, $\Delta_\partial = \partial\partial^* + \partial^*\partial$ and $\Delta_{\bar{\partial}} = \bar{\partial}^*\bar{\partial} + \bar{\partial}\bar{\partial}^*$ then*

$$\Delta_\partial = \Delta_{\bar{\partial}} = \frac{1}{2}\Delta_d.$$

We include here for later use some simple consequences of these identities.

**Corollary 1.4.11.** *Suppose $(M, g, J)$ is a Kähler manifold. Then we have the following identities.*

$$\mathbf{i}\Lambda(\partial\alpha) = -\bar{\partial}^*\alpha, \quad \forall \alpha \in \Omega^{0,1}(M), \tag{1.4.13}$$

$$\mathbf{i}\Lambda\bar{\partial}\beta = \partial^*\beta, \quad \forall \beta \in \Omega^{1,0}(M), \tag{1.4.14}$$

$$\mathbf{i}\Lambda(\partial\bar{\partial}f) = -\frac{1}{2}d^*df \quad \forall f \in \Omega^{0,0}(M). \tag{1.4.15}$$

**Proof** To prove (1.4.13) we use the commutator identity

$$[\Lambda, \partial] = \mathbf{i}\bar{\partial}^*.$$

We deduce

$$\Lambda\partial\alpha = \partial\Lambda\alpha + \mathbf{i}\bar{\partial}^*\alpha = \mathbf{i}\bar{\partial}^*\alpha$$

since $\Lambda\alpha = 0$ because $\alpha \in \Omega^{0,1}(M)$. The first identity is proved. The same method proves the second identity as well. The third identity follows from the first and the equality $\Delta_{\bar{\partial}} = \frac{1}{2}\Delta_d$. ■

The identities in Proposition 1.4.10 do not hold for almost Kähler manifolds but surprisingly the identities in Corollary 1.4.11 continue to hold on an arbitrary almost Kähler manifold. We will spend the remainder of this subsection proving this fact.

**Proposition 1.4.12.** *The identities (1.4.13) – (1.4.15) continue to hold for arbitrary almost Kähler manifolds.*

**Proof** We prove only (1.4.13) and

$$\frac{1}{2}d^* df = \bar{\partial}^* \bar{\partial} f, \ \ \forall f \in \Omega^{0,0}(M). \tag{1.4.16}$$

The identity (1.4.14) follows from (1.4.13) by complex conjugation while (1.4.15) follows from (1.4.13) and (1.4.16).

Denote by $\nabla$ the Chern connection of the almost Kähler structure and choose a local orthonormal frame $(e_i, f_i)$ as in Exercise 1.4.7. To prove (1.4.13) we use the identity $\partial\alpha = (d\alpha)^{1,1}$, that is,

$$\partial\alpha(\varepsilon_i, \bar{\varepsilon}_j) = (d\alpha)(\varepsilon_i, \bar{\varepsilon}_j), \ \ \forall i, j.$$

At this point we want to use the fact that the torsion of the Chern connection is $N$ and the identity (1.4.9)

$$(d\alpha)(\varepsilon_i, \bar{\varepsilon}_j) = (\nabla_{\varepsilon_i}\alpha)(\bar{\varepsilon}_j) - (\nabla_{\bar{\varepsilon}_j}\alpha)(\varepsilon_i) + \alpha(N(\varepsilon_i, \bar{\varepsilon}_j)).$$

$$= (\nabla_{\varepsilon_i}\alpha)(\bar{\varepsilon}_j) + \alpha\big(\, N(\varepsilon_i, \bar{\varepsilon}_j)\,\big)$$

because $\nabla_{\bar{\varepsilon}_j}\alpha \in \Omega^{0,1}(M)$.

To compute $\Lambda\partial\alpha$ we use the local description of $\Lambda$ in Exercise 1.4.7. We deduce

$$\mathbf{i}\Lambda\partial\alpha = \sum_k (\partial\alpha)(\varepsilon_k, \bar{\varepsilon}_k) = \sum_k \Big(\, (\nabla_{\varepsilon_k}\alpha)(\bar{\varepsilon}_k) + \alpha\big(N(\varepsilon_k, \bar{\varepsilon}_k)\big)\,\Big).$$

We need to analyze in greater detail the terms in the above sums. We will use the fact that for any $\beta \in \Omega^{0,1}(M)$ we have

$$\beta(JX) = -\mathbf{i}\beta(X), \ \ \forall X \in \mathrm{Vect}\,(M) \otimes \mathbb{C}.$$

This implies that

$$\beta(e_k) = \mathbf{i}\beta(f_k), \ \ \forall k. \tag{1.4.17}$$

Then

$$(\nabla_{\varepsilon_k}\alpha)(\bar{\varepsilon}_k) = \frac{1}{2}(\nabla_{e_k} - \mathbf{i}\nabla_{f_k})\alpha(e_k + \mathbf{i}f_k)$$

$$= \frac{1}{2}\Big(\nabla_{e_k}\alpha(e_k) + \nabla_{f_k}\alpha(f_k)\Big) + \frac{1}{2}\Big(-\mathbf{i}(\nabla_{f_k}\alpha)(e_k) + \mathbf{i}(\nabla_{e_k}\alpha)(f_k)\Big)$$

(use the fact that $\nabla_{e_k}\alpha, \nabla_{f_k}\alpha \in \Omega^{0,1}(M)$ and (1.4.17))

$$= (\nabla_{e_k}\alpha)(e_k) + (\nabla_{f_k}\alpha)(f_k).$$

Using the identity (1.4.12) we deduce

$$\sum_k (\nabla_{\varepsilon_k}\alpha)(\bar{\varepsilon}_k) = -\bar{\partial}^*\alpha.$$

To conclude the proof of (1.4.13) it suffices to show that

$$N(\varepsilon_k, \bar{\varepsilon}_k) = 0, \ \ \forall k. \tag{1.4.18}$$

We have

$$N(\varepsilon_k, \bar{\varepsilon}_k) = \frac{1}{2} N(e_k - \mathbf{i} f_k, e_k + \mathbf{i} f_k) = \mathbf{i} N(e_k, f_k) = \mathbf{i} N(e_k, J e_k)$$

$$= -\mathbf{i} J N(e_k, e_k) = 0.$$

The identity (1.4.13) is proved. Combining the above arguments with (1.4.11) one can easily obtain (1.4.17). The details are left to the reader. ■

**1.4.2. Cauchy-Riemann operators.** Suppose $(M, J)$ is an almost complex manifold and $E \to M$ is a complex Hermitian vector bundle over $M$. We denote by $\Omega^{p,q}(E)$ the space of smooth sections of the complex bundle $\Lambda^{p,q} T^* M \otimes E$ so that we have a decomposition

$$\Omega^k(E) = \bigoplus_{p+q=k} \Omega^{p,q}(E).$$

A *Cauchy-Riemann operator* (CR-operator for brevity) on $E$ is a first order p.d.o.

$$L : \Omega^{0,0}(E) \to \Omega^{0,1}(E)$$

such that

$$L(fu) = (\bar{\partial} f) \otimes u + f L u, \ \ \forall f \in C^\infty(M), \ \ u \in \Omega^{0,0}(E).$$

Let us remark that the above condition is simply a statement about the symbol of $L$. We denote by $\mathbf{CR}(E)$ the space of CR-operators on $E$ and by $\mathfrak{A}_h(E)$ the affine space of Hermitian connections on $E$. Denote by $P^{1,0}$ and $P^{0,1}$ the projectors associated to the decomposition

$$\Omega^1(E) = \Omega^{1,0}(E) \oplus \Omega^{0,1}(E).$$

Given a connection $A \in \mathfrak{A}_h(E)$ with covariant derivative

$$\nabla^A : \Omega^0(E) \to \Omega^1(E)$$

we obtain an operator

$$\bar{\partial}_A = P^{0,1} \circ \nabla^A : \Omega^{0,0}(E) \to \Omega^{0,1}(E).$$

We let the reader check that $\bar{\partial}_A$ is a CR-operator. We thus obtain a map

$$\bar{\partial}_\bullet : \mathfrak{A}_h \to \mathbf{CR}(E), \ \ A \mapsto \bar{\partial}_A.$$

**Proposition 1.4.13.** *The map $\bar{\partial}_\bullet$ is a bijection.*

**Proof** We first show that $\bar{\partial}_{\bullet}$ is injective. Suppose $A, B$ are two Hermitian connections such that $\bar{\partial}_A = \bar{\partial}_B$. Then

$$\delta = B - A$$

is a 1-form valued in the bundle of skew-Hermitian endomorphisms of $E$ such that

$$\delta^{0,1} = 0.$$

Note that

$$\delta^{0,1}(X) = \frac{1}{2}(\delta(X) + \mathbf{i}\delta(JX)), \ \ \forall X \in \mathrm{Vect}\,(M)$$

where $\delta(X)$ is a skew-Hermitian endomorphism and $\mathbf{i}\delta(JX)$ is Hermitian. This implies $\delta(X) = \delta(JX) = 0$ since any complex endomorphism decomposes uniquely as a sum of a skew-Hermitian and a Hermitian operator.

To prove the surjectivity we will construct a right inverse

$$\nabla^{\bullet} : \mathbf{CR}(E) \to \mathfrak{A}_h(E).$$

Fix a Hermitian connection $A_0$ on $E$ and denote by $L_0$ the associated CR-operator $\bar{\partial}_{A_0}$. If $L \in \mathbf{CR}(E)$ then

$$\beta = L - L_0 \in \Omega^{0,1}(\mathrm{End}\,(E)).$$

We have to construct a 1-form $\delta$ valued in the bundle of skew-Hermitian endomorphisms of $E$ such that

$$\delta^{0,1} = \beta.$$

In other words, $\delta$ satisfies the functional equation

$$\delta(X) + \mathbf{i}\delta(JX) = 2\beta(X), \ \ \forall X \in \mathrm{Vect}\,(M).$$

We deduce from the above equality that $\delta(X)$ is the skew-Hermitian part of the endomorphism $2\beta(X)$ so that

$$\delta(X) = \beta(X) - \beta(X)^*.$$

Now set

$$\nabla^{L_0+\beta}_{\cdot} = \nabla^{A_0}_{\cdot} + \beta(\cdot) - \beta(\cdot)^*.$$

The map

$$L_0 + \beta \mapsto \nabla^{L_0+\beta}$$

is a right inverse for $\bar{\partial}_{\bullet}$. ■

Suppose $L \in \mathbf{CR}\,(E)$. Then $L$ induces first order p.d.o.'s

$$L : \Omega^{p,q}(E) \to \Omega^{p,q+1}(E)$$

uniquely determined by

$$L(\alpha \otimes u) = \bar{\partial}\alpha \otimes u + (-1)^{p+q}\alpha \wedge Lu, \ \ \forall \alpha \in \Omega^{p,q}(M), \ u \in C^{\infty}(E).$$

If $A$ is Hermitian connection on $E$ we denote by the same symbol all the CR-operators

$$\bar{\partial}_A : \Omega^{p,q}(E) \to \Omega^{p,q+1}(E).$$

Then for every $u \in C^\infty(E)$ we have

$$\bar{\partial}_A^2 u = F_A^{0,2} u - (\partial_A u) \circ N \tag{1.4.19}$$

where $N$ denotes the Nijenhuis tensor of the almost complex structure on $N$.

**Exercise 1.4.8.** Use the arguments in the proof of Proposition 1.4.2 to prove the identity (1.4.19).

In the remaining part of this subsection we will assume the almost complex structure on $M$ is integrable. This means the manifold $M$ can be covered by (contractible) coordinate charts $U_\alpha \to \mathbb{C}^n$ such that the transition maps are holomorphic. A *holomorphic structure* on the rank-$r$ complex vector bundle $E$ is a collection of *holomorphic local trivializations*, i.e. a collection of local trivializations

$$\Psi_\alpha : E|_{U_\alpha} \to \underline{\mathbb{C}}^r_{U_\alpha}$$

such that the transition maps

$$g_{\beta\alpha} := \Psi_\beta(p) \circ \Psi_\alpha^{-1}(p) : U_{\alpha\beta} \to GL(r, \mathbb{C}) \subset \mathbb{C}^{r^2}$$

are *holomorphic*. A holomorphic vector bundle is a pair

(vector bundle, holomorphic structure).

Two holomorphic structures $\Psi = (\Psi_\alpha, g_{\beta\alpha} = \Psi_\beta \circ \Psi_\alpha^{-1})$ and $\Phi = (\Phi_\alpha, h_{\beta\alpha} = \Phi_\beta \circ \Phi_\alpha^{-1})$ are isomorphic if there exist *holomorphic* maps

$$T_\alpha : U_\alpha \to GL(r, \mathbb{C})$$

such that

$$h_{\beta\alpha} = T_\beta g_{\beta\alpha} T_\alpha^{-1}.$$

We denote by $\mathbf{Hol}\,(E)$ the set of isomorphism classes of holomorphic structures on $E$. (To be completely rigorous, one has to include in the definition of equivalence the gluing cocycles subordinated to different covers.)

**Exercise 1.4.9.** Prove that any holomorphic structure on $E$ induces an integrable complex structure on the total space of the bundle such that the canonical projection $E \to M$ is a holomorphic map. Moreover, two equivalent isomorphic holomorphic structures induce biholomorphic complex structures on the total space.

Fix a holomorphic structure on $E$ given by the local holomorphic trivialization $\Psi_\alpha$. There is a canonically associated sheaf of holomorphic sections. If $V$ is an open subset of $M$ and $V_\alpha = V \cap U_\alpha$ then a section $\psi$ of $E$ over $V$ is called holomorphic if the functions

$$\psi_\alpha := \Psi_\alpha \circ \psi|_{V_\alpha} \colon V_\alpha \to \mathbb{C}^r$$

are holomorphic. We denote by $\mathcal{O}_M(E)$ the sheaf of holomorphic local sections of $E$. The manifold $M$ is equipped with a fundamental sheaf $\mathcal{O}_M$, the sheaf of local holomorphic functions on $M$. Then $\mathcal{O}_M(E)$ is a sheaf of $\mathcal{O}_M$-modules. It is a locally free sheaf, i.e. it is locally isomorphic to the sheaf $\mathcal{O}_M^r$.

**Exercise 1.4.10.** Prove that two holomorphic structures on $E$ are isomorphic iff the associated sheaves of holomorphic sections are isomorphic as sheaves of $\mathcal{O}_M$-modules.

Denote by $e_i$ the canonical spanning sections of the trivial vector bundle $\underline{\mathbb{C}}^r_{U_\alpha}$ and define

$$\Psi_\alpha^{-1}(e_1) = u_1^\alpha, \cdots, \Psi_\alpha^{-1}(e_r) = u_r^\alpha \in \mathcal{O}_M(E, U_\alpha).$$

These sections span the fibers of $E_\alpha$. Any section $u \in C^\infty(E_\alpha)$ can be uniquely written as

$$u = \sum_i f_i u_i^\alpha, \quad f_i \in C^\infty(U_\alpha) \otimes \mathbb{C}.$$

Define $\bar\partial_\alpha \in \mathbf{CR}(E_\alpha)$ by

$$\bar\partial_\alpha u = \sum (\bar\partial f_i) \otimes u_i^\alpha.$$

Since the identifications $E_\alpha \cong E_\beta$ over $U_{\alpha\beta}$ are given by holomorphic maps we deduce

$$\bar\partial_\alpha = \bar\partial_\beta \text{ over } U_{\alpha\beta}.$$

Thus the operators $\bar\partial_\alpha$ glue together to form a CR-operator on $E$. It depends on the choice of the trivializations $\Psi_\alpha$. We will denote it by $\bar\partial_\Psi$.

**Exercise 1.4.11.** Show that

$$\bar\partial_\Psi \circ \bar\partial_\Psi = 0.$$

**Definition 1.4.14.** A CR-operator $L$ on a complex vector bundle $E$ over a *complex* manifold $M$ is called *integrable* if $L^2 = 0$. We will denote by $\mathbf{CR}_i(E)$ the space of complex integrable CR-operators.

Suppose $\Big(\Psi = (\Psi_\alpha)\Big)$ and $\Big(\hat\Psi = (\hat\Psi_\alpha)\Big)$ define two isomorphic holomorphic structures on $E$. Thus, there exist holomorphic maps

$$\gamma_\alpha : U_\alpha \to GL(r, \mathbb{C})$$

such that

$$\hat{\Psi}_\beta \circ \hat{\Psi}_\alpha^{-1}\gamma_\alpha = \gamma_\beta \Psi_\beta \circ \Psi_\alpha^{-1}.$$

Define

$$\Phi_\alpha := \gamma_\alpha^{-1}\hat{\Psi}_\alpha.$$

Observe that

$$\Phi_\beta \circ \Phi_\alpha^{-1} = \Psi_\beta \circ \Psi_\alpha^{-1}.$$

Thus, the collections $\big(\Psi = (\Psi_\alpha)\big)$ and $\big(\Phi = (\Phi_\alpha)\big)$ lead to the same holomorphic gluing cocycle. Moreover, since the maps $\gamma_\alpha$ are holomorphic we have

$$\bar{\partial}_{\hat{\Psi}} = \bar{\partial}_\Phi.$$

The collections $\Psi$ and $\Phi$ are cohomologous, i.e. there exist smooth maps

$$T_\alpha : U_\alpha \to GL(r, \mathbb{C})$$

such that

$$\Psi_\alpha = T_\alpha \circ \Phi_\alpha.$$

Clearly

$$T_\beta = g_{\beta\alpha} T_\alpha g_{\alpha\beta}$$

so that $T$ defines a complex automorphism of the bundle $E$. Thus, two collections of local trivializations which lead to the same (holomorphic) gluing cocycle differ by an automorphism of $E$.

Suppose now that $T \in \mathfrak{G}(E)$ is a complex (not necessarily holomorphic) automorphism of $E$. Using the trivializations $\Psi_\alpha$ it can be described as a collection of smooth maps $T_\alpha : U_\alpha \to GL(r, \mathbb{C})$ satisfying the gluing rules

$$T_\beta = g_{\beta\alpha} T_\alpha g_{\beta\alpha}^{-1} \Longleftrightarrow T_\beta^{-1} g_{\beta\alpha} T_\alpha = g_{\beta\alpha}.$$

It defines new trivializations

$$\Phi_\alpha : E_\alpha \to \underline{\mathbb{C}}^r, \quad \Phi_\alpha = T_\alpha^{-1} \circ \Psi_\alpha.$$

Notice that

$$\Phi_\beta \circ \Phi_\alpha^{-1} = T_\beta^{-1} \Psi_\beta \Psi_\alpha^{-1} T_\alpha = T_\beta^{-1} g_{\beta\alpha} T_\alpha = g_{\beta\alpha}$$

so that $\Phi_\alpha$ are compatible with the gluing cocycle $g_{\beta\alpha}$. We will denote $\Phi = \Psi \circ T$. We obtain a new CR-operator $\bar{\partial}_\Phi$.

If $s$ is a section of $E_\alpha$ then we can write

$$s = \sum_i s_i \Psi_\alpha^{-1}(e_i) \text{ and } Ts = \sum_i s_i \Psi_\alpha^{-1} T_\alpha(e_i).$$

Note that

$$\bar{\partial}_\Phi Ts = \sum_i (\bar{\partial} s_i)\Psi_\alpha^{-1} T_\alpha(e_i) = T\sum_i (\bar{\partial} s_i)\Psi_\alpha^{-1}(e_i) = T\bar{\partial}_\Psi s.$$

In other words

$$\bar{\partial}_{\Psi\circ T} = T\bar{\partial}_\Psi T^{-1}.$$

The group $\mathfrak{G}(E)$ of complex automorphisms of $E$ acts on $\mathbf{CR}_i(E)$ as above, by conjugation. We thus have a well defined map

$$\mathbf{Hol}\,(E) \to \mathbf{CR}_i(E)/\mathfrak{G}(E)$$

which associates to each holomorphic structure $\Psi$ on $E$ the $\mathfrak{G}(E)$-orbit in $\mathbf{CR}_i(E)$ of the CR-operator $\bar{\partial}_\Psi$. Observe that the sheaf $\mathcal{O}_M(E, \Psi)$ of local sections of $E$ holomorphic with respect $\Psi$ coincides precisely with the sheaf of local solutions of the partial differential equation

$$\bar{\partial}_\Psi u = 0, \quad u \text{ local smooth section of } E.$$

If $\Psi_1$ and $\Psi_2$ are two holomorphic structures such that the associated CR-operators lie in the same orbit of $\mathfrak{G}(E)$ then clearly the associated sheaves of holomorphic sections are isomorphic as sheaves of $\mathcal{O}_M$-modules and, according to Exercise 1.4.10, the two holomorphic structures are isomorphic. This means that the map

$$\mathbf{Hol}\,(E) \to \mathbf{CR}_i(E)$$

is one-to-one. This map is also *surjective* and we refer to [**29**, Chap. 2] or [**63**, Chap. I] for a proof of this nontrivial fact. The following results summarizes the above observations.

**Proposition 1.4.15.** *The map* $\mathbf{Hol}\,(E) \to \mathbf{CR}_i(E)/\mathfrak{G}(E)$, $(E, \Psi) \mapsto \bar{\partial}_\Psi$ *described above is a bijection.*

In view of this proposition, we can reconsider the manner in which we regard holomorphic bundles. In the sequel, by a holomorphic bundle over a complex manifold we will understand a pair $(E, L)$ where $E$ is a complex bundle and $L$ is an integrable CR-operator.

Suppose now that $E$ is equipped with a Hermitian metric $h$. As we have seen we have a bijection

$$\bar{\partial}_\bullet : \mathfrak{A}_h(E) \to \mathbf{CR}(E), \quad A \mapsto \bar{\partial}_A.$$

Set

$$\mathfrak{A}_h^{1,1} = \bar{\partial}_\bullet^{-1}(\mathbf{CR}_i(E)).$$

**Lemma 1.4.16.** *The space* $\mathfrak{A}_h^{1,1}(E)$ *consists of Hermitian connections* $A$ *such that* $F_A^{2,0} = F_A^{0,2} = 0$.

**Proof** Suppose $A \in \mathfrak{A}_h^{1,1}(E)$. Then using (1.4.19) we deduce $F_A^{0,2} = \bar{\partial}_A^2 = 0$. On the other hand, since the connection $A$ is compatible with the metric $h$, the curvature $F_A$ is skew-Hermitian so that $F_A^{2,0} = -(\overline{F^{2,0}}_A)^t = 0$. ■

There is an action of $\mathfrak{G}(E)$ on $\mathfrak{A}_h(E)$ induced by the isomorphism $\mathfrak{A}_h^{1,1}(E) \cong \mathbf{CR}_i(E)$. More precisely, given $T \in \mathfrak{G}(E)$ and $A \in \mathfrak{A}_h^{1,1}(E)$ we define $T \cdot A$ by the equality

$$\bar{\partial}_{T \cdot A} = T \bar{\partial}_A T^{-1}.$$

We have thus proved the following result.

**Proposition 1.4.17.** *Any Hermitian metric $h$ on a complex vector bundle $E$ over a complex manifold defines a bijection*

$$\mathbf{Hol}(E) \cong \mathfrak{A}_h^{1,1}(E)/\mathfrak{G}(E).$$

*Moreover, any integrable CR-operator $\bar{\partial}$ on $E$ induces a unique holomorphic structure $\Psi$ on $E$ and a unique Hermitian connection $A$ such $\bar{\partial}_A = \bar{\partial} = \bar{\partial}_\Psi$.*

**Remark 1.4.18.** The above identification has profound consequences. For example, in **[58]** it is shown that, modulo some topological identifications, it contains as a special case the classical Abel-Jacobi theorem.

**Example 1.4.19.** Suppose $L \to M$ is a complex line bundle over a complex manifold $M$ equipped with a Hermitian metric $h$. The group $\mathfrak{G}(L)$ can be identified with the group of smooth maps

$$f : M \to \mathbb{C}^*.$$

Suppose we are given an integrable CR-operator $\bar{\partial}$ on $L$. This induces a holomorphic structure on $L$ and a Hermitian connection $A$ such that

$$\bar{\partial}_A = \bar{\partial} \text{ and } F_A \in \Omega^{1,1}(M).$$

To find an explicit local description of $A$ we choose a local trivializing patch $U$ and a nowhere vanishing holomorphic section $s$ of $L$ over $U$. Set

$$\rho = h(s, s) = |s|_h^2.$$

The connection $A$ is locally described by a $(1,0)$-form $\theta$ determined by the conditions

$$\nabla^A s = \theta s,$$

$$d\rho = \theta h(s, s) + \bar{\theta} h(s, s) = \rho(\theta + \bar{\theta})$$

from which we deduce

$$\theta = \frac{\partial \rho}{\rho} = \partial \log \rho.$$

The curvature of $A$ is given by the 2-form

$$d\theta = \bar{\partial}\partial \log \rho.$$

Suppose now that $f \in \mathfrak{G}(L)$. We get a new CR-operator $\bar{\partial}_f$ on $L$:

$$\bar{\partial}_f = f\bar{\partial}f^{-1} = \bar{\partial} - \frac{\bar{\partial}f}{f}$$

defining the same holomorphic structure on $L$ as $\bar{\partial}$. Its associated Chern connection, denoted by $A_f$, can be determined as in the proof of Proposition 1.4.13 using the equality

$$A_f - A = -\frac{\bar{\partial}f}{f} + \frac{\partial \bar{f}}{\bar{f}}.$$

This formula describes the action of $\mathfrak{G}(L)$ on $\mathfrak{A}_h^{1,1}(L)$.

Suppose that instead of the metric $h$ we work with the metric

$$h_u = \exp(2u)h$$

where $u$ is a smooth real valued function on $M$. Denote by $A_u$ the Chern connection associated to the CR-operator $\bar{\partial}$ and the metric $h_u$. Then

$$\partial_{A_u} s = \theta_u s, \quad \theta_u = \partial \log |s|^2_{h_u} = \theta + 2\partial u$$

so that $A_u - A = 2\partial u$.

$$F_{A_u} = F_A + 2\bar{\partial}\partial u. \ \blacksquare$$

**Example 1.4.20.** Supppose $L_\ell \to S^2$ is a complex line bundle of degree $\ell \in \mathbb{Z}$ over $S^2 \cong \mathbb{P}^1$. Observe that any CR-operator on $L_\ell$ is automatically integrable since $\Omega^{0,2}(\mathbb{P}^1) = 0$.

Thus, for any Hermitian metric $h$ on $L$ we have

$$\mathfrak{A}_h(L) = \mathfrak{A}_h^{1,1}(L)$$

and we have a bijection

$$\mathfrak{A}_h(L) \to \mathbf{CR}_i(L) = \mathbf{CR}(L), \quad A \mapsto \bar{\partial}_A.$$

Fix a CR-operator $\vartheta : \Omega^{0,0}(L) \to \Omega^{0,1}(L)$. Then, for every metric $h$ on $L$ denote by $A_h$ the Chern connection determined by $\vartheta$ and $h$. If we change $h \to h_u := e^{2u}h$, $u : S^2 \to \mathbb{R}$ then, using the computations the previous example, the curvature of $A_h$ changes according to

$$F_{A_h} \to F_{A_h} + 2\bar{\partial}\partial u.$$

Suppose additionally that $S^2 := \mathbb{P}^1$ is equipped with a Kähler metric $g_0$. (All Riemannian metrics on a Riemann surface are automatically Kähler.)

Denote by $\omega_0$ the Kähler form. Then, using the Kähler-Hodge identities in Corollary 1.4.11 we deduce

$$2\bar{\partial}\partial u = 2\Lambda(\bar{\partial}\partial u)\omega_0 = (-\mathbf{i}\Delta_d u)\omega_0.$$

Let

$$c := -\frac{2\pi \deg(L)}{\mathrm{vol}_{g_0}(S^2)}$$

so that

$$\int_{S^2}\Big(\mathbf{i}c\omega_0 - F_{A_h}\Big) = 0.$$

Thus, the 2-form $\mathbf{i}c\omega_0 - F_{A_h}$ is *exact*, and there exists a smooth function $u : S^2 \to \mathbb{R}$, unique up to an additive constant, such that

$$2\bar{\partial}\partial u = \mathbf{i}c\omega_0 - F_{A_h}.$$

The curvature of $A_{h_u}$ is the *harmonic* 2-form

$$F_{A_{h_u}} := -\frac{2\pi \deg(L)}{\mathrm{vol}_{g_0}(S^2)}\mathbf{i}\omega_0.$$

The metric $h_u$ is determined by $(\vartheta, g_0)$, *uniquely* up to a positive multiplicative constant. ■

Suppose $(M, g, J)$ is a Kähler manifold and $E \to M$ is a holomorphic, Hermitian line bundle. Denote by $A$ the associated Chern connection and by $\bar{\partial}_A$ the family of operators

$$\bar{\partial}_A : \Omega^{p,q}(E) \to \Omega^{p,q+1}(E).$$

There is a Hodge $*$-operator

$$*_E : \Omega^{p,q}(E) \to \Omega^{n-p,n-q}(E^*) \tag{1.4.20}$$

defined as the the tensor product (over $\mathbb{C}$) of the complex conjugate-linear bundle morphisms

$$* : \Lambda^{p,q}T^*_{\mathbb{C}}M \to \Lambda^{n-p,n-q}T^*_{\mathbb{C}}M$$

and the metric duality

$$\mathcal{D}_E : E \to \bar{E} \cong E^*.$$

We have the following generalization of Proposition 1.4.10. For a proof we refer to [**49**].

**Proposition 1.4.21.** *Let $E \to M$ and $A$ be as above. Then*

$$\partial_A^2 = \bar{\partial}_A^2 = 0, \quad \partial_A\bar{\partial}_A + \bar{\partial}_A\partial_A = e(F_A)$$

*where $e(F_A)$ denotes the exterior multiplication by $F_A \in \Omega^{1,1}(M)$. Additionally, the Hodge identities continue to hold:*

$$\bar{\partial}_A^* = -*_E\,\bar{\partial}_A *_E, \quad \partial_A^* = -*\,\partial_A *,$$

$$[L,\Lambda]=H,$$

$$[\partial_A,L]=[\bar\partial_A,L]=[\partial_A^*,\Lambda]=[\bar\partial_A^*,\Lambda]=0,$$

$$[L,\partial_A^*]=\mathbf{i}\bar\partial_A,\quad [L,\bar\partial_A^*]=-\mathbf{i}\partial_A,$$

$$[\Lambda,\partial_A]=\mathbf{i}\bar\partial_A^*,\quad [\Lambda,\bar\partial_A]=-\mathbf{i}\partial_A^*.$$

We conclude with a Weitzenböck type identity we will need in 3.3.4.

**Proposition 1.4.22.** *Suppose* $(M,g,J)$ *is an* almost Kähler *manifold and* $E$ *is a Hermitian line bundle equipped with a Hermitian connection* $A$. *Then for every smooth section* $s$ *of* $E$ *we have the equality*

$$2\bar\partial_A^*\bar\partial_A s=(\nabla^A)^*\nabla^A s-\mathbf{i}(\Lambda F_A)s.$$

**Proof** Fix a local orthonormal frame $(e_i,f_i)$ as in Exercise 1.4.7. Then

$$\bar\partial_A=\sum_k \bar\varepsilon^k\wedge\nabla^A_{\bar\varepsilon_k}=\frac{1}{2}\sum_k(e^k-\mathbf{i}f^k)\wedge(\nabla^A_{e_k}+\mathbf{i}\nabla^A_{f_k})$$

$$=\frac{1}{2}\sum_k(e^k\wedge\nabla^A_{e_k}+f^k\wedge\nabla^A_{f_k})+\frac{\mathbf{i}}{2}\sum_k(e^k\wedge\nabla^A_{f_k}-f^k\wedge\nabla^A_{e_k})$$

$$=\frac{1}{2}\nabla^A+\frac{\mathbf{i}}{2}\sum_k(e^k\wedge\nabla^A_{f_k}-f^k\wedge\nabla^A_{e_k}).$$

For $s\in\Omega^{0,0}(E)$ we have $\bar\partial_A^*\bar\partial_A s=(\nabla^A)^*\bar\partial_A$ so that

$$\bar\partial_A^*\bar\partial_A s=\frac{1}{2}(\nabla^A)^*\nabla^A+\frac{\mathbf{i}}{2}(\nabla^A)^*\sum_k(e^k\wedge\nabla^A_{f_k}-f^k\wedge\nabla^A_{e_k})s.$$

For any vector field $X$ on $M$ we denote by $i(X)$ the contraction by $X$. Then

$$(\nabla^A)^*=\sum_j\big((\nabla^A_{e_j})^*i(e_j)+(\nabla^A_{f_j})^*i(f_j)\big).$$

Since

$$\big((\nabla^A_{e_j})^*i(e_j)+(\nabla^A_{f_j})^*i(f_j)\big)(e^k\wedge\nabla^A_{f_k}-f^k\wedge\nabla^A_{e_k})s$$

$$=\delta_j^k\big((\nabla^A_{e_j})^*\nabla^A_{f_k}-(\nabla^A_{f_j})^*\nabla^A_{e_k}\big)s$$

we deduce

$$\bar\partial_A^*\bar\partial_A s=\frac{1}{2}(\nabla^A)^*\nabla^A s+\frac{\mathbf{i}}{2}\sum_k\big((\nabla^A_{e_k})^*\nabla^A_{f_k}-(\nabla^A_{f_k})^*\nabla^A_{e_k}\big)s.$$

Using the identities

$$(\nabla^A_{e_k})^* = -\nabla^A_{e_k} - \mathbf{div}\,(e_k),\ \ (\nabla^A_{e_k})^* = -\nabla^A_{e_k} - \mathbf{div}\,(e_k)$$

and

$$-[\nabla^A_{e_k}, \nabla^A_{f_k}] = -F_A(e_k, f_k) - \nabla^A_{[e_k, f_k]}$$

we deduce

$$\bar\partial_A^* \bar\partial_A s = \frac{1}{2}(\nabla^A)^*\nabla^A s - \frac{\mathbf{i}}{2}\sum_k F_A(e_k, f_k)s$$

$$-\frac{\mathbf{i}}{2}\sum_k\Big(\nabla^A_{[e_k, f_k]} + \mathbf{div}(e_k)\nabla^A_{f_k} - \mathbf{div}(f_k)\nabla^A_{e_k}\Big)s$$

$$= \frac{1}{2}\Big(\,(\nabla^A)^*\nabla^A - \mathbf{i}\Lambda(F_A)\Big)s - \frac{\mathbf{i}}{2}\sum_k\Big(\nabla^A_{[e_k, f_k]} + \mathbf{div}(e_k)\nabla^A_{f_k} - \mathbf{div}(f_k)\nabla^A_{e_k}\Big)s$$

Hence, to conclude the proof of the proposition it suffices to prove the following identity:

$$\sum_k [e_k, f_k] = \sum_k (\mathbf{div}(f_k)e_k - \mathbf{div}(e_k)f_k). \tag{1.4.21}$$

The proof of this identity relies on the following elementary facts:

$$\omega = \sum_k e^k \wedge f^k,\ \ \omega^n = n!dv_g,\ \ d\omega = 0.$$

Let us now supply the details. First note that (1.4.21) is equivalent to

$$\sum_k e^j([e_k, f_k]) = \mathbf{div}(f_j)\ \text{ and }\ \sum_k f^j([e_k, f_k]) = -\mathbf{div}(e_k). \tag{1.4.22}$$

Next, observe that

$$\mathbf{div}(e_j) = *d*e^j = \frac{1}{(n-1)!} * d(f^j \wedge \omega^{n-1})$$

$$= \frac{1}{(n-1)!} * (df^j \wedge \omega^{n-1}) = \frac{1}{(n-1)!}(df^j \wedge \omega^{n-1})(e_1, f_1, \cdots, e_n, f_n)$$

and, similarly,

$$\mathbf{div}(f_j) = *d*f^j = -\frac{1}{(n-1)!}(de^j \wedge \omega^{n-1})(e_1, f_1, \cdots, e_n, f_n).$$

Thus (1.4.22) is equivalent to

$$\frac{1}{(n-1)!}(df^j \wedge \omega^{n-1})(e_1, f_1, \cdots, e_n, f_n) = -\sum_k f^j([e_k, f_k]) \tag{1.4.23}$$

$$\frac{1}{(n-1)!}(de^j \wedge \omega^{n-1})(e_1, f_1, \cdots, e_n, f_n) = -\sum_k e^j([e_k, f_k]). \tag{1.4.24}$$

Now introduce the operators

$$C_k = i(f_k)i(e_k), \quad k = 1, \cdots, n.$$

They enjoy some nice elementary properties.

$$C_k^2 = 0 \text{ and } [C_k, C_j] = 0, \quad \forall k, i. \tag{1.4.25}$$

$$C_k(\eta \wedge C_k\varphi) = C_k\eta \wedge C_k\varphi, \quad \forall \eta, \varphi \in \Omega^*(M), \quad \forall k. \tag{1.4.26}$$

Define $P := \prod_k C_k$, $P_k := \prod_{j\neq k} C_j$ and $S := \sum_k C_k$. Observe that

$$\frac{1}{(n-1)!}\omega^{n-1} = \frac{1}{n!}S\omega^n.$$

Thus

$$\frac{1}{(n-1)!}(df^j \wedge \omega^{n-1})(e_1, f_1, \cdots, e_n, f_n) = P(\, df^j \wedge \frac{1}{(n-1)!}\omega^{n-1})$$

$$= \frac{1}{n!}P(df^j \wedge S\omega^n)$$

(use the identities (1.4.25), (1.4.26))

$$= \frac{1}{n}\sum_k P_k(C_k(df^j) \wedge C_k\omega^n) = \frac{1}{n!}\sum_k C_k(df^j)P\omega^n = \sum_k C_k df^j$$

$$= \sum_k df^j(e_k, f_k) = -\sum_k f^j([e_k, f_k]).$$

This proves the equality (1.4.23). (1.4.24) is proved similarly. The proof of Proposition 1.4.22 is complete. ■

**Exercise 1.4.12.** Suppose $(M, g, J)$ is an almost Kähler *4-manifold* and $E \to M$ is a Hermitian line bundle equipped with a Hermitian connection. Denote by $A$ the Hermitian connection induced on the *line bundle* $\Lambda^{0,2}T^*M$ by the Chern connection. Show that for every section $\beta \in \Omega^{0,2}(E)$ we have the following Weitzenböck type identity:

$$\bar{\partial}_B\bar{\partial}_B^*\beta = \frac{1}{2}(\,(\nabla^{A\otimes B})^*\nabla^{A\otimes B} + \mathbf{i}\Lambda(F_A + F_B)).$$

**1.4.3. Dirac operators on almost Kähler manifolds.** Suppose $(M, g, J)$ is an almost Kähler manifold of dimension $2n$. We denote by $D$ the Levi-Civita connection of $g$ and by $\nabla$ the Chern connection of this almost Kähler structure. Recall that if $M$ is *Kähler* then $D = \nabla$.

The almost complex structure defines a canonical *spin*$^c$ structure $\sigma_0$ on $M$. We have seen that the line bundle associated to this structure is $K_M^{-1} = \Lambda^{0,n}T^*M$. The **Fundamental Fact** in §1.3.1 shows that the associated bundle of spinors is

$$\mathbb{S}_c \cong \Lambda^{0,*}T^*M, \;\; \mathbb{S}_c^{\pm} \cong \Lambda^{0,even/odd}T^*M.$$

The Chern connection induces Hermitian connections on $\Lambda^{0,p}T^*M$, $\forall p$ and in particular, a Hermitian connection on $K_M^{-1}$. In this manner we obtain a *geometric* Dirac operator

$$\mathfrak{D}_c : \Lambda^{0,even}T^*M \to \Lambda^{0,odd}T^*M.$$

We say that $\mathfrak{D}_c$ is the *canonical Hermitian Dirac operator* associated to the almost Kähler structure.

On the other hand, the Chern connection induces CR-operators

$$\bar{\partial} : \Lambda^{p,q}T^*M \to \Lambda^{p,q+1}$$

and we can form the first order p.d.o.

$$\bar{\partial} + \bar{\partial}^* : \Lambda^{0,even}T^*M \to \Lambda^{0,odd}T^*M.$$

**Proposition 1.4.23.** *Let $(M, g, J)$ be as above. Then*

$$\mathfrak{D}_c = \sqrt{2}(\bar{\partial} + \bar{\partial}^*).$$

**Proof** Choose a local orthonormal frame $(e_i, f_i)$ of $TM$ such that $f_i = Je_i$. Set $e_{i+n} = f_i$ and define $\varepsilon_i$, $\bar{\varepsilon}^j$ as usual. Denote by $\hat{D}$ the connection on $\mathbb{S}_c$ induced by the Levi-Civita connection on $TM$ and the Chern connection on $K_M^{-1}$. Then

$$\mathfrak{D}_c = \sum_i c(\varepsilon^i)\hat{D}_{\varepsilon_i} + \sum_i c(\bar{\varepsilon}^i)\hat{D}_{\bar{\varepsilon}_i}.$$

To proceed further we need to use the explicit description of the Clifford multiplication explained in the proof of the **Fundamental Fact**. We have to be careful about conventions because the description $\mathbb{S}_c \cong \Lambda^{0,*}T^*M$ uses the isomorphism $TM^{1,0} \cong T^*M^{0,1}$ given by

$$\varepsilon_i \longleftrightarrow \bar{\varepsilon}^i.$$

We deduce

$$c(\bar{\varepsilon}^j) = \sqrt{2}e(\bar{\varepsilon}^j), \;\; c(\varepsilon^k) = -\sqrt{2}i(\bar{\varepsilon}_k).$$

If we continue to denote by $\nabla$ the connection on $\Lambda^{0,*}T^*M$ induced by the Chern connection then, using Exercise 1.4.5, we deduce

$$\sqrt{2}(\bar{\partial} + \bar{\partial}^*) = \sum_i \mathbf{c}(\varepsilon^i)\nabla_{\varepsilon_i} + \sum_i \mathbf{c}(\bar{\varepsilon}^i)\nabla_{\bar{\varepsilon}_i}.$$

Next, note that since all the computations are local we can assume that, topologically, $M$ is the open ball in $\mathbb{R}^{2n}$. It has a spin structure and we denote by $\mathbb{S}_0$ the associated bundle of complex spinors. This spin structure also defines a square root $K^{1/2}$ of the canonical line bundle and we can write $\mathbb{S}_c \cong \mathbb{S}_0 \otimes K^{-1/2}$. As in Remark 1.3.19 the Chern connection induces a Dirac structure $(\mathbb{S}_0, \mathbf{c}, \hat{\nabla}, \nabla)$, where the connection $\hat{\nabla}$ on $\mathbb{S}_0$ satisfies

$$\hat{\nabla}_X \mathbf{c}(\alpha) = \mathbf{c}(\nabla_X \alpha), \ \forall X \in \mathrm{Vect}\,(M), \forall \alpha \in \Omega^1(M). \tag{1.4.27}$$

Using the Chern connection on $K_M^{-1}$ we obtain by twisting, as in §1.3.4, a connection on $\mathbb{S}_c$, which we continue to denote by $\hat{\nabla}$, satisfying the same compatibility relation (1.4.27). We can now define a new Dirac operator

$$\mathfrak{D}_h = \sum_i \mathbf{c}(\varepsilon^i)\hat{\nabla}_{\varepsilon_i} + \sum_i \mathbf{c}(\bar{\varepsilon}^i)\hat{\nabla}_{\bar{\varepsilon}_i}.$$

We have thus obtained three first order p.d.o.'s $\mathfrak{D}_c$, $\mathfrak{D}_h$, $\sqrt{2}(\bar{\partial} + \bar{\partial}^*)$ which have the same symbol. The proposition will be proved once we show these three operators actually coincide. The proof of this more refined statement will be carried out in two steps.

**Step 1**

$$\mathfrak{D}_c = \mathfrak{D}_h.$$

Set $S = \hat{\nabla} - \hat{D} \in \Omega^1(\mathrm{End}\,(\mathbb{S}_c))$. Then

$$\mathfrak{D}_h - \mathfrak{D}_c = \sum_{i=1}^{2n} \mathbf{c}(e^i)S(e_i) = \sum_{i=1}^{n} \mathbf{c}(\varepsilon^i)S(\varepsilon_i) + \sum_{i=1}^{n} \mathbf{c}(\bar{\varepsilon}^i)S(\bar{\varepsilon}_i).$$

Thus we have to show that

$$\sum_{i=1}^{n} \mathbf{c}(\varepsilon^i)S(\varepsilon_i) + \sum_{i=1}^{n} \mathbf{c}(\bar{\varepsilon}^i)S(\bar{\varepsilon}_i) = 0. \tag{1.4.28}$$

Using Proposition 1.4.4 we deduce

$$\nabla_i e_j = D_i e_j - \sum_{k=1}^{2n} N^i_{jk} e_k, \ \forall i, j = 1, \cdots, 2n$$

and

$$\nabla_i e^j = D_i e^j - \sum_{k=1}^{2n} N^i_{jk} e^k, \ \forall i, j = 1, \cdots, 2n$$

where $N = \sum_{i,j,k=1}^{2n} N^i_{jk} e_i \otimes e^j \otimes e^k$ denotes the Nijenhuis tensor. We will regard $N$ as a $T^*M$-valued 2-form using the metric duality

$$N(X,Y,Z) = g(X, N(Y,Z)), \quad \forall X,Y,Z \in \mathrm{Vect}\,(M).$$

Thus $N \in C^\infty(T^*M^{\otimes 3})$ and is skew symmetric in the last two variables. We can extend it by complex multilinearity to an element of $C^\infty(T^*M^{\otimes 3}) \otimes \mathbb{C}$. Using Exercise 1.4.5 (b), (c) we deduce that

$$N \in C^\infty((T^*M^{1,0})^{\otimes 3}) \oplus C^\infty((T^*M^{0,1})^{\otimes 3}).$$

From Remark 1.3.19 we deduce

$$S = \hat{\nabla} - \hat{D} = \frac{1}{4}\sum_{i,j,k=1}^{2n} e^i \otimes N(e_i, e_j, e_k)\mathbf{c}(e^j)\mathbf{c}(e^k)$$

$$= \frac{1}{4}\sum_{i,j,k=1}^{n} \varepsilon^i \otimes N(\varepsilon_i, \varepsilon_j, \varepsilon_k)\mathbf{c}(\varepsilon^j)\mathbf{c}(\varepsilon^k) + \frac{1}{4}\sum_{i,j,k=1}^{n} \bar{\varepsilon}^i \otimes N(\bar{\varepsilon}_i, \bar{\varepsilon}_j, \bar{\varepsilon}_k)\mathbf{c}(\bar{\varepsilon}^j)\mathbf{c}(\bar{\varepsilon}^k)$$

and therefore

$$\sum_{i=1}^{n} \mathbf{c}(\varepsilon^i)S(\varepsilon_i) = \frac{1}{4}\sum_{i,j,k=1}^{n} N(\varepsilon_i, \varepsilon_j, \varepsilon_k)\mathbf{c}(\varepsilon^i)\mathbf{c}(\varepsilon^j)\mathbf{c}(\varepsilon^k)$$

$(\mathbf{c}(\varepsilon^i)\mathbf{c}(\varepsilon^j) = -\mathbf{c}(\varepsilon^j)\mathbf{c}(\varepsilon^i),\ \mathbf{c}(\varepsilon^i)^2 = 0, \ \forall i,j)$

$$= \frac{1}{2}\sum_{1\le i<j<k\le n} (N(\varepsilon_i, \varepsilon_j, \varepsilon_k) + N(\varepsilon_k, \varepsilon_i, \varepsilon_j) + N(\varepsilon_j, \varepsilon_k, \varepsilon_i))\mathbf{c}(\varepsilon^i)\mathbf{c}(\varepsilon^j)\mathbf{c}(\varepsilon^k)$$

$$= \frac{1}{2}\sum_{1\le i<j<k\le n} (\mathfrak{b}N)(\varepsilon_i, \varepsilon_j, \varepsilon_k)\cdot \mathbf{c}(\varepsilon^i)\mathbf{c}(\varepsilon^j)\mathbf{c}(\varepsilon^k) \overset{(1.4.5)}{=} 0.$$

Similarly, one proves that

$$\sum_{i=1}^{n} \mathbf{c}(\bar{\varepsilon}^i)S(\bar{\varepsilon}_i) = 0.$$

The equality (1.4.28) is proved.

**Step 2**

$$\not{\mathfrak{D}}_h = \sqrt{2}(\bar{\partial} + \bar{\partial}^*).$$

Set $S = \hat{\nabla} - \nabla \in \Omega^1(\mathrm{End}\,(\mathbb{S}_c))$. Note that both connections $\nabla$ and $\hat{\nabla}$ satisfy the compatibility condition (1.4.27), so that

$$[S(X), \mathbf{c}(\alpha)] = 0, \quad \forall X \in \mathrm{Vect}\,(M),\ \alpha \in \Omega^1(M).$$

This means that $\forall x \in M$ the operator $S(X)_x$ commutes with every element in $\mathsf{Cl}(T^*_xM)\otimes\mathbb{C} = \mathrm{End}_{\mathbf{C}}(\mathbb{S}_c\,|_x)$. Using Schur's lemma we deduce that $S(X)_x$

is a multiple of the identity. In other words, there exists a purely imaginary 1-form $\mathfrak{a}$ such that

$$S = \mathfrak{a} \otimes \mathbf{id}.$$

We want to prove $\mathfrak{a} \equiv 0$. Note that the constant function $\mathbf{1}$ can be viewed as a section of $\Lambda^{0,0}T^*M \hookrightarrow \mathbb{S}_c$ so that

$$\mathfrak{a} = (\hat{\nabla} - \nabla)\mathbf{1} = \hat{\nabla}\mathbf{1}$$

so that it suffices to show $\hat{\nabla}\mathbf{1} \equiv 0$.

Locally we have

$$\nabla e_j = \sum_{i,k=1}^{2n} \Omega^i_{kj} e_i$$

and

$$\nabla e^j = \sum_{i,k=1}^{2n} \Omega^i_{kj} e^i.$$

Using the metric duality we can regard the End $(TM)$-valued 1-form $\Omega$ as a $T^*M$-valued 2-form

$$\Omega(e_k, e_i, e_j) = g(\nabla_k e_j, e_i).$$

We can extend it by complex linearity to an element of $C^\infty(T^*M^{\otimes 3}) \otimes \mathbb{C}$. Note that since $\nabla$ is compatible with the complex structure it preserves the splitting $TM \otimes \mathbb{C} = TM^{1,0} \oplus TM^{0,1}$. This implies that $\forall X \in$ Vect $(M)$ the 2-form $\Omega(X, \cdot, \cdot)$ has type $(1,1)$, i.e.

$$\Omega(X, \varepsilon_i, \varepsilon_j) = \Omega(X, \bar{\varepsilon}_j, \bar{\varepsilon}_j) = 0, \ \ \forall\, i, j = 1, \cdots, n.$$

Moreover, $\forall X \in$ Vect $(M)$

$$\nabla_X \varepsilon_j = \sum_i \Omega(X, \bar{\varepsilon}_i, \varepsilon_j)\varepsilon_j, \ \ \nabla_X \bar{\varepsilon}_j = \sum_i \Omega(X, \varepsilon_i, \bar{\varepsilon}_j)\bar{\varepsilon}_j.$$

The connection $\hat{\nabla}^0$ induced by $\nabla$ on $\mathbb{S}_0$ has the local description

$$\hat{\nabla}^0 = d - \frac{1}{4}\sum_{i,j,k} e^k \otimes \Omega(e_k, e_i, e_j)\mathbf{c}(e^i)\mathbf{c}(e^j)$$

$$= d - \frac{1}{4}\sum_{i,j,k} \varepsilon^k \otimes \Omega(\varepsilon_k, \varepsilon_i, \bar{\varepsilon}_j)\mathbf{c}(\varepsilon^i)\mathbf{c}(\bar{\varepsilon}^j) - \frac{1}{4}\sum_{i,j,k} \varepsilon^k \otimes \Omega(\varepsilon_k, \bar{\varepsilon}_j, \varepsilon_i)\mathbf{c}(\bar{\varepsilon}^j)\mathbf{c}(\varepsilon^i)$$

$$-\frac{1}{4}\sum_{i,j,k} \bar{\varepsilon}^k \otimes \Omega(\bar{\varepsilon}_k, \varepsilon_i, \bar{\varepsilon}_j)\mathbf{c}(\varepsilon^i)\mathbf{c}(\bar{\varepsilon}^j) - \frac{1}{4}\sum_{i,j,k} \bar{\varepsilon}^k \otimes \Omega(\bar{\varepsilon}_k, \bar{\varepsilon}_j, \varepsilon_i)\mathbf{c}(\bar{\varepsilon}^j)\mathbf{c}(\varepsilon^i).$$

Now define $\delta \in \Omega^1(M)$ by

$$\nabla(\bar{\varepsilon}^1 \wedge \cdots \wedge \bar{\varepsilon}^n) = \delta \otimes \bar{\varepsilon}^1 \wedge \cdots \wedge \bar{\varepsilon}^n.$$

The connection $\hat{\nabla}$ on $\mathbb{S}_c$ induced by $\nabla$ is

$$\hat{\nabla} = \hat{\nabla}^0 + \frac{1}{2}\delta.$$

Since $\mathbf{c}(\varepsilon^i)\mathbf{1} = 0$ we deduce

$$\hat{\nabla}_{\varepsilon_k}\mathbf{1} = -\frac{1}{4}\sum_{i,j}\Omega(\varepsilon_k, \varepsilon_i, \bar{\varepsilon}_j)\mathbf{c}(\varepsilon^i)\mathbf{c}(\bar{\varepsilon}^j) + \frac{1}{2}\delta(\varepsilon_k)$$

$(\mathbf{c}(\varepsilon^i)\mathbf{c}(\bar{\varepsilon}^j)\mathbf{1} = -2\delta^{ij})$

$$= \frac{1}{2}\sum_i \Omega(\varepsilon_k, \varepsilon_i, \bar{\varepsilon}_i) + \frac{1}{2}\delta(\varepsilon_k).$$

On the other hand, if we denote by $g_c$ the complexification of the metric $g$ (by complex linearity) we deduce

$$(\nabla_{\varepsilon_k}\bar{\varepsilon}^j)(\bar{\varepsilon}_i) = -\bar{\varepsilon}^j(\nabla_{\varepsilon_k}\bar{\varepsilon}_i) = -g_c(\varepsilon_j, \nabla_{\varepsilon_k}\bar{\varepsilon}_i)$$

$$= -\sum_\ell g_c(\varepsilon_j\,,\,\Omega(\varepsilon_k, \varepsilon_l, \bar{\varepsilon}_i)\bar{\varepsilon}_\ell\,) = -\sum_\ell \delta_{j\ell}\Omega(\varepsilon_k, \varepsilon_\ell, \bar{\varepsilon}_i) = -\Omega(\varepsilon_k, \varepsilon_j, \bar{\varepsilon}_i)$$

so that

$$\nabla_{\varepsilon_k}\bar{\varepsilon}_j = -\sum_i \Omega(\varepsilon_k, \varepsilon_j, \bar{\varepsilon}_i)\bar{\varepsilon}^i.$$

This implies immediately that

$$\delta(\varepsilon_k) = -\sum_i \Omega(\varepsilon_k, \varepsilon_i, \bar{\varepsilon}_i)$$

so that $\hat{\nabla}_{\varepsilon_k}\mathbf{1} = \mathfrak{a}(\varepsilon_k) = 0$. Similarly we have $\mathfrak{a}(\bar{\varepsilon}_k) = 0$ which shows that $\mathfrak{a} = 0$ and completes the proof of the proposition. ■

**Remark 1.4.24.** For an alternate proof of Proposition 1.4.23 we refer to [**119**].

The following result now follows immediately from the above. Its proof is left to the reader.

**Proposition 1.4.25.** *Supose $(M, g, J)$ is an almost Kähler manifold of dimension $2n$, $L \to M$ is a Hermitian line bundle and $B$ is a Hermitian connection on $L$. $L$ defines a spin$^c$ structure $\sigma_L = \sigma_c \otimes L$, where $\sigma_c$ is the spin$^c$ structure induced by $J$. Moreover,* $\det(\sigma_L) = K_M^{-1} \otimes L^2$. *Using the Chern connection $A_0$ on $M$ and the connection $B$ on $L$ we obtain a connection $A = A_0 \otimes B^{\otimes 2}$ on* $\det(\sigma_L)$ *and thus a geometric Dirac operator $\mathfrak{D}_A$ on $\mathbb{S}_{\sigma_L} = \Omega^{0,*}(L)$. Then*

$$\mathfrak{D}_A = \sqrt{2}(\bar{\partial}_B + \bar{\partial}_B^*).$$

## 1.5. Fredholm theory

When defining the Seiberg-Witten invariants one relies essentially on the fact that the various operators involved are Fredholm. In this section we discuss some important topological features of Fredholm operators.

**1.5.1. Continuous families of elliptic operators.** Suppose $(M, g)$ is a smooth, closed, compact, oriented Riemannian manifold and $E^0, E^1 \to M$ are *real* vector bundles equipped with a metric $\langle \cdot, \cdot \rangle$ and $\mathfrak{D}_0 : C^\infty(E^0) \to C^\infty(E^1)$ is a first order elliptic operator. Suppose $X$ is a smooth, compact, connected manifold. Using the natural projection $X \times M \to M$ we obtain by pullback a bundle $E_X \to X \times M$. Now consider

- a section $T$ of $\mathrm{Hom}\,(E^0_X, E^1_X)$.

We can regard $T$ as a smooth family $(T_x)_{x \in X}$ of morphisms of $E^0 \to E^1$. We can now form the family of elliptic operators

$$\mathfrak{D}_x : C^\infty(E^0) \to C^\infty(E^1)$$

described by

$$\mathfrak{D}_x = \mathfrak{D}_0 + T_x.$$

These operators have symbols independent of $x \in X$ and define closed, unbounded, Fredholm linear operators $L^2(E^0) \to L^2(E^1)$ with common domain $L^{1,2}(E)$. Moreover the map

$$\mathrm{ind}\,(\mathfrak{D}_\bullet) : X \to \mathbb{Z}, \quad x \mapsto \mathrm{ind}\,(\mathfrak{D}_x)$$

is constant since $X$ is connected.

Suppose $\dim \ker \mathfrak{D}_x$ is independent of $x$. Then $\dim \ker \mathfrak{D}^*_x = \dim \ker \mathfrak{D}_x - \mathrm{ind}\,(\mathfrak{D}_x)$ is also independent of $x$. We then get two smooth vector bundles $\ker \mathfrak{D}$ and $\ker \mathfrak{D}^*$ and a real line bundle

$$\det(\mathfrak{D}) = \det \ker \mathfrak{D} \otimes (\det \ker \mathfrak{D}^*)^*$$

called the *determinant line bundle* of the family $\mathfrak{D}$. Remarkably, one can still define such a line bundle even if the dimension of the kernels of $\mathfrak{D}_x$ jumps. To explain the construction we first recall a couple of facts proven in [**105**], Sec. 9.4.1. First, set for simplicity $H_i = L^2(E^i)$, $i = 0, 1$. For every closed subspace $V \subset H_1$ define the unbounded operator

$$\mathfrak{D}_{V,x} : H_0 \oplus V \to H_1$$

with domain $L^{1,2}(E^0) \oplus v$ acting according to

$$\mathfrak{D}_{V,x}(h \oplus v) = \mathfrak{D}_x h + v, \quad \forall u \in L^{1,2}(E^0),\ v \in V.$$

A *stabilizer* of the family $(\mathfrak{D}_x)_{x \in X}$ is a finite-dimensional subspace $V \subset H_1$ such that $\mathfrak{D}_{V,x}$ is surjective for all $x \in X$. We will denote by $\mathcal{S}(\mathfrak{D})$ the set of stabilizers.

**Example 1.5.1.** The cokernel of a single operator $\mathfrak{D}$, $\cong \ker \mathfrak{D}^*$, is a stabilizer for the one-member family $\mathfrak{D}$ so that $\mathcal{S}(\mathfrak{D}) \neq \emptyset$. In fact, any finite dimensional subspace of $H_1$ containing the cokernel will be a stabilizer. Observe that if we denote $V_0 = \ker \mathfrak{D}^*$ then

$$\ker \mathfrak{D}_{V_0} = \{u \oplus 0\,;\quad u \in \ker \mathfrak{D}\}$$

so that there is a natural isomorphism $\ker \mathfrak{D} \cong \ker \mathfrak{D}_{V_0}$.

If $V \in \mathcal{S}(\mathfrak{D})$ then for every $x \in X$ we have a natural short exact sequence of Hilbert spaces

$$0 \to \ker \mathfrak{D}_{V,x} \to H_0 \oplus V \xrightarrow{\mathfrak{D}_{V,x}} H_1 \to 0.$$

It admits a canonical splitting in the form of the bounded, right inverse

$$\mathsf{R}_{V,x} : H_1 \to (\ker \mathfrak{D}_{V,x})^{\perp} \subset V \oplus H_0$$

where $\mathsf{R}_V h_1 = h_0 \oplus v$ if and only if

$$(v \oplus h_0) \in (\ker \mathfrak{D}_V)^{\perp}, \quad \mathfrak{D} h_0 + v = h_1.$$

**Remark 1.5.2.** For any stabilizer $V$ of a family $\mathfrak{D}$ we could define $\mathfrak{D}_V^-$ by the equality

$$\mathfrak{D}_V^-(v + h_0) = -v + h_0.$$

This operator is onto and it has a right inverse $\mathsf{R}_V^-$ defined by the conditions

$$\mathsf{R}_V h_1 = v \oplus h_0 \Leftrightarrow (v \oplus h_0) \in (\ker \mathfrak{D}_V^-)^{\perp}, \quad \mathfrak{D} h_0 - v = h_1.$$

In this book we will consistently work with the first convention, $\mathfrak{D}_V$ and $\mathsf{R}_V$.

The following results can be deduced immediately from the considerations in [**105**, §9.4.1].

**Fact 1** $\mathcal{S}(\mathfrak{D}) \neq \emptyset$. Moreover, if $V \in \mathcal{S}(\mathfrak{D})$ and $W \supseteq V$ then $W \in \mathcal{S}(\mathfrak{D})$.

**Fact 2** For any $V \in \mathcal{S}(\mathfrak{D})$ the bounded linear operators $\mathsf{R}_{V,x}$ depend smoothly upon $x$ and the family $x \mapsto \ker \mathfrak{D}_{V,x}$ defines a smooth vector bundle $\ker \mathfrak{D}_V$ over $X$.

Suppose $V, W \in \mathcal{S}(\mathfrak{D})$, $V \subseteq W$. The short exact sequence

$$0 \to V \to W \to W/V \to 0 \tag{1.5.1}$$

admits a natural metric induced splitting by identifying $W/V$ with the orthogonal complement in $W$. We also have a natural dual split exact sequence

$$0 \to V^* \to W^* \to (W/V)^* \to 0. \tag{1.5.2}$$

Then there is a natural exact sequence

$$0 \to \ker \mathfrak{D}_V \hookrightarrow \ker \mathfrak{D}_W \to W/V \to 0 \tag{1.5.3}$$

where the first arrow is induced by the inclusion $V \hookrightarrow W$ and the second arrow is given by orthogonal projection. This sequence admits a natural splitting

$$s_{W/V} : W/V \to \ker \mathfrak{D}_W, \quad (-R_V(w/v)) \oplus (w/v).$$

Taking the direct sum of the split exact sequences (1.5.3) and (1.5.2) (*in this order*) we obtain the split exact sequence

$$0 \to \ker \mathfrak{D}_V \oplus V^* \to \ker \mathfrak{D}_W \oplus W^* \to W/V \oplus (W/V)^* \to 0 \tag{1.5.4}$$

which leads to an isomorphism

$$\ker \mathfrak{D}_V \oplus V^* \oplus (W/V) \oplus (W/V)^* \to \ker \mathfrak{D}_W \oplus W^*.$$

By passing to determinants we obtain a natural isomorphism

$$I_{W/V} : \det \ker \mathfrak{D}_V \otimes \det V^* \to \det \ker \mathfrak{D}_W \otimes \det W^*$$

defined by the commutative diagram below.

$$\begin{array}{ccc} \det \ker \mathfrak{D}_V \otimes \det V^* & \xrightarrow{\cong} & \det \ker \mathfrak{D}_V \otimes \det V^* \otimes \det(W/V) \otimes \det(W/V)^* \\ & \searrow^{I_{W/V}} & \downarrow \cong \\ & & \det \ker \mathfrak{D}_W \otimes \det W^* \end{array}$$

Set $\mathcal{L}_V := \det \ker \mathfrak{D}_V \otimes \det V^*$ so that $I_{W/V}$ is a line bundle isomorphism $\mathcal{L}_V \to \mathcal{L}_W$. Thus, isomorphism class of the real line bundle $\mathcal{L}_V$ is independent of $V \in \mathcal{S}(\mathfrak{D})$.

**Definition 1.5.3.** The isomorphism class of the line bundles $\mathcal{L}_V$ is called the *determinant line bundle of the family* $\mathfrak{D}$ and will be denoted by $\det \mathfrak{D}$.

The above construction has a built-in coherence, explicitly described in the next result.

**Proposition 1.5.4.** *If $V_1 \subset V_2 \subset V_3$ are stabilizers of the family $\mathfrak{D}_\bullet$ then*

$$I_{V_3/V_1} = I_{V_3/V_2} \circ I_{V_2/V_1}.$$

**Proof** We begin by making a few useful conventions. For any ordered basis $b$ of a vector space $E$ we will denote by $b^*$ the dual ordered basis of $E^*$, by $\det(b)$ the element it defines in $\det E$ and by $\det(b)^*$ the corresponding element in $\det E^*$.

If $b_1$ and $b_2$ are ordered bases in $E_1$ and $E_2$ we denote by $b_1 \cup b_2$ the ordered basis in the ordered direct sum $E_1 \oplus E_2$. Observe that

$$\det(b_1^* \cup b_2^*) = \det(b_1 \cup b_2)^*.$$

There is a natural isomorphism

$$\mathbb{R} \to \det(E \oplus E^*)$$

defined by $1 \mapsto \det(b \cup b^*)$, where $b$ is an arbitrary ordered basis of $E$. It is easy to see that this isomorphism is independent of $b$.

For $1 \leq i < j \leq 3$ denote by $s_{ij} : V_j/V_i \to \ker \mathfrak{D}_{V_j}$ the natural splitting $s_{V_j/V_i}$ of the exact sequence

$$(S_{ij}) \quad 0 \to \ker \mathfrak{D}_{V_i} \to \ker \mathfrak{D}_{V_j} \to V_j/V_i \to 0.$$

Fix an ordered basis $b_1$ of $V_1$, an ordered basis $\beta_1$ of $\ker \mathfrak{D}_{V_1}$ and ordered bases $b_2/b_1$, $b_3/b_2$ of $V_2/V_1$ and $V_3/V_2$. We get bases $b_2 = b_1 \cup (b_2/b_1)$ of $V_2$ and $b_3 = b_2 \cup b_3/b_2$ of $V_3$. Set $b_3/b_1 = b_2/b_1 \cup b_3/b_2$ so that $b_3 = b_1 \cup b_3/b_1$.

Using the split sequence $(S_{12})$ we obtain an ordered basis

$$\beta_2 = \beta_1 \cup s_{12}(b_2/b_1)$$

of $\ker \mathfrak{D}_{V_2}$ and similarly, from $(S_{23})$, an ordered basis

$$\beta_3 = \beta_2 \cup s_{23}(b_3/b_2) = \beta_1 \cup s_{12}(b_2/b_1) \cup s_{23}(b_3/b_2).$$

From the explicit description of $s_{ij}$ we deduce immediately that

$$s_{13}(b_2/b_1 \cup b_3/b_2) = s_{12}(b_2/b_1) \cup s_{23}(b_3/b_2).$$

This implies

$$\beta_3 = \beta_1 \cup s_{13}(b_3/b_1).$$

The above identities can be written succinctly as

$$\beta_j = \beta_i \cup s_{ij}(b_j/b_i).$$

The isomorphism $I_{ji}$ can now be described as follows:

$$\det(\beta_i \cup b_i^*) \mapsto \det(\beta_i \cup b_i^* \cup (b_j/b_i) \cup (b_j/b_i)^*) \mapsto$$

$$\mapsto \det(\beta_i \cup s_{ij}(b_j/b_i) \cup (b_i)^* \cup (b_j/b_i)^*)$$

$$= \det(\beta_i \cup b_i^* \cup (b_j/b_i)^*) = \det(\beta_j \cup b_j^*).$$

The proposition is now obvious. ■

**Exercise 1.5.1.** Suppose $\mathfrak{D}$ is a family such that $\dim \ker \mathfrak{D}_x$ is independent of $x \in X$. Show that

$$\det \mathfrak{D}_x \cong \det \ker \mathfrak{D}_x \otimes (\det \ker \mathfrak{D}_x^*)^*.$$

Suppose now that we have two families $(T_0)$, $(T_1)$ of morphisms parameterized by $X$. They are said to be *homotopic* if there exists a morphism $\tilde{T} : E^0_{[0,1]\times X} \to E^1_{[0,1]\times X}$ such that

$$\tilde{T}\,|_{\{i\}\times X} = T_i, \quad i = 0, 1.$$

**Proposition 1.5.5.** *Two homotopic families* $(T_i)$, $i = 0, 1$, *have isomorphic determinant line bundles*

$$\det \mathfrak{D}_0 \cong \det \mathfrak{D}_1.$$

**Proof** We denote by $\tilde{\mathfrak{D}}$ the family of operators parameterized by $[0,1] \times X$ generated by the homotopy $(\tilde{T})$. Fix $U \in \mathcal{S}(\tilde{\mathfrak{D}})$. Then $U \in \mathcal{S}(\mathfrak{D}^0) \cap \mathcal{S}(\mathfrak{D}^1)$. To prove the proposition it suffices to construct an isomorphism

$$\ker \mathfrak{D}^0_U \to \ker \mathfrak{D}^1_U.$$

To do this, consider the bundle $\ker \tilde{\mathfrak{D}}_U \to [0,1] \times X$, fix a connection on it and denote by $\mathcal{T}_x$ the parallel transport from $\ker \tilde{\mathfrak{D}}_{U,(0,x)}$ to $\ker \tilde{\mathfrak{D}}_{U,(1,x)}$ along the path $[0,1] \ni t \mapsto (t,x) \in [0,1] \times X$. Then $\mathcal{T}$ induces the corresponding isomorphism. Observe that the homotopy class of the isomorphism is *independent* of the choice of the connection on $\ker \tilde{\mathfrak{D}}_U$. ■

**Definition 1.5.6.** (a) The family $(\mathfrak{D}_x)_{x\in X}$ is called *orientable* if $\det \mathfrak{D}$ is trivial.

(b) An *orientation* on a real line bundle $L \to X$ is a homotopy class of isomorphisms $\phi : \det \mathfrak{D} \to \underline{\mathbb{R}}$. Two oriented line bundles $\phi_i : L_i \to \underline{\mathbb{R}}$, $i = 1, 2$, are said to be *equivalent* if there exists an isomorphism $\delta : L_1 \to L_2$ such that $\phi_2 \circ \delta$ and $\phi_1$ are homotopic through isomorphisms.

From Proposition 1.5.5 we deduce immediately the following consequence.

**Corollary 1.5.7.** *Suppose* $(T_i)$, $i = 0, 1$, *are two homotopic families. Then* $\det \mathfrak{D}^0$ *is orientable iff* $\det \mathfrak{D}^1$ *is orientable. Moreover, any orientation on* $\det \mathfrak{D}^0$ *canonically induces an orientation on* $\det \mathfrak{D}^1$.

In practice one is often led to ask the following question.

*How can one construct orientations on a given oriented family* $\mathfrak{D}$*?*

We will address two aspects of this issue.

**Step 1** Describe special cases when there is a canonical way of assigning orientations.

**Step 2** Describe how to transport orientations via homotopies.

**Step 1** To construct an orientation on $\det \not{\mathfrak{D}}$ it suffices to construct coherent orientations on the line bundles $\mathcal{L}_V$. The coherence means that the natural isomorphisms $I_{WV}$ are *orientation preserving*. We describe below several situations when such an approach is successful.

Suppose the family $(\not{\mathfrak{D}})_x$ is *nice*, i.e. satisfies the following two conditions:

(i) $\dim \ker \not{\mathfrak{D}}_x$ is independent of $x$.

(ii) The real vector bundles $\ker \not{\mathfrak{D}}$ and $\ker \not{\mathfrak{D}}^*$ are equipped with orientations.

For example, if $\operatorname{ind}(\not{\mathfrak{D}}_x) \equiv 0$ and all the operators $\not{\mathfrak{D}}_x$ are one-to-one (and hence also onto) then both the above conditions are satisfied. If $\not{\mathfrak{D}}_x$ is a family of *complex* operators satisfying (i) then the condition (ii) is automatically satisfied since the bundles in question are equipped with complex structures and thus canonical orientations.

To proceed further we need the following elementary fact.

**Exercise 1.5.2.** There exists a finite-dimensional subspace $V \subset H_1$ such that $\ker \not{\mathfrak{D}}^*$ is a subbundle of the trivial bundle $\underline{V}$.

We denote by $\hat{\mathcal{S}}(\not{\mathfrak{D}})$ the set of **oriented** finite-dimensional subspaces of $H_1$ such that the bundle $V_0 := \ker \not{\mathfrak{D}}^*$ is a subbundle of $\underline{V}$.

To proceed further we will need to make an orientation convention.

**Convention** *Consider a split exact sequence of finite-dimensional vector spaces*

$$0 \to E_0 \to E_1 \to E_2 \to 0.$$

*If any of the two spaces above is oriented then the third space is given the orientation determined by the splitting induced isomorphism*

$$E_0 \oplus E_2 \cong E_1.$$

*More precisely*

$$\mathbf{or}(E_0) \wedge \mathbf{or}(E_2) = \mathbf{or}(E_1).$$

Now let $V \in \hat{\mathcal{S}}(\not{\mathfrak{D}})$. Denote by $\hat{V}$ the orthogonal complement of the bundle $V_0 := \ker \not{\mathfrak{D}}_*$ inside the trivial bundle $\underline{V}$. To orient $\mathcal{L}_V = \ker \not{\mathfrak{D}}_V \otimes \det \underline{V}^*$ we equip $\ker \not{\mathfrak{D}}_V$ with a *compatible orientation*. This is done as follows.

### Orientation Recipe

◇ Orient $\hat{V} := V/V_0$ using the canonical split exact sequence of Hilbert spaces

$$0 \to V_0 \to \underline{V} \to \hat{V} \to 0.$$

where the second arrow denotes the orthogonal projection. Observe that $\ker \mathfrak{D}_{V_0}$ is **canonically isomorphic** to $\ker \mathfrak{D}$.

◇ Equip $\ker \mathfrak{D}_V$ with the orientation induced by split exact sequence (1.5.3)

$$0 \to \ker \mathfrak{D}_{V_0} \to \ker \mathfrak{D}_V \to V/V_0 = \hat{V} \to 0.$$

The orientation on $V$ and the above orientation on $\ker \mathfrak{D}_V$ induce an orientation on $\mathcal{L}_V$. Now observe that we have the following sequence of isomorphisms of *oriented* line bundles:

$$I_V := \det \ker \mathfrak{D} \otimes (\det \ker \mathfrak{D}_*)^* \cong \det \ker \mathfrak{D}_{V_0} \otimes \det V_0^* = \mathcal{L}_{V_0} \stackrel{I_{V/V_0}}{\longrightarrow} \mathcal{L}_V.$$

*Exactly* as in the proof of Proposition 1.5.4 we see that for any oriented stabilizers $V \subset W$ we deduce that $I_W = I_{WV} \circ I_V$ which shows that $I_{WV}$ is orientation preserving. This coherence allows us to equip $\det \mathfrak{D}$ with an orientation.

**Proposition 1.5.8.** *Suppose* $(\mathfrak{D}_x)_{x \in X}$ *is a* nice *family. Then* $\det \mathfrak{D}$ *admits a natural orientation which can be concretely described as follows.*

- *Pick* $V \in \hat{\mathcal{S}}(\mathfrak{D})$.
- *Equip the bundle* $\ker \mathfrak{D}_V$ *with the* compatible *orientation.*
- *Orient* $\det \underline{V}^* \otimes \det \mathfrak{D} \cong \det \mathfrak{D}_V$ *using the orientation on* $\underline{V}$ *and the compatible orientation on* $\ker \mathfrak{D}_V$.

There is another situation when one can canonically assign orientations. Suppose the vector bundles $E^0$ and $E^1$ are equipped with complex structures and the operators $\mathfrak{D}_0$ and $T_x$ are complex. Then the stabilizers can be chosen to be *complex subspaces* so that the bundles $\ker \mathfrak{D}_U$ are *complex*, hence equipped with canonical orientations. Arguing exactly as above we can deduce that the orientations thus obtained on the determinant line bundles are independent of the choice of complex stabilizers. We summarize the results proved so far in the following proposition.

**Proposition 1.5.9.** *If the family* $(\mathfrak{D}_x)$ *is the direct sum of a nice family and a complex one then its determinant line bundle can be given a canonical orientation.*

**Remark 1.5.10.** (a) The above observations extend to more general situations. Suppose that $\mathcal{H}^0, \mathcal{H}^1 \to X$ are two, smooth, real Hilbert vector bundles over a compact smooth manifold $X$ and $\mathfrak{D} : \mathcal{H}^0 \to \mathcal{H}^1$ is a Fredholm morphism. This means $\mathfrak{D}$ is a smooth morphism of Hilbert bundles such that for every $x \in X$ the induced map $\mathfrak{D}_x : \mathcal{H}^0_x \to \mathcal{H}^1_x$ is Fredholm. To such a morphism one can attach a determinant line bundle. Moreover, Proposition 1.5.9 continues to hold in this more general context.

(b) The construction in this section which associates to each continuous family of elliptic operators a line bundle on the parameter space has its origins in $K$-theory. Each continuous family of Fredholm operators parameterized by a compact $CW$-complex $X$ defines an element in $K(X)$, a certain abelian group naturally associated to $X$, which is a homotopy invariant of $X$. We recommend [**3**] for a beautiful introduction to this subject.

**Exercise 1.5.3.** Prove the claims in the above remark. (**Hint:** Consult [**3**].)

**Step 2** Suppose we have two homotopic *nice* families, $(\mathfrak{D}^0_x)_{x\in X}$ and $(\mathfrak{D}^1_x)_{x\in X}$. Using the canonical orientation on $\det \mathfrak{D}^0$ and the connecting homotopy we can produce another orientation on $\det \mathfrak{D}^1$. Naturally, one wonders what is the relationship between this transported orientation and the canonical orientation on $\det \mathfrak{D}^1$. It is natural to expect that the comparison between these orientations depends on the given homotopy.

We will consider only one situation, which suffices for most applications in Seiberg-Witten theory. Suppose $X$ consists of one point and $(\nabla^i, T_i)$, $i = 0, 1$, are two pairs (connection on $E$, morphism $E^0 \to E^1$). We get two Dirac operators

$$\mathfrak{D}_i : C^\infty(E^0) \to C^\infty(E^1).$$

Fix orientations on $\ker \mathfrak{D}_i$ and $\ker \mathfrak{D}_i^*$. Clearly the two families $(\nabla^i, T_i)$ satisfy the conditions (i) and (ii) and we thus get two oriented lines

$$\phi_i : \det \mathfrak{D}_i \to \mathbb{R}, \quad i = 0, 1.$$

Each homotopy $h(s) = \mathfrak{D}_s$ determines a homotopy class of isomorphisms

$$\psi : \det \mathfrak{D}_0 \to \det \mathfrak{D}_1$$

and we obtain an induced orientation on $\det \mathfrak{D}_1$ defined by the composition

$$\psi_1 : \det \mathfrak{D}_1 \stackrel{\psi^{-1}}{\to} \det \mathfrak{D}_0 \stackrel{\phi_0}{\to} \mathbb{R}.$$

We thus obtain a linear isomorphism

$$\psi_1 \circ \phi_1^{-1} : \mathbb{R} \stackrel{\phi_1^{-1}}{\to} \det \mathfrak{D}_1 \stackrel{\psi_1}{\to} \mathbb{R}$$

whose homotopy class is determined by a number $m \in \{-1, 1\}$. This real number is called the *orientation transport along the given homotopy.* We will denote it by $\varepsilon(\mathfrak{D}_1, h, \mathfrak{D}_0)$. We want to emphasize that this number depends on the chosen orientations on $\ker \mathfrak{D}_i$ and $\ker \mathfrak{D}_i^*$ and on the chosen homotopy $h(s)$.

**Example 1.5.11.** To understand the subtleties of the above construction we present in detail the following simple example. Consider the map

$$L : \mathbb{R}^n \to \mathbb{R}^k, \quad v \mapsto \sum_{i=1}^{k} \langle v, e_i \rangle e_i$$

where $n > k$, $(e_i)$ denotes the canonical basis of $\mathbb{R}^n$ and $\langle \bullet, \bullet \rangle$ denotes the usual inner product. The kernel of $L$ is precisely the subspace spanned by $e_{k+1}, \cdots, e_n$. We choose this ordered basis to orient $\ker L$. Observe two things.

**1.** $\operatorname{coker}(L) = 0$ so that an orientation of the line $\det L$ uniquely defines an orientation of $\ker L$.

**2.** The map $L$ is homotopic to the trivial map $\mathbb{R}^n \to \mathbb{R}^k$ whose kernel and cokernel are naturally oriented. This homotopy induces another orientation on $\det L$. The difference between these two orientations is precisely the orientation transport along the path $tL$, $t \in [0,1]$ defined above. We want to describe this explicitly since it is very similar to the situation we will encounter in Seiberg-Witten theory.

Consider the family $L_t : \mathbb{R}^n \to \mathbb{R}^k$, $v \mapsto tLv$, $t \in [0,1]$, and set

$$V := \operatorname{span}(e_1, \cdots, e_k) \subset \mathbb{R}^n.$$

$V$ is a stabilizer for the family $L_t$.

For $t = 0$ we have $V = \ker L_0^*$ and the compatible orientation of $V$ given by the rules above is the natural one, determined by the oriented basis $e_1, \cdots, e_k$. $\ker L_{V,0}$ is oriented using the natural isomorphism

$$\ker L_0 \oplus V \cong \ker L_{V,0}, \quad (\ker L_0 \oplus V) \ni (u \oplus v) \mapsto u \oplus 0.$$

Hence

$$e_1 \oplus 0, \cdots, e_n \oplus 0 \tag{1.5.5}$$

is an oriented basis of $\ker L_{V,0}$.

Observe that for each $t \geq 0$ the collection of vectors in $\mathbb{R}^n \oplus V$

$$\begin{gathered} v_1(t) := e_1 \oplus (-te_1), \cdots, v_k(t) := e_k \oplus (-te_k), \\ v_{k+1}(t) = e_{k+1} \oplus 0, \cdots v_n(t) := e_n \oplus 0 \end{gathered} \tag{1.5.6}$$

forms a basis of $\ker L_{V,t}$. When $t = 0$ it coincides with the basis (1.5.5). Thus for $t = 1$ it defines an oriented basis of $\ker L_{V,1}$.

The orientation on $\ker L$ which induces the above orientation is determined from the natural split exact sequence

$$0 \to \ker L \to \ker L_V \to V \to 0.$$

This leads to the isomorphism

(1.5.7)
$$\begin{gathered}\ker L \oplus V \cong \ker L_V,\\ \ker L \oplus V \ni (u \oplus v) \mapsto u \oplus v - \mathsf{R}v \oplus 0 \in \ker L_V\end{gathered}$$

where $\mathsf{R}$ denotes the canonical right inverse of $L$ which in this case is the natural inclusion $V \subset \mathbb{R}^n$.

The natural basis of

$$e_{k+1} \oplus 0, \cdots, e_n \oplus 0, \quad 0 \oplus e_1, \cdots, 0 \oplus e_k$$

of $\ker L \oplus V$ determines via the isomorphism (1.5.7) the following basis of $\ker L_V$:

$$e_{k+1} \oplus 0, \cdots, e_n \oplus 0, \quad (-e_1) \oplus e_1, \cdots, (-e_k) \oplus e_k.$$

The orientation defined by this basis differs from the positive orientation defined by the basis (1.5.6) by $(-1)^{k(n-k)+k}$. Thus $\ker L$ is oriented by the element $(-1)^{k(n-k+1)} e_{k+1} \wedge \cdots \wedge e_n$ of $\det \ker L$.

Returning to the general situation, let us additionally assume

$$\operatorname{ind} \mathfrak{D}_0 = \operatorname{ind} \mathfrak{D}_1 = 0. \tag{1.5.8}$$

The orientation transport has a couple of important properties.

**P0** *Fix* $\mathfrak{D}_0$ and $\mathfrak{D}_1$. Then $\varepsilon(\mathfrak{D}_1, h, \mathfrak{D}_0)$ depends only on the homotopy class of $h$.

**P1** If along the homotopy the operators $\mathfrak{D}_s$ are invertible then $\varepsilon(\mathfrak{D}_1, h, \mathfrak{D}_0) = 0$.

**Proof** Note that the trivial subspace is a stabilizer for the family $\mathfrak{D}_s$. This property now follows from the proof of Proposition 1.5.5. ■

**P2** Suppose $h_0$, resp. $h_1$, is a homotopy connecting $\mathfrak{D}_0$ to $\mathfrak{D}_1$, resp. $\mathfrak{D}_1$ to $\mathfrak{D}_2$. Denote by $h$ the resulting homotopy connecting $\mathfrak{D}_0$ to $\mathfrak{D}_2$. Then

$$\varepsilon(\mathfrak{D}_2, h, \mathfrak{D}_0) = \varepsilon(\mathfrak{D}_2, h_1, \mathfrak{D}_1) \cdot \varepsilon(\mathfrak{D}_1, h_0, \mathfrak{D}_0).$$

**Definition 1.5.12.** Suppose $h(s) = \mathfrak{D}_s$ is a homotopy connecting two operators $\mathfrak{D}_0$ and $\mathfrak{D}_1$.

(a) The *resonance set* of the homotopy is

$$\mathcal{Z}_h = \{ s \in [0,1] \; ; \; \ker \mathfrak{D}_s = \{0\} \, \}.$$

For each $s \in \mathcal{Z}_h$ we denote by $P_s$ the orthogonal projection onto $\ker(\mathfrak{D}_s)^*$.

(b) Set $C_s = \mathfrak{D}_s - \mathfrak{D}_0$. $C_s$ is a zeroth order p.d.o., i.e. a bundle morphism. Define $\dot{C}_s = \frac{d}{ds}C_s$. The homotopy is called *regular* if the resonance set is finite and $\forall s \in [0,1]$ the *resonance operator*

$$R_s : \ker \mathfrak{D}_s \stackrel{\dot{C}_s}{\to} L^2(E^1) \stackrel{P_s}{\to} \ker \mathfrak{D}_s^*$$

is a linear isomorphism.

**P3**([**119**]) Suppose $h$ is a regular homotopy connecting $\mathfrak{D}^0$ to $\mathfrak{D}^1$. Set $d_s = \dim \ker \mathfrak{D}_s = \dim \ker \mathfrak{D}_s^*$. Then

$$\varepsilon(\mathfrak{D}_1, h, \mathfrak{D}_0) = \operatorname{sign}(R_1)\operatorname{sign}(R_0) \prod_{s\in[0,1)} (-1)^{d_s} \tag{1.5.9}$$

where $\operatorname{sign}(R_i) = \pm 1$ $(i = 0,1)$ according to whether $R_i : \ker \mathfrak{D}_i \to \ker \mathfrak{D}_i^*$ preserves or reverses the chosen orientations.

**Proof** Using the product formula **P2** we can reduce the proof of (1.5.9) to two cases.

**Case 1** $\mathcal{Z}_h = \{0\}$. Set

$$\sigma_+ = \lim_{s\searrow 0} \varepsilon(\mathfrak{D}_s, h, \mathfrak{D}_0).$$

Using **P1** and **P2** we deduce $\varepsilon(\mathfrak{D}_1, h, \mathfrak{D}_0) = \sigma_+$. We have to show

$$\sigma_+ = (-1)^{d_0}\operatorname{sign}(R_0).$$

Set $V_0 = \ker \mathfrak{D}_0^*$ and fix an *oriented* basis $(f_1, \cdots, f_n)$ $(n = d_0 = \dim V_0)$ of $V_0$. Then $V_0$ is a stabilizer for $\mathfrak{D}_s$ for all sufficiently small $s \in [0, \varepsilon]$ and

$$\det \mathfrak{D}_s = \det \mathfrak{D}_{V_0,s} \otimes V_0^*.$$

For $s \neq 0$ the operator $\mathfrak{D}_s$ is invertible and for each $f_k$ there exists a unique $x_k \in L^{1,2}(E_0)$ such that

$$\mathfrak{D}_s x_k + f_k = 0. \tag{1.5.10}$$

Then $x_1 \oplus f_1, \cdots, x_n \oplus f_n$ is a basis of $\ker \mathfrak{D}_{V_0,s}$ and we see that the orientation of $V_0$ induces an orientation on $\ker \mathfrak{D}_{V_0,s}$. These orientations on $\ker \mathfrak{D}_{V_0,s}$ and $V_0$ are compatible (in the sense described at Step 1) and define according to Proposition 1.5.8 the canonical orientation on the line $\det \mathfrak{D}_s$, $s > 0$. For $s = 0$ we orient $\det \mathfrak{D}_0$ using the oriented bases $(e_1, \cdots, e_n)$ of $\ker \mathfrak{D}_0$ and $(f_1, \cdots, f_n)$ of $V_0$.

Denote by $Q_s$ the orthogonal projection onto $\ker \mathfrak{D}_{V_0,s} \subset L^2(E^1)$. The trivial connection $\frac{d}{ds}$ on the trivial bundle $L^2(E^1) \times [0,\varepsilon] \to [0,\varepsilon]$ induces a connection $Q_s \frac{d}{ds}$ on the bundle $\ker \mathfrak{D}_{V_0,\bullet} \to [0,\varepsilon]$. It produces a parallel transport map

$$\mathcal{T}_s : \ker \mathfrak{D}_{V_0,0} \to \ker \mathfrak{D}_{V_0,s}.$$

$\ker \mathfrak{D}_{V_0,0}$ is oriented by the oriented basis $e_1 \oplus 0, \cdots, e_n \oplus 0$ while $\ker \mathfrak{D}_{V_0,s}$ is oriented by the oriented basis $x_1 \oplus f_1, \cdots, x_n \oplus v_n$. Set

$$y_k(s) \oplus v_k(s) := \mathcal{T}_s(e_k \oplus 0) \in \ker \mathfrak{D}_{V_0,s}.$$

The vectors $y_k(s) \oplus v_k(s)$ determine a smoothly varying basis of $\ker \mathfrak{D}_{V_0,s}$ described by the initial value problem

$$\left\{ \begin{array}{rcl} \mathfrak{D}_s y_k(s) + v_k(s) & = & 0 \\ y_k(0) & = & e_k \\ v_k(0) & = & 0 \\ (\dot{v}_k, \dot{y}_k) & \in & (\ker \mathfrak{D}_{V_0,s})^{\perp} \end{array} \right. \tag{1.5.11}$$

Observe that $\sigma_+$ is $\pm 1$ depending on whether $\mathcal{T}_s$ preserves/reverses the above orientations for $s$ very small. In other words, to decide the sign of $\sigma_+$ we have to compare the orientations defined by the bases

$$(x_k(s) \oplus f_k) \ \text{ and } \ (y_k(s) \oplus v_k(s))$$

of $\ker \mathfrak{D}_{V_0,s}$. We cannot pass to the limit as $s \searrow 0$ since the vectors $x_k(s)$ "explode" near $s = 0$. The next result makes this statement more precise and will provide a way out of this trouble.

**Lemma 1.5.13.**

$$\|s x_k(s) + R_0^{-1} f_k\| = O(s) \ \text{ as } \ s \searrow 0 \tag{1.5.12}$$

*where* $\|\cdot\|$ *denotes the* $L^2$*-norm.*

**Proof of the lemma** First observe that we have an asymptotic expansion

$$\mathfrak{D}_s = \mathfrak{D}_0 + s\dot{C}_0 + O(s^2) \ \text{ as } s \searrow 0 \tag{1.5.13}$$

where $O(s^2)$ denotes a morphism $E^0 \to E^1$ whose norm as a bounded operator $L^2(E^0) \to L^2(E^1)$ is $\leq \text{const} \cdot s^2$ as $s \searrow 0$. Set

$$z_k(s) = \left\{ \begin{array}{ccc} s x_k(s) & if & s \neq 0 \\ -R_0^{-1} f_k & if & s = 0 \end{array} \right. .$$

We want to prove that

$$\|z_k(s) - z_k(0)\| = O(s) \ \text{ as } \ s \searrow 0.$$

Using the equalities (1.5.10) and (1.5.13) we deduce

$$(\mathfrak{D}_0 + s\dot{C}_0 + O(s^2)) z_k + s f_k = 0$$

so that

$$\mathfrak{D}_0 z_k = -s\dot{C}_0 z_k - s f_k + O(s^2) z_k. \tag{1.5.14}$$

We decompose $z_k$ following the orthogonal decomposition

$$L^2(E^0) = \ker \mathfrak{D}_0 \oplus (\ker \mathfrak{D}_0)^{\perp} \longrightarrow z_k = z_k^0 + z_k^{\perp}.$$

Recall that $P_0$ denotes the orthogonal projection onto $\ker \mathfrak{D}_0^* = \text{Range}(\mathfrak{D}_0)^\perp$. We can now rewrite (1.5.14) as

$$\begin{cases} \mathfrak{D}_0 z_k^\perp = (\mathbf{1} - P_0)(-s\dot{C}_0 z_k - sf_k + O(s^2)z_k) \\ \\ P_0\dot{C}_0 z_k + f_k = P_0 O(s) z_k. \end{cases} \tag{1.5.15}$$

From the first equation we deduce

$$\|z_k^\perp\| \leq Cs(\|z_k\| + \|f_k\|)$$

so that

$$\|z_k^\perp\| \leq Cs(\|z_k^0\| + \|f_k\|). \tag{1.5.16}$$

We can now rewrite the second equation in (1.5.15) as

$$R_0 z_k^0 = P_0 \dot{C}_0 z_k^0 = P_0 O(s)(z_k^0 + z_k^\perp) - f_k - P_0 \dot{C}_0 z_k^\perp$$

so that

$$z_k^0 = z_k(0) + R_0^{-1} O(s)(z_k^0 + z_k^\perp) - R_0^{-1} P_0 \dot{C}_0 z_k^\perp$$

and using (1.5.16) we deduce

$$\|z_k^0 - z_k(0)\| \leq Cs(\|z_k^0\| + 1).$$

The equality (1.5.12) is now obvious. ■

Notice that the bases $z_k(s) \oplus sf_k$ and $x_k(s) \oplus f_k$ define the same orientations on $\ker \mathfrak{D}_{V_0,s}$, for all $s > 0$ sufficiently small. Thus, in order to find the sign of $\sigma_+$ we have to compare the orientations determined by the bases $z_k(s) \oplus f_k$ and $y_k(s) \oplus v_k$ as $s \searrow 0$. The advantage now is that we can pass to the limit in *both* bases. Thus we need to compare the orientations determined by the bases $(-R_0^{-1} f_k) \oplus 0$ and $e_k \oplus 0$. They differ exactly by $(-1)^n \text{sign}\,(R_0)$ where $n = \dim \ker \mathfrak{D}_0 = d_0$.

**Case 2** $\mathcal{Z}_h = \{1\}$. Set $\sigma_- = \lim_{s \nearrow 1} \varepsilon(\mathfrak{D}_1, h, \mathfrak{D}_s)$. We have to show

$$\sigma_- = \text{sign}\,(R_1).$$

The proof is identical to the one in **Case 1**. The equality (1.5.12) has to be replaced with

$$\|sx_k(1-s) - R_1^{-1} f_k\| = O(s), \quad \text{as } s \searrow 0$$

because instead of (1.5.13) we have

$$\mathfrak{D}_{1-s} = \mathfrak{D}_1 - s\dot{C}_1 + O(s^2) \;\; \text{as } s \searrow 0 \tag{1.5.17}$$

In the end we have to compare the bases $R_1^{-1} f_k$ and $e_k$. Property **P3** is proved. ■

**Remark 1.5.14.** For a different proof of **P3** we refer to [**119**].

In Section 2.3 we will need the following technical result.

**Proposition 1.5.15.** *Suppose* $\mathfrak{D}_t$, $t \in [0,1]$, *is a continuous family of* real *first order elliptic operators*

$$\mathfrak{D}_t : L^{1,2}(E^0) \subset H_0 := L^2(E^0) \to H_1 := L^2(E^1)$$

*with the following properties.*

*(a)* $\operatorname{ind} \mathfrak{D}_t = 0$.

*(b)* $\mathfrak{D}_t$ *is invertible for* $t$ *close to* 0 *and* 1.

*(c) There exists a smooth family of continuous linear maps* $L_t : \mathbb{R} \to H_1$ *such that*

*(c0)* $L_t = 0$ *for* $t = 0, 1$.

*(c1) The map* $S_t := L_t + \mathfrak{D}_t : H_0 \oplus \mathbb{R} \to H_1$, $h_0 \oplus \mu \to L_t\mu + \mathfrak{D}_t h_0$ *is onto.*

*(c2) The real line bundle* $\mathcal{L} := \ker(S_\bullet) \to [0,1]$ *is oriented.*

*Observe that the fibers of* $\mathcal{L}$ *over* $i = 0, 1$ *can be identified with* $\mathbb{R}$ *via the natural isomorphisms*

$$\omega_i : \mathbb{R} \to \mathcal{L}_i, \quad \mu \mapsto (0, \mu). \tag{1.5.18}$$

*On the other hand, the orientation of* $\mathcal{L}$ *defines orientations* $\phi_i : \mathcal{L}_i \to \mathbb{R}$, $i = 0, 1$. *The homotopy class of the isomorphism* $\phi_i\omega_i : \mathbb{R} \to \mathbb{R}$ *is uniquely determined by a sign* $\varepsilon_i \in \{\pm 1\}$.

*Then the orientation transport along the path* $\mathfrak{D}_t$ *is* $\varepsilon_0/\varepsilon_1$.

**Proof** Recall how one computes the parallel transport. Fix an arbitrary oriented stabilizer $V$ for the family $\mathfrak{D}_t$. We get a vector bundle

$$\ker \mathfrak{D}_{V,\bullet} \to [0,1].$$

Once we fix a connection $\nabla$ on this bundle we get a parallel transport

$$T = T_\nabla : \ker \mathfrak{D}_{V,0} \to \ker \mathfrak{D}_{V,1}. \tag{1.5.19}$$

Using condition (b) we obtain isomorphisms $\ker \mathfrak{D}_{V,i} = 0 \oplus V$, $i = 0, 1$, defined explicitly by

$$\tau_i : 0 \oplus V \oplus \ni (0, v) \mapsto (-\mathfrak{D}_i^{-1} v, v) \in \ker \mathfrak{D}_{V,i}. \tag{1.5.20}$$

*Via these isomorphisms* we can regard $T$ as a map $\tau_1 \circ T \circ \tau_0^{-1} : V \to V$. The orientation transport is then the sign of its determinant. For $t \in [0,1]$ define $U_t : H_0 \oplus V \oplus \mathbb{R} \to H_1$ by

$$h_0 \oplus v \oplus \mu \mapsto S_t(h_0 \oplus \mu) + v = L_t\mu + \mathfrak{D}_{V,t}(h_0 \oplus v) = L_t\mu + v + \mathfrak{D}_t h_0.$$

There exist natural isomorphisms

$$I_t := \ker \mathfrak{D}_{V,t} \oplus \mathbb{R} \to \ker U_t$$

defined by

$$I_t := \ker \not{\mathfrak{D}}_{V,t} \oplus \mathbb{R} \ni (h_0, v, \mu) \mapsto (h_0, v, \mu) - \mathsf{R}_{V,t}(L_t \mu) \oplus 0.$$

On the other hand, we have isomorphisms

$$J_t : V \oplus \mathcal{L}_t \to \ker U_t^1$$

defined by

$$V \oplus \mathcal{L}_t \ni (v, h_0, \mu) \mapsto (h_0, v, \mu) - (h_0^t(v), 0, \mu^t(v))$$

where $(h_0^t(v), \mu^t(v))$ is the element in $H_0 \oplus \mathbb{R}$ uniquely determined by

$$(h_0^t(v), \mu^t(v)) \in (\ker S_t)^{\perp}, \quad L_t \mu^t(v) + \not{\mathfrak{D}}_t h_0^t(v) = v.$$

Using (c0) we deduce that for $t = 0, 1$ $\ker S_t = \mathbb{R} \oplus 0$ and we can be more explicit, namely

$$\mu^t(v) = 0, \quad h_0^t(v) = \not{\mathfrak{D}}_t^{-1} v.$$

Thus, for $t = 0, 1$ we have

$$J_t^{-1}(h, v, \mu) = (v, 0, \mu).$$

We thus get isomorphisms

$$I_t^{-1} \circ J_t : V \oplus \mathcal{L}_t \to \ker \not{\mathfrak{D}}_{V,t} \oplus \mathbb{R}$$

depending smoothly upon $t$. Now look at the following diagram.

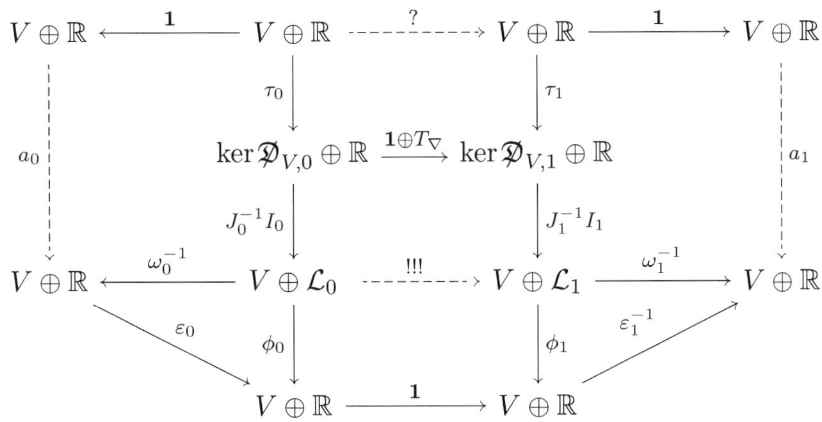


The maps $\tau_i$ are defined by (1.5.20) and $T_\nabla$ denotes the parallel transport defined in (1.5.19). The dashed arrows are defined tautologically, to make the diagram commutative. We are interested in the sign of the determinant of the ?-arrow. The maps $\phi_i$ are determined by the orientation (trivialization) of the fibers $\mathcal{L}_i$ induced by the orientation (a.k.a. trivialization) of $\mathcal{L}$.

The connection $\nabla$ induces via $J_t^{-1} I_t$ a connection $\nabla'$ on $\underline{V} \oplus \mathcal{L}$ with parallel transport $T'$. The (!!!)-arrow is precisely $T'$.

On the other hand, the orientation (trivialization) of $\mathcal{L}_t$ defines a canonical connection $\nabla^0$ on $\underline{V} \oplus \mathcal{L}$ with parallel transport $T^0$. Since $\nabla'$ and $\nabla^0$ are homotopic we deduce $T^0$ is homotopic to $T'$ so that in the above diagram the (!!!)-arrow is also equivalent to $T^0$. With respect to the trivializations $\phi_i$ the map $T^0$ is the **identity**, thus explaining the bottom arrow.

The isomorphism $a_i : V \oplus \mathbb{R} \to V \oplus \mathbb{R}$ is the identity. To see this observe that (for $i = 0$) we have

$$a_0(v \oplus \mu) = \omega_0^{-1} J_0^{-1} I_0 \tau_0(v \oplus \mu) = \omega_0^{-1} J_0^{-1} I_0((-\not{\mathfrak{D}}_0^{-1} v) \oplus v \oplus \mu)$$

$$= \omega_0^{-1} J_0^{-1}(\,(-\not{\mathfrak{D}}_0^{-1} v) \oplus v \oplus \mu) = \omega_0^{-1}(v \oplus \mu \oplus 0) = v \oplus \mu.$$

The proposition is now obvious from the diagram and the above explicit description of the maps $a_i$. ■

**Exercise 1.5.4.** Formulate and prove a generalization of the above proposition where instead of maps $L_t : \mathbb{R} \to H_1$ we have linear maps $L_t : E \to H_1$ in which $E$ is a finite-dimensional *oriented* space.

**1.5.2. Genericity results.** Suppose $X$, $Y$ and $\Lambda$ are Hilbert manifolds and

$$F : \Lambda \times X \to Y, \quad (\lambda, x) \mapsto y = F(\lambda, x)$$

is a smooth map. Fix $y_0 \in Y$. We are interested in studying the dependence upon the parameter $\lambda$ of the solution sets

$$S_\lambda = \{x \in X \ ; \ F(\lambda, x) = y\}.$$

More precisely, we are interested whether there exist values of the parameter $\lambda$ for which the solution sets $S_\lambda$ are smooth submanifolds. According to the implicit function theorem this will happen provided $y_0$ is a *regular value* of the map

$$F_\lambda : X \to Y, \quad x \mapsto F(\lambda, x),$$

that is, for every $x_0 \in S_\lambda$ the differential

$$\frac{\partial F_\lambda}{\partial x} : T_{x_0} X \to T_{y_0} Y$$

is a bounded linear surjection. We will say that $\lambda$ is a *good* parameter if $y_0$ is a regular value of $F_\lambda$. In this subsection we will address the following question.

*Is it possible that "most" parameters are good?*

A result providing a positive answer to this question is usually known as a genericity result.

Note first of all that if we expect genericity results it is natural to assume the parameter space $\Lambda$ is "sufficiently large". More precisely, we will assume that $y_0$ is a regular value of $F$. To understand why this is a statement about the size of $\Lambda$ introduce the "master space"

$$\mathcal{S} = \{(\lambda, x) \in \Lambda \times X \; ; \; F(\lambda, x) = y_0\}.$$

Since $y_0$ is a regular value of $F$ this means that for all $(\lambda, x) \in \mathcal{S}$ the differential

$$DF : T_{(\lambda,x)}\Lambda \times X \to T_{y_0}Y$$

is a bounded linear surjection. In particular, $\mathcal{S}$ is a smooth Hilbert manifold. We see that if $\Lambda$ is "too small" the above operator may not be surjective.

Denote by $\pi$ the natural projection $\Lambda \times X \to \Lambda$. We obtain a smooth map

$$\pi : \mathcal{S} \hookrightarrow \Lambda \times X \to \Lambda$$

and the solution sets $\mathcal{S}_\lambda$ can be identified with the fibers $\pi^{-1}(\lambda)$ of $\pi$. We see that any regular value of $\pi$ is necessarily a good parameter. Thus, if "most" parameters are regular values of $\pi$ then "most" of them must be good and we have a genericity result. This looks more and more like Sard's theorem but there is one aspect we have quietly avoided so far: the manifolds $X, Y, \Lambda$ may be infinite dimensional and thus out of the range of the standard Sard theorem. Fortunately, S. Smale [**124**] has shown that under certain conditions, the Sard theorem continues to hold in infinite dimensions as well. To formulate his result we need to introduce the notion of *nonlinear Fredholm maps.*

**Definition 1.5.16.** A smooth map $F : M \to N$ between Hilbert manifolds is said to be *Fredholm* if for every $m \in M$ the differential

$$D_mF : T_mM \to T_{F(m)}N$$

is a bounded, linear Fredholm operator. If $M$ is connected, the indices of the operators $D_mF$ are independent of $M$ and their common value is called the *index* of $F$ and is denoted by $\mathrm{ind}\,(F)$.

A subset in a topological space is said to be *generic* if it contains the intersection of an at most countable family of dense, open sets. Baire's theorem states that the generic sets in complete metric spaces or locally compact spaces are necessarily dense. The expression *"most $x$ satisfy the property ..."* will mean that the set of $x$ satisfying that property is generic.

**Theorem 1.5.17. (Sard-Smale)** *Suppose $F : M \to N$ is a smooth Fredholm map between paracompact Hilbert manifolds, where $M$ is assumed connected.*

*(a) If* $\operatorname{ind}(F) < 0$ *then* $F^{-1}(n) = \emptyset$ *for most* $n$.

*(b) If* $\operatorname{ind}(F) \geq 0$ *then most* $n \in N$ *are regular values of* $F$ *and for these* $n$ *the fibers* $F^{-1}(n)$ *are finite dimensional (possibly empty) smooth manifolds of dimension* $\operatorname{ind}(F)$.

Let us now return to the original problem. We want to apply the Sard-Smale theorem to the map $\pi : \mathcal{S} \to \Lambda$, so that we have to assume it is Fredholm. The following result describes a condition on $F$ which guarantees that $\pi$ is Fredholm.

**Lemma 1.5.18.** *Suppose that both* $\Lambda$ *and* $X$ *are connected,* $y_0$ *is a regular value of* $F$ *and for each* $\lambda \in \Lambda$ *the map* $F_\lambda : X \to Y$ *is Fredholm. Then* $\pi : \mathcal{S} \to \Lambda$ *is Fredholm and*

$$\operatorname{ind}(\pi) = \operatorname{ind}(F_\lambda), \quad \forall \lambda \in \Lambda.$$

**Exercise 1.5.5.** Prove the above lemma.

The final result of this subsection summarizes the above considerations.

**Theorem 1.5.19.** *Consider smooth, paracompact, connected Hilbert manifolds* $X, Y, \Lambda$, *a smooth map* $F : \Lambda \times X \to Y$ *and a point* $y_0 \in Y$ *satisfying the following conditions.*

*(i)* $y_0$ *is a regular value of* $F$.

*(ii) The maps* $F_\lambda : X \to Y$ *are Fredholm for all* $\lambda \in \Lambda$.

*Then the following hold.*

*(a) If* $\operatorname{ind}(F_\lambda) < 0$ *then* $S_\lambda = \emptyset$ *for most* $\lambda$.

*(b) If* $\operatorname{ind}(F_\lambda) \geq 0$ *then* $S_\lambda$ *is a smooth (possible empty) manifold of dimension* $\operatorname{ind}(F_\lambda)$ *for most* $\lambda \in \Lambda$.

*Chapter 2*

# The Seiberg-Witten Invariants

*Get your facts first, and then distort them as much as you please.*

Mark Twain

## 2.1. Seiberg-Witten monopoles

This section finally introduces the reader to the central objects of these notes, namely, the Seiberg-Witten monopoles. They are solutions of a nonlinear system of partial differential equations called the *Seiberg-Witten equations.* We will discuss several basic features of these objects.

**2.1.1. The Seiberg-Witten equations.** First we need to introduce the geometric background. It consists of a connected, oriented, Riemannian four dimensional manifold $(M, g)$ equipped with a *spin*$^c$ structure $\sigma$. There are two bundles naturally associated to this datum.

- The bundle of complex spinors $\mathbb{S}_\sigma = \mathbb{S}_\sigma^+ \oplus \mathbb{S}_\sigma^-$;
- The associated line bundle $\det(\sigma)$ which is equipped with an $U(1)$-structure.

*Fix* a Hermitian metric on $\det(\sigma)$ inducing this $U(1)$-bundle and denote by $\mathfrak{A}_\sigma = \mathfrak{A}_\sigma(M)$ the space of Hermitian connections on $\det(\sigma)$. Also, denote by $c_\sigma$ the first Chern class of $\det(\sigma)$, $c_\sigma = c_1(\det(\sigma))$.

We can now define the *configuration space*

$$\mathcal{C}_\sigma = \mathcal{C}_\sigma(M) = C^\infty(\mathbb{S}_\sigma^+) \times \mathfrak{A}_\sigma.$$

Observe that this is an affine space. We will denote its elements by the symbol $\mathsf{C} = (\psi, A)$ and by $\mathfrak{G}_\sigma = \mathfrak{G}_\sigma(M)$ the group of smooth maps $M \to S^1$. Given $A \in \mathfrak{A}_\sigma$ we obtain a *geometric Dirac structure* $(\mathbb{S}_\sigma, \mathbf{c}, \nabla^A, \nabla)$, where $\nabla$ denotes the Levi-Civita connection while $\nabla^A$ is the connection induced by $A$ on $\mathbb{S}_\sigma$ which is compatible with the Clifford multiplication, the Levi-Civita connection and the splitting $\mathbb{S}_\sigma^+ \oplus \mathbb{S}_\sigma^-$. As usual, we will denote by $\mathfrak{D}_A$ the Dirac operator $\Gamma(\mathbb{S}_\sigma^+) \to \Gamma(\mathbb{S}_\sigma^-)$ induced by this geometric Dirac structure.

We can now conjugate $\nabla^A$ with any element $\gamma \in \mathfrak{G}_\sigma$ and, as shown in Exercise 1.3.21, the connection $\gamma\nabla^A\gamma^{-1}$ is induced by the connection $A - 2(d\gamma)\gamma^{-1} \in \mathfrak{A}_\sigma$, that is,

$$\gamma\nabla^A\gamma^{-1} = \nabla^{A-2d\gamma/\gamma}.$$

We can regard the correspondence

$$\mathfrak{G}_\sigma \times \mathcal{C}_\sigma \ni (\gamma; \psi, A) \mapsto (\gamma\psi, A - 2d\gamma/\gamma) \in \mathcal{C}_\sigma$$

as a *left* action of $\mathfrak{G}_\sigma$ on $\mathcal{C}_\sigma$, $(\gamma, \mathsf{C}) \mapsto \gamma \cdot \mathsf{C}$. For each $\mathsf{C} \in \mathcal{C}_\sigma$ we denote by $\mathbf{Stab}(\mathsf{C})$ the stabilizer of $\mathsf{C}$ with respect to the above action

$$\mathbf{Stab}(\mathsf{C}) := \Big\{\gamma \in \mathfrak{G}_\sigma \; ; \; \gamma \cdot \mathsf{C} = \mathsf{C}\Big\}.$$

**Definition 2.1.1.** A configuration $\mathsf{C}$ is said to be *irreducible* if

$$\mathbf{Stab}(\mathsf{C}) = \{1\}.$$

Otherwise, it is said to be *reducible*. We will denote by $\mathcal{C}_{\sigma,irr}$ the set of irreducible configurations and by $\mathcal{C}_{\sigma,red}$ the set of reducible ones.

**Proposition 2.1.2.**

$$\mathcal{C}_{\sigma,red} = \Big\{\mathsf{C} = (\psi, A) \; ; \; \psi \equiv 0\Big\}.$$

*Moreover, if $\mathsf{C} = (\psi, A)$ is a reducible configuration, then $\mathbf{Stab}(\mathsf{C})$ is isomorphic to the subgroup $S^1 \subset \mathfrak{G}_\sigma$ consisting of constant maps.*

**Exercise 2.1.1.** Prove the above proposition.

The quadratic map $q$ introduced in Example 1.3.3 defines a map

$$q : C^\infty(\mathbb{S}_\sigma^+) \to \mathrm{End}_0(\mathbb{S}_\sigma^+), \;\; q(\psi) = \bar{\psi} \otimes \psi - \frac{1}{2}|\psi|^2\mathbf{id}.$$

$\mathrm{End}_0(\mathbb{S}_\sigma^+)$ denotes the space of traceless, symmetric endomorphisms of $\mathbb{S}_\sigma^+$. More precisely,

$$C^\infty(\mathbb{S}_\sigma^+) \ni \phi \overset{q(\psi)}{\mapsto} \langle \phi, \psi \rangle \psi - \frac{1}{2}|\psi|^2\phi \in C^\infty(\mathbb{S}_\sigma^+).$$

We want to emphasize one working convention.

*We will always assume that a Hermitian metric $\langle\bullet,\bullet\rangle$ on a complex vector space is assumed to be complex linear in the first variable and complex conjugate-linear in the second variable.*

**Definition 2.1.3.** Fix a closed, real 2-form $\eta \in \Omega^2(M)$. Then a $(\sigma,\eta)$-*monopole* is a configuration $\mathsf{C} = (\psi, A)$ satisfying the *Seiberg-Witten equations*

$$(2.1.1) \qquad (SW_{\sigma,\eta}) \left\{ \begin{array}{rcl} \mathfrak{D}_A\psi & = & 0 \\ \mathbf{c}(F_A^+ + \mathbf{i}\eta^+) & = & \frac{1}{2}q(\psi) \end{array} \right.$$

where the superscript "+" denotes the self-dual part of a 2-form and $\mathbf{c}$ denotes the Clifford multiplication by a form. The 0-monopoles will be called simply monopoles. The closed 2-form $\eta$ is called the *perturbation parameter*.

A few comments are in order.

• Note first that the Seiberg-Witten equations (2.1.1) depend on the metric $g$ in several ways: the symbol of the Dirac operator depends on the metric, the connection $\nabla^A$ depends on the Levi-Civita connection of the metric and the splitting $\Omega^2(M) = \Omega^2_+(M) \oplus \Omega^2_-(M)$ is also dependent on the metric.

• Notice also that the second equation in (2.1.1) is consistent with the isomorphism $\mathbf{i}\Omega^2_+(M) \cong \mathrm{End}_0(\mathbb{S}^+_\sigma)$ induced by the Clifford multiplication $\mathbf{c}$.

We denote by $\mathfrak{Z}_\sigma = \mathfrak{Z}_\sigma(g,\eta)$ the set of solutions of the Seiberg-Witten equations and set

$$\mathfrak{Z}_{\sigma,irr} = \mathfrak{Z}_\sigma \cap \mathcal{C}_{\sigma,irr}.$$

Observe next that if $\mathsf{C} \in \mathfrak{Z}_\sigma$ and $\gamma \in \mathfrak{G}_\sigma$ then $\gamma \cdot \mathsf{C} \in \mathfrak{Z}_\sigma$. Thus, $\mathfrak{Z}_\sigma$ is a $\mathfrak{G}_\sigma$-invariant subset of $\mathcal{C}_\sigma$. We set

$$\mathfrak{M}_\sigma = \mathfrak{M}_\sigma(g,\eta) = \mathfrak{Z}_\sigma/\mathfrak{G}_\sigma$$

and

$$\mathfrak{M}_{\sigma,irr} = \mathfrak{Z}_{\sigma,irr}/\mathfrak{G}_\sigma.$$

$\mathfrak{M}_\sigma$ is known as the *Seiberg-Witten moduli space*.

Besides the huge $\mathfrak{G}$-symmetry, the Seiberg-Witten equations are equipped with another special type of symmetry. The involution $\sigma \mapsto \bar{\sigma}$ on $Spin^c(M)$ defines a bijection $\hat{\vartheta} : \mathcal{C}_{\sigma,\eta} \to \mathcal{C}_{\bar{\sigma},-\eta}$ induced by the isomorphisms

$$\vartheta : \bar{\mathbb{S}}^+_\sigma \to \mathbb{S}^+_{\bar{\sigma}}, \ \ \det(\bar{\sigma}) \cong \overline{\det(\sigma)} \cong \det(\sigma)^*.$$

More precisely, $\hat{\vartheta}(\psi, A) = (\vartheta(\bar{\psi}), A^*)$ where for any connection $A$ on $\det(\sigma)$ we have denoted by $A^*$ the connection it induces on $\det(\sigma)^*$. The results in Exercise 1.3.23 coupled with the equality $F_{A^*} = -F_A$ show that if $\mathsf{C}$

is a $(\sigma, \eta)$-monopole then $\hat{\vartheta}(\mathsf{C})$ is a $(\bar{\sigma}, -\eta)$-monopole. Also observe that $\mathfrak{G}_\sigma = \mathfrak{G}_{\bar{\sigma}} = \mathfrak{G}$ and, for all $\gamma \in \mathfrak{G}$, we have

$$\hat{\vartheta}(\gamma \cdot \mathsf{C}) = \gamma^{-1} \cdot \hat{\vartheta}(\mathsf{C}). \tag{2.1.2}$$

This shows that we have a bijection

$$\hat{\vartheta} : \mathfrak{M}_\sigma(g, \eta) \to \mathfrak{M}_{\bar{\sigma}}(g, -\eta). \tag{2.1.3}$$

*In the remainder of this chapter $M$ will be assumed to be compact, connected, oriented and without boundary.*

The Seiberg-Witten equations are first order equations and thus cannot be the Euler-Lagrange equations of any action functional. However, the monopoles do have a variational nature.

**Proposition 2.1.4.** *Define $\mathfrak{E}_\eta : \mathcal{C}_\sigma \to \mathbb{R}$ by*

$$\mathfrak{E}_\eta(\psi, A) = \int_M \Big( |\nabla^A \psi|^2 + \frac{s}{4}|\psi|^2 + \frac{1}{2}\Big|\frac{1}{2}q(\psi) - \mathbf{c}(\mathbf{i}\eta^+)\Big|^2 + |F_A + 2\mathbf{i}\eta^+|^2 \Big) dv_g$$

*where $s$ denotes the scalar curvature of the metric $g$ and for any endomorphism $T : \mathbb{S}_\sigma^+ \to \mathbb{S}_\sigma^+$ we have denoted $|T|^2 := \mathrm{tr}(TT^*)$. Then*

$$\mathfrak{E}_\eta(\psi, A) = \int_M \Big( |\mathfrak{D}_A \psi|^2 + \frac{1}{2}|\mathbf{c}(F_A^+ + \mathbf{i}\eta^+) - \frac{1}{2}q(\psi)|^2 \Big) dv_g$$

$$+4 \int_M |\eta^+|^2 dv_g - 4\pi^2 \int_M c_\sigma^2$$

*where $c_\sigma = c_1(\det(\sigma))$. In particular, we deduce that*

$$\mathfrak{E}_\eta(\psi, A) \geq 4 \int_M |\eta^+|^2 dv_g - 4\pi^2 \int_M c_\sigma^2$$

*with equality if and only if $(\psi, A)$ is an $\eta$-monopole.*

**Proof** The proof relies on the following elementary identities.

**Lemma 2.1.5.** *Let $\alpha \in \mathbf{i}\Omega_+^2(M)$, $\psi \in C^\infty(\mathbb{S}_\sigma^+)$ and $T \in \mathrm{End}_0(\mathbb{S}_\sigma^+)$. Then we have the following pointwise identities:*

$$|q(\psi)|^2 := \mathrm{tr}(q(\psi)^2) = \frac{1}{2}|\psi|^4,$$

$$|\mathbf{c}(\alpha)|^2 = 4|\alpha|^2,$$

$$\langle T, q(\psi) \rangle \stackrel{def}{=} \mathrm{tr}(Tq(\psi)) = \langle T\psi, \psi \rangle.$$

**Proof of the lemma** All the computations are *pointwise* so it suffices to prove they hold at a given arbitrary point $x \in M$. Set $V = T_xM$. We now use the notations and the computations in Example 1.3.3. Then

$$|q(\psi)|^2 = \operatorname{tr}\left(\begin{bmatrix} \frac{1}{2}(|\alpha|^2 - |\beta|^2) & \alpha\bar{\beta} \\ \bar{\alpha}\beta & \frac{1}{2}(|\beta|^2 - |\alpha|^2) \end{bmatrix}^2\right)$$

$$= \frac{1}{2}(|\alpha|^2 - |\beta|^2)^2 + 2|\alpha|^2|\beta|^2$$

$$= \frac{1}{2}(|\alpha|^2 + |\beta|^2)^2 = \frac{1}{2}|\psi|^4.$$

The second equality follows from the identities

$$c(\sum_{k=0}^{s} x_k\eta_k)^2 = -2(\sum_{k=0}^{2} x_k^2)\mathbf{id}, \;\; \operatorname{tr}(\mathbf{id}) = 2.$$

To prove the third identity we observe it is linear in $T$ and since any $T \in \operatorname{End}_0(\mathbb{S}_\sigma^+ |_x)$ can be written as $T = \sum_{k=0}^{2} t_k \mathbf{c}(\mathbf{i}\eta_k)$, $t_k \in \mathbb{R}$, it suffices to prove it for $T = \mathbf{c}(\mathbf{i}\eta_k)$. The computations in Exercise 1.3.2 show that

$$\operatorname{tr}(q(\psi)\cdot \mathbf{c}(\mathbf{i}\eta_k)) = -\frac{\mathbf{i}}{4}\langle\psi, \mathbf{c}(\eta_k)\psi\rangle \cdot \operatorname{tr}(\mathbf{c}(\mathbf{i}\eta_k)^2)$$

$$= \frac{1}{4}\langle\psi, \mathbf{c}(\mathbf{i}\eta_k)\psi\rangle \cdot |\mathbf{c}(\eta_k)|^2 = \langle\psi, \mathbf{c}(\mathbf{i}\eta_k)\psi\rangle, \;\; k = 0, 1, 2.$$

The lemma is proved. ■

We can now continue the proof of the proposition. First, an integration by parts coupled with the Weitzenböck formula (1.3.16) gives

$$\int_M |\mathfrak{D}_A\psi|^2 dv_g = \int_M \langle\mathfrak{D}_A^*\mathfrak{D}_A\psi, \psi\rangle dv_g$$

$$= \int_M \Big(\langle(\nabla^A)^*\nabla^A\psi, \psi\rangle + \frac{s}{4}|\psi|^2 + \frac{1}{2}\langle\mathbf{c}(F_A^+)\psi, \psi\rangle\Big) dv_g$$

(use Lemma 2.1.5)

$$= \int_M \Big(|\nabla^A\psi|^2 + \frac{s}{4}|\psi|^2 + \frac{1}{2}\langle\mathbf{c}(F_A^+), q(\psi)\rangle\Big) dv_g.$$

Next observe that

$$\int_M |\mathbf{c}(F_A^+ + \mathbf{i}\eta^+) - \frac{1}{2}q(\psi)|^2 dv_g$$

$$= \int_M \Big(|\mathbf{c}(F_A^+)|^2 + |\frac{1}{2}q(\psi) - \mathbf{c}(\mathbf{i}\eta^+)|^2\Big) dv_g - 2\int_M \langle\mathbf{c}(F_A^+), \frac{1}{2}q(\psi) - \mathbf{c}(\mathbf{i}\eta^+)\rangle dv_g.$$

Hence

$$\int_M \Big( |\mathfrak{D}_A\psi|^2 + \frac{1}{2}|\mathbf{c}(F_A^+ + \mathbf{i}\eta^+) - \frac{1}{2}q(\psi)|^2 \Big) dv_g$$

(use Lemma 2.1.5)

$$= \int_M \Big( |\nabla^A\psi|^2 + \frac{s}{4}|\psi|^2 + \frac{1}{2}|\frac{1}{2}q(\psi) - \mathbf{c}(\mathbf{i}\eta^+)|^2 + 2|F_A^+|^2 + 4\langle F_A^+, \mathbf{i}\eta^+\rangle \Big) dv_g.$$

The last two terms can be rewritten as

$$2\int_M \Big( |F_A^+|^2 + 2\langle F_A^+, \mathbf{i}\eta^+\rangle \Big) dv_g$$

$$= 2\int_M \Big( \frac{1}{2}|F_A|^2 + 2\langle F_A, \mathbf{i}\eta^+\rangle + |F_A^+|^2 - \frac{1}{2}|F_A|^2 \Big) dv_g$$

$$= 2\int_M \Big( \frac{1}{2}|F_A + 2\mathbf{i}\eta^+|^2 - 2|\eta^+|^2 + \frac{1}{2}(|F_A^+|^2 - |F_A^-|^2) \Big) dv_g$$

$$= \int_M \Big( |F_A + 2\mathbf{i}\eta^+|^2 - 4|\eta^+|^2 \Big) dv_g - \int_M F_A \wedge F_A$$

$$= \int_M \Big( |F_A + 2\mathbf{i}\eta^+|^2 - 2|\eta^+|^2 \Big) dv_g + 4\pi^2 \int_M c_1(A) \wedge c_1(A).$$

Thus

$$\int_M \Big( |\mathfrak{D}_A\psi|^2 + \frac{1}{2}|\mathbf{c}(F_A^+ + \mathbf{i}\eta^+) - \frac{1}{2}q(\psi)|^2 \Big) dv_g$$

$$= \int_M \Big( |\nabla^A\psi|^2 + \frac{s}{4}|\psi|^2 + \frac{1}{2}|\frac{1}{2}q(\psi) - \mathbf{c}(\mathbf{i}\eta^+)|^2 + |F_A + 2\mathbf{i}\eta^+|^2 \Big) dv_g$$

$$+ \int_M \Big( 4\pi^2 c_\sigma^2 - 4|\eta^+|^2 \Big).$$

Proposition 2.1.4 is now obvious. ■

**2.1.2. The functional set-up.** So far we have worked exclusively in the smooth category. To define the Seiberg-Witten invariants we have to introduce additional structures on the moduli space $\mathfrak{M}_\sigma(g,\eta)$ and, in particular, we need to topologize it. The best functional framework for such purposes is supplied by the Sobolev spaces.

Pick a nonnegative integer $m$ and a real number $p \in (1,\infty)$ such that

$$m + 2 - \frac{4}{p} > 0.$$

This condition guarantees that the Sobolev spaces $L^{m+2,p}$ embed continuously in some Hölder space.

Now fix a *smooth* Hermitian connection $A_0$ on $\det(\sigma)$ and denote by $\mathfrak{A}_\sigma^{m+1,p}$ the space of $L^{m+1,p}$ connections on $\det(\sigma)$. More precisely,

$$\mathfrak{A}_\sigma^{m+1,p} = \Big\{ A = A_0 + \mathbf{i}a \ ; \ a \in L^{m+1,p}(T^*M) \Big\}, \ \ \mathfrak{A}_\sigma^{m+1} := \mathfrak{A}_\sigma^{m+1,2},$$

$$\mathcal{C}_\sigma^{m+1} := L^{k+1,2}(\mathbb{S}_\sigma^+) \times \mathfrak{A}_\sigma^{m+1}.$$

Next, define

$$\mathfrak{Y}^{m,p} = \mathfrak{Y}_\sigma^{m,p} = L^{m,p}(\mathbb{S}_\sigma^-) \oplus L^{m,p}(\mathbf{i}\Lambda_+^2 T^*M), \ \ \mathfrak{Y}^k := \mathfrak{Y}^{k,2}.$$

where $\Lambda_+^2 T^*M$ denotes the bundle of self-dual 2-forms. We want to emphasize that the Sobolev norms on the spaces of spinors are defined using the fixed reference connection $A_0$. Finally, define

$$\mathfrak{G}_\sigma^{m+2,p} = \Big\{ \gamma \in L^{m+2,p}(M, \mathbb{C}); \ |\gamma(m)| = 1, \ \ \forall m \in M \Big\},$$

$$\mathfrak{G}_\sigma^{k+2} := \mathfrak{G}_\sigma^{k+2,2}.$$

We see that since any $\gamma \in L^{m+2,p}(M, \mathbb{C})$ is continuous, the expression $|\gamma(m)|$ is well defined everywhere.

Using the isomorphism $\mathbf{c} : \mathbf{i}\Lambda_+^2 T^*M \to \mathrm{End}_0(\mathbb{S}_\sigma^+)$ we are free to identify $q(\psi) \in \mathrm{End}_0(\mathbb{S}_\sigma^+)$ with the self-dual 2-form $\mathbf{c}^{-1}(q(\psi))$. When no confusion is possible we will freely switch between the two interpretations of $q(\psi)$ writing $q(\psi)$ instead of $\mathbf{c}^{-1}(q(\psi))$.

**Lemma 2.1.6.** *For every $k \geq 1$ the correspondence $\psi \mapsto q(\psi)$ defines a $C^\infty$-map*

$$q : L^{k+1,2}(\mathbb{S}_\sigma^+) \to L^{k,2}(\mathbf{i}\Lambda_+^2 T^*M).$$

**Sketch of proof** We consider only the case $k = 1$ and we begin by showing that $q(\psi) \in L^{1,2}$, $\forall \psi \in L^{2,2}$.

Since $\psi \in L^{2,2}$ it follows from the Sobolev embedding that $\psi \in L^p$ for all $p \in (1, \infty)$ so that, using Lemma 2.1.5, we deduce

$$\int_M |q(\psi)|^2 dv_g = \frac{1}{2} \int_M |\psi|^4 dv_g \leq \infty.$$

Next observe that there exists a constant $C > 0$ such that

$$\int_M |\nabla q(\psi)|^2 dv_g \leq C \int_M |\nabla^{A_0}\psi|^2 |\psi|^2 dv_g.$$

Since $\psi \in L^{2,2}$ we deduce from the Sobolev inequality that $\nabla^{A_0}\psi \in L^q$ for some $q > 2$ restricted only by the inequality

$$0 = 2 - 4/2 > 1 - 4/q.$$

The Hölder inequality now implies

$$\int_M |\nabla q(\psi)|^2 dv_g$$

$$\leq C\left(\int_M |\nabla^{A_0}\psi|^{\frac{2}{q}} dv_g\right)^{q/2}\left(\int_M |\psi|^{2q/(q-2)} dv_g\right)^{(q-2)/q} < \infty.$$

The stated regularity follows from the identity

$$q(\psi_0+\dot{\psi}) = q(\psi_0) + \dot{\bar{\psi}}\otimes\psi_0 + \bar{\psi}_0\otimes\dot{\psi} - \mathfrak{Re}\langle\psi_0,\dot{\psi}\rangle\mathbf{id} + q(\dot{\psi}) \tag{2.1.4}$$

for all $\psi_0, \dot{\psi} \in L^{2,2}$. The details are left to the reader. ■

Suppose now that $\eta \in L^{k,2}(\Lambda^2 T^*M)$ is a fixed closed form (i.e. satisfies $d\eta = 0$ weakly). Arguing similarly we deduce the following result.

**Proposition 2.1.7.** *For every $k \geq 1$ the correspondence*

$$(\psi, A) \mapsto \mathfrak{D}_A\psi \oplus (F_A^+ + \mathbf{i}\eta^+ - q(\psi))$$

*induces a $C^\infty$-map $\mathbf{SW}_\eta : \mathcal{C}_\sigma^{k+1} \to \mathfrak{Y}^k$.*

**Exercise 2.1.2.** Prove the above proposition.

The group $\mathfrak{G}_\sigma^{k+2}$ also has a nice structure.

**Proposition 2.1.8.** *For every $k \geq 1$ the group $\mathfrak{G}_\sigma^{k+2}$ is a Hilbert-Lie group modeled by $L^{k+2,2}(M, \mathbf{i}\mathbb{R})$.*

**Proof** Again we consider only the case $k = 1$. Observe first that $\mathfrak{G}_\sigma^{3,2} \subset C^0(M, S^1)$. The space of continuous maps $M \to S^1$, topologized with the compact-open topology, is an Abelian topological group. Since the target $S^1$ is a $K(\mathbb{Z}, 1)$-space we deduce that the group of components of $C^0(M, S^1)$ is isomorphic to $H^1(M, \mathbb{Z})$. For any $\gamma \in C^0(M, S^1)$ we denote by $[\gamma] \in H^1(X, \mathbb{Z})$ the component containing $\gamma$. The identity component ($[\gamma] = 0$) consists of those maps $\gamma$ which can be written as $\gamma = \exp(\mathbf{i}f)$ for some continuous map $f : M \to \mathbb{R}$.

Define

$$\hat{\mathfrak{G}}_\sigma = \left\{\gamma \in \mathfrak{G}_\sigma^{3,2};\ [\gamma] = 0\right\} = \left\{\exp(\mathbf{i}f);\ \ f \in L^{3,2}(M, \mathbb{R})\right\}.$$

It is clear that it suffices to show that $\hat{\mathfrak{G}}_\sigma$ is a Hilbert-Lie group. This will be achieved in several steps.

• Observe first that

$$\hat{\mathfrak{G}}_\sigma \subset L^{3,2}(M, \mathbb{C}).$$

• Equip $\hat{\mathfrak{G}}_\sigma$ with the topology as a subset in the space of $L^{3,2}$-maps $M \to \mathbb{C}$.

• We now construct coordinate charts. . The coordinate chart at the origin is given by the *Cayley transform*

$$T : U_1 := \hat{\mathfrak{G}}_\sigma \setminus \{-1\} \to L^{3,2}(M, \mathbf{i}\mathbb{R}),$$

$$\exp(\mathbf{i}f) \mapsto T[e^{\mathbf{i}f}] = \frac{1-\exp(\mathbf{i}f)}{1+\exp(\mathbf{i}f)} = \frac{-2\mathbf{i}\sin(f)}{|1+\exp(\mathbf{i}f)|^2}.$$

Observe that $T$ is a bijection onto $L^{3,2}(M, \mathbf{i}\mathbb{R})$ since $T = T^{-1}$, i.e.

$$e^{\mathbf{i}f} = \frac{1-T[e^{\mathbf{i}f}]}{1+T[e^{\mathbf{i}f}]}.$$

For an arbitrary $\gamma \in \hat{\mathfrak{G}}_\sigma$ define

$$T_\gamma : U_\gamma := \gamma \cdot U_1 \to L^{3,2}(M, \mathbf{i}\mathbb{R})$$

by

$$T_\gamma(\varphi) = T(\gamma^{-1}\varphi).$$

To show that this is a smooth structure it suffices to show that the transition maps $T_\gamma \circ T_\beta^{-1}$ are smooth maps $L^{3,2}(M, \mathbf{i}\mathbb{R}) \to L^{3,2}(M, \mathbf{i}\mathbb{R})$. This follows immediately from the identity $T = T^{-1}$ so that

$$T_\gamma \circ T_\beta^{-1}(\mathbf{i}f) = T(\gamma \cdot \beta^{-1} \cdot T(\mathbf{i}f)).$$

We leave the details to the reader. ■.

**Exercise 2.1.3.** Finish the proof of the above proposition.

The tangent space of $\mathfrak{G}_\sigma^{k+2}$ at $\mathbf{1}$ is $L^{k+2,2}(M, \mathbf{i}\mathbb{R})$. The exponential map

$$\exp : T_1\mathfrak{G}_\sigma^{k+2} \to \mathfrak{G}_\sigma^{k+2}, \ \ \mathbf{i}f \mapsto e^{\mathbf{i}f}$$

is a local diffeomorphism, just as in the finite-dimensional case. Often, we will refer to the elements in this tangent space as *infinitesimal gauge transformations.*

Now observe that $\mathfrak{G}_\sigma^{k+2}$ acts on $\mathcal{C}_\sigma^{k+1}$ and $\mathfrak{Y}^k$ by

$$\mathcal{C}_\sigma^{k+1} \ni (\psi, A) \stackrel{\gamma}{\mapsto} (\gamma \cdot \psi, A - 2d\gamma/\gamma) \in \mathcal{C}_\sigma^{k+1},$$

$$\mathfrak{Y}^k = L^{k,2}(\mathbb{S}_\sigma^-) \oplus L^{k,2}(\mathbf{i}\Lambda_+^2 T^*M) \ni (\phi, \omega) \stackrel{\gamma}{\mapsto} (\gamma \cdot \phi, \omega) \in \mathfrak{Y}^k.$$

The following result should be obvious.

**Proposition 2.1.9.** *The above actions of $\mathfrak{G}_\sigma^{k+2}$ on $\mathcal{C}_\sigma^{k+1}$ and $\mathfrak{Y}_\sigma^k$ are smooth and, moreover, the map $\mathbf{SW}_\eta : \mathcal{C}_\sigma^{k+1} \to \mathfrak{Y}_\sigma^k$ is $\mathfrak{G}_\sigma^{k+2}$-equivariant, i.e.*

$$\mathbf{SW}_\eta(\gamma \cdot \mathsf{C}) = \gamma \cdot \mathbf{SW}_\eta(\mathsf{C}), \ \ \forall \mathsf{C} \in \mathcal{C}_\sigma^{k+1}, \ \gamma \in \mathfrak{G}_\sigma^{k+2}.$$

The above proposition shows that every $\mathsf{C} \in \mathcal{C}_\sigma^{k+1}$ defines a smooth map

$$\mathfrak{G}_\sigma^{k+2} \to \mathcal{C}_\sigma^{k+1}, \quad \gamma \mapsto \gamma \cdot \mathsf{C}.$$

Its differential at $\mathbf{1} \in \mathfrak{G}_\sigma^{k+2}$ is a linear map

$$\mathfrak{L}_{\mathsf{C}} : T_1\mathfrak{G}_\sigma^{k+2} \to T_{\mathsf{C}}\mathcal{C}_\sigma^{k+1}$$

explicitly described by

$$\mathfrak{L}_{\mathsf{C}} : \quad T_1\mathfrak{G}_\sigma^{k+2} \ni \mathbf{i}f \mapsto (\mathbf{i}f\psi, -2\mathbf{i}df)$$

where $\mathsf{C} = (\psi, A)$. We will often refer to $\mathfrak{L}_{\mathsf{C}}$ as the *infinitesimal action at* $\mathsf{C}$.

As in the smooth case the stabilizer of a configuration $\mathsf{C} = (\psi, A) \in \mathcal{C}_\sigma^{k+1}$ is either trivial

$$\mathbf{Stab}(\mathsf{C}) = \{1\} \iff \psi \not\equiv 0$$

or

$$\mathbf{Stab}(\mathsf{C}) = S^1 \iff \psi \equiv 0.$$

Set

$$\mathcal{C}_{\sigma,irr}^{k+1} = \left\{ \mathsf{C} \in \mathcal{C}_\sigma^{k+1};\ \mathbf{Stab}(\mathsf{C}) = \{1\} \right\}$$

and

$$\mathcal{C}_{\sigma,red}^{k+1} = \left\{ \mathsf{C} \in \mathcal{C}_\sigma^{k+1};\ \mathbf{Stab}(\mathsf{C}) \neq \{1\} \right\}$$

Observe that

$$T_1\mathbf{Stab}(\mathsf{C}) \cong \ker \mathfrak{L}_{\mathsf{C}}.$$

We have thus proved the following result.

**Proposition 2.1.10.** *The following statements are equivalent.*
*(i)* $\mathsf{C} = (\psi, A) \in \mathcal{C}_\sigma^{k+1}$ *is reducible.*
*(ii)* $\psi \equiv 0$.
*(iii)* $\mathbf{Stab}(\mathsf{C}) \cong S^1$.
*(iv)* $\ker \mathfrak{L}_{\mathsf{C}} \neq \{0\}$.

Define

$$\mathfrak{Z}_\sigma^{k+1}(g,\eta) = \mathbf{SW}_\eta^{-1}(0), \quad \mathfrak{M}_\sigma^{k+1} = \mathfrak{M}_\sigma^{k+1}(g,\eta), = \mathfrak{Z}_\sigma^{k+1}/\mathfrak{G}_\sigma^{k+2}$$

$$\mathfrak{Z}_{\sigma,irr}^{k+1}(g,\eta) = \mathfrak{Z}_\sigma^{k+1}(g,\eta) \cap \mathcal{C}_{\sigma,irr}^{k+1}, \quad \mathfrak{M}_{\sigma,irr}^{k+1}(g,\eta) = \mathfrak{Z}_{\sigma,irr}^{k+1}/\mathfrak{G}_\sigma^{k+2}.$$

**Proposition 2.1.11.** *Suppose* $\eta \in L^{k,2}(\Lambda^2T^*M)$, $k \geq 1$. *Then for every* $\mathsf{C} \in \mathfrak{Z}_\sigma^2(g,\eta)$ *there exists* $\gamma \in \mathfrak{G}_\sigma^{3,2}$ *such that* $\gamma \cdot \mathsf{C} \in \mathcal{C}_\sigma^{k+1}$. *In particular, if* $\eta$ *is smooth we deduce*

$$\mathfrak{M}_\sigma^2(g,\eta) \cong \mathfrak{M}_\sigma^k(g,\eta), \quad \forall k \geq 2,$$

*i.e. any $L^{2,2}$-solution $(\psi, A)$ of the Seiberg-Witten equations is gauge equivalent to a* smooth *solution.*

**Proof** The proof is a typical application of the *elliptic bootstrap* technique. Suppose $\mathsf{C} = (\phi, B) \in \mathcal{C}_\sigma^{2,2}$ satisfies the Seiberg-Witten equations $\mathbf{SW}_\eta(\mathsf{C}) = 0$. By definition $\mathbf{i}b = B - A_0 \in L^{2,2}(T^*M)$.

Using the Hodge decomposition of $\Omega^1(M)$ we can write

$$b = b_0 + df + d^*\beta$$

where $b_0$ denotes the harmonic part of $b$, $f \in L^{3,2}(M)$, $\beta \in L^{3,2}(\Lambda^2 T^*M)$. We now define

$$\gamma := \exp(\frac{\mathbf{i}}{2}f), \;\; (\psi, A) := \gamma \cdot \mathsf{C} = (\exp(\frac{\mathbf{i}}{2}f)\phi, A_0 + \mathbf{i}b_0 + \mathbf{i}d^*\beta).$$

Set $a = b_0 + d^*\beta$. The main point of this gauge transformation is that $d^*a = 0$. Using Exercise 1.3.22 we can rewrite the Seiberg-Witten equations for $(\psi, A)$ as

$$\left\{ \begin{array}{rcl} \mathfrak{D}_{A_0}\psi & = & -\frac{1}{2}\mathbf{c}(\mathbf{i}a)\psi \\ \mathbf{i}d^+a & = & \frac{1}{2}q(\psi) - \mathbf{i}\eta^+ - F^+_{A_0} \end{array} \right.$$

We can use the first equation to "boost" the regularity of $\psi$. Note that since $a, \psi \in L^{2,2}$ we deduce from the Sobolev embedding that $a, \psi \in L^p$ for all $p \in (1, \infty)$. This implies $\mathbf{c}(\mathbf{i}a)\psi$ in $L^p$ for all $p \in (1, \infty)$. Thus $\mathfrak{D}_{A_0\psi} \in L^p$, $\forall p \in (1, \infty)$ so that, by elliptic regularity $\psi \in L^{1,p}$, $\forall p < \infty$. In particular $\psi$ is Hölder continuous. As in the proof of the Lemma 2.1.6 we deduce $q(\psi) \in L^{1,p}$, $\forall p$.

To proceed further we need to use the following elementary fact.

**Exercise 2.1.4.** The operator $d^+ + d^* : \Omega^1(M) \to \Omega^2_+(M) \oplus \Omega^0(M)$ is elliptic.

We can now combine the second equation and the condition $d^*a = 0$ to obtain

$$(d^+ + d^*)a + \mathbf{i}\eta^+ \in L^{1,p}, \;\; \forall p < \infty.$$

Now observe that $L^{k,2}$ embeds continuously in $L^{k-1,4}$, $\forall k \geq 1$. Hence $\eta^+ \in L^{1,p}$, $\forall p < \infty$ and thus

$$(d + d^*)a \in L^{1,p}, \;\; \forall p < \infty.$$

Invoking the elliptic regularity results for the operator $d^+ + d^*$ we deduce $a \in L^{2,4}$. This implies immediately that $\mathbf{c}(\mathbf{i}a)\psi \in L^{1,p}$ for all $p < \infty$ and using this information back in the first equation we deduce $\psi \in L^{2,p}$, $\forall p < \infty$. This information improves the regularity of the right-hand side of the second equation and, arguing as above, we gradually deduce the conclusion of the proposition. ∎

The last result shows that by looking for monopoles (modulo gauge equivalence) in the larger class of Sobolev objects, we do not get anything new. However, the Sobolev setting is indispensable when dealing with structural issues.

## 2.2. The structure of the Seiberg-Witten moduli spaces

So far we have defined the moduli spaces as abstract sets of orbits of $\mathfrak{G}_\sigma$. In this section we show that these spaces, equipped with some natural Hausdorff topologies, are smooth, compact, oriented finite-dimensional manifolds.

**2.2.1. The topology of the moduli spaces.** Fix a closed form

$$\eta \in L^{k,2}(T^*M), \quad k \geq 1.$$

The moduli space $\mathfrak{M}_\sigma^{k+1}(g,\eta)$ is a subset of the set of orbits

$$\mathcal{B}_\sigma^{k+1} := \mathcal{C}_\sigma^{k+1}/\mathfrak{G}_\sigma^{k+2}.$$

If $\mathfrak{G}_\sigma^{k+2}$ were a compact Lie group then this quotient would have a natural *Hausdorff* topology. In our situation $\mathfrak{G}_\sigma^{k+2}$ is obviously noncompact. We cannot a priori exclude the possibility that two orbits of $\mathfrak{G}_\sigma^{k+2}$ on $\mathcal{C}_\sigma^{k+1}$ may have arbitrarily close points and thus the quotient topology on $\mathcal{B}_\sigma^{k+1}$ may not be Hausdorff.

In this subsection we will prove that a natural topology of $\mathcal{B}_\sigma^{k+1}$ is Hausdorff and $\mathfrak{M}_\sigma^{k+1}(g,\eta)$ is in fact a compact subset of $\mathcal{B}_\sigma^{k+1}$.

For any point $\mathsf{C} \in \mathcal{C}_\sigma^{k+1}$ we denote by $\mathfrak{O}_\mathsf{C}$ the orbit of $\mathfrak{G}_\sigma^{k+2}$ containing $\mathsf{C}$, that is,

$$\mathfrak{O}_\mathsf{C} = \Big\{\gamma\cdot\mathsf{C} \in \mathcal{C}_\sigma^{k+1};\ \gamma \in \mathfrak{G}_\sigma^{k+2}\Big\}.$$

Now define

$$\delta(\mathfrak{O}_{\mathsf{C}_1},\mathfrak{O}_{\mathsf{C}_2}) = \inf\{\|\gamma_1\cdot\mathsf{C}_1 - \gamma_2\cdot\mathsf{C}_2\|;\ \gamma_1,\gamma_2 \in \mathfrak{G}_\sigma^{k+2}\}$$

where for any configurations $\mathsf{C}_i = (\psi_i, A_i) \in \mathcal{C}_\sigma^{k+1}$, $i = 1,2$, we set

$$\|\mathsf{C}_1 - \mathsf{C}_2\|^2 := \int_M \Big(|\psi_1 - \psi_2|^2 + |A_1 - A_2|^2\Big)dv_g.$$

Note that

$$\|\gamma\cdot\mathsf{C}_1 - \gamma\cdot\mathsf{C}_2\| = \|\mathsf{C}_1 - \mathsf{C}_2\|$$

for all $\mathsf{C}_1, \mathsf{C}_2 \in \mathcal{C}_\sigma^{k+1}$ and $\gamma \in \mathfrak{G}_\sigma^{k+2}$ so that we can alternatively define

$$\delta(\mathfrak{O}_{\mathsf{C}_1},\mathfrak{O}_{\mathsf{C}_2}) = \inf\{\|\mathsf{C}_1 - \gamma\cdot\mathsf{C}_2\|;\ \gamma \in \mathfrak{G}_\sigma^{k+2}\}.$$

Clearly $\delta$ defines a map $\delta : \mathcal{B}_\sigma^{k+1} \times \mathcal{B}_\sigma^{k+1} \to \mathbb{R}_+$.

**Proposition 2.2.1.** *For $k \geq 1$ the pair $(\mathcal{B}_\sigma^{k+1}, \delta)$ is a metric space.*

**Proof** Again, we consider only the case $k = 1$. We only have to prove

$$\delta(\mathfrak{O}_{\mathsf{C}_1}, \mathfrak{O}_{\mathsf{C}_2}) = 0 \Rightarrow \mathfrak{O}_{\mathsf{C}_1} = \mathfrak{O}_{\mathsf{C}_2}.$$

Suppose $\delta(\mathfrak{O}_{\mathsf{C}_1}, \mathfrak{O}_{\mathsf{C}_2}) = 0$. Then there exists a sequence $\gamma_n \in \mathfrak{G}_\sigma^{3,2}$ such that

(2.2.1)

$$\int_M \Bigl( |\gamma_n(A_1 - A_2) + 2d\gamma_n|^2 + |\psi_2 - \gamma_n \cdot \psi_2|^2 \Bigr) dv_g = o(1) \text{ as } n \to \infty.$$

In particular, this implies

$$\int_M |d\gamma_n|^2 dv_g \leq const \cdot \int_M |A_1 - A_2|^2 dv_g + o(1) \text{ as } n \to \infty. \tag{2.2.2}$$

Since the sequence $\gamma_n$ is obviously bounded in $L^2$ we deduce from the above inequality that the sequence $\gamma_n$ is bounded in $L^{1,2}(M, \mathbb{C})$. We can now use the Sobolev embedding theorem to deduce that a subsequence of $\gamma_n$ (which we continue to denote by $\gamma_n$) converges weakly in $L^{1,2}$ and strongly in $L^p$, $1 \leq p < 4$, to a map $\gamma \in L^{1,2}$. Clearly $|\gamma| = 1$ almost everywhere on $M$.

Using the Sobolev embedding again we deduce that $\psi_2 \in L^q$ for all $q < \infty$ so that $\gamma_n \cdot \psi_2$ converges strongly in $L^2$ to $\gamma \cdot \psi_2$. By passing to the limit in the inequality

$$\int_M |\psi_1 - \gamma_n \cdot \psi_2|^2 dv_g = o(1) \text{ as } n \to \infty$$

we deduce $\psi_1 = \gamma \cdot \psi_2$.

On the other hand, since $A_1 - A_2 \in L^q$ for all $q < \infty$ the functional

$$\mathcal{F} : L^{1,2}(M, \mathbb{C}) \ni f \mapsto \int_M |f(A_1 - A_2) + 2df|^2 dv_g \in \mathbb{R}$$

is obviously convex and strongly continuous so that it is weakly lower semicontinuous (see [**19**, Chap. 1,3]) which implies

$$0 \leq \mathcal{F}(\gamma) \leq \liminf_n \mathcal{F}(\gamma_n) \stackrel{(2.2.1)}{=} 0.$$

Hence $\gamma$ is a weak solution of the partial differential equation

$$2d\gamma = \gamma(A_2 - A_1), \;\; \gamma \in L^{1,2}(\Lambda^0 T^*M \otimes \mathbb{C}). \tag{2.2.3}$$

Since the operator $d + d^*$ is elliptic and the right-hand side of the above equation is in any $L^q$, $q < \infty$, we deduce $\gamma \in L^{1,q}$ for all $q < \infty$. Using the Sobolev embedding $L^{2,2} \hookrightarrow L^{1,4}$ we can now deduce $\gamma(A_1 - A_2) \in L^{1,4}$. Plug this in (2.2.3) to deduce $\gamma \in L^{2,4}$. Sobolev inequalities again imply $\gamma(A_1 - A_2) \in L^{2,2}$ and putting this back in (2.2.3) we deduce $\gamma \in L^{3,2}$. Thus we have produced a $\gamma \in \mathfrak{G}_\sigma^{3,2}$ such that

$$A_1 = A_2 - 2d\gamma/\gamma, \;\; \psi_1 = \gamma \cdot \psi_2,$$

that is, $\mathsf{C}_1 = \gamma \cdot \mathsf{C}_2$ and $\mathfrak{O}_{\mathsf{C}_1} = \mathfrak{O}_{\mathsf{C}_2}$. The proposition is proved. ■

Clearly the canonical projection $\pi : \mathcal{C}_\sigma^{k+1} \to (\mathcal{B}_\sigma^{k+1}, \delta)$, $\mathsf{C} \mapsto \mathfrak{O}_\mathsf{C}$ is continuous since

$$\delta(\mathfrak{O}_{\mathsf{C}_1}, \mathfrak{O}_{\mathsf{C}_2}) \leq \|\mathsf{C}_1 - \mathsf{C}_2\|.$$

The moduli space $\mathfrak{M}_\sigma^{k+1}(g, \eta)$ is a subset in the *metric space* $\mathcal{B}_\sigma^{k+1}$ and thus it is equipped with a metric space structure as well. The induced topology has other remarkable features.

**Proposition 2.2.2.** *Fix the closed form* $\eta \in L^{m,2}(\Lambda^2 T^* M)$, $m = \max(k, 4)$, $k \geq 1$. *Then the metric space* $(\mathfrak{M}_\sigma^{k+1}(g, \eta), \delta)$ *is compact.*

**Proof** For simplicity we consider only the case $k = 1$. We have to show that given any sequence $\mathsf{C}_n \in \mathfrak{Z}_\eta^{2,2}$ there exist a sequence $\gamma_n \in \mathfrak{G}_\sigma^{3,2}$ and $\mathsf{C} \in \mathfrak{Z}_\eta^{2,2}$ such that

$$\|\gamma_n \mathsf{C}_n - \mathsf{C}\| = o(1) \text{ on a subsequence } n_k \to \infty.$$

To simplify the presentation we will denote the extracted subsequences by the same symbols as the original ones. Using Proposition 2.1.11 we see that modulo some gauge changes we can assume $\mathsf{C}_n = (\psi_n, A_n) \in \mathcal{C}_\sigma^5$. In particular, this means $\psi_n$ and $A_n$ are twice continuously differentiable.

Our next result presents the key estimate responsible for the compactness property of the moduli space.

**Lemma 2.2.3. (Key Estimate)** *Suppose* $\mathsf{C} = (\psi, A) \in \mathfrak{Z}_\sigma^5(g, \eta)$. *Then*

$$\|\psi\|_\infty^2 \leq 2\max(0, -\min s(x) + 4\|\eta^+\|_\infty).$$

**Proof of the lemma** Using the Kato inequality (see Exercise 1.2.1) we deduce that $\forall x \in M$

$$\Delta_M |\psi|^2(x) \leq 2\langle (\nabla^A)^* \nabla^A \psi, \psi \rangle_x$$

(use the Weitzenböck identity)

$$= 2\langle \mathfrak{D}_A^* \mathfrak{D}_A \psi, \psi \rangle_x - \frac{s(x)}{2} |\psi(x)|^2 - \langle \mathbf{c}(F_A^+)\psi, \psi \rangle_x$$

(use $\mathfrak{D}_A \psi = 0$, $\mathbf{c}(F_A^+) = \frac{1}{2} q(\psi) - \mathbf{i}\mathbf{c}(\eta^+)$ and Lemma 2.1.5)

$$= -\frac{s(x)}{2} |\psi(x)|^2 - \frac{1}{4} |\psi(x)|^4 - \langle \mathbf{c}(\mathbf{i}\eta^+)\psi, \psi \rangle_x$$

$$\leq -\frac{s(x)}{2} |\psi(x)|^2 - \frac{1}{4} |\psi(x)|^4 + 2\|\eta^+\|_\infty |\psi(x)|^2.$$

Set $u(x) = |\psi(x)|^2$. Thus $u$ is a nonnegative $C^2$-function satisfying the differential inequality

$$\Delta_M u + \frac{1}{4}u^2 + \frac{s - 4\|\eta^+\|_\infty}{2}u \leq 0.$$

If $x_0$ is a maximum point of $u$ then $\Delta_M u(x_0) \geq 0$ so that

$$\frac{u(x_0)}{2}\Big(\frac{1}{2}u(x_0) + s(x_0) - 4\|\eta^+\|\Big) \leq 0$$

so that

$$u(x_0) \leq \max(0, -2\min s(x) + 8\|\eta^+\|_\infty).$$

The lemma is proved. ■

To proceed further we need to introduce some notation.

- $\mathbf{H}^k(M,g) :=$ the space of harmonic $k$-forms on $(M,g)$.
- $\mathbf{H}^k(M,\mathbb{Z}) :=$ the lattice in $\mathbf{H}^k(M,g)$ defined by the morphism

$$H^k(M,\mathbb{Z}) \to H^k(M,\mathbb{R}).$$

Define

$$\rho(g) = \sup_u \inf_v \Big\{ \|u - v\|_2 /\ u \in \mathbf{H}^1(M,g),\ \ v \in \mathbf{H}^1(M,\mathbb{Z}) \Big\}.$$

In other words, $\rho(g)$ measures how far away from the vertices of the lattice $\mathbf{H}^1(M,\mathbb{Z})$ one can place a point in $\mathbf{H}^1(M,g)$. It is a finite quantity, bounded above by the diameter of the fundamental parallelepiped of the lattice.

We leave the reader to check the following consequence of Hodge theory.

**Exercise 2.2.1.**

$$\ker\Big( (d^+ + d^*) : \Omega^1(M) \to (\Omega^2_+ \oplus \Omega^0)(M)\Big) = \mathbf{H}^1(M,g).$$

Now write $A_n = A_0 + \mathbf{i}a_n$ and then use the Hodge decomposition

$$a_n = h_n + 2df_n + d^*\beta_n$$

where $h_n \in \mathbf{H}^1(M,g)$, $f_n \oplus \beta_n \in L^{6,2}((\Lambda^0 \oplus \Lambda^2)T^*M)$. Now pick $\chi_n \in 4\pi\mathbf{H}^1(M,\mathbb{Z})$ such that

$$\|\chi_n - h_n\|_2 = \inf\Big\{ \|\chi - h_n\|_2;\ \chi \in 4\pi\mathbf{H}^1(M,\mathbb{Z}) \Big\} \leq 4\pi\rho(g).$$

Such a choice is possible since $4\pi\mathbf{H}^1(M,\mathbb{Z})$ is a lattice in $\mathbf{H}^1(M,g)$.

**Lemma 2.2.4.** *There exists $\gamma_n \in C^\infty(M,S^1)$ such that*

$$\mathbf{i}\chi_n = 2d\gamma_n/\gamma_n.$$

**Proof of the lemma** Denote by $\tilde{\chi}_n$ the pullback of $\chi_n$ to the universal cover $\tilde{M}$ of $M$. Fix $\tilde{m}_0 \in \tilde{M}$ and for any $\tilde{m} \in \tilde{M}$ set

$$f_n(\tilde{m}) := \int_c \tilde{\chi}_n$$

where $c$ denotes an arbitrary smooth path connecting $\tilde{m}_0$ to $\tilde{m}$. Because the integrals of $\chi_n$ along the closed paths in $M$ belong to $4\pi\mathbb{Z}$ the map

$$\tilde{\gamma}_n := \exp(\mathbf{i}\tilde{f}_n/2) : \tilde{M} \to S^1$$

descends to a map $\gamma_n : M \to S^1$. Since $2d\tilde{\gamma}_n/\tilde{\gamma}_n = \mathbf{i}\tilde{\chi}_n$ we deduce $\mathbf{i}\chi_n = 2d\gamma_n/\gamma_n$. ■

Denote by $P : L^2(T^*M) \to L^2(T^*M)$ the orthogonal projection onto $\mathbf{H}^1(M,g)$. Replace the configurations $\mathsf{C}_n$ with

$$\mathsf{C}_n' := e^{\mathbf{i}f_n}\gamma_n \mathsf{C}_n = (\psi_n', A_0 + \mathbf{i}(h_n - \chi_n) + \mathbf{i}d^*\beta_n).$$

These satisfy the additional conditions

$$d^*a_n' = 0, \quad \|Pa_n'\|_2 \leq 4\pi\rho(g), \quad \forall n.$$

Since we are interested in gauge equivalence classes of configurations we could have assumed from the very beginning that $\mathsf{C}_n = \mathsf{C}_n'$. The Seiberg-Witten equations for $\mathsf{C}_n$ and the above additional conditions can be rewritten as

$$\text{(2.2.4)} \qquad \left\{ \begin{array}{rcl} \mathfrak{D}_{A_0}\psi_n & = & -\frac{1}{2}\mathbf{c}(\mathbf{i}a_n)\psi_n \\ \mathbf{i}(d^+ + d^*)a_n & = & \frac{1}{2}q(\psi_n) - \mathbf{i}\eta^+ - F_{A_0}^+ \\ \|Pa_n\|_2 & \leq & 4\pi\rho(g) \end{array} \right.$$

Using the Key Estimate we deduce that

$$\|(d^+ + d^*)a_n\|_\infty = O(1) \text{ as } n \to \infty.$$

Since $(d^+ + d^*)$ is elliptic and $\ker(d^+ + d^*) = \mathbf{H}^1(M,g)$ we deduce from Theorem 1.2.18 (v) that

$$\forall p < \infty : \ \|a_n - Pa_n\|_{1,p} = O(1) \text{ as } n \to \infty.$$

The space $\mathbf{H}^1(M,g)$ is finite dimensional so that all the Sobolev norms on it are equivalent. The third condition in (2.2.4) implies

$$\text{(2.2.5)} \qquad \forall m \in \mathbb{Z}_+, \ p < \infty : \ \|Pa_n\|_{m,p} = O(1)$$

so that

$$\text{(2.2.6)} \qquad \forall p < \infty : \ \|a_n\|_{1,p} = O(1).$$

Coupling the Sobolev embedding theorem with the Key Estimate and (2.2.6) we deduce

$$\|\mathbf{c}(\mathbf{i}a_n)\psi_n\|_\infty = O(1).$$

Using this in the first equation of (2.2.4) we deduce from the elliptic estimates

$$\forall p < \infty : \|\psi_n\|_{1,p} = O(1).$$

This implies

$$\forall p < \infty : \quad \|\mathbf{c}(\mathbf{i}a_n)\psi_n\|_{1,p} = O(1)$$

and using again the elliptic estimates for the first equation in (2.2.4) we deduce

$$\forall p < \infty : \quad \|\psi_n\|_{2,p} = O(1). \tag{2.2.7}$$

Using this in the second equation of (2.2.4) we deduce

$$\forall p < \infty : \quad \|(d^+ + d^*)a_n\|_{1,p} = O(1).$$

Finally we invoke Theorem 1.2.18 and (2.2.5) to conclude

$$\forall p < \infty : \quad \|a_n\|_{2,p} = O(1). \tag{2.2.8}$$

The inequalities (2.2.7), (2.2.8) and the Sobolev embedding theorem imply that a subsequence of $\mathsf{C}_n$ converges weakly in $L^{2,p}$ and strongly in $L^{1,q}$ to a configuration $\mathsf{C} \in \mathcal{C}_\sigma^{2,2}$. Clearly $\mathsf{C}$ is a solution of the Seiberg-Witten equations. The proposition is proved. ■

**Remark 2.2.5.** We could have continued the above proof a step further to conclude that the convergence $\mathsf{C}_n \to \mathsf{C}$ also takes place in the *strong* topology of $L^{k+1,2}$. We leave the reader to fill in the missing details.

The Key Estimate has an important immediate consequence.

**Corollary 2.2.6.** *Suppose the scalar curvature of $M$ is nonnegative, $s \geq 0$. If the closed 2-form $\eta \in L^{4,2}(\Lambda^2 T^*M)$ is such that*

$$\|\eta^+\|_\infty \leq \frac{1}{4} \min_{x \in M} s(x)$$

*then any $\eta$-monopole is reducible.*

**2.2.2. The local structure of the moduli spaces.** The space $\mathcal{B}_\sigma^{k+1}$ is the quotient of an infinite-dimensional affine space $\mathcal{C}_\sigma^{k+1}$ modulo the smooth action of $\mathfrak{G}_\sigma^{k+2}$. Moreover, the action of $\mathfrak{G}_\sigma^{k+2}$ on $\mathcal{C}_{\sigma,irr}^{k+1}$ is free so it is natural to expect that the quotient $\mathcal{B}_{\sigma,irr}^{k+1} := \mathcal{C}_{\sigma,irr}^{k+1}/\mathfrak{G}_\sigma^{k+2}$ is a Hilbert manifold.

To discuss the local structure of $\mathcal{B}_\sigma^{k+1}$ we need to introduce a stronger topology on $\mathcal{B}_\sigma^{k+1}$. Define

$$\delta_{k+1}(\mathfrak{O}_{\mathsf{C}_1}, \mathfrak{O}_{\mathsf{C}_2}) := \inf\Big\{ \|\gamma_1 \mathsf{C}_1 - \gamma_2 \mathsf{C}_2\|_{k+1,2};\ \ \gamma_1, \gamma_2 \in \mathfrak{G}_\sigma^{k+2} \Big\}.$$

Since $\delta \leq \delta_{k+1}$ we deduce that $\delta_{k+1}$ is indeed a metric on $\mathcal{B}_\sigma^{k+1}$. Remark 2.2.5 shows that $\mathfrak{M}_\sigma^{k+1}(g, \eta)$ is compact in this topology as well.

Suppose now that $\mathsf{C} = (\psi, A) \in \mathcal{C}^2_\sigma$. We can regard the infinitesimal action $\mathfrak{L}_{\mathsf{C}}$ as a *real* unbounded operator $L^2(M, \mathbf{i}\mathbb{R}) \to L^2(\mathbb{S}^+_\sigma \oplus \mathbf{i}T^*M)$ with domain $L^{1,2}(M, \mathbf{i}\mathbb{R})$. Its $L^2$-adjoint is the *real* unbounded operator

$$\mathfrak{L}^*_{\mathsf{C}} : L^2(\mathbb{S}^+_\sigma \oplus \mathbf{i}T^*M) \to L^2(M, \mathbf{i}\mathbb{R})$$

with domain $L^{1,2}(\mathbb{S}^+_\sigma \oplus \mathbf{i}T^*M)$, uniquely determined by

$$\Big\langle \mathfrak{L}_{\mathsf{C}}(\mathbf{i}f), \dot\psi \oplus \mathbf{i}\dot a \Big\rangle_{L^2} = \langle \mathbf{i}f, \mathfrak{L}^*_{\mathsf{C}}(\dot\psi \oplus \mathbf{i}\dot a)\rangle_{L^2},$$

$$\forall \mathbf{i}f \in L^2(M, \mathbf{i}\mathbb{R}),\ \dot\psi \oplus \mathbf{i}\dot a \in L^{1,2}(\mathbb{S}^+_\sigma \oplus \mathbf{i}T^*M).$$

More explicitly,

$$\Big\langle \mathfrak{L}_{\mathsf{C}}(\mathbf{i}f), \dot\psi \oplus \mathbf{i}\dot a \Big\rangle_{L^2} := \int_M f\,\mathfrak{Re}\langle \mathbf{i}\psi, \dot\psi\rangle - 2\langle df, \dot a\rangle \, dv_g$$

$$= -\int_M f\Big(\mathfrak{Im}\langle \psi, \dot\psi\rangle + 2d^*\dot a\Big)\, dv_g = \int_M \mathfrak{Re}\langle \mathbf{i}f, (-2\mathbf{i}d^*\dot a - \mathbf{i}\mathfrak{Im}\langle \psi, \dot\psi\rangle)\ \rangle dv_g.$$

On the other hand,

$$\langle \mathbf{i}f, \mathfrak{L}^*_{\mathsf{C}}(\dot\psi \oplus \mathbf{i}\dot a)\rangle_{L^2} := \int_M \mathfrak{Re}\langle \mathbf{i}f, \mathfrak{L}^*_{\mathsf{C}}(\dot\psi \oplus \mathbf{i}\dot a)\rangle dv_g.$$

Hence

(2.2.9) $$\mathfrak{L}^*_{\mathsf{C}}(\dot\psi \oplus \mathbf{i}\dot a) = -2\mathbf{i}d^*\dot a - \mathbf{i}\mathfrak{Im}\langle \psi, \dot\psi\rangle.$$

Now define the *local slice* at $\mathsf{C}$ as

$$\mathcal{S}_{\mathsf{C}} = \mathcal{S}^{k+1}_{\mathsf{C}} := \Big\{ \dot{\mathsf{C}} \in T_{\mathsf{C}}\mathcal{C}^2_\sigma;\ \ \mathfrak{L}^*_{\mathsf{C}}\dot{\mathsf{C}} = 0 \Big\}$$

$$= \Big\{ (\dot\psi, \mathbf{i}\dot a) \in L^{k+1,2}(\mathbb{S}_\sigma \oplus \mathbf{i}T^*M);\ \ \mathfrak{L}^*_{\mathsf{C}}(\dot\psi \oplus \mathbf{i}\dot a) = 0 \Big\}.$$

Observe that if $\mathsf{C}$ is reducible then

$$\mathcal{S}_{\mathsf{C}} = \Big\{ \dot\psi \oplus \mathbf{i}\dot a \in L^{k+1,2};\ \ d^*a = 0 \Big\}.$$

In this case $\mathbf{Stab}(\mathsf{C}) = S^1$ acts on $\mathcal{S}_{\mathsf{C}}$ by complex multiplication on the spinorial part

$$e^{\mathbf{i}t} \cdot (\dot\psi \oplus \mathbf{i}\dot a) = (e^{\mathbf{i}t}\dot\psi) \oplus \mathbf{i}\dot a.$$

The slice has a simple geometric interpretation. It consists of the vectors in $T_{\mathsf{C}}\mathcal{C}^2_\sigma$ which are $L^2$-orthogonal to the orbit $\mathfrak{O}_{\mathsf{C}}$.

Define an action of $\mathbf{Stab}(\mathsf{C})$ on $\mathfrak{G}^{k+2}_\sigma \times \mathcal{S}_{\mathsf{C}}$ by

$$h \cdot (\gamma, \dot{\mathsf{C}}) = (\gamma h^{-1}, h\dot{\mathsf{C}}).$$

This action commutes with the obvious left action of $\mathfrak{G}^{k+2}_\sigma$ on $\mathfrak{G}^{k+2}_\sigma \times \mathcal{S}_{\mathsf{C}}$ so that the quotient

$$(\mathfrak{G}^{k+2}_\sigma \times \mathcal{S}_{\mathsf{C}})/\mathbf{Stab}(\mathsf{C})$$

is equipped with a left $\mathfrak{G}_\sigma^{k+2}$-action.

**Proposition 2.2.7.** *Let* $\mathsf{C} = (\psi, A) \in \mathcal{C}_\sigma^{k+2}$, $k \geq 1$. *Then there exists a smooth map* $\mathcal{F} : \mathfrak{G}_\sigma^{k+2} \times \mathcal{S}_{\mathsf{C}} \to \mathcal{C}_\sigma^2$ *with the following properties.*

*(i)* $\mathcal{F}(1, 0) = \mathsf{C}$.

*(ii)* $\mathcal{F}$ *is* $\mathfrak{G}_\sigma^{k+2}$ *equivariant.*

*(iii)* $\mathcal{F}$ *is* $\mathbf{Stab}(\mathsf{C})$*-invariant.*

*(iv) There exists a* $\mathbf{Stab}(\mathsf{C})$*-invariant neighborhood of* $0 \in \mathcal{S}_{\mathsf{C}}$ *such that the induced map*

$$\hat{\mathcal{F}} : (\mathfrak{G}_\sigma^{k+2} \times U)/\mathbf{Stab}(\mathsf{C}) \to \mathcal{C}_\sigma^2$$

*is a diffeomorphism onto a* $\mathfrak{G}_\sigma^{k+2}$*-invariant open neighborhood of* $\mathsf{C}$ *in* $\mathcal{C}_\sigma^2$.

**Proof** Again, for simplicity, we consider only the case $k = 1$. The general case involves no new ideas. Define

$$\mathcal{F} : \mathfrak{G}_\sigma^3 \times \mathcal{S}_{\mathsf{C}} \to \mathcal{C}_\sigma^2,$$

$$(\gamma, \dot{\psi} \oplus \mathbf{i}\dot{a}) \mapsto (\gamma\psi + \gamma\dot{\psi}, A + \mathbf{i}\dot{a} - 2d\gamma/\gamma).$$

Clearly $\mathcal{F}$ is a smooth map. The conditions (i) – (iii) are obvious. To prove (iv) we will rely on the following result.

**Lemma 2.2.8.** *There exists a* $\mathbf{Stab}(\mathsf{C})$*-invariant neighborhood* $W$ *of* $(1, 0) \in \mathfrak{G}_\sigma^3 \times \mathcal{S}_{\mathsf{C}}$ *with the following properties.*

- **P1** *The restriction of* $\mathcal{F}$ *to* $W$ *is a submersion. In particular,* $\mathcal{F}(W)$ *is an open neighborhood of* $\mathsf{C} \in \mathcal{C}_\sigma^2$.
- **P2** *Each fiber of the map* $\mathcal{F} : W \to \mathcal{F}(W)$ *consists of a single* $\mathbf{Stab}(\mathsf{C})$*-orbit.*

**Proof of the lemma** We will use the implicit function theorem. The differential of $\mathcal{F}$ at $(1, 0) \in \mathfrak{G}_\sigma^{k+2} \times \mathcal{S}_{\mathsf{C}}$ $(k = 1)$ is the map

$$D_{(1,0)}\mathcal{F} : T_{(1,0)}(\mathfrak{G}_\sigma^3 \times \mathcal{S}_{\mathsf{C}}) = L^{3,2}(\mathbf{i}\Lambda^0 T^*M) \times \mathcal{S}_{\mathsf{C}} \to L^{2,2}(\mathbb{S}_\sigma^+ \oplus \mathbf{i}\Lambda^1 T^*M)$$

given by

$$(\mathbf{i}f, \underline{\dot{\psi}} \oplus \mathbf{i}\underline{\dot{a}}) \mapsto (\mathbf{i}f\psi + \underline{\dot{\psi}}) \oplus (\mathbf{i}\underline{\dot{a}} - 2\mathbf{i}df) = \mathfrak{L}_{\mathsf{C}}(\mathbf{i}f) + \underline{\dot{\psi}} \oplus \mathbf{i}\underline{\dot{a}}.$$

We will prove several facts.

**Fact 1** The kernel of $D_{(1,0)}\mathcal{F}$ is isomorphic to the kernel of $\mathfrak{L}_{\mathsf{C}}$.

**Fact 2** $D_{(1,0)}\mathcal{F}$ is surjective.

These two facts are elementary when $\mathsf{C} = (\psi, A)$ is reducible, $\psi \equiv 0$ and in this case they are left to the reader as an exercise.

**Exercise 2.2.2.** Prove **Fact 1** and **Fact 2** when $\mathsf{C}$ is reducible.

When $\psi \neq 0$ these facts require an additional analytical input.

**Fact 3** If $\psi \neq 0$ then the correspondence

$$f \overset{T}{\mapsto} 4\Delta f + |\psi|^2 f$$

defines a continuous bijection $L^{3,2}(M) \to L^{1,2}(M)$.

We now prove **Fact 1** and **Fact 2** when $\psi \neq 0$ assuming **Fact 3** which will be proved later on.

**Proof of Fact 1** We have to show that $D_{(1,0)}\mathcal{F}$ is injective, that is, the equation

$$\begin{cases} \mathbf{i}f\psi + \underline{\dot{\psi}} &= 0 \\ \mathbf{i}\underline{\dot{a}} - 2\mathbf{i}df &= 0 \\ \mathfrak{L}_{\mathsf{C}}^{*}(\underline{\dot{\psi}} \oplus \mathbf{i}\underline{\dot{a}}) &= 0 \end{cases}$$

has only the trivial solution $f = 0$, $\underline{\dot{\psi}} = 0$, $\underline{\dot{a}} = 0$. The first equation implies

$$\mathfrak{Im}\langle \psi, \underline{\dot{\psi}} \rangle = |\psi|^2 f.$$

Using the second and the third equations we deduce

$$0 = 2d^*\underline{\dot{a}} + \mathfrak{Im}\langle \psi, \underline{\dot{\psi}} \rangle = 4\Delta f + |\psi|^2 f.$$

**Fact 3** now implies that $f = 0$ and using this in the first and second equations we deduce $\underline{\dot{\psi}} = 0$ and $\underline{\dot{a}} = 0$.

**Proof of Fact 2** Let $\dot{\phi} \oplus \mathbf{i}\dot{b} \in T_{\mathsf{C}}\mathcal{C}_\sigma^2 = L^{2,2}(\mathbb{S}_\sigma^+ \oplus \mathbf{i}\Lambda^1 T^*M)$. Then the equality

$$D_{(1,0)}\mathcal{F}(\mathbf{i}f, \underline{\dot{\psi}} \oplus \mathbf{i}\underline{\dot{a}}) = \dot{\phi} \oplus \mathbf{i}\dot{b}, \;\; (\mathbf{i}f; \underline{\dot{\psi}}, \mathbf{i}\underline{\dot{a}}) \in T_{(1,0)}\mathfrak{G}_\sigma^3 \times \mathcal{S}_{\mathsf{C}}$$

is equivalent to

(2.2.10) $$\begin{cases} \mathbf{i}f\psi + \underline{\dot{\psi}} &= \dot{\phi} \\ \mathbf{i}\underline{\dot{a}} - 2\mathbf{i}df &= \mathbf{i}\dot{b} \\ \mathfrak{L}_{\mathsf{C}}^{*}(\underline{\dot{\psi}} \oplus \mathbf{i}\underline{\dot{a}}) &= 0 \end{cases}.$$

Using the Hodge decomposition of $\Omega^1(M)$ we can write $\underline{\dot{a}} = du + \dot{c}$ where $u \in L^{3,2}(M)$ and $\dot{c} \in L^{2,2}(T^*M)$ is co-closed. The second equality implies that $\dot{c}$ equals the co-closed part in the Hodge decomposition of $\dot{b}$. The exact part $du$ is uniquely determined by $\Delta u$ which, according to the second equation, is given by $2\Delta f + d^*\dot{b}$. Thus it suffices to determine $f$ and $\underline{\dot{\psi}}$. We claim that $f$ is the unique $L^{3,2}$-solution of the equation

(2.2.11) $$4\Delta f + |\psi|^2 f = -\mathfrak{Im}\langle \psi, \dot{\phi} \rangle - 2d^*\dot{b}$$

and

(2.2.12) $$\underline{\dot{\psi}} = \dot{\phi} - \mathbf{i}f\psi.$$

**Fact 3** guarantees that (2.2.11) has a unique solution. We see that with the above choices the first equation in (2.2.10) is automatically satisfied. The second equation is satisfied as soon as we choose $u$ as a solution of the equation

$$\Delta u = 2\Delta f + 2d^*\dot{b}.$$

This equation has a solution $u \in L^{3,2}(M)$ because the right-hand side has zero average, i.e. it is $L^2$-orthogonal to the kernel of the selfadjoint Fredholm operator $\Delta$. We only need to show that the third equation is satisfied as well, i.e.

$$2d^*\underline{\dot{a}} + \mathfrak{Im}\langle\psi, \underline{\dot{\psi}}\rangle = 0. \tag{2.2.13}$$

To show this, note that, according to the second equation in (2.2.10), we have

$$2d^*\underline{\dot{a}} = 4\Delta f + 2d^*b \overset{(2.2.11)}{=} -\mathfrak{Im}\langle\psi, \dot{\phi}\rangle - |\psi|^2 f \overset{(2.2.12)}{=} -\mathfrak{Im}\langle\psi, \underline{\dot{\psi}}\rangle.$$

**Fact 2** is proved.

**Proof of Fact 3** Arguing as in the proof of Lemma 2.1.6 we deduce that there exists a constant $C > 0$ such that

$$\|4\Delta f + |\psi|^2 f\|_{1,2} \leq C\|f\|_{3,2}, \;\; \forall f \in L^{3,2}(M)$$

so that $T$ does indeed define a bounded linear operator $L^{3,2} \to L^{1,2}$. Note also that if

$$4\Delta f + |\psi|^2 f = 0$$

then, multiplying both sides by $f$ and integrating by parts, we deduce

$$4\int_M |df|^2 dv_g + \int_M |\psi|^2 f^2 dv_g = 0$$

which shows that $df = 0$ and $f|\psi| = 0$. Since $\psi \neq 0$ we conclude that $f \equiv 0$ showing that $T$ is injective.

Now define $T_0 : L^{3,2}(M) \to L^{1,2}(M)$, $f \mapsto 4\Delta f$. $T_0$ is a Fredholm operator with index 0 since it is determined by a formally selfadjoint elliptic operator. The difference $T - T_0$ is the operator $f \mapsto |\psi|^2 f$ which, in view of Sobolev embedding theorems, is compact. Thus $T$ is Fredholm, injective and has index 0. Hence it must be surjective as well.

We now return to the proof of Lemma 2.2.8. Using the implicit function theorem we can find a $\mathbf{Stab}(\mathsf{C})$-invariant open neighborhood of $(1, 0) \in \mathfrak{G}^3_\sigma \times \mathcal{S}_{\mathsf{C}}$ such that $\mathcal{F}(W)$ is open. We are left to check **P2**. We distinguish two cases.

**A.** $\mathsf{C}$ *is irreducible.* In this case $\ker D_{(1,0)}\mathcal{F} = \ker \mathfrak{L}_{\mathsf{C}} = \{0\}$ and the assertion **P2** follows from the implicit function theorem.

**B.** $\mathsf{C}$ *is reducible,* $\mathsf{C} = (0, A)$. Denote by $\hbar$ the length of the shortest nonzero vector in the lattice $\mathbf{H}^1(M, 4\pi\mathbf{i}\mathbb{Z})$. Now fix $W$ small enough so that $\|\mathbf{i}\dot{a}\|_2 \leq \frac{\hbar}{4}$ for all $(\gamma, \dot{\psi} \oplus \mathbf{i}\dot{a}) \in W$.

Suppose $(\gamma_j, \dot{\psi}_j \oplus \mathbf{i}\dot{a}_j) \in W$ $(j = 1, 2)$ are such that

$$\mathcal{F}(\gamma_1, \dot{\psi}_1 \oplus \mathbf{i}\dot{a}_1) = \mathcal{F}(\gamma_2, \dot{\psi}_2 \oplus \mathbf{i}\dot{a}_2)$$

and if we set $\gamma = \gamma_2/\gamma_1$ we deduce

$$\dot{\psi}_1 = \gamma\dot{\psi}_2 \text{ and } \mathbf{i}\dot{a}_1 - \mathbf{i}\dot{a}_2 = -2d\gamma/\gamma.$$

The left-hand side of the second equality is co-closed while the right-hand side is closed. Thus, the right hand side represents a harmonic form, more precisely, an element in $\mathbf{H}^1(M, 4\pi\mathbf{i}\mathbb{Z})$. Since $\|\mathbf{i}\dot{a}_1 - \mathbf{i}\dot{a}_2\| \leq \frac{\hbar}{2}$ we conclude that $d\gamma/\gamma = 0$ so that $\mathbf{i}\dot{a}_1 = \mathbf{i}\dot{a}_2$ and there exists $t \in \mathbb{R}$ such that $\gamma = e^{\mathbf{i}t}$, that is, $\gamma_2 = e^{\mathbf{i}t}\gamma_1$. The lemma is proved. ■

Let us now prove (iv). Fix $W_0$ as in the statement of the lemma. The $\mathfrak{G}^3_\sigma$-invariant open set $\mathfrak{G}^3_\sigma \cdot W_0$ can be written as a product $\mathfrak{G}^3_\sigma \times U_0$ where $U_0$ is a $\mathbf{Stab}(\mathsf{C})$-invariant neighborhood of 0 in $\mathcal{S}_{\mathsf{C}}$. Denote by $\hbar$ the $L^2$-length of the shortest nonzero vector in the lattice $\mathbf{H}^1(M, 4\pi\mathbf{i}\mathbb{Z})$.

Now pick $V_r \subset U_0$ such that for all $\dot{\psi} \oplus \mathbf{i}\dot{a} \in V_r$ we have

$$\|\mathbf{i}\dot{a}\|_{2,2} + \|\dot{\psi}\|_{2,2} \leq r < \frac{\hbar}{2}. \tag{2.2.14}$$

Clearly $\mathcal{F}(\mathfrak{G}^3_\sigma \times V_r)$ is an open set because it coincides with $\mathfrak{G}^3_\sigma \cdot \mathcal{F}(V_r)$, which is open. We will show that if $r$ is sufficiently small the fibers of

$$\mathcal{F} : \mathfrak{G}^3_\sigma \times V_r \to \mathcal{C}^2_\sigma$$

are $\mathbf{Stab}(\mathsf{C})$-orbits. Consider $(\dot{\psi}_j \oplus \mathbf{i}\dot{a}_j) \in V_r$, $j = 1, 2$, and $\gamma \in \mathfrak{G}^{k+2}_\sigma$ such that

$$\mathcal{F}(\gamma, \dot{\psi}_1 \oplus \mathbf{i}\dot{a}_1) = \mathcal{F}(1, \dot{\psi}_2 \oplus \mathbf{i}\dot{a}_2),$$

This means

$$\psi + \dot{\psi}_2 = \gamma(\psi + \dot{\psi}_1) \text{ and } \mathbf{i}\dot{a}_2 = \mathbf{i}\dot{a}_1 - 2d\gamma/\gamma.$$

Denote by $\omega$ the harmonic part of the closed form $d\gamma/\gamma$, so that

$$d\gamma/\gamma = \omega + \mathbf{i}df, \quad f \in L^{3,2}(M).$$

Then

$$\frac{\hbar^2}{4} > \|\mathbf{i}\dot{a}_1 - \mathbf{i}\dot{a}_2\|_2^2 = 4\|\omega\|_2^2 + 4\|df\|_2^2.$$

From the definition of $\hbar$ we deduce that $\omega = 0$, so that $\gamma = e^{\mathbf{i}f}$ and

$$\dot{a}_2 = \dot{a}_1 - 2df. \tag{2.2.15}$$

The conditions $\mathfrak{L}_{\mathsf{C}}^*(\dot{\psi}_j \oplus \mathbf{i}\dot{a}_j) = 0$ imply

$$2d^*\dot{a}_j + \mathfrak{Im}\langle\psi, \dot{\psi}_j\rangle = 0.$$

If $\mathsf{C}$ is reducible ($\psi = 0$) then the above equality shows that $f = const.$ and the condition (iv) is proved. Suppose $\psi \neq 0$ and set

$$\nu := \|\psi\|_1.$$

Denote by $\hat{f}$ the $L^2$-orthogonal projection of $f$ onto the kernel of $\Delta_M$, more precisely

$$\hat{f} := \frac{1}{\mathrm{vol}(M)}\int_M f.$$

Since $f$ is defined only mod $2\pi\mathbb{Z}$ we can assume

$$\hat{f} \in [0, 2\pi].$$

The equality (2.2.15) yields

$$\|df\|_{2,2} \leq \frac{1}{2}(\|\dot{a}_1\|_{2,2} + \|\dot{a}_2\|_{2,2}).$$

Using Theorem 1.2.18 we deduce that there is a constant $C > 0$, depending only on the geometry of $M$, such that

$$\|f - \hat{f}\|_{3,2} \leq Cr.$$

Using the Sobolev embedding theorem we deduce

$$\|f - \hat{f}\|_\infty \leq Cr \tag{2.2.16}$$

where we use the same letter to denote the constants depending only on the geometric background.

On the other hand, from the equality $(1 - e^{\mathbf{i}f})\psi = e^{\mathbf{i}f}\dot{\psi}_1 - \dot{\psi}_2$ we deduce

$$Cr \stackrel{(2.2.14)}{\geq} \int_M |e^{\mathbf{i}f}\dot{\psi}_1 - \dot{\psi}_2| dv_g = \int_M |(1 - e^{\mathbf{i}f})\psi| dv_g = \int_M |e^{-\mathbf{i}\hat{f}}\psi - e^{\mathbf{i}(f-\hat{f})}\psi| dv_g$$

$$\geq \int_M |1 - \exp(-\mathbf{i}\hat{f})| \cdot |\psi| dv_g - \int_M |(\,1 - \exp(\mathbf{i}(f - \hat{f}))\,) \cdot \psi| dv_g$$

$$= |1 - \exp(-\mathbf{i}\hat{f})| \int_M |\psi| dv_g - \int_M |(\,1 - \exp(\mathbf{i}(f - \hat{f}))\,) \cdot \psi| dv_g$$

$$\stackrel{(2.2.16)}{\geq} |1 - \exp(-\mathbf{i}\hat{f})|\nu - Cr\nu$$

so that

$$|1 - \exp(-\mathbf{i}\hat{f})| \leq \frac{(C + \nu)r}{\nu}.$$

We conclude that

$$\|f\|_{3,2} \leq \frac{(C+\nu)r}{\nu}.$$

Suppose we fix $r$ at the very beginning such that $(e^{\mathbf{i}f}, \dot{\psi} \oplus \mathbf{i}\dot{a}) \in W_0$ as soon as

$$\|f\|_{3,2} \leq \frac{(C+\nu)r}{\nu}, \quad \|\dot{\psi}\|_{2,2} + \|\mathbf{i}\dot{a}\|_{2,2} \leq r.$$

This means

$$(1, \dot{\psi}_1 \oplus \mathbf{i}\dot{a}_1),\ (e^{\mathbf{i}f}, \dot{\psi}_2 \oplus \mathbf{i}\dot{a}_2) \in W_0$$

and

$$\mathcal{F}(1, \dot{\psi}_1 \oplus \mathbf{i}\dot{a}_1) = \mathcal{F}(e^{\mathbf{i}f}, \dot{\psi}_2 \oplus \mathbf{i}\dot{a}_2).$$

Then Lemma 2.2.8 (with $\psi \neq 0$) implies that $e^{\mathbf{i}f} = 1$. Proposition 2.2.7 is proved. ∎

Consider $\mathsf{C} = (\psi, A) \in \mathcal{C}^2_\sigma$ and a neighborhood of $0 \in \mathcal{S}_{\mathsf{C}}$ as in Proposition 2.2.7. Then the map $U \to (\mathcal{B}^{k+1}_\sigma, \delta_{k+1})$ given by

$$\dot{\psi} \oplus \mathbf{i}\dot{a} \mapsto \mathfrak{O}_{\mathsf{C}+(\dot{\psi}, \mathbf{i}\dot{a})}$$

is continuous, maps open sets to open sets and its fibers are the orbits of the $\mathbf{Stab}(\mathsf{C})$-action. Hence it induces a homeomorphism $\Phi$ of $U/\mathbf{Stab}(\mathsf{C})$ onto a neighborhood of $\mathfrak{O}_{\mathsf{C}}$ in $\mathcal{B}^{k+1}_\sigma$.

**Exercise 2.2.3.** Show that $\Phi$ is a bi-Lipschitzian map, i.e. there exists $C > 0$ such that

$$\frac{1}{C}\|(\dot{\psi}_1 - \dot{\psi}_2) \oplus (\dot{a}_1 - \dot{a}_2)\|_{k+1,2} \leq \delta_{k+1}(\Phi(\dot{\psi}_1 \oplus \mathbf{i}\dot{a}_1), \Phi(\dot{\psi}_2 \oplus \mathbf{i}\dot{a}_2)\ )$$

$$\leq C\|(\dot{\psi}_1 - \dot{\psi}_2) \oplus (\dot{a}_1 - \dot{a}_2)\|_{k+1,2},$$

$\forall \dot{\psi}_j \oplus \mathbf{i}\dot{a}_j \in U/\mathbf{Stab}(\mathsf{C})$.

From Proposition 2.2.7 we deduce the following important consequence. For any $\mathsf{C} \in \mathcal{C}^2_\sigma$ we denote by $[\mathsf{C}]$ the image of $\mathsf{C}$ in $\mathcal{B}^{k+1}_\sigma$.

**Corollary 2.2.9.** *The topological space $(\mathcal{B}^{k+1}_{\sigma,irr}, \delta_{k+1})$, $k \geq 2$, has a natural structure of smooth manifold. For every irreducible $\mathsf{C} \in \mathcal{C}^2_\sigma$, the tangent space to $\mathcal{B}^{k+1}_{\sigma,irr}$ at $[\mathsf{C}]$ can be naturally identified with $\mathcal{S}_{\mathsf{C}}$.*

Now fix the perturbation parameter $\eta \in L^{m,2}(\Lambda^2 T^*M)$, $m = \max(4, k)$ and an $\eta$-monopole $\mathsf{C} = (\psi, A)$. Modulo a gauge change, we can assume $\mathsf{C} \in \mathcal{C}^5_\sigma$ so that $\mathsf{C}$ is at least twice continuously differentiable. According to Proposition 2.2.7, to study the structure of a neighborhood of $[\mathsf{C}] \in \mathfrak{M}^{k+1}_\sigma(g, \eta)$ it suffices to understand the structure of a neighborhood of $\mathsf{C}$ in

$\mathfrak{Z}_\sigma^{k+1}(g,\eta)\cap\mathcal{S}_{\mathsf{C}}$. First, observe that the techniques in the proof of Proposition 2.1.11 imply the following result.

**Exercise 2.2.4.** Any $\mathsf{C}' \in \mathcal{S}_{\mathsf{C}}\cap\mathfrak{Z}_\sigma^{k+1}(g,\eta)$ has better regularity than $L^{k+1,2}$, namely, $\mathsf{C}' \in \mathcal{C}_\sigma^{m+1}$.

We have to understand the $L^{k+1,2}$-small solutions $\dot{\mathsf{C}} := (\dot{\psi}, \mathbf{i}\dot{a})$ of the equation

$$\left\{ \begin{array}{rcl} \mathbf{SW}_\eta(\mathsf{C}+\dot{\mathsf{C}}) & = & 0 \\ \mathfrak{L}_{\mathsf{C}}^*\dot{\mathsf{C}} & = & 0 \end{array} \right. . \tag{2.2.17}$$

We follow the well traveled path of perturbation theory and linearize this equation

$$\left\{ \begin{array}{rcl} D_{\mathsf{C}}\mathbf{SW}_\eta(\dot{\mathsf{C}}) & = & 0 \\ \mathfrak{L}_{\mathsf{C}}(\dot{\mathsf{C}}) & = & 0 \end{array} \right. .$$

At this point it helps to be more explicit. For $\psi, \dot{\psi} \in L^{k+1,2}(\mathbb{S}_\sigma^+)$ define

$$\dot{q}(\psi,\dot{\psi}) := \frac{d}{dt}|_{t=0}\, q(\psi + t\dot{\psi}) \overset{(2.1.4)}{=} \bar{\dot{\psi}} \otimes \psi + \bar{\psi} \otimes \dot{\psi} - \mathfrak{Re}\langle \psi, \dot{\psi}\rangle.$$

More precisely, $\dot{q}(\psi,\dot{\psi})$ is the traceless, selfadjoint endomorphism of $\mathbb{S}_\sigma$ given by

$$\phi \mapsto \dot{q}(\psi,\dot{\psi})\phi := \langle \phi, \dot{\psi}\rangle\psi + \langle \phi, \psi\rangle\dot{\psi} - (\mathfrak{Re}\langle \psi, \dot{\psi}\rangle)\phi.$$

We will identify it with a purely imaginary 2-form via the isomorphism induced by the Clifford multiplication. Then

$$D_{(\psi,A)}\mathbf{SW}_\eta(\dot{\psi} \oplus \mathbf{i}\dot{a}) = \left(\mathfrak{D}_A\dot{\psi} + \frac{1}{2}\mathbf{c}(\mathbf{i}\dot{a})\psi\right) \oplus \left(\, d^+\mathbf{i}\dot{a} - \frac{1}{2}\dot{q}(\psi,\dot{\psi})\,\right).$$

Thus, the linearized equations (2.2.17) define a bounded linear operator

$$\mathcal{T}_{\mathsf{C}} : L^{k+1,2}(\mathbb{S}_\sigma^+ \oplus \mathbf{i}T^*M) \to L^{k,2}(\mathbb{S}_\sigma^- \oplus \mathbf{i}\Lambda_+^2T^*M \oplus \mathbf{i}\Lambda^0T^*M)$$

described by

$$\begin{bmatrix} \dot{\psi} \\ \mathbf{i}\dot{a} \end{bmatrix} \xrightarrow{\mathcal{T}_{\mathsf{C}}} \begin{bmatrix} \mathfrak{D}_A\dot{\psi} + \frac{1}{2}\mathbf{c}(\mathbf{i}\dot{a})\psi \\ d^+\mathbf{i}\dot{a} - \frac{1}{2}\dot{q}(\psi,\dot{\psi}) \\ -2\mathbf{i}d^*\dot{a} - \mathbf{i}\mathfrak{Im}\langle \psi, \dot{\psi}\rangle \end{bmatrix}. \tag{2.2.18}$$

Observe that $\mathcal{T}_{\mathsf{C}} = \underline{\mathbf{SW}}_{\mathsf{C}} + \mathfrak{L}_{\mathsf{C}}^*$, where the underline signifies linearization.

**Lemma 2.2.10.** *The operator $\mathcal{T}_{\mathsf{C}}$ is Fredholm. Its* real *index is*

$$d(\sigma) := \frac{1}{4}(c_\sigma^2 - (2\chi + 3\tau))$$

*where $\chi$ denotes the Euler characteristic of $M$, $\tau := b_2^+ - b_2^-$ denotes the signature of $M$ and*

$$c_\sigma^2 := \int_M c_\sigma \wedge c_\sigma.$$

**Proof** Set $\mathsf{C}_0 := (0, A_0)$ where $A_0$ is the fixed, smooth reference connection on $\det(\sigma)$. The Sobolev embedding theorem shows that the difference $\mathcal{T}_{\mathsf{C}} - \mathcal{T}_{\mathsf{C}_0}$ is a compact operator $L^{k+1,2} \to L^{k,2}$ because it is a zeroth order p.d.o. Thus $\mathcal{T}_{\mathsf{C}}$ is Fredholm if and only if $\mathcal{T}_{\mathsf{C}_0}$ is Fredholm and both operators have the same index. On the other hand,

$$\mathcal{T}_{\mathsf{C}_0}\begin{bmatrix} \mathbf{i}\dot{\psi} \\ \mathbf{i}\dot{a} \end{bmatrix} = \begin{bmatrix} \mathfrak{D}_{A_0}\dot{\psi} \\ d^+\mathbf{i}\dot{a} \\ -2\mathbf{i}d^*\dot{a} \end{bmatrix}$$

which shows that $\mathcal{T}_{\mathsf{C}_0}$ is defined by the direct sum of two first order elliptic operators with smooth coefficients

$$\mathfrak{D}_{A_0} : \Gamma(\mathbb{S}_\sigma^+) \to \Gamma(\mathbb{S}_\sigma^-)$$

and

$$d^+ - 2d^* : \mathbf{i}\Omega^1(M) \to \mathbf{i}(\Omega_+^2 \oplus \Omega^0)(M).$$

Thus $\mathcal{T}_{\mathsf{C}_0}$ is Fredholm. We deduce

$$\mathrm{ind}_{\mathbf{R}}\mathcal{T}_{\mathsf{C}} = \mathrm{ind}_{\mathbf{R}}\mathcal{T}_{\mathsf{C}_0} = 2\mathrm{ind}_{\mathbf{C}}\mathfrak{D}_{A_0} + \mathrm{ind}_{\mathbf{R}}(d^+ - 2d^*)$$

(use the Atiyah-Singer index theorem)

$$= \frac{1}{4}(c_\sigma^2 - \tau) + (b_1 - b_2^+ - b_0) = \frac{1}{4}c_\sigma^2 + \frac{4b_1 - 4b_0 - 2(b_2^+ + b_2^-) - 3(b_2^+ - b_2^-)}{4}$$

$(\chi = 2(b_0 - b_1) + b_2)$

$$= \frac{1}{4}\big( c_\sigma^2 - (4b_0 - 4b_1 + 2b_2 + 3\tau)\big) = \frac{1}{4}\big(c_\sigma^2 - (2\chi + 3\tau)\big). \quad \blacksquare$$

It is reasonable to hope we could extract information about the local structure of $\mathfrak{M}_\sigma^{k+1}(g, \eta)$ near $[\mathsf{C}]$ using the implicit function theorem. This would require the surjectivity of $\mathcal{T}_{\mathsf{C}}$ and would imply that near $[\mathsf{C}]$ the moduli space is a smooth manifold of dimension $d(\sigma)$. Moreover, in this case, the tangent space at $[\mathsf{C}]$ could be identified with $\ker \mathcal{T}_{\mathsf{C}}$.

It is thus natural to investigate the surjectivity of $\mathcal{T}_{\mathsf{C}}$ and, in case this surjectivity is not there for us, to see how much of the implicit function argument we can salvage.

Consider the following sequence of operators:

$$(\mathcal{K}_{\mathsf{C}}) : \ 0 \to T_{\mathbf{1}}\mathfrak{G}_\sigma^{k+2} \xrightarrow{\mathfrak{L}_{\mathsf{C}}} T_{\mathsf{C}}\mathcal{C}_\sigma^2 \xrightarrow{\underline{\mathbf{SW}}_\eta} \mathfrak{Y}_\sigma^k \to 0.$$

Because $\mathbf{SW}_\eta$ is $\mathfrak{G}_\sigma^{k+2}$-equivariant and $\mathbf{SW}_\eta(\mathsf{C}) = 0$ we deduce

$$\frac{d}{dt}|_{t=0}\, \mathbf{SW}_\eta(e^{\mathbf{i}tf} \cdot \mathsf{C}) = 0,$$

that is, $\underline{\mathbf{SW}}_\eta \circ \mathfrak{L}_{\mathsf{C}} = 0$. Thus the sequence $(\mathcal{K}_{\mathsf{C}})$ is a cochain complex called *the deformation complex* at $\mathsf{C}$. Its cohomology will be denoted by $\mathcal{H}_{\mathsf{C}}^*$.

**Lemma 2.2.11.** *The deformation complex $\mathcal{K}_{\mathsf{C}}$ is Fredholm, that is, the coboundary maps have closed ranges and the cohomology spaces are finite dimensional. Moreover*

$$\mathcal{H}^0_{\mathsf{C}} \cong \ker \mathfrak{L}_{\mathsf{C}}, \;\; \mathcal{H}^1_{\mathsf{C}} \cong \ker \mathcal{T}_{\mathsf{C}}$$

*and*

$$\operatorname{coker} \mathcal{T}_{\mathsf{C}} \cong \mathcal{H}^0_{\mathsf{C}} \oplus \mathcal{H}^2_{\mathsf{C}}.$$

*In particular,*

$$d(\sigma) = \operatorname{ind}_{\mathbf{R}}(\mathcal{T}_{\mathsf{C}}) = -\chi_{\mathbf{R}}(\mathcal{H}^*_{\mathsf{C}}).$$

**Proof** Clearly $\mathcal{H}^0_{\mathsf{C}} = \ker \mathfrak{L}_{\mathsf{C}}$. Moreover, Hodge theory shows that the range of $\mathfrak{L}_{\mathsf{C}}$ is closed in $T_{\mathsf{C}}\mathcal{C}^2_\sigma$. We now regard $\mathfrak{L}_{\mathsf{C}}$ as an unbounded operator $L^2(-) \to L^2(-)$ with domain $L^{1,2}(\mathbf{i}\Lambda^0 T^*M)$. Its range is closed in $T_{\mathsf{C}}\mathcal{C}^0_\sigma = L^2(\mathbb{S}_\sigma \oplus \mathbf{i}\Lambda^1 T^*M)$ and we have an $L^2$-orthogonal decomposition

$$L^2(\mathbb{S}_\sigma \oplus \mathbf{i}\Lambda^1 T^*M) = \operatorname{Range}(\mathfrak{L}_{\mathsf{C}}) \oplus \ker \mathfrak{L}^*_{\mathsf{C}}.$$

Thus we have the isomorphism

$$\mathcal{H}^1_{\mathsf{C}} \cong \Big\{ \dot{\mathsf{C}} \in \ker \mathfrak{L}^*_{\mathsf{C}};\; \underline{\mathbf{SW}}_\eta(\dot{\mathsf{C}}) = 0 \Big\} \cong \ker \mathcal{T}_{\mathsf{C}}.$$

Since $\mathcal{T}_{\mathsf{C}}$ is Fredholm it maps $T_{\mathsf{C}}\mathcal{C}^2_\sigma$ onto a closed subspace of $\mathfrak{Y}^k_\sigma$. Since $\operatorname{Range}(\mathcal{T}_{\mathsf{C}}) = \operatorname{Range}(\underline{\mathbf{SW}}_\eta) \oplus \operatorname{Range}(\mathfrak{L}^*_{\mathsf{C}})$ we deduce that the range of $\underline{\mathbf{SW}}_\eta$ is $L^{k,2}$-closed. Moreover

$$\operatorname{coker} \mathcal{T}_{\mathsf{C}} \cong \operatorname{coker} \underline{\mathbf{SW}}_\eta \oplus \operatorname{coker} \mathfrak{L}^*_{\mathsf{C}} \cong \mathcal{H}^2_{\mathsf{C}} \oplus \ker \mathfrak{L}_{\mathsf{C}}.$$

This completes the proof of the lemma. ■

**Corollary 2.2.12.** *$\mathcal{T}_{\mathsf{C}}$ is surjective if and only if $\mathcal{H}^0_{\mathsf{C}} = \mathcal{H}^2_{\mathsf{C}} = 0$. In particular, $\mathcal{T}_{\mathsf{C}}$ can be surjective only if $\mathsf{C}$ is irreducible ($\Longleftrightarrow \mathcal{H}^0_{\mathsf{C}} = 0$).*

**Definition 2.2.13.** An $\eta$-monopole $\mathsf{C}$ is said to be *regular* if $\mathcal{H}^2_{\mathsf{C}} = 0$.

**Exercise 2.2.5.** Suppose $\mathsf{C} = (0, A)$ is a reducible $\eta$-monopole. Then $\mathsf{C}$ is regular iff the operator $\mathfrak{D}_A : L^{k+1,2}(\mathbb{S}^+_\sigma) \to L^{k,2}(\mathbb{S}^-_\sigma)$ is surjective and $b^+_2 = 0$.

**Corollary 2.2.14.** *If $\mathsf{C} \in \mathcal{C}^2_\sigma$ is a regular, irreducible $\eta$-monopole then a small neighborhood of $[\mathsf{C}]$ in $\mathfrak{M}^{k+1}_\sigma(g, \eta)$ can be given the structure of a smooth manifold of dimension $d(\sigma)$. The tangent space at $[\mathsf{C}]$ is naturally isomorphic to $\mathcal{H}^1_{\mathsf{C}}$.*

**Definition 2.2.15.** The integer $d(\sigma)$ is called the *virtual dimension* of the moduli space $\mathfrak{M}^{k+1}_\sigma(g, \eta)$.

We can provide some information about the structure of $\mathfrak{M}_\sigma^{k+1}(g,\eta)$ near irregular solutions as well. For simplicity set $U := \mathcal{H}^1_{\mathsf{C}}$ and denote by $V$ the $L^2$-orthogonal complement of $U$ in $\mathcal{S}_{\mathsf{C}}$. We need to understand the small solutions $\dot{\mathsf{C}}$ of the equation

$$\mathbf{SW}_\eta(\mathsf{C}+\dot{\mathsf{C}}) = 0, \quad \dot{\mathsf{C}} \in \mathcal{S}_{\mathsf{C}}. \tag{2.2.19}$$

Denote by $\mathbf{P}$ the $L^2$-orthogonal projection onto $U$ and by $\mathbf{Q}$ the $L^2$-orthogonal projection onto the $L^2$-closure of Range$(\underline{\mathbf{SW}}_\eta)$. We rewrite the equation $\mathbf{SW}_\eta(\mathsf{C}+\dot{\mathsf{C}}) = 0$ as

$$\begin{cases} \mathbf{QSW}_\eta(\mathsf{C}+\dot{\mathsf{C}}) &= 0 \\ (\mathbf{1}-\mathbf{Q})\mathbf{SW}_\eta(\mathsf{C}+\dot{\mathsf{C}}) &= 0 \end{cases}$$

Since $\mathbf{Q}\underline{\mathbf{SW}}_\eta : \mathcal{S}_{\mathsf{C}} \to \text{Range}\,(\underline{\mathbf{SW}}_\eta)$ is onto and $\mathbf{SW}_\eta$ is $\mathbf{Stab}(\mathsf{C})$-equivariant we deduce from the implicit function theorem that there exists a small $\mathbf{Stab}(\mathsf{C})$-invariant neighborhood $\mathcal{N}$ of 0 in $U = \ker \underline{\mathbf{SW}}_\eta|_{\mathcal{S}_{\mathsf{C}}}$ and a $\mathbf{Stab}(\mathsf{C})$-equivariant smooth map

$$f : \mathcal{N} \to V$$

so that the set

$$\Big\{\dot{\mathsf{C}};\ \mathbf{QSW}_\eta(\mathsf{C}+\dot{\mathsf{C}}) = 0, \ \ \|\dot{\mathsf{C}}\|_{k+1,2} \text{ is small}\Big\}$$

can be described as the graph of $f$. More precisely, this means that

$$\mathbf{QSW}_\eta(\mathsf{C}+u\oplus v) = 0, \ \ u \in \mathcal{N}, \ \ v \in V,$$

if and only if $v = f(u)$. The small solutions of (2.2.19) can be all obtained from the *finite-dimensional* equation

$$\kappa(u) = 0, \ \ u \in U,$$

where $\kappa : \mathcal{N} \to (\text{Range}\,\underline{\mathbf{SW}}_\eta)^\perp \cong \mathcal{H}^2_{\mathsf{C}}$,

$$u \mapsto (\mathbf{1}-\mathbf{Q})\mathbf{SW}_\eta(\mathsf{C}+u\oplus f(u)).$$

The map $\kappa$ is clearly $\mathbf{Stab}(\mathsf{C})$-equivariant. It is called the *Kuranishi map* at $\mathsf{C}$. If $\mathsf{C}$ is regular then the Kuranishi map is identically zero. We have thus proved the following result.

**Proposition 2.2.16.** *There exist a small* $\mathbf{Stab}(\mathsf{C})$*-invariant neighborhood* $\mathcal{N}$ *of* $0 \in \mathcal{H}^1_{\mathsf{C}}$ *and a* $\mathbf{Stab}(\mathsf{C})$*-equivariant smooth map*

$$\kappa : \mathcal{N} \to \mathcal{H}^2_{\mathsf{C}}$$

*such that a neighborhood of* $\mathsf{C}$ *on* $\mathfrak{M}_\sigma^{k+1}(g,\eta)$ *is homeomorphic to the quotient*

$$\kappa^{-1}(0)/\mathbf{Stab}(\mathsf{C}).$$

For more information on how to piece these local descriptions to a global picture we refer to the nice discussion in [**29**, Sec. 4.2.5] concerning the similar problem for Yang-Mills equations.

**2.2.3. Generic smoothness.** The considerations in the previous subsection lead naturally to the following question:

*Is it possible to choose the perturbation parameter* $\eta \in L^{m,2}$ *(*$m = \max(4, k)$*) so that for any* $\eta$*-monopole* $\mathsf{C}$ *we have* $\mathcal{H}^0_{\mathsf{C}} = \mathcal{H}^2_{\mathsf{C}} = 0$*?*

If this question had an affirmative answer then for such $\eta$'s the moduli space $\mathfrak{M}^{k+1}_\sigma(g, \eta)$ would be a compact smooth manifold of dimension $d(\sigma)$.

The vanishing of $\mathcal{H}^0_{\mathsf{C}}$ is easier to understand because $\mathcal{H}^0_{\mathsf{C}} = 0$ if and only if $\mathsf{C}$ is reducible. To formulate our next result we need to introduce some notation. For every form $\alpha$ on $M$ we denote by $[\alpha]$ its harmonic part in its Hodge decomposition.

**Proposition 2.2.17.** *The following conditions are equivalent.*
*(i) All* $\eta$*-monopoles are irreducible.*
*(ii)* $2\pi[c_\sigma]^+ \neq [\eta]^+$.

**Proof** *(ii)* $\Rightarrow$ *(i)* We argue by contradiction. Suppose there exists a reducible monopole $\mathsf{C} = (0, A)$. Then $F_A^+ + \mathbf{i}\eta^+ = 0$ so that $2\pi[c_\sigma]^+ = \mathbf{i}[F_A]^+ = [\eta]^+$. This contradicts (ii).

*(i)* $\Rightarrow$ *(ii)* We argue again by contradiction. Suppose $2\pi[c_\sigma]^+ = [\eta]^+$. Since $\eta$ is closed we can write

$$\eta = [\eta] + d\alpha, \quad \alpha \in L^{m+1,2}(\Lambda^1 T^* M).$$

Hence

$$\eta^+ = [\eta]^+ + \frac{1}{2}(d\alpha + *d\alpha) = [\eta]^+ + \frac{1}{2}(d\alpha - *d*(*\alpha)\,)$$

$$= [\eta]^+ + \frac{1}{2}(d\alpha + d^*(*\alpha)\,).$$

Similarly we have $F_A = [F_A] + d\beta$ where $[F_A] = -2\pi\mathbf{i}[c_\sigma]$ so that

$$F_A^+ = -2\pi\mathbf{i}[c_\sigma]^+ + \frac{1}{2}(d\beta + d^*(*\beta)\,).$$

Since $2\pi[c_\sigma]^+ = [\eta]^+$ we deduce $[F_A]^+ = -\mathbf{i}[\eta]^+$. Now pick a connection $A \in \mathfrak{A}^{k+1}_\sigma$ such that $F_A = [F_A] - \mathbf{i}d\alpha$. Then $F_A^+ + \mathbf{i}\eta^+ = 0$ so that $(0, A)$ is a reducible $\eta$-monopole. ■

Define

$$\mathcal{N}^k_\sigma = \mathcal{N}^k_{g,\sigma} = \Big\{\eta \in L^{k,2}(\Lambda^2 T^* M);\ d\eta = 0,\ [\eta]^+ \neq 2\pi[c_\sigma]^+\Big\}.$$

Observe that $\mathcal{N}_\sigma^k = \emptyset$ if $b_2^+ = 0$ while if $b_2^+ > 0$, $\mathcal{N}_\sigma^k$ is an open set in the space $\ker d \cap L^{k,2}(\Lambda^2 T^*M)$. We deduce the following consequence.

**Corollary 2.2.18.** *(a) If $b_2^+ = 0$ then for any perturbation parameter $\eta \in \ker d \cap L^{k,2}(\Lambda^2 T^*M)$ there exist reducible $\eta$-monopoles.*

*(b) If $b_2^+ > 0$ then $\mathcal{N}_\sigma^k \neq \emptyset$ and for any $\eta \in \mathcal{N}_\sigma^k$ there are no reducible $\eta$-monopoles.*

In the sequel, if $b_2^+ > 0$ the perturbation parameter will be assumed to belong to some $\mathcal{N}_\sigma^k$ where $k \geq 4$. The original question is then equivalent to the following

*Fix $k \geq 4$. Can we find $\eta \in \mathcal{N}_\sigma^k$ such that $\mathcal{H}^2_{\mathsf{C}} = 0$ for any $\eta$-monopole $\mathsf{C}$?*

This is where the genericity results come in. We will need to use them in a context slightly more general that the one in §1.5.2. We begin by presenting this context.

Note first that it suffices to look at the restriction of $\mathbf{SW}_\eta$ to $\mathcal{C}^{k+1}_{\sigma,irr}$. The map

$$\mathbf{SW}_\eta : \mathcal{C}^2_{\sigma,irr} \to \mathfrak{Y}^k$$

can be regarded as a section of the trivial vector bundle

$$\mathcal{U}^k : \mathfrak{Y}^k \times \mathcal{C}^2_{\sigma,irr} \to \mathcal{C}^2_{\sigma,irr}.$$

This bundle is equipped with a $\mathfrak{G}_\sigma^{k+2}$-action covering the $\mathfrak{G}_\sigma^{k+2}$-action on the base. More precisely, for every $\gamma \in \mathfrak{G}_\sigma^{k+2}$ and $(y, \mathsf{C}) \in \mathcal{U}^k$ we have

$$\gamma \cdot (y, \mathsf{C}) = (\gamma \cdot y, \gamma \cdot \mathsf{C}).$$

Observe that $\mathbf{SW}_\eta$ is a $\mathfrak{G}_\sigma^{k+2}$-equivariant section of this bundle. Thus $\mathbf{SW}_\eta$ descends to a section $[\mathbf{SW}_\eta]$ of

$$[\mathcal{U}]^k := \mathcal{U}^k / \mathfrak{G}_\sigma^{k+2} \to \mathcal{B}^{k+1}_{\sigma,irr}.$$

On the other hand, the trivial bundle is equipped with a $\mathfrak{G}_\sigma^{k+2}$-invariant connection $\tilde{\nabla}$ so that

$$\tilde{\nabla}_{\dot{\mathsf{C}}} \mathbf{SW}_\eta|_{\mathsf{C}} = (\underline{\mathbf{SW}}_\eta)(\dot{\mathsf{C}}), \ \ \forall \mathsf{C} \in \mathcal{C}^2_{\sigma,irr}, \ \ \dot{\mathsf{C}} \in T_{\mathsf{C}}\mathcal{C}^2_{\sigma,irr}.$$

Now observe that for every $\gamma \in \mathfrak{G}_\sigma^{k+2}$ we have

$$\gamma_* \mathcal{S}_{\mathsf{C}} = \mathcal{S}_{\gamma \cdot \mathsf{C}} \cong T_{[\mathsf{C}]} \mathcal{B}^{k+1}_{\sigma,irr}$$

where $\gamma_*$ denotes the differential of $\gamma : \mathcal{C}^2_{\sigma,irr} \to \mathcal{C}^2_{\sigma,irr}$.

The above observation show that $\tilde{\nabla}$ descends to a connection $\nabla$ on $T\mathcal{B}^{k+1}_{\sigma,irr}$ and its action can be read off from the action of $\tilde{\nabla}$ on $\mathcal{S}_{\mathsf{C}}$. For every

$\dot{\mathsf{C}} \in T_{\mathsf{C}}\mathcal{C}^2_\sigma$ we will denote by $[\dot{\mathsf{C}}]$ the $L^2$-orthogonal projection onto the $L^2$-closure of $\ker \mathfrak{L}_{\mathsf{C}}$. A priori $[\dot{\mathsf{C}}]$ is only an $L^2$-object but in fact we have the following result.

**Exercise 2.2.6.** Prove that if $\dot{\mathsf{C}} \in T_{\mathsf{C}}\mathcal{C}^2_\sigma$ then $[\dot{\mathsf{C}}] \in \mathcal{S}^{k+1}_{\mathsf{C}}$, that is, $[\dot{\mathsf{C}}] \in L^{k+1,2}(\mathbb{S}^+_\sigma \oplus \mathbf{i}\Lambda^1 T^*M)$.

The moduli space $\mathfrak{M}^{k+1}_\sigma(g,\eta)$ is precisely the zero set of the section $[\mathbf{SW}_\eta]$ of $[\mathcal{U}]^k$. We leave it to the reader to prove the following fact.

**Exercise 2.2.7.** (a) Suppose that for all $[\mathsf{C}] \in [\mathbf{SW}_\eta]^{-1}(0)$ the adjunction map

$$\mathfrak{a}_{\mathsf{C}} : T_{[\mathsf{C}]}\mathcal{B}^{k+1}_{\sigma,irr} \to \mathcal{V}^k_{[\mathsf{C}]}, \quad [\dot{\mathsf{C}}] \mapsto \nabla_{[\dot{\mathsf{C}}]}[\mathbf{SW}_\eta]$$

is surjective. Then $[\mathbf{SW}_\eta]^{-1}(0)$ is a smooth submanifold of $\mathcal{B}^{k+1}_{\sigma,irr}$.
(b) Let $\mathbf{SW}_\eta(\mathsf{C}) = 0$. Then the adjunction map $\mathfrak{a}_{\mathsf{C}}$ is surjective if and only if the map $D_{\mathsf{C}}\mathbf{SW}_\eta : T_{\mathsf{C}}\mathcal{C}^2_\sigma \to \mathfrak{Y}^k$ is surjective, i.e. $\mathcal{H}^2_{\mathsf{C}} = 0$.

**Definition 2.2.19.** The parameter $\eta \in \mathcal{N}^k_\sigma$ is said to be *good* if the adjunction map of every $\eta$-monopole is surjective.

We can rephrase the initial question as follows:

*Can we find good parameters?*

We follow the approach sketched in §1.5.2. In that case the bundle $[\mathcal{U}]^k$ was trivial. We can regard the family of sections $[\mathbf{SW}_\eta]$ as a section of the bundle

$$\mathcal{E} : [\mathcal{U}]^k \times \mathcal{N}^k_\sigma \to \mathcal{B}^{k+1}_{\sigma,irr} \times \mathcal{N}^k_\sigma, \quad (\mathsf{C},\eta) \mapsto \mathbf{SW}_\eta(\mathsf{C}).$$

The connection $\nabla$ on $[\mathcal{U}]^k$ induces by pullback a connection on $\mathcal{E}$ which we continue to denote by $\nabla$. Set

$$\mathcal{Z} = \Big\{([\mathsf{C}],\eta) \in \mathcal{B}^{k+1}_{\sigma,irr} \times \mathcal{N}^k_\sigma;\ \mathbf{SW}_\eta(\mathsf{C}) = 0\Big\}.$$

The space $\mathcal{Z}$ plays the same role as the "master space" introduced in §1.5.2. We will prove two things.

**Fact 1** For all $([\mathsf{C}],\eta) \in \mathcal{Z}$ the map

$$T_{([\mathsf{C}],\eta)}\mathcal{B}^{k+1}_{\sigma,irr} \times \mathcal{N}^k_\sigma \ni ([\dot{\mathsf{C}}],\dot{\eta}) \mapsto \nabla_{[\dot{\mathsf{C}}]}[\mathbf{SW}_\eta] + \nabla_{\dot{\eta}}[\mathbf{SW}_\eta] \in \mathcal{E}_{([\mathsf{C}],\eta)}$$

is surjective, so that $\mathcal{Z}$ is a smooth Banach manifold.

**Fact 2** The natural projection

$$\pi : \mathcal{Z} \to \mathcal{N}^k_\sigma, \quad (\mathsf{C},\eta) \mapsto \eta$$

is a Fredholm map with index $d(\sigma)$.

As shown in Lemma 1.5.18, **Fact 2** is implied by **Fact 1**. In particular, the regular values of $\pi$ are all good parameters. Thus we only need to prove **Fact 1**.

**Proof of Fact 1** Let $([\mathsf{C}], \eta) \in \mathcal{Z}$. Fix a representative $\mathsf{C} = (\psi, A) \in \mathcal{C}_\sigma^2$ of $\mathsf{C}$. Notice that since $\mathbf{SW}_\eta$ is $\mathfrak{G}_\sigma^{k+2}$-equivariant we have

$$\underline{\mathbf{SW}}_\eta(\dot{\mathsf{C}}) = \underline{\mathbf{SW}}_\eta([\dot{\mathsf{C}}]) = \nabla_{[\dot{\mathsf{C}}]}[\mathbf{SW}_\eta]\,|_{[\mathsf{C}]}$$

because the vector $\dot{\mathsf{C}} - [\dot{\mathsf{C}}]$ is tangent to the orbit of $\mathfrak{G}_\sigma^{k+2}$ through $\mathsf{C}$. Thus, to establish **Fact 1** it suffices to show that the map

$$S : T_{(\mathsf{C},\eta)}\mathcal{C}_\sigma^2 \times \mathcal{N}_\sigma^k \ni (\dot{\mathsf{C}}, \dot{\eta}) \mapsto D_{\mathsf{C}}\mathbf{SW}_\eta(\dot{\mathsf{C}}) + D_\eta \mathbf{SW}_\eta(\dot{\eta})$$

is onto. More explicitly,

$$S(\dot{\mathsf{C}}, \dot{\eta}) = \left[\begin{array}{c} \mathfrak{D}_A\dot{\psi} + \frac{1}{2}\mathbf{c}(\mathbf{i}\dot{a})\psi \\ d^+\mathbf{i}\dot{a} + \mathbf{i}\dot{\eta}^+ - \frac{1}{2}\dot{q}(\psi, \dot{\psi}) \end{array}\right] \in L^{k,2}\left(\begin{array}{c} \mathbb{S}_\sigma^- \\ \oplus \\ \mathbf{i}\Lambda_+^2 T^*M \end{array}\right).$$

Since the linear map $D_{\mathsf{C}}\mathbf{SW}_\eta : T_{\mathsf{C}}\mathcal{C}_\sigma^2 \to \mathfrak{Y}^k$ has closed range we deduce immediately that $S$ has closed range as well. To establish the surjectivity it suffices to show that if $\phi \oplus \mathbf{i}\omega \in \mathfrak{Y}^k$ is $L^2$-orthogonal to the range of $S$ then $\phi \equiv 0$ and $\omega \equiv 0$. Consider such a $(\phi \oplus \mathbf{i}\omega)$. This means

(2.2.20)

$$\int_M \langle \mathfrak{D}_A\dot{\psi} + \frac{1}{2}\mathbf{c}(\mathbf{i}\dot{a})\psi, \phi\rangle dv_g + \int_M \mathfrak{Re}\langle d^+\mathbf{i}\dot{a} + \mathbf{i}\dot{\eta}^+ - \frac{1}{2}\dot{q}(\psi, \dot{\psi}), \mathbf{i}\omega\rangle dv_g = 0$$

for all $\dot{\psi} \in L^{k+1,2}(\mathbb{S}_\sigma^+)$, $\dot{a} \in L^{k+1,2}(\Lambda^1 T^*M)$ and $\dot{\eta} \in \ker d \cap L^{k,2}(\Lambda_+^2 T^*M)$. Set $\dot{a} = 0$ and $\dot{\psi} = 0$ in the above equation. We conclude that

$$\int_M \langle \dot{\eta}^+, \omega\rangle dv_g = 0, \;\; \forall \dot{\eta} \in \ker d \cap L^{k,2}(\Lambda^2 T^*M).$$

On the other hand, there exists $\dot{\eta} \in \ker d \cap L^{k,2}(\Lambda^2 T^*M)$ such that $\dot{\eta}^+ = \omega$ (as in the proof of Proposition 2.2.17). This shows $\omega \equiv 0$. Now set $\dot{a} = 0$ in (2.2.20) so that

$$0 = \int_M \langle \mathfrak{D}_A\dot{\psi}, \phi\rangle dv_g = \int_M \langle \dot{\psi}, \mathfrak{D}_A^*\phi\rangle dv_g, \;\; \forall \dot{\psi} \in L^{k+1,2}(\mathbb{S}_\sigma^+).$$

This implies

$$\mathfrak{D}_A^*\phi = 0. \tag{2.2.21}$$

We can now conclude from (2.2.20) that

$$\int_M \langle \mathbf{c}(\mathbf{i}\dot{a})\psi, \phi\rangle dv_g = 0, \;\; \forall \dot{a} \in L^{k+1,2}(\Lambda^1 T^*M). \tag{2.2.22}$$

Above, by density, we can assume the equality holds for all $L^2$-forms $\dot{a}$. Fix a point $m_0 \in M$ such that $\psi(m_0) \neq 0$. Since $\psi$ is at least $C^2$ we deduce that

$\psi$ stays away from zero on an entire neighborhood of $m_0$. Using the explicit description of the Clifford multiplication given in §1.3.1 we deduce that the map

$$\Lambda^1 T_m^* M \ni \alpha \mapsto \mathbf{c}(\alpha)\psi(m_0) \in \mathbb{S}_\sigma^- \mid_m$$

is a bijection for any $m$ in a small neighborhood $U$ of $m_0$. We can use this map to produce a continuous 1-form $\dot{a}$ supported on $U$ such that

$$\mathbf{c}(\mathbf{i}\dot{a}(m))\psi(m) = \phi(m), \quad \forall m \in U.$$

Using this equality in (2.2.22) we deduce

$$\int_U |\psi(m)|^2 dv_g = 0.$$

Thus $\psi \equiv 0$ on $U$ and by unique continuation (see [**16**]) we deduce $\phi \equiv 0$ on $M$. **Fact 1** is proved.

Using the genericity theorem, Theorem 1.5.19, we now obtain the following important result.

**Theorem 2.2.20.** *Suppose $b_2^+ > 0$ and fix $k \geq 4$.*
*(a) If $d(\sigma) < 0$ then $\mathfrak{M}_\sigma^{k+1}(g, \eta) = \emptyset$ for generic $\eta$.*
*(b) If $d(\sigma) \geq 0$ then the set of good parameters $\eta \in \mathcal{N}_\sigma^k$ is generic. For such a parameter the moduli space $\mathfrak{M}_\sigma^{k+1}(g, \eta)$ is either a compact, smooth manifold of dimension $d(\sigma)$ or it is empty.*

The last result raises a natural question. Can the moduli spaces be empty if their virtual dimension is $\geq 0$? We will show that this is a frequent occurrence and in fact it happens for most *spin*$^c$ structures except possibly finitely many of them.

**Proposition 2.2.21.** *Fix $k \geq 4$ and $C_0 > 0$. Then there exists a finite set $F \subset Spin^c(M)$, depending on the metric $g$ and the constant $C_0$, such that for any $\sigma \in Spin^c(M) \setminus F$ and any perturbation parameter $\eta$ such that*

$$\|\eta\|_{k,2} \leq C_0$$

*the moduli space $\mathfrak{M}_\sigma^{k+1}(g, \eta)$ is empty.*

**Proof** Suppose $\sigma \in Spin^c(M)$ is such that $d(\sigma) \geq 0$ and $\eta$ is a perturbation parameter such that $\|\eta\|_{k,2} \leq C_0$. In the sequel we will use the same letter $C$ to denote constants depending only on $C_0$ and the geometry of $M$. The condition $d(\sigma) \geq 0$ implies

$$c_\sigma^2 \geq 2\chi + 3\tau. \tag{2.2.23}$$

If $\mathsf{C} = (\psi, A) \in \mathcal{C}^2_\sigma$ is an $\eta$- monopole then using the Key Estimate in Lemma 2.2.3 we deduce

$$\|\psi\|_\infty \leq C. \tag{2.2.24}$$

Since $\mathsf{C}$ is a minimum of the energy functional $\mathfrak{E}_\eta$ we deduce from Proposition 2.1.4 that

$$\mathfrak{E}_\eta(\psi, A) = 4\|\eta^+\|_2^2 - 4\pi^2 c_\sigma^2 \overset{(2.2.23)}{\leq} C.$$

Using the description of $\mathfrak{E}_\eta$ we deduce

$$\|F_A + 2\mathbf{i}\eta^+\|_2^2 \leq C + \frac{1}{4}\int_M |s| \cdot |\psi|^2 dv_g \overset{(2.2.24)}{\leq} C.$$

This implies

$$\|[F_A]\|_2 \leq C$$

where we recall that $[\alpha]$ denotes the harmonic part of the form $\alpha$. Thus the cohomology class $c_\sigma$ sits in a ball of radius $C > 0$ and in the lattice $\mathbf{H}^2(M, 2\pi\mathbf{i}\mathbb{Z})$. Thus $c_\sigma$ belongs to a finite set. Since only finitely many *spin*$^c$ structures $\sigma$ determine the same class $c_\sigma \in \mathbf{H}^2(M, 2\pi\mathbf{i}\mathbb{Z})$ the proposition is proved. ■

The bijection $\hat{\vartheta}$ introduced in (2.1.3) interacts nicely with the additional structures on the moduli spaces. Observe that if $\mathsf{C} \in \mathcal{C}_\sigma$ is a $(\sigma, \eta)$-monopole then we have an induced isomorphism between deformation complexes

$$\begin{array}{ccccccccc}
0 & \longrightarrow & T_1\mathfrak{G}_\sigma & \xrightarrow{\mathfrak{L}_\mathsf{C}} & T_\mathsf{C}\mathcal{C}_\sigma & \xrightarrow{\underline{\mathbf{SW}}_\eta} & \mathfrak{Y}_\sigma & \longrightarrow & 0 \\
 & & \downarrow{\scriptstyle -\mathbf{id}} & & \downarrow{\scriptstyle \underline{\hat{\vartheta}}} & & \downarrow{\scriptstyle \underline{\hat{\vartheta}}^\sharp} & & \\
0 & \longrightarrow & T_1\mathfrak{G}_{\bar{\sigma}} & \xrightarrow[\mathfrak{L}_{\hat{\vartheta}(\mathsf{C})}]{} & T_{\hat{\vartheta}(\mathsf{C})}\mathcal{C}_{\bar{\sigma}} & \xrightarrow[\underline{\mathbf{SW}}_{-\eta}]{} & \mathfrak{Y}_\sigma & \longrightarrow & 0
\end{array} \tag{2.2.25}$$

In particular, this proves the following.

**Proposition 2.2.22.** *If $\eta$ is a good parameter for the spin$^c$ structure then $-\eta$ is a good parameter for the spin$^c$ structure $\bar{\sigma}$ and the map*

$$\hat{\vartheta} : \mathfrak{M}_\sigma(g, \eta) \to \mathfrak{M}_{\bar{\sigma}}(g, -\eta)$$

*is a diffeomorphism.*

**2.2.4. Orientability.** Suppose now that $b_2^+ > 0$ and $\eta \in \mathcal{N}_\sigma^k$, $k \geq 4$, is a good parameter. For brevity, when no confusion is possible, we will write $\mathfrak{M}_\sigma(\eta)$ instead of $\mathfrak{M}_\sigma^{k+1}(\eta)$, etc. Then, if nonempty, the moduli space $\mathfrak{M}_\sigma(\eta)$ is a compact smooth manifold of dimension $d(\sigma)$. It is very natural to inquire whether it is orientable.

To understand what such a problem entails, observe that the family of finite-dimensional vector spaces

$$\ker \mathcal{T} := \Big\{ \ker \mathcal{T}_{\mathsf{C}};\ \ \mathsf{C} \in \mathfrak{Z}_\sigma(\eta) \Big\}$$

defines a smooth vector bundle over the infinite-dimensional Banach manifold $\mathfrak{Z}_\sigma(\eta)$ and more precisely, it is the pullback via the natural projection $\pi : \mathfrak{Z}_\sigma(\eta) \to \mathfrak{M}_\sigma(\eta)$ of the tangent bundle $T\mathfrak{M}_\sigma(\eta)$ . If we could prove that $\ker \mathcal{T}$ admits an orientation preserved by the action of $\mathfrak{G}_\sigma$ then the orientability of $\mathfrak{M}_\sigma(\eta)$ would be clear. Note first that the bundle $\det \ker \mathcal{T}$ can be formally identified with the determinant line bundle $\det \mathcal{T}$ because the elliptic operators $\mathcal{T}_{\mathsf{C}}$ are surjective for $\mathsf{C} \in \mathfrak{Z}_\sigma(\eta)$. This is only a formal identification because the base $\mathfrak{Z}_\sigma(\eta)$ is an infinite-dimensional manifold and determinant line bundles were defined only in a compact context. Fortunately Remark 1.5.10 provides a way out of this trouble.

Consider the space $\mathcal{M}_k$ of smooth maps

$$\mathfrak{Z}_\sigma^{k+1}(\eta) \to L^{k+1,2}(\mathrm{Hom}\,(\mathbb{S}_\sigma^+ \oplus \mathbf{i}T^*M\ ,\ \mathbb{S}_\sigma^- \oplus \mathbf{i}\Lambda_+^2 T^*M \oplus \mathbf{i}\Lambda^0 T^*M)\,).$$

We leave it to the reader to verify the following fact.

**Lemma 2.2.23.** *Each* $\Phi = (\Phi_{\mathsf{C}}) \in \mathcal{M}_k$ *defines a morphism of Hilbert vector bundles*

$$\mathcal{V}^k \to \mathcal{W}^k$$

*where* $\mathcal{V}^k$ *denotes the Hilbert vector bundle*

$$\Big(L^{k+1,2}(\mathbb{S}_\sigma^+ \oplus \mathbf{i}\Lambda^1 T^*M) \times \mathfrak{Z}_\eta^{k+1} \twoheadrightarrow \mathfrak{Z}_\eta^{k+1}\Big) \cong T\mathcal{C}_\sigma^{k+1}|_{\mathfrak{Z}_\eta^{k+1}}$$

*while* $\mathcal{W}^k$ *denotes the vector bundle*

$$(\mathfrak{Y}^k \oplus L^{k,2}(\mathbf{i}\Lambda^0 T^*M)\,) \times \mathfrak{Z}_\eta^{k+1} \twoheadrightarrow \mathfrak{Z}_\sigma^{k+1}(\eta).$$

*Moreover, for every* $\mathsf{C} \in \mathfrak{Z}_\sigma^{k+1}(\eta)$ *the linear operator*

$$\Phi_{\mathsf{C}} : \mathcal{V}_{\mathsf{C}}^k \to \mathcal{W}_{\mathsf{C}}^k$$

*is* compact.

The group $\mathfrak{G}_\sigma^{k+2}$ acts on $\mathcal{W}^k$, trivially on the factor $L^{k,2}(\mathbf{i}\Lambda^0 T^*M)$. We denote by $\hat{\mathcal{M}}_k$ the subspace of $\mathcal{M}_k$ consisting of $\mathfrak{G}_\sigma$-equivariant maps. For example the map $\mathcal{P} = \mathcal{P}_{\mathsf{C}}$, $\mathsf{C} = (\psi, A)$, defined by

$$\begin{bmatrix} \dot{\psi} \\ \mathbf{i}\dot{a} \end{bmatrix} \mapsto \begin{bmatrix} \frac{1}{2}\mathbf{c}(\mathbf{i}\dot{a})\psi \\ -\frac{1}{2}\dot{q}(\psi, \dot{\psi}) \\ -\mathbf{i}\mathfrak{Im}\langle \psi, \dot{\psi} \rangle \end{bmatrix} \tag{2.2.26}$$

belongs to $\hat{\mathcal{M}}_k$.

The bundles $\mathcal{V}^k$ and $\mathcal{W}^k$ descend to Hilbert vector bundles over $\mathfrak{M}_\sigma(\eta)$ which we denote by $[\mathcal{V}]^k$ and $[\mathcal{W}]^k$. The family $\mathcal{T}_{\mathsf{C}}$ descends to a morphism $\mathcal{T}_{[\mathsf{C}]}$ of these bundles over $\mathfrak{M}_\sigma(\eta)$. Moreover, for every $[\mathsf{C}] \in \mathfrak{M}_\sigma(\eta)$ the induced linear operator $\mathcal{T}_{[\mathsf{C}]} : [\mathcal{V}]^k_{[\mathsf{C}]} \to [\mathcal{W}]^k_{[\mathsf{C}]}$ is Fredholm. We can now use Remark 1.5.10 to deduce that there is a determinant line bundle $\det \mathcal{T}_{[\mathsf{C}]}$ satisfying

$$\det T\mathfrak{M}_\sigma(\eta) = \det(\mathcal{T}_{[\mathsf{C}]}).$$

To assign an orientation to $\mathfrak{M}_\sigma(\eta)$ (if any) we have to describe a trivialization of $\det(\mathcal{T}_{[\mathsf{C}]})$.

Now define $\mathcal{T}^0_{\mathsf{C}} := \mathcal{T}_{\mathsf{C}} - \mathcal{P}_{\mathsf{C}}$. More precisely

$$\mathcal{T}^0_{\mathsf{C}} \begin{bmatrix} \dot{\psi} \\ \mathbf{i}\dot{a} \end{bmatrix} = \begin{bmatrix} \mathfrak{D}_A\dot{\psi} \\ d^+\mathbf{i}\dot{a} \\ -2\mathbf{i}d^*\dot{a} \end{bmatrix}.$$

Because of equivariance we deduce that $\mathcal{T}^0_{\mathsf{C}}$ descends to a morphism from $[\mathcal{V}]^k \to [\mathcal{W}]^k$. Now set $\mathcal{T}^t_{[\mathsf{C}]} := \mathcal{T}^0_{[\mathsf{C}]} + t\mathcal{P}_{[\mathsf{C}]}$, $t \in [0,1]$. Note that for all $[\mathsf{C}] \in \mathfrak{M}_\sigma(\eta)$ and $t \in [0,1]$ the operator

$$\mathcal{T}^t_{[\mathsf{C}]} : [\mathcal{V}]^k_{[\mathsf{C}]} \to [\mathcal{W}]^k_{[\mathsf{C}]}$$

is Fredholm and $\mathcal{T}^1_{[\mathsf{C}]} = \mathcal{T}_{[\mathsf{C}]}$. The morphism $(\mathcal{T}^0_{[\mathsf{C}]})$ can be written as a direct sum

$$(\mathfrak{D}_\bullet) \oplus (d^+ - 2d^*).$$

The first summand is complex and thus it is equipped with a natural orientation. The second summand is *independent* of $[\mathsf{C}] \in \mathfrak{M}_\sigma(\eta)$ and thus an orientation is determined by fixing orientations on $\ker(d^+ - 2d^*)$ and $\operatorname{coker}(d^+ - 2d^*)$. Observe that

$$\ker(d^+ - 2d^*) \cong \mathbf{H}^1(M, g)$$

and

$$\operatorname{coker}(d^+ - 2d^*) \cong \mathbf{H}^2_+(M, g) \oplus \mathbf{H}^0(M, g).$$

Observe that $\mathbf{H}^0(M, g)$ is canonically isomorphic to $\mathbb{R}$. Thus, we can fix an orientation on $\det(d^+ - 2d^*)$ by fixing orientations on $H^2_+(M, \mathbb{R})$, $H^1(M, \mathbb{R})$

and then agreeing to equip coker $(d^+ - 2d^*)$ with the orientation induced by **ordered** direct sum decomposition

$$\text{coker}\,(d^+ - 2d^*) \cong \mathbf{H}^0(M) \oplus \mathbf{H}^2_+(M).$$

With these conventions in place, we obtain an orientation on $\det(\mathcal{T}^0_{[\mathsf{C}]})$ and, via the homotopy $\mathcal{T}^t_\bullet$, an orientation on $T\mathfrak{M}_\sigma(\eta)$.

**Definition 2.2.24.** If $M$ is a compact, closed, oriented smooth 4-manifold then a *homology orientation* on $M$ is a choice of orientations on $H^1(M,\mathbb{R}) \oplus H^2_+(M,\mathbb{R})$.

We have thus proved the following result.

**Proposition 2.2.25.** *There is a canonical procedure to assign to each homology orientation $\varepsilon$ on $M$ an orientation $o = o(\varepsilon)$ on $\mathfrak{M}_\sigma(\eta)$.*

Let us trace the effect of the involution $\hat{\vartheta}$ on the orientations. For each $\mathsf{C} = (\psi, A) \in \mathfrak{M}_\sigma$ it induces maps

$$\ker \mathcal{T}^0_{\mathsf{C}} \to \ker \mathcal{T}^0_{\hat{\vartheta}(\mathsf{C})} \;\text{ and }\; \text{coker}\, \mathcal{T}^0_{\mathsf{C}} \to \text{coker}\, \mathcal{T}_{\hat{\vartheta}(\mathsf{C})}.$$

These act as complex conjugation on $\ker \mathfrak{D}_\bullet$ and $\text{coker}\, \mathfrak{D}_\bullet$ while on $\ker(d^+ - 2d^*)$ and $\text{coker}(d^+ - 2d^*)$ they act as multiplication by $(-1)$. Thus the induced map $\det \mathcal{T}^0_{\mathsf{C}} \to \det \mathcal{T}^0_{\hat{\vartheta}(\mathsf{C})}$ changes the orientation by a factor $(-1)^{\nu_\sigma}$,

$$\nu_\sigma = \text{ind}_{\mathbb{C}} \mathfrak{D}_\bullet + \text{ind}(d^+ - 2d^*) = d_\sigma - \text{ind}_{\mathbb{C}} \mathfrak{D}_\bullet.$$

We have thus proved the following result.

**Proposition 2.2.26.** *The involution $\hat{\vartheta}$ induces an orientation preserving diffeomorphism*

$$\hat{\vartheta} : \mathfrak{M}_\sigma \to (-1)^{\nu_\sigma} \mathfrak{M}_{\bar{\sigma}}.$$

## 2.3. The structure of the Seiberg-Witten invariants

**2.3.1. The universal line bundle.** We have seen that if $b_2^+ > 0$ then, for generic $\eta \in \mathcal{N}^k_\sigma$, the moduli space $\mathfrak{M}_\sigma(\eta)$ is a smooth, compact, oriented submanifold of $\mathcal{B}_{\sigma,irr}$ of dimension $d(\sigma)$. The Banach manifold $\mathcal{B}_{\sigma,irr}$ is cohomologically nontrivial. More precisely we have the following result.

**Proposition 2.3.1.** *There exists an isomorphism of $\mathbb{Z}$-graded commutative rings with* **1**

$$H^*(\mathcal{B}_{\sigma,irr}, \mathbb{Z}) \cong \mathbb{Z}[u] \oplus \Lambda^* H^1(M, \mathbb{Z})$$

*where* $\deg u := 2$.

**Proof** Observe that $\mathcal{C}_{\sigma,irr}$ is a contractible space since it is the complement of an affine subspace of infinite codimension. Thus $\mathcal{B}_{\sigma,irr}$ is homotopically equivalent to the classifying space of the gauge group $\mathfrak{G}_\sigma$. Its topology is described in [**4**, Sect. 2]. More precisely $B\mathfrak{G}_\sigma$ is homotopically equivalent to one connected component of the space Map $(M, BS^1)$. Since

$$BS^1 \cong \mathbb{CP}^\infty \cong K(\mathbb{Z}, 2),$$

we deduce from a result of R. Thom that we have the homotopy equivalence

$$\mathrm{Map}(M, K(\mathbb{Z},2)) \cong \prod_{q=0}^{2} K(H^q(M,\mathbb{Z}); 2-q)$$

$$\cong H^2(M,\mathbb{Z}) \times K(\mathbb{Z}^{b_1}, 1) \times K(\mathbb{Z}, 2).$$

The components of this space are parameterized by the first Chern class $c_1 \in H^2(M,\mathbb{Z})$ and are all homotopic to

$$K(\mathbb{Z}^{b_1}, 1) \times K(\mathbb{Z}, 2)$$

The proposition is now obvious. ■

We will construct several integral cohomology classes on $\mathcal{B}_{\sigma,irr}$ which upon integration along the moduli space $\mathfrak{M}_\sigma(\eta)$ will lead to the Seiberg-Witten invariants.

First, recall that if $X$ and $Y$ are two metric spaces there is a natural operation

$$/ : H^n(X \times Y, \mathbb{Z}) \times H_k(X, \mathbb{Z}) \to H^{n-k}(Y, \mathbb{Z}), \quad (c, \alpha) \mapsto \alpha/c$$

called the *slant product*, defined dually by the equality

$$\langle \alpha/c, d\rangle = \langle \alpha, c \times d\rangle, \quad \forall \alpha \in H^*(X \times Y, \mathbb{Z}), \ (c,d) \in H_*(X,\mathbb{Z}) \times H_*(Y,\mathbb{Z}).$$

(Our definition differs by a sign, $(-1)^{k(n-k)}$ to be precise, from the definition in [**29**, Chap. 5] or [**126**, Chap. 6]. We prefer this choice since it agrees with the "fiber-first" convention in [**105**, §3.4.5] which has certain mnemonic advantages.)

Now consider the trivial line bundle $\underline{\mathbb{C}}$ over $M \times \mathcal{C}_{\sigma,irr}$. It is equipped with a natural *free* $\mathfrak{G}_\sigma$ action. More precisely, for any $(m, \mathsf{C}) \in M \times \mathcal{C}_{\sigma,irr}$ an element $\gamma \in \mathfrak{G}_\sigma$ defines a linear map

$$\gamma : \underline{\mathbb{C}}_{(m,\mathsf{C})} \to \underline{\mathbb{C}}_{(m,\gamma\cdot\mathsf{C})}, \quad z \mapsto \gamma(m)^{-1}z.$$

This $\mathfrak{G}_\sigma$-equivariant line bundle defines a complex line bundle on the quotient $M \times \mathcal{B}_{\sigma,irr}$. We call this the *universal Seiberg-Witten bundle* and we denote it by $\mathbb{U}_\sigma$.

We can now use the slant product to define the *$\mu$-map*

$$\mu : H_j(M \times \mathcal{B}_{\sigma,irr}, \mathbb{Z}) \to H^{2-j}(\mathcal{B}_{\sigma,irr}, \mathbb{Z}), \quad a \mapsto \mu(a) := c_1(\mathbb{U}_\sigma)/a.$$

Set $\Omega_\sigma := \mu(\mathbf{1}) \in H^2(\mathcal{B}_{\sigma,irr})$.

There are more intuitive ways of viewing these cohomology classes.

**1st interpretation** Fix $m_0 \in M$. Then $\mathbb{U}_\sigma$ defines by restriction a line bundle $\mathbb{U}_\sigma(m_0)$ over $\{m_0\} \times \mathcal{B}_{\sigma,irr}$. This bundle can be alternatively described as follows.

Consider the short exact sequence of Abelian groups

$$\mathbf{1} \hookrightarrow \mathfrak{G}_\sigma(m_0) \hookrightarrow \mathfrak{G}_\sigma \stackrel{\mathbf{ev}_{m_0}}{\rightarrow} S^1 \rightarrow \mathbf{1}$$

where $\mathbf{ev}_{m_0}$ is the **ev**aluation map

$$\mathfrak{G}_\sigma \ni \gamma \mapsto \gamma(m_0) \in S^1$$

and $\mathfrak{G}_\sigma(m_0)$ is the kernel of $\mathbf{ev}_{m_0}$. Then the quotient

$$\tilde{\mathcal{B}}_{\sigma,irr}(m_0) := \mathcal{B}_{\sigma,irr}/\mathfrak{G}_\sigma(m_0)$$

is equipped with a residual free $S^1 \cong \mathfrak{G}_\sigma/\mathfrak{G}_\sigma(m_0)$-action so that the projection $\tilde{\mathcal{B}}_{\sigma,irr}(m_0) \rightarrow \mathcal{B}_{\sigma,irr}$ defines a principal $S^1$-bundle. The bundle $\mathbb{U}_\sigma(m_0)$ is associated to this principal bundle via the tautological representation $S^1 \rightarrow \mathrm{Aut}(\mathbb{C})$. Then $\Omega_\sigma$ is the first Chern class of $\mathbb{U}_\sigma(m_0)$.

**2nd interpretation** The second interpretation adopts a dual point of view. In other words, we want to regard $c_1(\mathbb{U}_\sigma)$ as the "Poincaré dual" of the zero locus of a generic section of $\mathbb{U}_\sigma$. The Poincaré duality in this infinite-dimensional context should be understood as follows. A codimension 2 submanifold $Z$ of $M \times \mathcal{B}_{\sigma,irr}$ will be called a Poincaré dual of $c_1(\mathbb{U}_\sigma)$ if, for every finite-dimensional, compact, oriented smooth submanifold $X \hookrightarrow M \times \mathcal{B}_{\sigma,irr}$ which intersects $Z$ transversally, the restriction $c_1(\mathbb{U}_\sigma)\,|_X$ is the Poincaré dual of $Y := X \cap Z$ with respect to the duality on the *finite-dimensional* manifold $X$.

Clearly, to produce Poincaré duals to $c_1(\mathbb{U}_\sigma)$ is suffices to indicate a procedure for constructing large quantities of sections of $\mathbb{U}_\sigma$. The zero loci of these sections when smooth will be the sought for Poincaré duals.

To construct sections of $\mathbb{U}_\sigma$ it suffices to produce $\mathfrak{G}_\sigma$-equivariant sections of

$$\underline{\mathbb{C}} \rightarrow M \times \mathcal{C}_{\sigma,irr}.$$

These will be smooth functions $s : M \times \mathcal{C}_{\sigma,irr} \rightarrow \mathbb{C}$ such that

$$s(m, \gamma \cdot \mathsf{C}) = \gamma(m)^{-1} \cdot s(m, \mathsf{C}), \ \ \forall \gamma \in \mathfrak{G}_\sigma, \ (m, \mathsf{C}) \in M \times \mathcal{C}_{\sigma,irr}.$$

There exists a very cheap way of constructing such functions. For every $\phi \in C^\infty(\mathbb{S}_\sigma)$ define $s_\phi : M \times \mathcal{C}_{\sigma,irr} \rightarrow \mathbb{C}$ by

$$(m; \psi, A) \mapsto \langle \phi(m), \psi(m) \rangle_m.$$

It clearly satisfies the required equivariance properties since we agreed that a Hermitian metric will always be conjugate linear in the second variable.

Suppose there exists $m_0 \in M$ such that $s_\phi^{-1}(0)$ intersects a moduli space $\{m_0\} \times \mathfrak{M}_{\sigma,\eta}$ transversally along a codimension-two submanifold $Y_{\phi,m_0}$. We now see that the restriction of $\Omega_\sigma$ to the moduli space is the Poincaré dual of $Y_{\phi,m_0}$.

**Exercise 2.3.1.** Suppose $b_2^+ > 0$ and fix an integer $k \geq 5$. Show that for a generic choice of $m \in M$, $\phi \in L^{k+1,2}(\mathbb{S}_\sigma^+)$ and $\eta \in \mathcal{N}_\sigma$ the set

$$Y_{\phi,m} = s_\phi^{-1}(0) \cap \mathfrak{M}_\sigma^{k+1}(\eta)$$

is either empty or a submanifold of dimension $d(\sigma) - 2$.

The involution $\hat{\vartheta} : \mathcal{C}_\sigma \to \mathcal{C}_{\bar{\sigma}}$ reverses the $S^1$-action and we thus deduce

$$\hat{\vartheta}^* \Omega_{\bar{\sigma}} = -\Omega_\sigma. \tag{2.3.1}$$

**2.3.2. The case $b_2^+ > 1$.** Suppose now that $(M, g)$ is a compact, oriented Riemannian 4-manifold such that $b_2^+ > 1$. A *spin*$^c$ structure $\sigma$ is said to be *feasible* if $d(\sigma) \geq 0$. If $\sigma$ is *not feasible* we define the Seiberg-Witten invariant of the pair $(M, \sigma)$ by the equality

$$\mathbf{sw}_M(\sigma) := 0.$$

If $\sigma$ is feasible then the definition of this invariant requires additional work and we need to distinguish two cases.

**Case 1** $d(\sigma) = 0$. We want to mention here that this condition already imposes restrictions on the topological type of $M$. More precisely, this implies that the equation $x^2 = 2\chi + 3\tau$ has a solution $x \in H^2(M, \mathbb{Z})$ and, according to [**55**], this implies that the tangent bundle of $M$ can be equipped with an almost complex structure. In fact, all the *spin*$^c$ structures $\sigma$ such that $d(\sigma) = 0$ are the *spin*$^c$-structures determined by almost complex structures on $TM$. With this topological aside behind us, let us choose a generic $\eta \in \mathcal{N}_\sigma$ so that $\mathfrak{M}_\sigma(g, \eta)$ is a finite collection of irreducible solutions. We will show that a choice of orientations on $\mathbf{H}^1(M)$ and $\mathbf{H}_+^2(M)$ canonically determines a map

$$\epsilon : \mathfrak{M}_\sigma(g, \eta) \to \{\pm 1\}.$$

Here are the details.

For $[\mathsf{C}] = [(\psi, A)] \in \mathfrak{M}_\sigma(g, \eta)$ the operator $\mathcal{T}_\mathsf{C}$ is Fredholm, of index zero, with trivial kernel. Thus $\det \mathcal{T}_\mathsf{C}$ is equipped with a canonical orientation $\mathfrak{O}_{can}(\mathsf{C})$. Now, as in Sec. 2.2.4, set $\mathcal{T}_\mathsf{C}^0 := \mathcal{T}_\mathsf{C} - \mathcal{P}_\mathsf{C}$. Then

$$\ker \mathcal{T}_\mathsf{C}^0 \cong \ker \mathfrak{D}_A \oplus \mathbf{H}^1(M) \text{ and } \operatorname{coker} \mathcal{T}_\mathsf{C}^0 \cong \mathfrak{D}_A^* \oplus \mathbf{H}_+^2(M) \oplus \mathbf{H}^0(M).$$

Since $\ker \mathfrak{D}_A$ and $\ker \mathfrak{D}_A^*$ are complex spaces they are equipped with natural orientations. The space $\mathbf{H}^0(M)$ is canonically isomorphic to $\mathbb{R}$. Once we have *fixed* orientations on $\mathbf{H}^1(M)$ and $\mathbf{H}^2_+(M)$ we deduce that $\det \mathcal{T}^0_{\mathsf{C}}$ is equipped with a natural orientation.

*We want to remind the reader (see §2.2.4) that the space* $\mathbf{H}^2_+(M) \oplus \mathbf{H}^0(M)$ *is oriented by the ordered direct sum*

$$\mathbf{H}^0(M) \oplus \mathbf{H}^2_+(M).$$

*We will consistently use this ordering throughout the book.*

We now transport this orientation on $\det \mathcal{T}^0$ using the deformation

$$\mathcal{T}^s_{\mathsf{C}} := \mathcal{T}_{\mathsf{C}} + s\mathcal{P}_{\mathsf{C}}, \quad s \in [0,1],$$

to an orientation $\mathfrak{O}_{ind}(\mathsf{C})$ on $\det \mathcal{T}_{\mathsf{C}}$. The two orientations $\mathfrak{O}_{can}(\mathsf{C})$ and $\mathfrak{O}_{ind}(\mathsf{C})$ differ by a sign $\pm 1$ which we denote by $\epsilon(\mathsf{C})$. Observe that in the notation of §1.5.1 we have

$$\epsilon(\mathsf{C}) = \epsilon(\mathcal{T}_{\mathsf{C}}, \mathcal{T}_{\mathsf{C}} + s\mathcal{P}, \mathcal{T}^0_{\mathsf{C}}). \tag{2.3.2}$$

Now define

$$\mathbf{sw}_M(\sigma, g, \eta) = \sum_{\mathsf{C}} \epsilon(\mathsf{C}).$$

**Remark 2.3.2.** We want to point out an equivalent definition of $\epsilon(\mathsf{C})$. First observe that

$$\mathcal{T}^0_{\mathsf{C}} : \begin{bmatrix} \dot\psi \\ \mathbf{i}\dot a \end{bmatrix} \mapsto \begin{bmatrix} \mathfrak{D}_A \dot\psi \\ d^+ \mathbf{i}\dot a \\ -2\mathbf{i}d^* \dot a \end{bmatrix}$$

where $\mathsf{C} = (\psi, A)$ and

$$\mathcal{P}_{\mathsf{C}} : \begin{bmatrix} \dot\psi \\ \mathbf{i}\dot a \end{bmatrix} \mapsto \begin{bmatrix} \frac{1}{2}\mathbf{c}(\mathbf{i}\dot a)\psi \\ -\frac{1}{2}\dot q(\psi, \dot\psi) \\ -\mathbf{i}\mathfrak{Im}\langle \psi, \dot\psi \rangle \end{bmatrix}.$$

Both $\mathcal{T}^0_{\mathsf{C}}$ and $\mathcal{P}_{\mathsf{C}}$ are defined irrespective of whether $\mathsf{C}$ is a monopole or not. If we now pick an arbitrary configuration $\mathsf{C}' = (\psi', A')$ then the orientation transport along the affine path

$$(1-t)\mathcal{T}^0_{\mathsf{C}'} + t\mathcal{T}^0_{\mathsf{C}}$$

is always positive because the only fashion in which the kernels of these operators change is through the path of Dirac operators $\mathfrak{D}_{(1-t)A'+tA}$ which are complex and thus with no effect on the orientation issue. Thus we can define $\epsilon(\mathsf{C})$ as the orientation transport along an arbitrary path connecting an operator $\mathcal{T}^0_{\mathsf{C}'}$ to the operator $\mathcal{T}_{\mathsf{C}}$.

**Case 2** $d(\sigma) > 0$. Again we choose a generic $\eta \in \mathcal{N}_\sigma$ so that $\mathfrak{M}_\sigma(g, \eta)$ is a smooth, compact *orientable* manifold of dimension $d(\sigma)$. We can fix an orientation on the moduli space by choosing orientations on $\mathbf{H}^2_+(M)$ and $\mathbf{H}^1(M)$. Now define

$$\mathbf{sw}_M(\sigma, g, \eta) = \Big\langle (1 - \Omega_\sigma)^{-1}, [\mathfrak{M}_\sigma(g, \eta)] \Big\rangle$$

where $\langle \bullet, \ \bullet \rangle$ denotes the Kronecker pairing between cohomology and homology while $(1 - \Omega_\sigma)^{-1}$ stands for the formal series

$$(1 - \Omega_\sigma)^{-1} = 1 + \Omega_\sigma + \Omega_\sigma^2 + \cdots.$$

We see that $\mathbf{sw}_M(\sigma, g, \eta) = 0$ if $d(\sigma)$ is odd while if $d(\sigma) = 2k$ then

$$\mathbf{sw}_M(\sigma, g, \eta) = \int_{\mathfrak{M}_\sigma(g,\eta)} \Omega^k.$$

In the remainder of this subsection we will show that the quantity $\mathbf{sw}_M(\sigma, g, \eta)$ is in fact independent of the additional data $g$ and $\eta$ *provided that* $b_2^+(M) > 1$. Ultimately we will have to distinguish between the two cases $d(\sigma) = 0$ and $d(\sigma) > 0$ but we will begin by describing a general set-up, which applies to both situations.

Suppose we have two sets of parameters $(g_i, \eta_i)$, $i = 0, 1$, which are good with respect to the fixed *spin*$^c$ structure $\sigma$. Choose a smooth path of metrics $g(s)$ on $M$ such that

$$g(s) \equiv g_i \ \text{ for } \ |t - i| \leq \hbar, \ \ i = 0, 1,$$

where $\hbar$ is a fixed very small number. Fix the integer $k \geq 4$. We can organize the family

$$\{\mathcal{N}^k_{\sigma, g(s)}, \ \ s \in [0, 1]\}$$

as a bundle

$$\mathcal{N} \to [0, 1]$$

whose fibers are connected when $b_2^+ > 1$. In particular, the total space $\mathcal{N}$ is connected. A smooth path $s \mapsto \eta_s \in \mathcal{N}^k_{\sigma, g(s)}$ can be viewed as a smooth section of the bundle $\mathcal{N}$. Given such a section we get a family of moduli spaces

$$\widetilde{\mathfrak{M}} := \bigcup_s \mathfrak{M}_\sigma(g(s), \eta_s)$$

which can be thought of as defining a deformation of $\mathfrak{M}_\sigma(g_0, \eta_0)$ to $\mathfrak{M}_\sigma(g_1, \eta_1)$. Clearly, some of the spaces $\mathfrak{M}_s = \mathfrak{M}_\sigma(g(s), \eta(s))$ may not be smooth but the whole family may be organized as a smooth manifold with boundary $\mathfrak{M}_0 \cup \mathfrak{M}_1$ (see Figure 2.1). More rigorously, we hope the family $\widetilde{\mathfrak{M}}$ forms a cobordism from $\mathfrak{M}_0$ to $\mathfrak{M}_1$ inside $\mathcal{B}_{irr}$. We will show that we can choose the path $\eta_s$ wisely so that the family $\widetilde{\mathfrak{M}}$ does indeed form a cobordism. In fact,

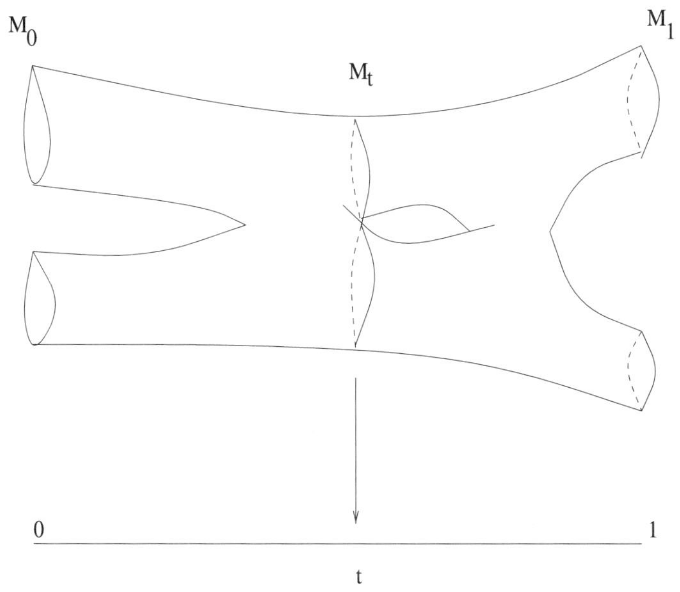


**Figure 2.1.** *A 2-dimensional cobordism*

this cobordism will be *oriented* and we will have an *orientation* preserving diffeomorphism

$$\partial\tilde{\mathfrak{M}} \cong \mathfrak{M}_1 \cup -\mathfrak{M}_0.$$

The existence of such a good path will be achieved using again the Sard-Smale transversality theorem. First we need to define an appropriate set of paths. We think of $\eta_s$ as an object over $I \times M$. More precisely it will be a $L^{k+1,2}$- section of $\pi^*\Lambda_+^2 T^*M$. Since $k+1 \geq 5$ we deduce from the Sobolev embedding theorem that such a section will be of class at least $C^2$ so that its restrictions $\eta_s$ to $\{s\} \times M$ are well defined and $C^2$ (in fact they are at least $L^{k,2}$ on $M$ according to the trace theorems of [**79**]). We will denote by $\mathcal{P}$ the subspace of such objects which additionally satisfy

$$\eta_s \equiv \eta_i \text{ for } |s-i| \leq \hbar, \;\; i = 0, 1,$$

and

$$\eta_s \in \mathcal{N}^k_{g(s),\sigma}, \;\; \forall t \in [0,1].$$

$\mathcal{P}$ is a Banach manifold modeled by the Banach space of $L^{k+1,2}$-sections of $\pi^*\Lambda_+^2 T^*M$ which are identically zero on the closed set $([0,\hbar] \cup [1-\hbar, 1]) \times M$.

Consider now the new configuration space $\widetilde{\mathcal{C}}^{k+1}_\sigma := [0,1] \times \mathcal{C}^{k+1}_\sigma$. Each path $\tilde{\eta} \in \mathcal{P}$ defines a new map $\widetilde{\mathbf{SW}} = \widetilde{\mathbf{SW}}_{\tilde{\eta}} : \widetilde{\mathcal{C}}^{k+1}_\sigma \to \mathfrak{Y}^k$ given by

$$\widetilde{\mathbf{SW}}(s, \mathsf{C}) = \mathbf{SW}_{g(s),\tilde{\eta}(s)}(\mathsf{C}).$$

The gauge group continues to act on $\widetilde{\mathcal{C}}_\sigma$ in an obvious fashion and the map $\widetilde{\mathbf{SW}}$ is $\mathfrak{G}_\sigma$-equivariant. The desired cobordism $\widetilde{\mathfrak{M}}$ can be alternatively described as

$$\widetilde{\mathfrak{M}} = \widetilde{\mathbf{SW}}^{-1}(0)/\mathfrak{G}_\sigma.$$

The structure problem for $\tilde{\mathfrak{M}}$ is very similar to that of $\mathfrak{M}$. It is in great measure determined by the deformation complexes at configurations $\tilde{\mathsf{C}} = (s, \mathsf{C})$ satisfying $\widetilde{\mathbf{SW}}(\tilde{\mathsf{C}}) = 0$. More explicitly, these are

$$(2.3.3) \qquad (\widetilde{\mathcal{K}}_{\tilde{\mathsf{C}}}): \quad 0 \to T_1\mathfrak{G}_\sigma \xrightarrow{\mathfrak{L}_{\tilde{\mathsf{C}}}} T_{\tilde{\mathsf{C}}}\widetilde{\mathcal{C}}_\sigma \xrightarrow{\underline{\widetilde{\mathbf{SW}}}} \mathfrak{Y}_\sigma \to 0$$

where the linearization $\underline{\widetilde{\mathbf{SW}}}$ is given by

$$\underline{\widetilde{\mathbf{SW}}}(\dot{s}, \dot{\mathsf{C}}) = \frac{d}{dt}|_{t=0}\, \mathbf{SW}_{g(s+t\dot{s}),\tilde{\eta}(s+t\dot{s})}(\mathsf{C} + t\dot{\mathsf{C}}).$$

This deformation complex is Fredholm because for every $(s, \mathsf{C}) \in \widetilde{\mathfrak{M}}$ we have an obvious short exact sequence of complexes

$$0 \to \mathcal{K}_{\mathsf{C}} \to \widetilde{\mathcal{K}}_{(s,\mathsf{C})} \to \mathcal{R} \to 0$$

where the residual complex $\mathcal{R}$ is finite dimensional and has index 1. The space $\widetilde{\mathfrak{M}}$ is a smooth manifold if $H^{even}(\widetilde{\mathcal{K}}_{\tilde{\mathsf{C}}}) = 0$ for all $\tilde{\mathsf{C}} \in \widetilde{\mathfrak{M}}$. Since $\tilde{\eta}(s) \in \mathcal{N}_{g(s),\sigma}$ we deduce $H^0(\widetilde{\mathcal{K}}_{s,\mathsf{C}}) = H^0(\mathcal{K}_{\mathsf{C}}) = 0$ so we only need to worry about $H^2$. To deal with this issue we use the same approach as in §2.2.3, based on the Sard-Smale transversality theorem.

Define

$$\widetilde{\mathcal{Z}} := \Big\{(\tilde{\eta}, \tilde{\mathsf{C}}) \in \mathcal{P} \times \widetilde{\mathcal{C}}_\sigma;\ \widetilde{\mathbf{SW}}_{\tilde{\eta}}(\tilde{\mathsf{C}}) = 0\Big\}.$$

Again, it suffices to prove that the map

$$\mathcal{P} \times \widetilde{\mathcal{C}}_\sigma \ni (\tilde{\eta}, \tilde{\mathsf{C}}) \mapsto \widetilde{\mathbf{SW}}_{\tilde{\eta}}(\tilde{\mathsf{C}}) \in \mathfrak{Y}_\sigma$$

is a submersion at the points in $\widetilde{\mathcal{Z}}$. Then the induced map

$$\pi : \widetilde{\mathcal{Z}}/\mathfrak{G}_\sigma \to \mathcal{P}$$

will be Fredholm of index $\operatorname{ind}_{\mathbb{R}}(\widetilde{\mathcal{K}}_{(s,\mathsf{C})}) = \operatorname{ind}_{\mathbb{R}}(\mathcal{K}_{\mathsf{C}}) + 1 = d(\sigma) + 1$.

To establish the submersion condition we have to show that if

$$\mathbf{SW}_{g(s),\tilde{\eta}(s)}(\mathsf{C}) = 0$$

then the linear map

$$(2.3.4) \qquad \begin{gathered} T_{(\tilde{\eta},s,\mathsf{C})}(\mathcal{P} \times I \times \mathcal{C}_\sigma) \ni (\underline{\tilde{\eta}}, \dot{s}, \dot{\mathsf{C}}) \mapsto \\ \mapsto \frac{d}{dt}|_{t=0}\, \mathbf{SW}_{g(s+t\dot{s}),(\tilde{\eta}+t\underline{\tilde{\eta}})(s+t\dot{s})}(\mathsf{C} + t\dot{\mathsf{C}}) \in T_0\mathfrak{Y}_\sigma \end{gathered}$$

is onto. Arguing **exactly** as in the proof of **Fact 1** in §2.2.3 one can show that a stronger statement is true, namely the map

$$(2.3.5)\quad T_{(\tilde{\eta},s,\mathsf{C})}(\mathcal{P}\times\mathcal{C}_\sigma)\ni(\underline{\dot{\eta}},\dot{\mathsf{C}})\mapsto\frac{d}{dt}|_{t=0}\,\mathbf{SW}_{g(s),(\tilde{\eta}+t\underline{\dot{\eta}})(s)}(\mathsf{C}+t\dot{\mathsf{C}})\in T_0\mathfrak{Y}_\sigma$$

is onto. Observe that (2.3.5) is obtained by setting $\dot{s}=0$ in (2.3.4).

**Remark 2.3.3.** The map in (2.3.5) has a major computational advantage over the map in (2.3.4). More precisely, the map in (2.3.4) requires an explicit understanding of how a Dirac operator and the Hodge operator vary with the metric. While these variations are known (see [**18, 37**]) their concrete descriptions are by no means pleasant. By setting $\dot{s}=0$ we have eliminated this computational nightmare and, remarkably, this restricted differential continues to be onto.

We conclude that for a generic choice of $\tilde{\eta}\in\mathcal{P}$ the parameterized moduli space $\tilde{\mathfrak{M}}_\sigma(\tilde{\eta})$ is a smooth manifold with boundary

$$\partial\tilde{\mathfrak{M}}_\sigma(\tilde{\eta})=\mathfrak{M}_\sigma(g_0,\eta_0)\sqcup\mathfrak{M}_\sigma(g_1,\eta_1).$$

To study the orientability of this parameterized moduli space we need to understand the family of Fredholm operators$\tilde{\mathcal{T}}_{(s,\mathsf{C})}$, $(s,\mathsf{C})\in\widetilde{\mathbf{SW}}_{\tilde{\eta}}^{-1}(0)$ described by

$$T_{(s,\mathsf{C})}(I\times\mathcal{C}_{\sigma,g(s)})\ni(\dot{s},\dot{\mathsf{C}})\mapsto\tilde{\mathcal{T}}_{(s,\mathsf{C})}(\dot{s},\dot{\mathsf{C}})$$

$$=\underline{\widetilde{\mathbf{SW}}}(\dot{s},\dot{\mathsf{C}})\oplus\mathfrak{L}_{\mathsf{C}}^{*_s}(\dot{\mathsf{C}})\in T_0\mathfrak{Y}_\sigma\oplus T_{\mathbf{1}}\mathfrak{G}_\sigma.$$

More explicitly, if $\mathsf{C}=(\psi,A)$ and $\dot{\mathsf{C}}=(\dot{\psi},\mathbf{i}\dot{a})$ then

(2.3.6)

$$\tilde{\mathcal{T}}_{(s,\mathsf{C})}:\begin{bmatrix}\dot{s}\\ \dot{\psi}\\ \mathbf{i}\dot{a}\end{bmatrix}\mapsto\begin{bmatrix}\mathfrak{D}_{A,g(s)}\dot{\psi}+\frac{1}{2}\mathbf{c}_{g(s)}(\mathbf{i}\dot{a})\psi\\ d^{+_{g(s)}}\mathbf{i}\dot{a}-\frac{1}{2}\dot{q}_{g(s)}(\psi,\dot{\psi})\\ -2\mathbf{i}d^{*_{g(s)}}\dot{a}-\mathbf{i}\mathfrak{Im}\langle\psi,\dot{\psi}\rangle_{g(s)}\end{bmatrix}+$$

$$\dot{s}\begin{bmatrix}(\frac{d}{dt}|_{t=0}\,\mathfrak{D}_{A,g(s+t)})\psi+\frac{1}{2}\frac{d}{dt}|_{t=0}\,\mathbf{c}_{g(s+t)}(\mathbf{i}\dot{a})\\ \frac{1}{2}(\frac{d}{dt}|_{t=0}\,*_{g(s+t)})F_A+(\frac{d}{dt}|_{t=0}\,\tilde{\eta}^{+_{g(s+t)}}(s+t))-\frac{1}{2}(\frac{d}{dt}|_{t=0}\,q_{g(s+t)})(\psi)\\ 0\end{bmatrix}$$

where a sub/superscript $g(s)$ attached to an object signifies that object is constructed in terms of the metric $g(s)$. The second term in the right-hand side of the above formula can be computed quite explicitly (see [**18, 37**]) but its exact expression is quite nasty. On the other hand, we will only use a few facts about this term. First of all, observe that this term vanishes for $|s-i| \leq \hbar$, $i=0,1$, since for such $s$ the metric $g(s)$ is independent of $s$. Second, this term involves *no derivatives* of $\mathbf{i}\dot{a}$ and $\dot{\psi}$ so that, as far as Fredholm properties are concerned, it is irrelevant. In fact, we will deform it to zero by considering the family $\tilde{\mathcal{T}}^{\tau}_{(s,\mathsf{C})}$, $0 \leq \tau \leq 1$, described by

$$\begin{bmatrix} \dot{s} \\ \dot{\psi} \\ \mathbf{i}\dot{a} \end{bmatrix} \mapsto \begin{bmatrix} \not{\mathfrak{D}}_{A,g(s)}\dot{\psi} \\ \\ d^{+_{g(s)}}\mathbf{i}\dot{a} \\ \\ -2\mathbf{i}d^{*_{g(s)}}\dot{a} \end{bmatrix} + \tau \begin{bmatrix} \frac{1}{2}\mathbf{c}_{g(s)}(\mathbf{i}\dot{a})\psi \\ \\ -\frac{1}{2}\dot{q}_{g(s)}(\psi,\dot{\psi}) \\ -\mathbf{i}\mathfrak{Im}\langle\psi,\dot{\psi}\rangle_{g(s)} \end{bmatrix} +$$

(2.3.7)
$$\tau\dot{s}\begin{bmatrix} (\frac{d}{dt}\mid_{t=0}\not{\mathfrak{D}}_{A,g(s+t)})\psi + \frac{1}{2}\frac{d}{dt}\mid_{t=0}\mathbf{c}_{g(s+t)}(\mathbf{i}\dot{a}) \\ \\ \frac{1}{2}(\frac{d}{dt}\mid_{t=0} *_{g(s+t)})F_A + \mathbf{i}(\frac{d}{dt}\mid_{t=0}\tilde{\eta}^{+_{g(s+t)}}(s+t)) + \frac{1}{2}(\frac{d}{dt}\mid_{t=0} q_{g(s+t)})(\psi) \\ \\ 0 \end{bmatrix}.$$

For $s$ fixed, the operator $\tilde{\mathcal{T}}^0_{(s,\mathsf{C})}$, *restricted to the subspace* $\dot{s}=0$, coincides with the operator $\mathcal{T}^0_{\mathsf{C}}$ considered in §2.2.4. More accurately, if we set

(2.3.8) $$H_0(s) := L^2(\mathbb{S}_{\sigma,g(s)} \oplus \mathbf{i}T^*M, g(s)) \; (\text{“} = \text{”} T_{\mathsf{C}}\mathcal{C}_\sigma)$$

and

(2.3.9) $$H_1(s) = L^2(\mathbb{S}_{\sigma,g(s)} \oplus \mathbf{i}\Lambda^2_{+_{g(s)}}T^*M, g(s)) \oplus L^2(\mathbf{i}\Lambda^0 T^*M, g(s))$$

then $\mathcal{T}^0_{\mathsf{C}}$ is an unbounded Fredholm operator $H_0(s) \to H_1(s)$ while $\tilde{\mathcal{T}}^0_{(s,\mathsf{C})}$ is an unbounded Fredholm operator $\mathbb{R} \oplus H_0(s) \to H_1(s)$. Moreover, we have the block decomposition

(2.3.10) $$\tilde{\mathcal{T}}^0 = [0 \;\; \mathcal{T}^0] : \mathbb{R} \oplus H_0(s) \to H_1(s).$$

Observe that if $|s-i| \leq \hbar$, $i=0,1$, then for every $\tau \in [0,1]$ we have a similar block decomposition

(2.3.11) $$\tilde{\mathcal{T}}^\tau = [0 \;\; \mathcal{T}^\tau] : \mathbb{R} \oplus H_0(s) \to H_1(s).$$

We have seen that the family $\det \mathcal{T}^0$ is orientable and we can specify an orientation by choosing an orientation in $H^1(M) \oplus H^2_+(M)$. Since $\ker \tilde{\mathcal{T}}^0_\bullet = \mathbb{R} \oplus \ker \mathcal{T}^0_\bullet$ we deduce that $\det \tilde{\mathcal{T}}^0$ is also orientable. The component $\mathbb{R}$ is naturally oriented and the positive orientation is given by the tangent vector $\frac{\partial}{\partial s}$. Thus, by fixing an orientation on $H^1(M) \oplus H^2_+(M)$ we induce an orientation on $\det \tilde{\mathcal{T}}^0_\bullet$ which induces an orientation on $\det \tilde{\mathcal{T}}_\bullet$ via the homotopy $\tilde{\mathcal{T}}^\tau$. This last orientation induces an orientation on $\tilde{\mathfrak{M}}_\sigma(\tilde{\eta})$. At this point we have to discuss separately the two situations $d(\sigma)=0$ and $d(\sigma)>0$.

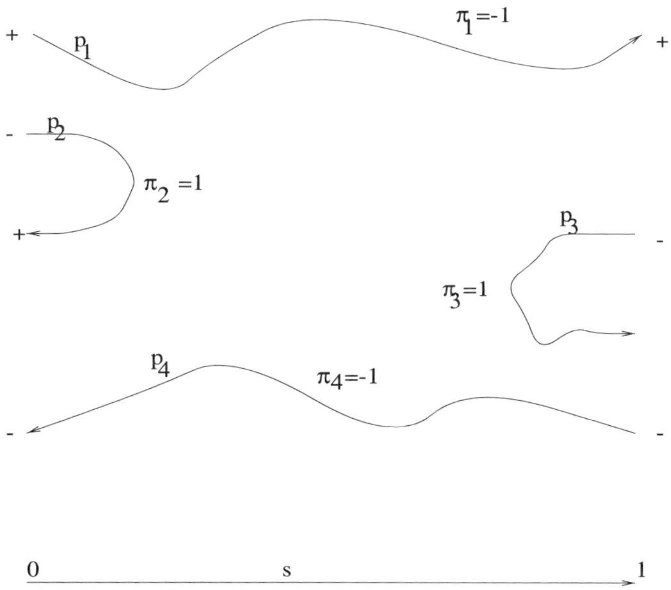


**Figure 2.2.** *A one-dimensional oriented cobordism*

• $d(\sigma) > 0$. The above considerations show that if we equip $\partial\tilde{\mathfrak{M}}_\sigma(\tilde{\eta})$ with the induced orientation (outer-normal-first convention) then $\partial\tilde{\mathfrak{M}}_\sigma(\tilde{\eta}) = \mathfrak{M}_\sigma(\eta_1) \sqcup -\mathfrak{M}_\sigma(\eta_0)$ as *oriented* manifolds. This follows from the fact that $\frac{\partial}{\partial s}$ coincides at $s = 1$ with the outer normal along $\mathfrak{M}_{\eta_1}$ while at $s = 0$ this vector field is the inner normal.

Now we can regard $\tilde{\mathfrak{M}}_\sigma(\tilde{\eta})$ as an *oriented* cobordism inside $\mathcal{B}_{\sigma,irr}$ between $\mathfrak{M}_\sigma(\eta_0)$ and $\mathfrak{M}_\sigma(\eta_1)$. From Stokes' theorem we deduce

$$\Big\langle (1-\Omega_\sigma)^{-1}, \mathfrak{M}_\sigma(\eta_1)\Big\rangle - \Big\langle (1-\Omega_\sigma)^{-1}, \mathfrak{M}_\sigma(\eta_0)\Big\rangle = \Big\langle (1-\Omega_\sigma)^{-1}, \partial\tilde{\mathfrak{M}}\Big\rangle$$

($d$ = exterior derivative)

$$= \int_{\tilde{\mathfrak{M}}} d(1-\Omega_\sigma)^{-1} = 0.$$

This shows that $\mathbf{sw}_M(\sigma, g_0, \eta_0) = \mathbf{sw}_M(\sigma, g_1, \eta_1)$.

• $d(\sigma) = 0$. In this case $\tilde{\mathfrak{M}}$ is a compact, oriented one-dimensional manifold with boundary so that it consists of a finite family of embeddings (see Figure 2.2)

$$p_j = p_j(t) : [0,1] \to \widetilde{\mathcal{B}}_{\sigma,irr} = \{(s, \mathsf{C});\;\; s \in [0,1],\;\; \mathsf{C} \in \mathcal{C}_{\sigma,irr}(g(s))/\mathfrak{G}\},$$

$j = 1, \cdots, \nu$, such that

$$s_j(0), s_j(1) \in \{0, 1\}, \;\; \forall j = 1, \cdots, \nu.$$

Above, $s_j$ denotes the composition

$$[0,1] \xrightarrow{p_j} \widetilde{\mathcal{B}}_{\sigma,irr} \xrightarrow{s} [0,1].$$

The integer $(-1)^{s_j(0)+s_j(1)} \in \{\pm 1\}$ is called the *parity* of the path $p_j$ and will be denoted by $\pi_j$. The path $p_j$ is called even/odd if $\pi_j = +/-$.

The end points of the path $p_j$ are irreducible monopoles $\mathsf{C}_j^0$, $\mathsf{C}_j^1$ and, as such, they come with signs attached $\epsilon_j^0 = \epsilon(C_j^0)$, $\epsilon_j^1 = \epsilon(\mathsf{C}_j^1) \in \{\pm 1\}$.

**Lemma 2.3.4.** *For every $j = 1, \cdots, \nu$ we have (see Figure 2.2)*

$$\epsilon_j^0 \epsilon_j^1 + \pi_j = 0.$$

Assume for the moment Lemma 2.3.4. Set $\mathbf{sw}_i := \mathbf{sw}(M, \sigma, g_i, \eta_i)$, $i = 0, 1$. Then (see Figure 2.2)

$$\mathbf{sw}_0 - \mathbf{sw}_1 = \sum_{j=1}^{\nu} (\,(-1)^{s_j(0)} \epsilon_j^0 + (-1)^{s_j(1)} \epsilon_j^1\,)$$

$$= \sum_{j=1}^{\nu} (-1)^{s_j(0)} \epsilon_j^1 (\,\epsilon_j^0 \epsilon_j^1 + \pi_j\,) = 0.$$

**Proof of Lemma 2.3.4** Fix $j = 1, \cdots, \nu$. Lift $p_j$ to a path $\tilde{p}_j(t) = \big(s_j(t), C_j(t)\big)$. $C_j(t) \in \mathcal{C}_\sigma$ is a $\big(g(s_j(t)), \eta(s_j(t))\big)$-monopole. Denote by $\mathcal{T}_t$ the operator

$$\mathcal{T}_t := \underline{\mathbf{SW}}_{g(s_j(t)), \mathsf{C}_j(t)} + \mathfrak{L}_{\mathsf{C}_j(t)}^{*s_j(t)}$$

described by (2.2.18) in §2.2. Denote by $\mathcal{T}_t^0$ the restriction of the operator $\tilde{\mathcal{T}}_{s_j(t), C_j(t)}^{\tau=0}$ (described in (2.3.7) with $\tau = 0$) to the subspace $\dot{s} = 0$. Clearly, the two families $\mathcal{T}_t$ and $\mathcal{T}_t^0$ are homotopic. The proof of the lemma will be carried out in two steps.

**Step 1**

$$\epsilon_j^0 \epsilon_j^1 = \epsilon(\mathcal{T}_1, \mathcal{T}_t, \mathcal{T}_0)$$

where on the right hand side we have the transport along the path $\mathcal{T}_t$ defined as in §1.5.1.

**Step 2**

$$\epsilon(\mathcal{T}_1, \mathcal{T}_t, \mathcal{T}_0) = -\pi_j.$$

**Proof of Step 1** For $t \in [0,1]$ $\mathcal{T}_0^0$ set $\mathcal{P}_t = \mathcal{T}_t - \mathcal{T}_t^0$. Then according to (2.3.2) we have

$$\epsilon_j^i = \epsilon(\mathcal{T}_i, \mathcal{T}_i^0 + u\mathcal{P}_i, \mathcal{T}_i^0, 0 \le u \le 1), \quad i = 0, 1.$$

Denote by $h$ the path of Fredholm operators which starts at $\mathcal{T}_0^0$, goes along $\mathcal{T}_t^0$ to $\mathcal{T}_1^0$ and then to $\mathcal{T}_1$ following the path $\mathcal{T}_1^0 + u\mathcal{P}_1$. Then

$$\epsilon(\mathcal{T}_1, h, \mathcal{T}_0^0) = \epsilon_j^1 \cdot \epsilon(\mathcal{T}_1^0, \mathcal{T}_t^0, \mathcal{T}_0^0).$$

The path $h$ is homotopic to the path $\lambda$ which starts at $\mathcal{T}_0^0$, goes along $\mathcal{T}_0^0 + u\mathcal{P}_0$ to $\mathcal{T}_0$ and then to $\mathcal{T}_1$ along $\mathcal{T}_t$:

$$\begin{array}{ccc} \mathcal{T}_0^0 & \xrightarrow{\mathcal{T}_t^0} & \mathcal{T}_1^0 \\ {\scriptstyle \epsilon_j^0}\big\downarrow & & \big\downarrow{\scriptstyle \epsilon_j^1} \\ \mathcal{T}_0 & \xrightarrow[\mathcal{T}_t]{} & \mathcal{T}_1 \end{array} \tag{2.3.12}$$

We have (see (2.3.12))

$$\epsilon(\mathcal{T}_1, h, \mathcal{T}_0^0) = \epsilon(\mathcal{T}_1, \lambda, \mathcal{T}_0^0) = \epsilon(\mathcal{T}_1, \mathcal{T}_t, \mathcal{T}_0) \cdot \epsilon_j^0.$$

Hence

$$\epsilon_j^0 \epsilon_j^1 = \epsilon(\mathcal{T}_1, \mathcal{T}_t, \mathcal{T}_0) \cdot \epsilon(\mathcal{T}_1^0, \mathcal{T}_t^0, \mathcal{T}_0^0).$$

Now observe that each operator $\mathcal{T}_t^0$ is the direct sum of the anti-self-duality operator of the metric $g(s_j(t))$ and a *complex spin*$^c$ Dirac operator. The anti-self-duality operators have *oriented* kernel and cokernel of constant dimensions so they have no contribution to the orientation transport. The Dirac components also have no contribution since we can use complex stabilizers for this family so that the parallel transport will be a *complex* map, thus preserving orientations. Hence $\epsilon(\mathcal{T}_1^0, \mathcal{T}_t^0, \mathcal{T}_0^0) = 1$ establishing **Step 1**.

**Proof of Step 2** We will use Proposition 1.5.15 of §1.5.1. First, for $t \in [0,1]$ define the operators

$$S_t : \mathbb{R} \oplus H_0(t) \to H_1(t), \quad L_t : \mathbb{R} \to H_1(t)$$

described by $(s = s_j(t))$

$$\begin{bmatrix} \mu \\ \dot{\psi} \\ \mathbf{i}\dot{a} \end{bmatrix} \mapsto \begin{bmatrix} \mathfrak{D}_{A,g(s)}\dot{\psi} \\ \\ d^{+_{g(s)}}\mathbf{i}\dot{a} \\ \\ -2\mathbf{i}d^{*_{g(s)}}\dot{a} \end{bmatrix} + \begin{bmatrix} \frac{1}{2}\mathbf{c}_{g(s)}(\mathbf{i}\dot{a})\psi \\ \\ -\frac{1}{2}\dot{q}_{g(s)}(\psi, \dot{\psi}) \\ \\ -\mathbf{i}\mathfrak{Im}\langle \psi, \dot{\psi} \rangle_{g(s)} \end{bmatrix} +$$

(2.3.13)

$$\mu \begin{bmatrix} (\frac{d}{dz}\mid_{z=0} \mathfrak{D}_{A,g(s+z)})\psi + \frac{1}{2}\frac{d}{dz}\mid_{z=0} \mathbf{c}_{g(s+z)}(\mathbf{i}\dot{a}) \\ \frac{1}{2}(\frac{d}{dz}\mid_{z=0} *_{g(s+z)})F_A + \mathbf{i}(\frac{d}{dz}\mid_{z=0} \tilde{\eta}^{+_{g(s+z)}}(s+z)) - \frac{1}{2}(\frac{d}{dz}\mid_{z=0} q_{g(s+z)})(\psi) \\ 0 \end{bmatrix}$$

and

$$\mathbb{R} \ni \mu \mapsto L_t(\mu) =$$

$$\begin{bmatrix} (\frac{d}{dz}\mid_{z=0} \mathfrak{D}_{A,g(s+z)})\psi + \frac{1}{2}\frac{d}{dz}\mid_{z=0} \mathbf{c}_{g(s+z)}(\mathbf{i}\dot{a}) \\ \frac{1}{2}(\frac{d}{dz}\mid_{z=0} *_{g(s+z)})F_A + \mathbf{i}(\frac{d}{dz}\mid_{z=0} \tilde{\eta}^{+_{g(s+z)}}(s+z)) - \frac{1}{2}(\frac{d}{dz}\mid_{z=0} q_{g(s+z)})(\psi) \\ 0 \end{bmatrix}.$$

Observe several things.

- $S_t = \tilde{\mathcal{T}}_{s,\mathsf{C}_j(t)}$(defined in (2.3.6) ).
- $S_t = L_t + \mathcal{T}_t$.
- $L_t = 0$ for $t$ near 0 and 1.
- The operators $S_t$ have index 1 and the bundle $\mathcal{L} = \ker S_\bullet$ is oriented as the tangent bundle of the oriented path $p_j(t)$.

The above observations show that we are precisely in the conditions of Proposition 1.5.15. We need to understand the orientations $\omega_i$ and $\phi_i$ in this special case.

Observe that $\ker S_i = \mathbb{R} \oplus 0 \subset T_{C_j(0)}\mathcal{C}_\sigma$ so that $\ker S_i$ is tautologically isomorphic to $\mathbb{R}$. The orientation $\omega_i$ is the tautological one, given by the vector $1 \in \mathbb{R}$. The orientation $\phi_i$ is the orientation induced from the orientation of $\ker S_\bullet$ as tangent bundle of the *oriented* path $p_j(t)$ and thus is given by the vector

$$\frac{ds_j}{dt}\mid_{t=i}.$$

Thus the parallel transport along the path $\mathcal{T}_t$ is

$$\operatorname{sign}\left(\frac{ds_j}{dt}\mid_{t=0} \cdot \frac{ds_j}{dt}\mid_{t=1}\right).$$

This number is clearly equal to $-\pi_j$. ■

The following theorem summarizes the results established so far.

**Theorem 2.3.5.** *Suppose $M$ is a compact, closed, oriented and homology oriented smooth 4-manifold such that $b_2^+(M) > 1$. Then the correspondence*

$$Spin^c(M) \ni \sigma \mapsto \mathbf{sw}_M(\sigma, g, \eta) =: \mathbf{sw}_M(\sigma) \in \mathbb{Z}$$

*is independent of the metric $g$ and the perturbation $\eta$ and is a diffeomorphism invariant of $M$. More precisely, for every orientation preserving diffeomorphism of $f$ we have*

$$\mathbf{sw}_M(\sigma) = \varepsilon(f)\mathbf{sw}_M(f^*\sigma)$$

*where $\varepsilon(f) = \pm 1$ depending on whether $f$ preserves/reverses the homology orientation of $M$.*

If $M$ is as in the above theorem then the Seiberg-Witten invariant is the map

$$\mathbf{sw}_M : \mathrm{Spin}^c(M) \to \mathbb{Z}.$$

Denote by $\mathbb{B}_M$ the support of **sw**. The elements of $\mathbb{B}_M$ are called *basic classes* . Observe that $\mathbb{B}_M$ is *finite* since, according to Proposition 2.2.21, for all but finitely many $\sigma \in Spin^c(M)$ the moduli space $\mathfrak{M}_\sigma$ is empty.

**Definition 2.3.6.** A smooth manifold $M$ with $b_2^+ > 1$ is said to be of *SW-simple type* if for every $\sigma \in \mathbb{B}_M$ we have $d(\sigma) = 0$.

All known examples of smooth 4-manifolds with $b_2^+ > 1$ are of *SW*-simple type. This prompted E. Witten ([**149**]) to state the following

**Conjecture.** *All smooth 4-manifolds with $b_2^+ > 1$ are of SW-simple type.*

Presently (January 2000) the validity of this conjecture has been established for very large families of 4-manifolds but a general argument is yet to be discovered.

Denote by $\Gamma_M$ the set of path components of the diffeomorphism group of $M$. $\Gamma_M$ is itself a group. It acts on $Spin^c(M)$ and $|\mathbf{sw}|$ is $\Gamma_M$-invariant. (**sw** may change signs under the action of $\Gamma_M$ which can affect the chosen orientations of $H^1(M)$ or $H^2_+$.) In particular, we deduce that $\mathbb{B}_M$ is a finite $\Gamma_M$-invariant set. Note that $\mathbb{B}_M$ is also invariant under the natural involution $\sigma \mapsto \bar{\sigma}$. Moreover, using Proposition 2.2.26 of §2.2.4 and (2.3.1) of §2.3.1 we deduce after some simple manipulations

$$\mathbf{sw}_M(\bar{\sigma}) = (-1)^{d(\sigma)/2+\nu_\sigma}\mathbf{sw}_M(\sigma) = (-1)^\kappa \mathbf{sw}_M(\sigma) \tag{2.3.14}$$

where $\kappa = \kappa_M := \frac{1}{2}(b_2^+ + 1 - b_1)$.

**Remark 2.3.7.** For many smooth manifolds $M$ (with $b_2^+ > 1$) the group $\Gamma_M$ is infinite and thus one expects that many of the orbits of $\Gamma_M$ on $Spin^c(M)$ are infinite. The above observations show that only the finite ones are potentially relevant in Seiberg-Witten theory. Observe that if $\sigma$

belongs to a finite orbit of $\Gamma_M$ then the stabilizers of $\sigma$ in $\Gamma_M$ must be very large (infinite) and thus we deduce that the basic classes live amongst very symmetric *spin*$^c$ structures.

Using Corollary 2.2.6 in §2.2.1 we deduce the following remarkable consequence.

**Corollary 2.3.8.** *Suppose $M$ is a smooth 4-manifold with $b_2^+ > 1$ which admits a metric $g_0$ with positive scalar curvature. Then $\mathbb{B}_M = \emptyset$, i.e.* $\mathbf{sw}_M(\sigma) = 0$ *for all $\sigma \in Spin^c(M)$.*

**Proof** To compute the Seiberg-Witten invariants we can use the metric $g_0$ and a small $\eta$ such that there are no reducible $(g_0, \eta)$-monopoles. According to Corollary 2.2.6 if $\eta$ is sufficiently small there are no irreducible ones as well. ■

The above corollary shows that in dimension four the Seiberg-Witten invariant is an obstruction to the existence of positive scalar curvature metrics. It is known (see [**50**], [**130**]) that in dimensions $\geq 5$ the existence of such a metric is essentially a homotopy theoretic problem. As we will see later, the Seiberg-Witten invariant is a smooth invariant, i.e. there exist (many) homeomorphic smooth four-manifolds with distinct Seiberg-Witten invariants (thus nondiffeomorphic). The corollary shows another "pathology" of the 4-dimensional world: the existence of a positive scalar curvature metric is decided not just by the homotopy type of the manifold but it depends in mysterious ways on the smooth structure.

**2.3.3. The case $b_2^+ = 1$.** Suppose now that $M$ is a compact smooth 4-manifold with $b_2^+ = 1$. In this case $\mathcal{N}^k_{\sigma,g}$ is not connected. Its connected components are easy to describe. Recall (see §2.2.3) that

$$\mathcal{N}^k_{\sigma,g} = \Big\{ \eta \in L^{k,2}(\Lambda^2 T^* M);\ d\eta = 0,\ \ [\eta]_g^+ \neq 2\pi[c_\sigma]_g^+ \Big\}$$

where $[\bullet]_g$ denotes the $g$-harmonic part of a differential form. When $b_2^+ = 1$ the space $\mathbf{H}^2_+(M, g)$ of harmonic, self-dual 2-forms is one-dimensional. Fix an orthonormal basis $\omega$ of this space. Then

$$[\eta]_g^+ = \langle \eta, \omega \rangle \omega$$

where

$$\langle \eta, \omega \rangle := \int_M (\eta, \omega)_g dv_g = \int_M \eta \wedge *\omega = \int_M \eta \wedge \omega.$$

Thus the condition $[\eta]_g^+ = 2\pi[c_\sigma]_g^+$ is equivalent to

$$\langle \eta, \omega \rangle = 2\pi \langle c_\sigma, \omega \rangle.$$

The above equation describes a hyperplane in the linear space of closed 2-forms and its complement is precisely $\mathcal{N}^k_{\sigma,g}$. We see that it consists of two connected components called *chambers*. The above hyperplane is called the *separating g-wall* and we will denote it by $W_{\sigma,g}$.

Fix a *spin*$^c$ structure $\sigma$ on $M$ and a Riemannian metric $g$. We can still pick a generic $\eta \in \mathcal{N}_{\sigma,g}$ such that $\mathfrak{M}_\sigma(g,\eta)$ is a smooth, compact, oriented manifold of dimension $d(\sigma)$ and define as usual

$$\mathbf{sw}_M(\sigma, g, \eta) = \langle (1-\Omega_\sigma)^{-1}, [\mathfrak{M}_\sigma(g,\eta)] \rangle$$

(or a signed count if $d(\sigma) = 0$). When trying to imitate the argument in §2.3.2 establishing the independence of this number on $(g,\eta)$ we encounter an obstacle. The correspondence

$$\mathcal{N}_{\sigma,g} \ni \eta \mapsto g \in Met(M) = \text{the space of Riemannian metrics on } M$$

defines a fibration

$$\tilde{\mathcal{N}}^k_\sigma := \bigcup_{g \in Met(M)} \mathcal{N}_{\sigma,g} \to Met(M).$$

Since the fibers $\mathcal{N}_{\sigma,g}$ are not connected the total space $\tilde{\mathcal{N}}_\sigma$ is not connected. It consists of two components separated by the *wall*

$$\tilde{W}_\sigma = \bigcup_{g \in Met(M)} W_{\sigma,g}.$$

This means that if we pick $(g_i, \eta_i) \in \tilde{\mathcal{N}}_\sigma$ $(i = 0, 1)$ in different connected components then any smooth path

$$[0,1] \ni t \mapsto (g_t, \eta_t) \in (\text{Metrics on } M) \times \{\eta \in \Omega^1(M);\ \ d\eta = 0\}$$

connecting the $(g_i, \eta_i)$ will, at certain instants $\tau$, cross the wall $\tilde{W}_\sigma$. This means there are reducible $(\sigma, g_\tau, \eta_\tau)$-monopoles and by putting together all the $(\sigma, g_t, \eta_t)$-monopoles for $t \in [0,1]$, as we did in the previous section, we can *never* get a smooth cobordism. The reducibles are at fault. To salvage something we need to understand how the wall crossing affects the cobordism. We will do this in a special yet quite general situation. More precisely, **in the remaining part of this subsection we will assume $M$ is simply connected.**

To define the Seiberg-Witten invariants we had to fix an orientation on $(H^1 \oplus H^2_+)(M)$. In this case this is equivalent to fixing an orientation on the one-dimensional space $\mathbf{H}^2_+(M,g)$. This orientation canonically determines an orthonormal basis.

**Remark 2.3.9.** Suppose $(M,\omega)$ is a symplectic 4-manifold satisfying $b_2^+ = 1$, and $g$ is a metric adapted to $\omega$ (see Exercise 1.4.2 of §1.4.1). Then $\omega$ is $g$-self-dual and since it is also closed it is harmonic. In particular, it defines an orientation on $\mathbf{H}^2_+(M,g)$. In the symplectic case *we will exclusively work with this orientation.*

Suppose we have fixed an orientation of $H^2_+(M)$. For any metric $g$ we denote by $\omega_g$ the oriented orthonormal basis of $\mathbf{H}^2_+(M,g)$. The two components of $\mathcal{N}_{\sigma,g}$ are

$$\mathcal{N}^{\pm}_{\sigma,g} = \{\eta \in L^{k,2}(\Lambda^2 T^*M)\ ;\ d\eta = 0,\ \ \pm\langle \eta - 2\pi c_\sigma, \omega_g\rangle > 0\}.$$

We will refer to them as the *positive/negative chambers*. We get a corresponding decomposition

$$\tilde{\mathcal{N}}_\sigma = \tilde{\mathcal{N}}^+_\sigma \cup \tilde{\mathcal{N}}^-_\sigma .$$

The above discussion shows that the map

$$\tilde{\mathcal{N}}_\sigma \ni (g,\eta) \mapsto \mathbf{sw}(\sigma, g, \eta) \in \mathbb{Z}$$

is continuous and thus it is constant on each of the two chambers. We will denote by $\mathbf{sw}^{\pm}_M(\sigma, g, \eta)$ the value on the $\pm$ chamber.

Before we enter into the details of wall crossing let us first observe that we can make certain simplifying assumptions. Suppose $(g_i, \eta_i)$, $i = 0, 1$, are in different chambers of $\tilde{\mathcal{N}}_\sigma$. To study what happens when crossing a wall we can assume $g_0 = g_1$ because we can find $\hat{\eta}_0$ such that the pairs $(g_0, \hat{\eta}_0)$ and $(g_1, \eta_1)$ live in the same chamber so that the corresponding Seiberg-Witten invariants are equal, as proved in the previous sections.

Let us now take $\eta_{\pm 1} \in \mathcal{N}^{\pm}_{\sigma,g}$. We will consider paths $(\eta(s))_{|s|\leq 1}$ such that $\eta(\pm 1) = \eta_{\pm 1}$, crossing the wall $W_g$ only once and we will study the singular cobordism

$$\tilde{\mathfrak{M}}_\sigma = \bigcup_s \mathfrak{M}_\sigma(g, \eta(s))$$

from $\mathfrak{M}_\sigma(g, \eta_{-1})$ to $\mathfrak{M}_\sigma(g, \eta_1)$. We can assume that $\eta_{\pm 1}$ are good perturbations so that $\mathfrak{M}_\sigma(g, \eta_{\pm 1})$ are compact, smooth oriented manifolds of dimension $d(\sigma) \geq 0$. In this case we have

$$\chi = b_0 + b_2 + b_4 = 3 + b_2^-,\quad \tau = 1 - b_2^-$$

so that

$$d(\sigma) = \frac{1}{4}(c_\sigma^2 - (2\chi + 3\tau)\,) = \frac{1}{4}(c_\sigma^2 - 9 + b_2^-).$$

Observe also that the index of a Dirac operator associated to the *spin*$^c$ structure $\sigma$ is

$$\text{(2.3.15)}\qquad \operatorname{ind}_{\mathbb{R}} \mathfrak{D}_\sigma = \frac{1}{4}(c_\sigma^2 - \tau) = \frac{1}{4}(c_\sigma^2 - 1 + b_2^-) = d(\sigma) + 2.$$

The local structure of the parameterized moduli space $\tilde{\mathfrak{M}}_\sigma$ at $\tilde{\mathsf{C}} = (s, \mathsf{C})$, $\mathsf{C} \in \mathfrak{M}_\sigma(g, \eta(s))$ is again described by the deformation complex (2.3.3)

$$(\widetilde{\mathcal{K}}_{\tilde{\mathsf{C}}}): \qquad 0 \to T_{\mathbf{1}}\mathfrak{G}_\sigma \xrightarrow{\mathfrak{L}_{\tilde{\mathsf{C}}}} T_{\tilde{\mathsf{C}}}\widetilde{\mathcal{C}}_\sigma \xrightarrow{\widetilde{\mathbf{SW}}} \mathfrak{Y}_\sigma \to 0.$$

Arguing exactly as in §2.3.2 we can slightly perturb the path $\eta(s)$ (keeping its endpoints fixed) such that for every $\tilde{\mathsf{C}} \in \tilde{\mathfrak{M}}_\sigma(\eta(s))$ the second cohomology group of this complex vanishes, that is,

$$H^2(\widetilde{\mathcal{K}}_{\tilde{\mathsf{C}}}) = 0, \ \ \forall \tilde{\mathsf{C}} \in \tilde{\mathfrak{M}}_\sigma(g, \eta(s)). \tag{2.3.16}$$

The perturbation of $\eta(s)$ (which we will continue to call $\eta(s)$ can be chosen so that it crosses the wall $W_g$ at a single point as well. Suppose for simplicity that this happens at $s = 0$. Since the path $\eta(s)$ goes from the negative chamber to the positive chamber we deduce

$$\frac{d}{ds}\Big|_{s=0} \langle \eta(s), \omega_g \rangle \geq 0. \tag{2.3.17}$$

At this point it is wise to break the flow of the argument to point out a significant fact. The above condition $H^2 = 0$ is equivalent to

$$\operatorname{coker} \tilde{\mathcal{T}}_{\tilde{\mathsf{C}}} \stackrel{def}{=} \operatorname{coker} (\widetilde{\mathbf{SW}} \oplus \mathfrak{L}^{*_s}_{\tilde{\mathsf{C}}}) = H^0(\widetilde{\mathcal{K}}_{\tilde{\mathsf{C}}})$$

where $\tilde{\mathcal{T}}$ is defined as in (2.3.6) with $g(s)$ independent of $s$, more precisely

$$\tilde{\mathcal{T}}_{(s,\mathsf{C})}: \begin{bmatrix} \dot{s} \\ \dot{\psi} \\ \mathbf{i}\dot{a} \end{bmatrix} \mapsto \begin{bmatrix} \mathfrak{D}_A\dot{\psi} + \frac{1}{2}\mathbf{c}(\mathbf{i}\dot{a})\psi \\ d^+\mathbf{i}\dot{a} - \frac{1}{2}\dot{q}(\psi, \dot{\psi}) \\ -2\mathbf{i}d^*\dot{a} - \mathbf{i}\mathfrak{Im}\langle \psi, \dot{\psi} \rangle \end{bmatrix}$$

$$+ \dot{s} \begin{bmatrix} 0 \\ \mathbf{i}(\frac{d}{dt}|_{t=0})\eta^+(s+t) \\ 0 \end{bmatrix}. \tag{2.3.18}$$

At a configuration $(s, \mathsf{C})$ with $\mathsf{C}$ reducible, $\mathsf{C} = (0, A)$ this has the form

$$\tilde{\mathcal{T}}_{(s,\mathsf{C})}: \begin{bmatrix} \dot{s} \\ \dot{\psi} \\ \mathbf{i}\dot{a} \end{bmatrix} \mapsto \begin{bmatrix} \mathfrak{D}_A\dot{\psi} \\ d^+\mathbf{i}\dot{a} \\ -2\mathbf{i}d^*\dot{a} \end{bmatrix} + \dot{s} \begin{bmatrix} 0 \\ +\mathbf{i}(\frac{d}{dt}|_{t=0})\eta^+(s+t) \\ 0 \end{bmatrix}.$$

We see that $H^2(\widetilde{\mathcal{K}}_{(s,0,A)}) = 0$ if and only if $\mathfrak{D}_A$ is surjective and the harmonic part of $(\frac{d}{dt}|_{t=0}\ \tilde{\eta}^+(s+t))$ is a **nonzero** multiple of the generator $\omega_g$ of $\mathbf{H}^2_+(M, g)$. This contrasts with the similar, unparametrized situation described in Exercise 2.2.5 of §2.2.2. That exercise shows that when $b_2^+ = 1$

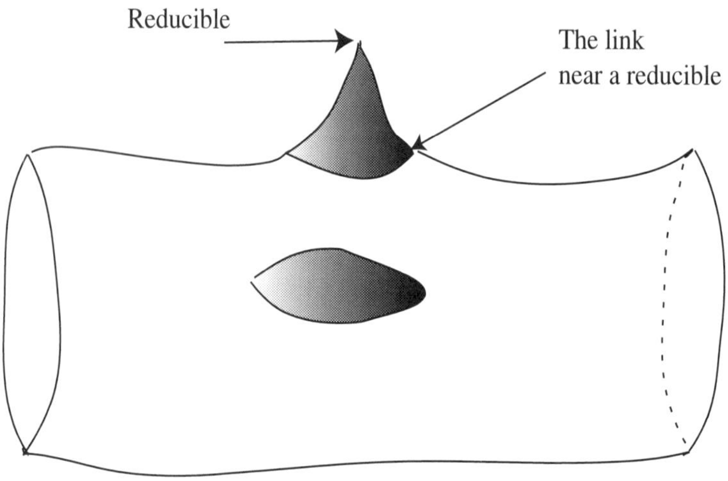


**Figure 2.3.** *A singular cobordism*

no reducible can be regular. However the reducibles can be regular in the parameterized moduli space!!! Observe that (2.3.17) can be improved to

$$\frac{d}{ds}|_{s=0} \langle \eta(s), \omega_g \rangle > 0. \tag{2.3.19}$$

If $(s, \mathsf{C}) \in \tilde{\mathfrak{M}}_\sigma$ and $s \neq 0$ then $\mathsf{C}$ is a $(g, \eta(s))$-monopole and, since $\eta(s) \in \mathcal{N}_{\sigma,g}$, it must be irreducible. This implies the 0-th cohomology of the complex $\widetilde{\mathcal{K}}_{(s,\mathsf{C})}$ is trivial and thus $(s, \mathsf{C})$ is a smooth point of the parameterized moduli space.

The configurations $(0, \mathsf{C}) \in \tilde{\mathfrak{M}}_\sigma$ arising when the wall is crossed require special considerations. If $\mathsf{C}$ is irreducible then, again, $(0, \mathsf{C})$ is a smooth point of the parameterized space. If $\mathsf{C}$ is reducible then using the Kuranishi local description as in Proposition 2.2.16 of §2.2.2 we deduce that a neighborhood of $(0, \mathsf{C})$ in $\tilde{\mathfrak{M}}_\sigma$ is homeomorphic to the quotient $B/S^1$, where $B$ is a small ball centered at the origin of $H^1(\widetilde{\mathcal{K}}_{(0,\mathsf{C})})$ and $S^1$ is the stabilizer of $\mathsf{C}$. The "cobordism" $\tilde{\mathfrak{M}}_\sigma$ has singularities, one for each reducible $(0, \mathsf{C})$. Figure 2.3 illustrates such a singular cobordism.

To proceed further we need to know some more about the structure of the singularities of the "cobordism" $\tilde{\mathfrak{M}}_\sigma$. Observe first that there exists a unique reducible point $(0, \mathsf{C}) = (0; (0, A)) \in \tilde{\mathfrak{M}}_\sigma$. Indeed $\mathsf{C} = (0, A)$ is a $(g, \eta(0))$-monopole iff

$$F_A^+ + \mathbf{i}\eta(0)^+ = 0. \tag{2.3.20}$$

Since $M$ is simply connected the group $\mathfrak{G}_\sigma$ is connected and thus every $\gamma \in \mathfrak{G}_\sigma$ can be written as $\exp(\mathbf{i}f)$, $f : M \to \mathbb{R}$. This means that, up to gauge equivalence, there exists a unique connection $A_0$ such that $F_{A_0} = -2\pi\mathbf{i}[c_\sigma]_g$. Arguing as in the proof of Proposition 2.2.17 of §2.2.3 we deduce that any connection satisfying (2.3.20) has the form

$$A = A_0 - \mathbf{i}\alpha$$

where $\alpha$ is any 1-form such that $\eta(0) = [\eta(0)]_g + d\alpha$. Again, since $M$ is simply connected $A$ is uniquely determined up to a gauge transformation.

The singularity of $\tilde{\mathfrak{M}}_\sigma$ at the unique reducible point $(0, \mathsf{C}) = (0; (0, A))$ is now easy to describe. Observe first that

$$H^1(\tilde{\mathcal{K}}_{(0,\mathsf{C})}) = \ker \tilde{\mathcal{T}}_{(0,\mathsf{C})} = V := \ker \mathfrak{D}_A.$$

It is a *complex* vector space of dimension

$$\operatorname{ind}_{\mathbb{C}} \mathfrak{D}_A \stackrel{(2.3.15)}{=} \frac{1}{2}d(\sigma) + 1.$$

The stabilizer $S^1 \subset \mathbb{C}$ acts on this complex vector space tautologically, by complex multiplication. If $B$ is a small ball in $V$ centered at the origin then $B/S^1$ is a cone on the projective space $\mathbb{CP}^{d(\sigma)/2}$, where $\mathbb{CP}^0 \stackrel{def}{=} \{pt.\}$. The boundary $L$ of $B/S^1$ is called the *link of the singularity* (see Figure 2.3).

Denote by $X$ the manifold $\tilde{\mathfrak{M}}_\sigma$ with a small neighborhood $N$ of the reducible point removed, $X = \tilde{\mathfrak{M}}_\sigma \setminus B/S^1$. The orientation on $(H^1 \oplus H^2_+)(M) = H^2_+(M)$ induces an orientation on $X$. As in the previous subsection, the induced orientation on the boundary component $\mathfrak{M}_\sigma(g, \eta_{\pm 1})$ of $X$ is $\pm$ the orientation as a moduli space. Understanding the induced orientation on the link $\partial X$ is a considerably more delicate issue. We have to distinguish two cases.

• $d(\sigma) > 0$. Let us first point out the source of complications when unraveling the orientation of the link. Denote by $(0, \mathsf{C}_0)$ the unique reducible point along the cobordism. As we have already indicated $\ker \tilde{\mathcal{T}}_{(0,\mathsf{C}_0)}$ is a complex space of dimension $d(\sigma)/2 + 1$ and the cokernel is the oriented one-dimensional space $\mathbf{H}^0(M, g)$. Thus

$$\mathcal{L}_0 := \det \tilde{\mathcal{T}}_{(0,\mathsf{C}_0)}$$

is naturally oriented. We will refer to this orientation as the *tautological orientation*. On the other hand, this line is a fiber of the line bundle

$$\left\{ \mathcal{L}_{(s,\mathsf{C})} := \det \tilde{\mathcal{T}}_{(s,\mathsf{C})}; \quad (s, \mathsf{C}) \in \tilde{\mathfrak{M}}_\sigma \right\}$$

and, as indicated in the previous subsection, this line bundle is equipped with a natural orientation, induced by an orientation on $(H^1 \oplus H^2_+)(M)$.

In turn, this induces an orientation on $\mathcal{L}_0$ which we will call the *Seiberg-Witten orientation.* We will denote by $\hat{\mathcal{L}}_0$ the line bundle equipped with the tautological orientation and by $\mathcal{L}_0^{sw}$ the line bundle $\mathcal{L}_0$ equipped with the Seiberg-Witten orientation. These two orientations differ by a sign $\epsilon \in \{\pm 1\}$.

Similarly, the neighborhood $N \cong B/S^1$ of $(0, \mathsf{C}_0)$ has two orientations: the *Seiberg-Witten orientation*, $\mathfrak{O}_{sw}$, as a subset of the moduli space, and the *tautological orientation*, $\hat{\mathfrak{O}}$, as a quotient of a complex vector space modulo the action of $S^1$. (To orient such quotients we use the **fiber-first convention**: orientation of total space = orientation orbit $\wedge$ orientation quotient.) These two orientations differ exactly by the same sign $\epsilon$.

Observe that the induced orientation on $L = \partial(N, \hat{\mathfrak{O}})$ is precisely the opposite orientation of $\mathbb{CP}^{d(\sigma)/2}$ as a complex manifold. (This follows after a little soul-searching using the fiber-first and outer normal-first orientation conventions.) Thus, the orientation of $L$ as a boundary component of $(X, \mathfrak{O}_{sw})$ is $\epsilon\times$ { the canonical orientation on $\mathbb{CP}^{d(\sigma)/2}$}. To compute $\epsilon$ we have to recall in detail the constructions of $\mathcal{L}_0^{sw}$ and $\hat{\mathcal{L}}_0$.

• **Constructing** $\mathcal{L}_0^{sw}$. Consider the one-parameter family of Fredholm operators

$$\tilde{\mathcal{T}}^\tau : \mathbb{R} \oplus \Gamma(\mathbb{S}_\sigma^+ \oplus T^*M) \to \Gamma(\mathbb{S}_\sigma^- \oplus \Lambda_+^2 \oplus \Lambda^0 T^*M), \quad \tau \in [0,1]$$

given by

$$(2.3.21) \qquad \begin{bmatrix} \dot{s} \\ \dot{\psi} \\ \dot{a} \end{bmatrix} \mapsto \begin{bmatrix} \mathfrak{D}_{A_0}\dot{\psi} \\ d^+\dot{a} \\ -2d^*\dot{a} \end{bmatrix} + \tau\dot{s}\begin{bmatrix} 0 \\ \dot{\eta}^+ \\ 0 \end{bmatrix}$$

where $\dot{\eta} := \frac{d}{ds}\,|_{s=0}\ \eta(s)$, and $\mathsf{C}_0 = (0, A_0)$. Notice that, up to an obvious factor of $\mathbf{i}$, we have $\tilde{\mathcal{T}}^1 = \tilde{\mathcal{T}}_{(0,\mathsf{C}_0)}$.

To obtain the Seiberg-Witten orientation on $\mathcal{L}_0$ we proceed as follows.

**1.** Orient $\ker \tilde{\mathcal{T}}^0 = \ker \mathfrak{D}_{A_0}$ and $\operatorname{coker} \tilde{\mathcal{T}}^0 = \ker \mathfrak{D}_{A_0}^* \oplus (\mathbf{H}_+^2 \oplus \mathbf{H}^0)(M, g)$ to obtain an orientation on $\det \tilde{\mathcal{T}}^0$. The spinor components are canonically oriented as complex vector spaces while $\mathbf{H}_+^2 \oplus \mathbf{H}^0$ is oriented by the ordered basis $\mathbf{1} \wedge \omega_g \in \det(\mathbf{H}_+^2 \oplus \mathbf{H}^0)$.

**2.** Transport the above orientation along the path $\tilde{\mathcal{T}}^\tau$ to obtain the Seiberg-Witten orientation on $\mathcal{L}_0 = \det \tilde{\mathcal{T}}^1$.

The orientation transport at Step 2 above is performed concretely as in Example 1.5.11 in §1.5.1. To begin with, observe the following fact.

$$\ker \tilde{\mathcal{T}}^0 = \mathbb{R} \oplus \ker \mathfrak{D}_{A_0}, \quad \ker \tilde{\mathcal{T}}^\tau = \ker \mathfrak{D}_{A_0}, \quad \forall \tau \in (0,1]$$

(the component $\mathbb{R}$ corresponds to $\dot{s}$) and

$$\operatorname{coker}\tilde{\mathcal{T}}^0 = \ker \mathfrak{D}^*_{A_0} \oplus \mathbf{H}^2(M) \oplus \mathbf{H}^0(M),$$

$$\operatorname{coker}\tilde{\mathcal{T}}^\tau = \ker \mathfrak{D}^*_{A_0} \oplus \mathbf{H}^0(M), \quad \forall \tau \in (0,1].$$

Since the components $\ker \mathfrak{D}_{A_0}$ and $\ker \mathfrak{D}^*_{A_0}$ are even-dimensional, oriented and stay unchanged along the deformation, they have no effect on the orientation transport so we can neglect them. To simplify the presentation we redefine $\tilde{\mathcal{T}}^\tau$ to denote the operator

$$\tilde{\mathcal{T}}^\tau : \mathbb{R} \oplus \Omega^1(M) \to \Omega^2_+(M) \oplus \Omega^0(M), \quad (\dot{s}, \dot{a}) \mapsto (d^+\dot{a} + \tau\dot{s}\dot{\eta}^+, -2d^*\dot{a}).$$

With this new convention we have

$$\ker \tilde{\mathcal{T}}^0 = \mathbb{R}, \quad \ker \tilde{\mathcal{T}}^\tau = \{0\}, \quad \tau \in (0,1],$$

$$\operatorname{coker}\tilde{\mathcal{T}}^0 = \mathbf{H}^2_+(M) \oplus \mathbf{H}^0(M), \quad \operatorname{coker}\tilde{\mathcal{T}}^\tau = \mathbf{H}^0(M), \quad \tau \in (0,1].$$

We can now perform the orientation transport.

**2a.** Choose an oriented stabilizer $V$ for the family $\tilde{\mathcal{T}}^\tau$. In this case $V = \mathbf{H}^0 \oplus \mathbf{H}^2_+$, with orientation $\mathbf{1} \wedge \omega_g$, will do the trick.

**2b.** Determine the compatible orientation on $\ker \tilde{\mathcal{T}}^0_V$ by describing an ordered basis. We follow the prescriptions in §1.5.1. In the notations of that section we have

$$V_0 = \operatorname{coker}\tilde{\mathcal{T}}^0 = \mathbf{H}^0 \oplus \mathbf{H}^2_+ = V$$

and $\hat{V}$ —the orthogonal complement of $V_0$ in $V$— is trivial. We have a natural isomorphism

$$\ker \tilde{\mathcal{T}}^0 \cong \ker \tilde{\mathcal{T}}^0_V, \quad v \mapsto (v, 0).$$

More precisely, the one-dimensional space $\ker \tilde{\mathcal{T}}^0$ is oriented by the vector

$$u_0 = (1, 0) \in \mathbb{R} \oplus \Omega^1$$

so that the one-dimensional space $\ker \tilde{\mathcal{T}}^0_{V_0}$ is oriented by the vector

$$\hat{u}_0 = (1,0,0,0) \in \mathbb{R} \oplus \Omega^1(M) \oplus \mathbf{H}^0 \oplus \mathbf{H}^2_+.$$

**2c.** We now parallel transport the orientation on $\ker \tilde{\mathcal{T}}^0_V$ to an orientation on $\det \mathcal{T}^1_V$. Observe that

$$\tilde{\mathcal{T}}^\tau_V : \mathbb{R} \oplus \Omega^1(M) \oplus V \to \Omega^2_+(M) \oplus \Omega^0(M)$$

is given by

$$(\dot{s}, \dot{a}, v, u\omega_g) \mapsto (d^+\dot{a} + \tau\dot{s}\dot{\eta}^+ + u\omega_g, v).$$

To determine the kernel of $\tilde{\mathcal{T}}^\tau_V$ observe that the harmonic part of $\dot{\eta}^+$ is a scalar multiple of $\omega_g$:

$$[\dot{\eta}^+]_g = \mu\omega_g.$$

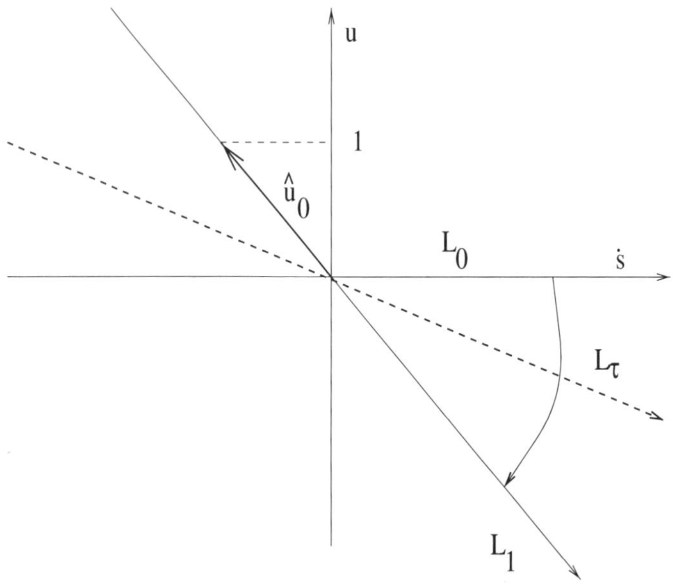


**Figure 2.4.** *Orientation transport*

According to (2.3.19) we have $\mu > 0$. Denote by $\dot{a}_0$ the unique 1-form such that

$$d^+\dot{a}_0 = -(\dot{\eta}^+ - [\dot{\eta}^+]_g), \quad d^*\dot{a}_0 = 0. \tag{2.3.22}$$

We can now describe

$$L_\tau := \ker \tilde{\mathcal{T}}_V^\tau = \{(\dot{s}, \tau\dot{s}\dot{a}_0, 0, u\omega_g) \in \mathbb{R} \oplus \Omega^1(M) \oplus \mathbf{H}^0 \oplus \mathbf{H}^2_+ ;\ \tau\mu\dot{s} + u = 0\}.$$

The orthogonal projections of these lines to the plane $\mathbb{R} \oplus \mathbf{H}^2_+$ can be visualized as a family of lines in the plane $(u, \dot{s})$ described by the equations

$$\tau\mu\dot{s} + u = 0$$

as in Figure 2.4. The line $L_{\tau=0}$ projects to the horizontal axis and the projection of the vector $\hat{u}_0$ induces the canonical positive orientation. The projection of the line $L_{\tau=1}$ has negative slope $-\mu$ and the parallel transport equips it with the "downhill" orientation.

• **Constructing** $\hat{\mathcal{L}}_0$. Recall that $\hat{\mathcal{L}}_0$ is the line $\det \tilde{\mathcal{T}}^1$ equipped with the natural orientation induced by the canonical orientations on $\ker \tilde{\mathcal{T}}^1 = \ker \mathfrak{D}_{A_0}$ and $\operatorname{coker} \tilde{\mathcal{T}}^1 = \ker \mathfrak{D}^*_{A_0} \oplus \mathbf{H}^0(M)$. To compare it with $\mathcal{L}_0^{sw}$ we need to describe the canonical orientation in terms of the stabilizer $V$ used above. Again we can neglect the spinor components in the definition of $\tilde{\mathcal{T}}^1$ and we will think of $\tilde{\mathcal{T}}^1$ as an operator

$$\tilde{\mathcal{T}}^1 : \mathbb{R} \oplus \Omega^1(M) \to \Omega^2_+(M) \oplus \Omega^0(M).$$

We use the notation and orientation construction in §1.5.1. In this case $V_0 := \operatorname{coker} \tilde{\mathcal{T}}^1 = \mathbf{H}^0$ and its orthogonal complement in $V = \mathbf{H}^0 \oplus \mathbf{H}^2_+$ is $\hat{V} = \mathbf{H}^2(M)$. We see that the orientation on $\hat{V}$ compatible with $\mathbf{1} \wedge \omega_g$ determined by the split exact sequence

$$0 \to V_0 \to V \to \hat{V} \to 0$$

is the orientation defined by the basis $\omega_g$. Denote by

$$\mathsf{R}_{V_0} : \Omega^2_+(M) \oplus \Omega^0(M) \to (\ker \tilde{\mathcal{T}}^1_{V_0})^\perp \subset \mathbb{R} \oplus \Omega^1(M) \oplus \mathbf{H}^0$$

the canonical right inverse of the surjective operator $\tilde{\mathcal{T}}^1_{V_0}$. The compatible orientation on $\ker \tilde{\mathcal{T}}^1_V$ is determined from the split exact sequence

$$0 \to \ker \tilde{\mathcal{T}}^1_{V_0} \to \tilde{\mathcal{T}}^1_V \to (V/V_0) = \hat{V} \to 0.$$

More explicitly, it is given by the basis

$$(0 \oplus 0 \oplus 0) \oplus \omega_g - \mathsf{R}_{V_0}(\omega_g) \oplus 0 \in (\mathbb{R} \oplus \Omega^1(M) \oplus \mathbf{H}^0) \oplus \mathbf{H}^2_+.$$

To determine $\mathsf{R}_{V_0}\omega_g$ observe that

$$\tilde{\mathcal{T}}^1_{V_0} : \mathbb{R} \oplus \Omega^1(M) \oplus \mathbf{H}^0 \to \Omega^2_+(M) \oplus \Omega^0(M)$$

is given by

$$(\dot{s}, \dot{a}, v) \mapsto (d^+a + \dot{s}\dot{\eta}^+, -2d^*a + v).$$

A simple computation shows that

$$\mathsf{R}_{V_0}\omega_g = (\frac{1}{\mu}, \frac{1}{\mu}\dot{a}_0, 0) \in \mathbb{R} \oplus \Omega^1(M) \oplus \mathbf{H}^0$$

where $\dot{a}_0$ is defined by (2.3.22). Thus, the oriented basis of $\ker \tilde{\mathcal{T}}^1_V$ is

$$\nu := (-\frac{1}{\mu}, -\frac{1}{\mu}\dot{a}_0, 0, \omega_g) \in \mathbb{R} \oplus \Omega^1(M) \oplus \mathbf{H}^0 \oplus \mathbf{H}^2_+.$$

By looking again at the projection onto the plane $\mathbb{R} \oplus \mathbf{H}^2_+$ we see that the canonical orientation of $L_{\tau=1}$, defined by the above vector, is the opposite of the Seiberg-Witten orientation discussed before. (The projection of $\nu$ is the "uphill" vector in Figure 2.4.) This shows $\epsilon = -1$.

Using Stokes' theorem we deduce

$$0 = \int_X d\Omega_\sigma^{d(\sigma)/2} = \int_{\partial X} \Omega_\sigma^{d(\sigma)/2}$$

$$= \int_{\mathfrak{M}_\sigma(g,\eta_1)} \Omega_\sigma^{d(\sigma)/2} - \int_{\mathfrak{M}_\sigma(g,\eta_{-1})} \Omega_\sigma^{d(\sigma)/2} + \epsilon \int_{\mathbb{CP}^{d(\sigma)/2}} \Omega_\sigma^{d(\sigma)/2}$$

$$= \mathbf{sw}_M(\sigma, g, \eta_1) - \mathbf{sw}_M(\sigma, g, \eta_0) + \langle \Omega_\sigma^{d(\sigma)/2}, \mathbb{CP}^{d(\sigma)/2} \rangle.$$

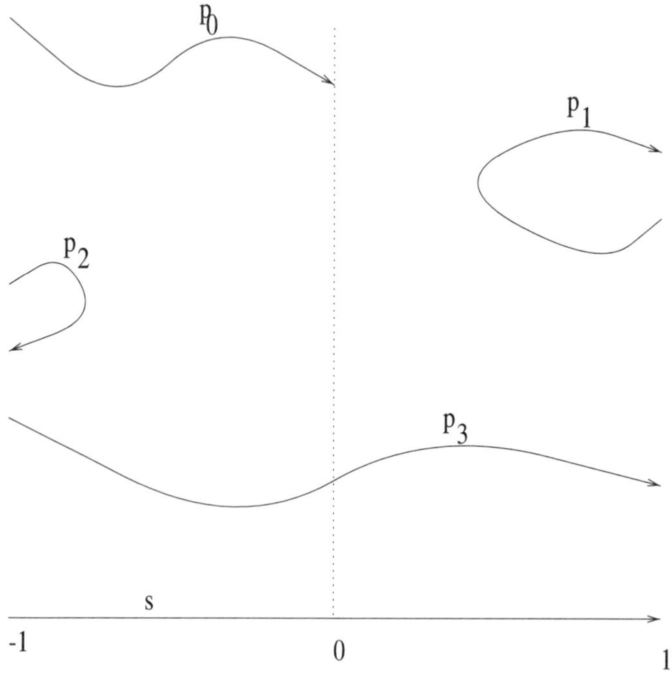


**Figure 2.5.** *A singular one-dimensional cobordism*

To compute the last integral observe that the restriction of $\mathbb{U}_\sigma$ to the link $L$ is the tautological line bundle over $\mathbb{CP}^{d(\sigma)/2}$. We conclude that

$$\mathbf{sw}_M^+(\sigma) - \mathbf{sw}_M^-(\sigma) = \mathbf{sw}_M(\sigma, g, \eta_1) - \mathbf{sw}_M(\sigma, g, \eta_0) = (-1)^{d(\sigma)/2}.$$

• $d(\sigma) = 0$. We make the simplifying assumption that $\eta_{\pm 1}$ are very close to the wall so that we have the approximation

$$\| \eta(s) - (\eta(0) + s\dot{\eta}) \|_{k,2} \ll s^2 \|\dot{\eta}(0)\|_{k,2}, \quad \forall s \in [-1, 1]. \tag{2.3.23}$$

The above inequality is a very fancy way of saying that, modulo negligible errors, we can assume the path $\eta(s)$ is affine, very very short and crosses the wall transversely only once, at $s = 0$, coming from the negative chamber and going to the positive one.

In this case, the singular cobordism $\tilde{\mathfrak{M}}_\sigma$ is a finite union of smooth oriented arcs in $\mathcal{B}_\sigma$

$$p_j : [-1, 1] \to [-1, 1] \times \mathcal{B}_\sigma, \quad t \mapsto (s_j(t), \mathsf{C}_j(t)) \;\; j = 0, 1, \cdots, n,$$

where

$$\mathsf{C}_j(t) \in \mathfrak{M}_\sigma(g, \eta(s_j(t))\,).$$

Again there is a unique reducible point $(0, \mathsf{C}_0)$ and a neighborhood $N$ is homeomorphic to $\mathbb{C}/S^1$ (see Figure 2.5).

Suppose that the path is $p_0$ so that $p_0(1) = \mathsf{C}_0$. As in the previous subsection we set

$$\epsilon_j^{\pm} = \epsilon(\mathsf{C}_j(\pm 1)),\ \ j = 1, \dots, n,$$

and

$$\epsilon_0 = \epsilon(\mathsf{C}_0(-1)).$$

We have

$$\mathbf{sw}_M(\sigma, g, \eta_1) - \mathbf{sw}_M(\sigma, g, \eta_{-1}) = \sum_{j=1}^{n}(s_j(-1)\epsilon_j^- + s_j(1)\epsilon_j^+) + s_0(-1)\epsilon_0.$$

The arguments in the previous subsection show that the first sum, corresponding to the smooth part of the cobordism, is zero. We claim that

$$\epsilon_0 s_0(-1) = 1. \tag{2.3.24}$$

The proof of this equality requires a refined perturbation analysis. Suppose $s_0(-1) = -1$ (the case $s_0(-1) = 1$ is analyzed in a similar fashion).Since

$$(s_0(t), \mathsf{C}_0(t)) \to (0, \mathsf{C}_0) \ \text{ as } \ t \to 1$$

then, modulo gauge transformations, we can write

$$\begin{aligned}(s_0(1-h), \mathsf{C}_0(1-h)) &= (0, \mathsf{C}_0) + h(\dot{s}, \dot{\mathsf{C}}_0) + h^2(\ddot{s}, \ddot{\mathsf{C}}_0) + O(h^3)\\ &= (0, \mathsf{C}_0) + h(\dot{s}, \dot{\psi}, \mathbf{i}\dot{a}) + h^2(\ddot{s}, \ddot{\psi}, \mathbf{i}\ddot{a}) + O(h^3)\end{aligned}$$

where $\dot{\mathsf{C}}_0$, $\ddot{\mathsf{C}}_0$ are vectors in the local slice at $\mathsf{C}_0$ and $\dot{s}$, $\ddot{s}$ are scalars. Moreover, we can assume $(\dot{s}, \dot{\mathsf{C}}_0) \neq 0$. Differentiating twice with the respect to $h$ (at $h = 0$) the equality

$$\mathbf{SW}_{\eta(s(1-h))}(\mathsf{C}_0(1-h)) = 0$$

we deduce

$$\mathfrak{D}_{A_0}\dot{\psi} = 0, \ \ \mathbf{i}d^+\dot{a} + \dot{s}\dot{\eta}^+ = 0, \tag{2.3.25}$$

$$\mathfrak{D}_{A_0}\ddot{\psi} + \frac{1}{2}\mathbf{c}(\mathbf{i}\dot{a})\dot{\psi} = 0, \ \ \mathbf{i}d^+\ddot{a} + \mathbf{i}\ddot{s}\dot{\eta}^+ + \frac{\mathbf{i}}{2}\dot{s}^2.\ddot{\eta}(0)^+ - \frac{1}{2}q(\dot{\psi}, \dot{\psi}) = 0 \tag{2.3.26}$$

Since $\dot{\mathsf{C}}_0$ and $\ddot{\mathsf{C}}_0$ belong to the local slice at $\mathsf{C}_0$ we deduce

$$d^*\dot{a} = d^*\ddot{a} = 0. \tag{2.3.27}$$

Recall that $\mathfrak{D}_{A_0}$ has index 1 and is onto. $\dot{\psi}$ is a vector in its one-dimensional kernel. On the other hand, since $[\dot{\eta}]^+ \neq 0$ the second equality in (2.3.25) is possible iff $\dot{s} = 0$ and $\dot{a} = 0$. (In drawing this conclusion we have used the fact that $\dot{a}$ is co-closed and $b_1(M) = 0$.) Thus $\dot{\psi}$ must be a nontrivial vector in $\ker \mathfrak{D}_{A_0}$. The equalities in (2.3.26) further simplify to

$$\mathfrak{D}_{A_0}\ddot{\psi} = 0, \ \ \mathbf{c}(\mathbf{i}d^+\ddot{a} + \mathbf{i}\ddot{s}\dot{\eta}^+(0)) - \frac{1}{2}q(\dot{\psi}, \dot{\psi}) = 0. \tag{2.3.28}$$

In particular, taking the inner product with $\mathbf{c}(\mathbf{i}\omega_g)$ we deduce

$$(\checkmark) \qquad 4\ddot{s}\mu = \int_M \langle\, \mathbf{c}(\mathbf{i}\omega_g)\,,\, q(\dot{\psi},\dot{\psi})\,\rangle dv_g = \int_M \langle\, \mathbf{c}(\mathbf{i}\omega)\dot{\psi},\dot{\psi}\,\rangle dv_g$$

where we recall that the positive number $\mu$ was determined by the equality $[\dot{\eta}(0)]^+ = \mu\omega_g$.

Observe that since we assumed the wall crossing takes place coming from the negative chamber and going towards the positive one, and since the *oriented* path $s_0(t)$ *ends* at the reducible we conclude $s_0(t) < 0$ for $t < 1$. This implies $\ddot{s} \le 0$. Using this in the last equality we conclude

$$\int_M \langle\, \mathbf{c}(\mathbf{i}\omega)\dot{\psi},\dot{\psi}\,\rangle dv_g \le 0$$

since $\mu > 0$.

At this point we need the following generic nondegeneracy result whose proof will be given later on.

**Lemma 2.3.10.** *In the very beginning we could have chosen the path $\eta(s)$ so that besides the conditions (2.3.16), $\eta(0) \in W_{\sigma,g}$, (2.3.19) and (2.3.23) it also satisfies*

$$\int_M \langle\, \mathbf{c}(\mathbf{i}\omega_g)\dot{\psi},\dot{\psi}\,\rangle dv_g < 0 \tag{2.3.29}$$

*where $(s = 0; \psi = 0, A_0)$ is the unique reducible on $\tilde{\mathfrak{M}}_\sigma(\eta(s))$ and $\dot{\psi} \in \ker \mathfrak{D}_{A_0} \setminus \{0\}$.*

From the lemma we deduce

$$-1 = s_0(-1) = \operatorname{sign} \ddot{s}. \tag{2.3.30}$$

Now consider the path of configurations

$$\mathsf{C}(t) = \mathsf{C}_0(-t), \quad t \in [-1,1].$$

Denote by $\mathcal{T}_t$ the linearization of $\mathbf{SW}_{g,\eta(s_0(-t))}$ at $\mathsf{C}(t)$, i.e.

$$\mathcal{T}_t = \underline{\mathbf{SW}}_{g,\eta(s_0(-t))} \oplus \mathfrak{L}^*_{\mathsf{C}(t)}.$$

The explicit form of $\mathcal{T}_t$ is

$$\tilde{\mathcal{T}}_t : \begin{bmatrix} \underline{\psi} \\ \\ \mathbf{i}\underline{a} \end{bmatrix} \mapsto \begin{bmatrix} \mathfrak{D}_A\underline{\psi} + \frac{1}{2}\mathbf{c}(\mathbf{i}\underline{a})\psi \\ \\ \mathbf{i}d^+\underline{a} - \frac{1}{2}\dot{q}(\psi,\underline{\psi}) \\ \\ -2\mathbf{i}d^*\underline{a} - \mathbf{i}\mathfrak{Im}\langle\psi,\underline{\psi}\rangle \end{bmatrix}$$

where $\underline{\psi} \in \Gamma(\mathbb{S}^+_\sigma)$, $\underline{a} \in \Gamma(\Lambda^1 T^*M)$ and

$$\mathsf{C}(t) = (\psi, A) = (\psi(t), A(t)) := \big(\psi(s_0(-t)), A(s_0(-t))\big).$$

Observe that with the above notation

$$\dot{\mathsf{C}}_0 = \frac{d}{dt}|_{t=-1} \mathsf{C}(t), \quad \ddot{\mathsf{C}}_0 = \frac{d^2}{dt^2}|_{t=-1} \mathsf{C}(t)$$

so that

$$\dot{\psi} = \frac{d}{dt}|_{t=-1} \psi(t), \quad \mathbf{i}\dot{a} := \frac{d}{dt}|_{t=-1} A(t) = 0.$$

We set

$$\underline{\mathsf{C}} = \begin{bmatrix} \underline{\psi} \\ \\ \mathbf{i}\underline{a} \end{bmatrix}$$

and we define

$$\dot{\mathcal{T}}\underline{\mathsf{C}} := \frac{d}{dt}|_{t=-1} \mathcal{T}_t\underline{\mathsf{C}}.$$

Observe that

$$\dot{\mathcal{T}}\underline{\mathsf{C}} = \begin{bmatrix} \frac{1}{2}\mathbf{c}(\mathbf{i}\dot{a})\underline{\psi} + \frac{1}{2}\mathbf{c}(\mathbf{i}\underline{a})\dot{\psi} \\ \\ -\frac{1}{2}\dot{q}(\dot{\psi}, \underline{\psi}) \\ \\ -\mathbf{i}\mathfrak{Im}\langle\dot{\psi}, \underline{\psi}\rangle \end{bmatrix}.$$

Let us now point out several things.

• The assumption that $\eta_{\pm 1}$ are very close to the wall so that (2.3.23) holds implies that the zero index operators $\mathcal{T}_t$ are actually nondegenerate (i.e. invertible) for $t \neq -1$.

• According to Remark 2.3.2 the sign $\epsilon_0$ is exactly the parallel transport $\epsilon(\mathcal{T}_1, \mathcal{T}_t, \mathcal{T}_{-1})$.

Using the above remarks and (1.5.9) of §1.5.1 we now deduce that

$$\epsilon_0 = (-1)^d \operatorname{sign} R$$

where $d = \dim_{\mathbb{R}} \ker \mathcal{T}_{-1}$ and if we denote by $P$ the orthogonal projection onto $\operatorname{coker} \mathcal{T}_{-1}$ then

$$R : \ker \mathcal{T}_{-1} \to \operatorname{coker} \mathcal{T}_{-1}, \ \ \underline{\mathsf{C}} \mapsto P\dot{\mathcal{T}} \,|_{t=-1} \underline{\mathsf{C}}.$$

Recall that $\operatorname{sign}(R) = \pm 1$ depending on whether $R$ preserves/reverses orientations.

Now observe that $\ker \mathcal{T}_{-1} = \ker \mathfrak{D}_{A_0}$ and an oriented *real* basis is given by $e_1 := \dot{\psi}$, $e_2 := \mathbf{i}\dot{\psi}$. Moreover, $\operatorname{coker} \mathcal{T}_{-1} = \mathbf{H}^0 \oplus \mathbf{H}^2_+$ and an oriented basis is given by $f_2 = \mathbf{i} \cdot \mathbf{1}$, $f_2 := \mathbf{i}\omega_g$.

Using (✓) we deduce

$$Re_1 = -\ddot{s}\mu\mathbf{i}\omega_g$$

and

$$Re_2 = -\mathbf{i}\mathfrak{Im}\langle\dot{\psi}, \mathbf{i}\dot{\psi}\rangle = \mathbf{i}\|\dot{\psi}\|^2.$$

Since $\ddot{s} < 0$ we deduce $\operatorname{sign}(R) = -1$. On the other hand, $d = 2$ so that $\epsilon_0 = 1$. Using the equality $s_0(-1) = -1$ we reach the desired conclusion that $\epsilon_0 s_0(-1) = 1$.

We can now formulate the main result of this section.

**Theorem 2.3.11. (Wall crossing formula)** *Suppose $M$ is a compact, oriented smooth 4-manifold such that $b_1 = 0$ and $b_2^+ = 1$. Then for every $\sigma \in Spin^c(M)$ such that $d(\sigma) \in 2\mathbb{Z}_+$ we have*

$$\mathbf{sw}_M^+(\sigma) - \mathbf{sw}_M^-(\sigma) = (-1)^{d(\sigma)/2}.$$

**Sketch of proof of Lemma 2.3.10** We will use the Sard-Smale theorem. Consider the smooth map

$$F : \mathcal{C}_\sigma^{k+1} \to L^{k,2}(\mathbb{S}_\sigma^-) \times \mathbb{R}, \;\; F(\psi, A) = \Bigl(\mathfrak{D}_A\psi, \int_M \langle \mathbf{c}(\mathbf{i}\omega_g)\psi\,,\,\psi\rangle dv_g\Bigr).$$

Now set

$$\mathcal{Z} = F^{-1}(0, -1).$$

Arguing as in §2.2.3 we deduce that for all $(\psi, A) \in \mathcal{Z}$ the differential

$$D_{(\psi,A)}F : T_{\psi,A}\mathcal{C}_\sigma^{k+1} \to T_{(0,-1)}L^{k,2}(\mathbb{S}_\sigma^-) \times \mathbb{R}$$

is onto, so that $\mathcal{Z}$ is a smooth manifold. Denote by $\pi$ the natural projection $\mathcal{C}_\sigma^{k+1} \to \mathfrak{A}_\sigma^{k+1}$. Its restriction

$$\pi : \mathcal{Z} \to \mathfrak{A}_\sigma^{k+1}$$

is Fredholm and has the same *real* index as the map

$$L^{k+1,2}(\mathbb{S}_\sigma) \ni \psi \mapsto (\mathfrak{D}_A\psi, \int_M \langle \mathbf{c}(\mathbf{i}\omega_g)\psi\,,\,\psi\rangle dv_g) \in L^{k,2}(\mathbb{S}_\sigma) \times \mathbb{R}.$$

The above map has real index 1. Thus by Sard-Smale for "most" $A \in \mathfrak{A}_\sigma$ the map $\pi$ is a Fredholm submersion along the fiber $\Psi_A := \pi^{-1}(A) \cap \mathcal{Z}$. In particular, this shows that the fiber $\Psi_A$ is a smooth one-dimensional manifold. If $(\psi, A) \in \Psi_A$ then $\dim_{\mathbb{C}} \mathfrak{D}_A = \operatorname{ind}_{\mathbb{C}} \mathfrak{D}_A = 1$ so that $\mathfrak{D}_A$ is onto. Moreover, $\Psi_A$ can be identified with the *circle*

$$\Bigl\{\psi \in \ker \mathfrak{D}_A\,;\;\; \int_M \langle \mathbf{c}(\mathbf{i}\omega_g)\psi, \psi\rangle dv_g = -1\Bigr\}.$$

Now pick $(\psi, A)$ as above and let $\eta_0 \in W_{g,\sigma}$ be defined by $F_A^+ + \mathbf{i}\eta_0^+ = 0$.

We will find the path $\eta(s)$ by looking amongst the paths

$$\eta = \eta(s) : (-\varepsilon, \varepsilon) \to \mathcal{N}_{\sigma,g},$$

at least $C^2$ in $s$, such that

$$\eta(s) \in \mathcal{N}^{\pm}_{\sigma,g} \text{ if } \pm s > 0,$$

$$\eta(0) = \eta_0$$

and

$$\int_M \dot{\eta}(0) \wedge \omega_g > \delta$$

where $\delta$ is a fixed small positive constant. The path is detected using the Sard-Smale theorem, where as space of parameters we take the space of paths $\eta(s)$ with the properties listed above. ■

**Remark 2.3.12.** There is a wall crossing formula in the case $b_1(M) > 0$ as well. However, both the formulation and its proof are much more involved. For more details we refer to [**23, 76, 112, 119**].

**2.3.4. Some examples.** We interrupt in this subsection the flow of general theoretical results to illustrate on two simple but revealing examples the power and the limitations of the wall crossing formula. The importance of these examples is not just purely academic.

**Example 2.3.13. (Seiberg-Witten invariants of $\mathbb{CP}^2$)** The complex projective plane $\mathbb{CP}^2$ is a complex manifold, so that its tangent bundle is naturally equipped with an integrable almost complex structure. In particular, this canonically defines a *spin*$^c$ structure $\sigma_0$ whose associated line bundle $\det(\sigma_0)$ is isomorphic to $K^{-1} = K^*$ – the dual of the canonical line bundle of $\mathbb{CP}^2$. Any other *spin*$^c$ structure $\sigma$ on $\mathbb{CP}^2$ has the form

$$\sigma = \sigma_0 \otimes L$$

where $L$ is a complex line bundle. Moreover

$$\det(\sigma) = 2L - K$$

where we use additive notation for the tensor product operation on line bundles and where $-K := K^{-1} = K^*$. In this case

$$\mathrm{Pic}^{\infty}(\mathbb{CP}^2) \cong H^2(\mathbb{CP}^2, \mathbb{Z}) \cong \mathbb{Z}$$

so that $Spin^c(\mathbb{CP}^2)$ is a $\mathbb{Z}$-torsor. To determine the chamber structure we need to understand the cohomology class $c_1(K)$. Since we will need it later and it requires no additional effort, we will solve this problem for all projective spaces $\mathbb{CP}^n$. We will follow the approach in [**17**].

We will freely use Poincaré duality to identify

$$H^2(\mathbb{CP}^n, \mathbb{Z}) = H_{2n-2}(\mathbb{CP}^n, \mathbb{Z}).$$

The positive generator of $H^2(\mathbb{CP}^n, \mathbb{Z})$ is represented by the homology class carried by a hyperplane in $\mathbb{CP}^n$ and we will denote it by $H$. Denote by $\tau$ the tautological line bundle over $\mathbb{CP}^n$. Since any hyperplane can be represented as the zero set of a holomorphic section of $\tau^*$ we deduce

$$c_1(\tau) = -H.$$

To follow the tradition of algebraic geometry we will denote $\tau^*$ by $H$ when no confusion is possible. (This amounts to identifying $\tau^*$ with $c_1^{top}(\tau^*) = H$.) Observe that we have the following exact sequence of complex vector bundles:

$$0 \to \underline{\mathbb{C}} \to H^{\oplus(n+1)} \to T\mathbb{CP}^n \to 0. \tag{2.3.31}$$

To see this, consider the exact *Euler sequence*

$$0 \to \tau \to \underline{\mathbb{C}}^{n+1} \to Q := \underline{\mathbb{C}}^{n+1}/\tau \to 0. \tag{2.3.32}$$

The tangent space to $\mathbb{CP}^n$ at $\ell \in \mathbb{CP}^n$ consists of infinitesimal deformations of the line $\ell \subset \mathbb{C}^{n+1}$, which can be described as linear maps $\ell \to \mathbb{C}^{n+1}/\ell$. Thus

$$T\mathbb{CP}^n \cong \mathrm{Hom}(\tau, Q) \cong \tau^* \otimes Q = H \otimes Q.$$

Thus, by tensoring (2.3.32) with $H$ we obtain (2.3.31). This implies

$$c_t(H^{\oplus(n+1)}) = c_t(\underline{\mathbb{C}})c_t(T\mathbb{CP}^n) = c_t(T\mathbb{CP}^n)$$

where $c_t(E)$ denotes the Chern polynomial $1 + c_1(E)t + c_2(E)t^2 + \cdots$. Hence

$$c_t(T\mathbb{CP}^n) = (c_t(H))^{n+1} = (1 + Ht)^{n+1}, \quad H^{n+1} = 0. \tag{2.3.33}$$

Hence

$$c_1(K) = c_1(-\det_{\mathbb{C}} T\mathbb{CP}^n) = -c_1(T\mathbb{CP}^n) = -(n+1)H. \tag{2.3.34}$$

In particular, we deduce

$$d(\sigma_0) = \frac{1}{4}(c(\sigma_0)^2 - (2\chi + 3\tau)) = \frac{1}{4}(9 - (6+3)) = 0.$$

Now consider $\mathbb{CP}^2$ with the Fubini-Study metric $g_0$. This metric has positive scalar curvature and moreover, up to a positive constant, the symplectic form $\omega_0$ associated to the Kähler structure on $\mathbb{CP}^2$ is harmonic and carries the cohomology class of $H$.

Thus

$$W_{\sigma_0, g_0} = \{\eta \in \mathcal{N}_{\sigma_0}\ ;\ \int_{\mathbb{CP}^2} (\eta - 2\pi c(\sigma_0)) \wedge \omega_0 = 0\}$$

and since $c(\sigma_0) = -K = 3H$ we deduce

$$\mathcal{N}^{\pm}_{\sigma_0, g_0} = \{\eta \in \mathcal{N}_{\sigma_0}\ \ \pm \int_{\mathbb{CP}^2} \eta \wedge \omega_0 > \pm 6\pi\}.$$

In particular $\eta = 0$ belongs to the negative chamber. Since $g_0$ has positive scalar curvature the $(g_0, \eta = 0)$ monopoles must be reducible and since $\eta = 0$ belongs to the negative chamber there are no such monopoles. Hence $\mathfrak{M}_{\sigma_0}(g_0, \eta = 0) = \emptyset$ so that

$$\mathbf{sw}_-(\sigma_0) = \mathbf{sw}(\sigma_0, g_0, \eta = 0) = 0.$$

Using the wall crossing formula we deduce

$$\mathbf{sw}_+(\sigma_0) = 1.$$

If $L_n$ denotes the line bundle with $c_1(L_n) = nH$ $(n \in \mathbb{Z})$ and $\sigma_n = \sigma_0 \otimes L_n$ then

$$c(\sigma_n) = c(\det \sigma_n) = (2n+3)H$$

and

$$d(\sigma_n) = n^2 + 3n \in 2\mathbb{Z}.$$

We have to exclude the cases $n = -1, -2$ which lead to negative virtual dimensions and thus to trivial invariants.

Next observe that

$$W_{\sigma_n, g_0} = \Big\{ \eta \in \mathcal{N}_{\sigma_n};\ \int_{\mathbb{CP}^2} \eta \wedge H = 2(2n+3)\pi \int_M H \wedge H \Big\}$$

Thus

$$\eta = 0 \in \begin{cases} \mathcal{N}^-_{\sigma_n} & \text{if} \quad n \geq -1 \\ \mathcal{N}^+_{\sigma_n} & \text{if} \quad n < -1 \end{cases}.$$

Arguing as before we deduce

$$\mathbf{sw}_+(\sigma_n) = \begin{cases} 0 & \text{if} \quad n \leq -1 \\ (-1)^{n(n+1)/2} & \text{if} \quad n > -1 \end{cases}.$$

**Example 2.3.14. (Seiberg-Witten invariants of $\mathbb{CP}^2 \# k\overline{\mathbb{CP}}^2$)** The smooth manifold

$$M = \mathbb{CP}^2 \# k\overline{\mathbb{CP}}^2$$

is a smooth realization of the algebraic construction known as the blow-up at $k$ points (see the next chapter). It is simply connected and $b_2 = k+1$. If we denote by $H$ the generator of $H_2(\mathbb{CP}^2, \mathbb{Z}) \cong H^2(\mathbb{CP}^2, \mathbb{Z})$ and by $E_i$ the generator of $H_2$ of the $i$-th copy of $\overline{\mathbb{CP}}^2$ in $M$ then the collection

$$\Big\{ H, E_i,\ i = 1, \dots, k \Big\}$$

is a $\mathbb{Z}$-basis of $H^2(M, \mathbb{Z})$. Observe that

$$H \cdot H = 1,\ \ H \cdots E_i = 0,\ \ E_i \cdot E_j = -\delta_{ij}$$

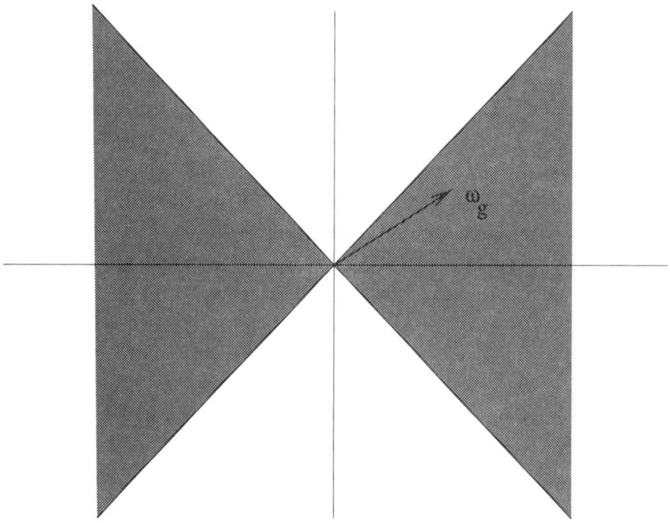


**Figure 2.6.** *The cone of vectors of nonnegative self-intersection in* $H^2(M,\mathbb{R})$

so that the intersection form has signature $(1,k)$. The intersection form defines a cone $C$ in $H^2(M,\mathbb{R})$ consisting of real cohomology classes of non-negative self-intersection. The space $C\setminus\{0\}$ has two connected components. An orientation on $H^2_+(M,\mathbb{R})$ is equivalent to declaring one of the components as the *positive cone*, $C_+$. In this case we denote by $C_+$ the connected components containing the class $H$.

A metric $g$ on $M$ produces two things on $H^2(M,\mathbb{R})$. First, it equips it with a Euclidean metric via the isomorphism with $\mathbf{H}^2(M,g)$. Second, it selects a linear subspace $\mathbf{H}^2_+(M,g)\subset\mathbf{H}^2(M,g)$. The form $\omega_g$ is defined as the unique vector of length 1 in $\mathbf{H}^2_+(M)\cap C_+$ (see Figure 2.6).

In contrast to $\mathbb{CP}^2$, there is no natural, unique way of defining a metric on $M$ but there are a few metric choices which we would like to discuss because of their future relevance.

• **The 1st choice.** Think of $\mathbb{CP}^2$ and each copy of $\overline{\mathbb{CP}}^2$ as equipped with the Fubini-Study metric. Now delete a small ball from each copy of $\overline{\mathbb{CP}}^2$ and $k$ small balls from $\mathbb{CP}^2$ and connect the resulting holed manifolds by short, thin tubes (see Figure 2.7, $k=2$). As explained in [**50**], this construction leads to a metric $g_1$ of positive scalar curvature.

Denote by $\omega_1$ the unique self-dual harmonic form of length 1 in $C_+$. If we let the sizes of the connecting necks go to zero then in the limit $\omega_1$ will converge to self-dual harmonic forms on the summands of $M$. Since $\overline{\mathbb{CP}}^2$ does not support such forms we see that the part of $\omega_1$ on the summands $\overline{\mathbb{CP}}^2$ is very small. Hence we can approximate $\omega_1$ with the restriction to

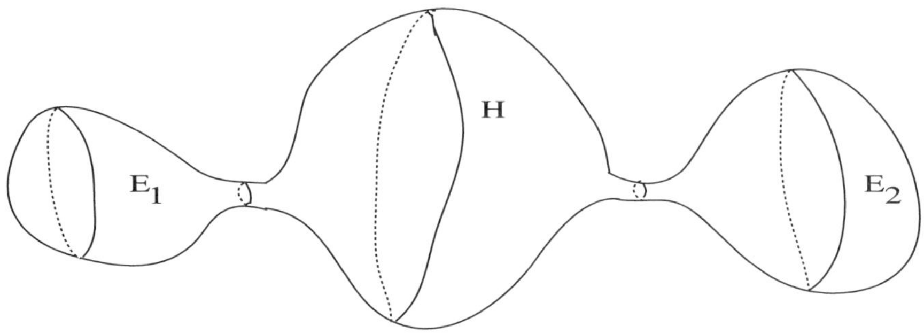


**Figure 2.7.** $\mathbb{CP}^2\#2\overline{\mathbb{CP}}^2$

$H$ on $\mathbb{CP}^2$ which is the symplectic form supported on $\mathbb{CP}^2$ induced by the Fubini-Study metric. Hence in cohomology we have

$$\omega_1 \approx H. \tag{2.3.35}$$

The manifold $M$ is equipped with a complex structure (which is by no means compatible with the above metric). Again this defines a *spin*$^c$ structure $\sigma_0$ with $\det(\sigma_0) = -K_M$, where again $-K_M$ denotes the dual of the canonical line bundle on $M$. One can show that (see Exercise 3.1.1)

$$K_M = -3H + \sum_i E_i.$$

Since $\chi_M = 3k+3$, $\tau_M = 1-k$ and $K_M \cdot K_M = 9-k$ we deduce

$$d(\sigma_0) = 0.$$

Using (2.3.35) we deduce

$$\int_M c(\sigma_0) \wedge \omega_1 \approx (3H - \sum_i E_i) \cdot H = 3 > 0$$

which shows that $\eta = 0 \in \mathcal{N}^-_{\sigma_0, g_1}$. Arguing as in the previous example we deduce

$$\mathbf{sw}_-(\sigma_0) = 0, \;\; \mathbf{sw}_+(\sigma_0) = 0. \tag{2.3.36}$$

• **2nd choice ([71]).** Let us assume $k$ is a perfect square $k = d^2$ and $d > 3$. Consider first a smooth embedded curve

$$\Sigma \hookrightarrow \mathbb{CP}^2$$

such that $[\Sigma] = dH$ in $H_2(\mathbb{CP}^2, \mathbb{Z})$. Hence

$$\Sigma \cdot \Sigma = d^2 = k.$$

Now blow-up $\mathbb{CP}^2$ in $k$ points. The surface $\Sigma$ sits in $M$. Each of the homology classes $-E_i$ is represented by an embedded 2-sphere which we continue to denote by $-E_i$. Denote by $\tilde{\Sigma}$ the surface obtained by connecting $\Sigma$ with each of the $-E_i$ by very thin tubes carrying no homology so that in $H_2(M, \mathbb{Z})$ we have the equality

$$[\tilde{\Sigma}] = dH - \sum_i E_i.$$

In particular we deduce

$$\tilde{\Sigma} \cdot \tilde{\Sigma} = 0$$

so there exists a small tubular neighborhood $U$ of $\tilde{\Sigma} \hookrightarrow M$ diffeomorphic to $D^2 \times \tilde{\Sigma}$ where $D^2$ denotes the unit disk in $\mathbb{R}^2$. Hence

$$N := \partial U \cong S^1 \times \tilde{\Sigma}.$$

Now choose a metric $g_L$ on $M$ $(L \gg 1)$ so that a tubular neighborhood of $N \hookrightarrow M$ is *isometric* with

$$[-L, L] \times S^1 \times (\tilde{\Sigma}, h)$$

where $h$ is a constant curvature metric on $\tilde{\Sigma}$. Denote by $\omega_L$ the unique $g_L$-harmonic self-dual form in $C_+$ such that

$$\omega_L \cdot H = \int_M \omega_l \wedge H = 1.$$

Observe that

$$\|\omega_L\|_{L^2(g_L)} \leq 1.$$

Indeed, if we pick an orthonormal basis $\omega_0, \omega_1, \cdots, \omega_k$ with $\omega_0$ self-dual, of norm 1 and in $C_+$ then

$$\omega_L = x_0\omega_0, \quad H = h_0\omega_0 + \sum_i h_i\omega_i, \quad x_0, h_0, h_i \in \mathbb{R}.$$

Then

$$\omega_L \cdot H = x_0 h_0 = 1$$

so that $\|\omega_L\| = x_0 = 1/h_0$. On the other hand, $1 = H \cdot H = h_0^2 - \sum_i h_i^2$ so that $h_0 \geq 1$.

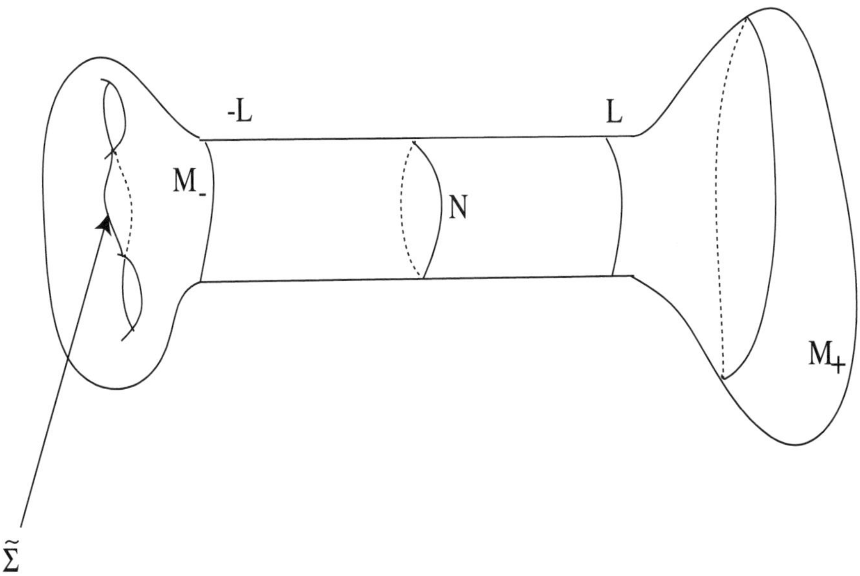


**Figure 2.8.** *Stretching the neck*

We want to figure out the sign of

$$\epsilon(L) := \int_M \omega_L \wedge c(\sigma_0)$$

for $L \to \infty$. First observe that

$$c(\sigma_0) = 3H - \sum_i E_i = \tilde{\Sigma} - (d-3)H.$$

The hypersurface $N$ divides $M$ into two parts $M_\pm$ as in Figure 2.8 where $M_-$ is the part containing the surface $\tilde{\Sigma}$ (hence $M_- \cong U$). Denote by $\omega_\pm(L)$ the restriction of $\omega_L$ to $M_\pm$. As $L \to \infty$, since $\|\omega_L\|_{L^2(g_L)} \leq 1$, the form $\omega_-(L)$ converges to a $L^2$-harmonic, self-dual form $\omega_-(\infty)$ on $M_+$ with a half-infinite cylinder attached. According to the results of [**6**] (see also Section 4.1), the cohomology class carried by $\omega_-(\infty)$ belongs to the image of the morphism

$$H^2(U, \partial U; \mathbb{R}) \to H^2(U, \mathbb{R}).$$

This image is trivial since $H^2(U, \partial U; \mathbb{R}) \cong \mathbb{R}$ is generated by the Thom class of the *trivial* line bundle $\mathbb{C} \times \tilde{\Sigma} \to \tilde{\Sigma}$. In particular, $\omega_+(\infty) = 0$ and

$$\lim_{L\to\infty} \omega_L \cdot [\tilde{\Sigma}] = \int_{\tilde{\Sigma}} \omega_-(\infty) = 0.$$

We conclude that

$$\lim_{L\to\infty} c(\sigma_0)\omega_L = \lim_{L\to\infty} ([\tilde{\Sigma}] \cdot \omega_L - (d-3)H \cdot \omega_L) = -(d-3) < 0.$$

Hence, for large $L$, the trivial closed 2-form lives in the positive chamber $\mathcal{N}_{\sigma_0, g_L}$ because

$$\langle 0 - \frac{1}{\|\omega_L\|}\omega_L, 2\pi c(\sigma_0)\rangle > 0.$$

Since $\mathbf{sw}_+(\sigma_0) \neq 0$ the above conclusion implies that for all large $L$ there exist $(g_L, 0)$-monopoles.

## 2.4. Applications

The theory developed so far is powerful enough to produce nontrivial topological and geometric applications. The goal of this section is to present some of them. More precisely we will present Kronheimer and Mrowka's proof of the Thom conjecture [**71**] for the projective plane and a proof of Donaldson's Theorem A on smooth, negative definite 4-manifolds [**28, 29**]. Because of its relevance in this section and later on as well, we have also included a separate technical subsection describing a few properties of the Seiberg-Witten equations on cylinders.

**2.4.1. The Seiberg-Witten equations on cylinders.** Suppose $(N, g)$ is a compact, oriented, Riemannian 3-manifold. We want to describe a few particular features of the Seiberg-Witten equations on the 4-manifold $\hat{N} = [a, b] \times N$ equipped with the product metric.

Some conventions are in order for this subsection. We will denote by $t$ the longitudinal coordinate on $\hat{N}$ and we will identify $N$ with the slice $\{b\} \times N$ of the cylinder $\hat{N}$. To distinguish objects of similar nature on $N$ and $\hat{N}$ we will use a hat "$\hat{\ }$" to denote the objects on the 4-manifold. Thus $d$ will denote the exterior derivative on $N$ while

$$\hat{d} = dt \wedge \partial_t + d$$

will denote the exterior derivative on $\hat{N}$. The metric on $\hat{N}$ will be denoted by $\hat{g}$ and the corresponding Hodge operator by $\hat{*}$. Denote by $\lrcorner_t$ the contraction by the tangent vector $\partial_t$.

Any differential form $\omega$ on $\hat{N}$ can be uniquely written as

$$\omega = dt \wedge f + a, \quad f := \lrcorner_t \omega, \quad a := \omega - dt \wedge f.$$

Above, $f$ and $a$ are paths of forms on $N$. Observe that

$$\hat{d}(dt \wedge f^0 + a^1) = dt \wedge (\dot{a}^1 - df^0) + da^1 \tag{2.4.1}$$

and

$$\hat{*}\omega^2 := \hat{*}(dt \wedge f^1 + a^2) = dt \wedge *a^2 + *f^1 \tag{2.4.2}$$

where the dot stands for $t$-differentiation. Then

$$\hat{d}_+(dt \wedge f^0 + a) = \frac{1}{2}(\hat{d} + \hat{*}\hat{d})(dt \wedge f^0 + a^1)$$

$$= \frac{1}{2}dt \wedge (\dot{a}^1 - df^0 + *da^1) + \frac{1}{2} * (\dot{a}^1 - df^0 + da^1)$$

and

$$\hat{d}^*(dt \wedge f^0 + a^1) = -\hat{*}\hat{d}\hat{*}(dt \wedge f^0 + a^1) = -(\dot{f}^0 - d^*a^1).$$

Fix a *spin*$^c$ structure on $N$. It induces by pullback a *spin*$^c$ structure $\hat{\sigma}$ on $\hat{N}$ with associated bundle of complex spinors

$$\hat{\mathbb{S}}_\sigma = \hat{\mathbb{S}}_\sigma^+ \oplus \hat{\mathbb{S}}_\sigma^-.$$

Denote by $\hat{\mathbf{c}}$ the Clifford multiplication on $\hat{\mathbb{S}}_\sigma$. We set $J := \hat{\mathbf{c}}(dt) : \hat{\mathbb{S}}_\sigma^+ \to \hat{\mathbb{S}}_\sigma^-$. Observe that $J$ produces an isomorphism between the restrictions of $\hat{\mathbb{S}}_\sigma^\pm$ to $N$. We set

$$\mathbb{S}_\sigma := \hat{\mathbb{S}}_\sigma^+ \mid_N \cong \hat{\mathbb{S}}_\sigma^- \mid_N .$$

The bundle $\mathbb{S}_\sigma$ is equipped with a Clifford structure given by the Clifford multiplication

$$\mathbf{c}(\alpha) = J\hat{\mathbf{c}}(\alpha) : \hat{\mathbb{S}}_\sigma^+ \mid_N \to \hat{\mathbb{S}}_\sigma^+ \mid_N .$$

$\mathbb{S}_\sigma$ is precisely the bundle of complex spinors associated to the *spin*$^c$ structure on the *odd*-dimensional manifold $N$.

For any 2-form $\hat{\alpha}$ on $\hat{N}$ we have $\hat{\mathbf{c}}(\hat{\alpha} - *\hat{\alpha}) = 0$ on $\hat{\mathbb{S}}_{\hat{\sigma}}^+$ so that, using (2.4.2), we deduce

(2.4.3) $$\mathbf{c}(\alpha) = \mathbf{c}(*\alpha), \ \forall \alpha \in \Omega^1(N)$$

and

(2.4.4) $$\mathbf{c}(dv(g)) = -\mathbf{1}.$$

Set $\det(\sigma) = \det \mathbb{S}_\sigma = \det(\hat{\sigma}) \mid_N$ and fix a smooth Hermitian connection $A_0$ on $\det(\sigma)$. It induces by pullback a Hermitian connection on $\det(\hat{\sigma})$ which we denote by $\hat{A}_0$. A Hermitian connection $\hat{A}$ on $\det(\hat{\sigma})$ is called *temporal* if

$$\lrcorner_t(\hat{A} - \hat{A}_0) = 0,$$

that is,

$$\hat{A} = \hat{A}_0 + \mathbf{i}a(t)$$

where $a(t)$ is a path of 1-forms on $A$. We set $A(t) = A_0 + \mathbf{i}a(t)$ so that $\hat{A}$ can be regarded as a path of Hermitian connections on $\det(\sigma)$. Using the identities (2.4.1) and (2.4.2) we deduce

(2.4.5) $$F_{\hat{A}} = \mathbf{i}dt \wedge \dot{a} + F_{A(t)}$$

and

$$2F^+_{\hat{A}} = dt \wedge (\mathbf{i}\dot{a} + *F_{A(t)}) + *(\mathbf{i}\dot{a} + *F_{A(t)}). \tag{2.4.6}$$

**Lemma 2.4.1.** *If $\hat{A}$ is a smooth Hermitian connection on $\det(\hat{\sigma})$ then there exists a smooth map*

$$\hat{f} : \hat{N} \to \mathbb{R}$$

*such that the connection $\exp(\mathbf{i}\hat{f}) \cdot \hat{A} := \hat{A} - 2\mathbf{i}d\hat{f}$ is temporal.*

**Proof** We write

$$\hat{A} = \hat{A}_0 + \mathbf{i}dt \wedge g(t) + \mathbf{i}a(t)$$

where $g(t) \oplus a(t)$ is a path of sections of $(\Lambda^0 \oplus \Lambda^1)T^*N$. Any function $\hat{f} : \hat{N} \to \mathbb{R}$ can be viewed as a path $f(t)$ of 0-forms on $N$. The condition

$$\lrcorner_t(\exp(\mathbf{i}\hat{f})(\hat{A} - \hat{A}_0)) = 0$$

is equivalent to

$$\mathbf{i}(g(t) - 2\dot{f}(t)) = 0.$$

We can define

$$\hat{f}(t,x) = \frac{1}{2}\int_a^t g(s,x)ds, \quad \forall t \in [a,b],\ x \in N. \ \blacksquare$$

Suppose now that $\hat{\mathsf{C}} = (\hat{\psi}, \hat{A})$ is a $\hat{g}$-monopole on $\hat{N}$. Modulo a $\widehat{\mathfrak{G}}_{\hat{\sigma}}$-change we can assume $\hat{A}$ is temporal so we can identify it with a path $A(t)$ of connections on $\det(\sigma)$. The spinor $\hat{\psi}$ can be viewed as a path $\psi(t)$ of sections in $\mathbb{S}_\sigma$. The connection $\hat{\nabla}^{\hat{A}}$ induced by $\hat{A}$ on $\hat{\mathbb{S}}_{\hat{\sigma}}$ has the form

$$\hat{\nabla}^{\hat{A}} = dt \otimes \partial_t + \nabla^{A(t)}$$

where $\nabla^{A(t)}$ is the connection induced by $A(t)$ on $\hat{\mathbb{S}}_{\hat{\sigma}}|_N \cong \mathbb{S}_\sigma \oplus \mathbb{S}_\sigma$. If $(e_i)$ is a local orthonormal frame on $N$ and $(e^i)$ denotes is dual coframe then we have

$$\hat{\mathfrak{D}}_{\hat{A}} = \hat{\mathbf{c}} \circ \hat{\nabla}^{\hat{A}} = \hat{\mathbf{c}}(dt)\partial_t + \sum_i \hat{\mathbf{c}}(e^i)\nabla^{A(t)}_{e_i} = J\left(\partial_t - \sum_i \mathbf{c}(e^i)\nabla^{A(t)}_{e_i}\right)$$

$$= J\left(\partial_t - \mathfrak{D}_{A(t)}\right)$$

where $\mathfrak{D}_{A(t)}$ denotes the geometric Dirac operator induced by the connection $A(t)$. Using the above identity, (2.4.3) and (2.4.6) we deduce that $\hat{\mathsf{C}} = (\psi(t), A(t) = \hat{A}_0 + \mathbf{i}a(t))$ satisfies the "evolution" equations

$$\begin{cases} \frac{d\psi}{dt} & = & \mathfrak{D}_{A(t)}\psi(t) \\ \mathbf{i}\dot{a} & = & \frac{1}{2}\mathbf{c}^{-1}(q(\psi(t))) - *F_{A(t)} \end{cases}. \tag{2.4.7}$$

To proceed further we imitate the four-dimensional situation and consider

$$\mathcal{C}_\sigma = \Gamma(\mathbb{S}_\sigma) \times \mathfrak{A}_\sigma$$

where $\mathfrak{A}_\sigma$ denotes the affine space of Hermitian connections on $\det(\sigma)$. Now define

$$\mathcal{E}_\sigma : \mathcal{C}_\sigma \to \mathbb{R},$$

by

$$\mathcal{E}_\sigma(\psi, A) = \frac{1}{2}\int_N (A - A_0) \wedge (F_A + F_{A_0}) + \frac{1}{2}\int_N \mathfrak{Re}\langle \mathfrak{D}_A\psi, \psi\rangle dv_g \tag{2.4.8}$$

We claim that the gradient of this functional (with respect to the $L^2$-metric on $\mathcal{C}_\sigma$) is given by precisely the right-hand side of (2.4.7).

The proof of this claim relies on the following technical result.

**Exercise 2.4.1.** Prove that for any real 1-form $\alpha$ on $N$ we have

$$2|\alpha(x)|^2 = 2| * \alpha(x)|^2 = |\mathbf{c}(\alpha(x))|^2 := -\mathrm{tr}\,(\mathbf{c}(\alpha(x))^2), \ \ \forall x \in N.$$

(**Note** the factor of 2 and **compare** to the analogous identity in Lemma 2.1.5 in §2.1.1 concerning self-dual forms.)

To verify this claim set $\mathbf{i}a := A - A_0 \in \mathbf{i}\Omega^1(N)$ (so that $\mathfrak{D}_A = \mathfrak{D}_{A_0} + \frac{1}{2}\mathbf{c}(\mathbf{i}a)$) and write $\mathcal{E}_\sigma(\psi, a)$ instead of $\mathcal{E}_\sigma(\psi, A)$. We have

$$\frac{d}{dt}|_{t=0}\, \mathcal{E}_\sigma(\psi + t\dot{\psi}, a + t\dot{a}) = \frac{1}{2}\int_N \mathbf{i}\dot{a} \wedge (\mathbf{i}da + 2F_{A_0}) + \frac{1}{2}\int_N \mathbf{i}a \wedge d\mathbf{i}\dot{a}$$

$$+\frac{1}{2}\int_N \Big(\frac{1}{2}\langle \mathbf{c}(\mathbf{i}\dot{a})\psi, \psi\rangle + 2\mathfrak{Re}\,\langle \mathfrak{D}_A\psi, \dot{\psi}\rangle\Big) dv_g$$

(use Stokes' theorem in the second integral)

$$= \frac{1}{2}\int_N \mathbf{i}\dot{a} \wedge (\mathbf{i}da + 2F_{A_0}) + \frac{1}{2}\int_N \mathbf{i}\dot{a} \wedge \mathbf{i}da$$

$$+\int_N \mathfrak{Re}\langle \mathfrak{D}_A\psi, \dot{\psi}\rangle dv_g + \frac{1}{4}\int_N \langle \mathbf{c}(\mathbf{i}\dot{a})\psi, \psi\rangle dv_g$$

(use $\langle \mathbf{c}(\mathbf{i}\dot{a})\psi, \psi\rangle = \mathfrak{Re}\,\mathrm{tr}\,(\mathbf{c}(\mathbf{i}\dot{a})q(\psi)) := \langle q(\psi), \mathbf{c}(\mathbf{i}\dot{a})\rangle$)

$$= \int_N \mathbf{i}\dot{a} \wedge F_A + \int_N \mathfrak{Re}\langle \mathfrak{D}_A\psi, \dot{\psi}\rangle dv_g + \frac{1}{4}\int_N \langle \mathbf{c}(\mathbf{i}\dot{a}), q(\psi)\rangle dv_g$$

$$= -\int_N \langle \mathbf{i}\dot{a}, *F_A\rangle dv_g + \int_N \mathfrak{Re}\langle \mathfrak{D}_A\psi, \dot{\psi}\rangle dv_g + \frac{1}{4}\int_N \langle \mathbf{c}(\mathbf{i}\dot{a}), q(\psi)\rangle dv_g$$

($*$ denotes the *complex linear* Hodge operator, and we use Exercise 2.4.1 in the last integral above)

$$= \int_N \langle \mathbf{i}\dot{a}, \frac{1}{2}\mathbf{c}^{-1}(q(\psi)) - *F_A\rangle dv_g + \int_N \mathfrak{Re}\langle \mathfrak{D}_A\psi, \dot{\psi}\rangle dv_g.$$

The functional $\mathcal{E}_\sigma$ is not $\mathfrak{G}_\sigma = \mathrm{Map}\,(N, S^1)$-invariant. In fact $\forall \gamma \in \mathfrak{G}_\sigma$ and $\mathsf{C} \in \mathcal{C}_\sigma$ we have

$$\mathcal{E}_\sigma(\gamma \cdot \mathsf{C}) = \mathcal{E}_\sigma(\mathsf{C}) - \int_N \frac{d\gamma}{\gamma} \wedge (F_A + F_{A_0})$$

$$= \mathcal{E}_\sigma(\mathsf{C}) - 4\pi^2 \int_M \frac{1}{2\pi \mathbf{i}} \frac{d\gamma}{\gamma} \wedge \frac{\mathbf{i}}{2\pi}(F_A + F_{A_0})$$

$$= \mathcal{E}_\sigma(\mathsf{C}) - 8\pi^2 \int_N \deg \gamma \wedge c_1(\det(\sigma)) \tag{2.4.9}$$

where $\deg \gamma \in H^1(N, \mathbb{Z})$ is the cohomology class $\gamma^*(\frac{1}{2\pi} d\theta)$. In particular, we deduce that $\mathcal{E}_\sigma$ is $\mathfrak{G}_\sigma$-invariant if and only if $c_1(\det \sigma)$ is a torsion class.

**Definition 2.4.2.** The critical points of the functional $\mathcal{E}_\sigma$ are called *g-monopoles* on $N$ corresponding to the *spin*$^c$ structure $\sigma$.

**Remark 2.4.3.** We want to point out a curious and somewhat confusing fact. More precisely, observe that the energy functional $\mathcal{E}_\sigma$ is *orientation sensitive.* By changing the orientation of $N$ respecting the normalization (2.4.4) the energy function changes to $-\mathcal{E}_\sigma$.

Inspired by the results in §2.1.1 we define the *energy* of a configuration $\hat{\mathsf{C}} = (\hat{\psi}, \hat{A})$ on $\hat{N}$ by

$$E(\hat{\mathsf{C}}) := \int_{\hat{N}} (|\hat{\nabla}^{\hat{A}} \hat{\psi}|^2 + \frac{\hat{s}}{4}|\hat{\psi}|^2 + \frac{1}{8}|q(\hat{\psi})|^2 + |F_{\hat{A}}|^2) dv(\hat{g})$$

where $\hat{s}$ denotes the scalar curvature of $\hat{g}$. If $\hat{A}$ is temporal, $\hat{A} = A(t) = \hat{A}_0 + \mathbf{i}a(t)$ then using (2.4.5) and the identity $|q(\psi)|^2 = \frac{1}{2}|\psi|^4$ we deduce

$$E(\hat{\psi}, \hat{A}) = \int_a^b dt \int_N (|\dot{\psi}|^2 + |\dot{a}|^2) dv(g)$$

$$+ \int_a^b \int_N (|\nabla^{A(t)} \psi(t)|^2 + \frac{s}{4}|\psi(t)|^2 + \frac{1}{16}|\psi(t)|^4 + |F_{A(t)}|^2) dv(g)$$

where $s$ denotes the scalar curvature of $g$. (Observe that on the cylinder $\hat{N}$ we have $s = \hat{s}$.)

**Lemma 2.4.4. (Main energy identity)** *Suppose $\hat{\mathsf{C}} = (\hat{\psi}, \hat{A})$ is a monopole on $\hat{N}$ such that $\hat{A}$ is temporal, $\hat{A} = A(t) = A_0 + \mathbf{i}a(t)$. Then*

$$\int_a^b dt \int_N \Big(|\dot{\psi}(t)|^2 + |\dot{a}(t)|^2\Big) dv(g)$$

$$= \int_a^b dt \int_N \Big(|\nabla^{A(t)} \psi|^2 + \frac{s}{4}|\psi|^2 + \frac{1}{16}|\psi|^4 + |F_{A(t)}|^2\Big) dv(g)$$

$$= \frac{1}{2} E(\hat{\psi}, \hat{A}).$$

**Proof** For brevity, we will write $A$ instead of $A(t)$ and $\psi$ instead of $\psi(t)$. Using the first equation in (2.4.7) we deduce

$$\int_N |\dot{\psi}|^2 dv(g) = \int_N |\mathfrak{D}_A \psi|^2 dv(g)$$

(use the Weitzenböck formula for $\mathfrak{D}_A$ and integration by parts )

$$= \int_N \Big( |\nabla^A \psi|^2 + \frac{s}{4}|\psi|^2 + \frac{1}{2}\mathfrak{Re}\langle \mathbf{c}(F_A)\psi, \psi \rangle \Big) dv(g).$$

Using the second equation in (2.4.7) and Exercise 2.4.1 we deduce

$$2\int_N |\dot{a}|^2 dv(g) = \int_N |\mathbf{c}(\dot{a})|^2 dv(g) = \int_N |\frac{1}{2}q(\psi) - \mathbf{c}(F_A)|^2 dv(g)$$

$$= \int_N \Big( |\frac{1}{2}q(\psi)|^2 + |\mathbf{c}(F_A)|^2 - \mathfrak{Re}\langle q(\psi), \mathbf{c}(F_A) \rangle \Big) dv(g)$$

(use Exercise 2.4.1 again )

$$= \int_N \Big( \frac{1}{8}|\psi|^4 + 2|F_A|^2 - \langle \mathbf{c}(F_A)\psi, \psi \rangle \Big) dv(g).$$

The energy identity is now obvious. ■

**Remark 2.4.5.** We want to point out a nice feature of the main energy identity. Its right-hand side is manifesly gauge independent while the left-hand side is apparently gauge dependent since the configuration $(\hat{\psi}, \hat{A})$ was chosen so that $\hat{A}$ is temporal.

The functional $\mathcal{E}_\sigma$ has nice variational properties, reminiscent of the Palais-Smale condition.

**Proposition 2.4.6.** *Suppose $\mathsf{C}_n = (\psi_n, A_n)$ is a sequence of smooth configurations such that*

$$\|\psi_n\|_\infty = O(1), \quad as \ n \to \infty \tag{2.4.10}$$

*and*

$$\|\nabla \mathcal{E}_\sigma(\mathsf{C}_n)\|_{L^2} = o(1), \quad as \ n \to \infty. \tag{2.4.11}$$

*Then there exists a sequence $\gamma_n \in \mathfrak{G}_\sigma$ such that $\gamma_n \cdot \mathsf{C}_n$ converges in any Sobolev norm to a critical point $\mathsf{C}_\infty$ of $\mathcal{E}_\sigma$*

$$\nabla \mathcal{E}_\sigma(\mathsf{C}_\infty) = 0.$$

**Proof** The condition (2.4.11) implies

$$\|\mathfrak{D}_{A_n}\psi_n\|_2 = o(1) \tag{2.4.12}$$

and

$$\|F_{A_n}\|_2 = \|\frac{1}{2}q(\psi_n)\|_2 + o(1). \tag{2.4.13}$$

Using the sup-bound on $\psi_n$ in the last inequality we deduce

$$\|F_{A_n}\|_2 = O(1).$$

Modulo changes of gauge, which can be used to reduce the size of the harmonic part of $F_{A_n}$ below a fixed, geometrically determined constant, the last inequality leads to $L^{1,2}$-bounds for $\mathbf{i}a_n := A_n - A_0$.

Throw this information back in (2.4.12) to obtain

$$\mathfrak{D}_{A_0}\psi_n = -\mathbf{c}(\mathbf{i}a_n)\psi_n + o(1).$$

The elliptic estimates coupled with the sup-bound on $\psi_n$ and the $L^{1,2}$-bound on $a_n$ lead to $L^{1,2}$-bounds on $\psi_n$. Bootstrap to obtain bounds on $(a_n, \psi_n)$ in arbitrary norms. These coupled with compact Sobolev embeddings allows us now to conclude that a subsequence of $\mathsf{C}_n$ converges in any Sobolev norm to some smooth $\mathsf{C}_\infty \in \mathcal{C}_\sigma$. The conclusion in the proposition now follows using (2.4.11) once again. ■

The last proposition has an important consequence.

**Corollary 2.4.7.** *Suppose $\hat{\mathsf{C}} = (\hat{\psi}, \hat{A})$ is a smooth finite energy monopole on $\hat{N}_\infty := \mathbb{R} \times N$ such that $\hat{A}$ is temporal and*

$$\|\hat{\psi}\|_\infty < \infty.$$

*Then there exists a sequence $t_n \to \infty$ such that, modulo $\mathfrak{G}_\sigma$, the configurations $(\psi(t_n), A(t_n))$ converge in any Sobolev norm to a critical point of $\mathcal{E}_\sigma$.*

**Proof** Using the main energy identity we deduce

$$\int_{-\infty}^{\infty} dt \int_N \Big( |\dot{\psi}(t)|^2 + |\dot{a}(t)|^2 \Big) dv(g) < \infty$$

so that there exists a sequence $t_n \to \infty$ such that

$$\Big\| \nabla \mathcal{E}_\sigma \big(\psi(t_n), A(t_n)\big) \Big\|_{L^2}^2 = \int_N \Big( |\dot{\psi}(t_n)|^2 + |\dot{a}(t_n)|^2 \Big) dv(g) = o(1).$$

The desired conclusion now follows from Proposition 2.4.6. ■

**2.4.2. The Thom conjecture.** To put the Thom conjecture in the proper context we begin by recalling a classical algebraic-geometry result. We will denote the tensor multiplication of line bundles additively, by $+$.

**Proposition 2.4.8. (Adjunction formula)** *Suppose $(X, J)$ is an almost complex manifold of dimension $2n$ and $Y \subset X$ is a submanifold of dimension $2(n-1)$ such that the natural inclusion*

$$TY \hookrightarrow TX\,|_Y$$

*is a morphism of complex bundles. Then*

$$K_Y = K_X\,|_Y + N_Y$$

*where $N_Y$ denotes the complex normal line bundle, $N_Y := TX\,|_Y / TY$ determined by the embedding $Y \hookrightarrow X$, and $K$ denotes the canonical line bundle, $K_M = \det(T^*M)^{1,0} = \det(T^{0,1}M)$.*

**Proof** Along $Y \hookrightarrow X$ we have the isomorphism of *complex* vector bundles

$$TX^{1,0}\,|_Y \cong TY^{1,0} \oplus N_Y.$$

By passing to determinants we deduce

$$-K_X\,|_Y = -K_Y + N_Y. \quad \blacksquare$$

Suppose now that $(X, \omega)$ is a Kähler manifold of complex dimension two and $\Sigma \hookrightarrow X$ is a *smooth complex curve on $X$* , i.e. a compact, connected, complex submanifold of $X$. Using the adjunction formula we deduce

$$K_\Sigma = K_X\,|_\Sigma + N_\Sigma.$$

Again we identify the complex line bundles with their first Chern class $c_1^{top}$. Integrating (=Kronecker pairing) the above equality over $\Sigma$ we deduce

$$\langle K_\Sigma, \Sigma\rangle = \langle K_X, \Sigma\rangle + \Sigma \cdot \Sigma$$

since, according to the Gauss-Bonnet theorem, the pairing $\langle N_\Sigma, \Sigma\rangle$ is the self-intersection of $\Sigma \hookrightarrow X$. Using Gauss-Bonnet again we deduce

$$\langle K_\Sigma, \Sigma\rangle = 2g(\Sigma) - 2$$

where $g(\Sigma)$ is the genus of the Riemannian surface $\Sigma$. This yields the *genus formula*

$$g(\Sigma) = 1 + \frac{1}{2}(K_X \cdot \Sigma + \Sigma \cdot \Sigma). \tag{2.4.14}$$

We specialize further and we assume $X = \mathbb{CP}^2$ and $\Sigma \to \mathbb{CP}^2$ is a smooth complex curve of degree $d$, i.e.

$$[\Sigma] = dH, \text{ in } H_2(\mathbb{CP}^2, \mathbb{Z}).$$

Using the equality $K_{\mathbb{CP}^2} = -3H$ established in §2.3.4 we deduce

$$g(\Sigma) = 1 + \frac{d(d-3)}{2}. \tag{2.4.15}$$

Kervaire and Milnor (see [**56, 62**]) have shown that if the homology class $dH \in H_2(\mathbb{CP}^2, \mathbb{Z})$ is characteristic for the intersection form (i.e. $d$ is odd) and can be represented by an embedded sphere then

$$1 = \tau(\mathbb{CP}^2) \equiv d^2 \mod 16.$$

In particular this shows that the class $3H$ cannot be represented by an embedded sphere.

To connect this fact with the genus formula (2.4.15) we introduce

$$g_{min} : H_2(\mathbb{CP}^2, \mathbb{Z}) \to \mathbb{Z}_+$$

where $g_{min}(dH)$ denotes the minimum of the genera of smoothly embedded Riemann surfaces $\Sigma \hookrightarrow \mathbb{CP}^2$ carrying the homology class $dH$. The above result of Kervaire and Milnor implies

$$g_{min}(dH) \geq 1, \quad d = 3.$$

The equality is optimal for $d = 3$ since according to (2.4.15) the curves of degree 3 on $\mathbb{CP}^2$ have genus 1. In particular this shows that

$$g_{min}(dH) = 1 + \frac{d(d-3)}{2}, \quad d = 1, 2, 3.$$

A famous conjecture, usually attributed to R. Thom, states that the above equality holds for all $d \geq 0$. Using the genus formula we can rephrase this by saying that the complex curves are genus minimizing amongst the smoothly embedded surfaces within a given homology class. The methods developed so far are powerful enough to offer a solution to this conjecture.

**Theorem 2.4.9.** *For every $d \geq 0$ we have the equality*

$$g_{min}(dH) = 1 + \frac{d(d-3)}{2}.$$

**Proof** We follow closely the ideas of Kronheimer and Mrowka [**71**]. The above observations show that it suffices to consider only the case $d > 3$.

Suppose $\Sigma \hookrightarrow \mathbb{CP}^2$ is a smoothly embedded surface such that $[\Sigma] = dH$, $d > 3$. Then

$$\Sigma \cdot \Sigma = k := d^2.$$

We blow up $\mathbb{CP}^2$ $k$ times $\mathbb{CP}^2 \dashrightarrow \mathbb{CP}^2 \# k\overline{\mathbb{CP}}^2$ and denote by $\pi$ the natural projection

$$M := \mathbb{CP}^2 \# k\overline{\mathbb{CP}}^2 \to \mathbb{CP}^2.$$

As in Example 2.3.36 denote by $E_i$, $i = 1, \cdots, k$ the homology classes carried by the exceptional divisors. Consider the proper transform $\tilde{\Sigma}$ in the blow-up in the sense of algebraic geometry. Topologically this means $\tilde{\Sigma}$ is the connected sum with the $k$ spheres representing the classes $-E_i$. Thus

$$\tilde{\Sigma} \cdot \tilde{\Sigma} = 0.$$

We now follow closely the geometric situation in Example 2.3.14. Denote by $U$ a small tubular neighborhood of $\tilde{\Sigma} \hookrightarrow M$ diffeomorphic to $D^2 \times \tilde{\Sigma}$ and set $N = \partial U \cong S^1 \times \tilde{\Sigma}$. Equip $\tilde{\Sigma}$ with a metric $g_0$ of constant scalar curvature $s_0$. The Gauss-Bonnet theorem implies

$$\frac{1}{4\pi} \int_{\tilde{\Sigma}} s_0 dv(g_0) = 2 - 2g(\tilde{\Sigma}) = 2 - 2g(\Sigma)$$

so that

$$s_0 = \frac{8\pi}{\mathrm{vol}_{g_0}(\tilde{\Sigma})} (1 - g(\Sigma)). \tag{2.4.16}$$

When no confusion is possible we will continue to denote by $g_0$ the product metric on $N = S^1 \times \tilde{\Sigma}$.

Now consider again the metric $g_n$, $n \gg 1$, of Example 2.3.14 so that a tubular neighborhood of $N \hookrightarrow M$ is isometric to the metric $dt^2 + d\theta^2 + g_0$ on $[-n, n] \times S^1 \times \tilde{\Sigma}$. Set $\hat{N}_n := [-n, n] \times N$. Again denote by $\hat{\sigma}_0$ the *spin*$^c$ structure induced by the natural complex structure on $M$ so that $\det(\hat{\sigma}_0) = -K_M = 3H - \sum_i E_i$. Denote by $\sigma_0$ the restriction of $\hat{\sigma}_0$ to $N$. We saw in that example that there exist (smooth) $(\hat{\sigma}_0, g_n, 0)$-monopoles $\hat{\mathsf{C}}_n = (\hat{\psi}_n, \hat{A}_n)$ for all $n \gg 1$.

**Lemma 2.4.10.** *There exists a constant $C > 0$, such that $\forall n \gg 1$ we have*

$$\|\hat{\psi}_n\|_{L^\infty(M)} < C \tag{2.4.17}$$

*and*

$$E(\hat{\mathsf{C}}_n \,|_{\hat{N}_n}) < C. \tag{2.4.18}$$

**Proof** Denote by $s_n(x)$ the scalar curvature of the metric $g_n$. Along the long neck $s_n(x)$ is comparable to $s_0$ while away from the neck it is bounded above by a constant independent of $n$ since the metric $g_n$ varies very little in that region. The inequality (2.4.17) is thus a consequence of the **Key Estimate** in §2.2.1.

To prove the second inequality denote by $R$ the complement of the neck in $M$ and let $E_n$ denote the energy of $\hat{\mathsf{C}}_n$ on $M$. Since $\hat{\mathsf{C}}_n$ is a $(\sigma_0, g_n, 0)$-monopole we deduce from Proposition 2.1.4 that

$$E_n = -2\pi^2 \int_M c_{\hat{\sigma}_0}^2 = -2\pi^2 K_M^2 = 2\pi^2(k - 9).$$

We deduce

$$E(\hat{\mathsf{C}}_n\,|_{\hat{N}_n}) = E_n - E(\hat{\mathsf{C}}_n\,|_R) \leq E_n - \int_R \frac{s_n(x)}{4}|\hat{\psi}_n(x)|^2 dv(g_n).$$

Since $s_n(x)$ and $|\hat{\psi}_n(x)|$ are bounded independent of $n$ and $R$ has finite volume, independent of $n$, we deduce that the right-hand side of the above inequality is bounded from above by a constant independent of $n$. This concludes the proof of the lemma. ■

Modulo a gauge transformation we can assume $\hat{\mathsf{C}}_n = (\hat{\psi}_n, \hat{A}_n)$ is temporal so that we can write

$$\hat{\psi}_n\,|_{\hat{N}_n} = \psi_n(t) \text{ and } \hat{A}_n = \hat{A}_0 + \mathbf{i}a_n(t).$$

Since

$$E(\hat{\mathsf{C}}_n\,|_{\hat{N}_n}) < C$$

there exists $|k_n| < n$ such that

$$E(\hat{\mathsf{C}}_n\,|_{[k_n,k_n+1]\times N}) < C/2n.$$

Using the main energy identity we deduce

$$\int_{k_n}^{k_n+1} dt \int_N |\dot{\psi}_n(t)|^2 + |\dot{a}_n(t)|^2 dv(g_0) < C/n.$$

Thus there exists $t_n \in [k_n, k_n+1]$ such that

$$\int_N |\dot{\psi}_n(t_n)|^2 + |\dot{a}_n(t_n)|^2 dv(g_0) < C/n. \tag{2.4.19}$$

Set

$$\mathsf{C}_n = \hat{\mathsf{C}}_n(t_n) = (\psi_n(t_n), A_0 + \mathbf{i}a_n(t_n)).$$

Lemma 2.4.10 and (2.4.19) show that the sequence $\mathsf{C}_n$ satisfies all the assumptions in Proposition 2.4.6. This leads to the conclusion that

◇ *there exist* $g_0$*-monopoles on* $N = S^1 \times \tilde{\Sigma}$ *corresponding to the spin*$^c$ *structure* $\sigma_0 = \hat{\sigma}_0\,|_N$.

To conclude the proof of Theorem 2.4.9 we will show that the existence of monopoles on $N$ imposes restrictions on $g(\tilde{\Sigma})$.

Observe first that any *spin*$^c$ structure $\sigma$ on $\tilde{\Sigma}$ induces by pullback via $p : N \to \tilde{\Sigma}$ a *spin*$^c$ structure $p^*\sigma$ on $N$. Next observe that

$$\sigma_0 = \hat{\sigma}_0\,|_N = p^*\hat{\sigma}_0\,|_{\tilde{\Sigma}}$$

so that

$$\det(\sigma_0) = p^*\big(\det(\hat{\sigma}_0\,|_{\tilde{\Sigma}})\big) = p^*(-K_M\,|_{\tilde{\Sigma}}).$$

The surface $\tilde{\Sigma}$ can be naturally viewed as a submanifold in $N$ which is the total space of a *trivial* $S^1$-bundle over $\tilde{\Sigma}$. The above equality implies

(2.4.20)

$$\int_{\tilde{\Sigma}} c_{\sigma_0} = -K_M \cdot \tilde{\Sigma} = (3H - \sum_{i=1}^{k} E_i) \cdot (dH - \sum_{i=1}^{k} E_i) = 3d - k = d(3-d).$$

If $\mathsf{C} = (\psi, A)$ is a $g_0$-monopole on $N$

$$\begin{cases} \mathfrak{D}_A\psi &= 0 \\ \mathbf{c}(*F_A) &= \frac{1}{2}q(\psi) \end{cases} \tag{2.4.21}$$

then arguing **exactly** as in the proof of the **Key Estimate** in §2.2.1 we deduce

$$\|\psi\|_\infty^2 \leq -2 \min_{x \in N} \bar{s}_0(x)$$

where $\bar{s}_0(x)$ denotes the scalar curvature of the metric $g_0$ on $N$. Now observe that since $N = S^1 \times \tilde{\Sigma}$ is equipped with the product metric the scalar curvature $\bar{s}_0$ at $(\theta, z) \in S^1 \times \tilde{\Sigma}$ is equal to $s_0(z)$ and using (2.4.16) we conclude

$$\|\psi\|_\infty^2 \leq \frac{16\pi}{\mathrm{vol}_{g_0}(\tilde{\Sigma})}(g(\Sigma) - 1). \tag{2.4.22}$$

Using Exercise 2.4.1 and (2.4.22) in the second equation of (2.4.21) we deduce

$$\sqrt{2}|F_A| = |\mathbf{c}(*F_A)| = \frac{1}{2}|q(\psi)| = \frac{1}{2\sqrt{2}}|\psi|^2 \leq \frac{4\sqrt{2}\pi}{\mathrm{vol}_{g_0}(\tilde{\Sigma})}(g(\Sigma) - 1)$$

so that

$$|F_A| \leq \frac{4\pi}{\mathrm{vol}_{g_0}(\tilde{\Sigma})}(g(\Sigma) - 1). \tag{2.4.23}$$

Using (2.4.20) and the assumption $d > 3$ we deduce

$$d(d-3) = \left| \int_{\tilde{\Sigma}} c_{\sigma_0} \right| \leq \frac{1}{2\pi} \int_{\tilde{\Sigma}} |F_A| dv(g_0) \overset{(2.4.23)}{\leq} 2(g(\Sigma) - 1).$$

This is exactly the content of Theorem 2.4.9. ■

**Remark 2.4.11.** (a) Presently the validity of the genus minimizing conjecture of Thom has been established in its full generality in the more general context of symplectic manifolds; see [**97, 114**] or the discussion at the end of §4.6.2. In this case the genus minimizing surfaces in a given homology class are precisely the symplectically embedded ones.

(b) In [**97, 101**] one can find a detailed and explicit description of the monopoles on $S^1 \times \Sigma$. For the more general case of circle bundles over a Riemann surface we refer to [**106**].

**2.4.3. Negative definite smooth 4-manifolds.** To help the reader better enjoy the beauty and the depth of the main result of this subsection we begin by surveying some topological facts. For more details we refer to [**29**, Chap. 1], [**51, 87**].

The world of topological 4-manifolds is very unruly and currently there is no best way to organize it, and not for lack of trying.

The fundamental group, which does wonders in dimension two and is sufficiently powerful in dimension three, is less effective in dimension four for a simple reason: every finitely presented group is the fundamental group of a smooth manifold (even symplectic, according to [**51**]). This shows that the algorithmic classification of 4-manifolds is more complicated than that of finitely presented groups, which is impossible. It is thus reasonable to try to understand first the simply connected 4-manifolds and in this dimension we have to be very specific whether we talk about topological or smooth ones.

The intersection form of simply connected topological 4-manifolds is a powerful invariant: it classifies them up to homotopy equivalence (according to J.H.C. Whitehead [**147**]) and almost up to a *homeomorphism* according to the award winning results of M. Freedman [**38**]. Recall that the intersection form of a closed 4-manifold is a symmetric, unimodular, bilinear map

$$q : \mathbb{Z}^n \times \mathbb{Z}^n \to \mathbb{Z}.$$

Unimodularity in this case means that the matrix describing $q$ with respect to some integral basis of $\mathbb{Z}^n$ has determinant 1.

To each intersection form one can associate three invariants: its rank, $n$ in this case, its signature and its type. The signature, $\tau(q)$, is defined as the difference between the number of positive eigenvalues and the number of negative eigenvalues of the symmetric matrix representing $q$ with respect to some basis of $\mathbb{Z}^n$. The intersection forms are of two types: even, if

$$q(x, x) \equiv 0 \mod 2, \ \ \forall x \in \mathbb{Z}$$

and odd, if it's not even. Observe that $q$ is even if and only if the matrix representing $q$ with respect to an arbitrary basis of $\mathbb{Z}^n$ has even diagonal entries. A quadratic form $q$ is called *positive/negative* if $\tau(q) = \pm \operatorname{rank} q$ and indefinite otherwise.

Two integral quadratic forms $q_1$, $q_2$ of the same rank $n$ are *isomorphic* if there exists $T \in GL(n, \mathbb{Z})$ such that

$$q_1(Tx, Tx) = q_2(x, x), \ \ \forall x \in \mathbb{Z}^n.$$

The quadratic forms over $\mathbb{Q}$ or $\mathbb{R}$ are completely classified up to isomorphism by their rank and signature. The situation is considerably more complicated in the integral case.

**Example 2.4.12.** The *diagonal definite form* of rank $n$ is the quadratic form $q = \langle \mathbf{1} \rangle_n$ whose matrix with respect to the canonical basis of $\mathbb{Z}^n$ is the identity matrix. More generally, a quadratic form is said to be *diagonal*(izable) if it is isomorphic to the direct sum $\langle \mathbf{1} \rangle_n \oplus \langle -\mathbf{1} \rangle_m$. The form $E_8$ is the positive definite quadratic form of rank 8 given by the symmetric matrix

$$E_8 = \begin{pmatrix} 2 & 1 & 0 & 0 & 0 & 0 & 0 & 0 \\ 1 & 2 & 1 & 0 & 0 & 0 & 0 & 0 \\ 0 & 1 & 2 & 1 & 0 & 0 & 0 & 0 \\ 0 & 0 & 1 & 2 & 1 & 0 & 0 & 0 \\ 0 & 0 & 0 & 1 & 2 & 1 & 0 & 1 \\ 0 & 0 & 0 & 0 & 1 & 2 & 1 & 0 \\ 0 & 0 & 0 & 0 & 0 & 1 & 2 & 1 \\ 0 & 0 & 0 & 0 & 1 & 0 & 1 & 2 \end{pmatrix}. \tag{2.4.24}$$

A more efficient and very much used way of describing this matrix is through its *Dynkin diagram* (see Figure 2.9). The $\bullet$'s describe a basis $v_1, \cdots, v_8$ of $\mathbb{Z}^8$.

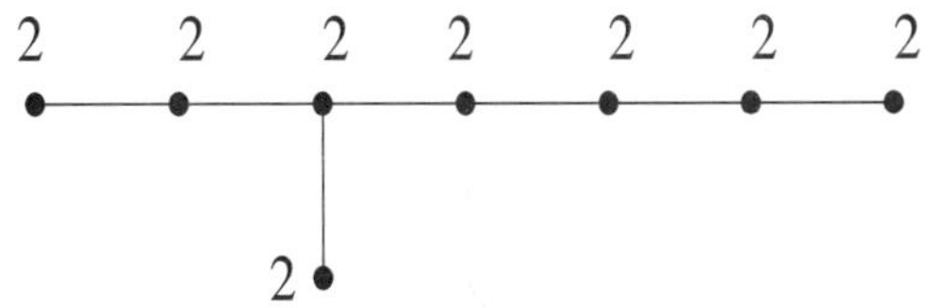


**Figure 2.9.** *The Dynkin diagram of* $E_8$

The 2's indicate that $q(v_i, v_i) = 2$ and the edges indicate that $q(v_i, v_j) = 1$ if and only if $v_i$ and $v_j$ are connected by an edge. $E_8$ is even and positive definite. $E_8$ is not diagonalizable over $\mathbb{Z}$. We also want to point out that often $E_8$ is described by a matrix very similar to the one in (2.4.24) where the 1's are replaced by $-1$'s. The two descriptions are equivalent and correspond to the change of basis $v_i \to (-1)^i v_i$.

Another important example of quadratic form is the *hyperbolic* form $H$ given by the matrix

$$H = \begin{pmatrix} 0 & 1 \\ 1 & 0 \end{pmatrix}.$$

It is even, indefinite, has zero signature and it is not diagonalizable.

The examples presented above generate a large chunk of the set of isomorphism classes of integral, unimodular, quadratic forms. More precisely, we have the following result, whose proof can be found in [**121**].

**Theorem 2.4.13.** *(a) Any odd,* indefinite *quadratic form is diagonalizable.*

*(b) Suppose $q$ is an* even *form. Then*

$$\tau(q) \equiv 0 \mod 8.$$

*(c) If $q$ is* even, indefinite *and $\tau(q) \geq 0$ then*

$$q \cong aE_8 \oplus bH := (\underbrace{E_8 \oplus \cdots \oplus E_8}_{a}) \oplus (\underbrace{H \oplus \cdots \oplus H}_{b})$$

*where $\tau(q) = 8a$ and $8a + 2b =$* rank $(q)$. *(When $\tau(q) < 0$ use $-q$ instead.)*

The classification of definite forms is a very complicated problem. It is known that the number of nonisomorphic definite quadratic, unimodular forms of rank $n$ goes very rapidly to $\infty$ as $n \to \infty$ (see [**121**]). The diagonal one however plays a special role. To describe one of its special features we need to introduce a new concept.

Suppose $q$ is a quadratic unimodular form of rank $n$. A vector $\mathbf{x}_0 \in \mathbb{Z}^n$ is called a *characteristic vector* of $q$ if

$$q(\mathbf{x}_0, y) \equiv q(y, y) \mod 2, \ \forall y \in \mathbb{Z}^n.$$

If we represent $q$ by a symmetric matrix $S$ using a basis of $\mathbb{Z}^n$ then a vector $\mathbf{x}$ is characteristic if its coordinates $(x_i)$ with respect to the chosen basis have the same parity as the diagonal elements of $S$, i.e.

$$x_i \equiv s_{ii} \mod 2, \ \forall i = 1, \cdots, n.$$

We see that $q$ is even if and only if $0$ is a characteristic vector.

**Example 2.4.14. (Wu's formula)** Suppose $M$ is a closed, compact oriented smooth 4-manifold with intersection form $q_M$. A special case of Wu's formula (see [**93**]) shows that the mod 2 reduction of any characteristic vector $\mathbf{x}$ of $q_M$ is precisely the second Stiefel-Whitney class $w_2(M)$. In particular, this implies that any smooth 4-manifold admits *spin*$^c$ structures (since any such structure corresponds to an integral lift of $w_2(M)$) and moreover,

$$\langle w_2(M), \alpha \rangle \equiv q_M(\alpha, \alpha) \mod 2, \ \forall \alpha \in H_2(M, \mathbb{Z}).$$

As explained in [**51**, Sec. 1.4], the last identity should be regarded as a mod 2 version of the adjunction formula.

The congruence (b) in Theorem 2.4.13 admits the following generalization (see [**121**]).

**Proposition 2.4.15.** *If $q$ is an integral, unimodular, quadratic form and* $\mathbf{x}$ *is a characteristic vector of $q$ then*

$$q(\mathbf{x}, \mathbf{x}) \equiv \tau(q) \mod 8.$$

Following [**32**] we introduce the *Elkies invariant* $\Theta(q)$ of a *negative definite* quadratic form $q$ as

$$\Theta(q) := \operatorname{rank}(q) + \max\{q(\mathbf{x},\mathbf{x});\ \ \mathbf{x}\ \text{a characteristic vector}\}.$$

Observe that since $q$ is negative definite $\Theta(q) \leq \operatorname{rank}(q) = -\tau(q)$ with equality if and only if $q$ is even. Moreover, by Proposition 2.4.15 we have $\Theta(q) \in 8\mathbb{Z}$. We have the following nontrivial result.

**Theorem 2.4.16. (Elkies, [32])** *For any negative definite quadratic form $q$ we have*

$$\Theta(q) \geq 0$$

*with equality if and only if $q$ is diagonal.*

Roughly speaking, this theorem says that if $q$ is not diagonal then the positive form $-q$ has short characteristic vectors.

We now return to topology. Michael Freedman's classification theorem states that given any *even* quadratic form there exists a unique, up to homeomorphism, simply connected (s.c.) topological 4-manifold with this intersection form. Moreover he showed that given any *odd* quadratic form there exist exactly *two* nonhomeomorphic topological s.c. 4-manifolds with this intersection form and at most one of them is smoothable (that is it admits smooth structures). We deduce the following remarkable consequence.

**Corollary 2.4.17.** *Two simply connected* smooth 4-*manifolds are* homeomorphic *if and only if they have isomorphic intersection forms.*

In the early 50's, Vladimir Rohlin ([**118**]) has showed that if the even form $q$ is the intersection form of a *smooth* s.c. 4-manifold then

$$\tau(q) \equiv 0 \mod 16.$$

According to Michael Freedman's classification, there exists a unique s.c. topological 4-manifold with intersection form $E_8$. The signature of $E_8$ is $8 = \operatorname{rank}(E_8)$. This topological 4-manifold cannot support smooth structures!!!

In the early 80's, Simon Donaldson ([**28**]) showed that this surprising fact is not singular.

**Theorem 2.4.18. (Donaldson, [28, 29])** *If $M$ is a smooth, compact, oriented 4-manifold with negative definite intersection form $q_M$ then $q_M$ is diagonal.*

This theorem shows that of the infinitely many negative definite quadratic forms only the diagonal ones can be the intersection forms of some *smooth* 4-manifold. Thus any negative definite topological 4-manifold with nondiagonalizable intersection form does not admit smooth structures !!!

**Proof of Theorem 2.4.18** We will argue by contradiction. Assume $q_M$ is not diagonal. We distinguish two cases.

• Assume first that $b_1(M) = \dim H^1(M, \mathbb{R}) = 0$. Then $\chi(M) = 2 + b_2$, $\tau(M) = -b_2$ so that for all $\sigma \in Spin^c(M)$ we have

$$d(\sigma) = \frac{1}{4}(c_\sigma^2 + b_2 - 4) = \frac{1}{4}(q_M(c_\sigma, c_\sigma) + \text{rank}\,(q_M)) - 1.$$

By Wu's formula $c_\sigma$ is a characteristic vector. Since $q_M$ is not diagonal we deduce from Elkies' theorem that $\Theta(q_M) > 0$ and we can find $\sigma \in Spin^c(M)$ such that $d(\sigma) = \frac{1}{4}\Theta(q_M) - 1 > 0$. Since $\Theta(q_M) \in 8\mathbb{Z}$ we deduce $d(\sigma) \in 2\mathbb{Z} + 1$.

For any closed 2-form $\eta$ on $M$ and any metric $g$ there exist reducible $(g, \eta)$-monopoles corresponding to the $\sigma$. They are determined by the condition

$$F_A^+ + \mathbf{i}\eta^+ = 0. \tag{2.4.25}$$

As in §2.2.3 we write $\eta = [\eta] + d\alpha$ and fix a connection $A_0$ such that

$$[F_{A_0}] = -2\pi\mathbf{i}[c_\sigma].$$

Any solution of (2.4.25) can be written as $A = A_0 - \mathbf{i}\alpha + \mathbf{i}\beta$ where $\beta$ is a closed 1-form. (Observe that such an $A$ satisfies $F_A = F_{A_0} - \mathbf{i}d\alpha$. Since $M$ is negative definite it automatically satisfies (2.4.25) because there are no self-dual harmonic 2-forms.) On the other hand, since $b_1(M) = 0$ any closed 1-form is exact so that $\beta = -2df$. This shows that all the solutions of (2.4.25) are $\mathfrak{G}_\sigma$ equivalent.

Using the Sard-Smale theorem as in §2.2.3 we can pick $\eta$ so that any $(g, \eta)$-monopole $\mathsf{C}$ is regular, i.e. the second cohomology group $\mathcal{H}^2_{\mathsf{C}}$ of the deformation complex at $\mathsf{C}$ is trivial. Denote by $\mathsf{C}_0 = (0, A_0)$ the unique (mod $\mathfrak{G}_\sigma$) reducible $(g, \eta)$-monopole. In this case, using the Kuranishi picture we deduce that away from $\mathsf{C}_0$ the moduli space is a smooth manifold while a neighborhood of $\mathsf{C}_0$ in the moduli space $\mathfrak{M}_\sigma(g, \eta)$ is homeomorphic to

$$\mathcal{H}^1_{\mathsf{C}_0}/S^1.$$

In this case $\mathcal{H}^1_{\mathsf{C}_0} \cong \ker \mathfrak{D}_{A_0}$. Since $\text{coker}\,\mathfrak{D}_{A_0} = \mathcal{H}^2_{\mathsf{C}_0} = 0$ we deduce

$$\dim_{\mathbb{C}} \mathfrak{D}_{A_0} = \text{ind}_{\mathbb{C}} \mathfrak{D}_{A_0} = \frac{1}{8}(c_\sigma^2 - \tau(M)) = \frac{1}{8}\Theta(q_M) = \frac{d(\sigma) + 1}{2}.$$

Thus, if $d(\sigma) = 1$ near $\mathsf{C}_0$ the moduli space is homeomorphic to the segment $[0, 1)$ while if $d(\sigma) > 1$ it looks like a cone over $\pm\mathbb{CP}^{\frac{d(\sigma)-1}{2}}$.

If we chop out a small neighborhood of $\mathsf{C}_0$ in $\mathfrak{M}_\sigma(g, \eta)$ we obtain a *smooth, compact, orientable* manifold $X$ with boundary $\pm\mathbb{CP}^{\frac{d(\sigma)-1}{2}}$.

If $d(\sigma) = 1$ then $X$ is a smooth, compact oriented one-dimensional manifold with boundary consisting of only one component. This is plainly impossible.

If $d(\sigma) > 1$ observe that the restriction of the universal line bundle $\mathbb{U}_\sigma$ to $\partial X$ is $\pm$ the tautological line bundle over $\pm\mathbb{CP}^{\frac{d(\sigma)-1}{2}}$ and thus is nontrivial. More precisely ($\Omega_\sigma = c_1(\mathbb{U}_\sigma)$)

$$\int_{\partial X} \Omega_\sigma^{\frac{d(\sigma)-1}{2}} = \pm 1.$$

The last equality is impossible since $\mathbb{U}_\sigma$ extends over $X$ and by Stokes' theorem we have

$$\int_{\partial X} \Omega_\sigma^{\frac{d(\sigma)-1}{2}} = \int_X d\Omega_\sigma^{\frac{d(\sigma)-1}{2}} = 0.$$

This contradiction completes the proof of Theorem 2.4.18 in the case $b_1(M) = 0$.

• $b_1(M) > 0$. We will reduce this case to the previous situation by a simple topological trick.

Choose a basis $c_1, \cdots, c_{b_1}$ of $H_1(M, \mathbb{Z})/\mathrm{Tors}$ and represent each of these cycles by smoothly embedded $S^1$'s. We can "kill" the homology class carried by each of these cycles by surgery (see [**51**]). This operation can be briefly described as follows.

Observe first that a tubular neighborhood $N$ of a smoothly embedded $S^1 \hookrightarrow M$ is diffeomorphic to $D^3 \times S^1$ where $D^k$ denotes the unit ball in $\mathbb{R}^k$. Fix such a diffeomorphism so that $\partial N \cong S^2 \times S^1$. Now remove $N$ to obtain a manifold with boundary $S^2 \times S^1$ to which we attach the handlebody $H = S^2 \times D^2$ (which has $\partial H = S^2 \times S^1$). This operation will kill each of the classes $c_i$ but will not affect $H_2/\mathrm{Tors}$ and the intersection form of $M$ since the classes $c_i$ are not torsion classes (use the Poincaré duals of $c_i$ to see this). In the end we obtain a smooth manifold with the same intersection form but with $b_1 = 0$. This places us in the previous situation. The proof of Theorem 2.4.18 is now complete. ∎

**Exercise 2.4.2.** Prove that the above sequence of surgeries does not affect the intersection form, as claimed.

**Remark 2.4.19.** Donaldson's theorem states that a smooth, simply connected, negative definite 4-manifold $X$ cannot be too complicated arithmetically: its intersection form is the simplest possible.

If we remove the negativity assumption, so that the intersection form $q_X$ is indefinite, then $q_X$ has a much simpler from. If $X$ is not spin then $q_X$

is odd and thus diagonal.[1] If $X$ is spin then $q_X$ is even and thus it has the form

$$q_X = aE_8 + bH, \quad a = \frac{1}{8}\tau(q), \quad 8|a| + 2b = \text{rank}\,(q).$$

In this case the integers $(a, b)$, $b > 0$, represent a measure of the complexity of $q_X$. Rohlin's theorem states there are restrictions on $(a, b)$. More precisely, $a$ must be an even integer. The celebrated 11/8-th conjecture states that there are even more drastic restrictions in this case, more precisely

$$11|a| = \frac{11}{8}|\tau(q)| \leq \text{rank}\,(q_X) = 8|a| + 2b.$$

This inequality is optimal because equality is achieved when $X$ is the $K3$ surface (see the next chapter). Using Seiberg-Witten theory M. Furuta has proved a 10/8-th theorem (see [**45**], or the simpler approach in [**19**]). More precisely, he showed that

$$10|a| + 1 \leq \text{rank}\,(q) = 8|a| + 2b.$$

[1]The example $m\mathbb{CP}^2\#n\overline{\mathbb{CP}}^2$ shows that any odd form is the intersection form of a smooth, s.c. 4-manifold.

*Chapter 3*

# Seiberg-Witten Equations on Complex Surfaces

*Anybody who is not shocked by this subject has failed to understand it.*

Niels Bohr

The Seiberg-Witten equations are very sensitive to the background geometry. In this chapter we study some of the effects a complex structure has on the Seiberg-Witten equations and, in particular, on the Seiberg-Witten invariants.

We will see that, very often, the complex structure leads to information so detailed about monopoles that we will be able to *explicitly* describe all of them and, in particular, count them.

## 3.1. A short trip in complex geometry

This section surveys some basic facts of complex geometry which are absolutely necessary in our study of monopoles. This survey is by no means complete or balanced but it is targeted to the applications we have in mind. It should motivate the reader not familiar with this subject to consult the references [**9, 10, 39, 49, 54, 59**] which served as sources of inspiration.

**3.1.1. Basic notions.** Suppose $M$ is a *compact* complex $n$-dimensional manifold without boundary and $E \to M$ is a holomorphic vector bundle as defined in Section 1.4. We denote by $\mathcal{O}_M(E)$ the sheaf of local holomorphic sections of $E$, by $\mathcal{O}_E^p$ the sheaf of holomorphic local sections of $\Lambda^{p,0}T^*M \otimes E$ and by $H^*(M, \mathcal{O}_M^p(E))$ the Čech cohomology of the sheaf $\mathcal{O}_M^p(E)$. When $p = 0$ we will write $H^q(M, E)$ instead of $H^q(M, \mathcal{O}_M(E))$ and when $E$ is the trivial holomorphic line bundle we will drop $E$ from the notation.

A *divisor* on $M$ is intuitively a codimension-1 complex subvariety. More rigorously a divisor is defined by an open cover $(U_\alpha)$ of $M$ and *nontrivial* meromorphic functions $f_\alpha : U_\alpha \dashrightarrow \mathbb{C}$ (i.e. holomorphic maps $f_\alpha : U_\alpha \to \mathbb{CP}^1$) such that $f_\alpha/f_\beta$ is a *nowhere vanishing holomorphic function on* $U_{\alpha\beta}$. The loci $\operatorname{ord}(f_\alpha) := f_\alpha^{-1}(\{0, \infty\})$ patch-up to a codimension-1 subvariety in $M$ called the *support* of the divisor and denoted by $\operatorname{supp}(D)$.

We consider two descriptions $(U_\alpha, f_\alpha)$ and $(V_a, g_a)$ to be equivalent if there is a cover $(W_i)$ finer then both covers $(U_\bullet)$ and $(V_\bullet)$ with the following property. For every $i, \alpha, a$ such that $W_i \subset U_\alpha \cap V_a$ there exists a nowhere vanishing holomorphic function $h : W_i \to \mathbb{C}$ so that $f_\alpha = h \cdot g_a$. We denote by $\operatorname{Div}(M)$ the space of divisors on $M$.

The previous definition captures the subtle notion of multiplicity. For example, if the divisor $D$ is given by the collection $(f_\alpha)$ then the collection $(f_\alpha^2)$ defines (in general a different) divisor, denoted by $2D$, which has identical support.

A divisor described by the cover of $M$ by itself and a (nontrivial) meromorphic function $f : M \dashrightarrow \mathbb{C}$ is called *principal.* We will denote this divisor by $(f)$ and by $\operatorname{PDiv}(M)$ the subspace of principal divisors.

If $D$ is a divisor given by a collection $(U_\alpha, f_\alpha)$ then we can regard the collection of holomorphic functions

$$g_{\beta\alpha} = f_\beta/f_\alpha : U_{\alpha\beta} \to \mathbb{C}^*$$

as a gluing cocycle for a holomorphic line bundle over $M$. Two equivalent descriptions of the divisor $D$ lead to isomorphic line bundles. We will denote this isomorphism class by $[D]$. With this interpretation, we can regard the collection $(f_\alpha)$ as a meromorphic section $f_D$ of $[D]$. Two equivalent descriptions lead to meromorphic sections which differ by a nonzero multiplicative constant. We see that the converse statement is true: any divisor can be viewed as described by a meromorphic section of a holomorphic line bundle.

We can define an operation on $\operatorname{Div}(M)$ as follows. If $D_i$, $i = 1, 2$, are divisors given by the same cover $(U_\alpha)$ (this can always be arranged by passing to finer covers) and meromorphic functions $f_{\alpha,i} : U_\alpha \dashrightarrow \mathbb{C}$ then $D_1 + D_2$ is the divisor given by the cover $U_\alpha$ and functions $f_{\alpha,1} f_{\alpha,2}$. We let the reader check that $(\operatorname{Div}(M), +)$ is an Abelian *group.*

One can give a more geometric description of the notion of divisor. First define a hypersurface of $M$ to be a *closed* subset $V$ locally defined as the zero set of a holomorphic function. A hypersurface may or may not be a smooth

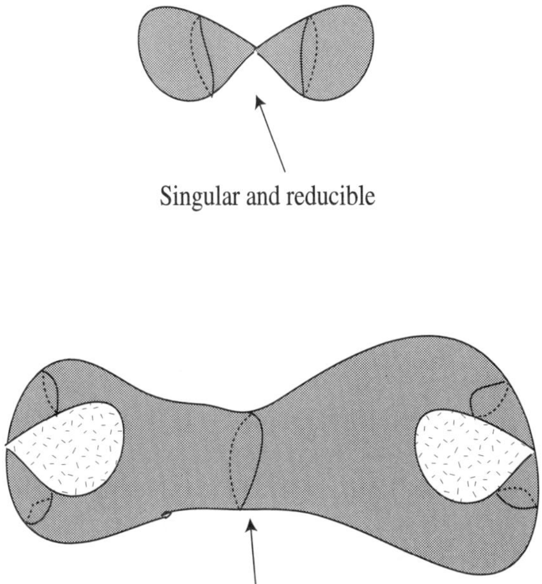


**Figure 3.1.** *Singular hypersurfaces*

manifold. A point $p$ on a hypersurface $V$ is called *smooth* if there exists a holomorphic function $f$ defined in a neighborhood $U$ of $p$ such $df(p) \neq 0$ and $U \cap V = f^{-1}(0)$. We denote by $V^*$ the set of smooth points of $V$. $V$ is said to be irreducible if $V^*$ is *connected* (see Figure 3.1).

Let us point out a subtlety of this definition. The line $z_2 = 0$ in $\mathbb{C}^2$ can be defined by many equations: $z_2 = 0$, $z_2^3 = 0$ etc. These equations define different divisors. The origin $(0,0)$ is not a smooth point for the defining equation $z_2^3 = 0$ but according to the definition it is a smooth point of this hypersurface since *there exists* a defining equation, $z_2 = 0$, for which the origin is a smooth point. In modern language, when we think of a hypersurface as a subscheme, we assume it is reduced. In less rigorous terms, we do not consider defining equations of the type

$$f^n = 0.$$

We will always *"reduce"* them to $f = 0$. For more details we refer to [**31, 49**]. The hypersurfaces behave in many respects like smooth submanifolds: the compact ones carry nontrivial homology classes and have finite volume. Moreover, we have the following important fact ([**75**]) .

**Proposition 3.1.1.** *Suppose $V$ is a hypersurface in a compact Kähler manifold $M$ of complex dimension $n$. Then $V$ defines a nontrivial homology class* *in* $H_{2n-2}(M, \mathbb{Z})$ *which is not torsion and, moreover,*

$$\langle \omega^{n-1}, V \rangle = \int_V \omega^{n-1} = (n-1)! \operatorname{vol}(V) > 0.$$

Putting together the (reduced) local equations of $V$ we obtain a divisor on $M$ which we continue to denote by $V$. We have the following result (see [**49**]).

**Proposition 3.1.2.** *The group* $\operatorname{Div}(M)$ *is isomorphic to the free abelian group generated by the irreducible hypersurfaces in* $M$.

Thus we can think of a divisor as a collection of irreducible hypersurfaces with attached multiplicities. The divisors on a curve (complex dimension 1) are finite collections of points with multiplicities while on a surface the divisors are finite collections of curves with multiplicities. (A *curve* on a surface is by definition an irreducible hypersurface.)

If $f : M \dashrightarrow \mathbb{C}$ is a meromorphic function then the divisor associated to the hypersurface $f^{-1}(0)$ (resp. $f^{-1}(\infty)$) is called the zero divisor (resp. the polar divisor) of $f$ and is denoted by $(f)_0$ (resp. $(f)_\infty$). The difference $(f) := (f)_0 - (f)_\infty$ is called the divisor determined by $f$. All principal divisors have the form $(f)$ for some meromorphic function $f$.

Two divisors $D_1$ and $D_2$ are said to be *linearly equivalent*, and we write this $D_1 \sim D_2$, if the corresponding holomorphic line bundles $[D_1]$ and $[D_2]$ are isomorphic. We let the reader check that this agrees with the classical definition $D_1 \sim D_2 \Leftrightarrow D_1 - D_2 \in \operatorname{PDiv}(M)$.

If we introduce the *Picard group* $\operatorname{Pic}(M)$ of isomorphism classes of holomorphic line bundles over $M$ we see that we have constructed an *injective* morphism of Abelian groups

$$\operatorname{Div}(M)/\operatorname{PDiv}(M) \to \operatorname{Pic}(M).$$

For a proof of the following result we refer to [**49**].

**Proposition 3.1.3.** *If* $M$ *is* algebraic, *i.e. it is a complex submanifold of a projective space* $\mathbb{CP}^N$ *then the morphism*

$$\operatorname{Div}(M)/\operatorname{PDiv}(M) \to \operatorname{Pic}(M)$$

*is an* isomorphism.

The elements of $\operatorname{Pic}(M)$ are described by holomorphic gluing cocycles and thus can be identified with the Čech cohomology group $H^1(M, \mathcal{O}^*)$ where $\mathcal{O}^*$ denotes the *multiplicative* sheaf of nowhere vanishing holomorphic functions. The short exact sequence of sheaves

$$0 \to \mathbb{Z} \to \mathcal{O} \to \mathcal{O}^* \to 0$$

leads to a long exact sequence

$$\cdots \to \operatorname{Pic}(M) \cong H^1(M, \mathcal{O}^*) \xrightarrow{\delta} H^2(M, \mathbb{Z}) \to \cdots.$$

For any holomorphic line bundle $L$ the class $\delta(L)$ is precisely the topological first Chern class $c_1(L)$.

A divisor $D$ is called *effective* (and we write this $D \geq 0$) if the corresponding section $f_D$ of $[D]$ is *holomorphic.* Equivalently, this means that $D$ is described by an open cover $(U_\alpha)$ and *holomorphic* functions $f_\alpha : U_\alpha \to \mathbb{C}$. Any effective divisor can be written as a sum $\sum_i n_i V_i$ where $n_i$ are *nonnegative* integers and $V_i$ are divisors associated to irreducible hypersurfaces.

**Example 3.1.4.** Suppose $V$ is a hypersurface. Continue to denote by $V$ the homology class in $H_{2n-2}(M, \mathbb{Z})$ determined by $V$. The divisor $V$ canonically defines a holomorphic section $f_V$ of $[V]$ satisfying $(f_V) = (f_V)_0 = V$. The Gauss-Bonnet-Chern theorem shows that the homology class carried by $V$ is the Poincaré dual of $c_1([V])$. That is why when no confusion is possible we will simultaneously denote by $V$ both the line bundle $[V]$ and the cohomology class $c_1([V])$.

For any divisor $D$ on $M$ we denote by $\mathcal{L}(D)$ the space of meromorphic functions $f$ such that $(f) + D \geq 0$. (By definition the identically zero function is included in $\mathcal{L}(D)$.) Observe that we have a map

$$i_D : \mathcal{L}(D) \to H^0(M, [D]) (= \text{the space of holomorphic sections of } [D])$$

described by

$$f \mapsto f \cdot f_D.$$

This map is injective, on account of the unique continuation principle. It is also surjective because for every holomorphic section $s$ of $[D]$ the ratio $s/f_D$, defined in the obvious way, is a meromorphic section of the trivial line bundle (hence a meromorphic function). Now observe that

$$(s/f_D) + D = (s) - (f_D) + D = (s) \geq 0.$$

We denote by $|D|$ the projective space $\mathbb{P}(\mathcal{L}(D))$. Equivalently,

$$|D| = \mathbb{P}(\,H^0(M, [D])\,).$$

$|D|$ is called the *complete linear system* generated by $D$. A projective subspace of $|D|$ is called a *linear system.* A linear system of dimension 1 is called a *pencil.* The complete linear system can be geometrically described as the space of effective divisors linearly equivalent to $D$. The *base locus* of a linear system $L \subset |D|$ consists of all points $p \in M$ which belong to the supports of all divisors in $L$. Equivalently, if we think of $L$ as a subspace of $\mathbb{P}(\,H^0(M, [D])\,)$ then the base locus is the intersection of the zero loci of the sections in $L$. We will denote the base locus by $B(L)$.

Any point $p \in M \setminus B(L)$ defines a hyperplane $H_p$ in $L$ consisting of the divisors containing $p$, or equivalently, of the holomorphic sections in $L$ which vanish at $p$. The correspondence $p \to H_p$ defines a holomorphic map

$$i_L : M \setminus B(L) \to L^* = \text{the dual of the projective space } L.$$

**Definition 3.1.5.** A divisor $D$ on a complex manifold $M$ is called *very ample* if $B(|D|) = \emptyset$ and the map $i_{|D|} : M \to |D|^*$ is an embedding. $D$ is called *ample* if $kD$ is very ample for $k \gg 0$.

**Example 3.1.6.** Consider a hyperplane $H$ in $\mathbb{CP}^N$. Its associated line bundle $[H]$ is the dual of the tautological line bundle. For every positive integer $d$, the holomorphic sections of $d[H]$ can be viewed as homogeneous complex polynomials of degree $d$ in $N+1$ variables. Thus

$$\dim H^0(M, d[H]) = \binom{d+N}{d}$$

so that

$$\dim |dH| = \binom{d+N}{d} - 1.$$

We can construct a pencil in $|dH|$ by choosing two linearly independent homogeneous polynomials $A$, $B$ of degree $d$. The pencil is the projective line $L$ defined by the linear space

$$\{\alpha A + \beta B;\ \ \alpha, \beta \in \mathbb{C}\}.$$

The pair $[\alpha : \beta]$ defines projective coordinates on $L^*$. The base locus is the variety

$$A^{-1}(0) \cap B^{-1}(0) \subset \mathbb{CP}^N.$$

The map

$$i_L : \mathbb{CP}^N \setminus B(L) \to \mathbb{CP}^1$$

is described explicitly as follows: $i_L(p) = [\alpha : \beta]$ if and only if $\alpha A(p) + \beta B(p) = 0$. We can visualize the pencil as a "fibration" $\mathbb{CP}^N \dashrightarrow \mathbb{CP}^1$.

Suppose $V$ is a codimension-1 submanifold of $M$. The associated holomorphic section $f_V$ of $[V]$ vanishes in a nondegenerate fashion precisely along $V$. If $\nabla$ is a connection on $[V]$ then we get an adjunction map

$$\mathfrak{a} : TM\,|_V \to [V]\,|_V, \quad X \mapsto \nabla_X f_V$$

vanishing precisely along the tangent bundle of $V$ because $f_V$ is nondegenerate so that $\mathfrak{a}$ induces an isomorphism of *real* bundles

$$\mathfrak{a} : N_V \to [V]\,|_V$$

where $N_V$ denotes the normal bundle to $V \hookrightarrow M$. Since $f_V$ is *holomorphic* the adjunction map preserves the complex structures so that we have an isomorphism of *holomorphic line bundles*

$$[V]\,|_V \cong N_V. \tag{3.1.1}$$

We can now rewrite the adjunction formula of §2.4.2 as

$$K_V \cong (K_M \otimes [V])\,|_V\,. \tag{3.1.2}$$

where $K_M$ denotes the canonical line bundle of $M$, $K_M = \Lambda^{n,0}T^*M$.

A large amount of information about the embedding $V \hookrightarrow M$ is contained in the following *structural short exact sequence*:

$$0 \to \mathcal{O}_M \stackrel{f_V\otimes}{\longrightarrow} \mathcal{O}_M([V]) \stackrel{r}{\to} \mathcal{O}_V([V]\,|_V) \to 0$$

where the last arrow is the restriction map. If $L$ is a holomorphic line bundle we can take the tensor product of the above sequence with the line bundle $L \otimes [-V]$ and we obtain

$$0 \to \mathcal{O}_M(L \otimes [-V]) \stackrel{f_V\otimes}{\longrightarrow} \mathcal{O}_M(L) \stackrel{r}{\to} \mathcal{O}_V(L\,|_V) \to 0 \tag{3.1.3}$$

As in Sec. 1.4 set

$$\Omega^{p,q}(E) := C^\infty(\Lambda^{p,q}T^*M \otimes E).$$

We can form the *Dolbeault complex*

$$0 \to \Omega^{p,0}(E) \stackrel{\bar\partial_E}{\longrightarrow} \Omega^{p,1}(E) \stackrel{\bar\partial_E}{\longrightarrow} \cdots \stackrel{\bar\partial_E}{\longrightarrow} \Omega^{p,n}(E) \to 0$$

whose cohomology is denoted by $H^{p,*}_{\bar\partial}(M, E)$.

**Theorem 3.1.7. (Dolbeault)** *There exist natural isomorphisms*

$$H^q(M, \mathcal{O}^p_M(E)) \cong H^{p,q}_{\bar\partial}(M, E), \quad q = 0, 1, \cdots, n.$$

Fix a Hermitian metric $g = g_M$ on $TM$ and a Hermitian metric $h = h_E$ on $E$. Then we can form the formal adjoints of the operators

$$\bar\partial_E : \Omega^{p,q}(E) \to \Omega^{p,q+1}(E).$$

The formal adjoint can be explicitly described in terms of the *conjugate linear* Hodge operator

$$*_E : \Omega^{p,q}(E) \to \Omega^{n-p,n-q}(E^*)$$

defined as in (1.4.20) of §1.4.2. More precisely we have (see [**49**])

$$\bar\partial_E^* = -\, *_E\, \bar\partial_{E^*}\, *_E\,.$$

We can form the Laplacian

$$\Delta_{\bar\partial} := \Delta_{\bar\partial_E} := \bar\partial_E \bar\partial_E^* + \bar\partial_E^* \bar\partial_E.$$

Since $\bar{\partial}_E^2 = (\bar{\partial}_E^*)^2 = 0$ we have

$$\Delta_{\bar{\partial}} = (\bar{\partial}_E + \bar{\partial}_E^*)^2$$

and a simple integration by parts shows that

$$\Delta_{\bar{\partial}}\omega = 0 \Leftrightarrow \bar{\partial}_E\omega = \bar{\partial}_E^*\omega = 0, \quad \omega \in \Omega^{*,*}(M).$$

A differential form satisfying one of the equivalent conditions above is called $\bar{\partial}$-harmonic. We will denote by $\mathbf{H}_{\bar{\partial}}^{p,q}(M,E)$ the space of $\bar{\partial}$-harmonic $E$-valued $(p,q)$-forms. We want to emphasize that this space *depends* on the metrics $g_M$ and $g_E$. However, its dimension depends *only on the complex structure of M!* More precisely, we have the following important result.

**Theorem 3.1.8. (Hodge)** *All the spaces* $\mathbf{H}_{\bar{\partial}}^{p,q}(M,E)$ *are finite-dimensional and the natural maps*

$$\mathbf{H}_{\bar{\partial}}^{p,q}(M,E) \to H_{\bar{\partial}}^{p,q}(M,E)$$

*are isomorphisms. In particular, the space of holomorphic global sections of* $E$ *is finite-dimensional since it is isomorphic to* $H_{\bar{\partial}}^{0,0}(M,E)$.

We set

$$h^{p,q}(E) = h_M^{p,q}(E) := \dim_{\mathbb{C}} \mathbf{H}_{\bar{\partial}}^{p,q}(M,E), \;\; h^p(E) := \dim_{\mathbb{C}} \mathbf{H}_{\bar{\partial}}^{0,p}(M,E)$$

and

$$\chi_p(E) := \sum_q (-1)^q h_M^{p,q}(E).$$

When $p=0$ we write $\chi(E)$ instead of $\chi_0(E)$. When $E$ is the trivial holomorphic line bundle, we write $h_M^{p,q}$ instead of $h_M^{p,q}(E)$ and we set

$$\chi_{hol}(M) := \chi_0(M,E) = \sum_{q=0}^n (-1)^q h_M^{0,q}.$$

The integer $h_M^{0,1}$ is denoted by $q(M)$ and is called the *irregularity.* The integer $(-1)^n(\chi_{hol}(M) - 1)$ is called the *arithmetic genus* and is denoted by $p_a(M)$.

The numbers $P_k(M) = h^0(M, K_M^k)$ are called the *plurigenera* of $M$. $P_1(M)$ is usually denoted by $p_g(M)$ and is called the *geometric genus* of $M$.

Observe that

$$p_g(M) = h^{n,0}(M).$$

**Theorem 3.1.9. (Riemann-Roch-Hirzebruch)**

$$\chi(E) = \int_M \mathbf{td}(M) \wedge \mathbf{ch}(E)$$

*where* $\mathbf{td}(M)$ *denotes the Todd class of the complex bundle* $TM$ *while* $\mathbf{ch}(E)$ *denotes the Chern character of* $E$.

In the above integral only the degree $2n$ part of the nonhomogeneous form $\mathbf{td}(M) \wedge \mathbf{ch}(E)$ is relevant. We present a few examples particularly important in the sequel. We consider only the case when $E$ is a complex line bundle. We will use additive notation for the tensor products and the duals of line bundles and we will frequently identify a line bundle with its (topological) Chern class or its Poincaré dual.

• $\dim_{\mathbb{C}} M = 1$. Thus $M$ is a Riemann surface of genus $g$. Then

$$\mathbf{td}(M) = 1 + \frac{1}{2}c_1(M) = 1 - \frac{1}{2}K_M, \ \ \mathbf{ch}(E) = 1 + c_1(E)$$

so that

$$\chi_0(M, E) = \int_M c_1(E) + \frac{1}{2}\int_M c_1(M).$$

The first integral is an integer called the *degree of* $E$ and denoted by $\deg E$ and the second integral is equal to $(2 - 2g)$ by the Gauss-Bonnet theorem. We conclude

$$\chi_0(M, E) = \deg E + 1 - g. \tag{3.1.4}$$

• $\dim_{\mathbb{C}} M = 2$. In this case

$$\mathbf{td}(M) = 1 + \frac{1}{2}c_1(M) + \frac{1}{12}(c_1(M)^2 + c_2(M)),$$

$$\mathbf{ch}(E) = 1 + c_1(E) + \frac{1}{2}c_1(E)^2.$$

Identifying $c_i(M)$ with $-K_M$ and $c_1(E)$ with $E$ we deduce

$$\chi_0(M, E) := \frac{1}{2}E(E - K_M) + \frac{1}{12}\int_M c_1(M)^2 + c_2(E).$$

Using the Gauss-Bonnet-Chern formula

$$\int_M c_2(M) = \chi_M (= \text{Euler characteristic of } \ M),$$

the Hirzebruch signature formula

$$\tau_M = \frac{1}{3}\int_M p_1(M)$$

and the universal identity

$$p_1(M) = c_1(M)^2 - 2c_2(M),$$

we conclude that

$$K_M^2 = 2\chi_M + 3\tau_M \tag{3.1.5}$$

and

$$\chi_0(M, E) = \frac{1}{2}E(E - K_M) + \chi_{hol}(M)$$

$$(3.1.6)\quad = \frac{1}{2}E(E-K_M) + \frac{1}{12}(K^2+\chi_M) = \frac{1}{2}E(E-K_M) + \frac{1}{4}(\chi_M+\tau_M).$$

(Above, the multiplication denotes the intersection pairing on the 4-manifold $M$.)

Observe that there is a natural, *complex bilinear* pairing

$$\langle \bullet, \bullet \rangle : \Omega^{0,q}(E) \times \Omega^{n,n-q}(E^*) \to \mathbb{C}$$

defined by

$$\langle \eta \otimes u, \omega \otimes v \rangle = \int_M \langle u, v \rangle \eta \wedge \omega,$$

$\forall u \in C^\infty(E)$, $v \in C^\infty(E^*)$, $\eta \in \Omega^{0,q}(M)$, $\omega \in \Omega^{n,n-q}(M)$. The above pairing can be regarded as a pairing

$$\langle \bullet, \bullet \rangle : \Omega^{0,q}(E) \times \Omega^{0,n-q}(K_M \otimes E^*) \to \mathbb{C}.$$

Clearly this map induces a bilinear pairing

$$(3.1.7)\qquad \langle \bullet, \bullet \rangle : \mathbf{H}^{0,q}_{\bar\partial}(M,E) \times \mathbf{H}^{0,n-q}_{\bar\partial}(M, K_M \otimes E^*) \to \mathbb{C}$$

and thus natural complex linear maps

$$(3.1.8)\qquad \begin{cases} \mathbf{H}^{0,q}_{\bar\partial}(M,E) \to \mathbf{H}^{0,n-q}_{\bar\partial}(M, K_M \otimes E^*)^* \\ \\ \mathbf{H}^{0,n-q}_{\bar\partial}(M, K_M \otimes E^*) \to \mathbf{H}^{0,q}_{\bar\partial}(M,E)^*. \end{cases}$$

**Theorem 3.1.10. (Serre duality)** *The pairing (3.1.7) is a duality, i.e. the natural maps (3.1.8) are isomorphisms.*

Using the natural metric on $\mathbf{H}^{p,q}_{\bar\partial}$ to identify

$$\overline{\mathbf{H}^{0,n-q}_{\bar\partial}(M, K_M \otimes E^*)} \cong \mathbf{H}^{0,n-q}_{\bar\partial}(M, K_M \otimes E^*)^*,$$

$$\overline{\mathbf{H}^{0,q}_{\bar\partial}(M,E)} \cong \mathbf{H}^{0,q}_{\bar\partial}(M,E)^*$$

we observe that the maps in (3.1.8) are precisely the complex linear maps induced by $*_E$,

$$*_E : \mathbf{H}^{0,q}_{\bar\partial}(M,E) \to \overline{\mathbf{H}^{0,n-q}_{\bar\partial}(M, K_M \otimes E^*)} \text{ etc.}$$

Observe that Serre duality implies

$$(3.1.9)\qquad h^{0,q}_M(E) = h^{0,n-q}_M(K_M \otimes E^*).$$

If $E$ is the trivial line bundle the above equality becomes

$$(3.1.10)\qquad h^{0,q}_M = h^{0,n-q}_M(K_M) = h^{n,n-q}_M$$

and in particular

$$p_g(M) = h^{n,0}_M = h^{0,n}_M.$$

Instead of the Cauchy-Riemann operators $\bar{\partial}_E : \Omega^{p,q}(E) \to \Omega^{p,q+1}(E)$ we can use their conjugates

$$\partial_E : \Omega^{p,q}(E) \to \Omega^{p+1,q}(E).$$

We can form similar complexes

$$0 \to \Omega^{0,q}(E) \xrightarrow{\partial_E} \Omega^{1,q}(E) \xrightarrow{\partial_E} \cdots \xrightarrow{\partial_E} \Omega^{n,q}(E) \to 0.$$

Their cohomology spaces are denoted by $H^{p,q}_{\partial}(M, E)$. Again, by choosing Hermitian metrics on $TM$ and $E$ we can form the Laplacian

$$\Delta_{\partial_E} = \partial_E \partial_E^* + \partial_E^* \partial_E = (\partial_E + \partial_E^*)^2$$

whose kernel we denote by $\mathbf{H}^{p,q}_{\partial}(M, E)$.

*In the remainder of this section we will assume the metric on $TM$ is Kähler unless otherwise indicated.*

Assume $E$ is the trivial line bundle equipped with the trivial Hermitian metric. Using the Kähler identities of Sec. 1.4 we deduce

$$\Delta_{\partial} = \Delta_{\bar{\partial}} \text{ on } \Omega^{p,q}(M)$$

which implies

$$\mathbf{H}^{p,q}_{\bar{\partial}}(M) = \mathbf{H}^{p,q}_{\partial}(M) = \overline{\mathbf{H}^{q,p}_{\bar{\partial}}(M)}$$

so that

$$h^{p,q}_M = h^{q,p}_M, \ \forall p, q. \tag{3.1.11}$$

If $\Delta_d$ denotes the Hodge-deRham Laplacian on (complex valued) forms on $M$ then

$$\frac{1}{2}\Delta_d = \Delta_{\bar{\partial}}$$

so that any $\bar{\partial}$-harmonic $(p, q)$-form on $M$ is also a $d$-harmonic form of degree $(p + q)$. This implies

$$\mathbf{H}^k_d(M) \otimes \mathbb{C} = \bigoplus_{p+q=k} \mathbf{H}^{p,q}_{\bar{\partial}}(M). \tag{3.1.12}$$

If $b_k(M)$ denotes the $k$-th Betti number of $M$ then the last identity implies

$$b_k(M) = \sum_{p+q=k} h^{p,q}_M. \tag{3.1.13}$$

The identities (3.1.10) and (3.1.11) lead to the *Hodge diamond* of a Kähler manifold. We describe it only in the case $\dim_{\mathbb{C}} M = 2$.

$$\begin{array}{ccccc} & & h^{0,0} & & \\ & h^{0,1} & \vdots & h^{1,0} & \\ h^{0,2} & \cdots & h^{1,1} & \cdots & h^{2,0} \\ & h^{2,1} & \vdots & h^{1,2} & \\ & & h^{2,2} & & \end{array}$$

The above configuration is symmetric with respect to the two diagonals, vertical and horizontal.

The Kähler identities discussed in Sec. 1.4 introduce additional, finer structure on the spaces $\mathbf{H}^{p,q}_{\bar{\partial}}(M)$. Instead of discussing the general situation, presented beautifully in [**54**], we will consider only the case of interest to us, namely $\dim_{\mathbb{C}} M = 2$.

Fix a point $p \in M$. Since $M$ is Kähler we can choose normal coordinates $(x^1, y^1, x^2, y^2)$ near $p$ so that $d\bar{z}^i = dx^i - \mathbf{i}dy^i$ form a local *holomorphic* frame of $\Lambda^{0,1}T^*_pM$. Denote by $\omega$ the symplectic form determined by the Kähler metric $g = g_M$, i.e.

$$\omega(X, Y) = -\mathfrak{Im}\, g(X, Y), \quad X, Y \in \mathrm{Vect}\,(M).$$

As shown in Example 1.3.3 the range of the restriction map

$$\mathbf{H}^2_+(M, \mathbb{R}) \otimes \mathbb{C} \to \Lambda^2 T^*_pM \otimes \mathbb{C}$$

is contained in the subspace $\mathbb{C}\omega_p \oplus \Lambda^{2,0}T^*_pM \oplus \Lambda^{0,2}T^*_pM$ while the range of the restriction map

$$\mathbf{H}^2_-(M, \mathbb{R}) \otimes \mathbb{C} \to \Lambda^2 T^*_pM$$

is contained in the orthogonal complement of $\mathbb{C}\omega_p$ in $\Lambda^{1,1}T^*_pM$. This orthogonal complement can be defined as the kernel of the contraction map (the dual of $L$ - the exterior multiplication by $\omega$)

$$\Lambda_p : \Lambda^{1,1}T^*_pM \to \Lambda^0 T^*_pM.$$

The Kähler identities in Sec. 1.4 show that the direct sum

$$\bigoplus_{p,q} \mathbf{H}^{p,q}_{\bar{\partial}}(M)$$

is an invariant subspace of $\Lambda$ so that these pointwise inclusions lead to global ones

$$\mathbf{H}^2_-(M, \mathbb{R}) \otimes \mathbb{C} \subset \mathbf{H}^{1,1}_{\omega}(M) := \ker(\ \Lambda : \mathbf{H}^{1,1}_{\bar{\partial}}(M) \to \mathbf{H}^{0,0}_{\bar{\partial}}(M)\ )$$

and

$$\mathbf{H}^2_+(M, \mathbb{R}) \otimes \mathbb{C} \subset L\mathbf{H}^{0,0}_{\bar{\partial}}(M) \oplus \mathbf{H}^{2,0}_{\bar{\partial}}(M) \oplus \mathbf{H}^{0,2}_{\bar{\partial}}(M)$$

$$= \mathbb{C}\omega \oplus \mathbf{H}^{2,0}_{\bar\partial}(M) \oplus \mathbf{H}^{0,2}_{\bar\partial}(M).$$

From the identity

$$\mathbf{H}^2(M,\mathbb{R}) \otimes \mathbb{C} = \mathbb{C}\omega \oplus \mathbf{H}^{1,1}_{\omega}(M) \oplus \mathbf{H}^{2,0}_{\bar\partial}(M) \oplus \mathbf{H}^{0,2}_{\bar\partial}(M)$$

we deduce that the above inclusions are equalities:

$$\mathbf{H}^2_+(M,\mathbb{R}) \otimes \mathbb{C} = \mathbb{C}\omega \oplus \mathbf{H}^{2,0}_{\bar\partial}(M) \oplus \mathbf{H}^{0,2}_{\bar\partial}(M), \tag{3.1.14}$$

$$\mathbf{H}^2_-(M,\mathbb{R}) \otimes \mathbb{C} = \mathbf{H}^{1,1}_{\omega}(M). \tag{3.1.15}$$

Observing that $p_g(M) = h^{2,0}(M) = h^{0,2}(M)$ we deduce from (3.1.14) that

$$b_2^+(M) = 2p_g(M) + 1. \tag{3.1.16}$$

The identities (3.1.14), (3.1.15) have another important consequence. Observe that the space of $(1,1)$-forms is invariant under conjugation and we can speak of real, harmonic $(1,1)$-forms.

**Corollary 3.1.11. (Hodge index theorem)** *The restriction of the intersection pairing on the space of real, harmonic $(1,1)$-forms on a Kähler surface has signature $(1, b_2^-)$.*

In the case of algebraic surfaces the Hodge index theorem can be formulated equivalently in more geometric terms.

According to the results of §1.4.2, given a Hermitian line bundle $L \to M$, we can describe the holomorphic structures on $L$ in terms of Hermitian connections $A$ such that $F_A^{2,0} = F_A^{0,2} = 0$. Thus the first Chern class of a holomorphic line bundle over a Kähler surface is a real $(1,1)$-class.

On the other hand, if $M$ is also algebraic then the holomorphic line bundles can also be described in terms of divisors, so that we have a map

$$\operatorname{Div}(M) \to \mathbf{H}^{1,1}_{\bar\partial}(M)_{\mathbb{R}}, \quad D \mapsto c_1([D]). \tag{3.1.17}$$

Suppose now that $c \in H^2(M,\mathbb{Z})$ is such that its harmonic part lies in $\mathbf{H}^{1,1}_{\bar\partial}(M)$. Then there exists a Hermitian line bundle $L \to M$ such that $c_1^{top}(L) = c$. Now we can find a Hermitian connection on $L$ whose curvature is harmonic and thus must be a $(1,1)$-class. This shows that the image of the map (3.1.17) is the lattice $\mathbf{H}^{1,1}_{\bar\partial}(M) \cap H^2(M,\mathbb{Z})$. Its rank, denoted by $\rho$, is called the Picard number of $M$. Observe that $\rho \leq h^{1,1}_M$.

According to the Hodge index theorem the restriction of the intersection form to this lattice has signature $(1, \rho - 1)$. This implies the following.

**Corollary 3.1.12. (Geometric version of the Hodge index theorem)** *Suppose $M$ is an algebraic surface. If $D, E$ are divisors on $M$ such that*

$$D^2 := D \cdot D > 0 \text{ and } D \cdot E = 0$$

*then either*

$$E^2 < 0$$

*or* $E \cdot D' = 0$ *for any divisor* $D'$.

**Definition 3.1.13.** A divisor $D$ on an algebraic surface is said to be *numerically equivalent* to 0, and we write $D \sim_n 0$, if $D \cdot E = 0$ for any divisor $E$. Two divisors $D_1$, $D_2$ are called *numerically equivalent* if $D_1 - D_2 \sim_n 0$. We denote by $\mathrm{Num}\,(M)$ the space of numerical equivalence classes of divisors.

Observe that the principal divisors are numerically equivalent to zero. The Hodge index theorem shows that the intersection form restricts to a nondegenerate quadratic form on $\mathrm{Num}\,(M)$. Observe that $\mathrm{Num}\,(M)$ is a free Abelian group. It coincides with $H^{1,1}_{\bar{\partial}}(M) \cap H^2(M, \mathbb{Z})$ and thus its rank is the *Picard* number of $M$. The restriction of the intersection form to $\mathrm{Num}\,(M)$ has signature $(1, \rho - 1)$.

Unraveling the structure of algebraic surfaces requires a good understanding of the "cone" $\mathrm{Num}_+(X)$, consisting of those divisors with positive self-intersection.

**Definition 3.1.14.** A divisor $D$ on an algebraic surface is called *big* if $D^2 > 0$.

A big divisor is not far from being effective. In fact, we have the following result.

**Proposition 3.1.15.** *If $D$ is a big divisor then there exists a positive integer such that either $nD$ or $-nD$ is effective.*

We present the proof (borrowed from [**59**]) since it relies on a simple but frequently used argument in the theory of algebraic surfaces.

**Proof** For every integer $n$ we have

$$\chi(nD) = h^0(nD) + h^{0,2}(nD) - h^{0,1}(nD)$$

$$= \frac{1}{2} nD \cdot (nD - K) + \frac{1}{4}(\chi_M + \tau_M).$$

Since $D^2 > 0$ we deduce $\chi(nD) \to \infty$ as $|n| \to \infty$ so that, using Serre duality, we deduce

$$h^0(nD) + h^0(K_M - nD) \to \infty.$$

If $nD$ is not effective for any $n \neq 0$ we deduce from the above that

$$h^0(K_M \pm nD) \to \infty, \text{ as } |n| \to \infty, \tag{3.1.18}$$

is effective for any $n \gg 0$. Choose a nontrivial holomorphic section $s_n$ of $K_M - nD$. This leads to an injection

$$H^{0,0}(M, K_M + nD) \stackrel{\otimes s_n}{\to} H^{0,0}(M, 2K_M)$$

so that

$$\dim_{\mathbb{C}} H^{0,0}(K_M + nD) \leq \dim H^{0,0}(2K), \quad \forall n \gg 0.$$

This is clearly impossible in view of (3.1.18). ■

We see that there is a built-in positivity in the notion of effectiveness. The reason behind it is essentially explained in the following simple observation: if the smooth complex curves $C_1$, $C_2$ embedded in an algebraic surface $M$ intersect transversely then they have positive intersection number

$$C_1 \cdot C_2 > 0.$$

A similar result is true without the smoothness and/or the transversality assumption. More precisely we have the following result (see [**10, 39**]).

**Proposition 3.1.16.** *Suppose $D_1$ and $D_2$ are two effective divisors on an algebraic surface such that their supports intersect in finitely many points. Then*

$$D_1 \cdot D_2 \geq 0$$

*with equality iff their supports are disjoint.*

To proceed further we need to introduce new notions.

**Definition 3.1.17.** A holomorphic Hermitian line bundle $L \to M$ on a complex manifold $M$ is called *positive* if there exists a Hermitian metric $g$ on $TM$ such that

$$\mathbf{i}F_A = -\mathfrak{Im}\, g$$

where $F_A$ denotes the curvature of the Chern connection on $L$. $L$ is called *negative* if $-L$ is positive.

**Theorem 3.1.18. (Kodaira vanishing theorem)** *Suppose $L$ is a negative line bundle on a complex manifold $M$. Then*

$$h^{0,q}(L) = 0, \quad \forall 0 \leq q < n.$$

**Theorem 3.1.19. (Kodaira embedding theorem)** *A complex manifold $M$ admits positive line bundles if and only if it is algebraic. More precisely, $L$ is a positive line bundle if and only if there exists an ample divisor $D$ such that $L = [D]$.*

It follows from the Kodaira embedding theorem that the self-intersection number of an ample divisor $E$ on an algebraic surface $M$ is always positive. In fact, given any effective divisor $D$ we have

$$D \cdot E > 0$$

To see this observe that the divisor $nE$ is very ample for $n \gg 0$ and so it defines an embedding

$$f : M \to |nE|^*.$$

Then $f(\mathrm{supp}\,(D))$ contains at most finitely many lines in $|nE|^*$. Now pick a hyperplane $H \subset |nE|^*$ not containing any of these lines but containing a point in $f(\mathrm{supp}\,(D))$. This hyperplane intersects $f(\mathrm{supp}\,(D))$ in finitely many points. This hyperplane corresponds to a nontrivial section $s$ of $[nE]$ whose zero set intersects $D$ in finitely many points. This implies

$$(s) \cdot D > 0.$$

Now observe that $(s) \sim_n nE$ so that

$$n(E \cdot D) = nE \cdot D > 0.$$

This extreme positivity of ample divisors characterizes them. More precisely, we have the following result.

**Theorem 3.1.20. (Nakai-Moishezon)** *A divisor $D$ on an algebraic surface $M$ is ample if and only if $D^2 > 0$ and $D \cdot E > 0$ for any effective divisor $E$.*

For a proof we refer to [**53**].

**Definition 3.1.21.** A divisor $D$ on an algebraic surface $M$ is said to be *numerically effective* (or *nef*) if $D \cdot E \geq 0$ for any effective divisor $E$.

Thus the ample divisors are both big and nef. However not all big and nef divisors are ample.

Algebraic geometers are interested in a rougher classification of complex manifolds, that given by bimeromorphisms. We present this notion only in the case of interest to us.

**Definition 3.1.22.** Suppose $M_1$ and $M_2$ are compact complex surfaces. A *bimeromorphic map*

$$f : M_1 \overset{\sim}{\dashrightarrow} M_2$$

is a surjective holomorphic map

$$f : M_1 \to M_2$$

such that there exist analytic proper subsets $S_i \subset M_i$, $i = 1, 2$, so that

$$f : M_1 \setminus S_1 \to M_2 \setminus S_2$$

is biholomorphic. Two surfaces are called bimeromorphic if there exists a bimeromorphic map between them. A surface is called *rational* if it is bimeromorphic to $\mathbb{CP}^2$.

**Example 3.1.23. (Complex blow-up)** Suppose $M$ is a complex surface. Fix a point $p \in M$ and local coordinates $(z_1, z_2)$ in a neighborhood $U$ of $p$ so that we can identify $p$ with the origin of $\mathbb{C}^2$ and $U$ with the unit disk $D \subset \mathbb{C}^2$ centered at the origin. We can regard $U \setminus \{p\}$ as an open subset of

$$\tilde{U} := \{(z, \ell) \in U \times \mathbb{CP}^1; \ \ z \in \ell\} \subset \tilde{\mathbb{C}}^2$$

where $\tilde{\mathbb{C}}^2$ is the total space of the tautological line bundle over $\mathbb{CP}^1$ and $\tilde{U}$ is an open neighborhood of the zero section. There is a natural holomorphic map

$$\pi : \tilde{U} \to U \setminus \{p\}, \ \ (z, \ell) \mapsto z$$

such that $E := \pi^{-1}(0)$ coincides with the zero section. Moreover $\pi : \tilde{U} \setminus E \to U \setminus \{p\}$ is *biholomorphic*. The blow-up of $M$ at $p$, denoted by $\tilde{M}_p$, is the manifold obtained by gluing $\tilde{U} \setminus E$ to $M \setminus \{p\}$ using the map $\pi$. Observe that $\pi$ extends to a natural surjection

$$\pi : \tilde{M}_p \to M.$$

This map is bimeromorphic and it is called the *blow-down map*. Its inverse (defined only on $M \setminus \{p\}$) is called the blow-up map. The zero section $E$ is a smooth *rational curve* (i.e. a holomorphically embedded $\mathbb{CP}^1 \hookrightarrow \tilde{M}_p$) with self-intersection $-1$. $E$ is called the *exceptional divisor* of the blow-up.

If $C$ is a complex curve on $M$ then the closure of $\pi^{-1}(C \setminus \{p\})$ is called the *proper transform* of $C$ and is denoted by $\pi^*(C)$. One can show that

$$\pi^*(C)^2 = C^2 - \mathrm{mult}_p(C).$$

The nonnegative integer $\mathrm{mult}_p(C)$ is called the multiplicity of $C$ at $p$. It is 0 if $p \notin C$, it is 1 if $C$ is smooth at $p$ and, in general, it is equal to the order of vanishing at $p$ of a defining equation for $C$ near $p$.

The blown-up manifold $\tilde{M}_p$ can itself be blown-up and so on. Iterating this procedure we obtain an *iterated blow-up* manifold $X$ and a natural surjection

$$\pi : X \to M$$

called the *iterated blow-down map*.

**Exercise 3.1.1.** Suppose $M$ is a complex manifold and $\tilde{M}$ is the blow-up of $M$ at some point. If $\sigma : \tilde{M} \to M$ denotes the natural projection then

$$K_{\tilde{M}} = \sigma^* K_M + [E]$$

where $E \subset \tilde{M}$ denotes the exceptional divisor.

In some sense, the above example captures the structure of any bimeromorphic map. More precisely, we have the following important result (see [**10, 49**]).

**Theorem 3.1.24. (Zariski)** *If $M_1 \to M_2$ is a bimeromorphic map between algebraic surfaces then there exist an algebraic surface $X$ and surjective holomorphic $\pi_i : X \to M_i$ with the following properties.*

*(i) The diagram below is commutative.*

$$\begin{array}{ccc} & X & \\ {}^{\pi_1}\swarrow & & \searrow^{\pi_2} \\ M_1 & \xrightarrow{f} & M_2 \end{array}$$

*(ii) $X$ is an iterated blow-up of both $M_1$ and $M_2$ and both maps $\pi_1$ and $\pi_2$ are iterated blow-down maps.*

The above result shows that the blow-up operation plays a special role in the theory of algebraic surfaces. It is therefore important to know if a given surface is a blow-up of another. Example 3.1.23 shows that for an algebraic surface to be a blow-up it is necessary that there exists a holomorphically embedded $\mathbb{CP}^1 \hookrightarrow X$ with self-intersection $-1$. The next remarkable result shows that this condition is also necessary. For a proof we refer to [**10, 49**].

**Theorem 3.1.25. (Castelnuovo-Enriques)** *Suppose $X$ is an algebraic surface containing a smooth rational curve with self-intersection $-1$. Denote by $E'$ the image of this embedding. Then there exist an algebraic surface $M$, a point $p \in M$ and holomorphic maps*

$$F : X \to \tilde{M}_p, \quad f : X \to M$$

*such that the following hold.*

*(i) The diagram below is commutative.*

$$\begin{array}{ccc} X & \xrightarrow{F} & \tilde{M}_p \\ & {}_f\searrow & \downarrow \pi \\ & & M \end{array} .$$

*(ii) $F$ is biholomorphic and $f^{-1}(p) = E'$.*

*The manifold $M$ is called the* blow-down *of $X$.*

**Definition 3.1.26.** A complex surface is called *minimal* if it contains no smooth rational curves (i.e. holomorphically embedded $\mathbb{CP}^1$'s) with self-intersection $(-1)$.

Thus, an algebraic surface is minimal if it cannot be blown down, i.e. it is not the blow-up of any surface.

We conclude our short survey in complex geometry with an important topological result due to S. Lefschetz.

**Theorem 3.1.27. (Lefschetz hypersurface theorem)** *Suppose $M \hookrightarrow \mathbb{CP}^N$ is an algebraic manifold of (complex) dimension $n$ and $F$ is a hypersurface in $\mathbb{CP}^N$ intersecting $M$ transversely. Then the inclusion induced morphisms*

$$H_q(M \cap F, \mathbb{Z}) \to H_q(M, \mathbb{Z}), \quad \pi_q(M \cap F) \to \pi_q(M)$$

*are isomorphisms for $i < n-1$ and surjections for $q = n-1$.*

For a very nice presentation of this theorem we refer to [**73**].

**Corollary 3.1.28.** *Any smooth hypersurface in $\mathbb{CP}^n$, $n \geq 3$, is simply connected.*

**Exercise 3.1.2.** Suppose $X$ is Kähler manifold of dimension $n \geq 3$ and $L \to X$ is an ample line bundle. Suppose there exists a holomorphic section $u$ of $L$ with transversal zero set $Y = u^{-1}(0)$. Show that the inclusion

$$Y \hookrightarrow X$$

induces isomorphisms $H_k(Y, \mathbb{Z}) \cong H_k(X, \mathbb{Z})$ and $\pi_k(Y) \cong \pi_k(X)$ for $k \leq n-2$.

**3.1.2. Examples of complex surfaces.** To give the reader a feeling about the general notions discussed in the previous subsection, we will, for a while, take a side road and present some beautiful algebraic geometric landscapes. In the sequel we write $\mathbb{P}^n$ for $\mathbb{CP}^n$.

So far, the only examples of complex surfaces we know are the projective plane $\mathbb{P}^2$, its iterated blow-ups and the products of pairs of Riemann surfaces. There is another unlimited source of examples: complex surfaces as zero sets of families of homogeneous polynomials.

**Example 3.1.29. (Quadrics in $\mathbb{P}^3$)** The space of quadratic homogeneous polynomials in four variables has dimension $\binom{5}{2} = 10$ and each such polynomial can be viewed as a holomorphic section of the line bundle $2H$ on $\mathbb{P}^3$.

If $Q(z_0, \cdots, z_3)$ is such a polynomial, the implicit function theorem implies that the zero set $Q = 0$ is a smooth submanifold of $\mathbb{P}^3$ if and only if $Q$ is nondegenerate as a quadratic form. On the other hand, all complex nondegenerate quadratic forms in four variables have the same canonical (diagonal) form. This implies that all quadrics in $\mathbb{P}^3$ are projectively equivalent, meaning that any two are related by a projective isomorphism of the ambient space $\mathbb{P}^3$. We thus have the freedom of choosing $Q$ in any way we want. Let

$$Q = z_0 z_3 - z_1 z_2.$$

The zero set $S$ of $Q$ is the image of the Segre embedding

$$\mathbb{P}^1 \times \mathbb{P}^1 \to \mathbb{P}^3, \quad ([s_0 : s_1], [t_0 : t_1]) \mapsto [s_0 t_0 : s_0 t_1 : s_1 t_0 : s_1 t_1]$$

which shows that the quadric $Q = 0$ is biholomorphic to $\mathbb{P}^1 \times \mathbb{P}^1$. This is a special example of a ruled surface. Observe that $S$ is spanned by two families of lines: the $A$-lines

$$A_{[t_0:t_1]} = \mathbb{P}^1 \times [t_0 : t_1], \quad [t_0 : t_1] \in \mathbb{P}^1,$$

and the $B$-lines

$$B_{[s_0:s_1]} = [s_0 : s_1] \times \mathbb{P}^1, \quad [s_0 : s_1] \in \mathbb{P}^1.$$

These lines have a nice intersection pattern. No two distinct lines of the same type meet while any $A$-line intersects any $B$-line in a unique point.

The quadrics are rational surfaces. To see this consider again the above quadric $S \subset \mathbb{P}^3$ and $p = [1 : 0 : 0 : 0] \in S$. The projective tangent plane to $S$ at $p$ intersects the quadric $S$ along the lines

$$\ell_1 := [s_0 : 0 : s_1 : 0] = A_{[1:0]}, \quad [s_0 : s_1] \in \mathbb{P}^1,$$

and

$$\ell_2 := B_{[1:0]} = [t_0 : t_1 : 0 : 0], \quad [t_0 : t_1] \in \mathbb{P}^1.$$

Now project $S$ from $p$ onto a plane $H \subset \mathbb{P}^3$. This means that to each $q \neq p$ we associate the point $\pi(q) \in H$, the intersection of the line $pq$ with $H$. The map $\pi : S \setminus \{p\} \to H$ is holomorphic but does not extend as a holomorphic map $S \to H$. Denote by $q_i$ the point where the line $\ell_i$ intersects $H$.

If we blow up $S$ at $p$ the points on the exceptional divisor correspond to the lines through $p$ tangent to $S$ and each of these lines intersects $H$ in a unique point. This shows that the projection $S \setminus \{p\} \to H$ leads to a well defined holomorphic map

$$\tilde{\pi} : \tilde{S}_p \to H.$$

Denote by $\hat{\ell}_i$ the proper transform of $\ell_i$ in the blow-up. Observe that $\hat{\ell}_i$ are smooth rational curves of self-intersection $-1$. The restriction

$$\tilde{\pi} : \tilde{S}_p \setminus (\hat{\ell}_1 \cup \hat{\ell}_2) \to H$$

is one-to-one while $\tilde{\pi}(\hat{\ell}_i) = q_i$. Using the Castelnuovo-Enriques theorem we can blow down the curves $\hat{\ell}_i$. Denote by $X$ the resulting surface. $\tilde{\pi}$ descends to a biholomorphism $X \to H$. Thus we arrived at $H \cong \mathbb{P}^2$ by blowing up once and blowing down twice, which shows that $S$ is rational.

**Exercise 3.1.3.** Show that any line on a quadric is either an $A$- or a $B$-line.

**Example 3.1.30. (Hirzebruch surfaces)** We have seen that a quadric can be viewed as the total space of a holomorphic family of lines ($\mathbb{P}^1$'s) parameterized by $\mathbb{P}^1$. The Hirzebruch surfaces $\mathbf{F}_n$, $n \geq 0$, are twisted versions of such families.

Define $\mathbf{F}_0 := \mathbb{P}^1 \times \mathbb{P}^1$ and $\mathbf{F}_1 = \mathbf{F}_1(\ell)$ as the graph of the projection from a point $p_0 \in \mathbb{P}^2$ to a line $\ell \subset \mathbb{P}^2$ not containing $p_0$. More precisely

$$\mathbf{F}_1 = \{(x, y) \in \mathbb{P}^2 \times \ell;\ \ x \in p_0 y\}$$

where $p_0 y$ denotes the line determined by the points $p_0$ and $y$. Observe that $\mathbf{F}_1$ coincides with the blow-up of $\mathbb{P}^2$ at $p_0$. We denote by $E \hookrightarrow \mathbf{F}_1$ the exceptional divisor. There is a natural map

$$\pi : \mathbf{F}_1(\ell) \cong \tilde{\mathbb{P}}^2_{p_0} \to E \cong \mathbb{P}^1$$

defined as follows. If $p \in E$ then set $\pi(p) = p$. If $p$ is not on the exceptional divisor then it corresponds to a unique point on $\mathbb{P}^2$ not equal to $p_0$; we continue to denote by $p$ this point on $\mathbb{P}^2$. The line $p_0 p$ defines a unique point on $E$ which we denote by $\pi(p)$. $\pi$ is holomorphic and its fibers are all lines, more precisely, the proper transforms of the lines through $p_0$. The proper transform of $\ell$ is a line $\tilde{\ell}$ on $\mathbf{F}_1$ with self-intersection 1. We will say that $E$ is the 0-section of the fibration $\pi : \mathbf{F}_1 \to \mathbb{P}^1$ and that $\tilde{\ell}$ is the $\infty$-section.

More generally, for $n \geq 0$ consider the line bundle $-nH \to \mathbb{P}^1$. We denote by $\mathbf{F}_n$ the projectivization of the rank-2 vector bundle

$$E_n = \underline{\mathbb{C}} \oplus (-nH) \to \mathbb{P}^1$$

meaning the bundle over $\mathbb{P}_1$ whose fiber over $p \in \mathbb{P}^1$ is the projective line $\mathbb{P}(E_n(p))$. By definition, $\mathbf{F}_n$ is equipped with a holomorphic map

$$\pi_n : \mathbf{F}_n \to \mathbb{P}^1$$

whose fibers are projective lines. The section $1 \oplus 0$ of $E_n$ defines a section of $\mathbf{F}_n$ called the 0-section and denoted by $D_0$. Observe that if $s$ is a section of $nH$ it defines a section of

$$\mathbb{P}((nH) \oplus \underline{\mathbb{C}}) \cong \mathbb{P}(\underline{\mathbb{C}} \oplus (-nH))$$

called the $\infty$-section and denoted by $D_\infty$. $D_0$ and $D_\infty$ are divisors and we will denote the classes they determine in $H_2(\mathbf{F}_n, \mathbb{Z})$ by the same symbols. Also, we denote by $F$ the cohomology class carried by a fiber. Since $D_0$ and $D_\infty$ are sections we have

$$D_0 \cdot F = D_\infty \cdot F = 1.$$

Clearly

$$F \cdot F = 0.$$

Since $D_0$ comes from the zero section of $-nH$ which has degree $-n$ we have

$$D_0^2 = -n.$$

The homotopy exact sequence of a fibration shows that $\mathbf{F}_n$ is simply connected while Gysin's exact sequence shows that $H^2(\mathbf{F}_n, \mathbb{Z}) = \mathbb{Z}F \oplus \mathbb{Z}D_0$, so that the intersection form of $\mathbf{F}_n$ is

$$q_n = \begin{bmatrix} 0 & 1 \\ 1 & -n \end{bmatrix}.$$

The intersection form is even iff $n$ is even, so that $\mathbf{F}_n$ is spinnable iff $n$ is even.

From a differentiable point of view the Hirzebruch surfaces are $S^2$-bundles over $S^2$ and these bundles are classified by $\pi_1(SO(3)) = \mathbb{Z}_2$. This shows that $\mathbf{F}_n$ is diffeomorphic to $\mathbf{F}_m$ if and only if $n$ and $m$ have the same parity.

It is easy to compute the canonical class $K$ of $\mathbf{F}_n$. It can be written as

$$K = xF + yD_0$$

so that

$$K \cdot F = y, \quad K \cdot D_0 = x - ny.$$

Using the adjunction formula we deduce

$$0 = g(F) = 1 + \frac{1}{2}F \cdot (F + K) = 1 + \frac{y}{2},$$

$$0 = g(D_0) = 1 + \frac{1}{2}D_0 \cdot (D_0 + K) = 1 + \frac{x - ny - n}{2}.$$

This shows $y = -2$ and $x = n - 2$ so that

$$K = (n-2)F - 2D_0.$$

Let us observe that the zero section $D_0$ is the unique smooth irreducible curve on $\mathbf{F}_n$ with negative self-intersection. Indeed, if $D$ were another such curve, $D \neq D_0$,

$$D = aF + bD_0$$

then

$$0 \leq D \cdot D_0 = a - nb, \quad 0 \leq D \cdot F = b$$

and

$$0 > D \cdot D = -nb^2 + 2ab = b(2a - nb).$$

The above inequalities are clearly impossible. Thus the Hirzebrúch surfaces $\mathbf{F}_n$ are minimal for $n \geq 2$ and $\mathbf{F}_n$ is not biholomorphic to $\mathbf{F}_m$ if $m \neq n$.

If we now blow up $\mathbf{F}_n$ at a point $p$ not situated on $D_0$ we obtain a surface

$$\tilde{\mathbf{F}}_n \to \mathbf{F}_n.$$

The proper transform of the fiber $F$ through $p$ is a rational curve $\tilde{F}$ of self-intersection $-1$ which can be blown down and we get a new surface $\mathbf{F}$. The pencil of fibers of $\mathbf{F}_n$ is transformed into a pencil of smooth rational curves of self-intersection 0 which cover each point of $\mathbf{F}$ exactly once. This shows that $\mathbf{F}$ is also a ruled surface, i.e. a holomorphic fiber bundle over $\mathbb{P}^1$ with fibers $\mathbb{P}^1$. On the other hand, the curve $D_0$ in $\mathbf{F}_n$ is mapped to a smooth rational curve $R$ in $\mathbf{F}$ with self-intersection $R^2 = D_0^2 + 1 = -n + 1$. This shows that $\mathbf{F}$ is biholomorphic to $\mathbf{F}_{n-1}$ and all Hirzebruch surfaces are bimeromorphic, and thus rational. One can show (see [**49**]) that any *minimal* rational surface is biholomorphic to either $\mathbb{P}^2$ or one of the Hirzebruch surfaces $\mathbf{F}_n$, $n \geq 2$.

**Example 3.1.31. (Cubics)** Consider six points $p_1, \cdots, p_6$ in general position in $\mathbb{P}^2$, meaning

◇ no three are collinear and

◇ no five are on the same conic.

The space of homogeneous cubic polynomials in three variables $z_0, z_1, z_2$ is $\binom{5}{3} = 10$-dimensional. The above six points define a four-dimensional subspace $V$ consisting of polynomials vanishing at the $p_i$. Each $P \in V$ defines a cubic curve $\{P = 0\} \subset \mathbb{P}^2$ containing all these six points.

Any point $q \in \mathbb{P}^2 \setminus \{p_1, \cdots, p_6\}$ determines a hyperplane

$$H_q = \{P \in V;\ \ P(q) = 0\}$$

so we get a holomorphic map

$$f : \mathbb{P}^2 \setminus \{p_1, \cdots, p_6\} \ni q \mapsto H_q \in \mathbb{P}(V^*).$$

This map can be equivalently described as follows. Fix a basis $Z_0, \cdots, Z_3$ of $V$. Then $f$ is the map

$$q \mapsto [Z_0(q) : \cdots : Z_3(q)] \in \mathbb{P}^3.$$

This map has singularities at the points $p_i$ but, by blowing up at these points we hope to obtain a well defined map,

$$\tilde{f} : \tilde{\mathbb{P}}^2_{p_1, \cdots, p_6} \to \mathbb{P}^3.$$

We refer the reader to [**10**] or [**49**] where it is shown that this map is well defined, its image is a smooth degree-3 surface $S$ in $\mathbb{P}^3$ and $f$ is a biholomorphic map $\tilde{\mathbb{P}}^2_{p_1, \cdots, p_6} \to S$. Conversely, one can show that any smooth cubic in $\mathbb{P}^3$ is biholomorphic to the blow-up of $\mathbb{P}^2$ at six points, not necessarily in general position. For details we refer to [**49**].

The surfaces presented so far were all rational and it took some ingenuity to establish that. Fortunately there is a very general method of deciding the rationality of a surface.

**Theorem 3.1.32. (Castelnuovo)** *If $M$ is an algebraic surface such that $q(M) := h^{0,1}(M) = 0$ and $p_2(M) := h^0(2K_M) = 0$ then $M$ can be obtained by iterated blow-up from $\mathbb{P}^2$ or one of the Hirzebruch surfaces. In particular, $M$ is rational.*

For a proof we refer to **[10]** or **[49]**.

**Example 3.1.33. (Hypersurfaces in $\mathbb{P}^3$)** The homogeneous polynomials of degree $d \geq 1$ in the variables $z_0, \cdots, z_3$ form a vector space $V_d$ of dimension

$$\dim_{\mathbb{C}} V_d := \binom{d+3}{3} = \frac{(d+3)(d+2)(d+1)}{6}.$$

For a generic $F \in V_d$ the zero locus $\{F = 0\}$ is a smooth hypersurface $X = X_d$ of degree 3 in the projective space $\mathbb{P}^3$. According to Lefschetz' theorem $X_d$ is simply connected for each $d$. Hence

$$q(X) = \frac{1}{2}b_1(X) = 0.$$

To compute the main invariants of $X_d$ we will rely on the adjunction formula. $X_d$ can be viewed as the zero set of a section of the line bundle $dH \to \mathbb{P}^3$. The adjunction formula holomorphically identifies $(dH)\mid_{X_d}$ with the normal bundle of $X_d \hookrightarrow \mathbb{P}^3$ from which we deduce

$$T\mathbb{P}^3 \mid_{X_d} = TX \oplus (dH)\mid_X,$$

$$c_t(\mathbb{P}^3)\mid_X = c_t(X)(1 + (dH)t)\mid_X$$

where $c_t$ denotes the Chern polynomial. Using the computations in §2.3.4 we deduce

$$(1 + tH)^4 \mid_X = c_t(TX)\Big(1 + (dH)t\Big)\mid_X, \quad H^4 = 0.$$

By setting $H_X := H\mid_X$ and observing that $H_X^2 = d$ (= the number of intersection points of a line with $X$) and $H_X^3 = 0$ we obtain

$$1 + c_1(TX)t + c_2(TX)t^2 = (1 + H_X t)^4\Big(1 + (dH_X)t\Big)^{-1}$$

$$= (1 + H_X t)^4\Big(1 - (dH_X)t + (d^2 H_X^2)t^2\Big)$$

$$= \Big(1 + (4H_X)t + (6H_X)t^2\Big)\Big(1 - (dH_X)t + d^3t^2\Big)$$

$$= 1 + (4 - d)H_X t + (d^3 - 4d^2 + 6)t^2.$$

Thus

$$K_X = -c_1(TX) = (d-4)H_X$$

and

$$K_X^2 = (d-4)^2 H_X^2 = d(d-4)^2.$$

On the other hand, $c_2(TX)$ is the Euler class of $TX$ and thus

$$\chi = d(d^2 - 4d + 6)$$

where $\chi$ denotes the Euler characteristic of $X$. Using the signature formula

$$K_X^2 = 2\chi + 3\tau$$

($\tau$ = signature) we deduce

$$\tau = \frac{d(4-d^2)}{3} = b_2^+ - b_2^-.$$

In this case $\chi = b_1(X) + b_2(X) + b_4(X) = 2 + b_2(X)$ so that

$$b_2(X) = d^3 - 4d^2 + 6d - 2 = b_2^+ + b_2^-.$$

Hence

$$b_2^+ = \frac{1}{2}(b_2 + \tau) = \frac{(d-1)(d-2)(d-3)}{3} + 1$$

and

$$p_g = \frac{b_2^+ - 1}{2} = \binom{d-1}{3}.$$

Observe that $K_X = w_2(X) \bmod 2$ and since $X$ is simply connected we deduce that the intersection form of $X$ is even iff $d$ is even. Equivalently, this means $X_d$ is spinnable iff $d$ is even. Using the Classification Theorem 2.4.13 of §2.4.3 we can now describe explicitly the intersection form of $X$.

Observe that for $d > 4$ the line bundle $K_X$ is ample so that according to the Kodaira vanishing theorem

$$H^j(X, nK_X) = 0, \;\; \forall k, j > 0.$$

Thus, using the Riemann-Roch-Hirzebruch formula we deduce

$$P_n(X) = h^0(nK_X) = \chi_0(nK_X) = \frac{1}{4}(\chi + \tau) + \frac{1}{2}n(n-1)K_X^2$$

$$= \frac{d(d-4)^2}{2}n(n-1) + \frac{1}{4}(\chi + \tau).$$

For $d < 4$ we deduce that $K_X = (d-4)H_X$ is negative, as the dual of the positive line bundle $(4-d)H\,|_X$. Using the Kodaira vanishing theorem we deduce that the line bundles $nK_X$, $n > 0$, do not admit holomorphic sections. Hence $q(X)$ and $P_2(X) = 0$. Castelnuovo's Theorem 3.1.32 once again shows that the hypersurfaces of degree $< 4$ in $\mathbb{P}^3$ are rational.

The case $d = 4$ deserves special consideration and will be discussed in a more general context in the next example. Observe only that

$$P_n(X_4) = 1, \quad \forall n > 0.$$

**Example 3.1.34. (K3 surfaces)** A $K3$ surface is a compact complex Kähler surface $X$ such that $b_1(X) = 0$ and whose canonical line bundle is topologically trivial.

Suppose $X$ is a $K3$ surface. Then

$$q(X) = \frac{1}{2}b_1 = 0.$$

Also

$$p_g = \dim H^0(K_X) = 1 = h^{2,0}(X) = h^{0,2}(X)$$

so that

$$b_2^+ = 2p_g + 1 = 3.$$

Using the signature formula we deduce

$$2\chi + 3\tau = K_X^2 = 0$$

so that

$$2(2 + b_2^+ + b_2^-) = 3(b_2^- - b_2^+).$$

Since $b_2^+ = 3$ we deduce $b_2^- = 19$ so that $\tau = -16$. The intersection form $q_X$ of $X$ is even since $w_2(X) \equiv K_X \mod 2$ so that, according to the Classification Theorem 2.4.13, we deduce that

$$q_X \cong 3H \oplus -2E_8.$$

M. Freedman's theorem shows that all $K3$ surfaces are homeomorphic to each other.

The smooth quartics (degree 4) in $\mathbb{P}^3$ are $K3$ surfaces. The space of degree-4 homogeneous polynomials in variables $z_0, \cdots, z_3$ form a space of dimension 35 and thus we get a 34-dimensional family of $K3$ surfaces. Not all quartics in this family are different. The group $PGL_4(\mathbb{C})$ (which has dimension $15 = 16-1$) acts by change of variables on this space of polynomials leading to isomorphic surfaces. If we mod out this action we are left with a 19-dimensional family of $K3$-surfaces. We only want to mention that not all $K3$ surfaces can be obtained in this manner (they form a 20-dimensional family).

**Remark 3.1.35.** All $K3$ surfaces are diffeormorphic to each other although not biholomorphic. In particular, all are simply connected. For more details we refer to [**9, 59**].

**Exercise 3.1.4.** Suppose $X$ is a $K3$ surface. Then $K_X$ is also *holomorphically* trivial.

**Example 3.1.36. (Elliptic surfaces)** An *elliptic surface* is a triple $(X, f, C)$ where $X$ is a complex surface, $C$ is a smooth complex curve (i.e. Riemann surface) and $f : X \to C$ is a holomorphic map such that there exists a finite set $F \subset C$ with the following properties:

◇ $f : X \setminus f^{-1}(F) \to C \setminus F$ is a submersion.

◇ For any $x \in C \setminus F$ the fiber $f^{-1}(x)$ is biholomorphic to a smooth elliptic curve (i.e. biholomorphic to a smooth cubic in $\mathbb{P}^2$).

We want to present two fundamental examples of elliptic surfaces. For a detailed presentation of this important class of complex surfaces we refer to [**40**].

**A.** Consider two smooth cubic curves $C_1, C_2 \subset \mathbb{P}^2$ intersecting in nine distinct points, $p_1, \cdots, p_9$. Thus $C_i$ are described as the zero sets of two homogeneous polynomials $P_i$, $i = 1, 2$, in the variables $(z_0, z_1, z_2)$. We get a map

$$f : \mathbb{P}^2 \setminus \{p_1, \cdots, p_9\} \to \mathbb{P}^1, \quad p \mapsto [P_1(p), P_2(p)].$$

Observe that $f(p) = [\lambda : \mu]$ if and only if $\mu P_1(p) + \lambda P_2(p) = 0$. This map induces a well defined map

$$F : X = \tilde{\mathbb{P}}^2_{p_1, \cdots, p_9} \to \mathbb{P}^1$$

whose generic fiber is a smooth elliptic curve (i.e. a biholomorphic to a smooth cubic on $\mathbb{P}^2$). The *discriminant locus* $\Delta_F \subset \mathbb{P}^1$, i.e. the set of critical values of $F$, is finite. In fact, the polynomials $P_1$, $P_2$ can be generically chosen so that the critical points of $F$ are nondegenerate, i.e. near such a point $F$ behaves like the function $z_1 z_2$ near $0 \in \mathbb{C}^2$. Such singular fibers have a *node* singularity and look like Figure 3.2. The Euler characteristic of such a singular fiber is 1 (see Figure 3.3 for a Mayer-Vietoris based proof).

It is an elementary exercise in topology to prove that if $F : S \to C$ is a holomorphic map whose fibers, except for finitely many $F_1, \cdots, F_\nu$, are smooth complex curves of genus $g$ then

$$\chi(S) = \chi(C)\chi(F) + \sum_{i=1}^{\nu} (\chi(F_i) - \chi(F)) \tag{3.1.19}$$

where $F$ denotes a generic fiber. In our case $\chi(F) = 0$ since the generic fibers are tori, so that

$$\chi(X) = \sum_{i=1}^{\nu} (\chi(F_i) - \chi(F)) = \nu$$

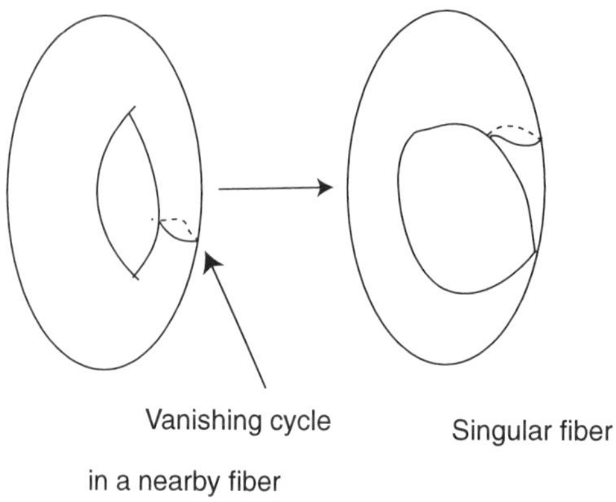


**Figure 3.2.** *A node singularity*

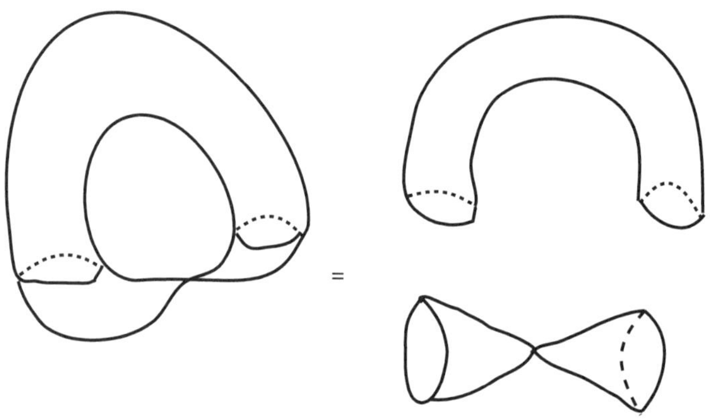

**Figure 3.3.** *Chopping the node*

where $\nu$ is the number of singular fibers of the fibration $F : X \to \mathbb{P}^1$. Thus

$$\nu := \chi(X) = 12.$$

The canonical class of $X$ is $K_X = -3H + \sum_{i=1}^{9} E_i$ so that, using the Riemann-Roch theorem, we deduce

$$\chi_{hol}(X) = \frac{1}{12}(K_X^2 + \chi(X)) = 1.$$

Observe that each of the nine exceptional divisors intersects each of the fibers of $F$ in exactly one point and thus they can be regarded as sections of the fibration $F : X \to \mathbb{P}^1$. Notice that the self-intersection numbers of these sections are all equal to $-1$. We will denote by $E(1)$ the smooth 4-manifold supporting the complex manifold $X$.

**B.** Consider two homogeneous cubic polynomials $A_0$ and $A_1$ in the variables $(z_0, z_1, z_2)$. The equation

$$t_0^n A_0(z_0, z_1, z_2) + t_1^n A_1(z_0, z_1, z_2) = 0$$

defines a hypersurface $V_n$ in $X = \mathbb{P}^1 \times \mathbb{P}^2$. For generic $A_0, A_1$ this is a smooth hypersurface. The natural projection

$$\mathbb{P}^1 \times \mathbb{P}^2 \longrightarrow \mathbb{P}^1$$

defines a holomorphic map $F_n : V_n \to \mathbb{P}^1$. Its fiber over the point $[t_0 : t_1]$ is the cubic

$$C_{[t_0:t_1]} = \{[z_0 : z_1 : z_2] \in \mathbb{P}^2;\ \ t_0^n A_0(z_0, z_1, z_2) + t_1^n A_1(z_0, z_1, z_2) = 0\}.$$

Hence $V_n$ is equipped with a structure of elliptic fibration. To compute some of its invariants we will use the adjunction formula. Denote by $H_i$ the hyperplane class in $H^2(\mathbb{P}^i, \mathbb{Z})$, $i = 1, 2$. The classes define by pullback classes in $H^2(X, \mathbb{Z})$ which we continue to denote by $H_i$. The Künneth formula shows that

$$H^2(X, \mathbb{Z}) = \mathbb{Z}H_1 \oplus \mathbb{Z}H_2,\ \ H^4(X, \mathbb{Z}) = \mathbb{Z}H_1 \cdot H_2 \oplus \mathbb{Z}H_2^2$$

and

$$H_1^2 = 0 = H_2^3,\ \ H_1 \cdot H_2^2 = 1.$$

We have

$$c_t(TX) = c_t(T\mathbb{P}^1)c_t(T\mathbb{P}^2) = (1 + H_1 t)^2 (1 + H_2 t)^3.$$

The normal bundle $N_{V_n}$ to $V_n \hookrightarrow X$ is $(nH_1 + 3H_2)\,|_{V_n}$ and thus it has Chern polynomial

$$c_t(N_{V_n}) = \Big(1 + (nH_1 + 3H_2)t\Big)\,|_{V_n}\,.$$

Hence

$$\begin{aligned}
c_t(TV_n) &= (1 + H_1 t)^2\,|_{V_n}\,(1 + H_2 t)^3\,|_{V_n}\,\Big(1 + (nH_1 + 3H_2)t\Big)^{-1}|_{V_n} \\
&= \Big(1 + (2H_1)t\Big)\,|_{V_n}\,\Big(1 + (3H_2)t + (3H_2^2)t^2\Big)\,|_{V_n} \\
&\quad \times\Big(1 - (nH_1 + 3H_2)t + (nH_1 + 3H_2)^2 t^2\Big)\,|_{V_n} \\
&= \left(1 + (2H_1 + 3H_2)t + (6H_1H_2 + 3H_2^2)t^2\right)|_{V_n} \\
&\quad \times\left(1 - (nH_1 + 3H_2)t + (6nH_1H_2 + 9H_2^2)t^2\right)|_{V_n} \\
&= 1 + (2 - n)H_1\,|_{V_n}\, t
\end{aligned}$$

$$+\left( (6n+6)H_1 \cdot H_2 + 12H_2^2 - (2H_1+3H_2)(nH_1+3H_2) \right)|_{V_n} t^2$$

$$= 1 + (2-n)H_1|_{V_n} t + \big(3nH_1H_2 + 3H_2^2\big)|_{V_n} t^2.$$

Thus

$$c_2(TV_n) = (3nH_1H_2 + 3H_2^2)|_{V_n} = \big(3nH_1H_2 + 3H_2^2\big) \cdot (nH_1 + 3H_2) = 12n.$$

Moreover

$$K_{V_n} = (n-2)H_1|_{V_n}$$

so that

$$K_{V_n}^2 = 0.$$

Observe that the Poincaré dual of the cohomology class $H_1|_{V_n} \in H^2(V_n, \mathbb{Z})$ is precisely the homology class carried by a fiber of $F_n : V_n \to \mathbb{P}^1$. Using the Riemann-Roch formula we deduce

$$\chi_{hol}(V_n) = n.$$

Let us now notice that $V_1$ is precisely the surface we considered in **A** since the natural projection $V_n \to \mathbb{P}^2$ has 9 singular fibers $F_i = \mathbb{P}^1 \times \{p_i\}$, $i = 1, \cdots, 9$, corresponding to the intersection points of the cubics $A_0 = 0$ and $A_1 = 0$ on $\mathbb{P}^2$. Each of these fibers has self-intersection $-1$ (*why?*) in $V_n$ and thus can be blown down.

Denote by $f_n : \mathbb{P}^1 \to \mathbb{P}^1$ the natural branched cyclic $n$-cover given by

$$[t_0 : t_1] \mapsto [t_0^n : t_1^n].$$

The map $f_n \times \mathbf{1} : \mathbb{P}^1 \times \mathbb{P}^2 \dashrightarrow \mathbb{P}^1 \times \mathbb{P}^2$ induces a biholomorphic map $g_n : V_n \to V_1$ such that the diagram below is commutative

$$\begin{array}{ccc} V_n & \xrightarrow{g_n} & V_1 \\ {\scriptstyle F_n}\downarrow & & \downarrow{\scriptstyle F_1} \\ \mathbb{P}^1 & \xrightarrow{f_n} & \mathbb{P}^1 \end{array}$$

Thus, we can regard the fibration $F_n : V_n \to \mathbb{P}^1$ as a pullback of the fibration $F_1 : V_1 \to \mathbb{P}^1$. A simple argument involving Lefschetz' hypersurface theorem implies $\pi_1(V_n) = 0$ (see [**40**, Sec. 2.2.1] for a different explanation). In particular, this shows $V_2$ is a $K3$ surface. Moreover, using the equality

$$\chi_{hol}(V_n) = 1 + p_g(V_n) \quad (q(V_n) = \frac{b_1}{2} = 0)$$

we deduce

$$p_g(V_n) = n - 1$$

so that

$$b_2^+(V_n) = 2p_g(V_n) + 1 = 2n - 1.$$

Using any section of $F_1 : V_1 \to \mathbb{P}^1$ we obtain by pullback a section $S_n : \mathbb{P}^1 \to V_n$ which defines a holomorphic embedding of $\mathbb{P}^1$ in $V_n$, that is, a smooth rational curve $S_n$ on $V_n$. Using the genus formula we deduce

$$0 = g(S_n) = 1 + \frac{1}{2}S_n \cdot (K_{V_n} + S_n).$$

On the other hand, we have $K_{V_n} = (n-2)F$ where $F$ denotes the Poincaré dual of the homology class of a fiber of $F_n : V_n \to \mathbb{P}^1$. Observe that $S_n \cdot F = 1$ since $S_n$ is a holomorphic section. Hence

$$0 = 1 + \frac{1}{2}(n - 2 + S_n \cdot S_n)$$

so that

$$S_n \cdot S_n = -n.$$

In particular, on the $K3$ surface $V_2$ we have $S_2 \cdot S_2 = -2$. We will denote by $E(n)$ the smooth 4-manifold $V_n$. We refer to [**51**, Chap. 3,7] for different $C^\infty$-descriptions of these important examples.

**Exercise 3.1.5.** Prove the identity (3.1.19).

**Exercise 3.1.6.** Show that the homology class $F$ carried by a fiber of $F_n : V_n \to \mathbb{P}^1$ is primitive , i.e. it cannot be written as $nF'$, $n > 1$, $F' \in H_2(E(n), \mathbb{Z})$. Use this information to describe the intersection form of $E(n)$ and then to conclude that $E(n)$ is spin if and only if $n$ is even.

**Exercise 3.1.7.** Prove that $V_n$ is simply connected using Lefschetz' hypersurface theorem.

**Exercise 3.1.8.** Suppose $X$ is an algebraic $K3$ surface which contains a smooth complex curve $C$ such that $C^2 = 0$. Prove the following:

(a) Show that $g(C) = 0$.

(b) Show that $\dim H^0([D]) = 2$ and the complete linear system determined by $D$ has no base points.

(c) Conclude that $X$ admits a natural structure of elliptic fibration.

(d) Show that a quartic $X \subset \mathbb{P}^3$ which contains a projective line $\ell$ also contains a curve $C$ as above. What is the self-intersection number of $\ell \hookrightarrow X$?

**3.1.3. Kodaira classification of complex surfaces.** The Riemann surfaces (i.e. complex curves) naturally split into three categories: rational (genus 0), elliptic (genus 1) and general type (genus $\geq 2$). This classification is natural from many points of view. From a metric standpoint these three types support different types of Riemannian metrics. From a complex analytic point of view, the canonical line bundles of these three classes display different behaviours.

A similar point of view can be adopted for complex surfaces as well. Recall that the plurigenera $P_n(X)$ of $X$ are the dimensions of the spaces of holomorphic sections of the line bundle $K_X^{\otimes n}$.

It can be shown that for any complex surface $X$ the sequence of integers $(P_n(X))$ displays one of the following asymptotic behaviors.

$-\infty$ $\quad P_n(X) = 0 \ \forall n \geq 1$.

**0** There exists $C > 0$ such that $P_n(X) < C \ \forall n \gg 1$ but $P_n(X)$ is not identically zero.

**1** There exists $C > 0$ such that

$$\frac{1}{C}n < P_n(X) < Cn, \quad \forall n \gg 1.$$

**2** There exists $C > 0$ such that

$$\frac{1}{C}n^2 < P_n(X) < Cn^2, \quad \forall n \gg 1.$$

Accordingly, the surface $X$ is said to have *Kodaira dimension* $-\infty, 0, 1$ or 2. The Kodaira dimension is denoted by $\operatorname{kod}(X)$. A complex surface of Kodaira dimension 2 is said to be of *general type.*

The plurigenera are invariant under blow-up, so that they are bimeromorphic invariants of a complex surface. In particular, the Kodaira dimension of a complex surface is a bimeromorphic invariant.

**Example 3.1.37.** (a) $\operatorname{kod}(\mathbb{P}^2) = \operatorname{kod}(V_1) = -\infty$. Since the Hirzebruch surfaces $\mathbf{F}_n$ are rational, they too have Kodaira dimension $-\infty$.

(b) $\operatorname{kod}(V_2) = 0$. More generally, any $K3$ surface has Kodaira dimension zero.

(c) $\operatorname{kod}(V_n) = 1$, $\forall n \geq 3$.

(d) Any hypersurface in $\mathbb{P}^3$ of degree $d \geq 5$ has Kodaira dimension 2.

**Exercise 3.1.9.** Prove the claims (c) and (d) in the above example.

In the remainder of this subsection we will focus our attention on algebraic surfaces. For the proofs of the following theorems and for more details we refer to [**39, 59**] and the references therein.

The Kodaira dimension contains a significant amount of information, as witnessed by the following result.

**Theorem 3.1.38.** *(a) If the algebraic surface $X$ has Kodaira dimension $-\infty$ then it is bimeromorphic to $\mathbb{P}^2$ or a geometrically ruled surface, i.e. a surface biholomorphic to a product $\mathbb{P}^1 \times C$, $C$ smooth complex curve.*

*(b) If an algebraic surface has Kodaira dimension $0$ then $P_n(X) \in \{0, 1\}$, $\forall n \geq 1$.*

*(c) An algebraic surface of Kodaira dimension $1$ is necessarily an elliptic surface.*

According to Theorems 3.1.24 and 3.1.25 each algebraic surface is bimeromorphic to a minimal one called a *minimal model.*

A bimeromorphism class of surfaces may contain several, minimal, nonbiholomorphic models. For example $\mathbb{P}^2$, $\mathbf{F}_n$, $n \geq 2$ are all minimal models of rational surfaces which are not biholomorphic.

The above example is in some sense an exception. More precisely, we have the following result.

**Theorem 3.1.39.** *An algebraic surface $X$ has a unique (up to biholomorphism) minimal model if and only if* $\mathrm{kod}\,(X) \geq 0$.

There is a simple intersection theoretic way of deciding which minimal surfaces have nonnegative Kodaira dimension. More precisely, we have the following result.

**Theorem 3.1.40.** *Suppose $X$ is a minimal algebraic surface. Then* $\mathrm{kod}\,(X) \geq 0$ *if and only if the canonical divisor $K_X$ is nef.*

Thus any minimal algebraic surface $X$ with $K_X$ nef can have Kodaira dimension $0, 1$ or $2$. The exact value of the Kodaira dimension is also decided by the intersection theoretic properties of the canonical divisor.

**Theorem 3.1.41.** *Suppose $X$ is a minimal algebraic surface with $K_X$ nef. Then $K_X^2 \geq 0$ and the following hold.*

*(a)* $\mathrm{kod}\,(X) = 0$ *if and only if $K_X$ is numerically equivalent to zero.*

*(b)* $\mathrm{kod}\,(X) = 1$ *if and only if $K_X^2 = 0$ but $K_X$ is not numerically equivalent to zero.*

*(c)* $\mathrm{kod}\,(X) = 2$ *if and only if $K_X$ is big, i.e. $K_X^2 > 0$. In this case*

$$P_n(X) = \frac{n(n-1)}{2} K_X^2 + \chi_{hol}(X).$$

## 3.2. Seiberg-Witten invariants of Kähler surfaces

The Seiberg-Witten equations simplify considerably in the presence of a Kähler metric. This section is devoted to the study of this interaction, Seiberg-Witten equations $\leftrightarrow$ Kähler metrics and some of its remarkable consequences.

**3.2.1. Seiberg-Witten equations on Kähler surfaces.** Consider a Kähler surface $M$ and denote by $\omega$ the associated symplectic form. Observe that the Kähler structure leads to several canonical choices on $M$.

• The complex structure on $M$ defines a canonical *spin*$^c$ structure $\sigma_0$ with associated line bundle $\det(\sigma_0) = K_M^{-1}$. $K_M^{-1}$ is naturally a holomorphic line bundle equipped with a natural Hermitian metric. Moreover

$$\mathbb{S}_0^+ = \Lambda^{0,0}T^*M \oplus \Lambda^{0,2}T^*M = \underline{\mathbb{C}} \oplus K_M^{-1}$$

and

$$\mathbb{S}_0^- = \Lambda^{0,1}T^*M.$$

This choice allows us to identify the *spin*$^c$ structures on $M$ with the space of complex line bundles via the correspondence

$$L \mapsto \sigma_0 \otimes L.$$

Observe that

$$\det(\sigma_0 \otimes L) = K_M^{-1} \otimes L^2.$$

Additionally, the associated bundles of complex spinors are

$$\mathbb{S}_L^+ := L \oplus L \otimes K_M^{-1}, \;\; \mathbb{S}_L^- = \Lambda^{0,1}T^*M \otimes L.$$

Thus, any even spinor $\psi \in \Gamma(\mathbb{S}_L^+)$ canonically splits as

$$\psi = \alpha \oplus \beta, \;\; \alpha \in \Gamma(L), \;\; \beta \in \Gamma(L \otimes K_M^{-1}). \tag{3.2.1}$$

In the new "coordinates" on $Spin^c(M)$ the involution $\sigma \mapsto \bar{\sigma}$ has the form

$$L \mapsto K_M - L.$$

• The Kähler structure on $M$ produces a Chern connection on $TM$ which induces a connection $A_0$ on $K_M^{-1}$ compatible both with the canonical metric and the canonical holomorphic structure.

• The metric and connection $A_0$ on $K_M^{-1}$ canonically define a Dirac operator $\mathfrak{D}_0 : \mathbb{S}_0^+ \to \mathbb{S}_0^-$ which, according to the computations in Sec. 1.4, is none other than the Dolbeault-Hodge operator

$$\sqrt{2}(\bar{\partial} + \bar{\partial}^*) : \Lambda^{0,even}T^*M \to \Lambda^{0,odd}T^*M.$$

Now observe that any Hermitian connection $A$ on $\det(\mathbb{S}_L^+)$ can be uniquely written as a tensor product

$$A := A_0 \otimes B^{\otimes 2} \tag{3.2.2}$$

where $B$ is a Hermitian connection on $L$. Since

$$F_A = F_{A_0} + 2F_B \tag{3.2.3}$$

we will use the less rigorous but more suggestive notation

$$A = A_0 \dot{+} 2B.$$

The computations in 1.4.3 show that the Dirac operator induced by $A$ is

$$\mathfrak{D}_A = \sqrt{2}(\bar{\partial}_B \oplus \bar{\partial}_B^*). \tag{3.2.4}$$

• Using the symplectic form $\omega$ we can associate to any complex line bundle $L \to M$ a real number $\deg_\omega(L)$ defined by

$$\deg_\omega(L) = \frac{\mathbf{i}}{2\pi}\int_M F_A \wedge \omega$$

where $A$ is an arbitrary Hermitian connection on $L$. Observe that the above integral is independent of $L$ because $\omega$ is *closed* and the cohomology class of $\frac{\mathbf{i}}{2\pi}F_A$ is independent of $A$.

• The deRham cohomology space $H^1(M, \mathbb{R})$ is naturally equipped with a complex structure.

To describe it recall that by Hodge duality there is a complex conjugate linear isomorphism

$$\mathbf{H}_{\bar{\partial}}^{0,1}(M) \to \mathbf{H}_{\bar{\partial}}^{1,0}(M), \quad \varphi \mapsto \bar{\varphi}.$$

Since

$$\mathbf{H}^1(M, \mathbb{R}) \otimes \mathbb{C} \cong_{\mathbb{C}} \mathbf{H}_{\bar{\partial}}^{0,1}(M) \oplus H_{\bar{\partial}}^{1,0}(M)$$

there exists an $\mathbb{R}$-linear isometry

$$T : \mathbf{H}_{\bar{\partial}}^{0,1}(M) \to \mathbf{i}\mathbf{H}^1(M, \mathbb{R})$$

defined by

$$\mathbf{H}_{\bar{\partial}}^{0,1}(M) \ni \varphi \mapsto \frac{\mathbf{i}}{\sqrt{2}}(\varphi + \bar{\varphi}) \in \mathbf{i}\mathbf{H}^1(M).$$

$T$ induces a natural orientation on $\mathbf{H}^1(M, \mathbb{R})$.

• The Kähler structure defines a natural orientation on $\mathbf{H}_+^2(M)$. More precisely, observe that we have a natural $\mathbb{R}$-linear isomorphism

$$\mathbf{i}\mathbb{R}\omega \oplus \mathbf{H}_{\bar{\partial}}^{0,2}(M) \to \mathbf{i}\mathbf{H}_+^2(M)$$

defined by the correspondences

$$\mathbf{i}\omega \mapsto \mathbf{i}\omega, \quad \mathbf{H}^{0,2}_{\bar\partial} \ni \beta \mapsto \frac{\mathbf{i}}{\sqrt{2}}(\beta + \bar\beta) \in \mathbf{i}\mathbf{H}^2_+(M).$$

The natural orientation on $\mathbb{R}\omega \oplus \mathbf{H}^{0,1}_{\bar\partial}(M)$ induces via the above isomorphism an orientation on $\mathbf{H}^2_+(M)$.

Let us point out a very confusing fact. Denote by $*_c$ the Hodge operator $\Omega^{p,q}(M) \to \Omega^{2-p,2-q}(M)$. Recall that $*_c$ is *conjugate linear.* A complex valued 2-form $\Omega$ on $M$ is said to be self-dual if

$$*_c\Omega = \bar\Omega$$

where the correspondence

$$\Omega^{p,q}(M) \ni \Omega \mapsto \bar\Omega \in \Omega^{q,p}(M)$$

is given by the Hermitian metric on $TM$. For example the 2-form $\Omega = \mathbf{i}\omega$ is self-dual but

$$*_c\Omega = -\mathbf{i} * \omega = \bar\Omega = -\Omega.$$

Now observe that any purely imaginary self-dual 2-form $\Phi$ decomposes as

$$\Phi = \Phi_0\omega + \Phi^{0,2} + \Phi^{2,0}$$

where

$$\Phi_0 \in \Omega^0(M, \mathbf{i}\mathbb{R}), \quad \Phi^{0,2} \in \Omega^{0,2}(M), \quad \Phi^{2,0} = -\overline{\Phi^{0,2}} \in \Omega^{2,0}(M)$$

and

$$\Phi_0 = \frac{1}{2}\Lambda\Phi. \tag{3.2.5}$$

Recall that $\Lambda$ is the adjoint of the exterior multiplication by $\omega$ and $\Lambda\omega = 2 = \dim_{\mathbb{C}} M$.

For any complex line bundle $L \to M$ and any $\psi = \alpha \oplus \beta \in \Gamma(\mathbb{S}^+_L)$ we can regard the endomorphism $q(\psi)$ of $\mathbb{S}^+_L$ as a purely imaginary self-dual 2-form, so that it has a decomposition

$$q(\psi) = q(\psi)_0\omega + q(\psi)^{0,2} + q(\psi)^{2,0}$$

as above. The identity (1.3.5) in Example 1.3.3 of §1.3.1 shows that

$$q(\psi)_0 = \frac{\mathbf{i}}{4}(|\alpha|^2 - |\beta|^2), \tag{3.2.6}$$

$$q(\psi)^{0,2} = \frac{1}{2}\bar\alpha\beta := \frac{1}{2}\bar\alpha \otimes \beta \in \Gamma(L^{-1} \otimes L \otimes K_M^{-1}) \cong \Omega^{0,2}(M). \tag{3.2.7}$$

• The Kähler form on $M$ also suggests a special family of perturbation parameters $\eta$. Fix $\mu \in \mathbf{H}^{0,2}_{\bar{\partial}}(M)$ so that $\bar{\mu}$ is a holomorphic section of $K_M$. For every $t \in \mathbb{R}$ define

$$\eta_t = \eta_t(\mu) := \mathbf{i}F_{A_0} + \frac{t}{8}\omega + 2(\mu + \bar{\mu}). \tag{3.2.8}$$

Now fix a *spin*$^c$ structure on $M$ or, equivalently, a complex Hermitian line bundle $L \to M$. Denote by $\mathcal{C}_L$ the space of configurations determined by this *spin*$^c$ structure. Using the identifications (3.2.1) and (3.2.2) we can alternatively describe $\mathcal{C}_L$ as

$$\mathcal{C}_L = \{(\alpha, \beta; B) \in \Gamma(L) \times \Gamma(L \otimes K_M^{-1}) \times \mathfrak{A}(L)\}$$

so that

$$\mathsf{C} = (\psi, A) = (\alpha \oplus \beta; A_0 \dot{+} 2B).$$

The $\eta_t$-perturbed Seiberg-Witten equations for $\mathsf{C}$

$$\left\{ \begin{array}{rcl} \mathfrak{D}_A\psi & = & 0 \\ \mathbf{c}(F_A^+ + \mathbf{i}\eta_t^+) & = & \frac{1}{2}q(\psi) \end{array} \right.$$

are equivalent to

$$\left\{ \begin{array}{rcl} \bar{\partial}_B\alpha + \bar{\partial}_B^*\beta & = & 0 \\ & & \\ \Lambda F_B & = & \frac{\mathbf{i}}{8}(|\alpha|^2 - |\beta|^2 - t) \\ & & \\ F_B^{0,2} + \mathbf{i}\mu & = & \frac{1}{8}\bar{\alpha}\beta \end{array} \right. . \tag{3.2.9}$$

The first equation in (3.2.9) is clear in view of (3.2.4). Let us explain the remaining two.

Observe first that

$$F_A^+ + \mathbf{i}\eta_t = 2F_B^+ + \frac{\mathbf{i}}{8}t\omega + 2\mathbf{i}(\mu + \bar{\mu})$$

and

$$\Lambda\Omega = \Lambda\Omega^+, \quad \forall \Omega \in \Omega^2(M) \otimes \mathbb{C}.$$

Thus

$$\Lambda(F_A^+ + \mathbf{i}\eta_t^+) = 2\Lambda F_B + \frac{\mathbf{i}}{4}t.$$

Using the identity (3.2.6) we deduce

$$\Lambda q(\psi) = \frac{\mathbf{i}}{2}(|\alpha|^2 - |\beta|^2).$$

The second equation in (3.2.9) is precisely the equality $\Lambda(F_A^+ + \mathbf{i}\eta_t) = \Lambda q(\psi)$.

Next observe that

$$(F_A + \mathbf{i}\eta_t)^{0,2} = F_{A_0}^{0,2} + 2F_B^{0,2} + 2\mathbf{i}\mu + \frac{\mathbf{i}}{8}t\omega^{0,2} = 2F_B^{0,2} + 2\mathbf{i}\mu$$

because $\omega$ is a $(1,1)$-form and $F_{A_0}^{0,2} = 0$ since $A_0$ is the Chern connection defined by a Hermitian metric and a **holomorphic structure** on $K_M^{-1}$. The last equality in (3.2.9) is now a consequence of (3.2.7).

The virtual dimension of the moduli space corresponding to the *spin*$^c$ structure $L$ is

$$d(L) = \frac{1}{4}\{(2L - K_M)^2 - (2\chi_M + 3\tau_M)\}$$

$$= \frac{1}{4}\{(4L^2 - 4L\cdot K_M + K_M^2) - K_M^2\} = L\cdot(L - K).$$

**Remark 3.2.1.** Suppose $b_2^+(M) = 1$ i.e. $p_g(M) = 0$. Then $\mu$ can only be $0$. To decide in which chamber $\eta_t$ lies we have to understand the sign of

$$\int_M (\eta_t - 2\pi c_1(\det\sigma\otimes L)\,)\wedge\frac{1}{\sqrt{2}}\omega$$

or, equivalently, the sign of

$$\frac{t}{8}\int_M \omega\wedge\omega + \mathbf{i}\int_M F_{A_0}\wedge\omega - 2\pi\deg_\omega(K_M^{-1}\otimes L^2).$$

Now observe that the second integral is precisely $2\pi\deg_\omega(K_M^{-1})$ so we have to decide the sign of

$$\frac{t\mathrm{vol}\,(M)}{4} - 4\pi\deg_\omega L.$$

We deduce that for $t > \frac{16\pi}{\mathrm{vol}\,(M)}\deg_\omega(L)$ the perturbation $\eta_t$ lies in the positive chamber with respect to the Kähler metric while for $t < \frac{16\pi}{\mathrm{vol}\,(M)}\deg_\omega(L)$ it lies in the negative chamber.

**Definition 3.2.2.** A complex line bundle $L\to M$ is said to have type $(1,1)$ with respect to the Kähler metric if its first Chern class is of type $(1,1)$ with respect to the Hodge decomposition

$$\mathbf{H}^2(M,\mathbb{C}) = \mathbf{H}_{\bar\partial}^{1,1}(M)\oplus\mathbf{H}_{\bar\partial}^{0,1}(M)\oplus\mathbf{H}_{\bar\partial}^{2,0}(M).$$

Observe that if $b_2^+(M) = 1$ then all classes have type $(1,1)$ since $p_g = \dim H_{\bar\partial}^{2,0}(M) = 0$.

We have the following vanishing result.

**Proposition 3.2.3.** *If $L\to M$ is a complex line bundle over $M$ which is not of type $(1,1)$ then the Seiberg-Witten invariant of $M$ corresponding to the spin$^c$ structure determined by $L$ is zero,*

$$\mathbf{sw}_M(L) = 0.$$

**Proof** We consider the equations (3.2.9) corresponding to $\mu = 0$ and $t = 0$. Applying $\bar{\partial}_B$ to the first equation we deduce

$$\bar{\partial}_B^2 \alpha + \bar{\partial}_B \bar{\partial}_B^* \beta = 0$$

so that

$$F_B^{0,2} \alpha + \bar{\partial}_B \bar{\partial}_B^* \beta = 0.$$

Take the inner product with $\beta$ and integrate by parts to obtain

$$\int_M \langle F_B^{0,2} \alpha, \beta \rangle dv + \int_M |\bar{\partial}_B^* \beta|^2 = 0.$$

Now use the third equation of (3.2.9) in the first integral above. We get

$$\frac{1}{8} \int_M |\bar{\alpha}\beta|^2 + \int_M |\bar{\partial}_B^* \beta| dv = 0.$$

This shows $\alpha \cdot \beta = 0$ so that $F_B^{0,2} = 0$. Since $F_B^{2,0} = \overline{F_B^{0,2}}$ we deduce $F_B$ is a $(1,1)$-class so that $L$ must be a $(1,1)$-line bundle. This shows that (3.2.9) has no solution in this case. ■

**3.2.2. Monopoles, vortices and divisors.** As was observed from the very beginning by Edward Witten in [**149**], the solutions of the equations (3.2.9) are equivalent to the complex analytic objects called *vortices*. These can then be described quite explicitly in terms of divisors on $M$. In particular, this opens the possibility of completely and explicitly describing the moduli spaces of monopoles.

Since we are interested only in Seiberg-Witten invariants then, according to Proposition 3.2.3, it suffices to consider only the case when $L$ has type $(1,1)$. To obtain further information about the solutions of (3.2.9) we will refine the technique used in the proof of Proposition 3.2.3. We follow closely the approach in [**13**].

Observe that since $L$ has type $(1,1)$ it follows from the third equation of (3.2.9) that $\mathbf{i}\mu$ is the $\bar{\partial}$-harmonic part of the $(0,2)$-form $\frac{1}{8}\bar{\alpha}\beta$. Denote by $[\bar{\alpha}\beta]$ the $\bar{\partial}$-harmonic part of $\bar{\alpha}\beta$. Again, applying $\bar{\partial}_B$ to the first equation in (3.2.9) we deduce as in the proof of Proposition 3.2.3

$$(\frac{1}{8}\bar{\alpha}\beta - \mathbf{i}\mu)\alpha = \bar{\partial}_B \bar{\partial}_B^* \beta = 0$$

or equivalently

$$\frac{1}{8}(\bar{\alpha}\beta - [\bar{\alpha}\beta])\alpha + \bar{\partial}_B \bar{\partial}_B^* \beta = 0.$$

Taking the inner product with $\beta$ and integrating by parts we get

$$\frac{1}{8}\int_M \langle \bar{\alpha}\beta - [\bar{\alpha}\beta], \bar{\alpha}\beta\rangle dv + \|\bar{\partial}_B^*\beta\|_{L^2}^2 = 0.$$

Since $[\bar{\alpha}\beta]$ is $L^2$-orthogonal to $\bar{\alpha}\beta - [\bar{\alpha}\beta]$ we deduce

$$\frac{1}{8}\|\bar{\alpha}\beta - [\bar{\alpha}\beta]\|_{L^2}^2 + \|\bar{\partial}_B^*\beta\|_{L^2}^2 = 0.$$

Thus

$$\bar{\partial}_B^*\beta = 0, \;\; \bar{\alpha}\beta = [\bar{\alpha}\beta] = 8\mathbf{i}\mu$$

and

$$F_B^{0,2} = \frac{1}{8}(\bar{\alpha}\beta - [\bar{\alpha}\beta]).$$

Using the equality $\bar{\partial}_B^*\beta = 0$ in the first equation of (3.2.9) we conclude that

$$\bar{\partial}_B\alpha = 0.$$

We have thus proved the following result.

**Proposition 3.2.4.** *Any solution $(\alpha, \beta, B)$ of (3.2.9) satisfies the conditions*

(3.2.10a) $$F_B^{0,2} = 0,$$

(3.2.10b) $$\bar{\partial}_B\alpha = \bar{\partial}_B^*\beta = 0,$$

(3.2.10c) $$\bar{\alpha}\beta = 8\mathbf{i}\mu,$$

(3.2.10d) $$\Lambda F_B = \frac{\mathbf{i}}{8}(|\alpha|^2 - |\beta|^2 - t).$$

**Definition 3.2.5.** The solutions of the system (3.2.10a) – (3.2.10d) are called $(\mu, t)$-*vortices*. When $\mu = 0$ we will call them simply vortices.

Obviously, any $(\mu, t)$-vortex is also an $\eta_t$-monopole.

The condition (3.2.10a) shows that $B$ induces an integrable complex structure on $L$. The equalities (3.2.10b) show that $\alpha$ is a holomorphic section of $L$ (with respect to the above holomorphic structure) and $\beta$ is an antiholomorphic section of $K_M^{-1} \otimes L = L - K_M$. Hence $\bar{\beta}$ is a holomorphic section of $K_M - L$. The equality (3.2.10c) can be rewritten as

(3.2.11) $$\alpha\bar{\beta} = -8\mathbf{i}\bar{\mu}.$$

In the above new formulation, $\bar{\mu}$ is a holomorphic section of $K_M$. To proceed further we have to distinguish two cases.

**A. The case** $\mu = 0$. Thus, $\alpha\bar{\beta} = 0$. Since both $\alpha$ and $\bar{\beta}$ are holomorphic sections the unique continuation principle implies that at least one of them must be identically zero.

Now let us observe that if a holomorphic line bundle $E \to M$ admits a nontrivial holomorphic section $s$ then $\deg_\omega(E) \geq 0$ because $\deg_\omega(E)$ can be interpreted as the integral of $\omega$ over the (possibly singular, possibly empty) complex curve $s^{-1}(0)$ on $M$. According to Proposition 3.1.1, this integral is none other than the *area* of this curve . Thus,

$$\alpha \neq 0 \Rightarrow \deg_\omega(L) \geq 0 \text{ and } \beta = 0$$

while

$$\beta \neq 0 \Rightarrow \deg_\omega(K_M - L) \geq 0 \text{ and } \alpha = 0.$$

On the other hand, observe that

$$\deg_\omega(L) = \frac{\mathbf{i}}{2\pi}\int_M F_B \wedge \omega = \frac{\mathbf{i}}{2\pi}\int_M \Lambda F_B \frac{1}{2}\omega^2$$

$$= \frac{\mathbf{i}}{2\pi}\int_M \Lambda F_B dv_M \overset{(3.2.10d)}{=} \frac{1}{16\pi}\int_M (|\beta|^2 + t - |\alpha|^2) dv_M.$$

If we fix $t$ such that

$$t \neq \frac{16\pi}{\operatorname{vol}(M)} \deg_\omega(L)$$

then the above equality shows that at least one of $\alpha$ or $\beta$ must be nontrivial. Moreover, when $t < \frac{16\pi}{\operatorname{vol}(M)} \deg_\omega(L)$ then $\alpha = 0$ and $\beta \neq 0$ because otherwise we would obtain

$$\beta = 0 \text{ and } \deg_\omega(L) = \frac{1}{16\pi}\int_M (t - |\alpha|^2) dv_M \leq \frac{t\operatorname{vol}(M)}{16\pi}.$$

Similarly, when $t > \frac{16\pi}{\operatorname{vol}(M)} \deg_\omega(L)$ we must have $\beta = 0$ and $\alpha \neq 0$. Using Remark 3.2.1 we obtain the following vanishing result.

**Proposition 3.2.6.** *(a) If* $b_2^+(M) > 1$ *and* $\mathbf{sw}_M(L) \neq 0$ *then*

$$0 \leq \deg_\omega L \leq \deg_\omega K_M.$$

*(b) If* $b_2^+(M) = 1$ *and* $\mathbf{sw}_M^+(L) \neq 0$ *then*

$$0 \leq \deg_\omega(L)$$

*while if* $\mathbf{sw}_M^-(L) \neq 0$ *then*

$$\deg_\omega(L) \leq \deg_\omega(K_M).$$

The above discussion also shows that for $t \gg 0$ the vortices are found amongst pairs $(E, \alpha)$ where $E$ is a holomorphic line bundle topologically isomorphic to $L$ and $\alpha$ is a holomorphic section. The metric on $L$ imposes an additional condition on $\alpha$ through (3.2.10d) in which $\beta = 0$. The pairs

*(holomorphic structure on L, holomorphic section of L)*

are precisely the effective divisors $D$ on $M$ such that

$$c_1([D]) = c_1(L).$$

Can we reverse this process? More precisely, given an effective divisor $[D]$ such that $c_1([D]) = c_1(L)$, can we find a solution $(\alpha, \beta = 0; B)$ of (3.2.10a) – (3.2.10d) such that $D$ is the divisor determined by $\alpha$, $D = \alpha^{-1}(0)$? To formulate an answer to this question let us first fix a Hermitian metric $h_0$ on $L$.

**Proposition 3.2.7.** *Suppose $L \to M$ has type $(1,1)$ and $\deg_\omega(L) \geq 0$. Fix*

$$t > \frac{16\pi}{\mathrm{vol}\,(M)} \deg_\omega(L). \tag{3.2.12}$$

*Given an integrable CR operator $\vartheta$ on $L$ and a $\vartheta$-holomorphic section $\alpha$ of $L$,*

$$\vartheta\alpha = 0,$$

*there exists a unique function $u \in C^\infty(M)$ such that the following hold.*

*(a) If $\vartheta_u := e^u \vartheta e^{-u}$ then $\alpha_u = e^u\alpha$ is $\vartheta_u$-holomorphic.*

*(b) If $B_u$ denotes the $h_0$-Hermitian connection on $L$ induced by the CR-operator $\vartheta_u$ then*

$$\Lambda F_{B_u} = \frac{\mathbf{i}}{8}(|\alpha_u|^2 - t), \tag{3.2.13}$$

*that is , $(\alpha_u \oplus 0; B_u)$ satisfies (3.2.10a)-(3.2.10d) with $\mu = 0$.*

**Proof** Observe first that, for any $u \in C^\infty(M)$, the CR-operators $\vartheta$ and $\vartheta_u$ define the same holomorphic structure on $L$ and that in fact the condition (a) above is tautological. Denote by $B_0$ the Chern connection determined by $\vartheta$ and $h_0$. Let $u \in C^\infty(M)$. As shown in Example 1.4.19 of §1.4.2 the Chern connection $B_u$ determined by $e^u\vartheta e^{-u}$ and $h_0$ is

$$B_u = B_0 + \partial u - \bar{\partial} u.$$

Its curvature is

$$F_{B_u} = F_{B_0} + \bar{\partial}\partial u - \partial\bar{\partial} u. \tag{3.2.14}$$

We have to find $u$ so that $(\alpha_u, B_u)$ satisfy (3.2.13), i.e.

$$\Lambda F_{B_u} = \frac{\mathbf{i}}{8}(|\alpha_u|^2 - t).$$

Using (3.2.14) we can rewrite this as an equation in $u$:

$$\Lambda(\bar{\partial}\partial u - \partial\bar{\partial}u) - \frac{\mathbf{i}}{8}|\alpha|^2_{h_0}e^{2u} = -\frac{\mathbf{i}t}{8} - \Lambda F_{B_0}. \tag{3.2.15}$$

On the other hand, according to Corollary 1.4.11 of §1.4.1 we have

$$\Lambda\bar{\partial}(\partial u) = -\mathbf{i}\bar{\partial}^*\partial u = -\frac{\mathbf{i}}{2}\Delta_d u$$

and

$$\Lambda\partial\bar{\partial}u = \mathbf{i}\partial^*\partial u = \frac{\mathbf{i}}{2}\Delta_d.$$

The equation (3.2.15) can now be rewritten as

$$\Delta_d u + \frac{1}{8}|\alpha|^2_{h_0}e^{2u} = (\frac{t}{8} - \mathbf{i}\Lambda F_{B_0}) =: f. \tag{3.2.16}$$

This equation was studied in great detail by J. Kazdan and F. Warner in [**61**] (see also [**105**] for a different approach). They proved the following result.

**Theorem 3.2.8. (Kazdan-Warner,** [**61**, Thm. 10.5]**)** *Suppose $k$ is a positive real number and $w(x)$ is a smooth function which is positive outside a set of measure zero in $M$. Then the equation*

$$\Delta_M u + w(x)e^{ku} = g \in C^\infty(M)$$

*has a solution (which is unique) if and only if*

$$\int_M g dv_M > 0.$$

Using the above existence theorem we deduce that the equation (3.2.16) has a solution (and no more than one) if and only if

$$\int_M f dv_M > 0.$$

In our case this means

$$t\mathrm{vol}\,(M) > 8\int_M \mathbf{i}\Lambda F_{B_0}dv_M = 16\pi\int_M \frac{\mathbf{i}}{2\pi}\Lambda F_B dv_M = 16\pi\deg_\omega(L)$$

which is precisely the condition (3.2.12). The proposition is proved. ■

We have the two-way correspondences

$t \gg 0$

$\eta_t(\mu = 0)$*-monopoles* $\leftrightarrow$ *effective divisors $D$ such that $c_1([D]) = c_1(L)$.*

$t \ll 0$

$$\eta_t(\mu = 0)\text{-}monopoles \leftrightarrow \text{ effective divisors } D \text{ such that}$$
$$c_1([D]) = c_1(K_M - L).$$

**Notation** *The symbol* $\mathbf{sw}_M^{(\pm)}(\sigma)$ *will denote* $\mathbf{sw}_M^{\pm}(\sigma)$ *if* $b_2^+(M) = 1$ *and* $\mathbf{sw}_M(\sigma)$ *if* $b_2^+(M) > 1$.

From the above correspondences we deduce immediately the following consequences.

**Corollary 3.2.9.** *Suppose $M$ is a Kähler surface and $L$ is a Hermitian line bundle.*
*(a) If* $\mathbf{sw}_M^{(+)}(L) \neq 0$ *then $L$ admits holomorphic structures with nontrivial holomorphic sections.*
*(a) If* $\mathbf{sw}_M^{(-)}(L) \neq 0$ *then $K_M - L$ admits holomorphic structures with nontrivial holomorphic sections.*

**Corollary 3.2.10.** *Suppose $M$ is a Kähler surface and $L$ is a Hermitian line bundle.*
*(a) If* $\deg_\omega L = 0$ *and* $\mathbf{sw}_M^{(+)}(L) \neq 0$ *then $L$ is the (topologically) trivial line bundle.*
*(b) If* $\deg_\omega(L) = \deg_\omega(K_M)$ *and* $\mathbf{sw}_M^{(-)}(L) \neq 0$ *then $L$ is (topologically) isomorphic to $K_M$.*

**Proof** We prove only (a). Part (b) follows from (a) using the involution $\sigma \mapsto \bar{\sigma}$ on $Spin^c(M)$.

We use the perturbation $\eta_t$, with $\mu = 0$ and $t \gg 0$. The condition $\mathbf{sw}_M^{(+)}(L) \neq 0$ implies that there exists a holomorphic structure on $L$ admitting holomorphic sections. If such a section does not vanish anywhere we deduce that $L$ is trivial. If it vanishes somewhere its zero locus defines an effective divisor $D$ and

$$\deg_\omega([D]) = \deg_\omega(L) = 0.$$

This contradicts Proposition 3.1.1, which states that $\deg_\omega([D])$ is a positive number expressible in terms of the area of supp $(D)$. The corollary is proved.

■

Clearly, gauge equivalent monopoles lead to identical divisors, so that the *set* of gauge equivalence classes of monopoles can be identified with the above *set* of divisors. This identification goes deeper. The set of effective divisors carrying the homology class Poincaré dual to $c_1(L)$ can be given a (Hilbert) scheme structure. This structure can be described in terms of the deformation complexes of the monopoles. If $M$ is algebraic this allows one to cast in an algebraic-geometric context the entire problem of computing the Seiberg-Witten invariants. We will not follow this approach but we refer the reader for details to [**21, 41, 42**].

**B. The case $\mu \neq 0$.** Suppose $(\alpha \oplus \beta, B)$ is a $(\mu, t)$-vortex. Thus $\alpha$ defines an effective divisor $D$ such that

$$c_1([D]) = c_1(L)$$

and

$$D \leq (\bar{\mu})$$

where $(\bar{\mu})$ denotes the effective divisor determined by the zeroes holomorphic section $\bar{\mu}$. More precisely, the effective divisor $D$ is the divisor determined by the holomorphic section $\bar{\beta}$. As in the case $\mu = 0$ we have the following result.

**Proposition 3.2.11. (O. Biquard, [13])** *Suppose $L$ is a complex line bundle over $M$ such that*

$$0 \leq \deg_\omega L \leq \deg_\omega(K_M).$$

*Fix a Hermitian metric $h_0$ on $L$. Suppose there exist an integrable CR operator $\vartheta$ on $L$ and holomorphic sections $\alpha \in \Gamma(L)$ and $\gamma \in \Gamma(K_M - L)$ such that*

$$\alpha\gamma = -8\mathbf{i}\bar{\mu}.$$

*Then there exists a unique function $u \in C^\infty(M)$ such that if $B_u$ denotes the Chern connection determined by $h_0$ and $\vartheta_u = e^u \vartheta e^{-u}$ then*

$$(\alpha_u, \beta_u, B_u) := (e^u \alpha, e^{-u}\bar{\gamma}, B_u)$$

*is a $(\mu, t)$-vortex.*

Observe that if $\vartheta^*$ is the CR operator induced by $\vartheta$ on $L^*$ then

$$(e^u \vartheta e^{-u})^* = e^{-u}\vartheta^* e^u.$$

This explains the definition of $\beta_u$.

**Proof** Clearly, for any smooth $u$ the collection $(\alpha_u, \beta_u, B_u)$ defined as in the statement of the propositions automatically satisfies the conditions (3.2.10a) – (3.2.10c) in the definition of a $(\mu, t)$-vortex. Thus, it suffices to find $u$ such that $(\alpha_u, \beta_u, B_u)$ satisfies (3.2.10d).

Denote by $B_0$ the Chern connection on $L$ determined by $h_0$ and $\vartheta$. Arguing exactly as in the proof of Proposition 3.2.7 we deduce that $u$ must be a solution of the equation

$$\Delta_d u + \frac{1}{8}|\alpha|^2_{h_0} e^{2u} - \frac{1}{8}|\gamma|^2_{h_0} e^{-2u} = f := \Big(\frac{t}{8} - \mathbf{i}\Lambda F_{B_0}\Big). \tag{3.2.17}$$

We have to show that the above equation admits a unique smooth solution.

**Existence** We will use the method of sub/supersolutions. For an approach based on the continuity method we refer to [**13**].

The method of sub/super-solutions is based on the following very general result.

**Theorem 3.2.12.** *Suppose $F : M \times \mathbb{R} \to \mathbb{R}$ is a smooth function and there exist two smooth functions $u, U : M \to \mathbb{R}$ such that*

$$u \leq U \ \text{ on } \ M, \tag{3.2.18}$$

$$\Delta_M u \leq F(x, u(x)), \ \ \forall x \in M, \tag{3.2.19}$$

*and*

$$\Delta_M U \geq F(x, U(x)) \ \ \forall x \in M. \tag{3.2.20}$$

*Then there exists a smooth solution $v$ of the partial differential equation*

$$\Delta_M v = F(x, v) \tag{3.2.21}$$

*such that $u \leq v \leq U$.*

The function $u$ (resp. $U$) is said to be a sub-(resp. super)-solution of (3.2.21). An outline of the proof of this theorem can be found in [**105**, §9.3.3]. For complete details we refer to [**1, 61**]. The proof is based on a very important principle in the theory of second order elliptic p.d.e.'s which will also play an important role in our existence proof.

**Comparison Principle** Suppose $g : M \times \mathbb{R} \to \mathbb{R}$ is a smooth function such that for all $x$ outside a set of measure zero the function

$$u \mapsto g(x, u)$$

is strictly increasing. Then

$$\Delta_M u + g(x, u) \geq \Delta_M v + g(x, v) \Longrightarrow u \geq v.$$

**Exercise 3.2.1.** Prove the comparison principle. (**Hint:** Consult [**105**, §9.3.3].)

Using Kazdan-Warner's Theorem 3.2.8 we deduce that for every $s \gg 0$ there exist smooth functions $U_s$ and $v_s$ on $M$ such that

$$\Delta_M U_s + \frac{1}{8}|\alpha|^2 e^{2U_s} = f + s,$$

$$\Delta_M v_s + \frac{1}{8}|\gamma|^2 e^{2v_s} = s$$

where $f$ is the function on the right-hand side of (3.2.17). Set

$$a = \frac{1}{8}\sup_{x\in M}|\alpha(x)|^2, \;\; b = \frac{1}{8}\sup_{x\in M}|\gamma(x)|^2,$$

$$f_{min} := \min_{x\in M} f(x).$$

Observe that if $c_s$ is the constant function defined by

$$ae^{2c_s} = f_{min} + s$$

then

$$\Delta_M c_s + |\alpha|^2 e^{2c_s} \leq f_s = \Delta U_s + |\alpha|^2 e^{2U_s}.$$

Using the comparison principle we deduce

$$U_s \geq c_s \to \infty \;\text{ as }\; s \to \infty. \tag{3.2.22}$$

In particular, this shows that for $s$ sufficiently large $U_s$ is a super-solution of (3.2.17) because

$$\Delta_M U_s + \frac{1}{8}|\alpha|^2 e^{2U_s} - \frac{1}{8}|\gamma|^2 e^{-2U_s} \geq f + s - be^{-2c_s} = f + s - \frac{ab}{f_{min}+s} > f$$

for $s \gg 0$. Similarly, if we denote by $d_s$ the constant function defined by

$$be^{2d_s} = s$$

we deduce

$$\Delta d_s + \frac{1}{8}|\gamma|^2 e^{2ds} \leq s$$

so that

$$v_s \geq d_s. \tag{3.2.23}$$

Set $u_s := -v_s$. Then

$$\Delta_M u_s + \frac{1}{8}|\alpha|^2 e^{2u_s} - \frac{1}{8}|\gamma|^2 e^{-2u_s} = -s + |\alpha|^2 e^{2u_s}$$

$$\leq -s + ae^{-2d_s} = -s + \frac{ab}{s} \leq f$$

for $s \gg 0$. Thus $u_s$ is a sub-solution of (3.2.17). Using (3.2.22) and (3.2.23) we deduce that for $s \gg 0$ we have

$$u_s \leq -d_s < c_s \leq U_s.$$

Using Theorem 3.2.12 we conclude that (3.2.17) has a smooth solution $u$ such that

$$u_s \leq u \leq U_s$$

for $s \gg 0$.

**Uniqueness** It follows immediately from the comparison principle in which $g(x,u) = \frac{1}{8}|\alpha(x)|^2 e^{2u} - \frac{1}{8}|\gamma(x)|^2 e^{-2u}$. The proof of Proposition 3.2.11 is now complete. ■

The above proposition has an immediate interesting geometric consequence.

**Proposition 3.2.13.** *Suppose $M$ is a Kähler surface such that $p_g(M) > 0$ and $K_M$ is not holomorphically trivial. Fix $\mu \in H^{2,0}_{\bar\partial}(M) \setminus \{0\}$ and denote by $(\mu)$ the effective divisor determined by this section. Then for all $t \in \mathbb{R}$ there exists a bijection between the set of orbits of $\eta_t(\mu)$-monopoles and the set $\mathcal{S}_\mu(M)$ of divisors $D$ on $M$ with the following properties.*
◇ $0 \leq D \leq (\mu)$.
◇ $c_1([D]) = c_1(L)$ *in* $H^2(M, \mathbb{Z})$.

**3.2.3. Deformation theory.** Now that we have an idea of the nature of monopoles we want to investigate whether the cohomology of the deformation complex associated to a monopole on a Kähler surface can be described in complex analytic terms.

Fix $\mu \in \mathbf{H}^{0,2}_{\bar\partial}(M)$, $t \in \mathbb{R}$ and $L \to M$ a type-$(1,1)$ Hermitian bundle over $M$. Suppose $(\alpha \oplus \beta, B)$ is a $(\mu, t)$-vortex corresponding to $L$.

The corresponding monopole is $\mathsf{C} = (\psi, A)$ where

$$\psi = \alpha \oplus \beta, \quad A := A_0 \dot{+} 2B.$$

The tangent space to $\mathcal{C}_L$ at $\mathsf{C}$ is

$$T_{\mathsf{C}}\mathcal{C}_L = \Gamma(\mathbb{S}_L \oplus \mathbf{i}\Lambda^1 T^* M)$$

where, for simplicity, we omitted the Sobolev labels. We will represent a tangent vector $\dot{\mathsf{C}} = (\dot\psi, \mathbf{i}\dot b)$ (where $\mathbf{i}\dot a = 2\mathbf{i}\dot b$) in complex analytic terms. Thus

$$\mathbf{i}\dot b = \frac{\mathbf{i}}{\sqrt{2}}(\varphi + \bar\varphi), \quad \varphi \in \Omega^{0,1}(M),$$

and

$$\dot\psi = \dot\alpha \oplus \dot\beta \in \Omega^{0,0}(L) \oplus \Omega^{0,2}(L).$$

Recall that (see §2.2.2)

$$\mathcal{T}_{\mathsf{C}} \begin{bmatrix} \dot\psi \\ \mathbf{i}\dot b \end{bmatrix} = \begin{bmatrix} \mathfrak{D}_A \dot\psi \\ 2d^+ \mathbf{i}\dot b \\ -4\mathbf{i}d^* \dot b \end{bmatrix} + \begin{bmatrix} \mathbf{c}(\mathbf{i}\dot b)\psi \\ -\frac{1}{2}\dot q(\psi, \dot\psi) \\ -\mathbf{i}\mathfrak{Im}\langle \psi, \dot\psi \rangle \end{bmatrix}.$$

We now proceed to express each of the objects in the above expression in terms of $\varphi$, $\dot\alpha$ and $\dot\beta$.

First, we have

$$\not{\mathfrak{D}}_A\dot{\psi} + \mathbf{c}(\mathbf{i}\dot{b})\psi = \sqrt{2}[\bar{\partial}_B \ \bar{\partial}_B^*] \cdot \begin{bmatrix} \dot{\alpha} \\ \dot{\beta} \end{bmatrix} + \frac{1}{\sqrt{2}}[\mathbf{c}(\mathbf{i}\varphi) \ \ \mathbf{c}(\mathbf{i}\bar{\varphi})] \cdot \begin{bmatrix} \alpha \\ \beta \end{bmatrix}$$

$$= \sqrt{2}(\bar{\partial}_B\dot{\alpha} + \bar{\partial}_B^*\dot{\beta}) + \frac{1}{\sqrt{2}}\mathbf{c}(\mathbf{i}\varphi)\alpha + \frac{1}{\sqrt{2}}\mathbf{c}(\mathbf{i}\bar{\varphi})\beta$$

(use the computations in Example 1.3.3 in §1.3.1)

$$= \sqrt{2}(\bar{\partial}_B\dot{\alpha} + \bar{\partial}_B^*\dot{\beta}) + \mathbf{i}(\varphi \wedge \alpha - \bar{\varphi} \lrcorner \beta)$$

where $\bar{\varphi}\lrcorner$ denotes the contraction by a $(1,0)$-form.

Next observe that the self-dual part of a complex 2-form $\theta$, defined by $\bar{\theta}^+ = *_c\theta^+$, is explicitly given by

$$\theta^+ = \theta_0\omega + \theta^{0,2} + \theta^{2,0} = \frac{1}{2}\Lambda\theta\omega + \theta^{0,2} + \theta^{2,0}.$$

In our case

$$\theta = 2\mathbf{i}d\dot{b} = \mathbf{i}\sqrt{2}d(\varphi + \bar{\varphi}) = \mathbf{i}\sqrt{2}(\partial + \bar{\partial})(\varphi + \bar{\varphi})$$

so that

$$2\mathbf{i}d^+\dot{b} = \frac{\mathbf{i}}{\sqrt{2}}\Lambda(\partial\varphi + \bar{\partial}\bar{\varphi})\omega + \mathbf{i}\sqrt{2}(\bar{\partial}\varphi + \partial\bar{\varphi}).$$

Since

$$q(\psi) = q(\alpha \oplus \beta) = \frac{\mathbf{i}}{4}(|\alpha|^2 - |\beta|^2)\omega + \frac{1}{2}(\bar{\alpha}\beta - \alpha\bar{\beta})$$

we deduce

$$\dot{q}(\psi, \dot{\psi}) = \frac{\mathbf{i}}{2}(\mathfrak{Re}\langle\alpha, \dot{\alpha}\rangle - \mathfrak{Re}\langle\beta, \dot{\beta}\rangle)\omega + \frac{1}{2}(\dot{\bar{\alpha}}\beta + \bar{\alpha}\dot{\beta} - \dot{\alpha}\bar{\beta} - \alpha\dot{\bar{\beta}}).$$

Next observe that

$$4d^*\dot{b} = 2\sqrt{2}(\partial + \bar{\partial})^*(\varphi + \bar{\varphi}) = 2\sqrt{2}(\bar{\partial}^*\varphi + \partial^*\bar{\varphi})$$

and

$$\mathfrak{Im}\langle\psi, \dot{\psi}\rangle = \mathfrak{Im}\langle\alpha, \dot{\alpha}\rangle + \mathfrak{Im}\langle\beta, \dot{\beta}\rangle.$$

Thus

$$(\dot{\alpha}, \dot{\beta}, \varphi) \longleftrightarrow (\dot{\psi}, \mathbf{i}\dot{b} = \frac{\mathbf{i}}{\sqrt{2}}(\varphi + \bar{\varphi})\ ) \in \ker \mathcal{T}_C$$

if and only if

(3.2.24a) $$\sqrt{2}(\bar{\partial}_B\dot{\alpha} + \bar{\partial}_B^*\dot{\beta}) + \mathbf{i}(\varphi \wedge \alpha - \bar{\varphi} \lrcorner \beta) = 0,$$

(3.2.24b) $$\Lambda(\partial\varphi + \bar{\partial}\bar{\varphi}) = \frac{1}{2\sqrt{2}}(\mathfrak{Re}\langle\alpha, \dot{\alpha}\rangle - \mathfrak{Re}\langle\beta, \dot{\beta}\rangle),$$

(3.2.24c) $$\mathbf{i}\bar{\partial}\varphi = \frac{1}{4\sqrt{2}}(\dot{\bar{\alpha}}\beta + \bar{\alpha}\dot{\beta}),$$

(3.2.24d) $$2\sqrt{2}(\bar{\partial}^*\varphi + \partial^*\bar{\varphi}) + \mathfrak{Im}\langle\alpha,\dot{\alpha}\rangle + \mathfrak{Im}\langle\beta,\dot{\beta}\rangle = 0.$$

These equations can be further simplified using the Kähler-Hodge identities in §1.4.1

$$\Lambda\partial\varphi = \mathbf{i}\bar{\partial}^*\varphi,\ \ \Lambda\bar{\partial}\bar{\varphi} = -\mathbf{i}\partial^*\bar{\varphi},\ \ \forall\varphi\in\Omega^{0,1}(M).$$

Using these identities in (3.2.24b) we deduce

$$\mathbf{i}\mathfrak{Im}\,\bar{\partial}^*\varphi = \frac{1}{2}(\bar{\partial}^*\varphi - \partial^*\bar{\varphi}) = -\frac{\mathbf{i}}{4\sqrt{2}}(\mathfrak{Re}\langle\alpha,\dot{\alpha}\rangle - \mathfrak{Re}\langle\beta,\dot{\beta}\rangle).$$

The equation (3.2.24d) can be rewritten as

$$\mathbf{i}\mathfrak{Re}\,\bar{\partial}^*\varphi = \frac{\mathbf{i}}{2}\Big(\bar{\partial}^*\varphi + \partial^*\bar{\varphi}\Big) = -\frac{\mathbf{i}}{4\sqrt{2}}(\mathfrak{Im}\langle\alpha,\dot{\alpha}\rangle + \mathfrak{Im}\langle\beta,\dot{\beta}\rangle).$$

Thus (3.2.24b) + (3.2.24d) are equivalent to a single equation

(3.2.25) $$\mathbf{i}\bar{\partial}^*\varphi = \frac{1}{4\sqrt{2}}\Big(\overline{\langle\alpha,\dot{\alpha}\rangle} - \langle\beta,\dot{\beta}\rangle\Big).$$

**Proposition 3.2.14.** $(\dot{\alpha},\dot{\beta},\varphi)\in\ker\mathcal{T}_{\mathbb{C}}$ *if and only if they satisfy the equations*

(3.2.26a) $$\bar{\partial}\varphi = 0,$$

(3.2.26b) $$\bar{\partial}_B\dot{\alpha} + \frac{\mathbf{i}}{\sqrt{2}}\varphi\wedge\alpha = 0,$$

(3.2.26c) $$\bar{\partial}_B^*\dot{\beta} - \frac{\mathbf{i}}{\sqrt{2}}\bar{\varphi}\,\lrcorner\,\beta = 0,$$

(3.2.26d) $$\dot{\alpha}\bar{\beta} + \alpha\dot{\bar{\beta}} = 0,$$

*and (3.2.25).*

**Proof** Clearly, if $(\dot{\alpha},\dot{\beta},\varphi)$ satisfy the equations (3.2.25), (3.2.26a)–(3.2.26d) then they satisfy (3.2.24a), (3.2.24c) and thus they must lie in the kernel of $\mathcal{T}_{\mathbb{C}}$. To prove the converse statement we follow the approach in [**13**].

Rewrite (3.2.24a) as

$$-(\sqrt{2}\bar{\partial}_B^*\dot{\beta} - \mathbf{i}\bar{\varphi}\,\lrcorner\,\beta) = \sqrt{2}\bar{\partial}_B\dot{\alpha} + \mathbf{i}\varphi\wedge\alpha$$

and observe that the operator $-\mathbf{i}\,\lrcorner\,\bar{\varphi}$ on $\Omega^{*,*}(L)$ is the adjoint of $\mathbf{i}\varphi\wedge$. We deduce

$$0 \geq -\|\sqrt{2}\bar{\partial}_B^*\dot{\beta} - \mathbf{i}\bar{\varphi}\,\lrcorner\,\beta\|_{L^2}^2 = \int_M \langle\sqrt{2}\bar{\partial}_B\dot{\alpha} + \mathbf{i}\varphi\wedge\alpha, \sqrt{2}\bar{\partial}_B^*\dot{\beta} + (\mathbf{i}\varphi\wedge)^*\beta\rangle dv_M$$

$$= \int_M \langle \sqrt{2}\bar{\partial}_B\dot{\alpha}, \sqrt{2}\bar{\partial}_B^*\dot{\beta}\rangle dv_M + \int_M \langle \mathbf{i}\varphi \wedge \alpha, (\mathbf{i}\varphi\wedge)^*\beta\rangle dv_M$$

$$+ \int_M \langle \sqrt{2}\bar{\partial}_B\dot{\alpha}, (\mathbf{i}\varphi\wedge)^*\beta\rangle dv_M + \int_M \langle \mathbf{i}\varphi \wedge \alpha, \sqrt{2}\bar{\partial}_B^*\dot{\beta}\rangle dv_M.$$

The first integral vanishes. This can be seen integrating by parts and using the equality $\bar{\partial}_B^2 = F_B^{0,2} = 0$ which follows from the fact that $(\alpha, \beta, B)$ is a vortex. We deduce similarly that the second integral vanishes because $(\mathbf{i}\varphi\wedge)^2 = 0$. We conclude that

$$0 \geq \int_M \langle \mathbf{i}\varphi \wedge \bar{\partial}_B\dot{\alpha}, \beta\rangle dv_M + \int_M \langle \bar{\partial}_B(\mathbf{i}\varphi \wedge \alpha), \dot{\beta}\rangle dv_M$$

$(\bar{\partial}_B\alpha = 0)$

$$= -\sqrt{2}\int_M \langle \bar{\partial}_B(\mathbf{i}\varphi \wedge \dot{\alpha}), \beta\rangle dv_M + \sqrt{2}\int_M \langle (\mathbf{i}\bar{\partial}\varphi)\dot{\alpha}, \beta\rangle dv_M + \int_M \langle (\mathbf{i}\bar{\partial}\varphi)\alpha, \dot{\beta}\rangle dv_M$$

$(\bar{\partial}_B^*\beta = 0)$

$$= \sqrt{2}\int_M \langle\, (\mathbf{i}\bar{\partial}\varphi)\dot{\alpha}, \beta\,\rangle dv_M + \sqrt{2}\int_M \langle\, (\mathbf{i}\bar{\partial}\varphi)\alpha, \dot{\beta}\,\rangle dv_M$$

$$\stackrel{(3.2.24c)}{=} \frac{1}{4}\int_M \langle\, (\dot{\bar{\alpha}}\beta + \bar{\alpha}\dot{\beta})\dot{\alpha}, \beta\,\rangle dv_M + \frac{1}{4}\int_M \langle\, (\dot{\bar{\alpha}}\beta + \bar{\alpha}\dot{\beta})\alpha, \dot{\beta}\,\rangle dv_M$$

$$= \frac{1}{4}\int_M |\dot{\bar{\alpha}}\beta + \bar{\alpha}\dot{\beta}|^2 dv_M.$$

Hence

$$\dot{\bar{\alpha}}\beta + \bar{\alpha}\dot{\beta} = 0 = \sqrt{2}\bar{\partial}_B\dot{\alpha} + \mathbf{i}\varphi \wedge \alpha = \sqrt{2}\bar{\partial}_B^*\dot{\beta} - \mathbf{i}\bar{\varphi} \lrcorner\, \beta$$

and using (3.2.24c) we deduce

$$\bar{\partial}\varphi = 0. \quad \blacksquare$$

## 3.3. Applications

The theory developed so far is powerful enough to allow the computation of the Seiberg-Witten invariants of many and wide classes of Kähler surfaces. In this section we will present such computations and some of their surprising topological consequences. We will conclude with a discussion of the Seiberg-Witten invariants of almost Kähler manifolds.

**3.3.1. A nonvanishing result.** Consider a Kähler surface $M$. We want to compute the Seiberg-Witten invariant determined by the canonical *spin*$^c$ structure $\sigma_0$ on $M$. In this case

$$\mathbb{S}_0 = \underline{\mathbb{C}} \oplus K_M^{-1}.$$

We will use the perturbation $\eta_t$ introduced in §3.2.1 in which $\mu = 0$ and $t = \lambda^2 \gg 0$ where $\lambda > 0$. If $b_2^+(M) = 1$ then, according to Remark 3.2.1 the perturbation parameter $\eta_t$ lies in the positive chamber defined by the Kähler metric.

In this case the $\eta_t$-monopoles are $t$-vortices $(\alpha \oplus \beta, B)$ where
$\alpha$ is a section of $\underline{\mathbb{C}}$,
$\beta$ is a section of $K_M^{-1}$ and
$B$ is a Hermitian connection on $\underline{\mathbb{C}}$.

The discussion in §3.2.2 shows that for $\lambda^2 \gg 0$ we have $\beta \equiv 0$ and $(\alpha, B)$ satisfy

$$F_B^{0,2} = 0, \tag{3.3.1a}$$

$$\Lambda F_B = \frac{\mathbf{i}}{8}(|\alpha|^2 - \lambda^2), \tag{3.3.1b}$$

$$\bar{\partial}_B \alpha = 0. \tag{3.3.1c}$$

Observe that if $B_0$ denotes the trivial connection on $\underline{\mathbb{C}}$ and $\alpha_0$ is the constant section $\alpha_0 \equiv \lambda$ of $\underline{\mathbb{C}}$ then $(\alpha_0, B_0)$ is a solution of (3.3.1a) – (3.3.1c). Notice also that the virtual dimension of the space of monopoles is 0 in this case.

**Proposition 3.3.1.** *Modulo $\mathfrak{G}_{\sigma_0}$ there is a unique $\eta_t$-monopole which is also nondegenerate.*

**Proof** To prove the uniqueness part we will rely on Proposition 3.2.7. The set of orbits of $\eta_t$-monopoles can be identified with the set of effective divisors $D$ such that

$$c_1([D]) = c_1(\underline{\mathbb{C}}) = 0.$$

There is only one such divisor, namely the trivial divisor since, according to Proposition 3.1.1 a nontrivial effective divisor carries a nontrivial homology class. This establishes the uniqueness claim in the proposition. Thus, modulo $\mathfrak{G}_{\sigma_0}$, the configuration

$$\mathsf{C}_0 = (\alpha_0 \oplus 0, A_0 \dot{+} 2B_0)$$

is the unique $\eta_t$-monopole. Observe that in this case we can write $\bar{\partial}$ instead of $\bar{\partial}_{B_0}$ Since the virtual dimension is 0 and $\mathsf{C}_0$ is nondegenerate (i.e. $H^2_{\mathsf{C}_0} = 0$) it suffices to show $H^1_{\mathsf{C}_0} = 0$, i.e.

$$\ker \mathcal{T}_{\mathsf{C}_0} = 0.$$

We will use Proposition 3.2.14.

Suppose $(\dot{\psi}, \mathbf{i}\dot{b}) = (\dot{\alpha} \oplus \dot{\beta}, \mathbf{i}(\varphi + \bar{\varphi}) \in \ker \mathcal{T}_{\mathsf{C}_0}$. Then $(\dot{\alpha}, \dot{\beta}, \varphi)$ satisfy the equations (3.2.25) – (3.2.26d). These further simplify because of the additional assumption $\beta(= \beta_0) = 0$. More precisely, we have

$$4\sqrt{2}\mathbf{i}\bar{\partial}^* \varphi = \lambda \dot{\bar{\alpha}}, \tag{3.3.2a}$$

$$\bar{\partial}\varphi = 0, \tag{3.3.2b}$$

$$\sqrt{2}\bar{\partial}\dot{\alpha} + \mathbf{i}\lambda\varphi = 0, \tag{3.3.2c}$$

$$\lambda\dot{\beta} = 0, \;\; \bar{\partial}^*\dot{\beta} = 0. \tag{3.3.2d}$$

Applying $\bar{\partial}^*_{B_0}$ to (3.3.2c) we obtain

$$0 = 2\bar{\partial}^*\bar{\partial}\dot{\alpha} + \mathbf{i}\sqrt{2}\lambda\bar{\partial}^*\varphi \overset{(3.3.2a)}{=} 2\bar{\partial}^*\bar{\partial}\dot{\alpha} + \frac{\lambda^2}{4}\dot{\alpha} = \Delta_M\dot{\alpha} + \lambda^2\dot{\alpha}.$$

Taking the inner product with $\dot{\alpha}$ and integrating by parts we deduce in standard fashion that $\dot{\alpha} = 0$. The equality (3.3.2c) now implies $\varphi = 0$. ■

The above proposition shows that $\mathbf{sw}_M(\sigma_0) = \pm 1$ if $b_2^+ > 1$ and $\mathbf{sw}^+_M(\sigma_0) = \pm 1$ if $b_2^+ = 1$. To decide which is the correct sign we will use its definition as an orientation transport. Form as usual

$$\mathcal{T}^{\tau}_{\mathsf{C}_0}\begin{bmatrix} \dot{\psi} \\ \mathbf{i}\dot{b} \end{bmatrix} = \begin{bmatrix} \mathfrak{D}_A\dot{\psi} \\ 2d^+\mathbf{i}\dot{b} \\ -4\mathbf{i}d^*\dot{b} \end{bmatrix} + \tau \begin{bmatrix} \mathbf{c}(\mathbf{i}\dot{b})\psi \\ -\frac{1}{2}\dot{q}(\psi, \dot{\psi}) \\ -\mathbf{i}\mathfrak{Im}\langle\psi, \dot{\psi}\rangle \end{bmatrix},$$

$$\tau \in [0, 1], \;\; A := B_0 \dot{+} 2A_0.$$

Then the sign is given by the orientation transport along the path $\mathcal{T}^{\tau}_{\mathsf{C}_0}$, $\epsilon(\mathcal{T}_{\mathsf{C}_0}, \mathcal{T}^{\tau}_{\mathsf{C}_0}, \mathcal{T}^0_{\mathsf{C}_0})$.

To compute the orientation transport we will rely on (1.5.9) in §1.5.1.

Arguing exactly as in the proof of Proposition 3.2.14 we deduce that $(\dot{\psi}, \mathbf{i}\dot{b}) = (\dot{\alpha} \oplus \dot{\beta}, \varphi) \in \ker \mathcal{T}^{\tau}_{\mathsf{C}_0}$ if and only if

$$4\sqrt{2}\mathbf{i}\bar{\partial}^* \varphi = \tau\lambda \dot{\bar{\alpha}}, \tag{3.3.3a}$$

$$\bar{\partial}\varphi = 0, \tag{3.3.3b}$$

$$\sqrt{2}\bar{\partial}\dot{\alpha} + \mathbf{i}\tau\lambda\varphi = 0, \tag{3.3.3c}$$

$$\tau\lambda\dot{\beta} = 0, \;\; \bar{\partial}^*\dot{\beta} = 0. \tag{3.3.3d}$$

To see this, replace $\mathbf{c}(\mathbf{i}\dot{b})$ with $\tau\mathbf{c}(\mathbf{i}\dot{b})$, $\dot{q}$ with $\tau\dot{q}$ and $\mathfrak{Im}\langle\psi, \dot{\psi}\rangle$ with $\tau\mathfrak{Im}\langle\psi, \dot{\psi}\rangle$ in the proof of Proposition 3.2.14 keeping in mind that $\alpha = \lambda$ and $\beta = 0$.

Arguing exactly as in the proof of Proposition 3.3.1 we deduce $\ker \mathcal{T}^{\tau}_{\mathsf{C}_0} = 0$ if $\tau > 0$. Moreover

$$\ker \mathcal{T}^0_{\mathsf{C}_0}$$

$$\cong \Big\{(\dot\alpha, \dot\beta, \varphi) \in \Gamma(\underline{\mathbb{C}}) \times \Gamma(K_M^{-1}) \times \Omega^{0,1}(M);\ \ \bar\partial\dot\alpha = 0 = \bar\partial^*\dot\beta, \varphi \in H^{0,1}_{\bar\partial}(M)\Big\}$$

$$\cong (\,\mathbb{C} \oplus \mathbf{H}^{0,2}_{\bar\partial}(M)\,) \oplus \mathbf{H}^{0,1}_{\bar\partial}(M).$$

The first summand corresponds to the spinor part of the kernel and the second summand corresponds to infinitesimal deformations of connections. The kernel is naturally oriented as a complex vector space.

To find the cokernel of $\mathcal{T}^0_{\mathsf{C}_0}$ we use the representation

$$\begin{bmatrix} \Omega^{0,0} \oplus \omega^{0,2}(M) \\ \oplus \\ \mathbf{i}\Omega^{0,1}(M) \cong \mathbf{i}\Omega^1(M) \end{bmatrix} \ni \begin{bmatrix} \dot\psi \\ \mathbf{i}\dot b \end{bmatrix}$$

$$\xrightarrow{\mathcal{T}^0_{\mathsf{C}_0}} \begin{bmatrix} \not{\mathfrak{D}}_A\dot\psi \\ 2d^+\mathbf{i}\dot b \\ -4\mathbf{i}d^*\dot b \end{bmatrix} \in \begin{matrix} \Omega^{0,1}(M) \cong \mathbb{S}_0^- \\ \oplus \\ \mathbf{i}\omega\otimes\Omega^0 \oplus \Omega^{0,2}(M) \cong \mathbf{i}\Omega^2_+(M) \\ \oplus \\ \mathbf{i}\Omega^0(M) \end{matrix}$$

and the computations in the beginning of §3.2.3. Recall that the isomorphism

$$\mathbf{i}\omega\otimes\Omega^0 \oplus \Omega^{0,2}(M) \cong \mathbf{i}\Omega^2_+(M)$$

is given by the *isometric* identifications

$$\mathbf{i}\omega \oplus \Phi \longleftrightarrow \mathbf{i}\omega + \frac{\mathbf{i}}{\sqrt{2}}(\Phi + \bar\Phi) = \frac{1}{\sqrt{2}}(\mathbf{i}\Phi - \overline{\mathbf{i}\Phi}).$$

This leads to the identification

$$\begin{aligned} \mathbf{i}\Omega^2_+(M) \ni \frac{1}{2}\dot q(\psi, \dot\psi) &= \frac{\mathbf{i}}{4}(\mathfrak{Re}\langle\alpha,\dot\alpha\rangle - \mathfrak{Re}\langle\beta,\dot\beta\rangle)\omega \\ &\quad + \frac{1}{4}(\dot{\bar\alpha}\beta + \bar\alpha\dot\beta - \dot\alpha\bar\beta - \alpha\dot{\bar\beta}) \end{aligned} \tag{3.3.4}$$

$$= \frac{1}{2}\dot q(\psi,\dot\psi) \longleftrightarrow \frac{\mathbf{i}}{4}(\mathfrak{Re}\langle\alpha,\dot\alpha\rangle - \mathfrak{Re}\langle\beta,\dot\beta\rangle)\omega \oplus \frac{1}{2\sqrt{2}\mathbf{i}}(\dot{\bar\alpha}\beta + \bar\alpha\dot\beta).$$

Consider a vector

$$\begin{bmatrix} \phi \\ \mathbf{i}u\omega \oplus \theta \\ \mathbf{i}f \end{bmatrix} \in \begin{matrix} \Omega^{0,1}(M) \cong \mathbb{S}_0^- \\ \oplus \\ \mathbf{i}\omega\otimes\Omega^0 \oplus \Omega^{0,2}(M) \cong \mathbf{i}\Omega^2_+(M) \\ \oplus \\ \mathbf{i}\Omega^0(M) \end{matrix}$$

in the cokernel of $\mathcal{T}^0_{\mathsf{C}_0}$. We deduce

$$\phi \in \operatorname{coker} \mathfrak{D}_A = \mathbf{H}^{0,1}_{\bar\partial}(M),$$

$$\mathbf{i}u\omega + \mathbf{i}(\theta + \bar\theta) \in \mathbf{i}\mathbf{H}^2_+(M)$$

and

$$\mathbf{i}f \in \mathbf{H}^0(M) \cong \mathbf{i}\mathbb{R}.$$

Thus $u$ must be constant and $\theta \in \mathbf{H}^{0,2}_{\bar\partial}(M)$. We conclude

$$\operatorname{coker} \mathcal{T}^0_{\mathsf{C}_0} = \mathbf{H}^{0,1}_{\bar\partial}(M) \oplus \mathbf{H}^{0,2}_{\bar\partial}(M) \oplus \mathbf{H}^0(M) \oplus \mathbb{R}\omega.$$

The vector space in the right hand-side of the above isomorphism is naturally oriented (here the order is essential) and it induces on $\operatorname{coker} \mathcal{T}_{\mathsf{C}_0}$ precisely the orientation discussed in §3.1.1.

To compute the orientation transport we need to determine the *resonance operator*

$$\mathbf{P}\frac{d}{d\tau}\,|_{\tau=0}\, \mathcal{T}^\tau_{\mathsf{C}_0} : \ker \mathcal{T}^0_{\mathsf{C}_0} \to \operatorname{coker} \mathcal{T}^0_{\mathsf{C}_0}$$

where $\mathbf{P}$ denotes the orthogonal projection onto $\operatorname{coker} \mathcal{T}^0_{\mathsf{C}_0}$. Observe that

$$\frac{d}{d\tau}\,|_{\tau=0}\, \mathcal{T}^\tau_{\mathsf{C}_0} \begin{bmatrix} \dot\psi \\ \mathbf{i}\dot b \end{bmatrix} = \begin{bmatrix} \mathbf{c}(\mathbf{i}\dot b)\psi \\ -\frac{1}{2}\dot q(\psi, \dot\psi) \\ -\mathbf{i}\,\mathfrak{Im}\langle \psi, \dot\psi\rangle \end{bmatrix}$$

where $\psi = \lambda \oplus 0$, $\mathbf{i}\dot b = \mathbf{i}(\varphi + \bar\varphi)$ and $\dot\psi = \dot\alpha \oplus \dot\beta$. Using the computations in §3.2.3 and (3.3.4) we deduce

$$\frac{d}{d\tau}\,|_{\tau=0}\, \mathcal{T}^\tau_{\mathsf{C}_0} \begin{bmatrix} \dot\psi \\ \mathbf{i}\dot b \end{bmatrix} = \begin{bmatrix} \lambda \mathbf{i}\varphi \\ -\frac{\mathbf{i}\lambda}{4}(\mathfrak{Re}\,\dot\alpha)\,\omega + \lambda\frac{\mathbf{i}}{2\sqrt{2}}\dot\beta \\ \mathbf{i}\lambda\,\mathfrak{Im}\,\dot\alpha \end{bmatrix}.$$

Clearly, $\frac{d}{d\tau}\,|_{\tau=0}\, \mathcal{T}^\tau_{\mathsf{C}_0}$ maps $\ker \mathcal{T}^0_{\mathsf{C}_0}$ bijectively onto $\operatorname{coker} \mathcal{T}^0_{\mathsf{C}_0}$ and it does so in an orientation preserving fashion. Formula (1.5.9) now shows that the orientation transport is 1. We have thus proved the following result.

**Theorem 3.3.2.** *Suppose $M$ is a Kähler surface and $\sigma_0$ is the canonical spin$^c$ structure. If $b_2^+ > 1$ we have*

$$\mathbf{sw}_M(\sigma_0) = 1$$

*while if $b_2^+ = 1$ we have*

$$\mathbf{sw}^+_M(\sigma_0) = 1.$$

The above nonvanishing result has immediate geometric consequences.

**Corollary 3.3.3.** *If $M$ is a $K3$ surface then $\sigma_0 = \bar{\sigma}_0$ is the only basic class of $M$ and $\mathbf{sw}_M(\sigma_0) = 1$.*

**Proof** Suppose $L$ is a Hermitian line bundle on $M$ such that $\mathbf{sw}_M(L) \neq 0$. Then

$$0 \leq \deg_\omega(L) \leq \deg_\omega(K_M) = 0$$

so that by Corollary 3.2.10 we deduce that $L$ is the trivial line bundle. ■

**Corollary 3.3.4.** *Suppose $M$ is a Kähler surface such that $p_g(M) > 0$. Then there exist no Riemannian metrics on $M$ with positive scalar curvature.*

Suppose $M$ is a Kähler surface such that $p_g(M) > 0$ (so that $b_2^+(M) > 1$). Using (2.3.14) of 2.3.2 we deduce

$$\mathbf{sw}_M(\bar{\sigma}_0) = \mathbf{sw}_M(K_M) = (-1)^\kappa \mathbf{sw}_M(0) = \mathbf{sw}_M(\sigma_0)$$

where

$$\kappa = \frac{1}{2}(b_2^+ + 1 - b_1) = \frac{1}{2}(2 - b_1 + 2p_g) = 1 - q + p_g = \chi_{hol}(M).$$

Thus $\sigma_0 (= 0)$ and $\bar{\sigma}_0 (= K_M)$ are basic classes of a Kähler surface with $p_g > 0$. If $M$ is an algebraic surface of general type we can be even more precise.

**Theorem 3.3.5.** *Let $M$ be a minimal algebraic surface of general type such that $p_g > 0$. Then $\sigma_0$ and $\bar{\sigma}_0$ are the only basic classes of $M$.*

**Proof** Suppose $L \to M$ is a Hermitian line bundle such that $\mathbf{sw}_M(L) \neq 0$. We want to show that (topologically) $L \cong \mathbb{C}$ or $L \cong K_M$. According to Corollary 3.2.10 it suffices to show

$$\deg_\omega(L) \in \{0, \deg_\omega K_M\}.$$

We argue by contradiction. This means $c_1(L)$ and $c_1(K_M)$ are linearly independent in $H^{1,1}_{\bar{\partial}}(M)$ and we denote by $V$ the two-dimensional space spanned by $K_M$ and $L$. We will show that the intersection form is positive definite on $V$, thus contradicting the Hodge index theorem.

Since $M$ is a minimal algebraic surface of general type we deduce

- $K_M$ is nef and
- $K_M^2 > 0$.

According to Corollary 3.2.9 the condition $\mathbf{sw}_M(L) \neq 0$ implies several things.

◇ The virtual dimension $d(L) = L \cdot (K_M - L) \geq 0$ so that $L^2 \geq K_M \cdot L$.

◇ There exists a holomorphic structure on $L$ which admits a nontrivial holomorphic section $u$.

$\diamond$ There exists a holomorphic structure on $K_M - L$ which admits a nontrivial holomorphic section $v$.

Observe that $D := u^{-1}(0) \neq \emptyset$ since $L$ is not the trivial line bundle. Hence $D$ is an effective divisor.

Since $K_M$ is nef we deduce

$$K_M \cdot D = K_M \cdot L \geq 0.$$

In fact

$$K_M \cdot L > 0.$$

Indeed, if $K_M \cdot L = 0$ then the conditions $K_M^2 > 0$ coupled with the Hodge index theorem would imply that $c_1(L) = c_1([D]) = 0$. This is impossible since $D$ is an effective divisor. Thus

$$L^2 \geq K_M \cdot L > 0. \tag{3.3.5}$$

Replacing $L \to K_M - L$ in the above arguments (which is equivalent to using the canonical involution $\sigma \mapsto \bar{\sigma}$ on $Spin^c(M)$) we deduce

$$K_M \cdot (K_M - L) > 0 \Longleftrightarrow K_M^2 > K_M \cdot L > 0. \tag{3.3.6}$$

We can represent the restriction of the intersection form to $V$ using the basis $(K_M, L)$. We obtain the $2 \times 2$ symmetric matrix

$$Q := \begin{bmatrix} K_M^2 & K_M \cdot L \\ K_M \cdot L & L^2 \end{bmatrix}.$$

Clearly $\operatorname{tr}(Q) = K_M^2 + L^2 > 0$ and, using (3.3.5) + (3.3.6) we deduce $\det(Q) > 0$. Thus $Q$ is positive definite, contradicting the Hodge index theorem. ■

The last proposition has a surprising topological consequence.

**Corollary 3.3.6.** *Suppose $M$ is a minimal algebraic surface of general type and $f : M \to M$ is a diffeomorphism. Then $f^*(K_M) = \pm K_M$.*

**Proof** It follows from the fact that the set of basic classes of $M$ is a diffeomorphism invariant of $M$: for any $\sigma \in \mathbb{B}_M$ we have $f^*\sigma \in \mathbb{B}_M$. ■

Thus the pair of *holomorphic objects* $(K_M, -K_M)$ of the minimal, general type surface $M$ is *a diffeomorphism invariant* of $M$!!!

**3.3.2. Seiberg-Witten invariants of simply connected elliptic surfaces.** The elliptic surfaces have a much richer structure than the surfaces of general type. They have more complex curves and thus we can expect a more sophisticated Seiberg-Witten theory.

We begin with a warm-up result showing that, as in the case of surfaces of general type, the basic classes of a minimal elliptic surfaces lie on the segment determined by the canonical classes $\sigma_0$ and $\bar{\sigma}_0$. If we use the language of line bundles this means the basic classes of such a surface lie on the segment in $H^2(M,\mathbb{Z})$ determined by the trivial line bundle and $K_M$.

**Definition 3.3.7.** A *proper* elliptic surface is a minimal algebraic elliptic surface $M$ such that $\operatorname{kod}(M) > 0$.

**Proposition 3.3.8.** *Suppose $M$ is a proper elliptic surface such that $p_g(M) > 0$. If $L$ is a $(1,1)$, Hermitian line bundle on $M$ such that $\mathbf{sw}_M(L) \neq 0$ then there exists $t \in [0,1]$ such that*

$$c_1(L) = tc_1(K_M) \;\; in \;\; H^{1,1}_{\bar{\partial}}(M).$$

**Proof** Since $M$ is a proper elliptic surface we deduce that $K_M$ is nef, nontrivial and $K_M^2 = 0$. Moreover, the metric $\omega$ is defined by an ample divisor $H$ and thus, for any line bundle $E$, we have $\deg_\omega(E) = H \cdot E$.

Suppose $L \not\cong \underline{\mathbb{C}}$, $K_M$. It suffices to prove $\hat{L} = K_M - L$ and $L$ are collinear, for then the inequality

$$0 < H \cdot L < H \cdot K_M$$

will force $L$ to lie on the segment going from 0 to $K_M$. We argue by contradiction. Suppose $c_1(L)$ and $c_1(K_M)$ are linearly independent (as classes in $H^{1,1}_{\bar{\partial}}(M)$).

Using Proposition 3.2.13 we deduce that there exist effective divisors $D'$ and $D''$ such that

$$[D'] + [D''] = K_M, \;\; c_1([D']) = c_1(L) \;\text{ in }\; H^2(M,\mathbb{Z}).$$

Since $K_M$ is nef we deduce

$$K_M \cdot L = K_M \cdot D' \geq 0, \;\; K_M \cdot \hat{L} = K_M \cdot D'' \geq 0$$

so that

$$K_M \cdot L = 0.$$

On the other hand, since $d(L) = d(\hat{L}) = L \cdot \hat{L} \geq 0$ we deduce

$$L^2 \geq K_M \cdot L \geq 0, \;\; \hat{L}^2 \geq K_M \cdot \hat{L} \geq 0$$

so that $L^2, \hat{L}^2 \geq 0$. From the identity

$$0 = K_M^2 = (L + \hat{L})^2 = L^2 + 2L \cdot \hat{L} + \hat{L}^2 \geq 0$$

we can now conclude $L^2 = \hat{L}^2 = L \cdot \hat{L} = 0$.

Set

$$t := (H \cdot L)/H^2 > 0,$$

$$s := (H \cdot \hat{L})/H^2 > 0$$

and

$$T := tH - L, \quad S := sH - \hat{L}.$$

Observe that

$$H \cdot T = H \cdot S = 0.$$

The vectors $H, S, T$ are linearly independent in $H^{1,1}_{\bar{\partial}}(M)$ and thus span a three-dimensional space $V$. We can now represent the restriction to $V$ of the intersection form as a symmetric $3 \times 3$ matrix using the basis $H$, $T$, $S$. An elementary computation shows this matrix is

$$Q = H^2 \begin{bmatrix} 1 & 0 & 0 \\ 0 & -t^2 & -(t^2 + s^2 + st) \\ 0 & -(t^2 + s^2 + st) & -s^2 \end{bmatrix}.$$

The $2 \times 2$ minor in the lower right hand corner has negative determinant and thus $Q$ has two positive eigenvalues. This contradicts the Hodge index theorem and completes the proof of the proposition. ■

To get more detailed information about the Seiberg-Witten invariants of an elliptic surface we need to have a deeper look into the structure of these surfaces. This is a very fascinating and elaborate subject. We want to present to the reader a few facts about elliptic surfaces which are needed in the computation of the Seiberg-Witten invariants. For more details we refer to [**9, 40**] or the original articles of K. Kodaira [**65**].

An important concept in the theory of elliptic surfaces is that of *multiple fiber*.

Suppose $\pi : M \to B$ is an algebraic elliptic surface over the smooth complex curve $B$. The fiber $F_b$ of $\pi$ at $b \in B$ is said to have multiplicity $m$ if there exists a holomorphic coordinate $w$ defined on a disk neighborhood $\Delta$ of $b$ such that

◇ $w(b) = 0$.

◇ There exists a holomorphic function $g : \pi^{-1}(\Delta) \to \Delta \subset \mathbb{C}$ such that $\pi = g^m$ on $\pi^{-1}(\Delta)$.

◇ The set $C_g$ of critical points of $g$ is finite.

The hypersurface $F_b = g^{-1}(0)$ is called the reduction of the fiber $\pi^{-1}(b)$. The multiple fiber is said to have smooth reduction if $C_g = \emptyset$ or, equivalently, if $F_b$ is smooth. Using the open cover $U_0 = \pi^{-1}(\Delta)$, $U_1 = M \setminus F_b$ and the holomorphic function

$$f_0 = \pi : U_0 \to \Delta \subset \mathbb{C}, \quad f_1 \equiv 1 : U_1 \to \mathbb{C}$$

we obtain a divisor $M_b$ on $M$. Observe that

$$M_b = mF_b.$$

The multiple fibers are not just theoretically possible. There is a simple way to construct elliptic surfaces with multiple fibers having smooth reductions. It relies on the *logarithmic transform*.

Let us first describe a simple procedure of constructing a smooth family of elliptic curves. Denote by $\mathbf{H}_+$ the half-plane $\{\mathfrak{Im}\,\tau > 0\} \subset \mathbb{C}$. Each $\tau \in \mathbf{H}_+$ defines a lattice

$$\Lambda_\tau = \{m + n\tau; \;\; m, n \in \mathbb{Z}\}.$$

It is known that any elliptic curve is biholomorphic to a quotient $C_\tau := \mathbb{C}/\Lambda_\tau$. If $X$ is a complex manifold and $\tau : X \to \mathbf{H}_+$ is a holomorphic map we can form a holomorphic family of smooth elliptic curves $C_\tau := (\mathbb{C}/\Lambda_{\tau(x)})_{x \in X}$. More precisely, $C_\tau$ is defined as the quotient

$$C_\tau := \mathbb{C} \times X/(\mathbb{Z} \oplus \mathbb{Z})$$

where $(m, n) \in \mathbb{Z} \oplus \mathbb{Z}$ acts on $(z, x) \in \mathbb{C} \times X$ by

$$(m, n)(z, x) = (z + m + n\tau(x), x).$$

We denote by $\pi_\tau$ the natural projection $C_\tau \to X$.

Suppose $\pi : M \to B$ is an elliptic surface and $b \in B$ is a regular value of $\pi$ so that the fiber $\pi^{-1}(b)$ is a smooth elliptic curve. Choose a small neighborhood $\Delta$ of $b \in B$ and a local coordinate $w$ on $\Delta$ such that $w(b) = 0$. For simplicity we assume that $w$ identifies $\Delta$ with the unit disk in $\mathbb{C}$. Then there exist[1] a holomorphic map $\tau : \Delta \to \mathbf{H}$ and a biholomorphic map

$$F : \pi^{-1}(\Delta) \to C_\tau$$

such that the diagram below is commutative.

$$\begin{array}{ccc} \pi^{-1}(\Delta) & \xrightarrow{F} & C_\tau \\ \Big\downarrow{\scriptstyle \pi} & & \Big\downarrow{\scriptstyle \pi_\tau} \\ \Delta & \xrightarrow{\mathbf{1}_\Delta} & \Delta \end{array}$$

Define $\Sigma \subset C_\tau \times \Delta$ by

$$\Sigma := \{(z, w, \zeta); \;\; w, \zeta \in \Delta, \;\; z \in C_{\tau(w)}, \;\; \zeta^m = w\}.$$

[1]This claim needs a proof and we refer to [**49**] for details.

More intuitively, $\Sigma$ is the pullback of the fibration $\pi_\tau : C_\tau \to \Delta$ via the $m$-fold branched cover

$$\Delta \to \Delta, \ \ \zeta \mapsto w := \zeta^m.$$

The natural map

$$\zeta : \Sigma \to \Delta, \ \ (z, w, \zeta) \mapsto \zeta \in \Delta$$

defines a structure of elliptic fibration on $\Sigma$. The fibers over $\zeta$ and $e^{2\pi \mathbf{i}/m}\zeta$ are biholomorphic to $C_{\tau(\zeta^m)} = C_{\tau(w)}$. This means we have a commutative diagram

$$\begin{array}{ccc} \Sigma & \overset{G}{\dashrightarrow} & C_\tau \\ {\scriptstyle \zeta}\downarrow & & \downarrow{\scriptstyle w=\pi_\tau} \\ \Delta & \xrightarrow{w=\zeta^m} & \Delta \end{array}$$

and we can also think of $\Sigma$ as the total space of the family of smooth elliptic curves $(C_{\tau(\zeta^m)})_{\zeta\in\Delta}$

We can now construct an automorphism $\phi : \Sigma \to \Sigma$

(3.3.7)

$$C_{\tau(\zeta^m)} \times \Delta \ni (z, \zeta) \mapsto \left( \ (z + \frac{\tau(\zeta^m)}{m}) \bmod \Lambda_{\tau(\zeta^m)}, \ e^{2\pi\mathbf{i}/m}\zeta \right) \in C_{\tau(\zeta^m)} \times \Delta.$$

Observe that the iterates of $\phi$ generate a cyclic group with $m$ elements which acts freely on $\Sigma$. We can form the quotient

$$\tilde{\Sigma} := \Sigma/(\phi).$$

The natural map $\zeta^m : \Sigma \to \Delta$ is invariant with respect to the action of this cyclic group and thus descends to a holomorphic map

$$u = \zeta^m : \tilde{\Sigma} \to \Delta.$$

It clearly induces a structure of elliptic fibration on $\tilde{\Sigma}$ and the fiber over $0 \in \Delta$ is multiple, with multiplicity $m$. Its reduction is smooth and is biholomorphic to $C_{\tau(0)}$. The fiber over $u \in \Delta \setminus \{0\}$ is smooth, it has multiplicity 1 and is biholomorphic to $C_{\tau(u)}$. Moreover, there is a biholomorphic map

$$L_m : \tilde{\Sigma} \setminus u^{-1}(0) \to C_\tau \setminus C_{\tau(0)}$$

induced by the $\phi$-invariant map

$$\Sigma \setminus \zeta^{-1}(0) \to C_\tau \setminus C_{\tau(0)}, \ \ (z, \zeta) \mapsto \left( \ (z - \frac{1}{2\pi\mathbf{i}} \log\zeta) \bmod \Lambda_{\tau(\zeta^m)}, \zeta^m \ \right).$$

Observe that the $2\pi\mathbf{i}\mathbb{Z}$-ambiguity of $\log\zeta$ vanishes when we mod out the $\Lambda_\tau$-action.

The logarithmic transform can now be described explicitly as follows.

▷ Remove the fibered neighborhood set $\pi^{-1}(\Delta_{1/2})$ of the fiber of $\pi$ over $w(b) = 0$ where $\Delta_{1/2^m}$ denotes the disk with the same center as $\Delta$ but with radius $1/2^m$.

▷ Glue back the elliptic fibration $\tilde{\Sigma}$ using the biholomorphism

$$L_m : \tilde{\Sigma}|_{\Delta\setminus\bar{\Delta}_{1/2}} \rightarrow \pi^{-1}(\Delta \setminus \bar{\Delta}_{1/2^m}).$$

We will denote the resulting manifold by $\mathbf{L}_m M$, or by $\mathbf{L}_m(b)M$ if the point $b$ where the logarithmic transform was performed is relevant. It is often useful to have a $C^\infty$-interpretation of this operation.

The fibered neighborhood $Y := \pi^{-1}(\Delta)$ is a 4-manifold with boundary diffeomorphic to $T^2 \times \Delta$. Its boundary is a three-dimensional torus $T^2 \times \partial\Delta$. We will denote by $w$ the complex coordinate on $\Delta$ and by $\xi_1$, $\xi_2$ the angular coordinates on $T^2$. When working in the $C^\infty$-category we can assume that the map $\tau : \Delta \rightarrow \mathbf{H}_+$ is constant $\tau(w) \equiv \mathbf{i}$.

Denote by $\hat{\Delta}$ another copy of $\Delta$ coordinatized by $\zeta = re^{\mathbf{i}\theta} \in \mathbb{C}$. We pull back this $T^2$-fibration using the $m$-fold branched cover

$$p_m : \hat{\Delta} \rightarrow \Delta, \quad \zeta \mapsto w = \zeta^m$$

and we obtain another $T^2$-fibration $\hat{Y} = p_m^* Y \rightarrow \hat{\Delta}$. Set $\omega := e^{2\pi\mathbf{i}/m}$ and identify the cyclic group $\mathbb{Z}_m$ with the subgroup of $S^1$ generated by $\omega$.

We can now define two $\mathbb{Z}_m$-actions on $\hat{Y}$:

$$\omega \star (\xi_1, \xi_2, \zeta) = (\xi_1, \xi_2, \omega\zeta)$$

and

$$\omega \circ (\xi_1, \xi_2, \zeta) = (\xi_1, \omega\xi_2, \omega\zeta).$$

The $\circ$-action corresponds to the holomorphic action described by the map $\phi$ in (3.3.7).

These two actions are not isomorphic and lead to two quotients

$$Y \cong \hat{Y}/(\omega, \star)$$

and

$$\tilde{Y} := \hat{Y}/(\omega, \circ).$$

On the other hand, the restrictions of these actions to $T^3 \cong \partial\hat{Y}$ are isomorphic. To see this pick a matrix $A \in SL(3, \mathbb{Z})$ such that

$$A \cdot \begin{bmatrix} 0 \\ 0 \\ 1 \end{bmatrix} = \begin{bmatrix} 0 \\ 1 \\ 1 \end{bmatrix}.$$

This means the last column of $A$ is the vector in the right-hand side of the above equality. For example, we can pick

$$A = \begin{bmatrix} 1 & 0 & 0 \\ 0 & 1 & 1 \\ 0 & 0 & 1 \end{bmatrix}.$$

Using the angular coordinates $(\xi_1, \xi_2, \theta)$ on $\partial\hat{Y}$ we can write the above two actions as

$$\omega \star \begin{bmatrix} \xi_1 \\ \xi_2 \\ \theta \end{bmatrix} = \frac{2\pi}{m} \begin{bmatrix} 0 \\ 0 \\ 1 \end{bmatrix} + \begin{bmatrix} \xi_1 \\ \xi_2 \\ \theta \end{bmatrix},$$

$$\omega \circ \begin{bmatrix} \xi_1 \\ \xi_2 \\ \theta \end{bmatrix} = \frac{2\pi}{m} \begin{bmatrix} 0 \\ 1 \\ 1 \end{bmatrix} + \begin{bmatrix} \xi_1 \\ \xi_2 \\ \theta \end{bmatrix}.$$

It is now clear that

$$A(\omega \star \vec{v}) = \omega \circ A\vec{v}, \quad \forall \vec{v} \in \mathbb{R}^3 \quad \bmod (2\pi\mathbb{Z})^3.$$

Thus $A$ induces a diffeomorphism

$$\bar{A} : \partial Y \to \partial\tilde{Y}.$$

This diffeomorphism *does not extend to a diffeomorphism* $\tilde{Y} \to Y$ although $\tilde{Y}$ and $Y$ are *diffeomorphic.*

We will produce a diffeomorphism $\tilde{Y} \to T^2 \times \Delta$ by constructing a map $T : \tilde{Y} \to T^2 \times \Delta$ whose fibers are precisely the orbits of the $(\omega, \circ)$ action. More precisely, set

$$T : \hat{Y} \to T^2 \times \Delta, \quad (\xi_1, \xi_2, \zeta) \mapsto (\xi_1, \xi_2^m, \xi_2^{-1}\zeta).$$

To understand the effect of $\bar{A}$ we need to introduce angular coordinates on $\partial Y$ and $\partial\tilde{Y}$.

On $\partial Y$ a natural choice is given by

$$(\xi_1, \xi_2, \xi_3) = (\xi_1, \xi_2, \zeta^m)$$

while on $\partial\tilde{Y}$ a natural choice is suggested by the definition of $T$

$$(\tilde{\xi}_1, \tilde{\xi}_2, \tilde{\xi}_3) = (\xi_1, \xi_2^m, \xi_2^{-1}\zeta).$$

The map $\bar{A}$ can be computed from the diagram

$$\begin{array}{ccc} (\xi_1, \xi_2, \zeta) & \xrightarrow{\ A\ } & (\xi_1, \xi_2\zeta, \zeta) \\ {\scriptstyle (\xi_1,\xi_2,\xi_3)} \downarrow & & \downarrow {\scriptstyle (\tilde{\xi}_1,\tilde{\xi}_2,\tilde{\xi}_3)} \\ (\xi_1, \xi_2, \zeta^m) & \overset{\bar{A}}{\dashrightarrow} & (\xi_1, (\xi_2\zeta)^m, \xi_2^{-1}) \end{array}$$

Thus $\bar{A}$ is given by

$$\tilde{\xi}_1 = \xi_1, \quad \tilde{\xi}_2 = \xi_2^m \xi_3, \quad \tilde{\xi}_3 = \xi_2^{-1}$$

or, in matrix notation,

$$\bar{A} = \begin{bmatrix} 1 & 0 & 0 \\ 0 & m & 1 \\ 0 & -1 & 0 \end{bmatrix} \in SL(3, \mathbb{Z}).$$

Its inverse is

$$G_m = \begin{bmatrix} 1 & 0 & 0 \\ 0 & 0 & 1 \\ 0 & -1 & m \end{bmatrix}.$$

Thus, in the $C^\infty$-category, the logarithmic transform is obtained by removing a fibered neighborhood $T^2 \times \Delta$ of a smooth fiber and then attaching it back in a new fashion, using the gluing map $G_m$.

We collect below some basic topological and geometric facts about elliptic surfaces admitting multiple fibers.

**Proposition 3.3.9.** *Suppose $\pi : M \to B$ is an elliptic surface with $r$ multiple fibers, with smooth reductions $F_1, \cdots, F_r$ and multiplicities $m_1, \cdots, m_r$. Then, there exists a holomorphic line bundle $L \to B$ of degree* $\deg L = 2g(B) - 2 + \chi_{hol}(M) = 2g(B) - 2 + \frac{1}{12}\chi_M$ *such that*

$$K_M \cong \pi^* L + \sum_{i=1}^{r} (m_i - 1) F_i.$$

For proofs of the above proposition we refer to [**9, 49**]. When $B \cong \mathbb{P}^1$ we can be more specific because in this case two holomorphic line bundles over $\mathbb{P}^1$ are holomorphically isomorphic if and only if they are topologically isomorphic, that is, they have the same degree. A holomorphic line bundle of degree $d$ over $\mathbb{P}^1$ can thus be described by any divisor $b_1 + \cdots + b_d$, where the points $b_i$ are pairwise distinct.

**Corollary 3.3.10.** *Suppose $\pi : M \to \mathbb{P}^1$ is an elliptic fibration with $r$ multiple fibers $F_1, \cdots, F_r$ with multiplicities $m_1, \cdots, m_r$. Then*

$$K_M = \left[ \sum_{j=1}^{\chi_{hol}(M)-2} M_{b_j} + \sum_{i=1}^{r} (m_i - 1) F_i \right]$$

*where the points $b_j \in \mathbb{P}^1$ are pairwise distinct* regular *values of $\pi$ and $M_{b_j} := \pi^{-1}(b_j)$.*

Denote by $E(n; m_1, \cdots, m_r)$ the smooth manifold obtained from the elliptic surfaces $E(n)$ by performing logarithmic transforms of multiplicities $m_1, \cdots, m_r$ on $r$ nonsingular fibers

$$E(n : m_1, \cdots, m_r) = \mathbf{L}_{m_1} \cdots \mathbf{L}_{m_r} V_n.$$

Denote by $F_1, \cdots, F_r$ the multiple fibers in $E(n; m_1, \cdots, m_r)$. For a proof of the following nontrivial result we refer to [**40**].

**Theorem 3.3.11.** *Suppose $\pi : M \to \mathbb{P}^1$ is an elliptic surface such that*

▶ $\chi_{hol}(M) = n > 0$.

▶ *There is no smooth rational curve $C \hookrightarrow M$ entirely contained in a fiber of $\pi$ and such that $C^2 = -1$.*

▶ *There are $r$ multiple fibers, with multiplicities $m_1, \cdots, m_r$ and smooth reductions $F_1, \cdots, F_r$.*

*Then the following hold.*

*(a) $M$ is diffeomorphic to $E(n; m_1, \cdots, m_r)$.*

*(b) $M$ is simply connected if and only if either $r \leq 1$ or $r = 2$ and the multiplicities $m_1, m_2$ are coprime.*

*(c) Denote by $m$ the least common multiple of $m_1, \cdots, m_r$ and by $F \in H_2(M, \mathbb{Z})/\mathrm{Tors}$ the homology class carried by a nonsingular fiber of $\pi$. Then there exists a primitive class $\mathbf{f} \in H_2(M, \mathbb{Z})/\mathrm{Tors}$ such that*

$$F = m\mathbf{f}, \;\; F_i = \frac{m}{m_i}\mathbf{f}, \;\; \forall i = 1, \cdots, r.$$

Using the above proposition we can now determine the homeomorphism type of the simply connected surfaces $E(n; m_1, m_2)$, where we allow $m_i = 1$.

In this case the least common multiple of $m_1, m_2$ is $m_1 m_2$. $H^2(M, \mathbb{Z})$ has no torsion and can be identified with $H_2(M, \mathbb{Z})$ via Poincaré duality. We deduce

$$\chi_M = 12n, \;\; b_2 = 12n - 2,$$

$$p_g = (n-1), \;\; b_2^+ = 2n - 1,$$

$$K_M = m\left\{(n-2) + \sum_i (1 - \frac{1}{m_i})\right\}\mathbf{f}. \tag{3.3.8}$$

Using Wu's formula we deduce that the intersection form of $M$ is even if and only if

$$\nu(n; m_1, m_2) = \left\{(n-2) + \sum_i (1 - \frac{1}{m_i})\right\}$$

is even. This happens if and only if

$$n \equiv m_1 + m_2 \equiv 0 \mod 2.$$

Using Corollary 2.4.17 we deduce the following result.

**Corollary 3.3.12.** *Two simply connected elliptic surfaces $E(n; m_1, m_2)$ and $E(n'; m_1', m_2')$ are* homeomorphic *if and only if*

$$n = n'$$

*and either*

$$n \equiv 0 \mod 2, \quad m_1 + m_2 \equiv m_1' + m_2' \mod 2.$$

*or,*

$$n \equiv 1 \mod 2.$$

We now have all the information we need to compute the Seiberg-Witten invariants of the elliptic surface $M = E(n; m_1, m_2)$, $(m_1, m_2) = 1$, $n \geq 3$. Denote by $F_1$ and $F_2$ the multiple fibers of $M$ and pick $(n-2)$ pairwise disjoint generic fibers, $M_{b_1}, \cdots, M_{b_{n-2}}$. The line bundle determined by the effective divisor

$$C_0 := \sum_j M_{b_j} + (m_1 - 1)F_1 + (m_2 - 1)F_2$$

is precisely the canonical line bundle $K_M$. $D$ determines a holomorphic section $s$ of $K_M$ such that $D$ coincides with the zero divisor determined by $s$. Using Proposition 3.2.13 we deduce that if the line bundle $L \to M$ determines a basic class of $M$ then there exists a divisor $D$ on $M$ such that

$$c_1([D]) = c_1(L) \text{ and } 0 \leq D \leq C_0.$$

This means $D$ must have the form

$$D = D(J, a_1, a_2) = \sum_{j \in J} M_{b_j} + a_1 F_1 + a_2 F_2$$

where $J \subset \{1, 2, \cdots, (n-2)\}$ and $0 \leq a_i < m_i$, $i = 1, 2$. Observe that with $D$ as above we have

$$c_1([D]) = m(|J| + \frac{a_1}{m_1} + \frac{a_2}{m_2})\mathbf{f}.$$

Since $m_1$ and $m_2$ are relatively prime we deduce

$$c_1([D(J, a_1, a_2)]) = c_1([D(J', a_1', a_2')]) \iff |J| = |J'|, \ a_1 = a_1', \ a_2 = a_2'.$$

Thus, if $L$ determines a basic class then $c_1(L)$ is collinear with $c_1(K_M)$ in $H^2(M, \mathbb{Z})$, the virtual dimension $D(L)$ is zero and moreover

$$c_1(L) = (mk + m_1 a_2 + m_2 a_1)\mathbf{f}, \ 0 \leq k \leq (n-2), \ 0 \leq a_i < m_i. \tag{3.3.9}$$

Thus the set of basic classes of $M$ has cardinality $\leq m_1m_2(n-1)$. We will denote by $L(k,a_1,a_2)$ the complex line bundle such that

$$c_1(L) = (mk + m_1a_2 + m_2a_1)\mathbf{f}.$$

Suppose $L = L(k,a_1,a_2)$. Then, according to Proposition 3.2.13, the set of orbits of monopoles corresponding to the *spin*$^c$ structure $\sigma_0 \otimes L$ and the perturbation $t\omega + s + \bar{s}$ can be identified with the set of effective divisors $D(J,a_1,a_2)$ such that $|J| = k$. There are exactly $\binom{n-2}{|J|}$ such divisors.

Given a divisor $D$ as above there exists a monopole

$$\mathsf{C} = \mathsf{C}_D = (\psi = \alpha \oplus \beta,\ \ A = A_0 \dot{+} 2B)$$

such that $B$ induces a holomorphic structure on $L$, $\alpha = \alpha_D$ is a holomorphic section of $L$, $\bar{\beta} = \bar{\beta}_D$ is a holomorphic section of $K_M - L$, $D$ coincides with the zero divisor determined by $\alpha$

$$\alpha\bar{\beta} = -8\mathbf{i}s,\ \ \Lambda F_B = \frac{\mathbf{i}}{8}(|\alpha|^2 - |\beta|^2 - t).$$

**Proposition 3.3.13. (O. Biquard [13])** *Each of the above monopoles* $\mathsf{C} = \mathsf{C}_D$ *is nondegenerate.*

**Proof** The idea of proof is inspired by [**13**]. Since the virtual dimension $d(L) = 0$ it suffices to show $\ker \mathcal{T}_{\mathsf{C}} = \{0\}$. Let

$$(\dot{\psi}, \mathbf{i}\dot{b}) \in \ker \mathcal{T}_{\mathsf{C}}.$$

As in §3.2.3 we write

$$\dot{\psi} = \dot{\alpha} \oplus \dot{\beta} \in \Omega^{0,0}(L) \oplus \Omega^{0,2}(L) = \Omega^{0,0}(L - K_M)$$

and

$$\mathbf{i}\dot{b} = \frac{\mathbf{i}}{\sqrt{2}}(\varphi + \bar{\varphi}),\ \ \varphi \in \Omega^{0,1}(M).$$

Then (see Proposition 3.2.14) $\dot{\alpha}$, $\dot{\beta}$ and $\varphi$ satisfy the equations

$$(3.3.10)\qquad \left\{\begin{array}{rcl} 4\sqrt{2}\mathbf{i}\bar{\partial}^*\varphi - \overline{\langle \alpha, \dot{\alpha}\rangle} + \langle \beta, \dot{\beta}\rangle & = & 0\\ \bar{\partial}\varphi & = & 0\\ \sqrt{2}\bar{\partial}_B\dot{\alpha} + \mathbf{i}\varphi\wedge\alpha & = & 0\\ \sqrt{2}\bar{\partial}_B^*\dot{\beta} - \mathbf{i}\bar{\varphi}\,\lrcorner\,\beta & = & 0\\ \dot{\alpha}\bar{\beta} + \alpha\dot{\bar{\beta}} & = & 0 \end{array}\right. .$$

The last equation shows $\dot{\alpha}/\alpha = -\dot{\bar{\beta}}/\bar{\beta}$ on $M \setminus \big(\alpha^{-1}(0) \cup \beta^{-1}(0)\big)$. We denote by $f$ this smooth function on $M \setminus \big(\alpha^{-1}(0) \cup \beta^{-1}(0)\big)$. Since $\alpha\bar{\beta} = -8\mathbf{i}s$ we deduce

$$(\bar{\beta}) = C_0 - (\alpha) = C_0 - D = \sum_{j\in\bar{J}} M_{b_j} + (m_1 - a_1 - 1)F_1 + (m_2 - a_2 - 1)F_2$$

where $\bar{J} := \{1, 2, \cdots, (n-2)\} \setminus J$. Since $\dot{\alpha} = \alpha f$ and $\dot{\bar{\beta}} = -\bar{\beta} f$ are smooth objects we deduce that $f$ extends to a smooth function on $M \setminus (F_1 \cup F_2)$.

**Lemma 3.3.14.** *The function $f$ extends to a smooth function on $M$.*

We will complete the proof of the proposition assuming the validity of the above lemma.

Observe that since $\bar{\partial}_B \alpha = 0$ we have (on $M \setminus \alpha^{-1}(0)$)

$$\bar{\partial} f = \bar{\partial}(\dot{\alpha}/\alpha) = (\alpha \bar{\partial}_B \dot{\alpha})/\alpha^2) = -\frac{\mathbf{i}}{\sqrt{2}}\varphi$$

where at the last step we used the third equation in (3.3.10). Since $M \setminus \alpha^{-1}(0)$ is dense in $M$ and $f$ is smooth we can conclude that the last equality is valid everywhere on $M$.

Using this identity in the first equation of (3.3.10) we obtain

$$0 = -8\bar{\partial}^* \bar{\partial} f - \overline{\langle \alpha, f\alpha \rangle} - \langle \beta, \bar{f}\beta \rangle = -(8\bar{\partial}^* \bar{\partial} + |\alpha|^2 + |\beta|^2) f.$$

Multiplying by $f$ and integrating by parts we deduce $f = 0$. This implies $\varphi = 0$, $\dot{\alpha} = f\alpha = 0$ and $\dot{\bar{\beta}} = -\bar{\beta} f = 0$. This concludes the proof of the proposition. ■

**Proof of Lemma 3.3.14** We will show that $f$ extends smoothly over $F_1$.

Suppose $F_1$ is the fiber of $\pi : M = E(n; m_1, m_2) \to \mathbb{P}^1$ over $0 \in \mathbb{C} \subset \mathbb{P}^1$. We denote by $w$ the coordinate on $\mathbb{C}$. Denote by $\Delta$ the unit disk centered at 0. By possibly rescaling we can assume that the restriction of $\pi$ to $\pi^{-1}(M)$ has the form

$$\pi = u^{m_1}$$

where $u : \pi^{-1}(\Delta) \to \Delta$ is a submersive holomorphic map.

Now fix a point $q \in F_1$ and a local holomorphic coordinate on $F_1$ near $q$. Then the pair of functions $(z, u)$ forms a local holomorphic coordinate system on a small neighborhood $U$ of $q$ in $M$. In this coordinate system $F_1$ is locally defined by $u = 0$ and the section $\alpha$ has the local description

$$\alpha = u^{a_1} \alpha_0$$

where $\alpha_0$ is a nowhere vanishing holomorphic function on $U$.

Since $\bar{\partial}\varphi = 0$ we can choose $U$ sufficiently small so that there exists $g \in C^\infty(U)$ such that $\varphi = \sqrt{2}\bar{\partial} g$. The second equation in (3.3.10) can be rewritten over $U$ as

$$\bar{\partial}(\dot{\alpha} + \mathbf{i}\alpha g) = 0.$$

Thus

$$h := \dot{\alpha} + \mathbf{i} u^{a_1} \alpha_0 g$$

is holomorphic on $U$. We now write

$$\dot{\alpha} = h - \mathbf{i}u^{a_1}\alpha_0 g \tag{3.3.11}$$

and use this in the last equation of (3.3.10). This yields

$$(h - \mathbf{i}u^{a_1}\alpha_0 g)\bar{\beta} + u^{a_1}\alpha_0\dot{\bar{\beta}} = 0$$

so that

$$h = u^{a_1}\alpha_0(\mathbf{i}g\bar{\beta} - \dot{\bar{\beta}}).$$

The last equality shows that the smooth function $h_0 = \alpha_0(\mathbf{i}g\bar{\beta} - \dot{\bar{\beta}})$ is holomorphic on $U \setminus F_1$ (where it equals $h/u^{a_1}$) and thus it must be holomorphic everywhere on $U$. This allows us to write

$$h = u^{a_1}h_0$$

where $h_0$ is holomorphic on $U$. Using this in (3.3.11) we deduce

$$\dot{\alpha} = u^{a_1}(h_0 - \mathbf{i}\alpha_0 g)$$

so that

$$f = \dot{\alpha}/\alpha = \frac{h_0 - \mathbf{i}\alpha_0 g}{\alpha_0}.$$

This proves that $f$ is bounded on $U$ since $\alpha_0$ does not vanish anywhere. ∎

We now know that if $L = L(k, a_1, a_2)$ then there are precisely $\binom{n-2}{k}$ $\mathfrak{G}$-orbits of nondegenerate irreducible monopoles corresponding to the *spin*$^c$ structure $\sigma_0 \otimes L$. To compute the Seiberg-Witten invariant we have to determine the signs attached to these monopoles.

Consider a monopole $\mathsf{C} = \mathsf{C}_D = (\alpha \oplus \beta, A_0 \dot{+} 2B)$ as in Proposition 3.3.13. We begin by rewriting the operator $\mathcal{T}_{\mathsf{C}}$ using the identifications

$$\Omega^{0,1} \ni \varphi \longleftrightarrow \mathbf{i}\dot{b} = \frac{\mathbf{i}}{\sqrt{2}}(\varphi + \bar{\varphi}) \in \mathbf{i}\Omega^1(M),$$

$$\Omega^{0,2}(M) \oplus \mathbf{i}\omega \otimes \Omega^0(M) \ni \theta \oplus u\omega \longleftrightarrow \frac{\mathbf{i}}{\sqrt{2}}(\theta + \bar{\theta}) \oplus \mathbf{i}u\omega \in \mathbf{i}\Omega^2_+(M),$$

$$\Gamma(\mathbb{S}^+_L) \ni \dot{\psi} \longleftrightarrow \dot{\alpha} \oplus \dot{\beta} \in \Omega^{0,0}(L) \oplus \Omega^{0,0}(L - K_M),$$

$$\Gamma(\mathbb{S}^-_L) = \Omega^{0,1}(L).$$

Using the computations in §3.2.3 and the identification (3.3.4) we deduce

$$\mathcal{T}_{\mathsf{C}}\begin{bmatrix} \dot{\alpha} \\ \dot{\beta} \\ \varphi \end{bmatrix} = \begin{bmatrix} \sqrt{2}(\bar{\partial}_B\dot{\alpha} + \bar{\partial}^*_B\dot{\beta}) \\ 2\bar{\partial}\varphi \\ -\mathbf{i}\sqrt{2}\mathfrak{Im}\,\bar{\partial}^*\varphi)\omega \\ -4\sqrt{2}\mathbf{i}\mathfrak{Re}\,\bar{\partial}^*\varphi \end{bmatrix} + \begin{bmatrix} \mathbf{i}(\varphi \wedge \alpha - \bar{\varphi} \lrcorner \beta) \\ \frac{\mathbf{i}}{2\sqrt{2}}(\dot{\bar{\alpha}}\beta + \bar{\alpha}\dot{\beta}) \\ -\frac{\mathbf{i}}{4}(\mathfrak{Re}\langle\alpha, \dot{\alpha}\rangle - \mathfrak{Re}\langle\beta, \dot{\beta}\rangle)\omega \\ -\mathbf{i}\mathfrak{Im}\langle\alpha, \dot{\alpha}\rangle - \mathbf{i}\mathfrak{Im}\langle\beta, \dot{\beta}\rangle \end{bmatrix}.$$

Define the isomorphism

$$\Upsilon : \Big(\mathbf{i}\Omega^0(M) \oplus \mathbf{i}\omega \otimes \Omega^0(M)\Big) \oplus \Omega^{0,2}(M) \to \Big(\Omega^0(M) \otimes \mathbb{C}\Big) \oplus \Omega^{0,2}(M),$$

$$\mathbf{i}f_0 \oplus \mathbf{i}f_1\omega \oplus \gamma \longleftrightarrow \Big(\frac{1}{4\sqrt{2}}f_0 + \frac{1}{\sqrt{2}}f_1\mathbf{i}\Big) \oplus \frac{1}{2}\gamma.$$

Using these last isomorphisms we can further rewrite $\mathcal{T}_{\mathsf{C}} \longleftrightarrow \Upsilon\mathcal{T}_{\mathsf{C}}$

$$\mathcal{T}_{\mathsf{C}}\begin{bmatrix} \dot{\alpha} \\ \dot{\beta} \\ \varphi \end{bmatrix} = \begin{bmatrix} \sqrt{2}(\bar{\partial}_B\dot{\alpha} + \bar{\partial}_B^*\dot{\beta}) \\ \bar{\partial}\varphi \\ -\bar{\partial}^*\varphi \end{bmatrix} + \begin{bmatrix} \mathbf{i}(\varphi \wedge \alpha - \bar{\varphi} \lrcorner \beta) \\ \\ \frac{\mathbf{i}}{4\sqrt{2}}(\dot{\bar{\alpha}}\beta + \bar{\alpha}\dot{\beta}) \\ \\ -\frac{\mathbf{i}}{4\sqrt{2}}(\ \langle\dot{\alpha}, \alpha\rangle - \langle\beta, \dot{\beta}\rangle\ ) \end{bmatrix}$$

$$= \begin{bmatrix} \sqrt{2}(\bar{\partial}_B\dot{\alpha} + \bar{\partial}_B^*\dot{\beta}) \\ \bar{\partial}\varphi \\ -\bar{\partial}^*\varphi \end{bmatrix} + \mathbf{i}\begin{bmatrix} \varphi \wedge \alpha \\ \\ \frac{1}{4\sqrt{2}}\bar{\alpha}\dot{\beta} \\ \\ -\frac{1}{4\sqrt{2}}\langle\dot{\alpha}, \alpha\rangle \end{bmatrix} + \mathbf{i}\begin{bmatrix} -\bar{\varphi} \lrcorner \beta \\ \\ \frac{1}{4\sqrt{2}}\dot{\bar{\alpha}}\beta \\ \\ \frac{1}{4\sqrt{2}}\langle\beta, \dot{\beta}\rangle \end{bmatrix}.$$

Observe that

$$\mathcal{T}_{\mathsf{C}}^0\begin{bmatrix} \dot{\psi} \\ \mathbf{i}\dot{b} \end{bmatrix} = \mathcal{T}_{\mathsf{C}}^0\begin{bmatrix} \dot{\alpha} \\ \dot{\beta} \\ \varphi \end{bmatrix} = \begin{bmatrix} \sqrt{2}(\bar{\partial}_B\dot{\alpha} + \bar{\partial}_B^*\dot{\beta}) \\ \bar{\partial}\varphi \\ -\bar{\partial}^*\varphi \end{bmatrix}$$

and our orientation conventions for $\ker \mathcal{T}_{\mathsf{C}}^0$ and $\operatorname{coker} \mathcal{T}_{\mathsf{C}}^0$ coincide with the orientations induced by the above identification of these spaces with complex spaces.

To determine the sign associated to the monopole $\mathsf{C}$ we will compute the orientation transport along a cleverly chosen deformation of $\mathcal{T}_{\mathsf{C}}^0$ to $\mathcal{T}_{\mathsf{C}}$, suggested by [**13**]. We will get the same result since it will be clear from the description of this deformation that it is homotopic to the deformation $\mathcal{T}_{\mathsf{C}}^{\tau}$ we have used so far.

The new deformation is a composite of two deformations. We first follow the path ($t \in [0, 1]$)

$$\mathcal{U}_t\begin{bmatrix} \dot{\alpha} \\ \dot{\beta} \\ \varphi \end{bmatrix} = \begin{bmatrix} \sqrt{2}(\bar{\partial}_B\dot{\alpha} + \bar{\partial}_B^*\dot{\beta}) \\ \bar{\partial}\varphi \\ -\bar{\partial}^*\varphi \end{bmatrix} + t\mathbf{i}\begin{bmatrix} \varphi \wedge \alpha \\ \\ \frac{1}{4\sqrt{2}}\bar{\alpha}\dot{\beta} \\ \\ -\frac{1}{4\sqrt{2}}\langle\dot{\alpha}, \alpha\rangle \end{bmatrix},$$

and then the path

$$\mathcal{V}_t \begin{bmatrix} \dot\alpha \\ \dot\beta \\ \varphi \end{bmatrix} = \begin{bmatrix} \sqrt{2}(\bar\partial_B\dot\alpha + \bar\partial_B^*\dot\beta) \\ \bar\partial\varphi \\ -\bar\partial^*\varphi \end{bmatrix} + \mathbf{i} \begin{bmatrix} \varphi\wedge\alpha \\ \\ \frac{1}{4\sqrt{2}}\bar\alpha\dot\beta \\ \\ -\frac{1}{4\sqrt{2}}\langle\dot\alpha,\alpha\rangle \end{bmatrix} + t\mathbf{i} \begin{bmatrix} -\bar\varphi\,\lrcorner\,\beta \\ \\ \frac{1}{4\sqrt{2}}\dot{\bar\alpha}\beta \\ \\ \frac{1}{4\sqrt{2}}\langle\beta,\dot\beta\rangle \end{bmatrix}.$$

Observe first that the operators $\mathcal{U}_t$ are complex linear so the orientation transport along this path is 1. Thus we only have to determine the orientation transport along $\mathcal{V}_t$. Let us first point out a very useful fact.

**Lemma 3.3.15.** $\ker\mathcal{V}_t = 0$ *for all* $t \in (0,1]$.

The proof is word for word the proof of Proposition 3.3.13 (which corresponds to $t = 1$) and can be safely left to the reader. Denote by $\mathbf{P}$ the orthogonal projection onto $\operatorname{coker}\mathcal{V}_0$ and set

$$R_0 = \mathbf{P}\frac{d}{dt}|_{t=0}\,\mathcal{V}_t : \ker\mathcal{V}_0 \to \operatorname{coker}\mathcal{V}_0.$$

Observe that

$$\dot{\mathcal{V}}_0 \begin{bmatrix} \dot\alpha \\ \dot\beta \\ \varphi \end{bmatrix} := \frac{d}{dt}|_{t=0}\,\mathcal{V}_t \begin{bmatrix} \dot\alpha \\ \dot\beta \\ \varphi \end{bmatrix} = \mathbf{i} \begin{bmatrix} -\bar\varphi\,\lrcorner\,\beta \\ \\ \frac{1}{4\sqrt{2}}\dot{\bar\alpha}\beta \\ \\ \frac{1}{4\sqrt{2}}\langle\beta,\dot\beta\rangle \end{bmatrix}$$

is *complex conjugate linear.* Thus $R_0$ is complex conjugate linear and if it is an $\mathbb{R}$-linear isomorphism, then the orientation transport will be

$$(-1)^{d_0}, \quad d_0 = \dim_{\mathbb{C}}\ker\mathcal{V}_0.$$

We will spend the remainder of this subsection proving that $R_0$ is indeed an isomorphism and determining $d_0$.

**Lemma 3.3.16.** *There exists a natural short exact sequence*

$$0 \to \mathbb{C} \to H^{0,0}([L,B]) = H^{0,0}([D(J,a_1,a_2)]) \to \ker\mathcal{V}_0 \to 0$$

*where* $[L,B]$ *denotes the line bundle* $L$ *equipped with the holomorphic structure defined by the Hermitian connection* $B$*. In particular,*

$$d_0 = h^0([D(J,a_1,a_2)]) - 1.$$

**Proof** Let $(\dot\alpha, \dot\beta, \varphi) \in \ker \mathcal{V}_0$, that is,

$$(3.3.12)\qquad \begin{cases} \bar\partial_B\dot\alpha + \bar\partial_B^*\dot\beta + \frac{\mathbf{i}}{\sqrt{2}}\varphi\wedge\alpha &= 0 \\ \bar\partial\varphi + \frac{\mathbf{i}}{4\sqrt{2}}\bar\alpha\dot\beta &= 0 \\ \bar\partial^*\varphi + \frac{\mathbf{i}}{4\sqrt{2}}\langle\dot\alpha,\alpha\rangle &= 0 \end{cases}.$$

We use the same strategy as in the proof of Proposition 3.2.14. Using the first equality in (3.3.12) we deduce

$$0 \geq -\|\bar\partial_B^*\dot\beta\|_{L^2}^2 = \langle \bar\partial_N\dot\alpha + \frac{\mathbf{i}}{\sqrt{2}}\varphi\wedge\alpha, \bar\partial_B^*\dot\beta\rangle_{L^2}$$

(use $\bar\partial_B\alpha = 0$)

$$= \langle\frac{\mathbf{i}}{\sqrt{2}}\bar\partial\varphi\wedge\alpha, \dot\beta\rangle_{L^2}$$

(use the second equation in (3.3.12)

$$= \frac{1}{8}\|\,|\alpha|\cdot|\dot\beta|\,\|_{L^2}^2.$$

This implies $\dot\beta \equiv 0$ and thus $\bar\partial\varphi = 0$, according to the second equation in (3.3.12). Since $h^{0,1}(M) = 0$ there exists a smooth complex valued function $f$ on $M$ such that $\sqrt{2}\bar\partial f = \varphi$.

The first equation in (3.3.12) can now be rewritten

$$\bar\partial_B(\dot\alpha + \mathbf{i}f\alpha) = 0$$

so that

$$h := \dot\alpha + \mathbf{i}f\alpha \in \mathbf{H}^{0,0}_{\bar\partial}([L,B])$$

and

$$\dot\alpha = h - \mathbf{i}f\alpha.$$

Using these last equalities in the third equation of (3.3.12) we deduce

$$(\bar\partial^*\bar\partial + \frac{1}{8}|\alpha|^2)f = -\frac{\mathbf{i}}{8}h\bar\alpha.$$

Since the positive operator $\bar\partial^*\bar\partial + \frac{1}{8}|\alpha|^2$ has bounded inverse we deduce

$$(3.3.13)\qquad f = f_\alpha(h) := -\frac{\mathbf{i}}{8}(\bar\partial^*\bar\partial + \frac{1}{8}|\alpha|^2)^{-1}(h\bar\alpha).$$

It is now clear that the correspondence

$$H^{0,0}_{\bar\partial}([L,B]) \ni h \mapsto (\dot\alpha, \dot\beta, \varphi) = (\,h - \mathbf{i}f_\alpha(h)\alpha, 0, \sqrt{2}\bar\partial f_\alpha(h)\,)$$

produces a $\mathbb{C}$-linear surjection

$$H^{0,0}_{\bar\partial}([L,B]) \to \ker \mathcal{V}_0.$$

Observe that its kernel is generated by

$$h_0 := \mathbf{i}\alpha.$$

Lemma 3.3.16 is proved. ∎

**Lemma 3.3.17.** *$R_0$ is a complex conjugate linear isomorphism.*

**Proof** Let

$$(\dot{\alpha}_0, \dot{\beta}_0 = 0, \varphi_0) \in \ker \mathcal{V}_0.$$

We will show that if

$$\dot{\mathcal{V}}_0 \begin{bmatrix} \dot{\alpha}_0 \\ 0 \\ \varphi_0 \end{bmatrix} = \mathbf{i} \begin{bmatrix} -\bar{\varphi}_0 \lrcorner \beta \\ \\ \frac{1}{4\sqrt{2}} \bar{\dot{\alpha}}_0 \beta \\ \\ 0 \end{bmatrix} \in \text{Range}\,(\mathcal{V}_0)$$

then $\dot{\alpha}_0 = 0$ and $\varphi_0 = 0$.

Suppose there exists $(\dot{\alpha}, \dot{\beta}, \varphi) \in \Omega^{0,0}(L) \times \Omega^{0,2}(L) \times \Omega^{0,1}(M)$ such that

$$\mathcal{V}_0 \begin{bmatrix} \dot{\alpha} \\ \dot{\beta} \\ \varphi \end{bmatrix} + \dot{\mathcal{V}}_0 \begin{bmatrix} \dot{\alpha}_0 \\ 0 \\ \varphi_0 \end{bmatrix} = \begin{bmatrix} 0 \\ 0 \\ 0 \end{bmatrix}.$$

This means

$$(3.3.14) \qquad \begin{cases} \bar{\partial}_B \dot{\alpha} + \bar{\partial}_B^* \dot{\beta} + \frac{\mathbf{i}}{\sqrt{2}} \varphi \wedge \alpha - \frac{\mathbf{i}}{\sqrt{2}} \bar{\varphi}_0 \lrcorner \beta &= 0 \\ \bar{\partial} \varphi + \frac{\mathbf{i}}{4\sqrt{2}} \bar{\alpha} \dot{\beta} + \frac{\mathbf{i}}{4\sqrt{2}} \bar{\dot{\alpha}}_0 \beta &= 0 \\ \bar{\partial}^* \varphi + \frac{\mathbf{i}}{4\sqrt{2}} \dot{\alpha} \bar{\alpha} &= 0 \end{cases}$$

and

$$(3.3.15) \qquad \begin{cases} \bar{\partial}_B \dot{\alpha}_0 + \frac{\mathbf{i}}{\sqrt{2}} \varphi_0 \wedge \alpha &= 0 \\ \bar{\partial} \varphi_0 &= 0 \\ \bar{\partial}^* \varphi_0 + \frac{\mathbf{i}}{4\sqrt{2}} \langle \dot{\alpha}_0, \alpha \rangle &= 0 \end{cases} .$$

Again we rely on the idea in the proof of Proposition 3.2.14. We have

$$0 \geq -\|\bar{\partial}_B^* \dot{\beta} - \frac{\mathbf{i}}{\sqrt{2}} \bar{\varphi}_0 \lrcorner \beta\|_{L^2}^2 = \langle \bar{\partial}_B \dot{\alpha} + \frac{\mathbf{i}}{\sqrt{2}} \alpha\varphi, \bar{\partial}_B^* \dot{\beta} - \frac{\mathbf{i}}{\sqrt{2}} \bar{\varphi}_0 \lrcorner \beta \rangle_{L^2}$$

$$= \langle \frac{\mathbf{i}}{\sqrt{2}} \bar{\partial}_B(\alpha\varphi), \dot{\beta} \rangle_{L^2} + \langle \frac{\mathbf{i}}{\sqrt{2}} \varphi_0 \wedge \bar{\partial}_B \dot{\alpha}, \beta \rangle_{L^2} + \langle (\frac{\mathbf{i}}{\sqrt{2}})^2 \alpha\varphi_0 \wedge \varphi, \beta \rangle_{L^2}$$

(use $\bar\partial\varphi_0 = 0$, $\bar\partial_B^*\beta = \bar\partial_B\alpha = 0$)

$$= \langle \frac{\mathbf{i}}{\sqrt{2}}\alpha\bar\partial\varphi, \dot\beta\rangle_{L^2} + \langle \frac{\mathbf{i}}{\sqrt{2}}\alpha\varphi_0 \wedge \frac{\mathbf{i}}{\sqrt{2}}\varphi, \beta\rangle_{L^2}$$

(use $\frac{\mathbf{i}}{\sqrt{2}}\alpha\varphi_0 = -\bar\partial_B\dot\alpha_0$)

$$= \frac{\mathbf{i}}{\sqrt{2}}\langle \alpha\bar\partial\varphi, \dot\beta\rangle_{L^2} - \frac{\mathbf{i}}{\sqrt{2}}\langle \bar\partial_B\dot\alpha_0 \wedge \varphi, \beta\rangle_{L^2}$$

$$= \frac{\mathbf{i}}{\sqrt{2}}\langle \alpha\bar\partial\varphi, \dot\beta\rangle_{L^2} - \frac{\mathbf{i}}{\sqrt{2}}\langle \bar\partial_B(\dot\alpha_0\varphi) - \dot\alpha_0\bar\partial\varphi, \beta\rangle_{L^2}$$

(use $\bar\partial_B^*\beta = 0$)

$$= \frac{\mathbf{i}}{\sqrt{2}}\langle \alpha\bar\partial\varphi, \dot\beta\rangle_{L^2} + \frac{\mathbf{i}}{\sqrt{2}}\langle \dot\alpha_0\bar\partial\varphi, \beta\rangle_{L^2}$$

$$= \langle \bar\partial\varphi, -\frac{\mathbf{i}}{\sqrt{2}}(\bar\alpha\dot\beta + \dot{\bar\alpha}_0\beta)\rangle_{L^2} \overset{(3.3.14)}{=} \frac{1}{8}\|\bar\alpha\dot\beta + \dot{\bar\alpha}_0\beta\|_{L^2}^2.$$

This shows

$$\text{(3.3.16)} \qquad \begin{cases} \bar\partial_B^*\dot\beta - \frac{\mathbf{i}}{\sqrt{2}}\bar\varphi_0 \lrcorner \beta &= 0 \\ \bar\partial_B\dot\alpha_0 + \frac{\mathbf{i}}{\sqrt{2}}\alpha\varphi_0 &= 0 \\ \bar\partial\varphi_0 &= 0 \\ \bar\partial^*\varphi_0 + \frac{\mathbf{i}}{4\sqrt{2}}\dot\alpha_0\bar\alpha &= 0 \\ \bar\alpha\dot\beta + \dot{\bar\alpha}_0\beta &= 0 \end{cases}$$

The above system of equations is very similar to (3.3.10). We can now conclude **exactly** as in the proof of Proposition 3.3.13 that the system (3.3.16) has only the trivial solution

$$\dot\alpha_0 = 0, \quad \dot\beta = 0, \quad \varphi_0 = 0.$$

This shows that $R_0$ is an isomorphism as claimed. ■

Observe that all divisors $D(J, a_1, a_2)$, $|J| = k$ are linearly equivalent. Indeed, for any two sets $J, J' \subset \{1, \cdots, n-2\}$ with $|J| = |J'|$ the divisors

$$C = \sum_{j\in J} M_{b_j}, \quad C' = \sum_{j\in J'} M_{b_j}$$

are linearly equivalent since the divisors

$$\sum_{j\in J} b_j, \quad \sum_{j\in J'} b_j$$

on $\mathbb{P}^1$ are linearly equivalent. Thus

$$d(J, a_1, a_2) := \dim_{\mathbb{C}} H^{0,0}_{\bar{\partial}}([D(J, a_1, a_2)])$$

depends only on $k = |J|$, $a_1$ and $a_2$. We will denote this dimension by $d(k, a_1, a_2)$. This shows that the Seiberg-Witten invariant of the *spin*$^c$ structure $\sigma_0 \otimes L(k, a_1, a_2)$ is nontrivial and more precisely

$$\mathbf{sw}_M = (-1)^{d(k,a_1,a_2)-1} \binom{n-2}{k}.$$

In particular, $M = E(n, m_1, m_2)$ has precisely $m_1 m_2 (n-1)$ basic classes. We can be even more precise.

**Proposition 3.3.18.** $d(k, a_1, a_2) = k + 1$.

**Proof** The key ingredient in the proof is the following fact concerning multiple fibers. Its proof can be found in [**49**].

**Lemma 3.3.19.** *Denote by $N_i$ the holomorphic normal bundle of $F_i \hookrightarrow M$, $i = 1, 2$. Then $N_i$ is an element of order $m_i$ in the group* $\mathrm{Pic}\,(F_i)$.

The proof of Proposition 3.3.18 will be completed in several steps. As in §3.1.1, for any effective divisor $D$ on $M$, we denote by $f_D$ one of the nontrivial holomorphic sections of $[D]$ canonically determined by $D$. Fix $k$ distinct regular fibers $M_{b_1}, \cdots, M_{b_k}$ and denote by $D_0$ the divisor

$$D_0 = \sum_{j=1}^{k} M_{b_j}.$$

We can identify $D_0$ with a smooth (reducible) curve on $M$. Now set $T = a_1 F_1 + a_2 F_2$ and $D = D_0 + T$.

**Step 1** *The proposition is true if* $a_1 = a_2 = 0$. To see this consider the structural sequence

$$0 \to \mathcal{O}_M \stackrel{f_{D_0}\cdot}{\to} \mathcal{O}_M([D_0]) \to \mathcal{O}_{D_0}([D_0]) \to 0$$

which leads to the long exact sequence

$$0 \to H^0(\mathcal{O}_M) \to H^0\Big(\mathcal{O}_M([D_0])\Big) \to H^0\Big(\mathcal{O}_{D_0}([D_0])\Big) \to H^1(\mathcal{O}_M) \to \cdots.$$

Since $M$ is simply connected we deduce $\dim_{\mathbb{C}} H^1(\mathcal{O}_M) = h^{0,1}_M = 0$. Thus we have the short exact sequence of complex vector spaces

$$0 \to H^0(\mathcal{O}_M) \to H^0\Big(\mathcal{O}_M([D_0])\Big) \to H^0\Big(\mathcal{O}_{D_0}([D_0])\Big) \to 0.$$

Hence

$$H^0\Big(\mathcal{O}_M([D_0])\Big) \cong H^0(\mathcal{O}_M) \oplus H^0\Big(\mathcal{O}_{D_0}([D_0])\Big)$$

$$\cong H^0(\mathcal{O}_M) \oplus \Big( \bigoplus_j H^0\Big(\mathcal{O}_{M_{b_j}}([D_0])\Big)\Big).$$

The holomorphic normal bundle to $M_{b_j} \hookrightarrow M$ is (holomorphically) trivial and, by the adjunction formula, it coincides with $[D_0]|_{M_{b_j}}$. Thus

$$H^0(\mathcal{O}_{M_{b_j}}([D_0])) \cong \mathbb{C}.$$

**Step 1** is now complete.

**Step 2** *If* $a_1 + a_2 > 0$ *then*

$$H^0\Big(\mathcal{O}_M([T])\Big) \cong \mathbb{C}, \quad H^1\Big(\mathcal{O}_M([T])\Big) \cong 0.$$

We will distinguish two cases: $a_1 + a_2 = 1$ and $a_1 + a_2 > 1$.

In the first case, assume $a_1 = 1$, $a_2 = 0$ so that $T = F_1$. Using the structural sequence

$$0 \to \mathcal{O}_M \to \mathcal{O}_M([F_1]) \to \mathcal{O}_{F_1}(N_1) \to 0$$

we obtain the long exact sequence

$$(*) \qquad \begin{aligned} 0 &\to H^0(\mathcal{O}_M) \to H^0\Big(\mathcal{O}_M([F_1])\Big) \to H^0\Big(\mathcal{O}_{F_1}(N_1)\Big) \\ &\to H^1(\mathcal{O}_M) \to H^1\Big(\mathcal{O}_M([F_1])\Big) \to H^1\Big(\mathcal{O}_{F_1}(N_1)\Big) \to \cdots . \end{aligned}$$

From Lemma 3.3.19 we deduce that the degree zero line bundle $N_1 \to F_1$ has no holomorphic sections so that

$$H^0\Big(\mathcal{O}_{F_1}(N_1)\Big) \cong 0.$$

The first portion of the long exact sequence now implies

$$H^0\Big(\mathcal{O}_M([F_1])\Big) \cong H^0(\mathcal{O}_M) \cong \mathbb{C}.$$

The Riemann-Roch theorem for the line bundle $N_1 \to F_1$ implies

$$\dim_{\mathbb{C}} H^0\Big(\mathcal{O}_M([F_1])\Big) - \dim_{\mathbb{C}} H^1\Big(\mathcal{O}_{F_1}(N_1)\Big)$$

$$= \chi(N_1) = \deg(N_1) + 1 - g(F_1) = 0$$

so that

$$H^1\Big(\mathcal{O}_{F_1}(N_1)\Big) \cong 0.$$

Using this in the second portion of the long exact sequence $(*)$ we deduce

$$H^1\Big(\mathcal{O}_M([F_1])\Big) \cong H^1(\mathcal{O}_M) \cong 0.$$

This completes **Step 2** in the case $a_1 + a_2 = 1$.

The general case follows by induction. Suppose $d := a_1 + a_2 > 1$ and assume $a_1 > 0$. Set $T_0 := T - F_1 = (a_1 - 1)F_1 + a_2 F_2$. We use the structural sequence

$$0 \to \mathcal{O}_M([T_0]) \to \mathcal{O}_M([T]) \to \mathcal{O}_{F_1}([T]) \to 0$$

with associated long exact sequence

$$(**)\qquad \begin{aligned} 0 \to H^0\Big(\mathcal{O}_M([T_0])\Big) &\to H^0\Big(\mathcal{O}_M([T])\Big) \to H^0\Big(\mathcal{O}_{F_1}([T])\Big) \\ \to H^1\Big(\mathcal{O}_M([T_0])\Big) &\to H^1\Big(\mathcal{O}_M([T])\Big) \to H^1\Big(\mathcal{O}_{F_1}([T])\Big) \to \cdots . \end{aligned}$$

The induction assumption implies

$$H^0\Big(\mathcal{O}_M([T_0])\Big) \cong \mathbb{C}, \;\; H^1\Big(\mathcal{O}_M([T_0])\Big) \cong 0.$$

Now observe that $[T]|_{F_1} \cong a_1 N_1$ and since $0 < a_1 < m_1$ we deduce from Lemma 3.3.19 that the degree zero line bundle $a_1 N_1$ is holomorphically nontrivial so that

$$H^0\Big(\mathcal{O}_{F_1}([T])\Big) \cong 0.$$

Invoking again the Riemann-Roch theorem for $a_1 N_1 \to F_1$ we deduce

$$H^1\Big(\mathcal{O}_{F_1}([T])\Big) \cong 0.$$

The conclusions of **Step 2** now follow from the sequence $(**)$.

**Step 3** *Conclusion.* Consider the structural sequence

$$0 \to \mathcal{O}_M([T]) \xrightarrow{f_{D_0}\cdot} \mathcal{O}_M([D]) \to \mathcal{O}_{D_0}([D]) \to 0$$

with associated long exact sequence

$$(***)\qquad \begin{aligned} 0 \to H^0\Big(\mathcal{O}_M([T])\Big) &\to H^0\Big(\mathcal{O}_M([D])\Big) \to H^0\Big(\mathcal{O}_{D_0}([D])\Big) \\ \to H^1\Big(\mathcal{O}_M([T])\Big) &\to H^1\Big(\mathcal{O}_M([D])\Big) \to H^1\Big(\mathcal{O}_{D_0}([D])\Big) \to \cdots . \end{aligned}$$

Observe that the restriction of $[D]$ to the disconnected curve $D_0$ is the holomorphically trivial line bundle. Thus

$$H^0\Big(\mathcal{O}_{D_0}([D])\Big) \cong \mathbb{C}^k.$$

Using **Step 2** we deduce $H^1\Big(\mathcal{O}_M([T])\Big) \cong 0$ so that the first part of $(***)$ reduces to a short exact sequence

$$0 \to H^0\Big(\mathcal{O}_M([T])\Big) \to H^0\Big(\mathcal{O}_M([D])\Big) \to H^0\Big(\mathcal{O}_{D_0}([D])\Big) \to 0.$$

Using **Step 2** again we deduce that the first space in the above sequence is one-dimensional. Proposition 3.3.18 is now clear. ■

The next theorem collects the results proved so far.

**Theorem 3.3.20.** *The simply connected elliptic surface* $M = E(n; m_1, m_2)$, $(m_1, m_2) = 1$, $n \geq 2$ *has exactly* $m_1 m_2(n-1)$ *basic classes*

$$\sigma(k, a_1, a_2) = \sigma_0 \otimes L_{k,a_1,a_2}$$

*where* $0 \leq k \leq n-2$, $0 \leq a_1 \leq m_1 - 1$, $0 \leq a_2 \leq m_2 - 1$ *and* $L_{k,a_1,a_2}$ *is the complex line bundle determined by*

$$c_1(L_{k,a_1,a_2}) = (m_1 m_2 k + m_1 a_2 + m_2 a_1)\mathbf{f}.$$

*Moreover,*

$$\mathbf{sw}_M(\sigma(k, a_1, a_2)) = (-1)^k \binom{n-2}{k}.$$

**Remark 3.3.21.** For different approaches to Theorem 3.3.20 we refer to [**21, 35, 42**].

The above theorem has a truly remarkable consequence.

**Corollary 3.3.22.** ([**82, 95, 129**]) *Two simply connected elliptic surfaces* $M = E(n; m_1, m_2)$ *and* $M' = E(n'; m_1', m_2')$ *are diffeomorphic if and only if*

$$n = n' \text{ and } \{m_1, m_2\} = \{m_1', m_2'\}. \tag{3.3.17}$$

**Proof** Clearly, (3.3.17) implies that the two surfaces are diffeomorphic. Conversely, suppose the two surfaces are diffeomorphic. In particular, they are homeomorphic and Corollary 3.3.12 implies

$$n = n'.$$

Since they are diffeomorphic they have the same number of basic classes so that

$$m_1 m_2 = m_1' m_2' := m.$$

Denote by $\mathbf{f}$ and $\mathbf{f}'$ the corresponding *primitive classes* on $M$ and $M'$. Since $\mathbb{B}_M = \mathbb{B}_{M'}$ we deduce that there exist $k_1, k_2, x_1, y_1, x_2, y_2 \in \mathbb{Z}$ such that

$$m_1'\mathbf{f}' = (mk_1 + m_1 x_2 + m_2 x_1)\mathbf{f}, \;\; m_2'\mathbf{f}' = (mk_2 + m_1 y_2 + m_2 y_1)\mathbf{f}$$

and

$$0 \leq k_1, k_2 \leq n-2, \;\; 0 \leq x_1, y_1 \leq m_1 - 1, \;\; 0 \leq x_2, y_2 \leq m_2 - 1.$$

We deduce

$$m_1' = mk_1 + (m_1 x_2 + m_2 x_1) \geq m_1 x_2 + m_2 x_1,$$

$$m_2' = mk_2 + (m_1 y_2 + m_2 y_1) \geq m_1 y_2 + m_2 y_1$$

and

$$m_1'|(m_1x_2+m_2x_1),\ \ m_2'|(m_1y_2+m_2y_1).$$

Thus,

$$m_1' = m_1x_2+m_2x_1,\ \ m_2' = m_1y_2+m_2y_1.$$

This implies

$$m_1m_2 = m_1'm_2' = (m_1x_2+m_2x_1)\cdot(m_1y_2+m_2y_1)$$

$$= m_1m_2(x_1y_2+x_2y_1) + m_1^2x_2y_2 + m_2^2x_1y_1.$$

We conclude

$$x_1y_1 = x_2y_2 = 0,\ \ x_1y_2+x_2y_1 = 1.$$

Some elementary manipulations now imply

$$\{m_1, m_2\} = \{m_1', m_2'\}. \ \blacksquare$$

Using Corollary 3.3.12 we can draw the following surprising conclusion.

**Corollary 3.3.23.** *There exist infinitely many smooth 4-manifolds* **homeomorphic** *to $E(n; m_1, m_2)$ but* **not diffeomorphic** *to it !!!*

**Proof** We can construct these manifolds of the form $E(n; m_1', m_2')$ such that

$$\{m_1', m_2'\} \neq \{m_1, m_2\}$$

but still $m_1+m_2 \equiv m_1'+m_2' \mod 2$ if $n \equiv 0 \mod 2$. $\blacksquare$

**Remark 3.3.24.** We have seen that the Seiberg-Witten invariants contain nontrivial information about the Kähler surfaces of Kodaira dimension $\geq 0$.

The Seiberg-Witten equations contain nontrivial information about the remaining case as well. C. Okonek and A. Teleman have used these equations in [**113**] to give a new, very short proof of van de Ven's conjecture stating that an algebraic surface diffeomorphic to a rational surface must in fact be rational. We refer to [**88, 113**] for more information.

### 3.3.3. The failure of the $h$-cobordism theorem in four dimensions.

Recall that two compact, closed, smooth manifolds $X_\pm$ are called *$h$-cobordant* if there exists a smooth manifold $W$ with boundary $\partial W = X_- \cup X_+$ such that the natural inclusions

$$X_\pm \hookrightarrow W$$

are homotopy equivalences. $W$ is also called an $h$-cobordism between $X_-$ and $X_+$. An $h$-cobordism $W$ is said to be trivial if it is diffeomorphic to a cylinder $[0,1] \times X$. The $h$-cobordism $W$ is said to be topologically trivial if it is *homeomorphic* to a cylinder.

In the award winning work [**125**], S. Smale has proved the following remarkable result.

**Theorem 3.3.25. (The $h$-cobordism theorem)** *Any $h$-cobordism between two simply connected smooth manifolds of dimension $n \geq 5$ is trivial. In particular, two smooth, compact, $h$-cobordant, simply connected manifolds of dimension $\geq 5$ are diffeomorphic.*

As explained in [**51**], the proof of Theorem 3.3.25 fails in dimension 4. Still, the $h$-cobordism relation is very restrictive.

**Theorem 3.3.26. (C.T.C. Wall, [145])** *(a) Any $h$-cobordism $W$ between two smooth, simply connected 4-manifolds $X$ and $Y$ induces an isomorphism*

$$f_W : (H^2(X,\mathbb{Z}), q_X) \to (H^2(Y,\mathbb{Z}), q_Y).$$

*(b) If $X$ and $Y$ are two smooth simply connected 4-manifolds and*

$$g : (H^2(X,\mathbb{Z}), q_X) \to (H^2(Y,\mathbb{Z}), q_Y)$$

*is an isomorphism then there exists an $h$-cobordism $W$ such that $g = f_W$.*

This theorem suggests the introduction of the following object. Suppose $X$ is a smooth, simply connected 4-manifold. Denote by $O(q_X)$ the group of automorphisms of the intersection form $q_X$. If $\Gamma_X$ denotes the group of components of the diffeomorphism group $\mathrm{Diff}\,(M)$ then there exists a natural map

$$\Gamma_X \to O(q_X)$$

with image $G_X$. Theorem 3.3.26 implies that if an $h$-cobordism $W$ is trivial then $f_W \in G_X$, i.e. the automorphism $f_W$ is induced by a diffeomorphism of $X$. This shows that the index

$$\delta_X := [O(q_X) : G_X]$$

is a measure of the "size" of the set of nontrivial $h$-self-cobordisms of $X$. In particular, if there exists a smooth manifold $X$ such that $\delta_X > 1$ then we can produce smoothly nontrivial cobordisms.

After considerable effort, M. Freedman succeeded in [**38**] in proving that a weaker version $h$-cobordism theorem continues to hold in four dimensions.

**Theorem 3.3.27. (M. Freedman)** *Any smooth cobordism between two, smooth, compact, simply connected* 4*-manifolds is* **topologically** *trivial.*

The weaker conclusion in the above theorem is **not** due to a limitation of the proof. It has deep and still mysterious roots. Yet, the mathematical world was taken completely by surprise when S. Donaldson announced the following result.

**Theorem 3.3.28.** *There exist* **smoothly** *nontrivial $h$-cobordisms.*

**Proof** We follow the approach in [**51**, Chap. 9]. Let $X$ be the $K3$ elliptic surface $E(2)$ . We will show that $\delta_X > 1$ by proving that the automorphism $(-1)$ of $q_X$ is not induced by any diffeomorphism. We argue by contradiction.

Suppose there exists such a diffeomorphism $f$. Since $X$ has a unique basic class $\sigma_0$ we deduce

$$f^*\sigma_0 = \sigma_0$$

and

$$\mathbf{sw}_X(f^*\sigma_0) = \mathbf{sw}_X(\sigma_0) = 1.$$

On the other hand, since $f$ acts as $-1$ on $H^2(M, \mathbb{Z})$ and $b_2^+(X) = 3$ we deduce that $f$ changes the orientation of $\mathbf{H}^2_+(X)$ by $-1$ and thus changes the Seiberg-Witten invariant by the same factor. ■

**3.3.4. Seiberg-Witten equations on symplectic 4-manifolds.** We hope that by now we have convinced the reader of the powerful impact of the Kähler condition on the Seiberg-Witten equations.

This condition can be relaxed in two ways. We can require the manifold to be complex but not Kähler or we can drop the integrability condition on the almost complex structure but preserve the symplectic form. Surprisingly, most of the consequences continue to hold under these weaker assumption.

The first situation was considered in great detail in [**13**] and involves no new analytical difficulties. By contrast, the symplectic situation is considerably more difficult. In a remarkable *tour de force*, C.H. Taubes has shown in [**134, 135, 136, 137, 138**] that the essential features of the Seiberg-Witten equations in the presence of a Kähler form survive when the Kähler condition is relaxed to a symplectic one.

It is beyond the scope of these notes to even attempt to survey Taubes' remarkable results. We have a much more modest goal in mind. We want to prove that the nonvanishing result of §3.3.1 has a symplectic counterpart. Our presentation will rely heavily on the results in Section 1.4.

Consider a symplectic 4-manifold $(M, \omega)$ equipped with a compatible metric $g$ and associated almost complex structure $J$ so that

$$\omega(X, Y) = g(JX, Y), \ \ \forall X, Y \in \mathrm{Vect}\,(M).$$

The almost complex structure canonically defines a *spin*$^c$ structure $\sigma_0$ with associated line bundle

$$\det(\sigma_0) \cong K_M^{-1}.$$

Any other *spin*$^c$ structure has the form

$$\sigma_L = \sigma_0 \otimes L, \ \ \det(\sigma_L) = K_M^{-1} \otimes L^2$$

where $L$ is a Hermitian line bundle. Moreover,

$$\Gamma(\mathbb{S}_L^+) = \Omega^{0,0}(L) \oplus \Omega^{0,2}(L), \ \ \Gamma(\mathbb{S}_L^-) \cong \Omega^{0,1}(L).$$

Thus, any spinor $\psi \in \Gamma(\mathbb{S}_L^+)$ naturally decomposes as

$$\psi = \alpha \oplus \beta \in \Omega^{0,0}(L) \oplus \Omega^{0,2}(L).$$

The Chern connection on $TM$ induces a connection $A_0$ on $K_M^{-1}$. Any Hermitian connection $A$ on $\det(\sigma_L)$ can be written as

$$A = A_0 \dot{+} 2B,$$

where $B$ is a Hermitian connection on $L$. From Proposition 1.4.25 we deduce that, exactly as in the Kähler case, we have

$$\mathfrak{D}_A = \sqrt{2}(\bar{\partial}_B + \bar{\partial}_B^*).$$

Imitating the situation in §3.2.1 we choose the perturbation parameter of the form

$$\eta_t := \mathbf{i}F_{A_0} + \frac{t}{8}\omega.$$

Again, we can rewrite the Seiberg-Witten equations in terms of $(\alpha, \beta, B)$ and, exactly as in §3.2.1 we deduce

$$(3.3.18) \qquad \begin{cases} \bar{\partial}_B\alpha + \bar{\partial}_B^*\beta & = & 0 \\ \Lambda F_B & = & \frac{\mathbf{i}}{8}(|\alpha|^2 - |\beta|^2 - t) \\ F_B^{0,2} & = & \frac{1}{8}\bar{\alpha}\beta \end{cases}.$$

The virtual dimension of the space of $\sigma_L$-monopoles is computed by the same formula as in 3.2.1

$$(3.3.19) \qquad d(\sigma_L) = L \cdot (K_M - L).$$

As in the Kähler case, for any Hermitian line $L \to M$, we denote by $\deg_\omega(L)$ the quantity

$$\deg_\omega(L) := \frac{\mathbf{i}}{2\pi}\int_M F_B \wedge \omega$$

where $B$ is an arbitrary Hermitian connection on $L$. Since $\omega$ is *closed* we deduce that the above expression is independent of $B$.

If $b_2^+(M) = 1$ then $\eta_t$ belongs to the $\pm$ chamber if

$$\pm\big(t - \frac{16\pi}{\mathrm{vol}\,(M)}\deg_\omega(L)\big) > 0.$$

**Theorem 3.3.29. (Taubes, [134, 135])** *(a)*

$$\mathbf{sw}_M^{(+)}(0) = \pm\mathbf{sw}_M^{(-)}(K_M) = \pm 1.$$

*(b) If* $\mathbf{sw}_M^{(+)}(L) \neq 0$ *then* $\deg_\omega(L) \geq 0$ *with equality if and only if* $L$ *is trivial.*

*(c) If* $\mathbf{sw}_M^{(-)}(L) \neq 0$ *then* $\deg_\omega(L) \leq \deg_\omega(K_M)$ *with equality if and only if* $L$ *is isomorphic to* $K_M$.

**Proof** We follow the approach in [**69**]. Using the involution $\sigma \mapsto \bar{\sigma}$ we see that it suffices to prove only that $\mathbf{sw}_M^{(+)}(\sigma_0) = \pm 1$ and (b).

Notice first that if $L$ is trivial then (3.3.18) has a nontrivial solution with $B$ the trivial connection, $\beta = 0$ and $\alpha = t^{1/2}$. Suppose now that $\mathbf{sw}_M^{(+)}(\sigma_L) \neq 0$. Fix $t \gg 0$ and consider an $\eta_t$-monopole

$$(\psi, A) = (\alpha, \beta, A = A_0 \dot{+} 2B)$$

corresponding to the *spin*$^c$ structure $\sigma_L$.

Using Proposition 1.4.22 we deduce

$$2\bar{\partial}_B^*\bar{\partial}_B\alpha = (\nabla^B)^*\nabla^B\alpha - \mathbf{i}\Lambda(F_B)\alpha$$

Taking the inner product with $\alpha$ and integrating by parts we deduce

$$\int_M |\nabla^B\alpha|^2 dv_M = \int_M \Big(2\langle\bar{\partial}_B^*\bar{\partial}_B\alpha, \alpha\rangle + \mathbf{i}\Lambda(F_B)|\alpha|^2\Big)dv_M. \tag{3.3.20}$$

Now use the first equation in (3.3.18) to deduce

$$\int_M \langle 2\bar{\partial}_B^*\bar{\partial}_B\alpha, \alpha\rangle dv_M = -2\int_M \langle\bar{\partial}_B^*\bar{\partial}_B^*\beta, \alpha\rangle dv_M = -2\int_M \langle\beta, \bar{\partial}_B^2\alpha\rangle dv_M$$

(use (1.4.19) in 1.4.2)

$$= -2\int_M \Big(\langle\beta, F_B^{0,2}\alpha\rangle - \langle\beta, (\partial_B\alpha)\circ N\rangle\Big)dv_M$$

(use the third equation in (3.3.18))

$$= \int_M \Big(-\frac{1}{4}|\alpha|^2|\beta|^2 + 2\langle\beta, (\partial_B\alpha)\circ N\rangle\Big)dv_M.$$

On the other hand, using the second equation in (3.3.18) we deduce

$$\int_M \mathbf{i}\Lambda(F_B)|\alpha|^2 dv_M = -\frac{1}{8}\int_M (|\alpha|^2 - |\beta|^2 - t)|\alpha|^2 dv_M.$$

Substituting this in (3.3.20) we obtain

$$\int_M \Big(|\nabla^B\alpha|^2 + \frac{|\alpha|^2}{8}(|\alpha|^2 + |\beta|^2 - t)|\alpha|^2\Big) dv_M = 2\int_M \langle \beta, (\partial_B\alpha)\circ N\rangle dv_M$$

or, equivalently,

$$\begin{aligned}\int_M \Big(|\nabla^B\alpha|^2 + \frac{1}{8}|\alpha|^2|\beta|^2 + \frac{1}{8}(|\alpha|^2 - t)^2 + \frac{t}{8}(|\alpha|^2 - t)\Big) dv_M \\ = 2\int_M \langle \beta, (\partial_B\alpha)\circ N\rangle dv_M.\end{aligned} \tag{3.3.21}$$

The right-hand side of (3.3.21) can be estimated using the interpolation inequality

$$|ab| \le \frac{\varepsilon}{2}a^2 + \frac{1}{2\varepsilon}b^2$$

and we obtain

$$\int_M \Big(|\nabla^B\alpha|^2 + \frac{1}{8}|\alpha|^2|\beta|^2 + \frac{1}{8}(|\alpha|^2 - t)^2 + \frac{t}{8}(|\alpha|^2 - t)\Big) dv_M$$

$$\le \frac{1}{2}\int_M |\nabla^B\alpha|^2 dv_M + C\int_M |\beta|^2 dv_M$$

where $C$ is some positive constant which depends only on the size of the Nijenhuis tensor $N$. Thus,

$$\begin{aligned}\int_M \Big(\frac{1}{2}|\nabla^B\alpha|^2 + \frac{1}{8}|\alpha|^2|\beta|^2 + \frac{1}{8}(|\alpha|^2 - t)^2 + \frac{t}{8}(|\alpha|^2 - t)\Big) dv_M \\ \le C\int_M |\beta|^2 dv_M.\end{aligned} \tag{3.3.22}$$

Now, using the identity

$$\deg_\omega(L) = \frac{\mathbf{i}}{2\pi}\int_M F_B \wedge \omega = \frac{\mathbf{i}}{2\pi}\int_M \Lambda F_B dv_M$$

$$= -\frac{1}{16\pi}\int_M \Big(|\alpha|^2 - |\beta|^2 - t\Big) dv_M$$

we deduce

$$\frac{t}{8}\int_M (|\alpha|^2 - t) dv_M = \frac{t}{8}\int_M |\beta|^2 dv_M - 2\pi t \deg_\omega(L).$$

Substituting this equality in (3.3.22) we obtain

(3.3.23)
$$\int_M \Big(\frac{1}{2}|\nabla^B\alpha|^2 + \frac{1}{8}|\alpha|^2|\beta|^2 + \frac{1}{8}(|\alpha|^2 - t)^2 + \frac{t}{8}|\beta|^2\Big)dv_M - 2\pi t \deg_\omega(L) \leq C\int_M |\beta|^2 dv_M.$$

Since $t \gg 0$ we can assume $t > 8C$. The last inequality then implies

$$2\pi t \deg_\omega(L)$$

$$\geq \int_M \Big(\frac{1}{2}|\nabla^B\alpha|^2 + \frac{1}{4}|\alpha|^2|\beta|^2 + \frac{1}{8}(|\alpha|^2 - t)^2 + \Big(\frac{t}{8} - C\Big)|\beta|^2\Big)dv_M \geq 0.$$

Hence

$$\deg_\omega(L) \geq 0.$$

Moreover, we see that $\deg_\omega(L) = 0$ if and only if $|\alpha| \equiv t^{1/2}$, $\nabla^B\alpha \equiv 0$ and $\beta \equiv 0$. This shows that $L$ must be trivial.

If $L$ is trivial the above inequality shows that for all $t > 4C$ there exists a unique (up to $\mathfrak{G}_{\sigma_0}$) $\eta_t$-monopole

$$\mathsf{C}_0 = (\alpha_0 = t^{1/2}, \beta_0 = 0, A_0).$$

In this case, the twisting connection $B$ on the trivial line bundle is the trivial connection. To complete the proof of Theorem 3.3.29 we only need to show $\mathsf{C}_0$ is nondegenerate. We follow a strategy very similar to the one employed in §3.3.1. Set $\lambda := t^{1/2}$.

As in §3.2.3 we can write

$$\dot{\mathsf{C}} = (\dot\alpha \oplus \dot\beta, \mathbf{i}\dot b = \frac{\mathbf{i}}{\sqrt{2}}(\varphi + \bar\varphi))$$

and we deduce $\dot{\mathsf{C}} \in \ker \mathcal{T}_{\mathsf{C}_0}$ if and only if

$$\sqrt{2}(\bar\partial\dot\alpha + \bar\partial^*\dot\beta) + \mathbf{i}(\varphi \wedge \alpha_0 - \bar\varphi \lrcorner \beta_0) = 0, \tag{3.3.24a}$$

$$\Lambda(\partial\varphi + \bar\partial\bar\varphi) = \frac{1}{2\sqrt{2}}(\mathfrak{Re}\langle\alpha_0, \dot\alpha\rangle - \mathfrak{Re}\langle\beta_0, \dot\beta\rangle), \tag{3.3.24b}$$

$$\mathbf{i}\bar\partial\varphi = \frac{1}{4\sqrt{2}}(\dot{\bar\alpha}\beta_0 + \bar\alpha_0\dot\beta), \tag{3.3.24c}$$

$$2\sqrt{2}(\bar\partial^*\varphi + \partial^*\bar\varphi) + \mathfrak{Im}\langle\alpha_0, \dot\alpha\rangle + \mathfrak{Im}\langle\beta_0, \dot\beta\rangle = 0. \tag{3.3.24d}$$

(Recall that above $\alpha_0 = \lambda$, $\beta_0 = 0$.) Using the Kähler-Hodge identities in Proposition 1.4.10 of §1.4.1 we deduce as §3.2.3 that (3.3.24b) and (3.3.24d) are equivalent to

$$\bar{\partial}^* \varphi = -\frac{\lambda \mathbf{i}}{4\sqrt{2}} \dot{\alpha}.$$

We deduce that $\dot{\mathsf{C}} \in \ker \mathcal{T}_{\mathsf{C}_0}$ if and only if

$$\sqrt{2}(\bar{\partial}\dot{\alpha} + \bar{\partial}^* \dot{\beta}) + \lambda \mathbf{i} \varphi = 0, \tag{3.3.25a}$$

$$\mathbf{i}\bar{\partial}\varphi = \frac{\lambda}{4\sqrt{2}} \dot{\beta}, \tag{3.3.25b}$$

$$\bar{\partial}^* \varphi = -\frac{\lambda \mathbf{i}}{4\sqrt{2}} \dot{\alpha}. \tag{3.3.25c}$$

Using the identities

$$\mathfrak{D}_{A_0} = \sqrt{2}(\bar{\partial} + \bar{\partial}^*) : \Omega^{0,0}(M) \to \Omega^{0,2}(M) \to \Omega^{0,1}(M)$$

and

$$\mathfrak{D}^*_{A_0} = \sqrt{2}(\bar{\partial}^* \oplus \bar{\partial}) : \Omega^{0,1}(M) \to \Omega^{0,0}(M) \oplus \Omega^{0,2}(M)$$

we can rewrite the above equalities as

$$\mathfrak{D}_{A_0} \dot{\psi} = -\frac{\lambda \mathbf{i}}{4} \varphi, \;\; \mathfrak{D}^*_{A_0} \varphi = -\frac{\lambda \mathbf{i}}{4} \dot{\psi}, \;\; \dot{\psi} = \begin{bmatrix} \dot{\alpha} \\ \dot{\beta} \end{bmatrix}.$$

Thus

$$\mathfrak{D}^*_{A_0} \mathfrak{D}_{A_0} \dot{\psi} = -\frac{\lambda^2}{16} \dot{\psi}.$$

Using the Weitzenböck presentation of the generalized Laplacian $\mathfrak{D}^*_{A_0} \mathfrak{D}_{A_0}$ we can rewrite the above equation as

$$(\nabla^* \nabla + \mathcal{R} + \frac{\lambda^2}{16}) \dot{\psi} = 0 \tag{3.3.26}$$

where $\mathcal{R}$ is a zeroth order operator independent of $\lambda$. If $\lambda$ is sufficiently large we deduce that the selfadjoint operator $\mathcal{R} + \lambda^2$ is positive definite so the only solution of (3.3.26) is $\dot{\psi} \equiv 0$. This forces $\varphi \equiv 0$ and thus

$$\ker \mathcal{T}_{\mathsf{C}_0} = 0, \;\; \forall \lambda \gg 0.$$

The proof of Theorem 3.3.29 is now complete. ■

**Remark 3.3.30.** We have not discussed if there is a natural way of determining the sign of the unique monopole $\mathsf{C}_0$. This issue is equivalent to the existence of natural orientations on $H^1(M)$ and $H^2_+(M)$. Such choices are still possible and lead to the conclusion that $\mathbf{sw}^{(+)}_M(\sigma_0) = 1$. For details we refer to [**57, 119**].

**Remark 3.3.31.** The above nonvanishing result implies that any symplectic (Kähler) 4-manifold admits almost complex structures which are not homotopic to an almost complex structure compatible with a symplectic (Kähler) structure; see [**27**].

**Remark 3.3.32.** One can use the information contained in Taubes' theorem to produce a very ingenious invariant of a symplectic 4-manifold, $(M,\omega)$.

Observe first that the symplectic structure determines a canonical *spin*$^c$ structure $\sigma_0$ which allows us to identify $Spin^c(M)$ with $H^2(M,\mathbb{Z})$. Using the morphism $H^2(N,\mathbb{Z}) \to H^2(M,\mathbb{Z})$ we can map the set of basic classes $\mathbb{B}_M$ to a finite collection of lattice points in $H^2(M,\mathbb{R})$. (The lattice is the image of $H^2(M,\mathbb{Z}) \to H^2(M,\mathbb{R})$.) The image of $\sigma_0$ is the origin of $H^2(M,\mathbb{R})$ while the image of $\bar{\sigma}_{can}$ coincides with the image of $c_1(K_M)$. For simplicity, we will denote by $K_M$ this image.

The symplectic form $\omega$ defines by integration a linear functional $L_\omega$ : $H^2(M,\mathbb{R}) \to \mathbb{R}$. Denote by $\mathcal{P}_{M,\omega}$ the convex hull of $\mathbb{B}_M \subset H^2(M,\mathbb{R})$. $\mathcal{P}_{M,\omega}$ is a convex polyhedron. Taubes' theorem imposes several restrictions on $\mathcal{P}_{M,\omega}$.

• Since $\sigma \in \mathbb{B}_M \Longleftrightarrow \bar{\sigma} \in \mathbb{B}_M$ we deduce that $\mathcal{P}_{M,\omega}$ is symmetric with respect to the point $\frac{1}{2}K_M$.

• The minimum (resp. maximum) of $L_\omega$ on $\mathcal{P}_{M,\omega}$ is achieved at precisely one point, 0 (resp. $K_M$) which must be a vertex of $\mathcal{P}_{M,\omega}$.

• The group $\Gamma_M$ = (group orientation preserving diffeomorphisms)/(subgroup of diffeomorphisms homotopic to 1) acts on $\mathbb{B}_M$ thus inducing an (affine) action on $\mathcal{P}_{M,\omega}$ which must leave invariant the finite set of vertices of $\mathcal{P}_{M,\omega}$.

Let us define a *special polyhedron* to be a $\Gamma_M$-invariant convex polyhedron $\mathcal{P}$ in the *affine* space $H^2(M,\mathbb{R})$ together with the following additional structure.

∘ The vertices of $\mathcal{P}$ are lattice points.

∘ $\mathcal{P}$ admits a center of symmetry $O$.

∘ There exist an affine map $L : \mathcal{P} \to \mathbb{R}$ and a pair of $O$-symmetric vertices $P_\pm$ of $\mathcal{P}$ such that $\pm L$ achieves its maximum exactly at $P_\pm$.

We will denote the special polynomials by $(\mathcal{P}, O, P_-, P_+, L)$. Clearly, $(\mathcal{P}_{M,\omega}, \frac{1}{2}K_M, \sigma_0, \bar{\sigma}_{can}, L_\omega)$ is a special polyhedron.

Two symplectic forms $\omega_0$ and $\omega_1$ are called isotopic if there exists a smooth path $\omega_t$ of symplectic forms connecting them. Two isotopic symplectic forms determine the same special polyhedron.

The group $\Gamma_M$ acts on the set of special polyhedra according to the rule

$$\gamma \cdot (\mathcal{P}, O, P_-, P_+, L) = (\gamma\mathcal{P}, \gamma O, \gamma P_-, \gamma P_+, \gamma L\gamma^{-1})$$

and two special polyhedra are said to be equivalent if they belong to the same $\Gamma_M$-orbit.

Two symplectic forms $\omega_0$ and $\omega_1$ are called *equivalent* if there exists an orientation preserving diffeomorphism $\varphi$ of $M$ such that $\varphi^*\omega_0$ is isotopic to $\omega_1$. Taubes' theorem implies that two equivalent symplectic forms determine equivalent special polyhedra.

It is very easy to construct invariants of equivalence classes of special polyhedra,

$$(\mathcal{P}, O, P_-, P_+, L).$$

More precisely, the number $\deg(P_-)$ of 1-faces of $\mathcal{P}$ which have $P_-$ as one end point is such an invariant. In particular, if $\omega$ is a symplectic form on $M$ then the integer

$$\nu(\omega) := \deg(\sigma_0(\omega))$$

is an invariant of the equivalence class of $\omega$. At a first glance, $\nu(\omega)$ may look like a very difficult to compute weak invariant.

In a recent stunning work [**90**], C.T. McMullen and C.H. Taubes have very elegantly constructed compact smooth 4-manifolds admitting symplectic structures with distinct $\nu$-invariant. They have thus given a positive answer to a longstanding question in symplectic topology: do there exist compact smooth manifolds admitting non-equivalent symplectic forms?

Theorem 3.3.29 has a nice topological consequence.

**Corollary 3.3.33.** *Suppose $M$ is a smooth, compact, closed oriented manifold such that $b_2^+(M) > 1$.*

*(a) If $\mathbf{sw}_M(\sigma) = 0$ for all $\sigma \in Spin^c(M)$ then $M$ cannot admit symplectic structures. In particular, if $M$ admits metrics of positive scalar curvature it cannot admit symplectic structures.*

*(b) If $|\mathbf{sw}_M(\sigma)| \neq 1$ for all $\sigma \in Spin^c(M)$ then $M$ cannot admit symplectic structures.*

**Remark 3.3.34.** Part (b) of Corollary 3.3.33, combined with some very ingenious topological constructions, was used in [**36, 131**] to produce many families of smooth 4-manifolds which admit no symplectic structures, and yet they have many of the known topological features of symplectic manifolds.

*Chapter 4*

# Gluing Techniques

*Treat nature in terms of the cylinder, the sphere, the cone, all in perspective.*

Paul Cézanne

## 4.1. Elliptic equations on manifolds with cylindrical ends

This section includes some basic analytic facts absolutely necessary in the understanding of the gluing problem. The main references for all of the following results are [**6, 74**]. We will follow the "ˆ" conventions of §2.4.1.

**4.1.1. Manifolds with cylindrical ends.** A *cylindrical* $(n+1)$-manifold is an oriented Riemannian $(n+1)$-manifold $(\hat{N}, \hat{g})$ with a cylindrical end modeled by $\mathbb{R}_+ \times N$ where $(N, g)$ is an oriented compact Riemannian $n$-manifold (see Figure 4.1). In more rigorous terms, this means that the complement of an open precompact subset of $\hat{N}$ is *isometric* in an orientation preserving fashion to the cylinder $\mathbb{R}_+ \times N$. This isometry is part of the structure of a cylindrical manifold. We will denote the canonical projection $\mathbb{R}_+ \times N \to N$ by $\pi$ while $t$ will denote the *outgoing* longitudinal coordinate along the neck. We will regularly denote the "slice" $N$ by $\partial_\infty \hat{N}$ and the metric $g$ by $\partial_\infty \hat{g}$. For each $t \geq 0$ we set $\hat{N}_t := \hat{N} \setminus (t, \infty) \times N$.

A *cylindrical structure* on a vector bundle $\hat{E} \to \hat{N}$ consists of a vector bundle $E \to N$ and a bundle isomorphism

$$\hat{\vartheta} : E|_{\mathbb{R}_+ \times N} \to \pi^* E.$$

We will use the notation $E := \partial_\infty \hat{E}$.

A cylindrical vector bundle will be a vector bundle together with a cylindrical structure $(\hat{\vartheta}, E)$. A section $\hat{u}$ of a cylindrical vector bundle is said to

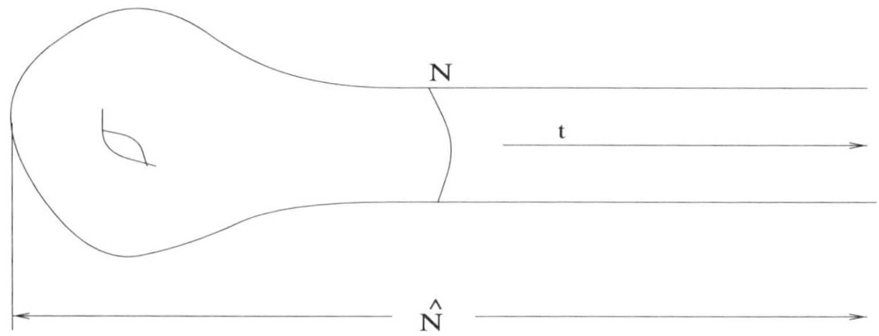


**Figure 4.1.** *Manifold with a cylindrical end*

be *cylindrical* if there exists a section $u$ of $\partial_\infty \hat{E}$ such that along the neck $\hat{u} = \pi^* u$. We will use the notation $u := \partial_\infty \hat{u}$.

Given any cylindrical vector bundle $(\hat{E}, \vartheta, E)$ there exists a canonical first order partial differential operator $P$, defined over the cylindrical end, uniquely determined by the conditions

$$P(fu) = \frac{df}{dt} u + fPu, \ \ \forall f \in C^\infty(\mathbb{R}_+ \times N), \ u \in \hat{E}\,|_{\mathbb{R}_+ \times N}$$

and $Pv = 0$ for any cylindrical section $v$ of $\hat{E}\,|_{\mathbb{R}_+ \times N}$. We will denote this operator by $\partial_t$.

**Example 4.1.1.** The cotangent bundle of a cylindrical manifold $(\hat{N}, \hat{g})$ has a natural cylindrical structure with $\partial_\infty T^*N = \underline{\mathbb{R}} \oplus T^*N$, where $\underline{\mathbb{R}}$ denotes the trivial real line bundle spanned by $dt$. The isomorphism $\hat{\vartheta}$ is given by

$$\hat{\vartheta}\alpha = \alpha(\partial_t) \oplus (\alpha - \alpha(\partial_t)dt), \ \ \forall \alpha \in \Omega^1(\hat{N}).$$

It is now clear that we can organize the set of cylindrical bundles over a given cylindrical manifold as a category. Moreover, we can perform all the standard tensorial operations in this category such as direct sums, tensor products, duals, etc.

**Exercise 4.1.1.** Formulate explicitly the exact definition of a cylindrical isomorphism of cylindrical vector bundles.

Denote by $\mathrm{VBUN}_{cyl}(\hat{N})$ the set of isomorphism classes of cylindrical vector bundles. We want to draw the reader's attention to one subtle fact. Two cylindrical vector bundles may be isomorphic as vector bundles but may **not** be isomorphic as *cylindrical* vector bundles. Define

$$\mathrm{Pic}^\infty_{cyl}(\hat{N}) \subset \mathrm{VBUN}_{cyl}(\hat{N})$$

as the space of isomorphism classes of cylindrical complex line bundles over $\hat{N}$. It is an Abelian group with respect to tensor multiplication. We have a forgetful morphism

$$\Phi : \mathrm{Pic}^{\infty}_{cyl}(\hat{N}) \to \mathrm{Pic}^{\infty}(\hat{N})$$

which is clearly onto. Its kernel consists of isomorphism classes of cylindrical structures on a trivial line bundle. We leave it to the reader to check the following fact.

**Exercise 4.1.2.**

$$\ker \Phi \cong H^1(N, \mathbb{Z})/H^1(\hat{N}, \mathbb{Z}) \cong \mathrm{Range}\Big(H^1(N, \mathbb{Z}) \stackrel{\delta}{\to} H^2(\hat{N}, N; \mathbb{Z})\Big).$$

A *cylindrical partial differential operator (p.d.o.)* will be a *first order* p.d.o. $\hat{L}$ between two cylindrical bundles $\hat{E}$, $\hat{F}$ such that along the neck $[T, \infty) \times N$ $(T \gg 0)$ it can be written as

$$\hat{L} = G\partial_t + L$$

where $L : C^\infty(E) \to C^\infty(E)$ is a first order p.d.o., $E = \hat{E}|_N$, $F = \hat{F}|_N$ and $G : E \to F$ is a cylindrical bundle morphism. We will use the notation

$$L := \partial_\infty \hat{L}.$$

If $\hat{\sigma}$ denotes the symbol of $\hat{L}$ then we see that $G = \hat{\sigma}(dt)$ and

$$\partial_\infty \hat{L} = \hat{L} - G\partial_t.$$

**Example 4.1.2.** If $\hat{E} \to \hat{N}$ is cylindrical then so is $T^*\hat{N} \otimes \hat{E}$. Any connection is a first order p.d.o. $C^\infty(\hat{E}) \to C^\infty(T^*\hat{N} \otimes \hat{E})$. A connection which is cylindrical as a p.d.o. is called *cylindrical.* Observe though the following "pathology". If $\hat{\nabla}$ is such a connection then along the neck it has the form

$$\hat{\nabla} = dt \otimes \partial_t + \partial_\infty \hat{\nabla}$$

where $\partial_\infty \hat{\nabla}$ is a first order p.d.o. $C^\infty(E) \to C^\infty(E) \oplus C^\infty(T^*N \otimes E)$. The component

$$C^\infty(E) \to C^\infty(T^*N \otimes E)$$

is a connection on $E$ while the component

$$A : C^\infty(E) \to C^\infty(E)$$

is a zeroth order operator, i.e. an endomorphism of $E$. Thus, $\partial_\infty \hat{\nabla}$ is no longer a connection. We define a *strongly cylindrical connection* to be a cylindrical connection such that the zeroth order component $A$ described above vanishes identically.

At this point it is illuminating to have another look at a notion we encountered in §2.4.1. Recall that a connection $\hat{\nabla}$ on a cylindrical bundle

$(\hat{E}, \hat{\vartheta})$ is called *temporal* if $\hat{\nabla}_t = \partial_t$. Thus, a connection is strongly cylindrical if it is both cylindrical and temporal.

A *cylindrical Hermitian bundle* is a cylindrical bundle $(\hat{E}, \hat{\vartheta})$ equipped with a cylindrical metric $\hat{h}$ and a strongly cylindrical connection $\hat{\nabla}^0$ compatible with $\hat{h}$.

Suppose $\hat{N}$ is an oriented cylindrical 4-manifold with $N := \partial_\infty \hat{N}$ and $\hat{\sigma}$ is a *spin*$^c$ structure on $\hat{N}$. We say that $\hat{\sigma}$ is a *cylindrical spin*$^c$-structure if there exist a *spin*$^c$ structure $\sigma$ on $N$ and an isomorphism

$$\varphi : \hat{\sigma} \,|_{\mathbb{R}_+ \times N} \rightarrow \mathbb{R}_+ \times \sigma$$

where $\mathbb{R}_+ \times \sigma$ denotes the natural *spin*$^c$ structure on $\mathbb{R} \times N$ induced by $\sigma$. ($\varphi$ has to be compatible in the obvious way with the cylindrical structure of $\hat{N}$.) We set $\sigma := \partial_\infty \hat{\sigma}$ and, whenever there is a potential ambiguity, we will denote a cylindrical *spin*$^c$ structure by a triple

$$\hat{\tau} := (\hat{\sigma}, \sigma, \varphi).$$

We set $\partial_\infty \hat{\tau} := \sigma$. Two such triples $\hat{\tau}_i = (\hat{\sigma}_i, \sigma_i, \varphi_i)$ are isomorphic if there exist isomorphisms

$$\hat{\Phi} : \hat{\sigma}_1 \rightarrow \hat{\sigma}_2, \quad \Phi : \sigma_1 \rightarrow \sigma_2$$

such that the diagram below is commutative.

$$\begin{array}{ccc} \hat{\sigma}_1 \,|_{\mathbb{R}_+ \times N} & \xrightarrow{\varphi_1} & \mathbb{R}_+ \times \sigma_1 \\ \Big\downarrow \hat{\Phi} & & \Big\downarrow \mathbb{R}_+ \times \Phi \\ \hat{\sigma}_2 \,|_{\mathbb{R}_+ \times N} & \xrightarrow{\varphi_2} & \mathbb{R}_+ \times \sigma_2 \end{array}$$

We denote by $Spin^c_{cyl}(\hat{N})$ the set of isomorphism classes of cylindrical *spin*$^c$-structures over $\hat{N}$. Observe that $\mathrm{Pic}^\infty_{cyl}(\hat{N})$ acts on $Spin^c_{cyl}(\hat{N})$ freely and transitively, so that $Spin^c_{cyl}(\hat{N})$ is a $\mathrm{Pic}^\infty_{cyl}(\hat{N})$-torsor.

**4.1.2. The Atiyah-Patodi-Singer index theorem.** Suppose now that $\hat{E}$ and $\hat{F}$ are cylindrical Hermitian bundles over $\hat{N}$. An **Atiyah-Patodi-Singer operator** (*APS* for brevity) is an *elliptic* cylindrical p.d.o. such that along the neck it has the form $\hat{L} = G\partial_t + L$ where

- $G$ is a homothety, i.e. there exists a positive constant $\lambda$ such that $GG^* = \lambda$;
- $\vec{\partial}_\infty \hat{L} := -G^{-1}\partial_\infty \hat{L} : C^\infty(E) \rightarrow C^\infty(E)$ is *formally selfadjoint*.

Traditionally, the *APS* operators are described in the form (see **[6]**):

$$\hat{L} = G\Big(\partial_t - A\Big).$$

The operator $A$ is none other than $\vec{\partial}_\infty \hat{L}$.

We will use the symbol $P(\hat{L})_\geq$ to denote the orthogonal projection onto the space spanned by the eigenvectors of $\vec{\partial}_\infty \hat{L}$ corresponding to eigenvalues $\geq 0$. $P(\hat{L})_>$ is defined similarly.

**Remark 4.1.3.** We want to draw attention to a confusing point. Consider an *oriented* Riemannian manifold $N$ and form the cylindrical manifold $\hat{N} = \mathbb{R} \times N$. $\partial_\infty \hat{N}$ has *two* components $N_{\pm\infty}$. The induced orientation on $N_{\pm\infty}$ is $\pm$ the orientation on $N$. Any bundle $E \to N$ and any selfadjoint Dirac-type operator $L : C^\infty(E) \to C^\infty(E)$ define in an obvious manner a cylindrical bundle $\hat{E} = \pi^* E$ and an $APS$ operator $\hat{L} = \partial_t - L$. Then $\vec{\partial}_\infty \hat{L}$ is a p.d.o. on the *disconnected* boundary $\partial_\infty \hat{N}$. On $N_{\pm\infty}$ we have

$$\vec{\partial}_\infty \hat{L}\,|_{N_{\pm\infty}} = \pm L.$$

To avoid confusion *always orient the manifold $\hat{N}$ first, and then give $\partial_\infty \hat{N}$ the induced orientation given by the outer-normal-first convention.* There is no room for variation around this rule since the orientability of a bordism implies the orientability of its boundary while the converse is certainly not true (think of the Möbius band).

Suppose $\hat{L} : C^\infty(\hat{E}) \to C^\infty(\hat{F})$ is an $APS$ operator between cylindrical Hermitian bundles. The $APS$ problem for $\hat{L}$ is the following boundary value problem:

$$(APS) \qquad \begin{cases} \hat{L}\hat{u} = 0 & \text{on } \hat{N}_r \\ P(\hat{L})_\geq \hat{u} = 0 & \text{on } \partial \hat{N}_r \end{cases}$$

where $r \gg 0$. If $\hat{L} = G\partial_t + L$ then the formal adjoint $\hat{L}^* = -G^*\partial_t + L^*$ is also an $APS$ operator. Indeed, using $G^* = \lambda G^{-1}$and $(\vec{\partial}_\infty \hat{L})^* = \vec{\partial}_\infty \hat{L}$ we deduce

$$\vec{\partial}_\infty \hat{L}^* = (G^*)^{-1} L^* = -(G^*)^{-1}(G\vec{\partial}_\infty \hat{L})^* = -G\vec{\partial}_\infty \hat{L} G^{-1} = -\frac{1}{\lambda} G\vec{\partial}_\infty \hat{L} G^*$$

so $\vec{\partial}_\infty \hat{L}^*$ is formally selfadjoint.

The formal adjoint $APS^*$ of the $APS$ boundary value problem is

$$(APS^*) \qquad \begin{cases} \hat{L}^*\hat{v} = 0 & \text{on } \hat{N}_r \\ P(\hat{L}^*)_> \hat{v} = 0 & \text{on } \partial \hat{N}_r \end{cases}.$$

**Remark 4.1.4.** As pointed out in [**6**], the solutions of $(APS)$ and $(APS^*)$ can be given an alternate description. For clarity, along the neck we write

$$\hat{L} := G(\partial_t - A), \;\; \hat{L}^* = -G^*(\partial_t - B), \;\; B := -GAG^{-1}.$$

$A$ and $B$ are first order selfadjoint elliptic operators and thus have discrete spectra, consisting only of eigenvalues of finite multiplicities. Denote by

$(\psi_{\lambda_m})_{\lambda_m\in\mathbb{R}}$ and $(\phi_{\mu_n})_{\mu_n\in\mathbb{R}}$, respectively, a complete orthonormal system of eigenfunctions of $A$ and $B$, respectively. Then

$$P(\hat{L})_{\geq}\hat{u} = 0 \Longleftrightarrow \hat{u}\,|_{\partial\hat{N}_r} \in \operatorname{span}_{L^2}\Big\{\psi_{\lambda_m};\ \ \lambda_m < 0\Big\},$$

$$P(\hat{L}^*)_{>}\hat{v} = 0 \Longleftrightarrow \hat{v}\,|_{\partial\hat{N}_r} \in \operatorname{span}_{L^2}\Big\{\phi_{\mu_n};\ \ \mu_n \leq 0\Big\}.$$

Suppose $\hat{u}$ and $\hat{v}$ are smooth solutions of $(APS)$ and $(APS^*)$, respectively. Along $\partial\hat{N}_r$, we can write

$$\hat{u} = \sum_{\lambda_m<0} u_{\lambda_m}\psi_{\lambda_m},\ \ u_{\lambda_m} \in \mathbb{C},\ \ \sum_{\lambda_m<0} |u_{\lambda_m}|^2 < \infty$$

and

$$\hat{v} = \sum_{\mu_n\leq 0} v_{\mu_n}\phi_{\mu_n},\ \ v_{\mu_n} \in \mathbb{C},\ \ \sum_{\mu_n\leq 0} |v_{\mu_n}|^2 < \infty.$$

Now extend $\hat{u}$ and $\hat{v}$ to $[r,\infty)\times N$ by setting

$$\hat{u}(t) = \sum_{\lambda_m<0} e^{\lambda_m(t-r)}u_{\lambda_m}\psi_{\lambda_m},\ \ \hat{v}(t) = \sum_{\mu_n\leq 0} e^{\mu_n(t-r)}v_{\mu_n}\phi_{\mu_n}$$

and continue to denote by $\hat{u}$ and $\hat{v}$ the sections thus produced over $\hat{N}$. One can show that $\hat{u}$ and $\hat{v}$ are smooth and

$$\hat{L}\hat{u} = 0,\ \ \hat{L}^*\hat{v} = 0.$$

These two sections also have nice behaviors as $t \to \infty$. $\hat{u}$ decays exponentially to zero (and thus it is an $L^2$-section on $\hat{N}$) while $\hat{v}(t)$ decays exponentially to

$$\hat{v}(\infty) := \sum_{\mu_n=0} v_{\mu_n}\phi_{\mu_n}.$$

The *Atiyah-Patodi-Singer index* of $\hat{L}$, denoted by $I_{APS}(\hat{L})$, is the quantity

$$I_{APS}(\hat{L}) = I_{APS}(L,\hat{N}_r) := \dim\ker(APS) - \dim\ker(\ APS^*).$$

A priori, this index may be infinite, or even worse, may not be well defined. The celebrated Atiyah-Patodi-Singer index theorem, [**6**], states that both $\dim\ker(APS)$ and $\dim\ker(APS^*)$ are finite and their difference can be explicitly expressed in terms of $\hat{L}$. To formulate this theorem we need to define the eta invariant.

The elliptic selfadjoint operators on closed compact manifolds behave in many respects as common finite-dimensional symmetric matrices. The eta invariant extends the notion of signature from finite-dimensional symmetric matrices to selfadjoint elliptic operators.

The signature of a finite-dimensional symmetric matrix $A$ is defined as

$$\mathrm{sign}\,(A) = \text{number of positive eigenvalues} - \text{number of negative eigenvalues}.$$

This definition however does not extend to infinite dimensions since the above terms are infinite. Following a strategy very dear to physicists one could try to "regularize" the definition. For each $s \in \mathbb{C}$ we set

$$\eta_A(s) = \sum_{\lambda \in \sigma^*(A)} \frac{\dim \ker(A - \lambda)}{\lambda |\lambda|^{s-1}} \tag{4.1.1}$$

$$= \sum_{\lambda > 0} \frac{\dim \ker(A - \lambda) - \dim \ker(A + \lambda)}{\lambda^s}$$

where $\sigma^*(A) = \mathrm{spec}\,(A) \setminus \{0\}$. Then one can define

$$\mathrm{sign}\,(A) = \eta_A(0).$$

The advantage of this new definition is that it is admirably suited for infinite-dimensional extensions. Assuming for simplicity that $A$ is invertible we can define

$$\eta_A(s) = \mathrm{tr}\,(A \cdot |A|^{-(s+1)}), \quad |A| = (A^2)^{1/2}.$$

Using the classical integral

$$\Gamma(\alpha) x^{-\alpha} = \int_0^\infty t^{\alpha-1} e^{-tx} dt, \quad x > 0, \ \alpha > 1,$$

we get $(x \mapsto A^2, \ \alpha \mapsto (s+1)/2)$

$$\eta_A(s) = \frac{1}{\Gamma(\,(s+1)/2\,)} \int_0^\infty t^{(s-1)/2} \mathrm{tr}\,(A e^{-tA^2}) dt.$$

The right-hand side of the above expression has two advantages. First of all, it makes sense even when $A$ is not invertible and on the other hand, it extends to infinite dimensions. We will denote the trace of an infinite-dimensional operator (when it exists) by "Tr" while "tr" is reserved for finite-dimensional operators. We have the following result.

**Proposition 4.1.5.** *(a) Consider a closed, oriented Riemannian manifold $(N, g)$ of dimension $d$, $E \to N$ a Hermitian vector bundle and*

$$A : C^\infty(E) \to C^\infty(E)$$

*a first order selfadjoint elliptic operator. Then*

$$\eta_A(s) := \frac{1}{\Gamma(\,(s+1)/2\,)} \int_0^\infty t^{(s-1)/2} \mathrm{Tr}\,(A e^{-tA^2}) dt \tag{4.1.2}$$

*is well defined for all $\mathfrak{Re}\, s \gg 0$ and extends to a meromorphic function on $\mathbb{C}$. Its poles are all simple and can be located only at $s = (d + 1 - n)/2$, $n = 0, 1, 2, \cdots$.*

*(b) For $|s| \gg 0$ the function $\eta_A(s)$ is described by the Dirichlet series (4.1.1).*

*(c) If $d$ is odd then the residue of $\eta_A(s)$ at $s = 0$ is zero so that $s = 0$ is a regular point.*

For a proof of this nontrivial result we refer to [**8**]. When $d$ is odd we define the eta invariant of $A$ by

$$\eta(A) := \eta_A(0).$$

**Remark 4.1.6.** (a) From the definition it follows directly that $\eta(-A) = -\eta(A)$ and $\eta(\lambda A) = \eta(A)$, $\forall \lambda > 0$.

(b) In [**14**] it is shown that if $A$ is an operator of Dirac type then one can define its eta invariant directly by setting $s = 0$ in (4.1.2). In other words, in this case

$$\eta(A) = \frac{1}{\sqrt{\pi}} \int_0^\infty t^{-1/2} \mathrm{Tr}\,(Ae^{-tA^2})dt. \blacksquare$$

**Example 4.1.7.** Let $N \cong S^1$ and $D_0 = \mathbf{i}\partial_\theta$. The spectrum of $D$ is $\mathbb{Z}$ and all its eigenvalues are simple. Thus, for $\mathfrak{Re}\, s \gg 1$ we have

$$\eta_{D_0}(s) = \sum_{n \neq 0} \frac{\operatorname{sign} n}{n^s} = 0.$$

By unique continuation we deduce that $\eta_{D_0}(0) = 0$. This simple equality reflects the symmetry of the spectrum of $D_0$. In general, the eta invariant should be regarded as a measure of the asymmetry (about the origin) of the spectrum.

More generally, define for each $a \in (0, 1)$ the operator

$$D_a := D_0 + a.$$

Its spectrum consists only of simple eigenvalues $\lambda_n(a) = n + a$, $n \in \mathbb{Z}$. Thus

$$\eta_{D_a}(s) = \sum_{n \geq 0} \frac{1}{(n+a)^s} - \sum_{n \geq 0} \frac{1}{(n+1-a)^s}$$

$$= \zeta(s, a) - \zeta(s, 1-a)$$

where

$$\zeta(s, a) := \sum_{n \geq 0} \frac{1}{(n+a)^s}$$

denotes the Riemann-Hurwitz function. Thus

$$\eta_{D_a}(0) = \zeta(0, a) - \zeta(0, 1-a)$$

and, according to [**148**, 13.21],

$$\zeta(0,a) = \frac{1}{2} - a.$$

We obtain the following identity (see [**7**]):

$$\eta_{D_a}(0) = 1 - 2a.$$

**Theorem 4.1.8. (Atiyah-Patodi-Singer, [6])**

$$I_{APS}(\hat{L}, \hat{N}_r) = \int_{\hat{N}_r} \rho(\hat{L}) dv_{\hat{g}} - \frac{1}{2}\Big(\dim\ker \vec{\partial}_\infty \hat{L} + \eta(\vec{\partial}_\infty \hat{L})\Big)$$

*where $\rho(\hat{L})$ denotes the local index density of $\hat{L}$, which depends only on the coefficients of $\hat{L}$ (see* [**12, 48, 117**] *for an exact definition) while $\eta(\vec{\partial}_\infty \hat{L})$ denotes the eta invariant of the operator $\vec{\partial}_\infty \hat{L}$. (The above integral is independent of $r \gg 0$.)*

Influenced by the above theorem we introduce the *$\xi$-invariant* (or *the reduced eta invariant*) of a selfadjoint elliptic operator $A$ by

$$\xi(A) := \frac{1}{2}(h(A) + \eta(A))$$

where $h(A) := \dim\ker A$. Note that $\xi(-A) = (h(A) - \eta(A))/2$ so that $A \mapsto \xi(A)$ is not an odd function.

**Exercise 4.1.3.** Let $\hat{L}_0$ and $\hat{L}_1$ be two $APS$ operators on $\hat{N}$ which differ by a zeroth order term. Suppose there exists $r_0 > 0$ such that $\hat{L}_0 = \hat{L}_1$ on $\hat{N} \setminus \hat{N}_{r_0}$. Prove that

$$I_{APS}(\hat{L}_0, \hat{N}_r) = I_{APS}(\hat{L}_1, \hat{N}_r), \;\; \forall r > r_0.$$

In many geometrically interesting situations the index density $\rho(\hat{L})$ has a very explicit description. We present below one such instance.

**Example 4.1.9.** Suppose $\hat{N}$ is a cylindrical 4-manifold equipped with a cylindrical *spin*$^c$ structure $\hat{\sigma}$ and $\hat{A}$ is a strongly cylindrical Hermitian connection on $\det(\hat{\sigma})$. Denote by $\sigma_\infty$ the induced *spin*$^c$ structure on $\partial_\infty \hat{N}$ and set $A = \partial_\infty \hat{A} = \hat{A}\,|_N$. Then, as shown in §2.4.1, the Dirac operator $\hat{\mathfrak{D}}_{\hat{A}}$ is an $APS$ operator and Theorem 4.1.8 takes the form

$$I_{APS}(\hat{\mathfrak{D}}_{\hat{A}}, \hat{N}_r) = \frac{1}{8}\int_{\hat{N}_r} \Big(-\frac{1}{3}p_1(\hat{\nabla}^{\hat{g}}) + c_1(\hat{A})^2\Big) - \xi(\mathfrak{D}_A) \tag{4.1.3}$$

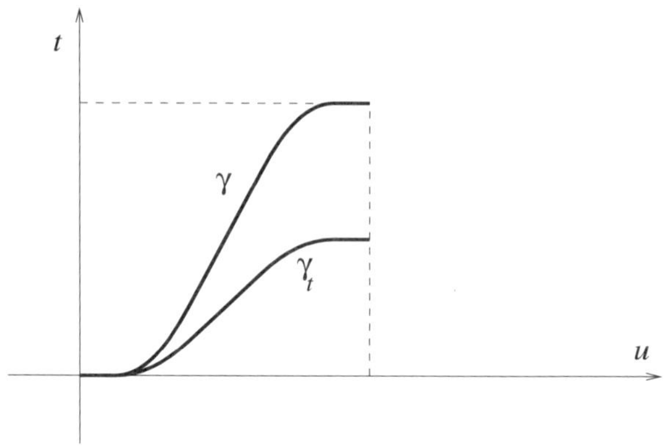


**Figure 4.2.** *The smoothing function* $\gamma$

where $p_1(\hat{\nabla}^{\hat{g}})$ denotes the first Pontryagin form of $TM$ determined from the Levi-Civita connection $\hat{\nabla}^{\hat{g}}$ on $T\hat{N}$ via the Chern-Weil construction. The 2-form $c_1(\hat{A})$ is defined similarly.

**4.1.3. Eta invariants and spectral flows.** While the eta invariant itself is a very complex object its deformation theory turns out to be a lot more tractable. More specifically, in this subsection we will address the following problem.

*Consider a smooth path of selfadjoint Dirac operators* $\mathfrak{D}_u$ *on an odd-dimensional manifold* $N$ $(\dim N = n)$. *Compute* $\xi(\mathfrak{D}_1) - \xi(\mathfrak{D}_0)$.

Set $\xi_t = \xi(\mathfrak{D}_t)$. We want to compute $\dot{\xi}_t = \frac{d\xi_t}{dt}$ although at this moment we have no guarantee that the map $t \mapsto \xi_t$ is differentiable.

Since the family of Dirac operators $(\mathfrak{D}_u)_{u\in[0,1]}$ may not be independent of $u$ near $u = 0, 1$ we need to smooth out the corners. To this end, consider a smooth, nondecreasing map $\gamma : [0,1] \to [0,1]$, $u \mapsto \gamma(u)$ such that $\gamma(0) = 0$, $\gamma(1)$ and $\gamma'(u) \equiv 0$ for $u$ near 0 and 1 (see Figure 4.2). Moreover, for each $0 < t \leq 1$ set $\gamma_t(u) = t\gamma(u)$ so that $\gamma_t$ connects 0 to $t$.

Denote by $u$ the longitudinal coordinate along $[0,1] \times N$. For every $0 < t \leq 1$ form the *APS* operator $\hat{L}_t$ on $[0,1] \times N$ defined by

$$\hat{L}_t = \partial_u - \mathfrak{D}_{t\gamma(u)}.$$

From Theorem 4.1.8 we get

$$i_t := I_{APS}(\hat{L}_t) = \rho_t - \frac{1}{2}(h_0 + h_t) + \frac{1}{2}(\eta_0 - \eta_t)$$

where $\rho_t$ denotes the integral of the index density of $\hat{L}_t$, $h_t = h(\mathfrak{D}_t)$, $\eta_t = \eta(\mathfrak{D}_t)$. The above formula can be rewritten as

$$\xi_t - \xi_0 = \rho_t + j_t \tag{4.1.4}$$

where $j_t = -(h_0 + i_t)$. The term $\rho_t$ depends smoothly on $t$ since the coefficients of $\hat{L}_t$ do. The term $j_t$ is $\mathbb{Z}$-valued so it cannot be smooth, unless it is constant. If $[\xi_t] = \xi_t \pmod{\mathbb{Z}}$ then the map $t \mapsto [\xi_t]$ is smooth and by (4.1.4)

$$\frac{d[\xi_t]}{dt} = \dot{\rho}_t. \tag{4.1.5}$$

We will deal with $\dot{\rho}_t$ a bit later. We first need to better understand the special nature of the discontinuities of $\xi_t$.

We see from (4.1.1) that the discontinuities of $\xi_t$ (and hence those of $j_t$) are due to jumps in $h_t$. We describe how the jumps in $h_t$ affect $\xi_t$ in a simple, yet generic situation. We assume $\mathfrak{D}_t$ is a regular family, i.e.

- The resonance set $\mathfrak{Z} = \{t \in [0,1]\ ;\ h_t \neq 0\}$ is finite.
- For every $t_0 \in \mathfrak{Z}$ and every sufficiently small $\varepsilon > 0$, there exist an open neighborhood $\mathcal{N}$ of $t_0$ in $[0,1]$ and smooth maps $\lambda_k : \mathcal{N} \to (-\varepsilon, \varepsilon)$, $k = 0, 1, \cdots, h_{t_0}$ such that for all $t \in \mathcal{N}$ the family $\{\lambda_k(t)\}_k$ describes *all* the eigenvalues of $\mathfrak{D}_t$ in $(-\varepsilon, \varepsilon)$ (including multiplicities) and, moreover, $\lambda_k(t_0) = 0$, $\dot{\lambda}_k(t_0) \neq 0$ for all $k = 1, 2, \cdots, h_{t_0}$.

Now for each $t \in \mathfrak{Z}$ set

$$\sigma_{\pm}(t) = \#\{k\ ;\ \pm\dot{\lambda}_k(t) > 0\}$$

and

$$\Delta_t\sigma = \begin{cases} -\sigma_-(0) & \text{if } t = 0 \\ \sigma_+(t) - \sigma_-(t) & \text{if } t \in (0,1) \\ \sigma_+(1) & \text{if } t = 1 \end{cases}.$$

If

$$\Delta_t\xi := \lim_{\varepsilon \to 0^+} (\xi_{t+\varepsilon} - \xi_{t-\varepsilon})$$

we see that $\Delta_t\xi = 0$ if $t \notin \mathfrak{Z}$ while for $t \in \mathfrak{Z}$ we have

$$\Delta_t\xi = \Delta_t\sigma. \tag{4.1.6}$$

(To understand the above formula it is convenient to treat $\mathfrak{D}_t$ as a finite-dimensional symmetric matrix and then keep track of the changes in its signature as the spectrum changes in the regular way described above.) Finally, define the spectral flow of the family $\mathfrak{D}_t$ by

$$SF(\mathfrak{D}_t) = \sum_{t \in [0,1]} \Delta_t\sigma. \tag{4.1.7}$$

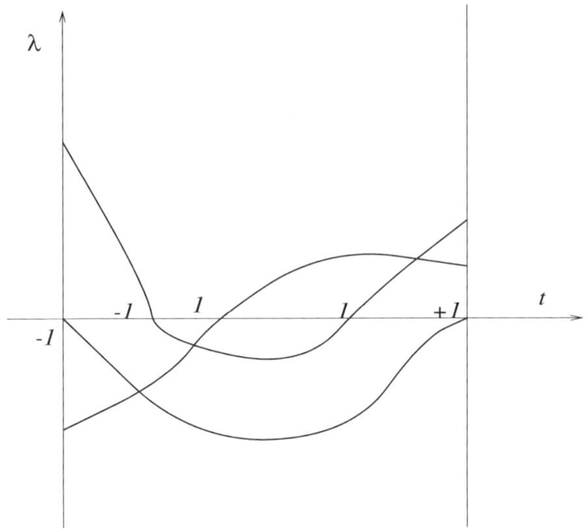


**Figure 4.3.** *Spectral flow*

For example, in Figure 4.3 we have represented those eigenvalues $\lambda_t$ of a smooth path of Dirac operators which vanish for some values of $t$. The $\pm 1$'s describe the jumps $\Delta_t \sigma$. Thus the spectral flow in Figure 4.3 is 1.

Intuitively, the spectral flow is the difference between the number of spectral curves $\lambda_k(t)$ which cross the axis $\lambda = 0$ going up and the number of spectral curves which cross this axis going down. The initial and final moments require separate consideration. At the initial moment only the going-down spectral curves contribute (with a nonpositive quantity), while at the final moment only the going-up spectral curves are relevant, contributing with a nonnegative quantity.

Using the equalities $j_1 - j_0 = \sum_t \Delta_t \xi$ and $j_0 = 0$ we deduce

$$j_1 - j_0 = -i_1 - h_0 = \sum_t \Delta_t \xi = \sum_{t \in [0,1]} \Delta_t \sigma = SF(\mathfrak{D}_t) \tag{4.1.8}$$

so that

$$i_1 = I_{APS}(\hat{L}_1) = -h_0 - SF(\mathfrak{D}_t). \tag{4.1.9}$$

From the equalities (4.1.4) and (4.1.8) we now conclude

$$\xi_1 - \xi_0 = SF(\mathfrak{D}_t) + \int_0^1 \frac{d[\xi_t]}{dt} dt. \tag{4.1.10}$$

**Remark 4.1.10.** In the above two equalities we have neglected the smoothing effect of $\gamma$. However, since $\gamma(u)$ is nondecreasing the crossing patterns of the eigenvalues of $t \mapsto \mathfrak{D}_t$ and $u \mapsto \mathfrak{D}_{\gamma(u)}$ are identical. This implies $SF(\mathfrak{D}_t) = SF(\mathfrak{D}_{\gamma(u)})$.

**Example 4.1.11.** To make sure our sign conventions are correct we test the equality (4.1.9) on a very simple example. Fix $\lambda \in \mathbb{R} \setminus \mathbb{Z}$ and for each $t \in [0,1]$ define

$$D_t = \mathbf{i}\partial_\theta + \lambda t : C^\infty(S^1) \to C^\infty(S^1).$$

$\operatorname{spec}(D_t) = t\lambda + \mathbb{Z}$ and all the eigenvalues are simple. The family $(D_t)$ is regular and its resonance set is

$$\mathfrak{Z}_\lambda = \{t \in [0,1]\ ;\ \lambda t \in \mathbb{Z}\}.$$

To compute the spectral flow note that when $\lambda > 0$ we have $\sigma_-(t) = 0$ and $\sigma_+(t) = 1$ for all $t \in \mathfrak{Z}_\lambda$ and thus

$$SF(D_t) = \#\mathfrak{Z}_\lambda - 1 = [\lambda].$$

When $\lambda < 0$ we have $\sigma_-(t) = 1$ and $\sigma_+(t) = 0$ for all $t \in \mathfrak{Z}_\lambda$ so that

$$SF(D_t) = -\#\mathfrak{Z}_\lambda = [\lambda].$$

We can form the operator $L_\lambda = \partial_t - D_t$ on $[0,1] \times S^1$. A separation of variables argument shows

$$I_{APS}(L_\lambda) = \#\{n \in \mathbb{Z}\ ;\ n > 0,\ n + \lambda < 0\} - \#\{n \in \mathbb{Z}\ ;\ n \leq 0,\ n + \lambda \geq 0\}$$

$$= \#\{n\ ;\ 0 < n < \lambda\} - \#\{n\ ;\ -\lambda \geq n \geq 0\}$$

$$= \begin{cases} [|\lambda|] & , \quad \lambda < 0 \\ -[\lambda] - 1 & , \quad \lambda > 0 \end{cases} = -[\lambda] - 1, \quad \forall \lambda \in \mathbb{R} \setminus \mathbb{Z}.$$

In our case $h_0 = 1$ and we see that $h_0 + \operatorname{ind}(L_\lambda) = -SF(D_t)$ which confirms (4.1.9). Again we have neglected the possible corners of the family $D_t$ near $t = 0, 1$ but the above computations stay the same if we work with the smoothed-out family $D_{\gamma(u)}$ instead.

It is now time to explain the continuous variation $\frac{d}{dt}[\xi]_t$. Formula (4.1.5) shows that this is a locally computable quantity. In fact, one can be more accurate than this.

Assume we have a family $(\mathfrak{D}_u)_{u \in [0,1]}$ of Dirac type operators on our $n$-dimensional manifold $N$ ($n$ is odd), acting on a Hermitian bundle $E \to N$. Observe that $\mathfrak{D}_u$ can be written as $\mathfrak{D}_0 + T_u$ where $T_u$ is a selfadjoint bundle endomorphism depending smoothly upon $u$. Set $\dot{T}_u = \frac{d}{du} T_u$ and $\xi_u = \xi(\mathfrak{D}_u)$. We then have the following result.

**Proposition 4.1.12.**

$$\frac{d}{du}[\xi_u] = -\frac{1}{\sqrt{\pi}} a_{n-1}(\dot{T}_u, \mathfrak{D}_u^2), \quad n := \dim N$$

*where* $a_j(\dot{T}_u, \mathfrak{D}_u)$ *is determined from the asymptotic expansion*

$$\operatorname{Tr}\left(\dot{T}_u \exp(-t\mathfrak{D}_u^2)\right) \sim \sum_{j\geq 0} a_j(\dot{T}_u, \mathfrak{D}_u^2) t^{(j-n)/2}, \quad t \to 0.$$

For a proof of a more general version of above result we refer to [**48**, Thm. 1.13.2]. (Watch out for an ambiguity in the statement of that theorem.)

The coefficients $a_j$ are local objects but apparently the above proposition replaces an abstract assertion with an impractical statement. In special situations though, the coefficients $a_j$ can be determined quite explicitly. Such is the case when $T_u$ is scalar, $T_u = u\lambda$ so that $\dot{T}_u = \lambda$. In this case $a_j(\dot{T}_u, \mathfrak{D}_u^2) = \lambda a_j(\mathfrak{D}_u^2)$ where the coefficient $a_j$ is determined from the asymptotic expansion

$$\operatorname{Tr} \exp(-t\mathfrak{D}_u^2) = \sum_{j\geq 0} a_j(\mathfrak{D}_u^2) t^{(j-n)/2}, \quad t \to 0.$$

For each $u$ the operator $\mathfrak{D}_u^2$ is a generalized Laplacian and so there exist a unique connection $\nabla_u$ and an endomorphism $\mathcal{R}_u$ such that

$$\mathfrak{D}_u^2 = \nabla_u^* \nabla_u + \mathcal{R}_u.$$

In [**48**, Chap. 4] it is shown that the coefficients $a_j$ can be expressed in terms of the metric $g$ on $N$ and the Weitzenböck remainder $\mathcal{R}_u$. As $j$ increases the actual description becomes more and more involved. However, for low $j$ the expression is quite manageable. For example (see [**48**, Chap. 4]) we have

$$a_0(\mathfrak{D}_u^2) = \frac{1}{(4\pi)^{n/2}} \int_N \operatorname{tr} \mathbf{id}_E \, dv_g = \frac{\operatorname{vol}_g(N) \cdot \operatorname{rk}(E)}{(4\pi)^{n/2}}, \tag{4.1.11}$$

$$a_2(\mathfrak{D}_u^2) = \frac{1}{(4\pi)^{n/2}} \int_N \operatorname{tr}\left(\mathcal{R}_u + \frac{s(g)}{6}\mathbf{id}_E\right) dv_g \tag{4.1.12}$$

where $s(g)$ denotes the scalar curvature of the metric $g$.

**Example 4.1.13.** We illustrate the strength of the above arguments on a simple example. Consider again the operators $D_u = \mathbf{i}\partial_\theta + u\lambda$ of Example 4.1.11. Assume $|\lambda| < 1/2$, $\lambda \neq 0$. In this case $n = 1$. Equip $S^1$ with the standard metric so that its length is $2\pi$. Using (4.1.11) we get

$$\frac{d}{du}[\xi_u] = -\frac{\lambda}{\sqrt{\pi}} a_0(D_u^2) = -\frac{\lambda}{\sqrt{\pi}} \cdot \frac{2\pi}{\sqrt{4\pi}} = -\lambda.$$

Note that our assumptions on $\lambda$ imply $h_1 = 0$. Since $h_0 = 0$ the variational formula (4.1.10) now yields

$$\xi_1 = \xi_0 - 1 - SF(D_u) + \int_0^1 \frac{d}{du}[\xi_u] du.$$

Since $\eta(D_0)=0$ we get

$$\eta(D_1)=1+2(SF(D_u)-\lambda).$$

From Example 4.1.11 we deduce $SF=0$ if $\lambda>0$ and $SF=-1$ if $\lambda<0$. Hence

$$\eta(\mathbf{i}\partial_\theta+\lambda)=\begin{cases}1-2\lambda & \text{if}\quad \lambda>0\\ -1-2\lambda & \text{if}\quad \lambda<0\end{cases}$$

This is in perfect agreement with the computation in [**7**] or Example 4.1.7.

For more general paths of Dirac operators the formula in Proposition 4.1.12 is for all intents and purposes useless. Fortunately, there is a geometric way out of this trouble supplied by Theorem 4.1.8.

We consider only a simple situation. Assume $N$ is an oriented Riemannian manifold of dimension 3 equipped with a *spin*$^c$ structure $\sigma$. Fix a smooth path of metrics $(g_u)_{u\in[0,1]}$ on $N$ such that $g_u\equiv g_i$ if $u$ is close to $i=0,1$. Next, choose a path $(A_u)_{u\in[0,1]}$ of Hermitian connections of $\det(\sigma)$ such that $A_u=A_i$ for $u$ close to $i=0,1$. For each $u$ denote by $\mathfrak{D}_u$ the associated Dirac operator on $N$ determined by $g_u$ and $A_u$. Consider now the manifold $\hat{N}=[0,1]\times N$ equipped with the metric $\hat{g}=du^2+g_u$. The Levi-Civita connection $\hat{\nabla}$ of $\hat{g}$ has the strongly cylindrical form

$$\hat{\nabla}=du\wedge\partial_u+\nabla^{g_u}$$

near $u=0,1$. The path of connections $(A_u)$ determines a connection $\hat{A}$ on the product *spin*$^c$ structure $\hat{\sigma}$ on $\hat{A}$. Denote by $\hat{\mathfrak{P}}_{\hat{A}}$ the geometric Dirac operator determined by $\hat{\nabla}$ and $\hat{A}$. This is an $APS$ operator on $\hat{N}$ and, more precisely, along $\hat{N}$ it has the form

$$\hat{\mathfrak{P}}_{\hat{A}}=\mathbf{c}(du)\big(\partial_u-\mathfrak{D}_{A_u}-T_u\big)$$

where $T_u$ are zeroth order operators such that

$$T_u\equiv 0,\quad \text{for } u \text{ near } 0 \text{ and } 1. \tag{4.1.13}$$

Set

$$\hat{\mathfrak{P}}^\natural_{\hat{A}}:=\mathbf{c}(du)\big(\partial_u-\mathfrak{D}_{A_u}\big).$$

Using (4.1.13), Exercise 4.1.3 and (4.1.9) we deduce

$$I_{APS}(\hat{\mathfrak{P}}_{\hat{A}})=I_{APS}(\hat{\mathfrak{P}}^\natural_{\hat{A}})=-h_0-SF(\mathfrak{D}_{A_u}).$$

Theorem 4.1.8 now implies

$$\xi_1-\xi_0=\frac{1}{8}\int_{\hat{N}}\Big(-\frac{1}{3}p_1(\hat{\nabla})+c_1(\hat{A})^2\Big)-I_{APS}(\hat{\mathfrak{P}}_{\hat{A}})-h_0$$

$$=SF(\mathfrak{D}_{A_u})+\frac{1}{8}\int_{\hat{N}}\Big(-\frac{1}{3}p_1(\hat{\nabla})+c_1(\hat{A})^2\Big).$$

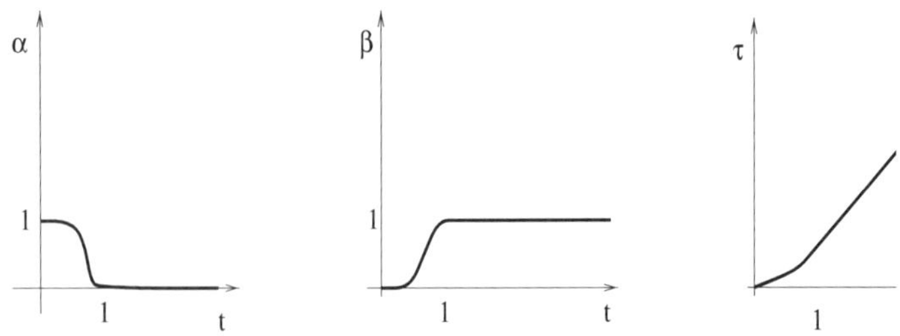


**Figure 4.4.** *Cutoff functions*

One can further simplify this formula by expressing the integral term as an integral over $N$ of some transgression forms. We refer to the beautiful paper [**7**] for more details.

**4.1.4. The Lockhart-McOwen theory.** Let us first introduce three important smooth cutoff functions $\alpha, \beta, \tau : \mathbb{R} \to \mathbb{R}_+$ satisfying the following conditions.

- $0 \leq \beta' \leq 4$.
- $\beta(t) \equiv 1$ on $[1, \infty)$ and $\equiv 0$ on $(-\infty, 1/2]$.
- $\alpha(t) = 1 - \beta(t)$.
- $\tau(t) = \int_0^t \beta(s)ds$.

The graphs of these three functions are depicted in Figure 4.4.

We can view $\tau$, first as a smooth function on the neck $\mathbb{R}_+ \times N$ and then, extending it by 0, as a smooth function on $\hat{N}$. In a similar way, we can regard $\alpha$ and $\beta$ as smooth functions on $\hat{N}$.

Fix a cylindrical Hermitian vector bundle $\hat{E} \to \hat{N}$. For each $\delta \in \mathbb{R}$, $k \in \mathbb{Z}_+$ and $p \in [1, \infty]$ we denote by $L^{k,p}_\delta(\hat{E})$ the space of $L^{k,p}_{loc}$-sections $\hat{u}$ of $\hat{E}$ such that

$$\|\hat{u}\|_{k,p;\delta} := \|e^{\delta\tau}\hat{u}\|_{k,p} < \infty$$

where $\|\cdot\|_{k,p}$ denotes the $L^{k,p}$-norm, defined in terms of the metric $\hat{g}$ and the *fixed* connection $\hat{\nabla}$. Notice that we have an isometry

$$\mathbf{m}_\delta : L^{k,p}_\delta(\hat{E}) \to L^{k,p}(\hat{E}), \quad \hat{u} \mapsto e^{\delta\tau}\hat{u}.$$

Much as in the compact case, these spaces are related by a series of Sobolev-type embeddings. For a proof of the following results we refer to [**84**, Sec. 3]. Set $n := \dim \hat{N}$.

**Theorem 4.1.14. (Continuous embeddings)** *There is a continuous embedding*

$$L^{k_0,p_0}_{\mu_0}(\hat{E}) \hookrightarrow L^{k_1,p_1}_{\mu_1}(\hat{E})$$

*if*

*(i)* $k_0 - k_1 \geq n(1/p_0 - 1/p_1)$,

*(ii)* $k_0 \geq k_1 \geq 0$ *and either*

*(iii)* $1 < p_0 \leq p_1 < \infty$ *with* $\mu_1 \leq \mu_0$ *or*

*(iii')* $1 < p_1 < p_0 < \infty$ *with* $\mu_1 < \mu_0$.

**Theorem 4.1.15. (Compact embeddings)** *If*

*(i)* $(k_0 - k_1) > n(1/p_0 - 1/p_1)$,

*(ii)* $k_0 > k_1$ *and*

*(iii)* $\mu_1 < \mu_0$

*then the embedding* $L^{k_0,p_0}_{\mu_0}(\hat{E}) \hookrightarrow L^{k_1,p_1}_{\mu_1}(\hat{E})$ *is compact.*

An $L^2_{loc}$-section $\hat{u}$ of a cylindrical bundle $\hat{E}$ is called *asymptotically cylindrical* (or *a-cylindrical*) if there exists an $L^2_{loc}$-cylindrical section $\hat{u}_0$ such that $\hat{u} - \hat{u}_0 \in L^2(\hat{E})$. We set $\partial_\infty \hat{u} := \partial_\infty \hat{u}_0$. Observe that $\hat{u}_0$ is uniquely determined by $\hat{u}$. (**N.B.** In [**6**] the asymptotically cylindrical sections were called extended $L^2$-sections. We use the new terminology only for coherence purposes.) The supremum of all $\mu \geq 0$ such that $\hat{u} - \hat{u}_0 \in L^2_\mu$ is called the *decay rate* of the a-cylindrical section $\hat{u}$.

We introduce a norm $\|\cdot\|_{ex}$ on the space of asymptotically cylindrical sections defined by

$$\|\hat{u}\|_{ex} = \|\hat{u} - \hat{u}_0\|_{L^2} + \|\partial_\infty \hat{u}\|_{L^2}$$

and we denote by $L^2_{ex}$ the resulting Hilbert space. It fits into an exact sequence of Hilbert spaces

$$0 \to L^2(\hat{E}) \hookrightarrow L^2_{ex}(\hat{E}) \stackrel{\partial_\infty}{\to} L^2(\partial_\infty \hat{E}) \to 0.$$

Using the cutoff function $\beta$ we can construct an entire family of splittings $i_r : L^2(\partial_\infty \hat{E}) \to L^2_{ex}(\hat{E})$, $r \in \mathbb{R}_+$, of this sequence described by

$$u(x) \mapsto (i_r u)(t,x) := \beta(t-r)u(x).$$

We will find it convenient to have a whole range of asymptotically cylindrical sections. Define $L^p_{ex}$ in the obvious way and then set

$$L^{k,p}_{\mu,ex}(\hat{E}) := \{\hat{u} \in L^{k,p}_{loc} \cap L^p_{ex}(\hat{E});\ \|\hat{u} - i_1\partial_\infty \hat{u}\|_{L^{k,p}_\mu(\hat{E})} + \|\partial_\infty \hat{u}\|_{L^{k,p}(E)} < \infty\}.$$

A p.d.o. $\hat{L} : C^\infty(E) \to C^\infty(E)$ is called *asymptotically cylindrical* if there exists $\mu > 0$ such that

$$\hat{A} \in L^{k,2}_\mu(\mathrm{Hom}\,(\hat{E}, \hat{F})),\ \ \forall k \in \mathbb{Z}_+$$

and $\hat{L} - \hat{A}$ is cylindrical. $\mu = \mu(\hat{L})$ is called the decay rate. A connection is called asymptotically (strongly) cylindrical if it differs from a (strongly) cylindrical one by zeroth order term in $\bigcap_{k\geq 0} L^{k,2}_{\mu}$. Its decay rate can be defined similarly.

An **asymptotic** *APS* **operator** (a-*APS* for brevity) is a first order operator which along the neck can be written as

$$\hat{L} = G(\partial_t - L_0) + \hat{A}$$

where $\hat{L}_0 := G(\partial_t - L_0)$ is an *APS* operator and $\hat{A} \in L^{k,2}_{\mu}(\mathrm{Hom}\,(\hat{E}, \hat{F}))$, $\forall k > 0$. The decay rate is defined exactly as before. We set $\vec{\partial}_{\infty}\hat{L} := L_0$. For later use define the *spectral gap*

$$\gamma(\hat{L}) := \mathrm{dist}\,(\,0\,,\,\mathrm{spec}\,(L_0) \setminus \{0\}\,).$$

Observe that if $\hat{L}$ is an a-*APS* operator then for every $r \gg 0$ we define ${}_r\hat{L}$ as the *APS* operator which along the neck is described by

$${}_r\hat{L} := G\Big(\partial_t - L_0\Big) + \alpha(t-r)\cdot\hat{A}.$$

If $\hat{L} : C^{\infty}(\hat{E}) \to C^{\infty}(\hat{F})$ is an a-*APS* operator on $\hat{N}$ then it defines a *bounded* operator

(4.1.14) $$\hat{L}_{\delta} = \hat{L}_{k,\delta} : L^{k+1,2}_{\delta}(\hat{E}) \to L^{k,2}_{\delta}(\hat{F}),\ \ k \in \mathbb{Z}_{+},$$

for any $\delta < \mu(\hat{L})$. Its formal adjoint with respect to the metric $L^2_{\delta}$ is denoted by $\hat{L}^{*\delta}$ and is given by

(4.1.15) $$\hat{L}^{*\delta} := \mathbf{m}_{-2\delta}\hat{L}^{*}\mathbf{m}_{2\delta}.$$

We can regard it either as a closed unbounded operator $L^2_{\delta} \to L^2_{\delta}$ or as a bounded operator $L^{1,2}_{\delta} \to L^2_{\delta}$.

The gluing construction uses the following spaces.

$$\ker_{\delta}\hat{L} := \ker\hat{L} \cap L^2_{\delta},\quad \ker_{ex}\hat{L} := \ker\hat{L} \cap L^2_{ex}.$$

The following result is proved in [**74**] .

**Theorem 4.1.16. (Lockhart-McOwen)** *Suppose $\hat{L}$ is an a-APS operator. Then for any $\delta < \mu(\hat{L})$ which is not an eigenvalue of $-\vec{\partial}_{\infty}\hat{L}$ the operator $\hat{L}_{k,\delta}$ is Fredholm and its index is independent of $k$.*

The following proposition is a slight generalization of [**6**, Prop. (3.11)].

**Proposition 4.1.17.** *Suppose $\hat{L}$ is an a-APS operator. Then the following hold.*

*(a)* $\ker_{\delta}\hat{L} = \ker\hat{L}_{k,\delta}$ , $\forall k \in \mathbb{Z}_{+},\ \delta < \mu(\hat{L})$.

*(b) The spaces* $\ker_{\delta}\hat{L}$, $\ker_{-\delta}\hat{L}^{*}$ *are independent of* $0 < \delta < \min(\mu(\hat{L}), \gamma(\hat{L}))$.

*(c) For every* $0 < \delta < \min(\mu(\hat{L}), \gamma(\hat{L}))$ *the continuous map* $\mathbf{m}_{2\delta} : L^2_\delta \to L^2_{-\delta}$ *induces an isomorphism*

$$\ker_\delta \hat{L}^{*_\delta} \cong \ker_{-\delta} \hat{L}^*.$$

*(d) For every* $0 < \delta < \min(\mu(\hat{L}), \gamma(\hat{L}))$ *we have the equality*

$$\ker_{-\delta} \hat{L} = \ker_{ex} \hat{L}.$$

*(e) For all* $r \gg 0$ *and for all* $0 < \delta < \min(\mu(\hat{L}), \gamma(\hat{L}))$ *the pullbacks by the inclusions* $\hat{N}_r \hookrightarrow \hat{N}$ *induce isomorphisms*

$$\ker({}_r\hat{L}, APS) \cong \ker_\delta({}_r\hat{L})$$

*and*

$$\ker({}_r\hat{L}^*, APS^*) \cong \ker_{-\delta}({}_r\hat{L}^*) = \ker_{ex}({}_r\hat{L}^*).$$

*(f)*

$$\operatorname{ind}(\hat{L}_\delta) = \lim_{r\to\infty} I_{APS}({}_r\hat{L}).$$

**Exercise 4.1.4.** Prove the above proposition.

The above results suggest the introduction of an $APS$ index for an a-$APS$ operator $\hat{L}$ by setting

$$I_{APS}(\hat{L}) := \lim_{r\to\infty} I_{APS}({}_r\hat{L}).$$

Using Proposition 4.1.17 and (4.1.9) we deduce that if $\hat{L} = G(\partial_t - L(t))$ is an a-$APS$ operator on $\mathbb{R} \times N$ then

$$\operatorname{ind}(\hat{L}_\delta) = I_{APS}(\hat{L}) = -\dim\ker L(-\infty) - SF(L(t)). \tag{4.1.16}$$

The remarks in §4.1.3 can be used to determine $i_\delta := \operatorname{ind}(\hat{L}_\delta)$ for arbitrary $\delta$. Assume for simplicity that $\hat{L}$ is an $APS$ operator (not just asymptotically). Set $A := \vec{\partial}_\infty \hat{L}$.

By definition, the map

$$\mathbf{m}_\delta : L^2_\delta \to L^2, \quad \psi \mapsto e^{\delta\tau(t)}\psi$$

is an isometry so that

$$i_\delta(\hat{L}) = i_0(\mathbf{m}_\delta \hat{L} \mathbf{m}_\delta^{-1}) = I_{APS}(\mathbf{m}_\delta \hat{L} \mathbf{m}_\delta^{-1}, \hat{N}_r).$$

A simple computation shows that

$$\hat{L}_\delta := \mathbf{m}_\delta \hat{L} \mathbf{m}_\delta^{-1} = \hat{L} - \delta\tau'(t)G.$$

Observe that $\vec{\partial}_\infty \hat{L}_\delta = A + \delta =: A_\delta$ and

$$i_\delta = I_{APS}(\hat{L}_\delta, \hat{N}_r).$$

Set $C_r := [r, r+1] \times N$. We have

$$I_{APS}(\hat{L}_\delta, \hat{N}_{r+1}) - I_{APS}(\hat{L}, \hat{N}_r) = -(\,\xi(A_\delta) - \xi(A)\,) + \int_{C_r} \rho(\hat{L}_\delta) dv_{\hat{g}}.$$

On the other hand, the above index density can be expressed as in (4.1.4) in terms of the $APS$ index of the operator $\hat{L}_\delta = \hat{L} - \delta G$ on $C_r$.

$$\int_{C_r} \rho(\hat{L}_\delta) dv_{\hat{g}} = \xi(A_\delta) - \xi(A) + h(A) + I_{APS}(\hat{L} - \delta G, C_r).$$

Finally, according to (4.1.9), the last term can be expressed as a spectral flow

$$I_{APS}(\hat{L} - \delta G, C_r) = -h(A) - SF(A + t\delta,\ t \in [0,1]). \tag{4.1.17}$$

Putting all of the above together we obtain the following useful equality:

$$i_\delta = I_{APS}(\hat{L}) - SF(A + t\delta,\ t \in [0,1]). \tag{4.1.18}$$

This is in perfect agreement with Theorem 1.2 in [**74**]. Note also that if $\delta$ is sufficiently small then there is no spectral flow correction in the above formula.

**Exercise 4.1.5. (Excision formula)** Consider two a-$APS$ operators

$$\hat{L}_0, \hat{L}_1 : \Gamma(\hat{E}_+) \to \Gamma(\hat{E}_-)$$

on $\hat{N}$ which have the *same principal symbol.* Set $A_i := \vec{\partial}_\infty \hat{L}_i$, $i = 0, 1$. Prove that

$$I_{APS}(\hat{L}_0) - I_{APS}(\hat{L}_1) = SF(A_0 \to A_1) \tag{4.1.19}$$

where $SF(A_0 \to A_1)$ denotes the spectral flow of the affine path of elliptic operators $A_t = A_0 + t(A_1 - A_0)$, $t \in [0,1]$.

**Remark 4.1.18.** The above exercise illustrates one of the many "anomalies" of the non-compact situation. The operators $\hat{L}_0$ and $\hat{L}_1$ are obviously homotopic via the affine homotopy

$$\hat{L}_t := (1-t)\hat{L}_0 + t\hat{L}_1.$$

However, for some values of $t$, the operator $\hat{L}_1$ *may not* define a Fredholm operator

$$L^{1,2}_\delta(\hat{E}_+) \to L^2_\delta(\hat{E}_-)$$

so that it is possible $\operatorname{ind}(\hat{L}_{0,\delta}) \neq \operatorname{ind}(\hat{L}_{1,\delta})$. The correction is given by precisely the spectral flow $SF(A_0 \to A_1)$.

**4.1.5. Abstract linear gluing results.** The main result of this subsection is a general gluing theorem of Cappell-Lee-Miller [**24**]. To formulate it in a more intuitive fashion we need to introduce the asymptotic notions in [**110**]. We begin with the notions of asymptotic map and asymptotic exactness. An *asymptotic map* is a sequence $(U_r, V_r, f_r)_{r>0}$ with the following properties:

• There exist Hilbert spaces $H_0$ and $H_1$ such that $U_r$ is a closed subspace of $H_0$ and $V_r$ is a closed subspace of $H_1$, $\forall r > 0$.

• $f_r$ is a densely defined linear map $f_r : U_r \to H_1$ with closed graph and range $R(f_r)$, $\forall r > 0$.

• $\lim_{r\to\infty} \hat{\delta}(R(f_r), V_r) = 0$ where, following [**60**], we set

$$\hat{\delta}(U, V) = \sup\Bigl\{\text{dist}\,(u, V)\ ;\ u \in U,\ |u| = 1\Bigr\}.$$

We will denote asymptotic maps by $U_r \stackrel{f_r}{\longrightarrow}{}^a V_r$.

**Example 4.1.19.** Suppose $H_0 = \mathbb{R} = U_r$, $H_1 = \mathbb{R}\oplus\mathbb{R}$ and $V_r = \mathbb{R}\oplus 0 \subset H_1$. Then the sequence of maps

$$f_r : H_0 \to H_1,\ \ t \mapsto (rt, t)$$

defines an asymptotic map $U_r \stackrel{f_r}{\longrightarrow}{}^a V_r$. Observe that $f_r$ does not converge in any reasonable sense to any linear map.

There is a super-version of this notion when $U_r$ and $V_r$ are $\mathbb{Z}_2$-graded and are closed subspaces in $\mathbb{Z}_2$-graded Hilbert spaces such that the natural inclusions are even operators.

Define the gap between two closed subspaces $U, V$ in a Hilbert space $H$ by

$$\delta(U, V) = \max\Bigl\{\hat{\delta}(U, V),\ \hat{\delta}(V, U)\Bigr\}.$$

The sequence of asymptotic maps

$$U_r \stackrel{f_r}{\longrightarrow}{}^a V_r \stackrel{g_r}{\longrightarrow}{}^a W_r,\ \ r \to \infty,$$

is said to be *asymptotically exact* if

$$\lim_{r\to\infty} \delta(R(f_r), \ker g_r) = 0.$$

The following result (proved in [**110**]) explains the above terminology.

**Proposition 4.1.20.** *If the sequence*

$$U_r \xrightarrow{f_r}{}^a V_r \xrightarrow{g_r}{}^a W_r, \quad r \to \infty,$$

*is asymptotically exact, $P_r$ denotes the orthogonal projection onto* $\ker g_r$ *and $Q_r$ the orthogonal projection onto $W_r$ then there exists $r_0 > 0$ such that the sequence*

$$U_r \overset{P_r \circ f_r}{\longrightarrow} V_r \overset{Q_r \circ g_r}{\longrightarrow} W_r$$

*is exact for all $r > r_0$.*

An asymptotic map $U_r \xrightarrow{f_r}{}^a V_r$ is said to be an *asymptotic isomorphism* if the sequence

$$0 \to U_r \xrightarrow{f_r}{}^a V_r \to 0$$

is asymptotically exact.

Two cylindrical manifolds $(\hat{N}_i, \hat{g}_i)$, $i = 1, 2$, are called *compatible* if there exists an orientation reversing diffeomorphism

$$\varphi : N_1 \to N_2$$

such that

$$g_1 = \varphi^* g_2.$$

Two cylindrical vector bundles $(\hat{E}_i, \vartheta_i, E_i = \partial_\infty \hat{E}_i) \to \hat{N}_i$ are said to be compatible if there exists a vector bundle isomorphism

$$\gamma : E_1 \to E_2$$

covering $\varphi$.

For simplicity, we will fix some (ghost) reference, orientation *reversing* diffeomorphism $\Phi_0 : N_1 \to N_2$. We set $N := N_1$ so that we can identify $\varphi$ with an orientation preserving self-diffeomorphism of $N$. It is very convenient to think of the end of $\hat{N}_2$ as the cylinder $(-\infty, 0) \times N$ so that the outgoing coordinate on $\hat{N}_2$ is $-t$. Note that the compatibility condition provides a way of identifying $\partial_\infty \hat{E}_1$ with $\partial_\infty \hat{E}_2$ so that we can compare a section of $\partial_\infty \hat{E}_1$ to a section of $\partial_\infty \hat{E}_2$.

The sections $\hat{u}_i$ of the compatible cylindrical bundles $\hat{E}_i$ are called compatible if $\partial_\infty \hat{u}_1 = \partial_\infty \hat{u}_2$. The cylindrical partial differential operators $\hat{L}_i$ on $\hat{N}_i$ , $i = 1, 2$, are compatible if along their necks they have the form

$$\hat{L}_1 = G_1 \partial_t - L_1, \;\; G_2 \partial_t - L_2, \;\; G_1 + G_2 = L_1 - L_2 = 0.$$

Consider two compatible cylindrical manifolds $\hat{N}_i$, $i = 1, 2$. For every orientation preserving diffeomorphism $\varphi : N \to N$ and every $r \gg 0$ we

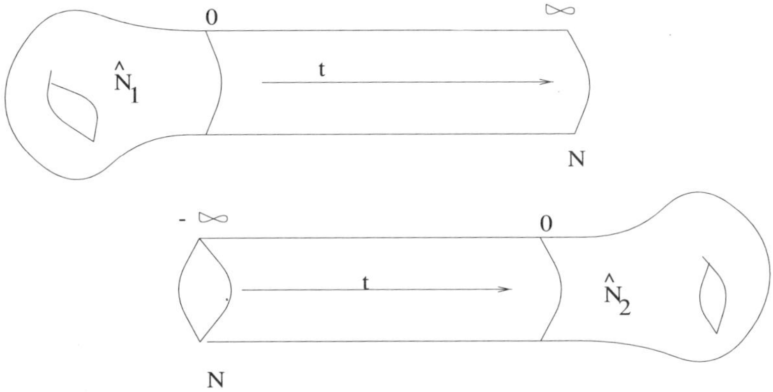


**Figure 4.5.** *Gluing two cylindrical manifolds*

denote by $\hat{N}(r) = \hat{N}(r, \varphi)$ the manifold obtained by attaching

$$\hat{N}_1(r) := \hat{N}_1 \setminus (r+1, \infty) \times N$$

to

$$\hat{N}_2(r) := \hat{N}_2 \setminus (-\infty, -r-1) \times N$$

(see Figure 4.5) using the obvious orientation preserving identification

$$\phi_r \times \Phi_0 \circ \varphi : [r, r+1] \times N_1 \to [-r-1, -r] \times N_2$$

where

$$\phi_r(t) := t - 2r - 1.$$

Two compatible cylindrical bundles $\hat{E}_i$ can be glued in an obvious way to form a bundle $\hat{E}(r) = \hat{E}_1 \#_r \hat{E}_2$ for all $r \gg 0$. We want to emphasize that the topological types of the resulting manifold $\hat{N}(r)$ and the bundle $\hat{E}(r)$ depend on the gluing isomorphisms $\gamma$. In the sequel, to simplify the presentation, we will drop $\varphi$ and $\gamma$ from our notations.

Given two compatible cylindrical sections $\hat{u}_i$ of $\hat{E}_i$, i.e $\partial_\infty \hat{u}_1 = \partial_\infty \hat{u}_2$, we can glue them together to a section $\hat{u}_1 \#_r \hat{u}_2$ of $\hat{E}_1 \#_r \hat{E}_2$. More generally, if $\hat{u}_i$ are only $L^2_{ex}$-sections with identical asymptotic values then we can approximate them by cylindrical sections

$$\hat{u}_i \approx \hat{u}_i(r) := \alpha_r(t)\hat{u}_i + \beta_r(t)\partial_\infty \hat{u}_i, \quad i = 1, 2,$$

where $\alpha_r(t) := \alpha(|t| - r)$ and $\beta_r(t) := \beta(|t| - r)$, $\forall t \in \mathbb{R}$, $r \gg 0$. Observe that if $\hat{u}_i$ are genuine cylindrical sections then $\hat{u}_i(r) = \hat{u}_i$ for all $r \gg 0$. Now define

$$\hat{u}_1 \#_r \hat{u}_2 := \hat{u}_1(r) \#_r \hat{u}_2(r), \quad r \gg 0. \tag{4.1.20}$$

The cylindrical partial differential operators $\hat{L}_i$ on $\hat{N}_i$ , $i = 1, 2$, are compatible if along their necks they have the form

$$\hat{L}_1 = G_1\partial_t - L_1, \;\; G_2\partial_t - L_2, \;\; G_1 + G_2 = L_1 - L_2 = 0.$$

Such pairs $\hat{L}_i$ of compatible cylindrical operators can be glued following the above pattern and we let the reader fill in the obvious details. Using the above cutoff trick we can extend the gluing construction to compatible asymptotic operators, i.e. pairs of operators which differ from a compatible cylindrical pair by zeroth order terms in $\bigcap_{k>0} L^{k,2}_\delta$. Cylindrical connections are special examples of cylindrical operators so the above gluing construction includes the gluing of compatible asymptotically cylindrical connections as a special case.

Suppose $\hat{D}_i : C^\infty(\hat{E}_i) \to C^\infty(\hat{E}_i)$ are compatible, formally selfadjoint a-$APS$ operators of Dirac type. Observe that the compatibility condition implies (on account of orientations) $\vec{\partial}_\infty \hat{D}_1 = -\vec{\partial}_\infty \hat{D}_2$ so we set $D := \vec{\partial}_\infty \hat{D}_1$.

We can now form the Dirac type operator

$$\hat{D}(r) := \hat{D}_1 \#_r \hat{D}_2 : C^\infty(\hat{E}(r)) \to C^\infty(\hat{E}(r)).$$

Fix $0 < \delta < \min\Big(\gamma(\hat{D}_i), \; \mu(\hat{D}_i)\Big)$ and a continuous function

$$c : \mathbb{R}_+ \to \mathbb{R}_+$$

satisfying

$$c(r) = o(1/r), \;\; \frac{1}{c(r)} = O(e^{\delta r}) \;\; \text{as } r \to \infty.$$

Define $\hat{\mathcal{K}}_{c(r)}$ as the finite-dimensional subspace of $L^2(\hat{E}(r))$ spanned by eigenvectors of $\hat{D}(r)$ corresponding to eigenvalues in the interval $[-c(r), c(r)]$. Observe that $\hat{\mathcal{K}}_{c(r)} \subset C^\infty(\hat{E}(r))$. One should think of this space as an approximation for the kernel of $\hat{D}(r)$ for $r \gg 0$.

The formulation of the main gluing result requires the introduction of some **splitting maps**

$$S_i^r : C^\infty(\hat{E}(r)) \to L^2_{ex}(\hat{E}_i), \;\; i = 1, 2.$$

We explain the construction for $i = 1$. First, regard $\hat{N}_{1,r}$ as a submanifold of $\hat{N}(r)$ in an obvious fashion. Thus any smooth section $\hat{u}$ of $\hat{E}(r) \to \hat{N}(r)$ defines by restriction a section $\tilde{u}_1(r)$ over $\hat{N}_{1,r}$. Denote by $z_r$ the midpoint of the overlapping interval $[r, r+1]$ and set

$$\partial_r \hat{u} := \tilde{u}_1(r)\,|_{z_r \times N} \;.$$

Now set

$$S_1^r \hat{u} = \alpha_r(t)\tilde{u}_1(r) + \beta_r(t)\partial_r \hat{u}.$$

Observe that $S_1^r\hat{u}$ is a cylindrical section of $\hat{E}_1$ and

$$\partial_\infty S_1^r\hat{u} := \partial_r\hat{u}.$$

With $S_2^r : C^\infty(\hat{E}(r)) \to L^2_{ex}(\hat{E}_2)$ defined in a symmetrical fashion we have the obvious equality

$$\partial_\infty S_i^r\hat{u} = \partial_\infty S_2^r\hat{u}.$$

We assemble these maps in a single *splitting map*

$$S^r := S_1^r \oplus S_2^r : C^\infty(\hat{E}(r)) \to L^2_{ex}(\hat{E}_1) \oplus L^2_{ex}(\hat{E}_2).$$

Denote by $L_i \subset L^2(E)$ the image of $\ker_{ex}\hat{D}_i$ via the map $\partial_\infty$. Observe that $L_i \subset \ker D$. The spaces $L_i$ have additional structure which we now proceed to describe.

The symbols of the operators $\hat{D}_i$ define Clifford multiplications on the bundles $\hat{E}_i$ and that is why we will denote them by the same symbol

$$\hat{\mathbf{c}} : T^*\hat{N}_i \to \operatorname{End}(\hat{E}_i).$$

Set $J := \hat{\mathbf{c}}(dt)$. The operator $J$ is skew-symmetric and satisfies $J^2 = -\mathbf{1}$ so that it induces a symplectic structure on $L^2(E)$ defined by

$$\omega(u,v) := \int_N (Ju, v)dv_g$$

Since $\{J, D\} := JD + DJ = 0$ we deduce that $\mathcal{H} := \ker D$ is a symplectic space. We have the following result (see [**16, 104**]).

**Lemma 4.1.21.** *The spaces $L_i$ are Lagrangian subspaces of $\mathcal{H}$ i.e.*

$$L_i^\perp = JL_i.$$

We get a *difference map*

$$\Delta : \ker_{ex}\hat{D}_1 \oplus \ker_{ex}\hat{D}_2 \to L_1 + L_2 \subset \ker D,\ \ (\hat{u}_1, \hat{u}_2) \mapsto \partial_\infty\hat{u}_1 - \partial_\infty\hat{u}_2.$$

The following result is due to Cappell-Lee-Miller [**24**]. For a shorter proof, in this asymptotic mappings context we refer to [**110**]. This result will be the key to understanding the monopole gluing problem.

**Theorem 4.1.22. (Linear Gluing Theorem)** *Using the above notation and hypotheses we have an asymptotically exact sequence*

$$0 \to \hat{\mathcal{K}}_{c(r)} \xrightarrow{S^r}{}^{a} \ker_{ex}\hat{D}_1 \oplus \ker_{ex}\hat{D}_2 \xrightarrow{\Delta} L_1 + L_2 \to 0. \tag{4.1.21}$$

We want to point out that the above sequence naturally splits. More precisely, the gluing map

$$\#_r : \ker\Delta \to L^2(\hat{E}(r))$$

defines an asymptotic map $\ker\Delta \longrightarrow {}^a\mathcal{H}_r$ which is an asymptotic right inverse for $S_r$.

The above result also shows that the cut off level $c(r)$ is somewhat artificial since $\mathcal{K}_{c(r)}$ is asymptotically independent of $c(r)$. This shows that as $r \to \infty$ the eigenvalues $\lambda_r$ of $\hat{D}(r)$ satisfying

$$|\lambda_r| = O(r^{-1-\varepsilon})$$

are subject to the sharper constraint

$$|\lambda_r| = O(r^{-n}), \quad \forall n \geq 1.$$

We conclude this discussion with a special case of Theorem 4.1.22 particularly relevant in Seiberg-Witten theory.

Suppose now the entire problem is supersymmetric. Thus, $\hat{E}_1$ splits as $\hat{E}_1^+ \oplus \hat{E}_1^-$ and $\hat{D}$ has the block decomposition

$$\hat{D} = \begin{bmatrix} 0 & \hat{\not{D}}^* \\ \hat{\not{D}} & 0 \end{bmatrix}.$$

The restriction $E$ of $\hat{E}_1$ to $N$ induces a splitting $E = E^+ \oplus E^-$ and we can write

$$J = \begin{bmatrix} 0 & -G^* \\ G & 0 \end{bmatrix}$$

where $G^*G = \mathbf{1}_{E^+}$, $GG^* = \mathbf{1}_{E^-}$. Moreover, $J(E^\pm) = JE^\mp$ and

$$D = \begin{bmatrix} \mathbf{D} & 0 \\ 0 & -J\mathbf{D}J^{-1} \end{bmatrix}.$$

The space $\mathcal{H}$ is $\mathbb{Z}_2$-graded,

$$\mathcal{H} = \mathcal{H}^+ \oplus \mathcal{H}^-$$

and $G\mathcal{H}^+ = \mathcal{H}^-$.

The bundle $\hat{E}_2$ is also $\mathbb{Z}_2$-graded and the compatibility assumptions must include the condition $\partial_\infty E_1^\pm = \partial_\infty \hat{E}_2^\pm$.

$$L_i = L_i^+ \oplus L_i^-, \quad L_i^\pm \subset \mathcal{H}^\pm$$

and the Lagrangian condition translates into

$$(L_i^+)^\perp = G^* L_i^-, \quad (L_i^-)^\perp = G L_i^+ \tag{4.1.22}$$

where $\perp$ denotes the orthogonal complements in $\mathcal{H}^\pm$.

All the spaces $\hat{\mathcal{K}}_{c(r)}$, $\ker_{ex}\hat{D}_i$ and $L_i$ in the statement of Theorem 4.1.22 are $\mathbb{Z}_2$-graded and in this case we can be more specific: all the asymptotic maps in (4.1.21) are *even*. Moreover, the spaces $\hat{\mathcal{K}}_{c(r)}$ have a particularly interesting description. To explain it we have to write $\hat{D}(r)$ is supersymmetric form

$$\hat{D}(r) = \begin{bmatrix} 0 & \hat{\not{D}}(r)^* \\ \hat{\not{D}}(r) & 0 \end{bmatrix}.$$

For every selfadjoint operator $A$ and any compact interval $I$ we denote by $\mathrm{Spec}(A; I)$ the spectral subspace corresponding to the part of the spectrum situated in $I$. Then

$$\hat{\mathcal{K}}^+_{c(r)} \cong \mathrm{Spec}(\hat{\not{D}}(r)^*\hat{\not{D}}(r);\ [0, c(r)^2])$$

and

$$\hat{\mathcal{K}}^-_{c(r)} \cong \mathrm{Spec}(\hat{\not{D}}(r)\hat{\not{D}}(r)^*;\ [0, c(r)^2]).$$

Observe that $\dim \hat{\mathcal{K}}^+_{c(r)} - \dim \hat{\mathcal{K}}^-_{c(r)}$ is a quantity *independent of* $r$ because it is equal to $\mathrm{ind}\, \hat{\not{D}}(r)$

**4.1.6. Examples.** We conclude this section with several examples which in our view best reveal the nature and the complexity of the objects involved in the gluing theorem. Moreover, we will need these computations later on in concrete gauge theoretic applications.

**Example 4.1.23.** Suppose $\hat{N}$ is a cylindrical manifold. The Hodge-de Rham operator

$$d + d^* : \Omega^*(\hat{N}) \to \Omega^*(\hat{N})$$

is a cylindrical *APS* operator. According to [**6**, Prop. 4.9], the $L^2$-kernel of this operator can be identified with the *"image of the relative in the absolute"*, i.e. with the image of the natural morphism

$$H^*(\hat{N}_t, \partial\hat{N}_t) \to H^*(\hat{N}_t)$$

(for some $t > 0$). To understand the extended kernel let us recall that we are working with the canonical cylindrical structure on $T^*\hat{N}$ and we have

$$\partial_\infty \Lambda^* T^* \hat{N} \cong \Lambda^* T^* \partial_\infty \hat{N} \oplus \Lambda^* T^* \partial_\infty \hat{N}.$$

Along the neck we have the isomorphisms

$$\Lambda^{even/odd} T^* \hat{N} = \Lambda^{even/odd} \pi^* T^* N \oplus dt \wedge \Lambda^{odd/even} \pi^* T^* N.$$

We see that the induced grading on $\partial_\infty \Lambda^* T^* \hat{N}$ is not the obvious one. The asymptotic boundary map

$$\partial_\infty : \ker_{ex}(d + d^*) \to \Omega^*(N) \oplus \Omega^*(N)$$

has two components. Given an a-cylindrical form $\hat{\alpha}$ on $\hat{N}$ we have

$$\partial_\infty \hat{\alpha} := \alpha_0 \oplus dt \wedge *\alpha_1$$

and we will set

$$\alpha_0 := \partial^0_\infty \hat{\alpha} \text{ and } \alpha_1 = \partial^1_\infty \hat{\alpha}.$$

Denote by $L_{an}$ the image of the morphism

$$\partial_\infty : \ker_{ex}(\hat{d} + \hat{d}^*) \to H^*(N) \oplus H^*(N)$$

and by $L_{top}$ the image of the morphism $H^*(\hat{N}) \to H^*(\partial\hat{N})$. We have the following isomorphisms:

$$L_{an} \cong \text{Range}\,(\partial^0_\infty) \oplus *\text{Range}\,(\partial^1_\infty) \cong L_{top} \oplus *L_{top}. \tag{4.1.23}$$

For the reader's convenience we include a short proof of this fact.

Observe first of all that $L_{top}$ is a Lagrangian subspace of $H^*(N)$, i.e. $*L_{top} = L_{top}^{\perp}$, so that $2\dim L_{top} = \dim H^*(N)$. Next, notice as in [**24**, Sect. 10] that if $\hat{\alpha} \in \ker_{ex}(\hat{d} + \hat{d}^*)$ then

$$\partial_\infty \hat{*} \hat{\alpha} = \pm\partial^1_\infty \hat{\alpha} \pm dt \wedge *\partial^0_\infty \hat{\alpha} \;\Leftrightarrow\; \partial^i_\infty \hat{*} = \pm * \partial^{1-i}_\infty, \;\; i = 0, 1.$$

This implies $\partial^i_\infty \hat{\alpha} \in L_{top}$ $(i = 0, 1)$ so that

$$L_{an} \subset L_{top} \oplus *L_{top}.$$

Both spaces above are Lagrangian and thus have the same dimension, $\dim H^*(N)$. Hence they must be equal.

By comparing the short exact sequences

$$0 \to \ker_0(d + d^*) \to \ker_{ex}(d + d^*) \to L_{top} \oplus (*L_{top}) \to 0$$

and

$$0 \to \ker_0(d + d^*) \hookrightarrow H^*(\hat{N}) \to L_{top} \to 0$$

we conclude that the natural map $\varphi : \ker_{ex}(d+d^*) \to H^*(\hat{N})$ *is not injective (!)* because we have

$$\dim \ker_{ex}(\hat{d} + \hat{d}^*) = \dim \ker_0(\hat{d} + \hat{d}^*) + 2\dim L_{top} = \dim H^*(\hat{N}) + \dim L_{top}.$$

On the other hand, $\varphi$ is surjective. Indeed, the isomorphism (4.1.23) shows that given the harmonic forms $\alpha_0, \alpha_1 \in L_{top}$ there exists a form $\hat{\alpha} \in \ker_{ex}(d + d^*)$ such that $\partial^i_\infty \hat{\alpha} = \alpha_i$. Its image in $H^*(\partial\hat{N})$ via the morphism

$$\upsilon : \ker_{ex}(\hat{d} + \hat{d}^*) \xrightarrow{\varphi} H^*(\hat{N}) \to L_{top}$$

is the form $\partial^0_\infty \hat{\alpha}$. Thus the above composition is onto and its kernel can be identified with the subspace of *a-cylindrical* harmonic forms $\hat{\alpha}$ such that $\partial^0_\infty \hat{\alpha} = 0$. It has dimension

$$\dim \ker \upsilon = \dim \ker_0(\hat{d} + \hat{d}^*) + \dim L_{top}.$$

On the other hand, $\ker(H^*(\hat{N}) \to L_{top}) = \ker_0(\hat{d} + \hat{d}^*) \subset \text{Range}\,(\varphi)$ so that

$$\dim \ker v = \dim \ker \varphi + \dim \ker(H^*(\hat{N}) \to L_{top})$$

$$= \dim \ker \varphi + \dim \ker_0(\hat{d} + \hat{d}^*).$$

Hence $\dim \ker \varphi = \dim L_{top} = \dim \ker_{ex}(\hat{d} + \hat{d}^*) - \dim H^*(\hat{N})$. This proves the surjectivity of $\varphi$. Its kernel is a subspace of $\ker \partial^0_\infty$. Moreover, the induced map

$$\partial^1_\infty : \ker \varphi \to L_{top}$$

is a bijection. Observe that if $\hat{\alpha} \in \ker \varphi \setminus \{0\}$ (i.e. $\hat{\alpha}$ is a nontrivial harmonic form representing $0 \in H^*(\hat{N})$) then $\partial^1_\infty \hat{\alpha} \neq 0$ so that $\partial^0_\infty \hat{*} \hat{\alpha} \neq 0$ which shows that the harmonic form $*\hat{\alpha}$ represents a *nontrivial* element in $H^*(\hat{N})$ !!!

These facts can be very clearly observed on the simplest situation. Suppose $\hat{N} = \mathbb{R} \times N$. Then for any harmonic form $\alpha$ on $N$ the form $dt \wedge \alpha$ is both harmonic and in $L^2_{ex}$ but its image in $H^*(\hat{N})$ is obviously trivial since $dt \wedge \alpha = d(t\alpha)$. On the other hand, $\hat{*}(dt \wedge \alpha) = \pm * \alpha$ is in $L^2_{ex}$ but it represents a nontrivial cohomology class.

**Exercise 4.1.6.** Fix $0 < \varepsilon \ll 1$. Use the results in the above example together with the Gluing Theorem 4.1.22 to prove that there exists $R = R_\varepsilon > 0$ such that for all $r > R_\varepsilon$ zero is the only eigenvalue in the interval $[-r^{-1-\varepsilon}, r^{-1-\varepsilon}]$ of the Hodge-de Rham operator $d + d^*$ on the closed manifold $\hat{N}(r)$ (introduced in §4.1.5).

**Example 4.1.24.** Suppose $\hat{N}$ is a cylindrical 4-manifold. We can then form the *anti-self-duality operator*

$$\mathbf{ASD} : \Omega^1(\hat{N}) \to (\Omega^2_+ \oplus \Omega^0)(\hat{N}), \;\; \hat{\alpha} \mapsto \sqrt{2}(\hat{d}\hat{\alpha})^+ \oplus -\hat{d}^*\hat{\alpha}.$$

**Remark 4.1.25.** Let us explain the two unusual features of this definition. The factor $\sqrt{2}$ guarantees that **ASD** is an *APS* operator. The choice of $-d^*$ instead of the regular $d^*$ is motivated by consistency reasons. When we investigated the linearization $\mathcal{T}_{\mathsf{C}}$ of the Seiberg-Witten equation we encountered the operator $d^+ \oplus -2d^*$. The negative sign appears because we worked with the left action of the gauge group. Changing this into a positive sign will affect all the orientation conventions.

Observe that if $\pi : \mathbb{R} \times N \to N$ denotes the natural projection then along the cylinder we have the bundle *isometries*

$$\Lambda^1 T^* \hat{N} \to (\Lambda^1 \oplus \Lambda^0)\pi^* T^* N, \;\; \omega^1 \mapsto (a, f) := (\omega - \lrcorner_t \omega \,,\, \lrcorner_t \omega),$$

$$\Lambda^2_+ T^* \hat{N} \to \Lambda^1 \pi^* T^* N, \;\; \eta \mapsto \sqrt{2} \lrcorner_t \eta$$

where $\lrcorner_t$ denotes the contraction by $\partial_t$. As in §2.4.1, any differential form $\omega$ on $\hat{N}$ can be uniquely written as

$$\omega = dt \wedge f + a, \quad f := \lrcorner_t \omega, \quad a := \omega - dt \wedge f.$$

Moreover,

$$\hat{d}(dt \wedge f^0 + a^1) = dt \wedge (\dot{a}^1 - df^0) + da^1,$$

$$\hat{*}\omega^2 := \hat{*}(dt \wedge f^1 + a^2) = dt \wedge *a^2 + *f^1,$$

$$\hat{d}_+(dt \wedge f^0 + a) = \frac{1}{2}(\hat{d} + \hat{*}\hat{d})(dt \wedge f^0 + a^1)$$

$$= \frac{1}{2}dt \wedge (\dot{a}^1 - df^0 + *da^1) + \frac{1}{2} * (\dot{a}^1 - df^0 + da^1)$$

and

$$\hat{d}^*(dt \wedge f^0 + a^1) = -\hat{*}\hat{d}\hat{*}(dt \wedge f^0 + a^1) = -(\dot{f}^0 - d^* a^1).$$

We can now regard the **ASD**-operator $\sqrt{2}\hat{d}_+ - \hat{d}^*$ as a p.d.o.

$$C^\infty(\,(\Lambda^1 \oplus \Lambda^0)\pi^* T^* N) \to C^\infty(\,(\Lambda^1 \oplus \Lambda^0)\pi^* T^* N),$$

$$\begin{bmatrix} a \\ f \end{bmatrix} \mapsto \begin{bmatrix} \dot{a} + *da - df \\ \dot{f} - d^* a \end{bmatrix}.$$

We see that **ASD** has the *APS* form

$$\mathbf{ASD} \begin{bmatrix} a \\ f \end{bmatrix} = \left( \frac{\partial}{\partial t} - \begin{bmatrix} -*d & d \\ d^* & 0 \end{bmatrix} \right) \begin{bmatrix} a \\ f \end{bmatrix}$$

$$\partial_\infty \Lambda^1 T^* \hat{N} \cong (\Lambda^1 \oplus \Lambda^0) T^* \partial_\infty \hat{N} \cong \partial_\infty (\Lambda^2_+ \oplus \Lambda^0) T^* \hat{N}$$

and

$$\vec{\partial}_\infty(\mathbf{ASD}) = \begin{bmatrix} -*d & d \\ d^* & 0 \end{bmatrix}.$$

The operator $-\vec{\partial}_\infty(\mathbf{ASD})$ is called the *odd signature operator* and we will denote it by **SIGN**. (The negative sign is due mostly to historical reasons but not solely.) It depends on the metric $g$ and its eta invariant will be denoted by $\eta_{sign}(g)$ so that the Atiyah-Patodi-Singer has the form

$$I_{APS}(\mathbf{ASD}) = \int_{\hat{N}} \rho(\mathbf{ASD}) + \frac{1}{2}(\eta_{sign}(g) - \dim\ker \mathbf{SIGN}).$$

**Remark 4.1.26.** If we define the "classical" **ASD**-operator by

$$\mathbf{ASD}_{cl} := \sqrt{2}\hat{d}^+ \oplus \hat{d}^*$$

then

$$\mathbf{ASD}_{cl}\begin{bmatrix} a \\ f \end{bmatrix} = \begin{bmatrix} 1 & 0 \\ 0 & -1 \end{bmatrix}\Big(\frac{\partial}{\partial t} - \begin{bmatrix} -*d & d \\ d^* & 0 \end{bmatrix}\Big)\begin{bmatrix} a \\ f \end{bmatrix}$$

and

$$\vec{\partial}_\infty(\mathbf{ASD}_{cl}) = \begin{bmatrix} -*d & d \\ d^* & 0 \end{bmatrix} = \vec{\partial}_\infty \mathbf{ASD}.$$

If we assume $\hat{N}$ is spin and $\mathbb{S} = \mathbb{S}_+ \oplus \mathbb{S}_-$ is the associated bundle of complex spinors then the Clifford multiplication map

$$\hat{\mathbf{c}} : \Lambda T^*\hat{N} \otimes \mathbb{C} \to \operatorname{End}(\mathbb{S})$$

induces isomorphisms ( *but not isometries*)

$$\Lambda^1 T^*\hat{N} \otimes \mathbb{C} \cong \operatorname{Hom}(\mathbb{S}_+, \mathbb{S}_-) \cong \mathbb{S}_+^* \otimes \mathbb{S}_- \cong \mathbb{S}_+ \otimes \mathbb{S}_- \tag{4.1.24}$$

and

$$(\Lambda^0 \oplus \Lambda^2_+)T^*\hat{N} \otimes \mathbb{C} \cong \operatorname{End}(\mathbb{S}_+) \cong \mathbb{S}_+^* \otimes \mathbb{S}_+ \cong \mathbb{S}_+ \otimes \mathbb{S}_+. \tag{4.1.25}$$

The operator $\mathbf{ASD}_{cl}$ can be regarded as an operator

$$\mathbf{ASD}_{cl} : C^\infty(\mathbb{S}_- \otimes \mathbb{S}_+) \to C^\infty(\mathbb{S}_+ \otimes \mathbb{S}_+).$$

If $\mathfrak{D} : C^\infty(\mathbb{S}_+) \to C^\infty(\mathbb{S}_-)$ denotes the canonical Dirac operator then we can identify $\mathbf{ASD}_{cl}$ with the geometric Dirac operator $\mathfrak{D}^*$ twisted by the bundle $\mathbb{S}_+$ equipped with the Levi-Civita induced connection (see [**5**, Sec. 6] and the references therein for details).

The operators **ASD** and $\mathbf{ASD}_{cl}$ have the same local index densities since

$$\mathbf{ASD}^* \cdot \mathbf{ASD} = \mathbf{ASD}_{cl}^* \cdot \mathbf{ASD}_{cl}, \;\; \mathbf{ASD} \cdot \mathbf{ASD}^* = \mathbf{ASD}_{cl} \cdot \mathbf{ASD}_{cl}^*.$$

This common index density is

$$\rho_{asd}(\hat{g}) = -\frac{1}{2}\Big(\mathbf{e}(\hat{\nabla}^{\hat{g}}) + \frac{1}{3}p_1(\hat{\nabla}^{\hat{g}})\Big) \tag{4.1.26}$$

where $\mathbf{e}(\hat{\nabla}\hat{g}) \in \Omega^4(\hat{N})$ is the Euler *form* associated to the Levi-Civita connection of $\hat{N}$ (via the Chern-Weil construction) and $p_1(\hat{\nabla}^{\hat{g}}) \in \Omega^4(\hat{N})$ is the first Pontryagin form associated to the Levi-Civita connection of $\hat{g}$. This follows essentially from the above identification of $\mathbf{ASD}_{cl}$ with a geometric Dirac operator (see [**5, 6**] for more details). Thus, as far as index computations are concerned, it makes no difference whether we work with **ASD** or $\mathbf{ASD}_{cl}$.

**Exercise 4.1.7.** Show that $D := \mathbf{ASD}_{cl}$ is a Dirac operator, i.e. both $D^*D$ and $DD^*$ are generalized Laplacians.

Suppose $\alpha \in \ker_{ex} \mathbf{ASD}$. Then

$$(\hat{d} + \hat{*}\hat{d})\hat{\alpha} = 0 \text{ and } \hat{d}\hat{*}\hat{\alpha} = 0.$$

We deduce that $\hat{d}^*\hat{d}\hat{\alpha} = 0$. Taking the inner product with $\hat{\alpha}$ and using the integration by parts formula of Sec. 1.2 over $\hat{N}_r$ $(r \gg 0)$ we deduce

$$\int_{\hat{N}_r} |\hat{d}\hat{\alpha}|^2 d\hat{v} = \pm \int_{\partial N_r} \hat{\alpha} \wedge * \lrcorner_t \hat{d}\hat{\alpha}.$$

The boundary term goes to zero as $r \to \infty$ since $\hat{\alpha} \in L^2_{ex}$ and we deduce $\hat{d}\hat{\alpha} = 0$. Thus $\hat{\alpha} \in \ker_{ex}(\hat{d} + \hat{d}^*)$ so that

$$\ker_{ex}(\mathbf{ASD}) = \ker_{ex}(\hat{d} + \hat{d}^*) \mid_{\Omega^1(\hat{N})} .$$

Arguing similarly we deduce

$$\ker_{ex} \mathbf{ASD}^* = P_+ \ker_{ex}(\hat{d} + \hat{d}^*) \mid_{\Omega^2(\hat{N})} \oplus \mathbb{R} \tag{4.1.27}$$

where $P_+$ denotes the projection $\Omega^2 \to \Omega^2_+$.

We can now determine $\partial_\infty \ker_{ex}(\mathbf{ASD})$ and $\partial_\infty \ker_{ex}(\mathbf{ASD}^*)$. To present this description observe that the spaces $L_{top}$ discussed in the previous example are graded by the degree. We denote by $L^i_{top}$ the degree-$i$ subspace. Since $L^3_{top} = 0$ we deduce

$$\partial_\infty \ker_{ex}(\mathbf{ASD}) = L^1_{an} = L^1_{top} \oplus (dt \wedge *L^3_{top}) = L^1_{top}. \tag{4.1.28}$$

Since $\partial_\infty \ker_{ex}(\mathbf{ASD}^*)^\perp = G\partial_\infty \ker_{ex}(\mathbf{ASD})$ (see (4.1.22)) we deduce

$$\partial_\infty \ker_{ex}(\mathbf{ASD}^*) = *L^2_{top} \oplus L^0_{top} \tag{4.1.29}$$

The above equality can also be seen directly from (4.1.27). We can use the above simple observations to compute the $APS$ index of $\mathbf{ASD}$. Let us assume for simplicity that both $\hat{N}$ and $N$ are connected.We have

$$I_{APS}(\mathbf{ASD}) = \dim \ker_{L^2}(\mathbf{ASD}) - \dim \ker_{ex}(\mathbf{ASD}^*).$$

The first space can be identified with the image of $H^1(\hat{N}, N)$ in $H^1(N)$. Using the long exact sequence of the pair $(\hat{N}, N)$ we deduce

$$\dim \ker_{L^2}(\mathbf{ASD}) = \dim H^1(\hat{N}, N) = \hat{b}^3$$

where $\hat{b}^k := \dim H^k(\hat{N})$. On the other hand,

$$\dim \ker_{ex}(\mathbf{ASD}^*) = \dim P_+ \ker_{ex}(\hat{d} + \hat{d}^*) + 1.$$

We want to identify the right-hand side of the above equality with known topological invariants. For a 2-form $\hat{\alpha} \in \ker(\hat{d} + \hat{d}^*)$ the condition $\hat{*}\hat{\alpha} = \hat{\alpha}$ implies

$$\partial^0_\infty \hat{\alpha} = \partial^1_\infty \hat{\alpha}$$

so that we have a natural map

$$P_+ \ker_{ex}(\hat{d} + \hat{d}^*) \to L^2_{top}, \quad \hat{\alpha} \mapsto \partial^0_\infty \hat{\alpha}.$$

From the isomorphism (4.1.23) we deduce the above map is onto. Its kernel is none other than the self-dual part of $\ker_{L^2}(\hat{d}+\hat{d}^*)$. Thus

$$\dim P_+ \ker_{ex}(\hat{d}+\hat{d}^*) = \dim P_+ \ker_{L^2}(\hat{d}+\hat{d}^*) + \dim L^2_{top}.$$

The radical of the intersection form on $H^2(\hat{N}, N)$ is precisely the kernel of the morphism

$$H^2(\hat{N}, N) \to H^2(\hat{N})$$

so that

$$\dim P_+ \ker_{L^2}(\hat{d}+\hat{d}^*) = \hat{b}_+$$

where $\hat{b}_\pm$ denotes the dimension of the positive/negative eigenspace of the intersection form. Thus

$$\dim P_+ \ker_{ex}(\hat{d}+\hat{d}^*) = \hat{b}_+ + l^2$$

where $l^k := \dim L^k_{top}$. Hence

$$I_{APS}(\mathbf{ASD}) = \hat{b}^3 - \hat{b}_+ - l^2 - 1.$$

On the other hand, we have the following identities which are either tautological or follow from the long exact sequence of the pair $(\hat{N}, N)$ coupled with the identity $l^k + l^{3-k} = \dim H^k(N)$:

$$\begin{cases} b^2 &= \hat{b}_+ + r + \hat{b}_- \\ r &= l^2 \\ \tau &= \hat{b}_+ - \hat{b}_- \\ \hat{b}^k - \hat{b}^{4-k} &= l^k - l^{4-k} \end{cases}$$

where $r$ is the dimension of the radical of the intersection form and $\tau$ is its signature. After some elementary manipulations involving the above identities we reach the conclusion

$$I_{APS}(\mathbf{ASD}) = -\frac{1}{2}(\chi + \tau + h) \tag{4.1.30}$$

where $\chi = \sum_k (-1)^k \hat{b}^k$ and $h = \dim(H^0 \oplus H^1)(N)$.

We conclude this section with a detailed discussion of a very special choice of $\hat{N}$ which will be needed for further applications.

**Example 4.1.27.** Supppose $L_\ell \to S^2$ is a Hermitian line bundle of degree $\ell \in \mathbb{Z}$ over the 2-sphere. We assume $S^2$ is equipped with a round metric $g_0$ so that its area is $\pi$. Thus its radius is $1/2$ so its Gauss (sectional) curvature is 4. Denote by $\omega_0$ the volume form on $S^2$.

The metric on $L$ determines a unit disk bundle $D_\ell \to S^2$ with boundary a principal $S^1$-bundle

$$\begin{array}{ccc} S^1 & \hookrightarrow & N_\ell \\ & & \downarrow \pi \\ & & S^2 \end{array}$$

Observe that $L_0$ is the trivial line bundle and $N_0 \cong S^1 \times S^2$ while $L_{-1}$ is the tautological line bundle over $\mathbb{P}^1 \cong S^2$ and in this case $N_{-1} \cong S^3$. Moreover, $D_{-1}$ can be identified with a tubular neighborhood of $\bar{\mathbb{P}}^1 \hookrightarrow \bar{\mathbb{P}}^2$.

$N_\ell$ is equipped with a free $S^1$-action whose orbits coincide with the fibers of $\pi$. We denote by $\zeta$ its canonical infinitesimal generator. A *global angular form* is an $S^1$-invariant 1-form $\varphi \in \Omega^1(N_\ell)$ such that $\zeta \lrcorner \varphi = 1$. Equivalently, this means that the restriction of $\varphi$ to any fiber of $\pi$ coincides with the angular form $d\theta$ on $S^1$. Using the language of principal $S^1$-bundles as in **[64]** we can say that $\mathbf{i}\varphi$ defines a connection on the principal bundle $N_\ell$. Notice that

$$L_\zeta d\varphi = 0, \quad \zeta \lrcorner (d\varphi) = L_\zeta \varphi - d(\zeta \lrcorner \varphi) = 0.$$

Thus $\mathbf{i}d\varphi$ is the pullback of an imaginary closed 2-form $\Omega$ on $S^2$, the curvature of the connection $\mathbf{i}\varphi$. Moreover

$$-\frac{1}{2\pi}\int_{S^2} \Omega = \int_{S^2} c_1(L) = \deg(L) = \ell. \tag{4.1.31}$$

The choice of global angular form is not unique. We can alter $\varphi$ by the pullback of a 1-form $\alpha$ on $S^2$. The curvature will change according to the rule

$$\mathbf{i}\Omega \to \mathbf{i}\Omega + \mathbf{i}d\alpha.$$

In particular, we can choose the global angular form so that its curvature is harmonic

$$\Omega = c\omega_0, \quad c \in \mathbb{R}.$$

Using this in the equality (4.1.31) we deduce

$$-\frac{c}{2\pi}\operatorname{area}(S^2) = -\frac{1}{2\pi}\int_{S^2} c\omega_0 = \ell$$

so that $c = -2\ell$. Thus with this choice we have

$$d\varphi = -2\ell\pi^*\omega_0.$$

Observe that $\ker\varphi$ determines a subbundle of $TN_\ell$ isomorphic to $\pi^*TS^2$. Thus

$$TN \cong \mathbb{R}\zeta \oplus \ker\varphi \cong \mathbb{R}\zeta \oplus \pi^*TS^2.$$

For each $r > 0$ we construct a metric $g_r$ on $TN_\ell$ uniquely determined by the conditions

$$g_r(\zeta, \zeta) = r^2, \quad g_r\,|_{\ker\varphi} = (\pi^* g_0)\,|_{\pi^* TS^2} \, .$$

The metric $g_r$ is the restriction of a natural metric $\hat{g}_r$ on $D_\ell$. Denote by $h$ the Hermitian metric on $L_\ell$.

To describe $\hat{g}_r$ observe that the angular form $\varphi$ induces a Hermitian connection $A_0$ on $L_\ell$. This produces a splitting of the tangent bundle $TL_\ell$ into vertical and horizontal parts.

$$TL_\ell := VTL_\ell \oplus HTL_\ell.$$

The vertical part is spanned by vectors tangent to the fibers of $\pi : L_\ell \to S^2$ and is isomorphic to $\pi^* L_\ell$. The horizontal part is generated by the locally covariant constant sections in the following sense. Choose local coordinates $z = (x, y)$ on a neighborhood $U$ of a point $p_0 \in S^2$ and a local unitary frame $\mathbf{f}$ of $L_\ell|_U$. Then a point $P \in \pi^{-1}(U)$ can be described by a pair of complex numbers $(\xi, z)$ uniquely determined by the conditions

$$P \in \pi^{-1}(z), \quad P = \xi \mathbf{f}_z.$$

A tangent vector $(\dot{\xi}, \dot{z}) \in T_P L_\ell$ is vertical if $\dot{z} = 0$. It is horizontal if

$$\dot{\xi} + \mathbf{i}a_z(\dot{z})\xi = 0$$

where $\mathbf{i}a \in \mathbf{i}\Omega^1(U)$ is the 1-form representing $A_0$ with respect to the unitary frame $\mathbf{f}$.

Consider the family of hypersurfaces $X_r \subset L_\ell$

$$X_r := \{(p, v);\ \ p \in S^2,\ v \in \pi^{-1}(p),\ \ h_p(v, v) = r^2\}.$$

$X_r$ is locally described by the equation

$$X_r = \{(\xi, z);\ \ |\xi|^2 = r^2\}.$$

Observe that all these hypersurfaces are diffeomorphic to $N_\ell$. Since $A_0$ is a Hermitian connection, the horizontal sub-bundle is tangent to the hypersurfaces $X_r$. If we choose polar coordinates $(r, \theta)$ (away from the zero section) in each fiber

$$\rho := |\xi|, \quad \xi \mathbf{f} := \rho e^{\mathbf{i}\theta} \mathbf{f}$$

then the horizontal distribution can be described by the equation

$$(\dot{\rho}, \dot{\theta}, \dot{z}) \in T_{\xi,z} L_\ell, \quad \dot{\rho} = 0, \quad \mathbf{i}\dot{\theta} + \mathbf{i}a(\dot{z}).$$

The 1-form $d\theta + a$ is precisely the global angular form expressed in the local coordinates $(r, \theta, z)$.

Now we can define a metric $\hat{g}_r$ on $TL_\ell := VTL_\ell \oplus HTL_\ell$ by

$$\hat{g}_r := r^2 h \oplus \pi^*(g_0).$$

The restriction of $\hat{g}_r$ to $X_1$ coincides with $g_r$. We want to prove that the scalar curvatures of $g_r$ and $\hat{g}_r$ are *everywhere positive* provided $r$ is sufficiently small. We will use Cartan's moving frame method. For more details concerning this method we refer to [**105**, Chap. 4].

Pick a local (oriented) orthonormal frame $\tau_1, \tau_2$ of $TS^2 \mid_U$, denote by $\theta^1, \theta^2$ the dual coframe and set

$$\vec{\theta} := \begin{bmatrix} \theta^1 \\ \theta^2 \end{bmatrix}.$$

Then the structural equations for the Riemann metric $g_0$ imply

$$d\vec{\theta} = \begin{bmatrix} 0 & \mu \\ -\mu & 0 \end{bmatrix} \cdot \vec{\theta}, \;\; \mu \in \Omega^1(U).$$

The $\underline{so}(2)$-valued 1-form

$$\begin{bmatrix} 0 & -\mu \\ \mu & 0 \end{bmatrix}$$

describes the Levi-Civita connection with respect to the frame $\tau_1, \tau_2$:

$$\nabla^{LC}\tau_1 = \mu\tau_2, \;\; \nabla^{LC}\tau_2 = -\mu\tau_1.$$

Then

$$d\mu = -\frac{1}{4}\omega_0$$

where $1/4$ is the sectional curvature of $S^2$. Set

$$\varphi^1 := r d\rho, \;\; \varphi^2 := r\varphi, \;\; \varphi^3 := \pi^*\theta^1, \;\; \varphi^4 = \pi^*\theta^2.$$

Observe that the metric $\hat{g}_r$ can be described as

$$\hat{g}_r = \left\{ (\varphi^1)^2 + (\varphi^2)^2 \right\} + \left\{ (\varphi^3)^2 + (\varphi^4)^2 \right\}$$

so that $(\varphi^1, \cdots, \varphi^4)$ is a local, oriented, $\hat{g}_r$-orthonormal coframe of $T^*L_\ell$. Set

$$\vec{\varphi} := \begin{bmatrix} \varphi^1 \\ \varphi^2 \\ \varphi^3 \\ \varphi^4 \end{bmatrix}.$$

The Cartan structural equations show that there exists a *unique* $4 \times 4$ matrix

$$\hat{S}_r = [\theta^i_j]_{1 \leq i,j \leq 4}, \;\; \theta^i_j \in \Omega^1(L_\ell),$$

such that

$$d\vec{\varphi} = \hat{S}_r \wedge \vec{\varphi}, \;\; \theta^i_j = -\theta^j_i.$$

Moreover, the curvature of the Levi-Civita connection of the metric $\hat{g}_r$ is given by

$$\hat{\Omega}_r = -d\hat{S}_r + \hat{S}_r \wedge \hat{S}_r.$$

If $(\zeta_1, \zeta_2, \zeta_3, \zeta_4)$ denotes the frame $\hat{g}_r$-dual to $\vec{\varphi}$ then the scalar curvature of the metric $\hat{g}_r$ is given by

$$\hat{s}_r = \sum_{i \neq j} \langle \hat{\Omega}_r(\zeta_i, \zeta_j)\zeta_j, \zeta_i \rangle.$$

We have

$$d\vec{\varphi} = \begin{bmatrix} 0 \\ -2r\ell\varphi^3 \wedge \varphi^4 \\ \pi^*\mu \wedge \varphi^4 \\ -\pi^*\mu \wedge \varphi^3 \end{bmatrix}$$

and

$$\theta^i_j = -\frac{1}{2}\sum_{k=2}^{4}\Big\{d\varphi^i(\zeta_j, \zeta_k) + d\varphi^j(\zeta_i, \zeta_k) - d\varphi^k(\zeta_i, \zeta_j)\Big\}\varphi^k.$$

We deduce

$$\theta^1_2 = -\frac{1}{2}\sum_{k=2}^{4}\Big\{d\varphi^2(\zeta_1, \zeta_k) - d\varphi^k(\zeta_1, \zeta_2)\Big\}\varphi^k = \frac{1}{2}\sum_{k=2}^{4} d\varphi^k(\zeta_1, \zeta_2)\varphi^k = 0,$$

$$\theta^1_3 = -\frac{1}{2}\sum_{k=2}^{4}\Big\{d\varphi^3(\zeta_1, \zeta_k) - d\varphi^k(\zeta_1, \zeta_3\Big\}\varphi^k = 0,$$

$$\theta^1_4 = 0,$$

$$\theta^2_3 = -\frac{1}{2}\sum_{k=2}^{4}\Big\{d\varphi^2(\zeta_3, \zeta_k) + d\varphi^3(\zeta_2, \zeta_k) - d\varphi^k(\zeta_2, \zeta_3)\Big\}\varphi^k = r\ell\varphi^4,$$

$$\theta^2_4 = -\frac{1}{2}\sum_{k=2}^{4}\Big\{d\varphi^2(\zeta_4, \zeta_k) + d\varphi^2(\zeta_3, \zeta_k) - d\varphi^k(\zeta_2, \zeta_4)\Big\} = -r\ell\varphi^3,$$

$$\theta^3_4 = -\frac{1}{2}\sum_{k=2} 4\Big\{d\varphi^3(\zeta_4, \zeta_k) + d\varphi^4(\zeta_3, \zeta_k) - d\varphi^k(\zeta_3, \zeta_4)\Big\}\varphi^k,$$

$$= -\frac{1}{2}\Big\{-d\varphi^2(\zeta_3, \zeta_4)\varphi^2 - 2d\varphi^3(\zeta_3, \zeta_4)\varphi^3 - 2d\varphi^4(\zeta_3, \zeta_4)\varphi^4)\Big\} = -r\ell\varphi^2 + \pi^*\mu.$$

Thus

$$\hat{S}_r = \begin{bmatrix} 0 & 0 & 0 & 0 \\ 0 & 0 & r\ell\varphi^4 & -r\ell\varphi^3 \\ 0 & -r\ell\varphi^4 & 0 & -r\ell\varphi^2 + \pi^*\mu \\ 0 & r\ell\varphi^3 & r\ell\varphi^2 - \pi^*\mu & 0 \end{bmatrix}.$$

The Riemann curvature tensor of $\hat{g}_r$ is

$$\hat{\Omega}_r = -d\hat{S}_r + \hat{S}_r \wedge \hat{S}_r$$

$$= \begin{bmatrix} 0 & 0 & 0 & 0 \\ 0 & 0 & r\ell\pi^*\mu \wedge \varphi^3 & r\ell\pi^*\mu \wedge \varphi^4 \\ 0 & -r\ell\pi^*\mu \wedge \varphi^3 & 0 & (\frac{1}{4} - 2r^2\ell^2)\varphi^3 \wedge \varphi^4 \\ 0 & -r\ell\pi^*\mu \wedge \varphi^4 & (2r^2\ell^2 - \frac{1}{4})\varphi^3 \wedge \varphi^4 & 0 \end{bmatrix}$$

$$+ \begin{bmatrix} 0 & 0 & 0 & 0 \\ 0 & 0 & * & * \\ 0 & -r^2\ell^2\varphi^2 \wedge \varphi^3 - r\ell\varphi^3 \wedge \pi^*\mu & 0 & * \\ 0 & -r^2\ell^2\varphi^2 \wedge \varphi^4 - r\ell\varphi^4 \wedge \pi^*\mu & r^2\ell^2\varphi^3 \wedge \varphi^4 & 0 \end{bmatrix}$$

$$= \begin{bmatrix} 0 & 0 & 0 & 0 \\ 0 & 0 & r^2\ell^2\varphi^2 \wedge \varphi^3 & r^2\ell^2\varphi^2 \wedge \varphi^4 \\ 0 & -r^2\ell^2\varphi^2 \wedge \varphi^3 & 0 & (\frac{1}{4} - 3r^2\ell^2)\varphi^3 \wedge \varphi^4 \\ 0 & -r^2\ell^2\varphi^2 \wedge \varphi^4 & (3r^2\ell^2 - \frac{1}{4})\varphi^3 \wedge \varphi^4 & 0 \end{bmatrix}.$$

The scalar curvature of $\hat{g}_r$ is

$$\hat{s}_r = 2\Big\{r^2\ell^2\varphi^2 \wedge \varphi^3(\zeta_2 \wedge \zeta_3) + r^2\ell^2\varphi^2 \wedge \varphi^4(\zeta_2 \wedge \zeta_4)$$
$$+(\frac{1}{4} - 3r^2\ell^2)\varphi^3 \wedge \varphi^4(\zeta_3, \zeta_4)\Big\} = \frac{1}{2} - 2r^2\ell^2.$$

We see that

(4.1.32) $$\hat{s}_r > 0, \quad \forall r < \frac{1}{2|\ell|}.$$

A similar computation shows that the scalar curvature of $g_r$ is

$$s_r = \frac{1}{2} - 2r^2\ell^2 = \hat{s}_r.$$

Observe that we can slightly perturb the metric $\hat{g}_r$ in a neighborhood $U$ of $\partial D_\ell$ so that the new metric continues to have *positive* scalar curvature and its restriction to a smaller tubular neighborhood $V \subset U$ of $N_\ell$ is isometric to the product metric $dt^2 + g_r$ on $(0, \varepsilon] \times N_\ell$.

More precisely, near $\partial D_\ell$, $\hat{g}_r$ has the form

$$\hat{g}_r = r^2 d\rho^2 + r^2 \rho^2 \varphi^2 + \pi^* g_0, \quad \rho \in (1-\varepsilon, 1].$$

Define the perturbed metric to be $\tilde{g}_r := r^2 d\rho^2 + r^2 \alpha(\rho)^2 \varphi^2 + \pi^* g_0$, where the cut off function $\alpha$ is depicted in Figure 4.6.

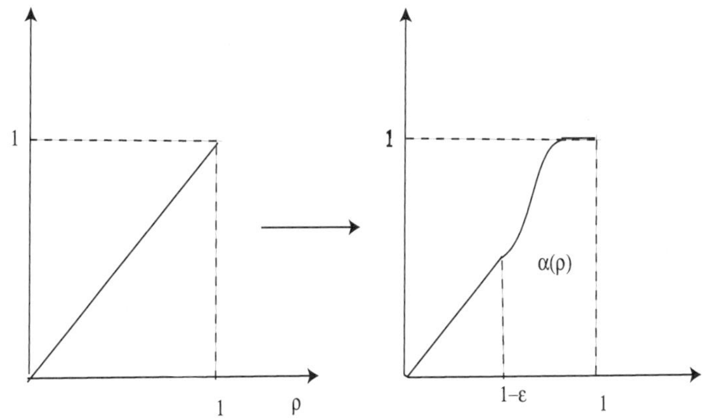


**Figure 4.6.** *Smoothing the linear function $\rho \to \rho$*

The scalar curvature of $\tilde{g}_r$ differs from the scalar curvature of $\hat{g}_r$ by a term bounded from above by $Cr^2 \|\alpha\|_{C^2}$ where $C$ is a universal constant. The scalar curvature $s(\tilde{g}_r)$ will be positive as soon as $r$ is sufficiently small.

The classical topological invariants of $N_\ell$, $\ell \neq 0$, are easy to compute. To determine its fundamental group observe that $N_\ell$ is a $\mathbb{Z}_{|\ell|}$-quotient of $N_{-1} \cong S^3$. To see this represent $S^3$ as the unit sphere in $\mathbb{C}^2$

$$S^3 = \{(z^1, z^2) \in \mathbb{C}^2;\ \ |z^1|^2 + |z^2|^2 = 1\}$$

and the cyclic group $\mathbb{Z}_{|\ell|}$ as the multiplicative subgroup of $S^1$ consisting $|\ell|$-th roots of 1. Then $\mathbb{Z}_{|\ell|}$ acts on $S^3$ by

$$\rho(z^1, z^2) = (\rho z^1, \rho z^2) \ \ (\rho^{|\ell|} = 1)$$

and this action commutes with the Hopf action of $S^1$

$$e^{\mathbf{i}t}(z^1, z^2) = (e^{\mathbf{i}t} z^1, e^{\mathbf{i}t} z^2).$$

This action descends to an $S^1$-action on the quotient $N = S^3/\mathbb{Z}_{|\ell|}$ and the stabilizer of each point with respect to this action is precisely $\mathbb{Z}_{|\ell|}$. Thus $N$ is equipped with a free $S^1 \cong S^1/\mathbb{Z}_{|\ell|}$-action and the natural projection

$$\pi : S^3 \to N$$

satisfies

$$\pi(e^{\mathbf{i}t}x) = e^{\ell \mathbf{i}t}\pi(x).$$

Thus $N \to N/S^1$ is a principal $S^1$-bundle and the $|\ell|$-fold cover $\pi : S^3 \to N$ maps the fibers of the Hopf bundle $S^3 \to S^2$ to the fibers of $N \to N/S^1$. Moreover the restriction to fibers is an $|\ell|$-fold cover. This shows that $N$ is a circle bundle of degree $-|\ell|$ over $S^2$, i.e. $N \cong N_{-|\ell|}$. (To obtain the bundles of positive degree we have to replace the Hopf action by its conjugate in the above arguments.) This shows that

$$\pi_1(N_\ell) = \mathbb{Z}_{|\ell|}$$

and the homotopy class of a fixed orbit is a generator of this cyclic group. Thus

$$H^1(N_\ell, \mathbb{Z}) \cong H_2(N_\ell, \mathbb{Z}) = 0, \quad H_1(N_\ell, \mathbb{Z}) \cong H^2(N_\ell, \mathbb{Z}) \cong \mathbb{Z}_{|\ell|}.$$

It is convenient to describe the isomorphism $H^2(N_\ell, \mathbb{Z}) \cong \mathbb{Z}_{|\ell|})$ from a different perspective.

The manifold $N_\ell$ bounds a disk bundle $D_\ell$ of degree $\ell$ and we have a long exact sequence

$$0 = H^1(N_\ell, \mathbb{Z}) \to H^2(D_\ell, N_\ell; \mathbb{Z}) \to H^2(D_\ell, \mathbb{Z}) \cong H^2(S^2, \mathbb{Z}) \to H^2(N_\ell)$$

$$\to H^3(D_\ell, N_\ell; \mathbb{Z}) \cong H_1 D_\ell, \mathbb{Z}) = 0$$

where at the last step we have used Poincaré duality. On the other hand, the Thom isomorphism theorem shows that the Poincaré dual $\tau \in H^2(D_\ell, N_\ell; \mathbb{Z})$ of $S^2 \hookrightarrow D_\ell$ satisfies

$$i^*\tau = \ell \times \text{generator of } H^2(S^2, \mathbb{Z})$$

and the map

$$H^0(S^2, \mathbb{Z}) \to H^2(D_\ell, N_\ell; \mathbb{Z}) \cong H_2(D_\ell, \mathbb{Z}), \quad u \mapsto \tau \wedge \pi^* u$$

is an isomorphism. Above, $\pi$ denotes the natural projection $D_\ell \to S^2$ while $i$ denotes the inclusion of $S^2$ in $D_\ell$ as the zero section. Thus, $\tau$ is a generator of $H^2(D_\ell, N_\ell; \mathbb{Z})$. The image of $\tau$ via the morphism

$$H^2(D_\ell, N_\ell; \mathbb{Z}) \to H^2(D_\ell, \mathbb{Z}) \overset{\cong i^*}{\to} H^2(S^2; \mathbb{Z})$$

is precisely $i^*\tau$. Thus, the image of $H^2(D_\ell, N_\ell; \mathbb{Z}) \to H^2(D_\ell, \mathbb{Z}) \cong \mathbb{Z}$ is the subgroup $\ell\mathbb{Z}$. The surjective morphism $H^2(D_\ell, N_\ell)$ is none other than the natural projection

$$H^2(D_\ell, \mathbb{Z}) \cong \mathbb{Z} \to \mathbb{Z} \to \mathbb{Z}/\ell\mathbb{Z} \cong H^2(N_\ell, \mathbb{Z}).$$

If we now identify $H^2(N_\ell, \mathbb{Z})$ with the Abelian group $\mathrm{Pic}^\infty(N_\ell)$ of isomorphism classes of smooth complex line bundles then the above observations show that the restriction map

$$\mathrm{Pic}^\infty(D_\ell) \to \mathrm{Pic}^\infty(N_\ell) \tag{4.1.33}$$

is a surjection, i.e. any complex line bundle over $N_\ell$ extends to a line bundle over $D_\ell$. Such extensions are not unique. The kernel of the morphism (4.1.33) is freely generated by the $VTD_\ell \cong \pi^* L_\ell =$ the pullback of $L_\ell \to S^2$ to the disk bundle $D_\ell \hookrightarrow L_\ell$.

Consider the operator **ASD** on $D_\ell$ determined by the metric $\tilde{g}_r$. Because of the cylindrical nature of $\tilde{g}_r$ near $\partial D_\ell$ we can attach a cylinder $[0,\infty) \times N_\ell$ and obtain a cylindrical manifold $\hat{N}_\ell$. We will continue to denote the cylindrical metric on $\hat{N}_\ell$ by $\tilde{g}_r$. Assume $\ell \neq 0$. Then

$$I_{APS}(\mathbf{ASD}) = -\frac{1}{2}(\chi(D_\ell) + \tau(D_\ell) + h(N_\ell))$$

$$= -\frac{1}{2}(2 + \tau(D_\ell) + 1) = -\frac{1}{2}(2 + 1 + \mathrm{sign}\,(\ell)).$$

Moreover,

$$\ker_{L^2}(\mathbf{ASD}) \cong H^1(D_\ell, N_\ell) = 0.$$

Observe that

$$\partial_\infty \ker_{ex}(\mathbf{ASD}) = *L^2_{top} \oplus L^0_{top} \cong 0 \oplus \mathbb{R}.$$

Thus

$$\dim \ker_{ex}(\mathbf{ASD}) = \dim \ker_{L^2}(\mathbf{ASD}) + 1 = 1$$

and

$$\dim \ker_{ex}(\mathbf{ASD}^*) = \frac{1}{2}(1 + \mathrm{sign}(\ell)) + 1.$$

This confirms the prediction

$$I_{APS}(\mathbf{ASD}) = -\frac{1}{2}(3 + \tau(D_\ell)).$$

We can now use the Atiyah-Patodi-Singer index theorem to conclude that

$$-\frac{1}{2}(3 + \tau(D_\ell)) = -\frac{1}{2}\int_{D_\ell}\Big(\mathbf{e}(\hat{\nabla}^{\tilde{g}_r}) + \frac{1}{3}p_1(\hat{\nabla}^{\tilde{g}_r})\Big) + \frac{1}{2}\Big(\eta_{sign}(g_r) - h\Big).$$

Since $h = b_0(N) + b_1(N) = 1$ we deduce

$$\eta_{sign}(g_r) = -2 - \tau(D_\ell) + \int_{D_\ell}\Big(\mathbf{e}(\hat{\nabla}^{\tilde{g}_r}) + \frac{1}{3}p_1(\hat{\nabla}^{\tilde{g}_r})\Big).$$

On the other hand, the Gauss-Bonnet theorem for manifolds with boundary (see [**48**, §2.7.6 – 7]) implies

$$\int_{D_\ell} \mathbf{e}(\hat{\nabla}^{\tilde{g}_r}) = \chi(D_\ell) = 2$$

so that

$$\eta_{sign}(g_r) = \frac{1}{3}\int_{D_\ell} p_1(\hat{\nabla}^{\tilde{g}_r}) - \tau(D_\ell). \tag{4.1.34}$$

The last equality is valid for any 4-manifold with boundary, not just the disk bundles $D_\ell$. It justifies the name *signature defect* used to refer to $\eta_{sign}(g)$ since the right-hand side of (4.1.34) would be zero if $D_\ell$ were a *closed* manifold. One of the main motivations for the research conducted in the beautiful papers [**6, 7, 8**] was the need to better understand the nature of this defect.

Let us now turn our attention to Dirac operators. Again we restrict to the case $\ell \neq 0$. Since the tangent bundle of any compact, oriented 3-manifold is trivializable we deduce $w_2(N_\ell) = 0$. Thus $N_\ell$ is spinnable. The universal coefficients theorem shows that

$$H^2(N_\ell, \mathbb{Z}_2) \cong \mathbb{Z}_\ell \otimes \mathbb{Z}_2 \cong \begin{cases} \mathbb{Z}_2 & \ell \equiv 0 \mod 2 \\ 0 & \ell \equiv 1 \mod 2 \end{cases}.$$

Hence, if $\ell$ is even there are precisely two nonisomorphic spin structures on $N_\ell$ while when $\ell$ is odd there is exactly one isomorphism class of spin structures.

If $\sigma \in Spin^c(N_\ell)$ then $c_1(\det \sigma) \equiv 0 \mod 2$. This implies that the range of correspondence

$$Spin^c(N_\ell) \ni \sigma \mapsto c_1(\det \sigma) \in H^2(N_\ell, \mathbb{Z})$$

is the subgroup $G_\ell$ of $\mathbb{Z}_\ell$ generated by $2 \mod \ell$. We will identify $G_\ell$ with a subset of $\{0, 1, \cdots, |\ell| - 1\}$.

Fix $\sigma \in Spin^c(N_\ell)$ and denote by $k$ the element in $G_\ell$ determined by $c_1(\det(\sigma))$. Since $c_1(\det \sigma)$ is a torsion class the line bundle $\det(\sigma)$ supports at least one flat connection $A_\sigma$. This connection is determined by its holonomy along the fibers (which generate $\pi_1(N_\ell)$) and is given by a complex number

$$\rho_\sigma := \exp\Big(\frac{2\pi k\mathbf{i}}{\ell}\Big).$$

As in [**106**, p. 369], we form the connection

$$B_\sigma := A_\sigma + \frac{\mathbf{i}k}{\ell}\varphi$$

so that

$$F_{B_\sigma} = \frac{\mathbf{i}k}{\ell} d\varphi = -2k\mathbf{i}\pi^*\omega_0.$$

The holonomy of $B_\sigma$ along any fiber is zero. (*Can you see why ?*) Since the curvature is the pullback of a form on the base of the fibration $N_\ell \to S^2$ we deduce that $B_\sigma$ is the pullback of a connection $B'_\sigma$ on a line bundle $L'_\sigma \to S^2$ such that $\pi^*(L'_\sigma) \cong L_\sigma$. The Chern class of $L'_\sigma$ is

$$c_1(L'_\sigma) - = \frac{\mathbf{i}}{2\pi} \cdot (-2k\mathbf{i}\omega_0) = \frac{k}{\pi}\omega_0.$$

Since

$$\int_{S^2} \omega_0 = \pi$$

this class corresponds to the element $k \in H^2(S^2, \mathbb{Z}) \cong \mathbb{Z}$. Since the pullback $\pi^* : H^2(S^2, \mathbb{Z}) \to H^2(N_\ell, \mathbb{Z})$ is given by the natural projection $\mathbb{Z} \to \mathbb{Z}/\ell\mathbb{Z}$ we deduce that $k \in G_\ell \mod \ell$ and $c_1(L_\sigma) = k$.

On $N_\ell$ there is a canonical *spin*$^c$ structure $\sigma_0$ induced from the natural *spin*$^c$ structure $\hat{\sigma}_0$ on $D_\ell$ determined by the complex structure. Observe that as a complex vector bundle we have

$$TD_\ell \cong \pi^* L_\ell \oplus \pi^* TS^2 \cong \pi^*(L_\ell \oplus K^{-1})$$

where $K$ denotes the canonical line bundle on $S^2 \cong \mathbb{P}^1$. Observe that $\deg K = -\chi(S^2) = -2$. Then

$$\det(\hat{\sigma}_0) \cong K^{-1}_{D_\ell} \cong \pi^*(L_\ell \otimes K^{-1}).$$

This induces a *spin*$^c$ structure $\sigma_0$ on $N_\ell$ satisfying

$$\det(\sigma_0) = \pi^*(L_\ell \otimes K^{-1}) \,|_{N_\ell} \cong \pi^* K^{-1} \,|_{N_\ell}$$

since $L_\ell \,|_{N_\ell} \cong \underline{\mathbb{C}}$. Thus $c_1(\sigma_0) \equiv 2 \mod \ell$. For every $n \in \mathbb{Z}$ denote by $L_n$ the degree $n$ line bundle over $S^2$ and set

$$\hat{\sigma}_n := \hat{\sigma}_0 \otimes L_n, \;\; \sigma_n := \hat{\sigma}_n \,|_{N_\ell} .$$

Observe that

$$c_1(\det(\hat{\sigma}_0)) = \pi^* L_{\ell+2}, \;\; c_1(\det(\hat{\sigma}_n)) = \pi^* L_{\ell+2+2n}.$$

Then $\sigma_n = \sigma_m \iff n \equiv m \mod \ell$ so that

$$Spin^c(N_\ell) = \{\sigma_n;\; n \in \mathbb{Z} \mod \ell\}.$$

Observe that $c_1(\det \sigma_n) \equiv (2n + \ell + 2) \mod \ell$. Following [**109**], for each $n \in \mathbb{Z}$ we define the canonical representative $L_{\sigma_n}$ of $\sigma_n$ to be the complex line bundle $L \to S^2$ uniquely determined by the requirements

$$\deg L \equiv n \mod \ell, \;\; -\frac{1 + \deg L}{\ell} \in [0, 1).$$

We set

$$h(\sigma_n) := -\frac{1 + \deg L_{\sigma_n}}{\ell}.$$

The rational number $h(\sigma_n)$ has a simple geometric interpretation namely, $\exp(-4\pi h(\sigma_n)\mathbf{i})$ is the holonomy along the fibers of $N_\ell \to S^2$ of the flat connections over $\det(\sigma_n)$.

The previous considerations show that a flat connection over $\det(\sigma_n)$ extends to a flat connection over $\det(\hat{\sigma}_n)$ if and only if $2n + 2 + \ell = 0$.

Fix a *spin*$^c$ structure $\sigma_n$ on $N_\ell$ and denote by $A_n$ a smooth flat connection on $\det(\sigma_n)$. (There is only one gauge equivalence class of such flat connections.) Suppose that there exists an asymptotic strongly cylindrical connection $\hat{A}_n$ on $\det(\hat{\sigma}_n) \to \hat{N}_\ell$, with positive decay rate $\mu$ and with anti-selfdual $L^2(\tilde{g}_r)$-curvature $F_{\hat{A}_n}$ ($F^+_{\hat{A}_n} = 0$). (We will see later that if $\ell < 0$ then there exist such connections $\hat{A}_n$.) The connection $\hat{A}_n$ determines an asymptotically cylindrical Dirac operator $\mathfrak{D}_{\hat{A}_n}$ with

$$\vec{\partial}_\infty \mathfrak{D}_{\hat{A}_n} = \mathfrak{D}_{A_n}.$$

The Weitzenböck formula implies that $\ker \mathfrak{D}_{A_n} = 0$ since $F_{A_n} = 0$ and the scalar curvature of $g_r$ is positive. This implies

$$\ker_{ex} \mathfrak{D}^*_{\hat{A}_n} = \ker_{L^2} \mathfrak{D}^*_{\hat{A}_n}$$

so that

$$I_{APS}(\mathfrak{D}_{\hat{A}_n}) = \dim_{\mathbb{C}} \ker_{L^2} \mathfrak{D}_{\hat{A}_n} - \dim_{\mathbb{C}} \ker_{L^2} \mathfrak{D}^*_{\hat{A}_n}.$$

We claim that $\ker_{L^2} \mathfrak{D}_{\hat{A}_n}$ is trivial.

For $T \gg 0$ set

$$\hat{N}_\ell(T) := \hat{N}_\ell \setminus (T, \infty) \times N_\ell.$$

Denote by $t \to \infty$ the longitudinal coordinate along the long neck of $\hat{N}_\ell$, $J := \hat{\mathbf{c}}(dt)$ and for each $T > 1$ set

$$\hat{N}_\ell(T) := \hat{N}_\ell \setminus (T, \infty) \times N_\ell.$$

Let $\hat{\psi} \in \ker_{L^2} \mathfrak{D}_1$. Observe that since $\ker \mathfrak{D}_{A_n} = 0$ we have

$$\|\hat{\psi}|_{\{t\}\times N_\ell}\|_{C^1} = o(1), \ \text{ as } t \to \infty. \tag{4.1.35}$$

Using the Weitzenböck formula (in which $F^+_{\hat{A}_n} = 0$) and the integration by parts formula in Exercise 1.2.2 of Sec. 1.2 we deduce

$$0 = \int_{\hat{N}_\ell(T)} \langle \mathfrak{D}^*_{\hat{A}_n} \mathfrak{D}_{\hat{A}_n} \hat{\psi}, \hat{\psi} \rangle dv(\tilde{g}_r)$$

$$= \int_{\hat{N}_\ell(T)} \left( \left\langle (\hat{\nabla}^{\hat{A}_n})^* \hat{\nabla}^{\hat{A}_n} \hat{\psi}, \hat{\psi} \right\rangle + \frac{s(\tilde{g}_r)}{4} |\hat{\psi}|^2 \right) dv(\tilde{g}_r)$$

$$= \int_{\hat{N}_\ell(T)} \Big( |\nabla^{\hat{A}_n} \hat{\psi}|^2 + \frac{s(\tilde{g}_r)}{4} |\hat{\psi}|^2 \Big) dv(\tilde{g}_r) - \int_{\partial \hat{N}_\ell(T)} \langle \hat{\nabla}^{\hat{A}_n}_{\partial_t} \hat{\psi}, \hat{\psi} \rangle dv(g_r).$$

The estimate (4.1.35) now implies

$$\int_{\hat{N}_\ell(T)} \Big( |\nabla^{\hat{A}_n} \hat{\psi}|^2 + \frac{s(\tilde{g}_r)}{4} |\hat{\psi}|^2 \Big) dv(\tilde{g}_r) = o(1) \text{ as } T \to \infty.$$

Now let $T \to \infty$. Since the scalar curvature of $\tilde{g}_r$ is positive we conclude that $\psi \equiv 0$. Thus

$$I_{APS}(\mathfrak{D}_{\hat{A}_n}) = -\dim \ker_{ex} \mathfrak{D}^*_{\hat{A}_n} = -\dim \ker_{L^2} \mathfrak{D}^*_{\hat{A}_n}.$$

Denote by $\eta_{dir}(\sigma_n, g_r)$ the eta invariant of the Dirac operator $\mathfrak{D}_{A_n}$. Formula (4.1.3) of §4.1.2 implies

$$-\dim \ker_{ex} \mathfrak{D}^*_{\hat{A}_n} = I_{APS}(\mathfrak{D}_1)$$

$$= -\frac{1}{24} \int_{\hat{N}_\ell} \frac{1}{3} p_1(\hat{\nabla}^{\tilde{g}_r}) + \frac{1}{8} \int_{\hat{N}_\ell} c_1(\hat{A}_n) \wedge c_1(\hat{A}_n) - \frac{1}{2} \eta_{dir}(\sigma, g_r).$$

Thus

$$4\eta_{dir}(\sigma, g_r) = -\frac{1}{3} \int_{D_\ell} p_1(\hat{\nabla}^{\tilde{g}_r}) + 8 \dim \ker_{ex} \mathfrak{D}^*_{\hat{A}_n} - \int_{\hat{N}_\ell} c_1(\hat{A}_n) \wedge c_1(\hat{A}_n).$$

Using the equation (4.1.34) we obtain

$$\begin{aligned} \mathbf{F}(\sigma_m, g_r) &:= 4\eta_{dir}(\sigma, g_r) + \eta_{sign}(g_r) \\ &= 8 \dim \ker_{ex} \mathfrak{D}^*_{\hat{A}_n} - \tau(D_\ell) + \int_{\hat{N}_\ell} c_1(\hat{A}_n) \wedge c_1(\hat{A}_n) \end{aligned} \tag{4.1.36}$$

In [**107, 109**] we showed that

$$\mathbf{F}(\sigma_n, g_r) = 4\ell h(\sigma_n)(h(\sigma_n) - 1) + \ell - \operatorname{sign}(\ell). \tag{4.1.37}$$

We deduce

$$8 \dim \ker_{ex} \mathfrak{D}^*_{\hat{A}_n} = 4\ell h(\sigma_n)(h(\sigma_n) - 1) + \ell - \int_{\hat{N}_\ell} c_1(\hat{A}_n) \wedge c_1(\hat{A}_n).$$

Suppose for example $\ell < -1$ and $-1 \leq n < |\ell| - 1$. Then

$$h(\sigma_n) = -\frac{n+1}{\ell}$$

so that

$$4\ell h(\sigma_n)(h(\sigma_n) - 1) = \frac{4(n+1)(n+1+\ell)}{\ell}.$$

To compute the integral term we use the intersection form on $H^2(N, \mathbb{Z})$ induced by the Poincaré duality

$$H^2(D_\ell, \partial D_\ell; \mathbb{Z}) \times H^2(D_\ell, \mathbb{Z}) \to \mathbb{Z}.$$

Then

$$\int_{\hat{N}_\ell} c_1(\hat{A}_n) \wedge c_1(\hat{A}_n) = \frac{(2n+2+\ell)^2}{\ell}$$

since $\det(\hat{\sigma}_n) = \pi^* L_{2n+2+\ell}$. We conclude that

$$\dim \ker_{ex} \mathfrak{D}^*_{\hat{A}_n} = 0.$$

## 4.2. Finite energy monopoles

This very technical section offers a glimpse into the analytical theory of the Seiberg-Witten equations on 4-manifolds with cylindrical ends. To keep the technical details within reasonable limits we will consider only some special, simpler situations required by the topological applications we have in mind. This choice has an academic advantage as well: it offers the reader a quite extensive picture of what to expect relying on a relatively moderate analytical machinery. For an exhaustive presentation of this type of problem in the Yang-Mills context we refer to [**96, 133**].

We tried to keep the presentation as self-contained as possible but, to keep the length of this section within reasonable limits, we had to appeal to certain basic facts about elliptic partial differential equations we did not include in this book. These can be found in [**47, 105**].

**4.2.1. Regularity.** Suppose $\hat{N}$ is an oriented cylindrical 4-manifold with $N := \partial_\infty \hat{N}$. Fix a cylindrical *spin*$^c$ structure $\hat{\tau} = (\hat{\sigma}, \sigma, \varphi)$ on $\hat{N}$ ($\sigma := \partial_\infty \hat{\sigma}$) (see §4.1.1 for precise definitions). Denote by $\hat{\mathbb{S}}_{\hat{\sigma}} = \hat{\mathbb{S}}^+_{\hat{\sigma}} \oplus \hat{\mathbb{S}}^-_{\hat{\sigma}}$ the bundle of complex spinors associated to $\hat{\sigma}$, and by $\mathbb{S}_\sigma$ the bundle of complex spinors associated to $\sigma$. $\hat{\mathbb{S}}_\sigma$ can be equipped with a cylindrical structure such that $\mathbb{S}_\sigma = \partial_\infty \hat{\mathbb{S}}^+_{\hat{\sigma}}$.

We denote by $\hat{\mathcal{C}}_{\hat{\sigma}}$ the configuration space consisting of pairs $\hat{\mathsf{C}} := (\hat{\psi}, \hat{A})$ where $\hat{\psi} \in L^{2,2}_{loc}(\hat{\mathbb{S}}^+_{\hat{\sigma}})$ and $\hat{A}$ is an $L^{2,2}_{loc}$ Hermitian connection of $\det(\hat{\sigma})$. Define $\widehat{\mathfrak{G}}_{\hat{\sigma}}$ as the space of $L^{3,2}_{loc}$-maps $\hat{\gamma} : \hat{N} \to S^1$. For every point $p_0 \in \hat{N}$ we define the subgroup $\widehat{\mathfrak{G}}_{\hat{\sigma}}(p_0) \subset \widehat{\mathfrak{G}}_\sigma$ consisting of maps $\hat{\gamma} : \hat{N} \to S^1$ such that $\hat{\gamma}(p_0) = 1$. (Such gauge transformations are said to be *based at* $p_0$.)

A *finite energy monopole* is a configuration $\hat{\mathsf{C}} = (\hat{\psi}, \hat{A}) \in \hat{\mathcal{C}}$ satisfying the Seiberg-Witten equations

$$\begin{cases} \hat{\mathfrak{D}}_{\hat{A}} \hat{\psi} & = & 0 \\ F^+_{\hat{A}} & = & \hat{\mathbf{c}}^{-1}\big(\frac{1}{2} q(\psi)\big) \end{cases}$$

and the growth condition

$$E(\hat{\mathsf{C}}) := \int_{\hat{N}} \Big( |\nabla^{\hat{A}}\hat{\psi}|^2 + \frac{1}{8}|q(\hat{\psi})|^2 + |F_{\hat{A}}|^2 + \frac{\hat{s}}{4}|\hat{\psi}|^2 \Big) dv(\hat{g}) < \infty.$$

We will denote by $\hat{\mathfrak{Z}}_{\hat{\sigma}}$ the set of finite energy monopoles on $\hat{N}$.

As in the closed case, we will need to use perturbation parameters. In this case they will take the form of closed, *compactly supported* 2-forms $\eta \in \Omega^2(\hat{N})$ of appropriate regularity.

**Proposition 4.2.1.** *Let $\hat{\mathsf{C}} = (\hat{\psi}, \hat{A}) \in \hat{\mathfrak{Z}}_{\hat{\sigma}}$. Then there exists $\hat{\gamma} \in \widehat{\mathfrak{G}}_{\hat{\sigma}}$ such that $\hat{\gamma} \cdot \hat{\mathsf{C}} \in C^\infty$.*

**Proof** The proof relies on the following technical result.

**Lemma 4.2.2.** *Suppose $M$ is a smooth, compact, oriented Riemannian 4-manifold with smooth boundary $\partial M = N$, $\sigma \in Spin^c(M)$ and $\mathsf{C} = (\psi, A)$ is a $L^{2,2}$-monopole corresponding to the spin$^c$-structure $\sigma$. Then there exists a $L^{3,2}_{loc}$-map $f : M \to \mathbb{R}$ such that $e^{\mathbf{i}f} \cdot \mathsf{C}$ is smooth in the interior of $M$.*

We will present the proof of this lemma after we explain why it implies Proposition 4.2.1.

Let $\hat{\mathsf{C}} = (\hat{\psi}, \hat{A})$ be a finite energy monopole on $\hat{N}$. Set $C_n := (n, n+2) \times N$, $n \in \mathbb{Z}_+$. Using Lemma 4.2.2 we can find $L^{3,2}_{loc}$-maps

$$f : \hat{N}_1 = \hat{N} \setminus (1, \infty) \times N \to \mathbb{R}, \quad f_n : C_n \to \mathbb{R}$$

such that

$$e^{\mathbf{i}f} \cdot \hat{\mathsf{C}}|_{\hat{N}_1} \in C^\infty(\hat{N}_1), \quad e^{\mathbf{i}f_n} \hat{\mathsf{C}}|_{C_n} \in C^\infty(C_n), \quad \forall n \in \mathbb{Z}_+.$$

Set $u_0 = f_0 - f$, $u_n := f_n - f_{n-1}$, $\forall n \geq 1$. Observe that $u_n$ is a smooth function on $(n, n+1) \times N$, $\forall n \in \mathbb{Z}_+$ because on this cylinder we have

$$-2\mathbf{i}du_n = e^{\mathbf{i}f_n} \cdot \hat{A} - e^{\mathbf{i}f_{n-1}} \cdot \hat{A} \in C^\infty\big((n, n+1) \times N\big).$$

Fix $0 < \varepsilon \ll \frac{1}{8}$. For each $n \in \mathbb{Z}_+$ define $\varphi_n \in C^\infty_{comp}(\hat{N})$ such that $\varphi_n \equiv u_n$ on $(n + \frac{1}{2} - \varepsilon, n + \frac{1}{2} + \varepsilon) \times N$ and $\varphi_n \equiv 0$ outside $(n + \frac{1}{2} - 2\varepsilon, n + \frac{1}{2} + 2\varepsilon) \times N$. Finally, set

$$h_n : (n - 1/2 - \varepsilon, n + 1/2 + \varepsilon) \times N \to \mathbb{R}, \quad h_n = f_{n-1} + \varphi_n, \quad n \geq 1,$$

and

$$h_0 \equiv f + \varphi_0 \;\text{ on }\; \hat{N} \setminus [1/2, \infty) \times N.$$

Observe that $h_{n-1} \equiv h_n$ on $(n + \frac{1}{2} - \varepsilon, n + \frac{1}{2} + \varepsilon) \times N$ so that the collection $(h_n)$ defines an $L^{3,2}_{loc}$-map

$$h : \hat{N} \to \mathbb{R}.$$

On the other hand, on the cylinder $(n-1/2-\varepsilon, n+1/2+\varepsilon)\times N$ we have

$$e^{\mathbf{i}h}\cdot\hat{\mathsf{C}} = e^{\mathbf{i}\varphi_n}\cdot\left(e^{\mathbf{i}f_{n-1}}\hat{\mathsf{C}}\right)\in e^{\mathbf{i}\varphi_n}\cdot C^\infty\subset C^\infty. \quad \blacksquare$$

**Proof of Lemma 4.2.2** Fix a Hermitian connection $A_0$ on $\det(\sigma)$ which is smooth up to the boundary of $M$ and set $\mathbf{i}a := A - A_0$. The Dirichlet problem

$$\begin{cases} \Delta_M u &= \frac{1}{2}d^*a \text{ in } M \\ u &= 0 \text{ on } \partial M \end{cases}$$

has a unique solution $u\in L^{3,2}(M)$ (see [**47**, Chap. 8]). Set $\gamma := e^{\mathbf{i}u}$ and $(\phi, B) := \gamma\cdot(\psi, A)$. If $\mathbf{i}b := B - A_0$ then

$$\mathbf{i}b = \mathbf{i}a - 2\mathbf{i}du$$

so that $d^*b = \mathbf{i}(d^*a - 2\Delta_M u) = 0$. The Seiberg-Witten equations for $(\phi, B)$ can be rewritten as an *elliptic* system

$$\mathfrak{D}_{A_0}\phi = -\frac{1}{2}\mathbf{c}(\mathbf{i}b)\phi, \tag{4.2.1a}$$

$$(d^+\oplus d^*)b = \left(\frac{1}{2}q(\phi) - F^+_{A_0}\right)\oplus 0. \tag{4.2.1b}$$

An elliptic bootstrap, identical to the one in the proof of Proposition 2.1.11 of §2.1.2 concludes the proof of Lemma 4.2.2. $\blacksquare$

Proposition 4.2.1 shows that there is no loss of generality by working only with *smooth* finite energy monopoles. Observe also that nowhere in the proof have we relied on the growth condition $E(\hat{\mathsf{C}}) < \infty$ to establish regularity modulo $\widehat{\mathfrak{G}}$. The growth condition affects only the asymptotic behavior. In particular, the considerations in 2.4.1 show that

$$\hat{\mathfrak{Z}}_{\hat{\sigma}}\neq\emptyset \Longrightarrow \textit{there exist three-dimensional } \sigma\textit{-monopoles on } N$$

In the next subsection we will have a closer look at three-dimensional monopoles.

**4.2.2. Three-dimensional monopoles.** Consider a closed, compact, oriented Riemannian manifold $(N, g)$ and a *spin*$^c$ structure $\sigma\in \mathit{Spin}^c(N)$. We want to define a functional set-up which closely follows the relationship between the four- and three-dimensional theory.

Define a configuration space $\mathcal{C}_\sigma$ consisting of pairs $(\psi, A)$ where $\psi\in L^{2,2}(\mathbb{S}_\sigma)$ and $A$ is an $L^{2,2}$-connection on $\det(\sigma)$. (Often we will need to consider configurations of different regularity, which will be indicated by Sobolev superscripts attached to $\mathcal{C}_\sigma$. E.g., $\mathcal{C}^r_\sigma$ refers to configurations in $L^{r,2}$. )

Denote by $\mathfrak{G}_\sigma$ the group of $L^{3,2}$-maps $\gamma : N \to S^1$. Observe that since $\dim N = 3$ the Sobolev-Morrey embedding theorem implies $L^{3,2}$ embeds in a Hölder space and, as in §2.1.2, we can conclude that $\mathfrak{G}_\sigma$ is a Hilbert-Lie group with commutative Lie algebra $T_1\mathfrak{G}_\sigma := L^{3,2}(N, \mathbf{i}\mathbb{R})$. For every $* \in N$ we set

$$\mathfrak{G}_\sigma(*) := \{\gamma \in \mathfrak{G}_\sigma;\ \ \gamma(*) = 1\}.$$

$\mathfrak{G}_\sigma(*)$ will be called the group of gauge transformations based at $*$. Observe that $\mathfrak{G}_\sigma(*)$ acts *freely* on $\mathcal{C}_\sigma$. Now set $\mathcal{B}_\sigma : \mathcal{C}_\sigma/\mathfrak{G}_\sigma$ and $\mathcal{B}_\sigma(*) := \mathcal{C}_\sigma(*)/\mathfrak{G}_\sigma(*)$. As in §2.2.2 we can equip $\mathcal{B}_\sigma$ and $\mathcal{B}_\sigma(*)$ with natural Sobolev metrics.

For every $\mathsf{C} \in \mathcal{C}_\sigma$ we denote by

$$\mathfrak{L}_\mathsf{C} : T_1\mathfrak{G}_\sigma \to T_\mathsf{C}\mathcal{C}_\sigma$$

the infinitesimal action at $\mathsf{C}$

$$\mathfrak{L}_\mathsf{C}(\mathbf{i}f) := \frac{d}{dt}|_{t=0}\, e^{\mathbf{i}tf}\cdot \mathsf{C} = (\mathbf{i}f\psi, A - 2\mathbf{i}df).$$

Its formal $(L^2)$ adjoint is

$$T_\mathsf{C}\mathcal{C}_\sigma \ni \dot{\mathsf{C}} \mapsto \mathfrak{L}_\mathsf{C}^*\dot{\mathsf{C}} = \mathfrak{L}_\mathsf{C}^*(\dot\psi, \mathbf{i}\dot a) = -2\mathbf{i}d^*\dot a - \mathbf{i}\mathfrak{Im}\,\langle\psi, \dot\psi\rangle.$$

As in the four-dimensional case, we can identify $\ker\mathfrak{L}_\mathsf{C}$ with the Lie algebra of the stabilizer $\mathbf{Stab}\,(\mathsf{C})$ with respect to the $\mathfrak{G}_\sigma$ action.

Since $\mathcal{C}_\sigma$ is an affine space we can identify the tangent space $T_\mathsf{C}\mathcal{C}_\sigma$ with $\mathcal{C}_\sigma$ via the map

$$\dot{\mathsf{C}} \mapsto \mathsf{C} + \dot{\mathsf{C}}.$$

Define the *slice* $\mathcal{S}_\mathsf{C} \subset T_\mathsf{C}\mathcal{C}_\sigma \cong \mathcal{C}_\sigma$ at $\mathsf{C}$ by

$$\mathcal{S}_\mathsf{C} := \ker\mathfrak{L}_\mathsf{C}^* \cap L^{2,2}.$$

More generally, we set $\mathcal{S}_\mathsf{C}^r := \ker\mathfrak{L}_\mathsf{C}^* \cap L^{r,2}$. The slice at $\mathsf{C}$ is equipped with a natural $\mathbf{Stab}\,(\mathsf{C})$-action and, exactly as in the four-dimensional case (see §2.2.2), we have the following result.

**Proposition 4.2.3.** *There exists a small* $\mathbf{Stab}\,(\mathsf{C})$*-invariant neighborhood* $U_\mathsf{C}$ *of* $\mathsf{C} \in \mathcal{S}_\mathsf{C}$ *such that every orbit of* $\mathfrak{G}_\sigma$ *which intersects* $U_\mathsf{C}$ *does so trasversally, along a single* $\mathbf{Stab}\,(\mathsf{C})$*-orbit. In particular, every* $\mathfrak{G}_\sigma(*)$*-orbit intersects* $U_\mathsf{C}$ *transversely in at most one point.*

From the above proposition we conclude that $\mathcal{B}_\sigma(*)$ is a Hilbert manifold while $\mathcal{B}_\sigma$ is smooth away from the reducible orbits.

A *three-dimensional monopole* is a configuration $\mathsf{C} = (\psi, A) \in \mathcal{C}_\sigma$ satisfying the Seiberg-Witten equations

$$\left\{\begin{array}{rcl} \mathfrak{D}_A\psi & = & 0 \\ \frac{1}{2}q(\psi) & = & \mathbf{c}(*F_A) \end{array}\right. .$$

Denote by $\mathfrak{Z}_\sigma \subset \mathcal{C}_\sigma$ the set of three-dimensional monopoles. Exactly as in the four-dimensional case we conclude that each three-monopole is $\mathfrak{G}_\sigma$-equivalent to a smooth one and $\mathfrak{M}_\sigma := \mathfrak{Z}_\sigma/\mathfrak{G}_\sigma$ is a compact subset of $\mathcal{B}_\sigma$.

**Remark 4.2.4.** Arguing exactly as in the proof of Lemma 2.2.3 one can prove that if $(\psi, A)$ is a 3-monopole then

$$\sup_{x\in N} |\psi(x)|^2 \leq 2 \sup_{x\in N} |s(x)|$$

where $s$ is the scalar curvature of $N$. We have already used this fact in the proof of the Thom conjecture in §2.4.2.

To describe the local structure of $\mathfrak{M}_\sigma$ we need to linearize the Seiberg-Witten equations along a slice. The monopoles are zeros of the smooth map

$$SW : \mathcal{C}_\sigma \rightarrow \mathcal{C}^1_\sigma \cong T_{\mathsf{C}}\mathcal{C}_\sigma, \;\; (\psi, A) \mapsto (\mathfrak{D}_A\psi, q(\psi) - \mathbf{c}(*F_A))$$

As explained in §2.4.1, the map $SW$ is the formal (i.e. $L^2$) gradient of the energy functional

$$\mathcal{E}_\sigma : \mathcal{C}_\sigma \rightarrow \mathbb{R},$$

$$\mathcal{E}_\sigma(\psi, A) = \frac{1}{2}\int_N (A - A_0)\wedge(F_A + F_{A_0}) + \frac{1}{2}\int_N \langle \mathfrak{D}_A\psi, \psi\rangle dv_g$$

where $A_0$ is a fixed, smooth reference Hermitian connection on $\det(\sigma)$. Since

$$\frac{d}{dt}|_{t=0}\, \mathcal{E}_\sigma(e^{t\mathbf{i}f}\cdot \mathsf{C}) = 0$$

we deduce

$$D_{\mathsf{C}}\mathcal{E}_\sigma(\mathfrak{L}_{\mathsf{C}}\mathbf{i}f) = 0 \Longleftrightarrow \Big\langle\, SW(\mathsf{C}),\, \mathfrak{L}_{\mathsf{C}}(\mathbf{i}f)\,\Big\rangle_{L^2} = 0, \;\; \forall \mathbf{i}f \in T_1\mathfrak{G}_\sigma$$

so that

$$SW(\mathsf{C}) \in \mathcal{S}^1_{\mathsf{C}}, \;\; \forall \mathsf{C} \in \mathcal{C}_\sigma.$$

Observe also that for every $\gamma \in \mathfrak{G}_\sigma$ we have

$$SW(\gamma\cdot(\psi, A)) = (\gamma\mathfrak{D}_A\psi, q(\psi) - \mathbf{c}(*F_A))$$

so that

$$\|SW(\gamma\cdot\mathsf{C})\|_{L^2} = \|SW(\mathsf{C})\|_{L^2}.$$

Hence $\mathsf{C} \rightarrow \|SW(\mathsf{C})\|_{L^2}$ is a well defined continuous function on $\mathcal{B}_\sigma$ which we denote by $\mathfrak{f}$. We can regard $SW(\mathsf{C})$ as an $S^1$-invariant tangent vector field on $\mathcal{B}_\sigma(*)$ or as a genuine tangent vector field on $\mathcal{B}_{\sigma,irr}$. For $\dot{\mathsf{C}} \in T_{\mathsf{C}}\mathcal{C}_\sigma$ and $\mathbf{i}f \in T_1\mathfrak{G}_\sigma$ define

$$\mathcal{T}_{\mathsf{C}}\begin{bmatrix} \dot{\mathsf{C}} \\ \mathbf{i}f \end{bmatrix} = \begin{bmatrix} \underline{SW} & -\frac{1}{2}\mathfrak{L}_{\mathsf{C}} \\ & \\ -\frac{1}{2}\mathfrak{L}^*_{\mathsf{C}} & 0 \end{bmatrix}\begin{bmatrix} \dot{\mathsf{C}} \\ & \\ \mathbf{i}f \end{bmatrix}$$

$$:= \begin{bmatrix} \frac{d}{dt}|_{t=0}\, SW(\mathsf{C}+t\dot{\mathsf{C}}) - \frac{1}{2}\mathfrak{L}_{\mathsf{C}}(\mathbf{i}f) \\ \\ -\frac{1}{2}\mathfrak{L}^*_{\mathsf{C}}\dot{\mathsf{C}} \end{bmatrix} \in \overline{T_{\mathsf{C}}\mathcal{C}_\sigma}^{L^2} \oplus L^2(N, \mathbf{i}\mathbb{R}).$$

More explicitly, if $\mathsf{C} := (\psi, A)$ and $\dot{\mathsf{C}} = (\dot{\psi}, \mathbf{i}\dot{a})$ then

$$(4.2.2)\quad \mathcal{T}_{\mathsf{C}} \begin{bmatrix} \dot{\psi} \\ \mathbf{i}\dot{a} \\ \mathbf{i}f \end{bmatrix} = \begin{bmatrix} \mathfrak{D}_A & 0 & 0 \\ 0 & -*d & d \\ 0 & d^* & 0 \end{bmatrix} \cdot \begin{bmatrix} \dot{\psi} \\ \mathbf{i}\dot{a} \\ \mathbf{i}f \end{bmatrix} + \begin{bmatrix} \frac{1}{2}\mathbf{c}(\mathbf{i}\dot{a})\psi - \frac{\mathbf{i}}{2}f\psi \\ \frac{1}{2}\dot{q}(\psi, \dot{\psi}) \\ \frac{\mathbf{i}}{2}\mathfrak{Im}\,\langle \psi, \dot{\psi}\rangle \end{bmatrix}.$$

Denote by $\mathcal{T}^0_{\mathsf{C}}$ the first operator on the right-hand side of (4.2.2) and set $\mathcal{P}_{\mathsf{C}} := \mathcal{T}_{\mathsf{C}} - \mathcal{T}^0_{\mathsf{C}}$. Notice that $\mathcal{P}_{\mathsf{C}}$ is a zeroth order operator while $\mathcal{T}_{\mathsf{C}}$ is a first order, formally selfadjoint elliptic operator.

**Exercise 4.2.1.** Prove directly that $\mathcal{T}_{\mathsf{C}}$ is formally selfadjoint.

Suppose $\mathsf{C}_0$ is a 3-monopole. To understand the local structure of $\mathfrak{M}_\sigma$ near $\mathsf{C}_0$ it suffices to understand the structure of the critical set of the restriction of $\mathcal{E}_\sigma$ to a small neighborhood $U$ of $\mathsf{C}_0 \in \mathcal{S}_{\mathsf{C}_0}$. For every $\mathsf{C} \in \mathcal{C}_\sigma$ we denote by $\Pi_{\mathsf{C}}$ the $L^2$-orthogonal projection

$$T_{\mathsf{C}}\mathcal{C}^0_\sigma \to \mathcal{S}^0_{\mathsf{C}_0}.$$

Since $T_{\mathsf{C}}\mathcal{C}^0_\sigma$ is independent of $\mathsf{C}$, $T_{\mathsf{C}}\mathcal{C}^0_\sigma \cong L^2(\mathbb{S} \oplus \mathbf{i}T^*N)$, we can write $\Pi$ instead of $\Pi_{\mathsf{C}}$.

**Exercise 4.2.2.** Show that $\Pi T_{\mathsf{C}}\mathcal{C}^r_\sigma \subset \mathcal{S}^r_{\mathsf{C}_0}$, $\forall r \geq 0$.

**Lemma 4.2.5.** *There exist a* **Stab** $(\mathsf{C}_0)$*-invariant neighborhood* $U = U_{\mathsf{C}_0}$ *of* $\mathsf{C}_0 \in \mathcal{S}_{\mathsf{C}_0}$ *and a constant* $\lambda > 0$ *such that*

$$\frac{1}{\lambda}\|SW(\mathsf{C})\|_{L^2} \leq \|\Pi SW(\mathsf{C})\|_{L^2} \leq \lambda\|SW(\mathsf{C})\|_{L^2},\ \ \forall \mathsf{C} \in U.$$

It is worth emphasizing the main point of the above result. Roughly speaking, it says that, for $\mathsf{C}$ sufficiently close to $\mathsf{C}_0$, the component of $SW(\mathsf{C})$ orthogonal to $\mathcal{S}_{\mathsf{C}_0}$ is small compared to the component along $\mathcal{S}_{\mathsf{C}_0}$. In particular, if $\mathsf{C} \in \mathcal{S}_{\mathsf{C}_0}$ is close to $\mathsf{C}_0$ then $SW(\mathsf{C})$ vanishes if and only if its component along $\mathcal{S}_{\mathsf{C}_0}$ vanishes.

**Proof** Observe that we always have

$$\|\Pi SW(\mathsf{C})\|_{L^2} \leq \|SW(\mathsf{C})\|_{L^2}$$

so it suffices to find a neighborhood $U$ of $\mathsf{C}_0 \in \mathcal{S}_{\mathsf{C}_0}$ and $\lambda > 1$ such that

$$\|SW(\mathsf{C})\|_{L^2} \leq \lambda\|\Pi SW(\mathsf{C})\|_{L^2},\ \ \forall \mathsf{C} \in U.$$

We will prove a slightly more general result. More precisely, we will show that there exists a neighborhood $U$ of $\mathsf{C}_0 \in \mathcal{S}_{\mathsf{C}_0}$ such that for any $\mathsf{C} \in U$ and any $\Psi \in \mathcal{S}_{\mathsf{C}}$ we have the equality

$$\|\Psi\|_{L^2} \leq \lambda \|\Pi\Psi\|_{L^2}.$$

Lemma 4.2.5 follows by setting $\Psi := SW(\mathsf{C})$ in the above inequality.

We argue by contradiction. Suppose there exist sequences $\mathsf{C}_n \in \mathcal{S}_{\mathsf{C}_0}$ and $\Psi_n \in \mathcal{S}_{\mathsf{C}_n}$such that

$$\mathsf{C}_n \stackrel{L^{2,2}}{\longrightarrow} \mathsf{C}_0, \;\; \|\Psi_n\|_{L^2} = 1, \;\; \|\Pi\Psi_m\|_{L^2} < \frac{1}{n}.$$

Set $\Upsilon_n := \Pi\Psi_n$ and $\Xi_n := (1-\Pi)\Psi_n$. Then

$$1 \geq \|\Xi_n\| > \Big(1 - \frac{1}{n}\Big). \tag{4.2.3}$$

Now observe that $\hat{\Xi}_n \perp \mathcal{S}_{\mathsf{C}_0}$ so there exists a unique $\mathbf{i}f_n \in (\ker \mathfrak{L}_{\mathsf{C}_0})^{\perp} \cong (T_1\mathbf{Stab}(\mathsf{C}_0))^{\perp} \subset T_1\mathfrak{G}_\sigma$ such that

$$\mathfrak{L}_{\mathsf{C}_0}(\mathbf{i}f_n) = \hat{\Xi}_n,$$

$$\mathfrak{L}_{\mathsf{C}_n}(\mathbf{i}f) = \mathfrak{L}_{\mathsf{C}_0} + R_n$$

where $R_n$ is a zeroth order p.d.o. (bundle morphism) such that $\|R_n\|_{2,2} = o(1)$ as $n \to \infty$. The condition

$$\mathfrak{L}^*_{\mathsf{C}_n}\psi_n = 0$$

can be rewritten as

$$0 = (\mathfrak{L}^*_{\mathsf{C}_0} + R^*_n)(\Upsilon_n + \Xi_n) = \mathfrak{L}^*_{\mathsf{C}_0}\Xi_n + R^*_n\Xi_n = \mathfrak{L}^*_{\mathsf{C}_0}\mathfrak{L}_{\mathsf{C}_0}(\mathbf{i}f_n) + R^*_n\Xi_n.$$

Thus $\mathbf{i}f_n \perp \ker \mathfrak{L}^*_{\mathsf{C}_0}\mathfrak{L}_{\mathsf{C}_0}$ and

$$\|\mathfrak{L}^*_{\mathsf{C}_0}\mathfrak{L}_{\mathsf{C}_0}(\mathbf{i}f_n)\|_{L^p} = \|R^*_n\Xi_n\|_{L^p}, \forall p \in (1,\infty).$$

Using the Sobolev inequalities we deduce that there exists $C > 0$ such that

$$\|R_n\|_{L^\infty} \leq C\|R_n\|_{2,2}.$$

Hence there exists $C > 0$ such that

$$\|R^*_n\Xi_n\|_{L^2} \leq C_q\|R_n\|_{2,2}\|\Xi_n\|_{L^2}, \;\; \forall n.$$

Using the elliptic estimate of Theorem 1.2.18 (v) for the generalized Laplacian $\mathfrak{L}^*_{\mathsf{C}_0}\mathfrak{L}_{\mathsf{C}_0}$ we deduce that there exists a constant $C > 1$ such that

$$\|f_n\|_{2,2} \leq C\|R_n\Xi_n\|_{L^2} = o(1) \;\; \text{as } n \to \infty.$$

This implies $f_n \to 0$ in $L^{2,2}$ and since $\mathfrak{L}_{\mathsf{C}_0}(\mathbf{i}f_n) = \Xi_n$ we deduce $\Xi_n \to 0$ in $L^2$. This contradicts the inequality (4.2.3). Lemma 4.2.5 is proved. ■

Fix a neighborhood $U$ of $\mathsf{C}_0 \in \mathcal{S}_{\mathsf{C}_0}$ as in the above lemma. The critical points of $\mathcal{E}_\sigma|_U$ are determined from the equation

$$\Pi SW(\mathsf{C}) = 0, \quad \mathsf{C} \in U.$$

Equivalently, this means there exists a unique $\mathbf{i}f \in T_1\mathfrak{G}_\sigma$ such that

$$\mathbf{i}f \perp \ker \mathfrak{L}_{\mathsf{C}_0}, \quad SW(\mathsf{C}) + \mathfrak{L}_{\mathsf{C}_0}(\mathbf{i}f) = 0.$$

Thus, the problem of understanding the structure of $\mathfrak{M}_\sigma$ near $\mathsf{C}_0$ boils down to understanding the local structure of the equation

$$SW(\mathsf{C}_0 + \dot{\mathsf{C}}) = 0 \tag{4.2.4}$$

where $\mathfrak{L}^*_{\mathsf{C}_0}\dot{\mathsf{C}} = 0$ and $\|\dot{\mathsf{C}}\|_{2,2}$ is very small.

Set

$$H^0_{\mathsf{C}_0} := \ker \mathfrak{L}_{\mathsf{C}_0}, \quad H^1_{\mathsf{C}_0} := \Big\{\dot{\mathsf{C}} \in \mathcal{C}_{\mathsf{C}};\ \underline{SW}(\dot{\mathsf{C}}) = 0, \ \ \mathfrak{L}^*_{\mathsf{C}_0}\dot{\mathsf{C}} = 0\Big\}$$

and denote by $\Pi_1 : \mathcal{S}_{\mathsf{C}_0} \to H^1_{\mathsf{C}_0}$ the $L^2$-orthogonal projection. Observe that

$$\ker \mathcal{T}_{\mathsf{C}_0} = H^1_{\mathsf{C}_0} \oplus H^0_{\mathsf{C}_0}.$$

For every $r > 0$ we set

$$B_{\mathsf{C}}(r) := \{\dot{\mathsf{C}} \in H^1_{\mathsf{C}};\ \ \|\dot{\mathsf{C}}\|_{L^2} < r\}.$$

The equation (4.2.4) is equivalent to the pair of equations

$$(1 - \Pi_1)\Big(SW(\mathsf{C}_0 + \dot{\mathsf{C}})\Big) = 0, \quad \dot{\mathsf{C}} \in \mathcal{S}_{\mathsf{C}_0}, \quad \|\dot{\mathsf{C}}\|_{2,2} \leq \varepsilon, \tag{$\dagger_\varepsilon$}$$

$$\Pi_1\Big(SW(\mathsf{C}_0 + \dot{\mathsf{C}})\Big) = 0, \quad \dot{\mathsf{C}} \in \mathcal{S}_{\mathsf{C}_0}, \quad \|\dot{\mathsf{C}}\|_{2,2} \leq \varepsilon. \tag{$\dagger\dagger_\varepsilon$}$$

The local structure of $(\dagger_\varepsilon)$ can be easily analyzed using the implicit function theorem. Our next result states that the solution set of $(\dagger_\varepsilon)$ can be represented as the graph of a $\mathbf{Stab}(\mathsf{C}_0)$-equivariant map

$$\Phi_1 : H^1_{\mathsf{C}_0} \to \ker \Pi_1$$

tangent to $H^1_{\mathsf{C}_0}$ at 0.

**Proposition 4.2.6.** *Suppose* $\mathsf{C}_0$ *is a smooth* 3*-monopole. There exist* $r_0 = r_0(\mathsf{C}_0) > 0$, $\varepsilon = \varepsilon(\mathsf{C}_0)$, $\nu = \nu(\mathsf{C}_0) > 0$ *and a smooth* $\mathbf{Stab}(\mathsf{C})$*-equivariant map*

$$\Phi_1 : B_{\mathsf{C}_0}(r_0) \to \ker(1 - \Pi_1)\mathcal{S}_{\mathsf{C}_0}$$

*satisfying the following requirements.*

*(i)* $\Phi_1(0) = 0$.

*(ii) Any solution* $\dot{\mathsf{C}}'$ *of* $(\dagger_\varepsilon)$ *decomposes as*

$$\dot{\mathsf{C}}' = \dot{\mathsf{C}} \oplus \Phi_1(\dot{\mathsf{C}})$$

*where* $\dot{\mathsf{C}} = \Pi_1 \dot{\mathsf{C}}' \in B_{\mathsf{C}_0}(r_0)$. *In particular,*

$$(\mathbf{1} - \Pi_1)\Big( SW\big( \mathsf{C} + \dot{\mathsf{C}} + \Phi_1(\dot{\mathsf{C}})\big) + \mathfrak{L}_{\mathsf{C}}\Phi_0(\dot{\mathsf{C}})\Big) = 0,$$

$\forall \dot{\mathsf{C}} \in B_{\mathsf{C}}(r)$.

*(iii)* $\|\Phi_1(\dot{\mathsf{C}})\|_{2,2} \leq \nu \|\dot{\mathsf{C}}\|^2$, $\|D_{\dot{\mathsf{C}}}\Phi_1(v)\|_{2,2} \leq C\|v\| \cdot \|\dot{\mathsf{C}}\|$, $\forall\, v, \dot{\mathsf{C}} \in H^1_{\mathsf{C}_0}$. *(*$H^1_{\mathsf{C}_0}$ *is a finite-dimensional space and thus all norms on it are equivalent.)*

The proof is a consequence of the implicit function theorem applied to the nonlinear equation

$$F(\dot{\mathsf{C}}) = 0$$

where $F$ is the $\mathbf{Stab}(\mathsf{C}_0)$-equivariant map

$$F : \mathcal{S}_{\mathsf{C}_0} \to (1 - \Pi_1)\mathcal{S}^1_{\mathsf{C}_0}, \;\; \dot{\mathsf{C}} \mapsto (1 - \Pi_1)\Pi SW(\mathsf{C}_0 + \dot{\mathsf{C}}).$$

The linearization of this map at $\dot{\mathsf{C}} = 0$ is $(1 - \Pi_1)\underline{SW}_{\mathsf{C}_0}$, which is onto and has kernel $H^1_{\mathsf{C}_0}$.

Set

$$Q_{\mathsf{C}_0} : B_{\mathsf{C}_0}(r_0) \to H^1_{\mathsf{C}_0}, \;\; \dot{\mathsf{C}} \mapsto \Pi_1 SW(\mathsf{C}_0 + \dot{\mathsf{C}} + \Phi_1(\dot{\mathsf{C}})).$$

$Q_{\mathsf{C}_0}$ is called the *Kuranishi map* at $\mathsf{C}_0$. It is a $\mathbf{Stab}(\mathsf{C}_0)$-equivariant map and the above discussion shows that $Q^{-1}_{\mathsf{C}_0}/\mathbf{Stab}(\mathsf{C}_0)$ is homeomorphic to a neighborhood of $\mathsf{C}_0$ in $\mathfrak{M}_\sigma$.

**Definition 4.2.7.** A 3-monopole $\mathsf{C}_0$ is called *regular* if $Q_{\mathsf{C}_0} \equiv 0$.

**Example 4.2.8.** Suppose $\mathsf{C}_0 = (\Psi_0, A_0)$ is a *smooth* reducible 3-monopole, i.e. $\psi_0 \equiv 0$. Then

$$\mathcal{S}_{\mathsf{C}_0} = \Big\{ \dot{\phi} \oplus \mathbf{i}\dot{b} \in L^{2,2}(\mathbb{S}_\sigma \oplus \mathbf{i}T^*N);\;\; d^*\dot{b} = 0 \Big\}$$

and

$$\mathcal{T}_{\mathsf{C}_0} = \mathcal{T}^0_{\mathsf{C}_0} = \mathfrak{D}_{A_0} \oplus \mathbf{SIGN}.$$

Thus

$$H^1_{\mathsf{C}_0} \cong \ker \mathfrak{D}\!\!\!/_{A_0} \oplus \mathbf{i}\mathbf{H}^1(N, g), \;\; H^0_{\mathsf{C}_0} \cong \mathbf{i}\mathbf{H}^0(N, g) \cong \mathbf{i}\mathbb{R}.$$

Fix $(\dot{\psi}, \mathbf{i}\dot{a}) \in B_{\mathsf{C}_0}(r_0)$. Then $(\dot{\phi}, \mathbf{i}\dot{b}) := \Phi_1(\dot{\psi}, \mathbf{i}\dot{a})$ is the solution of the equation

$$(\dot{\phi}, \mathbf{i}\dot{b}) \in (1 - \Pi_1)\mathcal{S}_{\mathsf{C}_0},$$
$$(1 - \Pi_1)\Big(\mathfrak{D}_{A_0 + \mathbf{i}\dot{a} + \mathbf{i}\dot{b}}(\dot{\psi} + \dot{\phi}), *F_{A_0 + \mathbf{i}\dot{a} + \mathbf{i}\dot{b}} - \tfrac{1}{2}q(\dot{\psi} + \dot{\phi})\Big) = 0$$

or equivalently,

$$(4.2.5)\qquad \begin{aligned} &(1-\Pi_1')\Big(\mathfrak{D}_{A_0+\mathbf{i}\dot{a}+\mathbf{i}\dot{b}}\dot{\phi}+\frac{1}{2}\mathbf{c}(\mathbf{i}\dot{a}+\mathbf{i}\dot{b})\dot{\psi})\Big)=0,\\ &(1-\Pi_1'')\Big(\mathbf{i}*d\dot{b}-\frac{1}{2}q(\dot{\psi}+\dot{\phi})\Big)=0 \end{aligned}$$

where $\Pi_1'$ denotes the orthogonal projection onto $\ker\mathfrak{D}_{A_0}$ and $\Pi_1''$ denotes the orthogonal projection onto $\mathbf{H}^1(N,g)$.

Suppose now that $\ker\mathfrak{D}_{A_0}=0$. Then $\Pi_1'\equiv 0$, $\dot{\psi}\equiv 0$ and thus (4.2.5) is equivalent to

$$(4.2.6)\qquad \mathfrak{D}_{A_0+\mathbf{i}\dot{a}+\mathbf{i}\dot{b}}\dot{\phi}=0,\ \ (1-\Pi_1'')\Big(\mathbf{i}*d\dot{b}-q(\dot{\phi})\Big)=0.$$

The map $\Phi_1$ of Proposition 4.2.6 is described by a pair of maps on

$$\dot{b}=\dot{b}(\dot{a}),\ \ \dot{\phi}=\dot{\phi}(\dot{a}),\ \ \dot{a}\in\mathbf{H}^1(N,g),\ \|\dot{a}\|_{L^2}\le r_0,\ \ \|b\|_{2,2}\le\nu\|a\|_2^2.$$

By making $r_0$ even smaller we can assume $\mathfrak{D}_{A_0+\mathbf{i}\dot{a}+\mathbf{i}\dot{b}(\dot{a})}$ is invertible, being very close to the invertible operator $\mathfrak{D}_{A_0}$. This shows that $\dot{\phi}\equiv 0$ and the second equation of (4.2.6) implies $\dot{b}\equiv 0$. Thus $\Phi_1\equiv 0$.

To compute the Kuranishi map at $\mathsf{C}_0$ we need to compute

$$\Pi_1''(*F_{A_0+\mathbf{i}\dot{a}}),\ \ \dot{a}\in\mathbf{H}^1(N,g).$$

Now observe that since $\mathsf{C}_0$ is reducible we have $F_{A_0}=0$. Thus $*F_{A_0+\mathbf{i}\dot{a}}=\mathbf{i}*d\dot{a}$, which clearly has trivial projection on the space of harmonic 1-forms. We have thus shown that if $A_0$ is a flat connection on $\det(\sigma)$ such that $\ker\mathfrak{D}_{A_0}=0$ then $(0,A_0)$ is a regular, reducible monopole.

The stabilizer of $\mathsf{C}_0$ is $S^1$ which acts trivially on $H^1_{\mathsf{C}_0}=\mathbf{i}\mathbf{H}^1(N,g)$ so that there exists an open neighborhood of $\mathsf{C}_0$ in $\mathfrak{M}_\sigma$ homeomorphic to an open ball in $\mathbb{R}^{b_1(N)}$ and consisting only of reducible monopoles.

**Definition 4.2.9.** A pair

$$(\sigma,g)=(spin^c\text{ structure on }N,\text{ Riemannian metric on }N)$$

is called *good* if all irreducible $(\sigma,g)$-monopoles are regular and for any flat connection $A$ on $\det(\sigma)$ the operator $\mathfrak{D}_A$ is invertible.

The discussion in the above example has the following consequence.

**Proposition 4.2.10.** *If $g$ is a positive scalar curvature metric on $N$ then $(\sigma,g)$ is good for every $\sigma\in Spin^c(N)$. Moreover, $\mathfrak{M}_\sigma$ is either empty or it is a compact smooth manifold diffeomorphic to a $b_1(N)$-dimensional torus consisting only of regular reducible monopoles.*

**Remark 4.2.11.** Suppose $(\sigma,g)$ is a good pair and $\mathsf{C}_0=(\psi_0,A_0)$ is a smooth monopole. If $\mathsf{C}_0$ is reducible then $H^1_{\mathsf{C}_0}\cong H^1(N,\mathbb{R})$ and the action of $\mathbf{Stab}\,(\mathsf{C}_0)$ on $H^1_{\mathsf{C}_0}$ is trivial. This proves that $T_{\mathsf{C}}\mathfrak{M}_\sigma\cong H^1_{\mathsf{C}}$, $\forall\mathsf{C}\in\mathfrak{M}_\sigma$.

For each smooth monopole $\mathsf{C}$ and $0 < \kappa \ll 1$ we define the *Kuranishi neighborhood* of $\mathsf{C}$

$$U_{\mathsf{C}}(\kappa) := \{\dot{\mathsf{C}} \in \mathcal{S}_{\mathsf{C}};\ \ \|\dot{\mathsf{C}}\|_{2,2} < \min(\kappa, \varepsilon(\mathsf{C}))\}$$

where $\varepsilon(\mathsf{C})$ is determined as in Proposition 4.2.6. After we factor out the action of $\mathbf{Stab}(\mathsf{C})$ it determines an open neighborhood of $\mathsf{C}$ in $\mathcal{B}_\sigma$.

**A word about notation** *When no serious confusion is possible, we will continue to denote by $U_{\mathsf{C}_0}(\kappa)$ the neighborhood of $[\mathsf{C}_0]$ in $\mathcal{B}$ determined by $U_{\mathsf{C}_0} \subset \mathcal{S}_{\mathsf{C}_0}$. For example, the statement $\mathsf{C} \in U_{\mathsf{C}_0}(\kappa)$ means $\mathsf{C} - \mathsf{C}_0 \in \mathcal{S}_{\mathsf{C}_0}$ and $\|\mathsf{C} - \mathsf{C}_0\|_{2,2} < \kappa$ while the statement $[\mathsf{C}] \in U_{\mathsf{C}_0}$ provides information only about the gauge equivalence class of $\mathsf{C}$ and not $\mathsf{C}$ itself.*

The family

$$\Big\{U_{\mathsf{C}}(\kappa);\ \ [\mathsf{C}] \in \mathfrak{M}_\sigma\Big\}$$

is then an open cover of the compact subset $\mathfrak{M}_\sigma \subset \mathcal{B}_\sigma$. We can extract a finite subcover

$$U_{\mathsf{C}_1}(\kappa), \cdots, U_{\mathsf{C}_m}(\kappa)$$

and we set

$$\kappa_0 := \min\{\kappa(\mathsf{C}_1), \cdots, \kappa(\mathsf{C}_m)\},$$

$$U_\kappa := \bigcup_{i=1}^{m} U_{\mathsf{C}_i}(\kappa),\ \ \forall \kappa < \kappa_0.$$

$U_\kappa$ is an open neighborhood of $\mathfrak{M}_\sigma$ in $\mathcal{B}_\sigma$ called a *Kuranishi neighborhood* of $\mathfrak{M}_\sigma$. Observe that for every $\mathsf{C} \in U_\kappa$

$$\mathrm{dist}_{2,2}([\mathsf{C}], \mathfrak{M}_\sigma) \le \kappa.$$

**4.2.3. Asymptotic behavior. Part I.** Consider a semi-infinite cylinder

$$\hat{N} := (\mathbb{R}_+ \times N, dt^2 + g)$$

and $\sigma$ a *spin*$^c$ structure on $N$. We will denote by $\hat{\sigma}$ the induced cylindrical *spin*$^c$ structure on $\hat{N}$. For every smooth configuration

$$\hat{\mathsf{C}} = (\hat{\psi}, \hat{A}) \in \Gamma(\hat{\mathbb{S}}_\sigma^+) \times \mathfrak{A}_\sigma$$

we define the scalar quantity called the *energy density* as

$$\rho_{\hat{\mathsf{C}}} := |\hat{\nabla}^{\hat{A}}\hat{\psi}|^2 + \frac{1}{8}|q(\hat{\psi})|^2 + |F_{\hat{A}}|^2 + \frac{\hat{s}}{4}|\hat{\psi}|^2.$$

Thus,

$$E(\hat{\mathsf{C}}) = \int_{\hat{N}} \rho_{\hat{\mathsf{C}}} dv_{\hat{g}}.$$

For every interval $I \subset \mathbb{R}_+$ and every $\varepsilon > 0$ we set

$$I_\varepsilon := \{t \in \mathbb{R}_+; \;\; \mathrm{dist}\,(t, I) \leq \varepsilon\}$$

$$E_{\hat{\mathsf{C}}}(I) := \int_{I\times N} \rho_{\hat{\mathsf{C}}} dv_{\hat{g}}.$$

Fix a Hermitian connection $A_0$ on $\det(\sigma) \to N$ and denote by $\hat{A}_0$ its pullback to $\det(\hat{\sigma}) \to \hat{N}$. Any smooth Hermitian connection $\hat{A}$ on $\det(\hat{\sigma})$ can be written as

$$\hat{A} = \hat{A}_0 + \mathbf{i}\varphi(t)dt + \mathbf{i}a(t)$$

where $\varphi(t)$ (resp. $a(t)$) is a smooth path of 0-forms (resp. 1-forms) on $N$. Set

$$A(t) := A_0 + \mathbf{i}a(t) = \hat{A}\,|_{\{t\}\times N}\,.$$

If $\hat{\gamma} := e^{\mathbf{i}f(t)}$ is a gauge transformation on $\hat{N}$ then

$$\hat{\gamma}\cdot\hat{A} = \hat{A}_0 + \mathbf{i}(\varphi(t) - 2\frac{df}{dt})dt + \mathbf{i}(a(t) - 2df(t))$$

where we recall that $d$ denotes the *three-dimensional* exterior derivative along $N$. If we regard $\hat{\gamma}$ as a smooth path of gauge transformations $\gamma_t$ on $N$ then the above computation shows

$$(\hat{\gamma}\cdot\hat{A})(t) = \gamma_t \cdot A(t).$$

In other words, the assignment $\hat{A} \mapsto A(t)$ defines a unique class $[A(t)] \in \mathfrak{A}_\sigma/\mathfrak{G}_\sigma$. This also implies that for any smooth configuration $\hat{\mathsf{C}}$ the assignment

$$\jmath_t : \hat{\mathsf{C}} \mapsto \mathsf{C}(t) := \hat{\mathsf{C}}\,|_{\{t\}\times N}$$

defines a unique gauge equivalence class $[C(t)] \in \mathcal{B}_\sigma = \mathcal{C}_\sigma/\mathfrak{G}_\sigma$. Clearly, the path $t \mapsto [C(t)]$ in $\mathcal{B}_\sigma$ is continuous. In particular, the quantity

$$\nu_{\hat{\mathsf{C}}}(t) := \mathfrak{f}(\mathsf{C}(t)) = \|SW(\mathsf{C}(t))\|_{L^2}$$

is well defined and independent of the gauge equivalence class of $\hat{\mathsf{C}}$.

Suppose now that $\hat{\mathsf{C}}$ is a 4-monopole. Modulo a smooth gauge transformation we can assume $\hat{\mathsf{C}}$ is temporal

$$\hat{\mathsf{C}} = (\psi(t), A(t)).$$

Then, for every interval $I \subset \mathbb{R}_+$ we have

$$\int_I \|SW(\mathsf{C}(t))\|^2_{L^2} dt = \int_t dt \int_N |\dot{\psi}(t)|^2 + |\dot{A}|^2 dv_g = \frac{1}{2} E_{\hat{\mathsf{C}}}(I)$$

so that

$$\|\nu_{\mathsf{C}}\|^2_{L^2(I)} = \frac{1}{2} E_{\hat{\mathsf{C}}}(I). \tag{4.2.7}$$

A simple application of Hölder's inequality shows that

$$\mathrm{dist}_{L^2}([\mathsf{C}(t_0)], [\mathsf{C}(t_1)]) \leq \frac{1}{2} E_{\hat{\mathsf{C}}}([t_0, t_1])^{1/2} (t_1 - t_0)^{1/2}. \tag{4.2.8}$$

Consider a finite interval $I = [t_0, t_1] \in \mathbb{R}_+$ and set

$$|s| := \max_{x \in N} |s_g(x)|.$$

Observe that

$$\frac{1}{16} \|\hat{\psi}\|^4_{L^4(I \times N)} = \frac{1}{16} \int_I dt \int_N |\hat{\psi}(t, x)|^4 dv_g$$

$$= \frac{1}{8} \int_I dt \int_N |q(\hat{\psi})|^2 \leq E_{\hat{\mathsf{C}}}(I) - \frac{1}{4} \int_I dt \int_N s|\hat{\psi}|^2 dv_g$$

$$\leq E_{\hat{\mathsf{C}}}(I) + \frac{|s|}{4} \int_I dt \int_N |\hat{\psi}|^2 dv_g$$

$$\leq E_{\hat{\mathsf{C}}}(I) + \frac{|s|}{4} (t_1 - t_0)^{1/2} \mathrm{vol}_g(N)^{1/2} \|\hat{\psi}\|^2_{L^4(I \times N)}$$

$$\leq E_{\hat{\mathsf{C}}}(I) + \frac{1}{32} \|\hat{\psi}\|^4_{L^4(I \times N)} + \frac{|s|^2}{2} (t_1 - t_0) \mathrm{vol}_g(N).$$

We have thus obtained the following $L^4$-estimate.

$$\|\hat{\psi}\|^4_{L^4(I \times N)} \leq 32 E_{\hat{\mathsf{C}}}(I) + 16 |s|^2 (t_1 - t_0) \mathrm{vol}_g(N). \tag{4.2.9}$$

We can build on this estimate to obtain a priori $L^\infty$-estimates for $\hat{\psi}$.

**Proposition 4.2.12.** *There exists a constant $C > 0$ which depends only on the metric $g$ such that*

$$\|\hat{\psi}\|^4_{L^\infty([T, T+1] \times N)} \leq C\Big(E_{\hat{\mathsf{C}}}([T-1, T+2]) + 1\Big), \quad \forall T > 1. \tag{4.2.10}$$

**Proof** We have

$$0 = \mathfrak{D}^*_{\hat{A}} \mathfrak{D}_{\hat{A}} \hat{\psi} = (\hat{\nabla}^{\hat{A}})^* \hat{\nabla}^{\hat{A}} \hat{\psi} + \frac{s}{4} \hat{\psi} + \frac{1}{2} \hat{\mathbf{c}}(F^+_{\hat{A}}) \hat{\psi}.$$

We can now use Kato's inequality and the equality $\hat{\mathbf{c}}(F^+_{\hat{A}}) = \frac{1}{2} q(\hat{\psi})$ to conclude that

$$\hat{\Delta}_{\hat{g}} |\hat{\psi}|^2 \leq 2 \Big\langle (\hat{\nabla}^{\hat{A}})^* \hat{\nabla}^{\hat{A}} \hat{\psi}, \hat{\psi} \Big\rangle = -\frac{s}{2} |\hat{\psi}|^2 - \frac{1}{4} |\hat{\psi}|^4.$$

Now set $u := |\hat{\psi}|^2$ so that we have

$$\hat{\Delta}_{\hat{g}} u + \frac{s}{2} u \leq -\frac{1}{4} u^2 \leq 0.$$

We can rewrite this as a differential inequality of the type

$$\hat{\Delta}_{\hat{g}} u + au \leq 0$$

where $a = \frac{s}{2} \in L^\infty([T-1, T+2] \times N)$. Using the DeGiorgi-Nash-Moser inequality (see [**11**] or [**47**, Thm. 8.17]) we deduce that there exists a constant $C > 0$ which depends only on $g$ such that

$$\sup_{[T,T+1]\times N} u \leq C\Big(|s| + \|u\|_{L^2([T-1,T+2]\times N)}\Big)$$

$$\overset{(4.2.9)}{\leq} C'\Big(E_{\hat{\mathsf{C}}}([t-1, T+2]) + 1\Big)^{1/2}. \blacksquare$$

**Corollary 4.2.13.** *If $\hat{\mathsf{C}} = (\hat{\psi}, \hat{A})$ is a finite energy monopole on $\hat{N} = \mathbb{R}\times N$ then there exists a constant $C > 0$ which depends only on the metric $g$ such that*

$$\|\hat{\psi}\|^4_{L^\infty(\hat{N})} \leq C\Big(E_{\hat{\mathsf{C}}}(\mathbb{R}_+) + 1\Big). \tag{4.2.11}$$

The next result, whose proof is deferred to §4.2.5, shows that if the total kinetic energy over a time period of length 4 is small enough, then the kinetic energy at each moment must be small. In other words, "bursts" of energy are prohibited.

**Lemma 4.2.14.** *Fix a smooth connection $A_0$ on $\det(\sigma)$. There exist $C_0 > 0$ and $0 < \omega_0 < 1$ such that for every smooth* **temporal** *monopole $\hat{\mathsf{C}}$ on $[-2, 2] \times N$ satisfying*

$$\hat{\mathsf{C}} = (\mathsf{C}(t)) = (\psi(t), A_0 + \mathbf{i}a(t)), \;\; a(t) \in \Omega^1(N),$$

$$E^2 := \int_{-2}^{2} dt \int_N \Big(|\dot{\psi}(t)|^2 + |\dot{a}(t)|^2\Big) dv_N \leq \omega_0$$

*we have*

$$\|SW(\mathsf{C}(t))\|^2_{L^2(N)} = \int_N \Big(|\dot{\psi}(t)|^2 + |\dot{a}(t)|^2\Big) dv_N \leq C_0 E^2, \;\; \forall t \in [-1, 1].$$

**Corollary 4.2.15.** *There exist $C > 0$ and $\omega_0 \in (0, 1)$ such that if $\hat{\mathsf{C}}$ is a smooth monopole on $[-2, 2] \times N$ satisfying*

$$E^2 := E_{\hat{\mathsf{C}}}([-2, 2]) \leq \omega_0$$

*then*

$$\|SW(\hat{\mathsf{C}}|_{t\times N})\|_{L^2(N)} \leq C_0 E, \;\; \forall t \in [-1, 1].$$

**Proof** Since the above inequality is invariant under gauge transformations on $[-2,2]\times N$ we can assume $\hat{\mathsf{C}}$ is in temporal gauge and then apply Lemma 4.2.14. ■

For every $\hbar > 0$ denote by $\mathfrak{f}^{\hbar}$ the level set of $\mathfrak{f}$

$$\mathfrak{f}^{\hbar} = \{\mathsf{C} \in \mathcal{C}_{\sigma};\ \ \mathfrak{f}(\mathsf{C}) < \hbar\}.$$

Observe that $\mathfrak{f}^{\hbar}$ is an open neighborhood of $\mathfrak{Z}_{\sigma}$ in $\mathcal{B}_{\sigma}$. The following result refines Proposition 2.4.6 of 2.4.1. We leave its proof to the reader.

**Proposition 4.2.16.** *There exists a function*

$$\hbar : (0,1) \to (0,\infty),\ \ \kappa \mapsto \hbar(\kappa)$$

*such that*

*(i)* $\lim_{\kappa\to 0}\hbar(\kappa) = 0$.

*(ii)If* $\mathsf{C} \in \mathfrak{f}^{\hbar(\kappa)}$ *then there exist a smooth monopole* $\mathsf{C}_0 \in \mathfrak{Z}_{\sigma}$ *and* $\gamma \in \mathfrak{G}_{\sigma}$ *such that*

$$\gamma\cdot\mathsf{C} \in U_{\mathsf{C}_0}(\kappa).$$

From the above proposition we deduce the following consequence.

**Corollary 4.2.17.** *If* $\mathfrak{M}_{\sigma} = \emptyset$ *there exists* $\hbar_0 > 0$ *such that* $\mathfrak{f}(\mathsf{C}) > \hbar_0$, $\forall\mathsf{C}$.

The above result, coupled with Corollary 4.2.15, leads to the following conclusion.

**Corollary 4.2.18.** *If* $\hat{\mathsf{C}}$ *is a finite energy monopole on* $\mathbb{R}_+\times N$ *then for any sequence* $t_n \to \infty$ *we can find a subsequence* $t_{n_k}$ *such that* $[\mathsf{C}(t_{n_k})]$ *converges to a point in* $\mathfrak{M}_{\sigma}$.

If $\mathfrak{M}_{\sigma}^{(1)},\cdots,\mathfrak{M}_{\sigma}^{(\ell)}$ are the connected components of $\mathfrak{M}_{\sigma}$ we can find $\kappa_0 > 0$ such that $U_{\kappa_0}$ consists of *disjoint* open neighborhoods $U_{\kappa_0}^{(j)}$ of $\mathfrak{M}_{\sigma}^{(j)}$, $j = 1,\cdots,\ell$. Set

$$d_0 = d_0(\kappa_0) := \min_{i\neq j}\operatorname{dist}_{L^2}\big(U_{\kappa_0}^{(i)}, U_{\kappa_0}^{(j)}\big).$$

**Exercise 4.2.3.** Show that

$$\liminf_{\kappa_0\searrow 0} d_0(k_0) > 0.$$

**Hint:** Show that if $\kappa_0$ is sufficiently small there exists a constant $C > 0$ depending only on the geometry of $N$ and $C(E_0)$ such that

$$\operatorname{dist}_{L^2}([\mathsf{C}],\mathfrak{M}_{\sigma}^{(i)}) \le C\kappa_0,\ \ \forall i,\ \forall[\mathsf{C}] \in U_{\kappa_0}^{(i)}.$$

Corollary 4.2.15 shows that if $\hat{\mathsf{C}}$ is a finite energy monopole and $T > 0$ is such that

$$E_{\hat{\mathsf{C}}}([T, \infty)) \ll \hbar(\kappa)$$

then $[\mathsf{C}(t)] \in U_\kappa$, $\forall t > T + 1$. Clearly, for large $t$ the path $t \mapsto [\mathsf{C}(t)]$ will wander inside a single component $U_\kappa^{(j)}$ of $U_\kappa$. We have thus proved the following result.

**Corollary 4.2.19.** *Suppose $\hat{\mathsf{C}}$ is a finite energy smooth monopole on $\hat{N}$. Then there exist a connected component $\mathfrak{M}_\sigma^{(j)}$ of $\mathfrak{M}_\sigma$ and, for all $\kappa > 0$, an instant of time $t = t(\kappa) > 0$, such that $[\mathsf{C}(t)] \in U_\kappa^{(j)}$ for all $t > t(\kappa)$.*

A priori, the path $[\mathsf{C}(t)]$ in the above corollary may wander around smaller and smaller neighborhoods $U_\kappa^{(j)}$ of $\mathfrak{M}_\sigma^{(j)}$ without converging to any specified 3-monopole so the limit set may consist of several points in $\mathfrak{M}_\sigma$. The results we proved so far show that the manner in which $[\mathsf{C}(t)]$ travels around $\mathfrak{M}_\sigma$ is quite constrained. More precisely, for every triple of arbitrarily small constants $a, b, c > 0$ there exists an instant of time $T = T(a, b, c) > 0$ such that for all $t > T$ the distance between $[\mathsf{C}(t)]$ and $\mathfrak{M}_\sigma$ is $< a$, the kinetic energy $\|\dot{\psi}(t)\|_{L^2}^2 + \|\dot{a}(t)\|_{L^2}^2$ at time $t$ is $< b$, and there is not much energy left, i.e.

$$E_{\hat{\mathsf{C}}}([T, \infty)) < c.$$

The energy functional $\mathcal{E}$ on $N$ (whose critical points are the 3-monopoles) may not descend to $\mathcal{C}_\sigma/\mathfrak{G}_\sigma$ so it may not induce a well defined function on $\mathfrak{M}_\sigma$. On the other hand, it descends to function on $\mathcal{C}_\sigma/\mathfrak{G}_\sigma^1$ where $\mathfrak{G}_\sigma^1$ denotes the identity component of $\mathfrak{G}_\sigma$. We denote by $\tilde{\mathfrak{M}}_\sigma$ the space of $\mathfrak{G}_\sigma^1$-orbits of 3-monopoles. $\mathcal{E}$ defines a continuous map from the discrete set of components of $\tilde{\mathfrak{M}}_\sigma$ to $\mathbb{R}$. $\mathfrak{M}_\sigma$ is a quotient of $\tilde{\mathfrak{M}}_\sigma$ modulo the action of the discrete group $H^1(N, \mathbb{Z})$. Since

$$\mathcal{E}(\mathsf{C}(t_1)) - \mathcal{E}(\mathsf{C}(t_0)) = E_{\hat{\mathsf{C}}}([t_0, t_1])$$

$\mathcal{E}(\mathsf{C}(t))$ has a well defined limit $\mathcal{E}_\infty$ as $t \to \infty$ so that the path $\mathsf{C}(t)$ "orbits" closer and closer around one of the components of $\tilde{\mathfrak{M}}_\sigma$ where $\mathcal{E} \equiv \mathcal{E}_\infty$.

In the next subsection we will show that these restrictions, coupled with the ellipticity of the Seiberg-Witten equations on cylinders, will force $[\mathsf{C}(t)]$ to converge to a specified monopole $[\mathsf{C}_0] \in \mathfrak{M}_\sigma$. To minimize the volume of technicalities we will make the simplifying assumption below which is satisfied in all concrete applications we have in mind. For a presentation of the general situation in the similar case of Yang-Mills equations we refer to [**96, 133**].

(N) *In the remainder of this chapter we will work exclusively with good pairs $(\sigma, g)$.*

**4.2.4. Asymptotic behavior. Part II.** Suppose $\hat{\mathsf{C}}$ is a finite energy monopole on $\hat{N}$. In the last subsection we have shown that for every $0 < \kappa \ll 1$ there exist a smooth monopole $\mathsf{C}_0$ and an interval $J = [t_0, t_1] \subset \mathbb{R}_+$ such that for every $t \in J$ the configuration $[\mathsf{C}(t)] \in U_{\mathsf{C}_0}(\kappa)$. We deduced this conclusion by taking advantage of the nice dynamical description of the Seiberg-Witten equations in temporal gauge. These arguments were however not powerful enough to deduce, for example, that once $[\mathsf{C}(t)]$ enters a neighborhood $U_{\mathsf{C}_0}(\kappa)$ of $[\mathsf{C}_0]$ it is then forced to stay inside it. From a technical point of view this is due essentially to a lack of estimates of the length of the path $[\mathsf{C}(t)]$, that is, estimating $L^1$-norms of $t$-derivatives on *long* time intervals. It is desirable to control the length of a portion of this path in terms of its energy. To obtain such estimates we need to modify $\hat{\mathsf{C}}$ by a gauge transformation which will capture the *elliptic* character of the Seiberg-Witten equations on a cylinder. Following [**96, 133**] we introduce the following notion.

**Definition 4.2.20.** Let $\kappa \in (0, 1)$ and $\mathsf{C}_0$ be a smooth monopole on $N$. A configuration $\hat{\mathsf{C}}$ on a cylinder $I \times N$ is said to be in *$\kappa$-standard gauge* with respect to $\mathsf{C}_0$ if there exist smooth paths

$$I \ni t \to (\mathbf{i}f(t), \mathsf{V}(t)) \in (\ker \mathfrak{L}_{\mathsf{C}_0})^{\perp} \times \mathcal{S}_{\mathsf{C}_0}, \;\; \mathsf{V}(t) = (\psi(t), \mathbf{i}a(t))$$

such that $\|\mathsf{V}(t)\|_{2,2} = \|\psi(t)\|_{2,2} + \|a(t)\|_{2,2} < \kappa$, $\forall t \in I$ and

$$\hat{\mathsf{C}} = (\psi_0 + \psi(t), A_0 + \mathbf{i}f(t)dt + \mathbf{i}a(t)).$$

For a proof of the following technical result we refer to [**96**, Lemma 2.4.3].

**Lemma 4.2.21.** *Assume $\hat{\mathsf{C}}$ is a smooth configuration on $I \times N$ and $\mathsf{C}_0$ is a smooth monopole on $N$ such that $\mathsf{C}(t)$ is gauge equivalent to a configuration in $U_{\mathsf{C}_0}(\kappa)$, $\forall t \in I$. Then there exists a smooth gauge transformation*

$$\hat{\gamma} : I \times N \to S^1$$

*such that $\hat{\gamma} \cdot \hat{\mathsf{C}}$ is in $\kappa$-standard gauge with respect to $\mathsf{C}_0$.*

Suppose now that $\hat{\mathsf{C}}$ is a smooth 4-monopole on $I \times N$ in $\kappa$-standard gauge with respect to the smooth 3-monopole $\mathsf{C}_0 = (\psi_0, A_0)$. Thus, we can write

$$\hat{\mathsf{C}} = (\hat{\psi} = \psi_0 + \psi(t), \hat{A} = A_0 + \mathbf{i}df(t)dt + \mathbf{i}a(t))$$

where, for any $t \in I$,

(4.2.12) $$\|a(t)\|_{\frac{3}{2},2} + \|\psi(t)\|_{\frac{3}{2},2} \leq \kappa,$$

$$\mathfrak{L}^*_{\mathsf{C}_0}(\psi(t), \mathbf{i}a(t)) = 0, \ \ \mathbf{i}f(t) \perp \ker \mathfrak{L}_{\mathsf{C}_0}.$$

Then, using the identities (2.4.1) and (2.4.2) in §2.4.1, we deduce

$$F_{\hat{A}} = F_{A_0} + \mathbf{i}dt \wedge (\dot{a}(t) - df(t)) + \mathbf{i}da(t),$$

$$F^+_{\hat{A}} = \frac{\mathbf{i}}{2}\Big(dt \wedge (\dot{a} + *F_{A_0} + *da(t) - df(t)) + *(\dot{a}(t) + F_{A_0} + da(t) - df(t))\Big)$$

$(J := \hat{\mathbf{c}}(dt),\ A(t) := A_0 + \mathbf{i}a(t))$,

$$\hat{\mathfrak{D}}_{\hat{A}} = J\Big(\partial_t - \mathfrak{D}_{A(t)} + \frac{\mathbf{i}}{2}f(t)\Big).$$

If we suppress the $t$ dependence in the above notation and we use the identity

$$\mathfrak{D}_A\psi_0 = \Big(\mathfrak{D}_{A_0} + \frac{1}{2}\mathbf{c}(\mathbf{i}a)\Big)\psi_0 = \frac{1}{2}\mathbf{c}(\mathbf{i}a)\psi_0$$

we can rewrite the Seiberg-Witten equations for $\hat{\mathsf{C}}$ as follows.

(4.2.13a) $$\frac{d}{dt}\psi = \Big(\mathfrak{D}_A - \frac{\mathbf{i}}{2}f\Big)(\psi + \psi_0) = \mathfrak{D}_{A_0}\psi + \frac{1}{2}(\mathbf{c}(\mathbf{i}a) - \mathbf{i}f)(\psi_0 + \psi),$$

(4.2.13b) $$\mathbf{i}\frac{d}{dt}a = \frac{1}{2}q(\psi_0 + \psi) - *\mathbf{i}da + \mathbf{i}df - *F_{A_0},$$

(4.2.13c) $$d^*a + \frac{1}{2}\mathfrak{Im}\langle\psi_0, \psi\rangle = 0.$$

One unpleasant feature of these equations is the apparent lack of information on the $t$-derivatives of $f$. Still, the size of $f$ can be controlled in terms of the sizes of $(\psi, A)$. To achieve this we will need an elementary identity whose proof is left to the reader.

**Exercise 4.2.4.** ([**107**]) Suppose $\psi$ is a smooth spinor on $N$ and $A$ is a smooth Hermitian connection on $\det(\sigma)$. Then

(4.2.14) $$d^*q(\psi) = -\mathbf{i}\mathfrak{Im}\langle\psi, \mathfrak{D}_A\psi\rangle. \ \blacksquare$$

For simplicity, in the sequel will denote the $t$-derivatives by dots. Also, we will denote by the same letter $C$ *all positive constants which depend only on* $\mathsf{C}_0$, *the total energy of* $\hat{\mathsf{C}}$ *and the metric* $g$.

Differentiating (4.2.13c) with respect to $t$ we get

$$\mathbf{i}d^*\dot{a} + \frac{\mathbf{i}}{2}\mathfrak{Im}\langle\psi_0, \dot{\psi}\rangle = 0.$$

Now use (4.2.13c) and (4.2.14) to obtain

$$0 = \frac{1}{2} d^* q(\psi + \psi_0) + \mathbf{i} d^* df + \frac{\mathbf{i}}{2} \mathfrak{Im} \langle \psi_0, \dot{\psi} \rangle$$

$$\stackrel{(4.2.13a)}{=} -\frac{\mathbf{i}}{2} \mathfrak{Im} \langle \psi_0 + \psi, \mathfrak{D}_A(\psi_0 + \psi) \rangle + \mathbf{i} d^* df + \frac{\mathbf{i}}{2} \mathfrak{Im} \langle \psi_0, \dot{\psi} \rangle$$

$$= -\frac{\mathbf{i}}{2} \mathfrak{Im} \langle \psi_0 + \psi, \dot{\psi} + \frac{\mathbf{i}}{2} f(\psi + \psi_0) \rangle + \mathbf{i} d^* df + \frac{\mathbf{i}}{2} \mathfrak{Im} \langle \psi_0, \dot{\psi} \rangle$$

$$= \mathbf{i} d^* df + \frac{\mathbf{i}}{4} |\psi_0 + \psi|^2 f - \frac{\mathbf{i}}{2} \mathfrak{Im} \langle \psi, \dot{\psi} \rangle \stackrel{(4.2.13a)}{=}$$

$$\mathbf{i} d^* df + \frac{\mathbf{i}}{4} \mathfrak{Re} \langle \psi_0 + \psi, \psi_0 + \psi \rangle f - \frac{\mathbf{i}}{2} \mathfrak{Im} \langle \psi, \mathfrak{D}_{A_0} \psi \rangle - \frac{\mathbf{i}}{4} \mathfrak{Re} \langle \psi, (\psi_0 + \psi) \rangle f$$

$$= \mathbf{i} d^* df + \frac{\mathbf{i}}{4} \mathfrak{Re} \langle \psi_0, \psi_0 + \psi \rangle f - \frac{\mathbf{i}}{2} \mathfrak{Im} \langle \psi, \mathfrak{D}_{A_0} \psi \rangle$$

$$= \mathbf{i} d^* df + \frac{\mathbf{i}}{4} |\psi_0|^2 f + \frac{\mathbf{i}}{4} \mathfrak{Re} \langle \psi_0, \psi \rangle f - \frac{\mathbf{i}}{2} \mathfrak{Im} \langle \psi, \mathfrak{D}_{A_0} \psi \rangle$$

$$= \frac{\mathbf{i}}{4} \mathfrak{L}^*_{\mathsf{C}_0} \mathfrak{L}_{\mathsf{C}_0} (\mathbf{i} f) + \frac{1}{4} \mathfrak{Re} \langle \psi_0, \psi \rangle \mathbf{i} f - \frac{\mathbf{i}}{2} \mathfrak{Im} \langle \psi, \mathfrak{D}_{A_0} \psi \rangle .$$

Hence

$$\mathfrak{L}^*_{\mathsf{C}_0} \mathfrak{L}_{\mathsf{C}_0} \mathbf{i} f = -\mathfrak{Re} \langle \psi_0, \psi \rangle \mathbf{i} f + 2\mathfrak{Im} \langle \psi, \mathfrak{D}_{A_0} \psi \rangle . \tag{4.2.15}$$

The proof of the following result is a simple application of Theorem 1.2.18 (v) and is left to the reader.

**Lemma 4.2.22.** *For each $\psi$ such that $\|\psi\|_{2,2} \leq \kappa$ consider the operator*

$$T_\psi : \ker \mathfrak{L}^{\perp}_{\mathsf{C}_0} \cap L^{3,2} \to L^{1,2}(N, \mathbf{i}\mathbb{R}), \ \ \mathbf{i} f \mapsto \mathfrak{L}^*_{\mathsf{C}_0} \mathfrak{L}_{\mathsf{C}_0} + \mathfrak{Re} \langle \psi_0, \psi \rangle \mathbf{i} f .$$

*Then, if $\kappa$ is sufficiently small the operator $T_\psi$ is invertible. Moreover for every $r \in \{0, 1\}$ and every $p \in (1, 2]$ there exists a constant $C > 0$ depending only on $p$, $r$ and the geometry of $N$ such that*

$$\|f\|_{2+r,p} \leq C \|T_\psi \mathbf{i} f\|_{r,p} .$$

Using the above lemma we deduce that there exists a constant $C > 0$ such that

$$\|f\|_{2,2} \leq C \|\mathfrak{Im} \langle \psi, \mathfrak{D}_{A_0} \psi \rangle \|_{L^2} .$$

The Sobolev embedding theorems show that we have continuous embeddings

$$L^{2,2}(N) \hookrightarrow L^\infty(N), \ \ L^{1,2}(N) \hookrightarrow L^6(N).$$

Using Hölder's inequality we deduce that there exists a constant $C > 0$ such that for every $a \in L^{2,2}(N)$ and $b \in L^{1,2}(N)$ we have

$$\|a \cdot b\|_{L^{1,2}} \leq C\|a\|_{2,2} \cdot \|b\|_{1,2}.$$

Hence

$$\|\mathfrak{Im}\langle\psi, \mathfrak{D}_{A_0}\psi\rangle\|_{L^{1,2}} \leq C\|\psi\|_{2,2}\|\mathfrak{D}_{A_0}\psi\|_{1,2} \leq C\|\psi\|_{2,2}^2.$$

We have thus established the estimate

$$\|f\|_{2,2} \leq C\|\psi\|_{2,2}^2 \overset{(4.2.12)}{<} C\kappa^2. \tag{4.2.16}$$

Since $\kappa$ is meant to be very small we deduce that $f(t)$ is very small as long as $\hat{\mathsf{C}}|_{I\times N}$ is in $\kappa$-standard gauge. Set

$$\mathsf{V}(t) := (\psi(t), \mathbf{i}a(t)).$$

The flow equations (4.2.13) can be rewritten as

$$\dot{\mathsf{V}} = SW(\mathsf{C}_0 + \mathsf{V}) + \begin{bmatrix} -\frac{\mathbf{i}f}{2}\psi_0 - \frac{\mathbf{i}f}{2}\psi \\ \mathbf{i}df \end{bmatrix} \tag{4.2.17}$$

where

$$\mathfrak{L}_{\mathsf{C}_0}^*\mathsf{V} = 0 \tag{4.2.18}$$

and

$$\mathbf{i}f = 2T_\psi^{-1}(\mathbf{i}\mathfrak{Im}\langle\psi, \mathfrak{D}_{A_0}\psi\rangle). \tag{4.2.19}$$

We will denote the second term on the right-hand side of (4.2.17) by $\mathfrak{N}(\mathsf{V})$. Observe that

$$\mathfrak{N}(\mathsf{V}) = -\frac{1}{2}\mathfrak{L}_{\mathsf{C}_0+\mathsf{V}}(\mathbf{i}f). \tag{4.2.20}$$

The estimate (4.2.16) shows that

$$\|\mathfrak{N}(\mathsf{V})\|_{2,2} \leq C\|\mathsf{V}\|_{2,2}^2. \tag{4.2.21}$$

**Remark 4.2.23.** One can show exactly as in [**96**, Chap. 2] that there exists a natural $L^2$-metric on $\mathcal{S}_{\mathsf{C}_0}$ such that in a neighborhood of $0 \in \mathcal{S}_{\mathsf{C}_0}$ the equations (4.2.17) have the form

$$\dot{\mathsf{V}} = \tilde{\nabla}\mathcal{E}|_{\mathcal{S}_{\mathsf{C}_0}}(\mathsf{C}_0 + \mathsf{V})$$

where the gradient $\tilde{\nabla}$ is computed with respect to this metric. ■

For every $0 < \kappa \ll 1$ we can find $T_0(\kappa) = T_0(\kappa, \hat{\mathsf{C}}) \gg 0$ such that for all $t_0 \geq T_0(\kappa)$ there exists a smooth monopole $\mathsf{C}_0 = \mathsf{C}_0(t_0) \in \mathfrak{M}_\sigma$ so that

$$
\left\{
\begin{array}{l}
[\mathsf{C}(t_0)] \in U_{\mathsf{C}_0}(\kappa^2) \\
\\
E_{\hat{\mathsf{C}}}([T_0(\kappa), \infty)) \leq \kappa^6 \quad , \quad \forall t \geq T_0(\kappa). \\
\\
\|SW([\mathsf{C}(t)])\|^2_{L^2} < \kappa^6
\end{array}
\right.
\tag{4.2.22}
$$

**Fix** $t_0 \geq T_0(\kappa)$ and define

$$
\begin{aligned}
T_\kappa(t_0) &:= \sup\Big\{\tau > 0;\ \ [\mathsf{C}(t_0 + t)] \in U_{\mathsf{C}_0(t_0)}(\kappa),\ \ \forall t \in [0, \tau]\Big\} \\
&= \sup\Big\{T > 0;\ \|\mathsf{V}(t_0 + t)\|_{2,2} \leq \kappa,\ \ \forall t \in [0, T]\Big\}
\end{aligned}
$$

where $\mathsf{V}(t)$ is determined as above by placing $\hat{\mathsf{C}}$ in $\kappa$-standard gauge at $\mathsf{C}_0$ over the time interval for which this is possible. Roughly speaking, $T_\kappa(t_0)$ is the length of the time interval, beginning at $t_0$, during which the orbit $[\mathsf{C}(t)]$ stays $\kappa$-close to $[\mathsf{C}_0] := [\mathsf{C}_0(t_0)]$. We want to get more precise information about the size of

$$
\mathrm{dist}_{2,2}\big(\,[\mathsf{C}(t_0 + t)], [\mathsf{C}_0]\,\big)
$$

for $0 \leq t \leq T_\kappa(t_0)$.

One of the main advantages of working in standard gauges comes from the fact that the 4-dimensional equations become "almost" *elliptic* and thus one can control stronger norms by weaker ones. More precisely, we have the following result.

**Lemma 4.2.24.** *There exist $\kappa_0 > 0$ and $C > 0$ with the following property. For any finite energy monopole $\hat{\mathsf{C}}$ on $\mathbb{R}_+ \times N$ and all*

$$
0 < \kappa < \kappa_0,\ \ t_0 > T_0(\kappa, \hat{\mathsf{C}}),\ \ t \in [t_0 + 1, T_\kappa(t_0)],\ \ [\mathsf{C}_0] \in \mathfrak{M}_\sigma
$$

*such that*

$$
\mathrm{dist}_{L^{2,2}}\big(\,[\mathsf{C}(t_0)], [\mathsf{C}_0]\,\big) < \kappa^2
$$

*we have*

$$
\begin{aligned}
&\mathrm{dist}_{2,2}([\mathsf{C}(t_0 + t)], [\mathsf{C}_0])^2 \\
&\leq C\Big(\mathrm{dist}_{L^2}([\mathsf{C}(t_0 + t)], [\mathsf{C}_0])^2 + E_{\hat{\mathsf{C}}}([t - 1, t + 1])\Big) \\
&\leq C\Big(\mathrm{dist}_{L^2}([\mathsf{C}(t_0 + t)], [\mathsf{C}_0])^2 + \kappa^6\Big).
\end{aligned}
\tag{4.2.23}
$$

In order to keep the flow of arguments uninterrupted we will defer the proof of the above lemma to the next subsection. This lemma roughly states that the $L^{2,2}$ -distance between $[\mathsf{C}(t_0+t)]$ and $[\mathsf{C}_0]$ can be controlled by the weaker metric $\mathrm{dist}_{L^2}$. This type of control immediately leads to nontrivial lower estimates on the duration $T_\kappa(t_0)$.

**Lemma 4.2.25.** *There exists a positive constant $C$ such that for all $0 < \kappa \ll 1$ we have*

$$T_\kappa(t_0) \geq C\Big(\frac{\kappa}{\delta+\kappa^2}\Big)^2.$$

**Proof** Let $T = T_\kappa(t_0)$. We rewrite

$$\mathsf{C}(t_0+t) = \mathsf{C}_0 + \mathsf{V}(t),\ \ \mathfrak{L}^*_{\mathsf{C}_0}\mathsf{V}(t) = 0,\ \ \|\mathsf{V}(t)\|_{2,2} \leq \kappa.$$

(*Note the time shift in the argument of* $\mathsf{V}$.) The maximality of $T$ implies

$$\|\mathsf{V}(T)\|_{2,2} = \kappa$$

so that using Lemma 4.2.24 we deduce

$$\|\mathsf{V}(T)\|_{L^2} \geq C\|\mathsf{V}(T)\|_{2,2} - \kappa^2 \geq C\kappa - \kappa^3. \tag{4.2.24}$$

The distance $\|\mathsf{V}(T) - \mathsf{V}(0)\|_{L^2}$ can be estimated using the flow equations (4.2.17). We have

$$\|\mathsf{V}(T) - \mathsf{V}(0)\|_{L^2} = \int_0^T \|\dot{\mathsf{V}}(t)\|_{L^2}dt$$

$$\leq \int_0^T \big(\|SW(\mathsf{C}(t_0+t))\|_{L^2} + \|\mathfrak{N}(\mathsf{V}(t))\|_{L^2}\big)dt$$

$$\stackrel{(4.2.21)}{\leq} C\Big(T^{1/2}E_{\hat{\mathsf{C}}}([t_0,t_0+T])^{1/2} + T\kappa^2\Big) \leq C(T^{1/2}\kappa^3 + \kappa^2 T) \leq CT\kappa^2.$$

Hence,

$$\|\mathsf{V}(T)\|_{L^2} \leq \|\mathsf{V}(0)\|_{L^2} + \|\mathsf{V}(T) - \mathsf{V}(0)\|_{L^2} \leq \kappa^2 + CT\kappa^2. \tag{4.2.25}$$

Lemma 4.2.25 now follows by comparing (4.2.24) and (4.2.25). ∎

Since the configurations $[\hat{\mathsf{C}}(t)]$ lie in a very small neighborhood of $\mathsf{C}_0$ it is natural to expect that the linearization of the flow (4.2.13) at $\mathsf{C}_0$ will contain information about the nonlinear situation. We now want to suitably decompose the flow (4.2.13) into a linear part and a small nonlinear perturbation, and analyze how much of the linear behavior is preserved under perturbation. At this stage the regularity assumption on $\mathsf{C}_0$ introduces substantial simplifications.

Consider again the **Stab** $(\mathsf{C}_0)$-equivariant map

$$\Phi_1 : U_{\mathsf{C}_0} \to (1 - \Pi_1)\mathcal{S}_{\mathsf{C}_0}$$

introduced in Proposition 4.2.6. Denote by $\mathcal{A}$ the linearization of $SW$ at $\mathsf{C}_0$:

$$\mathcal{A} := \underline{SW}_{\mathsf{C}_0}.$$

**Lemma 4.2.26.** *$\mathcal{A}$ defines a closed, densely defined linear operator*

$$\ker \mathfrak{L}^*_{\mathsf{C}_0} \cap L^2 \to \ker \mathfrak{L}^*_{\mathsf{C}_0}$$

*with domain*

$$\ker \mathfrak{L}^*_{\mathsf{C}_0} \cap L^{1,2}.$$

*This operator is selfadjoint with compact resolvent. Moreover* $\ker \mathcal{A} = H^1_{\mathsf{C}_0}$.

**Exercise 4.2.5.** Prove the above lemma.

The spectrum $\mathrm{spec}\,(\mathcal{A})$ of $\mathcal{A}$ is discrete, consisting of eigenvalues with finite multiplicities. We have an $L^2$-orthogonal decomposition

$$\mathcal{S}_{\mathsf{C}_0} = H^1_{\mathsf{C}_0} \oplus \mathcal{S}^+_{\mathsf{C}_0} \oplus \mathcal{S}^-_{\mathsf{C}_0}$$

corresponding to the partition

$$\mathrm{spec}\,(\mathcal{A}) = \{0\} \cup \mathrm{spec}\,(\mathcal{A}) \cap (0, \infty) \cup \mathrm{spec}\,(\mathcal{A}) \cap (-\infty, 0).$$

Correspondingly, any vector $\mathsf{U} \in \mathcal{S}_{\mathsf{C}_0}$ decomposes as

$$\mathsf{U} = \mathsf{U}_0 + \mathsf{U}^+ + \mathsf{U}^-.$$

Denote by $\mu_+ = \mu_+(\mathsf{C}_0)$ the smallest positive eigenvalue of $\mathcal{A}$, by $-\mu_- = -\mu_-(\mathsf{C}_0)$ the largest negative eigenvalue of $\mathcal{A}$ and

$$\mu := \min(\mu_-, \mu_+).$$

Now set

$$\mathsf{V}_0(t) := \Pi_1 \mathsf{V}(t),\ \ \xi(t) := \mathsf{V}_0(t) + \Phi_1(\mathsf{V}_0(t)),\ \ \mathsf{U}(t) := \mathsf{V}(t) - \xi(t).$$

Observe that $\mathsf{U}_0 = 0$. Since $\mathsf{C}_0$ is regular, the graph of the map $\Phi_1$ describes the critical points of $SW$ in $U_{\mathsf{C}_0}(\kappa)$. To proceed further observe that

$$SW(\mathsf{C}_0 + \mathsf{V}) = SW(\mathsf{C}_0 + \xi + \mathsf{U}) = SW(\mathsf{C}_0 + \xi + \mathsf{U}) - SW(\mathsf{C}_0 + \xi)$$

$$= \mathcal{A}(\xi + \mathsf{U}) - \mathcal{A}(\xi) + R(\xi + \mathsf{U}) - R(\xi) = \mathcal{A}\mathsf{U} + R(\xi + \mathsf{U}) - R(\xi)$$

where

$$\|R(\mathsf{X})\|_{1,2} \leq C\|\mathsf{X}\|^2_{2,2},\ \ \forall \mathsf{X} \in \mathcal{S}_{\mathsf{C}_0}.$$

Set

$$Q(\mathsf{V}) := R(\xi + \mathsf{U}) - R(\xi) + \mathfrak{N}(\mathsf{V}).$$

$Q$ satisfies a similar quadratic estimate as $R$:

$$\|Q(\mathsf{X})\|_{L^2} \le C\|\mathsf{X}\|_{2,2}^2, \quad \forall \mathsf{X} \in \mathcal{S}_{\mathsf{C}_0}. \tag{4.2.26}$$

We can be much more precise. The following estimates are proved in the next subsection.

**Lemma 4.2.27.** *There exists $C > 0$ such that $\forall t \in [0, T_\kappa]$ we have*

$$\|R(\xi(t) + U(t)) - R(\xi)\|_{L^2} \le C\|\mathsf{V}(t)\|_{2,2} \cdot \|\mathsf{U}(t)\|_{L^2}, \tag{4.2.27a}$$

$$\| \Pi_1 \mathfrak{N}(\mathsf{V}(t)) \|_{L^2} \le C\|\mathsf{V}(t)\|_{2,2} \cdot \|\mathsf{U}(t)\|_{L^2}, \tag{4.2.27b}$$

$$\left| \langle \mathfrak{N}(V), \mathsf{U}^{\pm} \rangle_{L^2} \right| \le C\|\mathsf{V}(t)\|_{2,2} \cdot \|\mathsf{U}(t)\|_{L^2}^2. \tag{4.2.27c}$$

The estimates in Lemma 4.2.27 can be used to provide a crucial *lower bound* for $\|SW(\mathsf{V}(t))\|_{L^2}$.

**Lemma 4.2.28.** *If $\kappa$ is sufficiently small we have*

$$\|SW(\mathsf{C}_0 + \mathsf{V}(t))\|_{L^2} \ge C\|\mathsf{U}(t)\|_{L^2}, \quad \forall t \in [0, T_\kappa(t_0)]. \tag{4.2.28}$$

**Proof** We have

$$\|SW(\mathsf{V}(t))\|_{L^2} = \|\mathcal{A}\mathsf{U} + R(\xi + \mathsf{U}) - R(\mathsf{U})\|$$

$$\ge \|\mathcal{A}\mathsf{U}\|_{L^2} - \|R(\xi + \mathsf{U}) - R(\mathsf{U})\|_{L^2}$$

$$\ge \mu\|\mathsf{U}\|_{L^2} - C\kappa\|\mathsf{U}\|_{L^2}. \ \blacksquare$$

The flow equations (4.2.17) now decompose as

$$\dot{\mathsf{V}}_0(t) = \Pi_1 Q(\mathsf{V}), \tag{4.2.29a}$$

$$\dot{\mathsf{U}}^+(t) = \mathcal{A}\mathsf{U}^+ + Q(\mathsf{V})^+ - \frac{d}{dt}\Big(\Phi_1(\mathsf{V}_0(t))\Big)^+, \tag{4.2.29b}$$

$$\dot{\mathsf{U}}^-(t) = \mathcal{A}\mathsf{U}^- + Q(\mathsf{V})^- - \frac{d}{dt}\Big(\Phi_1(\mathsf{V}_0(t))\Big)^-. \tag{4.2.29c}$$

Set

$$f_0(t) := \|\mathsf{V}_0(t)\|_{L^2}^2, \quad f_\pm(t) := \|\mathsf{U}^\pm(t)\|_{L^2}^2,$$

$$f(t) := f_+(t) + f_-(t) = \|\mathsf{U}(t)\|_{L^2}^2.$$

Since $\|\Phi_1(\mathsf{V}_0)\|_{L^2} \le \|\mathsf{V}_0\|_{2,2}^2 \le C\|\mathsf{V}_0\|_{L^2}^2$ we deduce that the problem of estimating $\|\mathsf{V}(t)\|_{L^2}$ is equivalent to the problem of estimating $f_0(t)$ and $f(t)$.

From (4.2.29a), (4.2.27a) and (4.2.27b) we get

$$\|\dot{\mathsf{V}}_0(t)\| \le Cf^{1/2}.$$

In particular,

$$\|\frac{d}{dt}\Phi_1(\mathsf{V}_0(t))\|_{L^2} = \|D_{\mathsf{V}_0(t)}\Phi_1\dot{\mathsf{V}}_0(t)\| \le \|D_{\mathsf{V}_0(t)}\Phi_1\|_{L^2}\|\dot{\mathsf{V}}_0(t)\|_{L^2}$$

$$\le C\kappa\|\mathsf{V}_0(t)\|_{L^2}f^{1/2} \le C\kappa f^{1/2}.$$

Thus,

$$\Big|\langle\frac{d}{dt}\Phi_1(\mathsf{V}_0(t)), \mathsf{U}^\pm\rangle_{L^2}\Big| \le C\kappa f. \tag{4.2.30}$$

Using (4.2.27a) and (4.2.27c) we deduce

$$\Big|\langle Q(V), \mathsf{U}^\pm\rangle_{L^2}\Big| \le C\kappa f. \tag{4.2.31}$$

Now, take the $L^2$-inner product of (4.2.29b) with $\mathsf{U}^+(t)$ and use (4.2.30), (4.2.31) and the inequality

$$\langle \mathcal{A}\mathsf{U}^+(t), \mathsf{U}^+(t)\rangle_{L^2} \ge \mu_+\|\mathsf{U}^+(t)\|^2_{L^2} = \mu_+ f_+(t).$$

We get

$$\dot{f}_+(t) \ge 2\mu_+ f_+(t) - C_+\kappa f. \tag{4.2.32}$$

Using the equality (4.2.29c) we deduce similarly that

$$\dot{f}_-(t) \le -2\mu_- f_i + C_-\kappa f. \tag{4.2.33}$$

By replacing $C_\pm$ with $\max(C_+, C_-)$ we can assume $C_+ = C_-$. Set $h := f_+ - f_-$. Notice that $h$ satisfies a differential inequality of the type

$$\dot{h} \ge 2\mu f \ge 2\mu h, \quad \forall t \in [0, T_\kappa]. \tag{4.2.34}$$

**Remark 4.2.29.** The trick in [**133**, Lemma 9.4] applies without change in this situation as well, allowing us to conclude that

$$f(t) \le 2\Big(f_+(0) + f_-(T)\Big)\Big(e^{-\mu t} + e^{\mu(t-T)}\Big), \quad \forall 0 < t < T < T_\kappa(t_0). \tag{4.2.35}$$

Observe that this estimate is valid for any monopole $\hat{\mathsf{C}}$ on a cylinder $[-1, T+1] \times N$ provided the total energy is sufficiently small and the path $[\mathsf{C}(t)]$ lies entirely in a Kuranishi neighborhood of a 3-monopole $\mathsf{C}_0$. ■

**Lemma 4.2.30.** *Suppose there exists $0 < \tau \le T_\kappa(t_0)$ such that $h(t) \le 0$ for all $0 \le t \le \tau$. Then there exist $c, C > 0$ such that for all $t \in [0, \tau]$ we have*

$$f(t) \le 2e^{-(2\mu_- - c\kappa)t}f(0),$$

$$\|\mathsf{V}_0(t)\| \le C\|\mathsf{V}_0(0)\| \le C\kappa^2,$$

$$\|\mathsf{V}(t)\|^2_{L^2} \le C\Big(\|\mathsf{V}(0)\|^2_{L^2} + \kappa^4 e^{-(2\mu_- - c\kappa)t}\Big)$$

*and*

$$\|\mathsf{V}(t)\|_{2,2}^2 \leq C\Big(\|\mathsf{V}(0)\|_{L^2}^2 + \kappa^6 + \kappa^4 e^{-(2\mu_- - c\kappa)t}\Big) \leq C\kappa^4(1 + \kappa^2 e^{-(2\mu_- - c\kappa)t}).$$

**Proof** The inequality $f_+(t) \leq f_-(t)$ implies $f(t) \leq 2f_-(t)$. Using this information in (4.2.33) we deduce that

$$\dot{f}_-(t) \leq -(2\mu_- - c\kappa)f_-$$

from which we obtain by integration

$$f(t) \leq 2f_-(t) \leq 2e^{-(2\mu_- - c\kappa)t})f(0).$$

Using (4.2.29a) we deduce

$$f_0^{1/2}(t) = \|\Pi_1\mathsf{V}(t)\| \leq \|\Pi_1\mathsf{V}(0)\| + \int_0^t \|\Pi_1\dot{\mathsf{V}}(s)\|ds$$

$$\leq \|\Pi_1\mathsf{V}(0)\| + C\int_0^t f^{1/2}(s)ds$$

$$\leq C(\|\Pi_1\mathsf{V}(0)\| + f(0)^{1/2}e^{-(\mu_- - c\kappa)t}) \leq C\|\mathsf{V}(0)\|_{2,1}.$$

We now conclude using Lemma 4.2.24. ■

Set

$$\tau_\kappa(t_0) := \sup\{\tau \in [0, T_\kappa(t_0)];\;\; f_+(t) \leq f_-(t),\;\; \forall 0 \leq t < \tau\}. \tag{4.2.36}$$

**Lemma 4.2.31.** *For every $\varepsilon > 0$ there exist $0 < \kappa < \varepsilon$ and $t_0 > T_0(\kappa) > 0$ such that $T_\kappa(t_0) = \infty$.*

**Proof** We argue by contradiction. Thus, assume there exists $\varepsilon_0 > 0$ such that for all $\kappa < \varepsilon_0$ and all $t_0 > T_0(\kappa)$ we have $T := T_\kappa(t_0) < \infty$. Taking into account the maximality of $T_\kappa(t_0)$ we deduce

$$\|\mathsf{V}(T)\|_{2,2} = \kappa$$

so that

$$\|\mathsf{V}(T)\|_{L^2} \leq \kappa. \tag{4.2.37}$$

Using Lemma 4.2.30 we now deduce $\tau := \tau_\kappa(t_0) < T$. Set $t_1 := t_0 + \tau$ and define $\chi = \chi(\kappa)$ by

$$\chi^2 := \max\Big\{\kappa^2, \operatorname{dist}_{2,2}([\mathsf{C}(t_0 + \tau)], [\mathsf{C}_0])\Big\}.$$

Lemma 4.2.30 shows that $\kappa \leq \chi = O(\kappa)$. Observe that for $t \geq t_1$ the configuration $[\mathsf{C}(t)]$ satisfies the conditions (4.2.22),

$$\begin{cases} [\mathsf{C}(t_1)] \in U_{\mathsf{C}_0}(\chi^2) \\ E_{\hat{\mathsf{C}}}([t_1,\infty))) \leq \kappa^6 \leq \chi^6 \\ \sup_{t>t_1} \|SW([\mathsf{C}(t)])\|^2_{L^2} < \chi^6 \end{cases} \tag{4.2.38}$$

so that

$$c\chi^{-1} < T_1 := T_\chi(t_1) < \infty.$$

Redefine $\mathsf{V}(t) := \mathsf{V}(t_1 + t)$, $t \in [0, T_\chi(t_1)]$ etc. Observe that by maximality

$$\|\mathsf{V}(T_1)\|_{2,2} = \chi. \tag{4.2.39}$$

From the definition of $t_1$ as $t_1 = t_0 + \tau_\kappa(t_0)$ and the maximality of $\tau_\kappa(t_0)$ we deduce

$$f_+(t) > f_-(t), \ \ \forall t \in (0, T_1].$$

Using the inequality (4.2.32) we deduce

$$\frac{1}{2}\|\mathsf{U}(t)\|^2_{L^2} \leq f_+(t) \leq f_+(T_1)e^{-(2\mu_+ - c\chi)(T_1 - t)}$$

$$\leq \|U(T_1)\|^2_{L^2} e^{-(2\mu_+ - c\chi)(T_1 - t)}, \ \ \forall t \in [0, T_1].$$

Then

$$\|\mathsf{V}_0(T_1)\| \leq \|\mathsf{V}_0(0)\| + \|\mathsf{V}_0(T_1) - \mathsf{V}_0(0)\| \leq \|\mathsf{V}_0(0)\| + \int_0^{T_1} \|\Pi_1 \dot{\mathsf{V}}(t)\| dt$$

$$\leq \|\mathsf{V}_0(0)\| + C\int_0^{T_1} \|\mathsf{U}(t)\|_{L^2} dt$$

$$\leq \|\mathsf{V}_0(0)\| + \|U(T_1)\|_{L^2} \int_0^{T_1} e^{-(\mu_+ - c\chi)(T_1 - t)} dt$$

$$\leq \|\mathsf{V}(0)\| + C\|U(T_1)\|_{L^2} \overset{(4.2.28)}{\leq} \|\mathsf{V}_0\| + C\|SW(\mathsf{C}_0 + \mathsf{V}(t))\|_{L^2}$$

$$\overset{(4.2.38)}{\leq} \|\mathsf{V}_0(0)\| + O(\kappa^3) \overset{(4.2.38)}{\leq} \chi^2 + O(\kappa^3) = O(\kappa^2). \tag{4.2.40}$$

Thus

$$\|U(T_1)\|_{L^2} \geq \|\mathsf{V}(T_1)\|_{L^2} - C\|\mathsf{V}_0(T_1)\|_{L^2}$$

$$\overset{(4.2.23)}{\geq} C(\|\mathsf{V}(T_1)\|_{2,2} - \kappa^6) - \|\mathsf{V}_0(T_1)\|$$

$$\overset{(4.2.40)}{\geq} C\|\mathsf{V}(T_1)\|_{2,2} - C\kappa^2 \overset{(4.2.39)}{\geq} C(\kappa - \kappa^2).$$

This contradicts the inequality (4.2.28) which, coupled with the last condition in (4.2.38), implies

$$\|U(T_1)\|_{L^2} = O(\kappa^3). \quad \blacksquare$$

The above lemma has an immediate consequence.

**Corollary 4.2.32.** *There exists* $[\mathsf{C}_0] \in \mathfrak{M}_\sigma$ *such that*

$$\lim_{t\to\infty} \mathrm{dist}_{2,2}([\mathsf{C}(t)], [\mathsf{C}_0]) = 0.$$

**Proof** Lemma 4.2.31 shows that for every limit point $[\mathsf{C}_0] \in \mathfrak{M}_\sigma$ and any neighborhood $U$ of $[\mathsf{C}_0]$ in $\mathcal{C}_\sigma/\mathfrak{G}_\sigma$ there exists an instant of time $t = t_U$ such that $[\mathsf{C}(t)] \in U$, $\forall t \geq t_U$. In particular, this shows there exists exactly one limit point. $\blacksquare$

We can now prove the main result of this section.

**Theorem 4.2.33.** *Suppose* $\hat{\mathsf{C}} = (\hat{\psi}, \hat{A})$ *is a smooth finite energy monopole on* $\mathbb{R}_+ \times N$. *Then there exist a smooth gauge transformation*

$$\hat{\gamma} : \mathbb{R}_+ \times N \to S^1$$

*and a smooth monopole* $\mathsf{C}_0 = (\psi_0, A_0)$ *on* $N$ *such that*

$$\hat{\gamma} \cdot \hat{\mathsf{C}} = (\psi(t), A_0 + \mathbf{i}a(t) + \mathbf{i}f(t)dt),$$

$$\mathfrak{L}^*_{\mathsf{C}_0}(\psi(t) - \psi_0, \mathbf{i}a(t)) = 0 \Longleftrightarrow (\psi(t), A_0 + a(t)) \in \mathcal{S}_{\mathsf{C}_0}, \quad \forall t \gg 0,$$

$$\lim_{t\to\infty} e^{\lambda t}\Big(\|\psi(t) - \psi_0\|_{L^{2,2}(N)} + \|a(t)\|_{L^{2,2}(N)} + \|f(t)\|_{L^{3,2}(N)}\Big) = 0,$$

$\forall 0 \leq \lambda < \mu_-(\mathsf{C}_0)$.

**Proof** Fix a smooth representative $\mathsf{C}_0$ of the limit of $[\mathsf{C}(t)]$ as $t \to \infty$. For all $\kappa$ sufficiently small we can find a smooth gauge transformation $\hat{\gamma}$ on $\mathbb{R}_+ \times N$ such that $\hat{\gamma} \cdot \hat{\mathsf{C}}$ is in $\kappa$-standard gauge with respect to $\mathsf{C}_0$ on a semi-cylinder $[T_0(\kappa), \infty) \times N$. Re-label $\hat{\mathsf{C}} := \hat{\gamma} \cdot \hat{\mathsf{C}}$. Then there exists a $t_0 \geq T_0(\kappa) > 0$ such that

$$E_{\hat{\mathsf{C}}}([t_0, \infty)) < \kappa^3,$$

$$\|\mathsf{C}(t_0), \mathsf{C}_0\|_{L^{2,2}(N)} := \|\psi(t) - \psi_0\|_{L^{2,2}(N)} + \|a(t)\|_{L^{2,2}(N)} \leq \kappa^2,$$

$$\|\mathsf{C}(t_0 + t) - \mathsf{C}_0\|_{L^{2,2}(N)} \leq \kappa,$$

$\forall 0 \leq t \leq T_\kappa(t_0)$. Observe that $\tau_\kappa(t_0)$ defined in (4.2.36) is infinite. Indeed, if $\tau_\kappa(t_0) < \infty$ then, arguing as in the proof of Lemma 4.2.31, we would deduce that $f_+(\tau_\kappa + t)$ increases exponentially. This is plainly impossible.

Using Lemma 4.2.30 we deduce

$$\|\mathsf{U}(t)\|_{L^2} \leq Ce^{-(\mu_- - c\kappa)t}, \ \forall t \geq T_0(\kappa)$$

and

$$\|\Pi_1 \mathsf{V}(t)\| = \|\Pi_1 \mathsf{V}(t) - \Pi_1 \mathsf{V}(\infty)\| \leq \int_t^\infty \|\Pi_1 \dot{\mathsf{V}}(s)\| ds$$

$$\leq C \int_t^\infty e^{-(\mu_- - c\kappa)s} ds \leq Ce^{-(\mu_- - c\kappa)t}.$$

This shows that

$$\|\mathsf{V}(t)\|_{2,2} \leq C\|\mathsf{V}(t)\|_{L^2} \leq Ce^{-(\mu_- - c\kappa)t}, \ \forall t \geq T_0(\kappa)$$

so that

$$\lim_{t\to\infty} e^{(\mu_- - c\kappa)t} \mathrm{dist}_{2,2}([\mathsf{C}(t)], [\mathsf{C}_0]) = 0, \ \forall \kappa \ll 1. \ \blacksquare$$

**Remark 4.2.34.** The gauge transformation $\hat{\gamma}$ postulated by the above theorem may not be in the identity component of the group of gauge transformations on $\mathbb{R}_+ \times N$. The group of components is parameterized by $H^1(N, \mathbb{Z})$. If $\hat{\gamma}$ lies in the component parameterized by $c \in H^1(N, \mathbb{Z})$ then we can find a smooth map

$$\gamma : N \to S^1$$

which belongs to the component of $\mathfrak{G}_\sigma$ corresponding to $c$. We can think of $\gamma$ as a $t$-independent gauge transformation on $\mathbb{R}_+ \times N$. Moreover $\hat{\gamma}_c := \hat{\gamma} \cdot \gamma^{-1}$ lies in the identity component of the group of gauge transformations on $\mathbb{R}_+ \times N$ and $\hat{\gamma}_c \cdot \hat{\mathsf{C}}$ will satisfy similar asymptotic behavior as $\hat{\gamma} \cdot \hat{\mathsf{C}}$ with $\mathsf{C}_0$ replaced by $\gamma_c^{-1} \cdot \mathsf{C}_0$. Thus we can strengthen the conclusion of Theorem 4.2.33 by adding the fact that $\hat{\gamma}$ can be chosen to be of the special form $\hat{\gamma} = e^{\mathbf{i}\hat{f}}$.

The above convergence result can be slightly strengthened.

**Proposition 4.2.35.** *With the above notation, for every nonnegative integer $m$ and every $0 \leq \lambda < \frac{\mu_-}{2}$ there exists a constant which depends only $m$ and $\lambda$ and the geometry of $N$ such that*

$$\|\mathsf{V}(t)\|_{L^{k,2}_\lambda([T_0(\kappa),\infty)\times N)} \leq C\kappa.$$

**Exercise 4.2.6.** Prove the above proposition.

**Proposition 4.2.36.** *Fix an instant of time $T_0 > 0$. Then there exists a constant $\kappa_0 > 0$ with the following property. For every $\kappa < \kappa_0$, and every monopole $\hat{\mathsf{C}}$ on $\mathbb{R}_+ \times N$ such that*

$$\|\rho_{\hat{\mathsf{C}}}\|^2_{L^2([T_0,\infty)\times N} = E_{\hat{\mathsf{C}}}([T_0,\infty)) < \kappa^6,$$

*and*

$$\mathrm{dist}_{L^{2,2}}([\mathsf{C}(T_0)], \mathfrak{M}_\sigma) \le \kappa^2$$

*we have*

$$\sup_{t > T_0+1} \|SW([\mathsf{C}(t)])\|^2_{L^2(N)} \le C\kappa^6$$

- $[\mathsf{C}(t)] \in U_\kappa$, $\forall t > T_0$.

- *There exist a monopole $\mathsf{C}_\infty$ on $N$ and a smooth gauge transformation $\hat{\gamma}$ on $\mathbb{R}_+ \times N$ such that*

$$\lim_{t\to\infty} \|\hat{\gamma}\mathsf{C}\,|_{t\times N} - \mathsf{C}_\infty\|_{L^{2,2}(N)}.$$

Proposition 4.2.36 is a simple consequence of the previous considerations and we leave its proof to the reader.

**Exercise 4.2.7.** Prove Proposition 4.2.36.

Proposition 4.2.35 can be roughly interpreted as saying that, if the total energy of the monopole $\hat{\mathsf{C}}$ is below a certain *capture level*, then its dynamics is constrained to a small Kuranishi neighborhood of some 3-monopole on $N$.

Up to now we have worked on a very special cylindrical manifold, $\hat{N} := \mathbb{R}_+ \times N$. The results we proved extend without difficulty to the case when $\hat{N}$ is a cylindrical manifold without boundary such that $\partial_\infty \hat{N} = N$. The next result summarizes all the facts proved so far.

**Theorem 4.2.37.** *Fix $T > 0$. There exists a constant $\hbar > 0$ with the following property. If $m \in \mathbb{Z}_+$, $0 \le \lambda < \mu_-(\mathsf{C}_0)$, there exists a constant $C$ depending on $m$, $\lambda$ and the geometry of $N$ such that for any smooth monopole $\hat{\mathsf{C}} = (\hat{\psi}, \hat{A})$ satisfying*

$$\int_{[T,\infty)\times N} \rho_{\hat{\mathsf{C}}} \le \hbar$$

*there exist a smooth function*

$$\hat{u} : \mathbb{R}_+ \times N \to S^1$$

*and a smooth monopole* $\mathsf{C}_0 = (\psi_0, A_0)$ *on* $N$ *such that along the neck*

$$e^{\mathbf{i}\hat{u}} \cdot \hat{\mathsf{C}} = (\psi(t), A_0 + \mathbf{i}a(t) + \mathbf{i}f(t)dt)$$

$$\mathfrak{L}^*_{\mathsf{C}_0}(\psi(t) - \psi_0, \mathbf{i}a(t)) = 0 \Longleftrightarrow (\psi(t), A_0 + a(t)) \in \mathcal{S}_{\mathsf{C}_0}, \ \ \forall t \geq T$$

*and*

$$\|\psi(t) - \psi_0\|_{L^{m,2}_\lambda([T,\infty)\times N)} + \|a(t)\|_{L^{m,2}_\lambda([T,\infty)\times N)} + \|f(t)\|_{L^{m,2}_\lambda([T,\infty)\times N)} < C.$$

**Remark 4.2.38.** We would like to say a few words about an alternate proof of Theorem 4.2.33 which works in the more general situation when (**N**) is not satisfied (see [**96**]). For simplicity we will describe it briefly in our nondegenerate context.

Observe that (4.2.15) can be rewritten as

$$T_\psi \mathbf{i}f = 2\mathfrak{Im}\langle\psi, \mathfrak{D}_A\psi\rangle = 2\mathfrak{Im}\langle\psi, \dot{\psi}\rangle$$

where

$$\|\mathfrak{Im}\langle\psi, \dot{\psi}\rangle\|_{L^2(N)} \leq C\|\psi\|_{2,2}\|\dot{\psi}\|_{L^2}$$

from which we deduce that

$$\|\mathfrak{N}(\mathsf{V})\|_{L^2(N)} \leq C\|\mathsf{V}\|_{L^{2,2}}\|\dot{\mathsf{V}}\|_{L^2}.$$

Next observe that there exists a constant depending only on the geometry of $N$ such that if $\mathsf{V} \in U_{\mathsf{C}_0}(\kappa)$ is sufficiently small in the $L^{2,2}$-norm then

$$|\mathcal{E}(\mathsf{C}_0 + \mathsf{V}) - \mathcal{E}(\mathsf{C}_0)|^{1/2} \leq C\|SW(\mathsf{C}_0 + \mathsf{V})\|_{L^2(N)},$$

$$\|SW(\mathsf{C}_0 + \mathsf{V})\|_{L^2} \geq C\mathrm{dist}_{L^2}\Big(\mathsf{C}_0 + \mathsf{V}, \mathfrak{M}_\sigma \cap U_{\mathsf{C}_0}(\kappa)\Big).$$

If $\kappa$ is sufficiently small then, following the proof of [**123**, Lemma 1, p. 541], we deduce that if $\mathsf{V}(t) \in U_{\mathsf{C}_0}(\kappa)$ for all $t \in [t_0, t_1]$ then

$$\int_{t_0}^{t_1} \|\dot{\mathsf{V}}(t)\|_{L^2(N)}dt \leq C\Big(E_{\hat{\mathsf{C}}}([t_0, \infty))^{1/2} - E_{\hat{\mathsf{C}}}([t_1, \infty))^{1/2}\Big) \tag{4.2.41}$$

$$\leq C' E_{\hat{\mathsf{C}}}([t_0, t_1])^{1/2}$$

where $C, C'$ are geometric constants. Using Corollary 4.2.15 it is now a relatively simple job to establish the existence of an asymptotic limit. We refer for details to [**96**, Chap. 4].

**4.2.5. Proofs of some technical results.** As promised, we include in this subsection some proofs which would have diverted the reader's attention had they been included in the middle of the flow of arguments in the previous subsections.

**Proof of Lemma 4.2.14** Set $C_T := [-T, T] \times N$ and denote by $\hat{A}_0$ the connection induced by $A_0$ on the cylinder $C_2$. There exists $t \in [-2, 2]$ such that

$$\|SW(\mathsf{C}(t_0))\|^2_{L^2(N)} < E/4 \leq \omega_0/4.$$

Now fix $\omega_0$ sufficiently small so that

$$\operatorname{dist}_{L^{2,2}}([\mathsf{C}(t_0)], \mathfrak{M}_\sigma) \leq 1/100$$

for some $t_0 \in [-1, 1]$. Set $\mathsf{C}_0 := (0, A_0)$ and

$$\delta := \sup\Big\{ \operatorname{dist}_{L^{2,2}}([\mathsf{C}_0], [\mathsf{C}]);\ [\mathsf{C}] \in \mathfrak{M}_\sigma \Big\}.$$

Observe that $\delta < \infty$ since $\mathfrak{M}_\sigma$ is compact. We can find a smooth gauge transformation such that

$$\|\mathsf{C}_0 - \gamma \cdot \mathsf{C}(t_0)\|_{L^2} \leq \delta + 1/50.$$

Now observe that both the hypotheses and the conclusion of Lemma 4.2.14 are invariant under the action of the group of smooth gauge transformations on $N$. Thus, modulo such a transformation we can assume that our monopole $\hat{\mathsf{C}}$ satisfies the additional restriction

$$\|a(t_0)\|_{L^{2,2}(N)} \leq \delta + 1/50$$

for some $t_0 \in [-1, 1]$. Hölder's inequality now implies

$$\begin{aligned} \|a(t)\|_{L^2(N)} &\leq \|a(t_0)\|_{L^2(N)} + \int_{t_0}^{t} \|\dot{a}(s)\|_{L^2(N)} ds \\ &\leq \delta + 1/50 + 2E^{1/2}. \end{aligned} \tag{4.2.42}$$

The Seiberg-Witten equations have the form

$$\begin{cases} \mathfrak{D}_{\hat{A}_0}\psi = -\frac{\mathbf{i}}{2}\hat{\mathbf{c}}(a(t))\psi \\ \mathbf{i}\dot{a} = \frac{1}{2}q(\psi) - *\mathbf{i}da - *F_{A_0} \end{cases}.$$

If we apply $d^*$ to the last equality we deduce

$$\mathbf{i}d^*\dot{a} = \frac{1}{2}d^*q(\psi) \overset{(4.2.14)}{=} -\frac{\mathbf{i}}{2}\mathfrak{Im}\langle \psi, \mathfrak{D}_{A_0+a(t)}\psi\rangle = -\frac{\mathbf{i}}{2}\mathfrak{Im}\langle \psi, \dot{\psi}(t)\rangle.$$

Now regard $\dot{a}$ as a 1-form on the four-dimensional cylinder. Since $\lrcorner_t \dot{a} = 0$ we deduce $\hat{d}^*\dot{a} = d^*\dot{a}$. Set $b := \dot{a}$, $\phi := \dot{\psi}$. By differentiating the Seiberg-Witten equations with respect to $t$ we deduce

$$\begin{cases} \mathfrak{D}_{\hat{A}_0}\phi = -\frac{\mathbf{i}}{2}\hat{\mathbf{c}}(a(t))\phi + \frac{\mathbf{i}}{2}\hat{\mathbf{c}}(b)\psi \\ \mathbf{i}\dot{b} = q(\psi,\phi) - *\mathbf{i}db \\ \hat{d}^*b = -\frac{\mathbf{i}}{2}\mathfrak{Im}\langle\psi,\phi\rangle \end{cases}$$

$$\Longleftrightarrow \begin{cases} \mathfrak{D}_{\hat{A}_0}\phi = -\frac{\mathbf{i}}{2}\hat{\mathbf{c}}(a(t))\phi + \frac{\mathbf{i}}{2}\hat{\mathbf{c}}(b)\psi \\ \mathbf{ASD}(\mathbf{i}b) = \Big(q(\psi,\phi)\Big) \oplus \Big(-\frac{\mathbf{i}}{2}\mathfrak{Im}\langle\psi,\phi\rangle\Big) \end{cases}.$$

According to (4.2.10) there exists a geometric constant $C > 0$ such that

$$\sup_{|t|\leq 1} \|\psi(t)\|_{L^\infty}(N) < C(1 + E^{1/4}) \leq C$$

so that

$$\|\mathbf{ASD}(\mathbf{i}b)\|_{L^2(C_2)} \leq CE.$$

Using interior elliptic estimates for the elliptic operator **ASD** we deduce

$$\|b\|_{L^{1,2}(C_{3/2})} \leq C\Big(E + \|b\|_{L^2(C_2)}\Big) \leq CE.$$

Thus, for all $t \in [-3/2, 3/2]$ we have

$$\|a(t)\|_{1,2} \leq \|a(t_0)\|_{1,2} + |t - t_0|^{1,2} \int_{t_0}^t \|b(s)\|_{L^{1,2}(N)} ds \leq C.$$

Using the Sobolev embedding

$$L^{1,2}(N) \hookrightarrow L^6(N)$$

we deduce

$$\|a(t)\|_{L^6(C_{3/2})} \leq C.$$

Thus

$$\|\hat{\mathbf{c}}(a)\phi\|_{L^{3/2}(C_{3/2})} \leq C\|\phi\|_{L^2(C_{3/2})} \leq CE.$$

Using interior elliptic estimates for

$$\mathfrak{D}_{\hat{A}_0}\phi = -\frac{\mathbf{i}}{2}\hat{\mathbf{c}}(a(t))\phi + \frac{\mathbf{i}}{2}\hat{\mathbf{c}}(b)\psi \tag{4.2.43}$$

on $C_{3/2}$ we deduce

$$\|\phi\|_{L^{1,3/2}(C_{4/3})}$$

$$\leq C\Big(\|\phi\|_{L^{3/2}(C_{3/2})} + \| -\frac{\mathbf{i}}{2}\hat{\mathbf{c}}(a(t))\phi + \frac{\mathbf{i}}{2}\hat{\mathbf{c}}(b)\psi\|_{L^{3/2}(C_{3/2})}\Big) \leq CE.$$

Using the Sobolev embedding $L^{1,3/2}(C_{4/3}) \hookrightarrow L^{12/5}(C_{4/3})$ and the Hölder inequality (with $1/6 + 5/12 = 7/12$) we deduce

$$\|\hat{\mathbf{c}}(a)\phi\|_{L^{12/7}(C_{4/3})} \leq CE$$

and we conclude as before using (4.2.43) that

$$\|\phi\|_{L^{1,12/7}(C_{5/4})} \leq CE.$$

Now use the Sobolev embedding $L^{1,12/7}(C_{5/4}) \hookrightarrow L^3(C_{5/4})$ and the Hölder inequality (with $1/6 + 1/3 = 1/2$) to deduce

$$\|\hat{\mathbf{c}}(a)\phi\|_{L^2(C_{5/4})} \leq CE.$$

Using (4.2.43) again we deduce

$$\|\phi\|_{L^{1,2}(C_{6/5})} \leq CE.$$

Thus

$$\|b\|_{L^{1,2}(C_{6/5})} + \|\phi\|_{L^{1,2}(C_{6/5})} \leq CE.$$

Using trace theorems (see [**79**]) we deduce

$$\|b(t)\|_{L^2} + \|\phi(t)\|_{L^2} \leq \|b\|_{L^{1,2}(C_{6/5})} + \|\phi\|_{L^{1,2}(C_{6/5})} \leq CE, \quad \forall t \in [-1,1].$$

The last inequality is precisely the content of Lemma 4.2.14. ■

**Proof of Lemma 4.2.24** Consider $\tau_0 > 0$ such that

$$\|\mathsf{V}(t)\|_{2,2} \leq C\kappa, \ \ \forall |t-\tau_0| \leq 1, \ \ E_{\hat{\mathsf{C}}}([t_0+\tau_0-1,\infty)) \leq \kappa^6.$$

Set $I_j = (\tau_0 - 1/2^j, \tau_0 + 1/2^j)$. We will first prove that there exists $j > 0$ such that

$$\|\mathsf{V}(t)\|_{L^{3,2}(I_j\times N)} \leq C\|\mathsf{V}(t)\|_{L^2(I_0\times N)} \tag{4.2.44}$$

where $\mathsf{V}(t) = \mathsf{C}(t) - \mathsf{C}_0$. We follow an approach similar to the one used in the proof of Lemma 4.2.14.

Rewrite equations (4.2.17) and (4.2.18) as an *elliptic system* over the 4-manifold $I_0 \times N$

$$(\partial_t - \mathfrak{D}_{A_0})\psi(t) = \frac{1}{2}\mathbf{c}(\mathbf{i}a(t))\psi - \frac{\mathbf{i}f}{2}(\psi(t)+\psi_0), \tag{4.2.45a}$$

$$\mathbf{ASD}\cdot\begin{bmatrix} \mathbf{i}a(t) \\ -\mathbf{i}f(t)\end{bmatrix} = \begin{bmatrix} \frac{1}{2}q(\psi_0+\psi) - *F_{A_0} \\ \\ -\frac{\mathbf{i}}{2}\mathfrak{Im}\langle\psi_0,\psi\rangle - \mathbf{i}\dot{f}\end{bmatrix}. \tag{4.2.45b}$$

The component $f$ is uniquely determined by $\psi$ via the differential equation on $N$

$$T_{\psi(t)}(\mathbf{i}f) := \mathfrak{L}^*_{\mathsf{C}_0}\mathfrak{L}_{\mathsf{C}_0}\mathbf{i}f + \mathfrak{Re}\langle\psi_0,\psi\rangle\mathbf{i}f = 2\mathbf{i}\mathfrak{Im}\langle\psi,\mathfrak{D}_{A_0}\psi\rangle. \tag{4.2.46}$$

Observe also that

$$\mathbf{ASD}\cdot\left[\ \mathbf{i}a(t)\ \right]=\left[\begin{array}{c}\frac{1}{2}q(\psi_0+\psi)-*F_{A_0}+\mathbf{i}df\\ \\ -\frac{\mathbf{i}}{2}\mathfrak{Im}\langle\psi_0,\psi\rangle\end{array}\right]. \tag{4.2.47}$$

Our strategy is very simple although the details are somewhat cumbersome. We will use the fact that (4.2.45a) + (4.2.45b) form an elliptic system and then, relying on interior elliptic estimates, we will gradually prove that stronger and stronger norms of the right-hand side, on gradually smaller subdomains of $I_0\times N$, can be estimated from above by the $L^2$-norm of $\mathsf{V}$ on $I_0\times N$.

Observe first that $L^{2,2}(N)$ embeds continuously in $L^\infty(N)$ because $N$ is three-dimensional. The $L^{1,2}$-norm of the right hand side of (4.2.46) is bounded from above by $C\|\psi\|_{2,2}$ and thus we have a bound

$$\|f\|_{L^{3,2}(N)}\leq C\|\psi\|_{L^{2,2}(N)}.$$

Using interior elliptic estimates for the elliptic equation (4.2.45a) on $I\times N$ we deduce

$$\|\psi(t)\|_{L^{1,2}(I_1\times N)}\leq C\Big(\|\psi(t)\|_{L^2(I_0\times N)}+\|c(\mathbf{i}a(t)\psi(t)\|_{L^2(I_0\times N)}$$
$$+\|\mathbf{i}f(\psi+\psi_0)\|_{L^2(I_0\times N)}\Big)$$

(use $\|\psi\|_\infty\leq C$)

$$\leq C(\|\psi(t)\|_{L^2(I_0\times N)}+\|a(t)\|_{L^2(I_0\times N)})=C\|\mathsf{V}(t)\|_{L^2(I_0\times N)}. \tag{4.2.48}$$

In particular, we deduce

$$\|\dot\psi(t)\|_{L^2(I_1\times N)}\leq C\|\mathsf{V}(t)\|_{L^2(I_0\times N)}. \tag{4.2.49}$$

Set $\phi(t):=\mathfrak{D}_{A_0}\psi$. Then

$$\dot\phi(t)-\mathfrak{D}_{A_0}\phi(t)=\frac{1}{2}[\mathfrak{D}_{A_0},\mathbf{c}(\mathbf{i}a)]\psi+\frac{1}{2}\mathbf{c}(\mathbf{i}a)\phi-\frac{\mathbf{i}}{2}\mathbf{c}(df)\psi-\frac{\mathbf{i}f}{2}\phi. \tag{4.2.50}$$

Thus, we have

$$\|\phi(t)\|_{L^{1,2}(I_2\times N)}\leq C\Big(\|\phi(t)\|_{L^2(I_1\times N)}+\big\|\ \|\mathbf{c}(\mathbf{i}df(t))\psi(t)\|_{L^2(N)}\ \big\|_{L^2(I_1)}$$
$$+\big\|\ \|\mathbf{c}(\mathbf{i}a)(t)\phi\|_{L^2(N)}\ \big\|_{L^2(I_1)}+\|\ [\mathfrak{D}_{A_0},\mathbf{c}(\mathbf{i}a)]\psi\ \|_{L^2(I_1\times N)}\Big).$$

Now use

$$\|df\|_{L^{2,2}(N)}+\|\psi\|_{L^{2,2}(N)}+\|a\|_{L^{2,2}(N)}\leq C\kappa,$$

$$L^{1,2}(N)\hookrightarrow L^6(N)\hookrightarrow L^4(N)$$

and

$$\|\mathbf{c}(\mathbf{i}a)\phi\|_{L^2(N)} \leq C\|a\|_{L^4(N)}\|\phi\|_{L^4(N)}$$

to deduce

$$\|\mathbf{c}(\mathbf{i}a)(t)\phi\|_{L^2(N)} + \|\mathbf{c}(\mathbf{i}df(t))\psi(t)\|_{L^2(N)} \leq C\kappa\big(\|\phi\|_{L^2(N)} + \|\psi\|_{L^2(N)}\big)$$

and

$$\|[\mathfrak{D}_{A_0}, \mathbf{c}(\mathbf{i}a)]\psi\|_{L^2(N)} \leq C\kappa\|\psi\|_{L^{1,2}(I_1\times N)} \leq C\kappa\|\mathsf{V}\|_{L^2(I_0\times N)}$$

Hence

$$\begin{aligned} &\|\phi(t)\|_{L^{1,2}(I_2\times N)} \leq C\Big(\|\phi(t)\|_{L^2(I_1\times N)} + \kappa\|\mathsf{V}(t)\|_{L^2(I_1\times N)})\Big) \\ &\leq C\Big(\|\psi(t)\|_{L^{1,2}(I_1\times N)} + \|\mathsf{V}(t)\|_{L^2(I_\times N)}\Big) \leq C\|\mathsf{V}(t)\|_{L^2(I_0\times N)} \end{aligned} \tag{4.2.51}$$

Differentiating (4.2.46) with respect to $t$ we deduce

$$\begin{aligned} T_{\psi(t)}(\mathbf{i}\dot{f}) &= F(t) \\ &:= -\mathbf{i}\mathfrak{Re}\langle\dot{\psi}(t),\psi_0\rangle - 2\mathbf{i}\mathfrak{Im}\langle\dot{\psi}(t),\phi\rangle + 2\mathbf{i}\mathfrak{Im}\langle\psi,\dot{\phi}\rangle. \end{aligned} \tag{4.2.52}$$

Since $\dot{f}\perp\ker\mathfrak{L}_{\mathsf{C}_0}$ and $\|\psi(t)\|_{L^{2,2}(N)}$ is small we deduce from Lemma 4.2.22 that for every $1<p\leq 2$ there exists a constant $C_p>0$ such that

$$\|\dot{f}\|_{L^{2,p}(N)} \leq C\|F(t)\|_{L^p(N)}.$$

Using the Sobolev embedding $L^{1,2}(N)\hookrightarrow L^6(N)$, Hölder's inequality (in the case $4/6 = 1/6 + 1/2$) and the estimates

$$\|\phi\|_{L^{1,2}(N)} \leq C\kappa, \quad \|\psi\|_\infty < C$$

we deduce

$$\begin{aligned} \|F(t)\|_{L^{3/2}(N)} &\leq C(\|\dot{\psi}(t)\|_{L^2(N)} + \|\dot{\psi}(t)\|_{L^2(N)}\|\phi(t)\|_{L^{1,2}(N)} + \kappa\|\dot{\phi}(t)\|_{L^2(N)}) \\ &\leq C\Big(\kappa\|\dot{\phi}(t)\|_{L^2(N)} + \|\dot{\psi}(t)\|_{L^2(N)}\Big). \end{aligned}$$

Invoking the Sobolev embedding

$$L^{2,3/2}(N)\hookrightarrow L^{1,2}(N)$$

we deduce

$$\|\dot{f}(t)\|_{L^{1,2}(N)} \leq C\Big(\|\dot{\psi}(t)\|_{L^2(N)} + \kappa\|\dot{\phi}(t)\|_{L^2(N)} + \|\dot{\psi}(t)\|_{L^2(N)}\Big)$$

so that we get

$$\|\dot{f}(t)\|_{L^{1,2}(N)} \leq C\|\dot{f}(t)\|_{L^2(N)} \leq C\Big(\|\dot{\psi}(t)\|_{L^2(N)} + \|\dot{\phi}(t)\|_{L^2(N)}\Big). \tag{4.2.53}$$

Integrating over $I_2$ and taking (4.2.49) and (4.2.51) into account we deduce

$$\begin{aligned} \|\dot{f}\|_{L^2(I_2\times N)} + \|d\dot{f}\|_{L^2(I_2\times N)} &\leq C(\|\dot{\psi}(t)\|_{L^2(I_2\times N)} + \|\dot{\phi}\|_{L^2(I_2\times N)}) \\ &\leq C\|\mathsf{V}(t)\|_{L^2(I_0\times N)}. \end{aligned} \tag{4.2.54}$$

To proceed further observe that

$$q(\psi_0 + \psi) = q(\psi_0) + 2q(\psi_0, \psi) + q(\psi)$$

where $q(u, v)$ is the symmetric bilinear map associated to the quadratic map $q(u)$,

$$q(u, v) := \frac{1}{4}(q(u + v) - q(u - v)).$$

Since $q(\psi_0) = 2 * F_{A_0}$ the equation (4.2.45b) can be rewritten as

$$\mathbf{ASD} \cdot \begin{bmatrix} \mathbf{i}a(t) \\ -\mathbf{i}f(t) \end{bmatrix} = \begin{bmatrix} \frac{1}{2}q(\psi) + q(\psi_0, \psi) \\ \\ -\frac{\mathbf{i}}{2}\mathfrak{Im}\langle\psi_0, \psi\rangle - \mathbf{i}\dot{f} \end{bmatrix}. \tag{4.2.55}$$

Using interior elliptic estimates we deduce

$$\|(a, f)\|_{L^{1,2}(I_3 \times N)} \leq C_p\Big(\|a(t)\|_{L^2(I_0 \times N)} + \|\psi\|_{L^2(I_2 \times N)} + \|\dot{f}\|_{L^2(I_2 \times N)}\Big) \leq C\|\mathsf{V}(t)\|_{L^2(I_0 \times N)}. \tag{4.2.56}$$

Putting together the estimates (4.2.48) and (4.2.56) we deduce

$$\|\mathsf{V}(t)\|_{L^{1,2}(I_3 \times N)} \leq C\|\mathsf{V}(t)\|_{L^2(I_0 \times N)}, \quad p \in (1, 2). \tag{4.2.57}$$

Thus, we have estimated the $L^{1,2}(I_3 \times N)$-norm of $\mathsf{V}(t)$ by a weaker one, $L^2(I_0 \times N)$. We iterate this procedure. Observe that the $L^{1,2}(I_3 \times N)$-norm of the right-hand side of (4.2.45a) is bounded from above by the $L^2(I_0 \times N)$-norm of $\mathsf{V}$ so, invoking the interior elliptic estimates, we deduce

$$\|\psi\|_{L^{2,2}(I_4 \times N)} \leq C\|\mathsf{V}\|_{L^2(I_0 \times N)}.$$

Using this estimate and estimate (4.2.53) in (4.2.47) we deduce that the $L^{1,2}(I_4 \times N)$-norm of the right-hand side of (4.2.47) is bounded from above by the $L^2(I_0 \times N)$-norm of $\mathsf{V}$. Using the interior elliptic estimates we deduce

$$\|a\|_{L^{2,2}(I_5 \times N)} \leq C\|\mathsf{V}\|_{L^2(I_0 \times N)}.$$

This shows

$$\|\mathsf{V}\|_{L^{2,2}(I_5 \times N)} \leq C\|\mathsf{V}\|_{L^2(I_0 \times N)}. \tag{4.2.58}$$

Differentiating (4.2.50) with respect to $t$ we deduce that $\dot{\phi}$ satisfies the elliptic equation

$$\begin{aligned} \partial_t\dot{\phi} - \mathfrak{D}_{A_0}\dot{\phi} &= \frac{1}{2}[\mathfrak{D}_{A_0}, \mathbf{c}(\mathbf{i}\dot{a})]\psi + \frac{1}{2}[\mathfrak{D}_{A_0}, \mathbf{c}(\mathbf{i}a)]\dot{\psi} \\ &\quad -\frac{\mathbf{i}}{2}\Big(\mathbf{c}(d\dot{f})\psi + \mathbf{c}(df)\dot{\psi} + \dot{f}\phi + f\dot{\phi}\Big). \end{aligned} \tag{4.2.59}$$

By trace results (see [**79**]) we deduce

$$\|\dot{a}(t)\|_{L^{1/2,2}(N)} \leq C\|\dot{a}(t)\|_{L^{1,2}(I_5\times N)}, \quad \|\dot{\psi}(t)\|_{L^{1/2,2}(N)} \leq C\|\dot{\psi}(t)\|_{L^{1,2}(I_5\times N)}.$$

Using the continuous Sobolev embeddings

$$L^{1/2,2}(N) \hookrightarrow L^3(N), \quad L^{1,2}(N) \hookrightarrow L^6(N)$$

and the Hölder inequality, which produces a bounded bilinear map

$$L^3(N) \times L^6(N) \to L^2(N), \quad (u,v) \mapsto uv,$$

we deduce

$$\Big\| \, [\mathfrak{D}_{A_0}, \mathbf{c}(\mathbf{i}\dot{a})]\psi \Big\|_{L^2(N)}$$

$$\leq C\Big( \|\dot{a}\|_{L^{1,2}(N)}\|\psi\|_{L^\infty(N)} + \|\dot{a}\|_{L^{1/2,2}(N)}\|\psi\|_{L^{1,2}(N)}\Big)$$

$$\leq C\kappa\Big( \|\mathsf{V}\|_{L^2(I_0\times N)} + \|\dot{a}\|_{L^{1,2}(N)}\Big)$$

so that

$$\Big\| \, [\mathfrak{D}_{A_0}, \mathbf{c}(\mathbf{i}\dot{a})]\psi \, \Big\|_{L^2(I_5\times N)} \leq C\kappa\Big( \|\mathsf{V}\|_{L^2(I_0\times N)} + \|a\|_{L^{2,2}(I_5\times N)}\Big).$$

Using (4.2.53) and the $L^\infty$-estimates on $f$ and $\psi$ we deduce

$$\Big\|\mathbf{c}(d\dot{f})\psi + \mathbf{c}(df)\dot{\psi} + \dot{f}\phi + f\dot{\phi}\Big\|_{L^2(I_5\times N)} \leq C\|\mathsf{V}\|_{L^2(I_0\times N)}.$$

Applying the interior elliptic estimates to (4.2.59) we deduce

$$\|\dot{\phi}\|_{L^{1,2}(I_6\times N)} \leq C\|\mathsf{V}\|_{L^2(I_0\times N)}.$$

Differentiating (4.2.52) with respect to $t$ we deduce

$$\mathfrak{L}^*_{\mathsf{C}_0}\mathfrak{L}_{\mathsf{C}_0}\mathbf{i}\ddot{f} + \mathbf{i}\mathfrak{Re}\langle\psi(t),\psi_0\rangle\ddot{f} = -\mathbf{i}\mathfrak{Re}\langle\ddot{\psi}(t),\psi_0\rangle\dot{f} + 4\mathbf{i}\mathfrak{Im}\langle\dot{\psi}(t),\dot{\phi}\rangle$$

$$+2\mathbf{i}\mathfrak{Im}\langle\ddot{\psi}(t),\phi(t)\rangle + 2\mathbf{i}\mathfrak{Im}\langle\psi,\ddot{\phi}\rangle.$$

We can rewrite the last equation as

$$T_{\psi(t)}(\mathbf{i}\ddot{f}) = -\mathbf{i}\mathfrak{Re}\langle\ddot{\psi}(t),\psi_0\rangle\dot{f} + 4\mathbf{i}\mathfrak{Im}\langle\dot{\psi}(t),\dot{\phi}\rangle$$

$$+2\mathbf{i}\mathfrak{Im}\langle\ddot{\psi}(t),\phi(t)\rangle + 2\mathbf{i}\mathfrak{Im}\langle\psi,\ddot{\phi}\rangle.$$

Since $\|\psi(t)\|_{L^{2,2}(N)}$ is small we deduce from Lemma 4.2.22 that for every $1 < p \leq 2$ we have

$$\|\ddot{f}\|_{L^{2,p}(N)}$$

$$\leq C_p\Big\|-\mathfrak{Re}\langle\ddot{\psi}(t),\psi_0\rangle\dot{f} + 4\mathfrak{Im}\langle\dot{\psi}(t),\dot{\phi}\rangle + 2\mathfrak{Im}\langle\ddot{\psi}(t),\phi(t)\rangle + 2\mathfrak{Im}\langle\psi,\ddot{\phi}\rangle\Big\|_{L^p(N)}.$$

Now observe that

$$\|\mathfrak{Re}\langle\ddot{\psi}(t),\psi_0\rangle\dot{f}\|_{L^{3/2}(N)} \leq C\|\ddot{\psi}\|_{L^2(N)}\|\dot{f}\|_{L^6(N)} \leq C\|\ddot{\psi}\|_{L^2(N)}\|\dot{f}\|_{L^{1,2}(N)}$$

(use 4.2.53) and trace results)

$$\leq C\|\mathsf{V}\|_{L^2(I_0\times N)}\|\ddot{\psi}\|_{L^2(N)}.$$

Similarly

$$\|\mathfrak{Im}\langle\ddot{\psi}(t),\phi(t)\rangle\|_{L^{3/2}(N)} \leq C\|\ddot{\psi}\|_{L^2(N)}\|\phi\|_{L^{1,2}(N)} \leq C\kappa\|\ddot{\psi}\|_{L^2(N)}.$$

Next observe

$$\|\mathfrak{Im}\langle\dot{\psi},\dot{\phi}\rangle\|_{L^{3/2}} \leq C\|\dot{\psi}\|_{L^3(N)}\|\dot{\phi}\|_{L^3(N)} \leq C\|\dot{\psi}(t)\|_{L^{1/2,2}(N)}\|\dot{\phi}\|_{L^{1/2,2}(N)}$$

(use trace results)

$$\leq C\|\dot{\psi}\|_{L^{1,2}(I_6\times N)}\|\dot{\phi}\|_{L^{1,2}(I_6\times N)}.$$

Finally

$$\|\mathfrak{Im}\langle\psi,\ddot{\phi}\rangle\|_{L^{3/2}(N)} \leq C\|\mathfrak{Im}\langle\psi,\ddot{\phi}\rangle\|_{L^2(N)} \leq C\|\ddot{\phi}\|_{L^2(N)}.$$

We conclude that

$$\|\ddot{f}\|_{L^{1,2}(N)} \leq C\|\ddot{f}\|_{L^{2,3/2}(N)}$$

$$\leq C\Big(\|\mathsf{V}\|_{L^2(I_0\times N)}\|\ddot{\psi}\|_{L^2(N)} + \|\dot{\psi}\|_{L^{1,2}(I_6\times N)}\|\dot{\phi}\|_{L^{1,2}(I_6\times N)} + \|\ddot{\phi}\|_{L^2(N)}\Big)$$

Integrating the last inequality over $I_6$ we deduce

$$\|\ddot{f}\|_{L^2(I_6\times N)} + \|d\ddot{f}\|_{L^2(I_6\times N)} \leq C\|\mathsf{V}\|_{L^2(I_0\times N)} \tag{4.2.60}$$

Now, look at the elliptic system (4.2.45a) + (4.2.47) in which the $L^{2,2}(I_6\times N)$-norm of the right hand side can be estimated from above by $\|\mathsf{V}\|_{L^2(I_0\times N)}$. Invoking the interior elliptic estimates once again we obtain (4.2.44).

Now using trace results (see [**79**]) we get

$$\|\mathsf{V}(\tau_0)\|^2_{L^{2,2}(N)} \leq C\|\mathsf{V}(t)\|^2_{L^{3,2}(I_j\times N)} \leq C\|\mathsf{V}(t)\|^2_{L^2(I_0\times N)}$$

$$= C\int_{\tau_0-1}^{\tau_0+1}\mathrm{dist}_{L^2}([\mathsf{C}(t_0+t)],[\mathsf{C}_0])^2dt \leq$$

$$C\int_{\tau_0-1}^{\tau_0+1}\Big(\mathrm{dist}_{L^2}([\mathsf{C}(t_0+t)],[\mathsf{C}(t_0+\tau_0)])^2 + \mathrm{dist}_{L^2}\big([\mathsf{C}(t_0+\tau_0)],[\mathsf{C}_0]\big)^2\Big)dt \overset{(4.2.8)}{\leq}$$

$$C\int_{\tau_0-1}^{\tau_0+1}\Big(\mathrm{dist}_{L^2}\big([\mathsf{C}(t_0+\tau_0)],[\mathsf{C}_0]\big)^2 + |t-\tau_0|E_{\hat{\mathsf{C}}}([t_0+\tau_0-1,t_0+\tau_0+1])\Big)dt$$

$$\leq C\Big(\mathrm{dist}_{L^2}([\mathsf{C}(t_0+\tau_0)],[\mathsf{C}_0])^2 + \kappa^6\Big).$$

The conclusion in Lemma 4.2.24 is now obvious. ■

**Proof of Lemma 4.2.27** Set

$$\xi := \begin{bmatrix} \mathbf{i}a_\xi \\ \psi_\xi \end{bmatrix}, \quad \mathsf{U} := \begin{bmatrix} \mathbf{i}a_u \\ \psi_u \end{bmatrix}.$$

The quadratic remainder $R(\mathsf{V}) = SW(\mathsf{C}_0 + \mathsf{V}) - \mathcal{A}\mathsf{V}$ can be expressed explicitly and, after some elementary manipulations left to the reader, we get

$$R(\xi + \mathsf{U}) = \begin{bmatrix} \frac{1}{2}\mathbf{c}(\mathbf{i}a_\xi + \mathbf{i}a_u)(\psi_\xi + \psi_u) \\ \\ \frac{1}{2}q(\psi_\xi) + q(\psi_\xi, \psi_u) + \frac{1}{2}q(\psi_u) - *F_{A_0} \end{bmatrix},$$

$$R(\xi) = \begin{bmatrix} \frac{1}{2}\mathbf{c}(\mathbf{i}a_\xi)\psi_\xi \\ \\ \frac{1}{2}q(\psi_\xi) - *F_{A_0} \end{bmatrix}.$$

Clearly

$$\|R(\xi + \mathsf{U}) - R(\xi)\|_{L^2} \leq C\|\mathsf{V}\|_{2,2}\|\mathsf{U}\|_2.$$

The term $\mathfrak{N}(\mathsf{V})$ requires a bit more work. We use the identity (4.2.20)

$$2\mathfrak{N}(\mathsf{V}) = -\mathfrak{L}_{\mathsf{C}_0+\mathsf{V}}(\mathbf{i}f) = \mathfrak{L}_{\mathsf{C}_0}(\mathbf{i}f) - \begin{bmatrix} \mathbf{i}f\psi \\ 0 \end{bmatrix} =: \mathfrak{L}_{\mathsf{C}_0}(\mathbf{i}f) + \varphi.$$

Now define $A_\xi := A_0 + \mathbf{i}a_\xi$, and observe that

$$F := \mathfrak{Im}\langle \psi, \mathfrak{D}_{A_0}\psi\rangle = \mathfrak{Im}\langle \psi, \mathfrak{D}_{A_\xi}\psi\rangle,$$

$$= \mathfrak{Im}\langle \psi, \mathfrak{D}_{A_\xi}\psi_\xi\rangle + \mathfrak{Im}\langle \psi, \mathfrak{D}_{A_\xi}\psi_u\rangle = \mathfrak{Im}\langle \psi, \mathfrak{D}_{A_\xi}\psi_u\rangle$$

We claim that

$$\|F\|_{L^{-1,2}(N)} \leq C\|\mathsf{V}\|_{2,2} \cdot \|\mathsf{U}\|_{L^2}, \tag{4.2.61}$$

that is,

$$|\langle F, \tau\rangle_{L^2}| \leq C\|\mathsf{V}\|_{2,2} \cdot \|\mathsf{U}\|_{L^2} \cdot \|\tau\|_{1,2}, \quad \forall \tau \in C^\infty(N).$$

Indeed, using the Sobolev embedding $L^{2,2}(N) \hookrightarrow L^\infty(N)$ we deduce

$$\left|\int_N \tau \mathfrak{Im}\langle \psi, \mathfrak{D}_{A_\xi}\psi_u\rangle dv_g\right| \leq C\|\psi\|_{L^\infty}\|\mathfrak{D}_{A_\xi}\psi_u\|_{L^{-1,2}}\| \; |\tau\psi| \; \|_{1,2}$$

$$\leq C\|\mathfrak{D}_{A_0}\psi_u + \mathbf{c}(\mathbf{i}a_\xi)\psi_u\|_{L^{-1,2}}\|\tau\|_{1,2}\|\psi\|_{2,2}$$

$$\leq C\|\tau\|_{1,2}\|\psi\|_{2,2}\Big(\|\mathfrak{D}_{A_0}\psi_u\|_{-1,2} + \|\mathbf{c}(\mathbf{i}a_\xi)\psi_u\|_{-1,2}\Big)$$

$$\leq C\|\tau\|_{1,2}\|\mathsf{V}\|_{2,2}\|\mathsf{U}\|_{L^2}.$$

The equality $T_\psi(\mathbf{i}f) = 2\mathbf{i}F$ now implies

$$\|f\|_{1,2} \leq C\|F\|_{-1,2} \leq C(\|\psi\|_\infty + \|\mathsf{V}\|_{2,2}) \cdot \|\mathsf{U}\|_{L^2} \tag{4.2.62}$$

so that

$$\|\mathfrak{L}_{\mathsf{C}_0+\mathsf{V}}(\mathbf{i}f)\|_{L^2} \leq C\|f\|_{1,2} \leq C\|\mathsf{V}\|_{2,2} \cdot \|\mathsf{U}\|_{L^2}.$$

This proves (4.2.27b). To prove (4.2.27c) observe that

$$2\Big|\langle \mathfrak{N}(\mathsf{V}), \mathsf{U}^{\pm}\rangle_{L^2}\Big| = \Big|\int_N \langle \mathfrak{L}_{\mathsf{C}_0}(\mathbf{i}f) + \varphi, \mathsf{U}^{\pm}\rangle dv_g\Big|$$

$(\mathfrak{L}^*_{\mathsf{C}_0}\mathsf{U}^{\pm} = 0)$

$$= \Big|\int_N \langle \varphi, \mathsf{U}^{\pm}\rangle dv_g\Big| \leq \|\varphi\|_{L^2}\|\mathsf{U}\|_{L^2} \leq \|f\|_{L^4}\|\psi\|_{L^4}\|\mathsf{U}^{\pm}\|_{L^2}$$

(use the Sobolev embedding $L^{1,2}(N) \hookrightarrow L^4(N)$)

$$\stackrel{(4.2.62)}{\leq} C\|\psi\|_{1,2}\|\mathsf{V}\|_{2,2} \cdot \|\mathsf{U}\|^2_{L^2}.$$

This concludes the proof of Lemma 4.2.27. ∎

## 4.3. Moduli spaces of finite energy monopoles: Local aspects

We have so far studied the internal structure of a single finite energy monopole. We now shift the emphasis to a different structural problem. Namely, we would like to describe some natural structures on the set of finite energy monopoles.

This problem encompasses both a local and a global aspect. The local aspect refers to the smoothness properties and the expected dimension of this moduli space. The global issues we will discuss are concerned with the compactness and orientability properties of this space.

**4.3.1. Functional set-up.** To analyze the possible structures on the set of gauge equivalence classes of finite energy monopoles on a 4-manifold with cylindrical ends we need to define an appropriate configuration space a priori containing the set of such monopoles. Consider a cylindrical 4-manifold $(\hat{N}, \hat{g})$ and a cylindrical *spin*$^c$ structure $\hat{\sigma}$ on $\hat{N}$. Set $\sigma := \partial_\infty\hat{\sigma}$. Again we will be working under the nondegeneracy assumption (**N**) in 4.2.3, that the pair $(g, \sigma)$ is good.

The asymptotic analysis in the previous section suggests that it is wise to restrict our attention to a special class of connections on $\det\hat{\sigma}$. We will follow an approach inspired by [**96, 99**]. Observe first the following consequence of the nondegeneracy assumption (**N**).

**Lemma 4.3.1.** *The quantity*

$$\mu_-(\sigma, g) := \inf\Big\{\mu_-([\mathsf{C}_\infty]);\ \ [\mathsf{C}_\infty] \in \mathfrak{M}_\sigma\Big\}$$

*is strictly positive.*

**Exercise 4.3.1.** Prove Lemma 4.3.1.

Proposition 4.2.35 shows that it is natural to restrict our attention only to configurations with stringent restrictions on their asymptotic behaviour. Fix $0 < \mu < \mu_-(\sigma, g)$ and denote by $\hat{\mathcal{C}}^\infty_{\mu,ex}$ the set of *smooth* configurations $\hat{\mathsf{C}}$ on $\hat{N}$ which differ from a *strongly cylindrical* configuration by an $L^{2,2}_\mu$-term. More precisely, along the neck $\hat{\mathsf{C}}$ has the form

$$\hat{\mathsf{C}} = (\hat{\psi}, \hat{A}) = (\psi(t), A_\infty + \mathbf{i}f(t)dt + \mathbf{i}a(t)), \ \ t \in \mathbb{R}_+, \ \ A_\infty \in \mathfrak{A}_\sigma$$

and there exist $\psi_\infty \in C^\infty(\mathbb{S}_\sigma)$, $a_\infty \in \Omega^1(N)$ such that

$$\|f\|_{L^{2,2}_\mu} + \|a(t) - a_\infty\|_{L^{2,2}_\mu} + \|\psi(t) - \psi_\infty\|_{L^{2,2}_\mu} < \infty.$$

We set

$$\partial_\infty \hat{\mathsf{C}} := \mathsf{C}_\infty = (\psi_\infty, A_\infty + \mathbf{i}a_\infty).$$

We thus have a natural projection

$$\partial_\infty : \hat{\mathcal{C}}^\infty_{\mu,ex} \to \mathcal{C}^\infty_\sigma = \text{smooth configurations on } N.$$

As in §4.1.4, for every $r \geq 0$ we can construct a right inverse

$$i_r : \mathcal{C}^\infty_\sigma \to \hat{\mathcal{C}}^\infty_{\mu,ex}$$

for $\partial_\infty$, $\partial_\infty \circ i_r = \mathbf{1}$. The space $\hat{\mathcal{C}}^\infty_{\mu,ex}$ is equipped with a natural metric

$$d_\mu(\hat{\mathsf{C}}_1, \hat{\mathsf{C}}_2) := \|\partial_\infty \hat{\mathsf{C}}_1 - \partial_\infty \hat{\mathsf{C}}_2\|_{2,2} + \Big\|(\hat{\mathsf{C}}_1 - i_1\partial_\infty \hat{\mathsf{C}}_1) - (\hat{\mathsf{C}}_2 - i_1\partial_\infty \hat{\mathsf{C}}_2)\Big\|_{L^{2,2}_\mu}.$$

We can now define[1] $\hat{\mathcal{C}}_{\mu,ex}$ as the completion of $\hat{\mathcal{C}}^\infty_{\mu,\mu}$ with respect to the metric $d_\mu$. It is naturally equipped with a structure of Banach manifold. Observe that $\partial_\infty$ extends to a smooth map

$$\partial_\infty : \hat{\mathcal{C}}_{\mu,ex} \to \mathcal{C}_\sigma.$$

$\partial_\infty$ is a surjective submersion.

Proposition 4.2.35 shows that for any smooth finite energy monopole $\hat{\mathsf{C}}$ there exists $\hat{\gamma} \in C^\infty(\hat{N}, S^1)$ such that $\hat{\gamma} \cdot \hat{\mathsf{C}} \in \hat{\mathcal{C}}^\infty_{\mu,ex}$. We want to prove that the converse statement is true: any monopole $\hat{\mathsf{C}} \in \hat{\mathcal{C}}^\infty_{\mu,ex}$ has finite energy.

**Proposition 4.3.2.** *Fix a smooth configuration* $\hat{\mathsf{C}}_0 = (\hat{\psi}_0, \hat{A}_0) \in \hat{\mathcal{C}}_{\mu,ex}$ *such that*

$$\int_{\hat{N}} F_{\hat{A}_0} \wedge F_{\hat{A}_0} < \infty.$$

[1]This a departure from the traditional functional set-up which involves fractional Sobolev spaces, [**96, 133**]. Our configurations have regularity slightly better than $L^{2,2}(\hat{N})$ because, by definition, their asymptotic traces are not in $L^{3/2,2}(\partial\hat{N})$ but in the more regular space $L^{2,2}(\partial\hat{N})$.

*Then* $\hat{\mathsf{C}} = (\hat{\psi}, \hat{A}) \in \hat{\mathcal{C}}_{\mu,ex}$ *has finite energy*

$$E(\hat{\mathsf{C}}) := \int_{\hat{N}} \Big( |\hat{\nabla}^{\hat{A}}\hat{\psi}|^2 + \frac{1}{8}|q(\hat{\psi})|^2 + |F_{\hat{A}}|^2 + \frac{\hat{s}}{4}|\hat{\psi}|^2 \Big) dv(\hat{g}) < \infty$$

*if and only if*

$$E(\hat{\mathsf{C}}) := \int_{\hat{N}} \Big( |\not{\mathfrak{D}}_{\hat{A}}\hat{\psi}|^2 + \frac{1}{2}|\hat{\mathbf{c}}(F^+_{\hat{A}}) - \frac{1}{2}q(\hat{\psi})|^2 \Big) dv(\hat{g})$$

$$+2\mathcal{E}_\sigma(\partial_\infty \hat{\mathsf{C}}) + \int_{\hat{N}} F_{\hat{A}_0} \wedge F_{\hat{A}_0} < \infty$$

*where* $\mathcal{E}_\sigma : \mathcal{C}_\sigma \to \mathbb{R}$ *is the energy functional described in (2.4.8) of §2.4.1, defined in terms of the reference connection* $A_0 := \partial_\infty \hat{A}_0$. *In particular, if* $\hat{\mathsf{C}} \in \hat{\mathcal{C}}^\infty_{\mu,ex}$ *is a monopole then*

$$E(\hat{\mathsf{C}}) := 2\mathcal{E}_\sigma(\partial_\infty \hat{\mathsf{C}}) + \int_{\hat{N}} F_{\hat{A}_0} \wedge F_{\hat{A}_0} = \int_{\hat{N}} F_{\hat{A}} \wedge F_{\hat{A}} < \infty.$$

**Proof** Set $\hat{N}_T := \hat{N} \setminus (T, \infty) \times N$. Using the integration by parts formulæ in Exercise 1.2.2 (in which all the inner products are real valued) we deduce

$$\int_{\hat{N}_T} |\not{\mathfrak{D}}_{\hat{A}}\hat{\psi}|^2 dv(\hat{g}) = \int_{\partial \hat{N}_T} B_{\not{\mathfrak{D}}_{\hat{A}}}(\hat{\psi}, \not{\mathfrak{D}}_A) dv(g) + \int_{\hat{N}_T} \langle \not{\mathfrak{D}}^*_{\hat{A}} \not{\mathfrak{D}}_{\hat{A}}\hat{\psi}, \hat{\psi}\rangle dv(\hat{g})$$

(use the Weitzenböck formula)

$$= \int_{\partial \hat{N}_T} B_{\not{\mathfrak{D}}_{\hat{A}}}(\hat{\psi}, \not{\mathfrak{D}}_A \hat{\psi}) dv(g)$$

$$+ \int_{\hat{N}_T} \Big( \langle (\hat{\nabla}^{\hat{A}})^* \hat{\nabla}^{\hat{A}}\hat{\psi}, \hat{\psi}\rangle + \frac{s}{4}|\hat{\psi}|^2 + \frac{1}{2}\langle \hat{\mathbf{c}}(F^+_{\hat{A}})\hat{\psi}, \hat{\psi}\rangle \Big) dv(\hat{g})$$

$$= \int_{\partial \hat{N}_T} \Big( B_{\not{\mathfrak{D}}_{\hat{A}}}(\hat{\psi}, \not{\mathfrak{D}}_{\hat{A}}\hat{\psi}) - B_{\hat{\nabla}^{\hat{A}}}(\hat{\psi}, \hat{\nabla}^{\hat{A}}\hat{\psi}) \Big) dv(g)$$

$$+ \int_{\hat{N}_T} \Big( |\hat{\nabla}^{\hat{A}}\hat{\psi}|^2 + \frac{s}{4}|\hat{\psi}|^2 + \frac{1}{2}\langle \hat{\mathbf{c}}(F^+_{\hat{A}}), q(\hat{\psi})\rangle \Big) dv(\hat{g}).$$

Denote the above boundary integral by $R_\partial(T)$. As in the proof of Proposition 2.1.4 we have

$$\frac{1}{2}\int_{\hat{N}_T} |\hat{\mathbf{c}}(F^+_{\hat{A}}) - \frac{1}{2}q(\hat{\psi})|^2 dv(\hat{g})$$

$$= \int_{\hat{N}_T} \Big( 2|F^+_{\hat{A}}|^2 + \frac{1}{8}|q(\hat{\psi})|^2 - \frac{1}{2}\langle \hat{\mathbf{c}}(F^+_{\hat{A}}), q(\hat{\psi})\rangle \Big) dv(\hat{g}).$$

By adding the above equalities we deduce

$$\int_{\hat{N}_T} \Big( |\hat{\not{\mathfrak{D}}}_{\hat{A}}\hat{\psi}|^2 + \frac{1}{2}|\hat{\mathbf{c}}(F_{\hat{A}}^+) - \frac{1}{2}q(\hat{\psi})|^2 \Big) dv(\hat{g})$$

$$= R_\partial(T) + \int_{\hat{N}_T} \Big( |\hat{\nabla}^{\hat{A}}\hat{\psi}|^2 + \frac{s}{4}|\hat{\psi}|^2 + 2|F_{\hat{A}}^+|^2 + \frac{1}{8}|q(\hat{\psi})|^2 \Big) dv(\hat{g})$$

$$= R_\partial(T) + \int_{\hat{N}_T} \Big( |\hat{\nabla}^{\hat{A}}\hat{\psi}|^2 + \frac{s}{4}|\hat{\psi}|^2 + |F_{\hat{A}}|^2 + \frac{1}{8}|q(\hat{\psi})|^2 \Big) dv(\hat{g})$$

$$- \int_{\hat{N}_T} F_{\hat{A}} \wedge F_{\hat{A}}.$$

Using Exercise 1.2.2 we deduce

$$R_\partial(T) = \int_{\partial\hat{N}_T} \Big( \langle J\hat{\psi}, \hat{\not{\mathfrak{D}}}_{\hat{A}}\hat{\psi}\rangle - \langle \psi, \hat{\nabla}_t^{\hat{A}}\hat{\psi}\rangle \Big) dv(g)$$

$(A(T) := \hat{A}\,|_{\partial\hat{N}_T})$

$$= \int_{\partial\hat{N}_T} \Big( \langle J\hat{\psi}, J(\hat{\nabla}_t^{\hat{A}} - \mathfrak{D}_{A(T)})\hat{\psi}\rangle - \langle \psi, \hat{\nabla}_t^{\hat{A}}\hat{\psi}\rangle \Big) dv(g)$$

$$= - \int_{\partial\hat{N}_T} \langle \psi, \mathfrak{D}_{A(T)}\hat{\psi}\rangle dv(g).$$

On the other hand, we can write $F_{\hat{A}} = F_{\hat{A}_0} + \hat{d}(\hat{A} - \hat{A}_0)$ so that

$$F_{\hat{A}} \wedge F_{\hat{A}} = F_{\hat{A}_0} \wedge F_{\hat{A}_0} + \hat{d}\Big( (\hat{A} - \hat{A}_0) \wedge (F_{\hat{A}} + F_{\hat{A}_0}) \Big).$$

Thus

$$\int_{\hat{N}_T} F_{\hat{A}} \wedge F_{\hat{A}} = \int_{\partial\hat{N}_T} (\hat{A} - \hat{A}_0) \wedge (F_{\hat{A}} + F_{\hat{A}_0}) + \int_{\hat{N}_T} F_{\hat{A}_0} \wedge F_{\hat{A}_0}$$

so that if we set $\mathsf{C}(T) := \hat{\mathsf{C}}\,|_{\partial\hat{N}_T}$ we deduce

$$\int_{\hat{N}_T} \Big( |\hat{\not{\mathfrak{D}}}_{\hat{A}}\hat{\psi}|^2 + \frac{1}{2}|\hat{\mathbf{c}}(F_{\hat{A}}^+) - \frac{1}{2}q(\hat{\psi})|^2 \Big) dv(\hat{g})$$

$$= \int_{\hat{N}_T} \Big( |\hat{\nabla}^{\hat{A}}\hat{\psi}|^2 + \frac{s}{4}|\hat{\psi}|^2 + |F_{\hat{A}}|^2 + \frac{1}{8}|q(\hat{\psi})|^2 \Big) dv(\hat{g})$$

$$-2\mathcal{E}_\sigma(\mathsf{C}(T)) - \int_{\hat{N}_T} F_{\hat{A}_0} \wedge F_{\hat{A}_0}.$$

The first part of the proposition now follows by letting $T \to \infty$.

The second part is an immediate consequence of the above proof and the fact that $\partial_\infty\hat{\mathsf{C}} = (\psi_\infty, A_\infty)$ is a monopole so that $\mathfrak{D}_{A_\infty}\psi_\infty = 0$. ■

We now need to define an appropriate gauge group. Set

$$\widehat{\mathfrak{G}}_{\mu,ex} := \Big\{ \hat{\gamma} \in L^{3,2}_{\mu,ex}(\hat{N}, \mathbb{C});\ \ |\hat{\gamma}(p)| = 1\ \ \forall p \in \hat{N} \Big\}.$$

Observe that

$$\widehat{\mathfrak{G}}_{\mu,ex} \cdot \hat{\mathcal{C}}_{\mu,ex} \subset \hat{\mathcal{C}}_{\mu,ex}.$$

We can now define a metric $d_\mu$ on $\widehat{\mathfrak{G}}_{\mu,ex}$ by setting

$$d_\mu(\hat{\gamma}_1, \hat{\gamma}_2) := \|\partial_\infty \hat{\gamma}_1 - \partial_\infty \hat{\gamma}_2\|_{3,2}$$

$$+\|(\hat{\gamma}_1(t) - i_0 \partial_\infty \hat{\gamma}_1) - (\hat{\gamma}_2 - i_0 \partial_\infty \hat{\gamma}_2)\|_{L^{3,2}_\mu(\mathbb{R}_+ \times N)}.$$

$\widehat{\mathfrak{G}}_{\mu,ex}$ equipped with the above metric becomes a topological group and we have a *continuous* group morphism

$$\partial_\infty : \widehat{\mathfrak{G}}_{\mu,ex} \to \mathfrak{G}_\sigma.$$

**Proposition 4.3.3.** *$\widehat{\mathfrak{G}}_{\mu,ex}$ is a Hilbert Lie group and $T_1 \widehat{\mathfrak{G}}_{\mu,ex} \cong L^{3,2}_{\mu,ex}(\hat{N}, \mathbf{i}\mathbb{R})$.*

**Exercise 4.3.2.** Prove the above proposition.

The group $\widehat{\mathfrak{G}}_{\mu,ex}$ may not be connected. Its group of components is isomorphic to $H^1(\hat{N}, \mathbb{Z})$. Since the map

$$H^1(\hat{N}, \mathbb{Z}) \to H^1(N, \mathbb{Z}) = \text{the group of components of } \mathfrak{G}_\sigma$$

may not be onto, the morphism

$$\partial_\infty : \widehat{\mathfrak{G}}_{\mu,ex} \to \mathfrak{G}_\sigma$$

may not be onto. It becomes onto if we restrict to the identity components of the two groups. We will indicate these components by the superscript 1.

**Lemma 4.3.4.** *The morphism $\partial_\infty : \widehat{\mathfrak{G}}^1_{\mu,ex} \to \mathfrak{G}^1_\sigma$ admits a natural right inverse*

$$E : \mathfrak{G}^1_\sigma \to \widehat{\mathfrak{G}}^1_{\mu,ex},\ \ \exp(\mathbf{i}f) \mapsto \exp(\mathbf{i} i_0 f).$$

We will denote by $\widehat{\mathfrak{G}}_\mu$ the kernel and by $\mathfrak{G}^\partial_\sigma$ the image of the morphism $\partial_\infty : \widehat{\mathfrak{G}}_{\mu,ex} \to \mathfrak{G}_\sigma$ so that

$$\mathfrak{G}_\sigma / \mathfrak{G}^\partial_\sigma \cong H^1(N, \mathbb{Z}) / H^1(\hat{N}, \mathbb{Z}).$$

Fix $\hat{\mathsf{C}}_0 = (\hat{\psi}, \hat{A}) \in \hat{\mathcal{C}}_{\mu,ex}$ and set $\mathsf{C}_\infty := \partial_\infty \hat{\mathsf{C}}_0$, $G_\infty := \mathbf{Stab}\,(\mathsf{C}_\infty)$. Define

$$\mathcal{S}_\infty := \{\dot{\mathsf{C}} \in T_{\mathsf{C}_\infty} \mathcal{C}_\sigma;\ \ \mathfrak{L}^*_{\mathsf{C}_\infty} \dot{\mathsf{C}} = 0\}.$$

Fix a tiny neighborhood $U_\infty$ of $0 \in \mathcal{S}_\infty$ such that every $\mathfrak{G}_\sigma$ orbit intersects $\mathsf{C}_\infty + U_\infty$ along at most one $G_\infty$-orbit. We deduce that any $\mathfrak{G}^\partial_\sigma$-orbit intersects $U_\infty$ along at most one $G_\infty$-orbit. Set

$$\hat{U}_\infty := \partial^{-1}_\infty(\mathsf{C}_\infty + U_\infty).$$

We see that any $\widehat{\mathfrak{G}}_\sigma$-orbit intersects $\hat{U}_\infty$ along at most one orbit of the group $\widehat{\mathfrak{G}}_{\mu,ex}(\mathsf{C}_\infty) := \partial_\infty^{-1}(G_\infty)$. Thus, the problem of understanding the local structure of $\hat{\mathcal{C}}_{\mu,ex}/\widehat{\mathfrak{G}}_{\mu,ex}$ is equivalent to the problem of understanding the local structure of

$$\hat{U}_\infty/\widehat{\mathfrak{G}}_{\mu,ex}(\mathsf{C}_\infty).$$

Observe that $\widehat{\mathfrak{G}}_{\mu,ex}(\mathsf{C}_\infty)$ is a commutative Hilbert Lie group with Lie algebra

$$T_1\widehat{\mathfrak{G}}_{\mu,ex}(\mathsf{C}_\infty) = \{\mathbf{i}f \in L^{3,2}_{\mu,ex};\ \partial_\infty(\mathbf{i}f) \in T_1G_\infty\}.$$

Observe that there is a natural action of $\widehat{\mathfrak{G}}_{\mu,ex}(\mathsf{C}_\infty)$ on $\partial_\infty^{-1}(\mathsf{C}_\infty)\times U_\infty$ defined by

$$\hat{\gamma}\cdot(\hat{\mathsf{C}}, \mathsf{C}_\infty + \dot{\mathsf{C}}) := (\hat{\gamma}\cdot\hat{\mathsf{C}}\,,\,\mathsf{C}_\infty + (\partial_\infty\hat{\gamma})\cdot\dot{\mathsf{C}}).$$

The following result should be obvious.

**Lemma 4.3.5.** *The natural map*

$$\partial_\infty^{-1}(\mathsf{C}_\infty)\times U_\infty \to \hat{U}_\infty,\quad (\hat{\mathsf{C}},\dot{\mathsf{C}}) \mapsto \hat{\mathsf{C}} + i_0\dot{\mathsf{C}}$$

*is a $\widehat{\mathfrak{G}}_{\mu,ex}(\mathsf{C}_\infty)$-equivariant diffeomorphism.*

The last lemma reduces the structure problem to understanding the quotient $\partial_\infty^{-1}(\mathsf{C}_\infty)/\widehat{\mathfrak{G}}_{\mu,ex}(\mathsf{C}_\infty)$. Observe now that $\partial_\infty^{-1}(\mathsf{C}_\infty)$ is a smooth Hilbert manifold modeled by $L^{2,2}_\mu(\mathbb{S}^+_{\hat{\sigma}} \oplus \mathbf{i}T^*\hat{N})$. The group $\widehat{\mathfrak{G}}_{\mu,ex}(\mathsf{C}_\infty)$ acts smoothly on this manifold and, as in the closed case, we can define the infinitesimal action

$$\mathfrak{L}_{\hat{\mathsf{C}}_0} : T_1\widehat{\mathfrak{G}}_{\mu,ex}(\mathsf{C}_\infty) \to T_{\hat{\mathsf{C}}_0}\partial_\infty^{-1}(\mathsf{C}_\infty),\quad \mathbf{i}f \mapsto \frac{d}{ds}|_{s=0}\, e^{s\mathbf{i}f}\cdot\hat{\mathsf{C}}_0.$$

Set

$$\hat{\mathcal{S}}_{\hat{\mathsf{C}}_0} := \{\underline{\hat{\mathsf{C}}} \in T_{\hat{\mathsf{C}}_0}\partial_\infty^{-1}(\mathsf{C}_\infty);\ \mathfrak{L}^{*_\mu}_{\hat{\mathsf{C}}_0}\underline{\hat{\mathsf{C}}} = 0\}$$

where $*_\mu$ denotes the $L^2_\mu$-adjoint as in §4.1.4. Set $\hat{G}_0 := \mathbf{Stab}\,(\hat{\mathsf{C}}_0)$. Notice that the induced map $\hat{G}_0 \to G_\infty$ is one-to-one.

Let us first observe an immediate consequence of the Lockhart-McOwen Theorem 4.1.16.

**Lemma 4.3.6.** *There exists $\mu_0 = \mu_0(\sigma, g) \in (0, \mu_-(\sigma, g)]$ such that the operator*

$$(\hat{d} + \hat{d}^{*_\mu}) : L^{1,2}_\mu(\Lambda T^*\hat{N}) \to L^{1,2}_\mu(\Lambda T^*\hat{N})$$

*is Fredholm for every $0 < \mu < \mu_0(\sigma, g)$.*

(4.3.1) *In the sequel we will always assume* $0 < \mu < \mu_0(\sigma, g)$.

**Proposition 4.3.7.** *There exists a small $\hat{G}_0$-invariant neighborhood $\hat{V}$ of $0 \in \hat{\mathcal{S}}_{\hat{\mathsf{C}}_0}$ such that every orbit of $\widehat{\mathfrak{G}}_{\mu,ex}(\mathsf{C}_\infty)$ intersects $\hat{\mathsf{C}}_0 + V$ along at most one $\hat{G}_0$-orbit.*

**Proof** We will follow the strategy used in the proof of Proposition 2.2.7 in §2.2.2. Consider

$$\mathcal{F} : \widehat{\mathfrak{G}}_{\mu,ex}(\mathsf{C}_\infty) \times \hat{\mathcal{S}}_{\hat{\mathsf{C}}_0} \to \partial_\infty^{-1}(\mathsf{C}_\infty)$$

defined by

$$\mathcal{F}(\hat{\gamma}; \hat{\psi}, \mathbf{i}\hat{a}) = (\hat{\gamma}(\hat{\psi}_0 + \hat{\psi}), \hat{A}_0 + \mathbf{i}\hat{a} - 2d\hat{\gamma}/\hat{\gamma}).$$

We have the following counterpart of Lemma 2.2.8.

**Lemma 4.3.8.** *There exists a $\hat{G}_0$-invariant neighborhood $W$ of $(1, 0) \in \widehat{\mathfrak{G}}_{\mu,ex}(\mathsf{C}_\infty) \times \hat{\mathcal{S}}_{\hat{\mathsf{C}}_0}$ with the following properties.*

- **P1** *The restriction of $\mathcal{F}$ to $W$ is a submersion. In particular, $\mathcal{F}(W)$ is an open neighborhood of $\hat{\mathsf{C}}_0$ in $\partial_\infty^{-1}(\mathsf{C}_\infty)$.*
- **P2** *Each fiber of the map $\mathcal{F} : W \to \mathcal{F}(W)$ consists of a single $\hat{G}_0$-orbit.*

**Proof of Lemma 4.3.8** We will use the implicit function theorem. The differential of $\mathcal{F}$ at $(1, 0)$ is the bounded linear map

$$D\mathcal{F} : T_1\widehat{\mathfrak{G}}_{\mu,ex}(\mathsf{C}_\infty) \times \hat{\mathcal{S}}_{\hat{\mathsf{C}}_0} \to T_{\hat{\mathsf{C}}_0}\partial_\infty^{-1}(\mathsf{C}_\infty)$$

described by

$$(\mathbf{i}\hat{f}, \underline{\hat{\psi}}, \mathbf{i}\underline{\hat{a}}) \mapsto (\mathbf{i}\hat{f}\hat{\psi}_0 + \underline{\hat{\psi}}) \oplus (\mathbf{i}\underline{\hat{a}} - 2\mathbf{i}d\hat{f}) = \mathfrak{L}_{\hat{\mathsf{C}}_0}(\mathbf{i}\hat{f}) + \underline{\hat{\psi}} \oplus \mathbf{i}\underline{\hat{a}}.$$

We want to prove that $D\mathcal{F}$ is surjective and $\ker D\mathcal{F} \cong T_1\hat{G}_0$.

- $\ker D\mathcal{F} \cong T_1\hat{G}_0$. If $(\mathbf{i}\hat{f}, \underline{\hat{\psi}}, \mathbf{i}\underline{\hat{a}}) \in \ker D\mathcal{F}$ then $\mathfrak{L}^{*\mu}_{\hat{\mathsf{C}}_0}(\underline{\hat{\psi}} \oplus \mathbf{i}\underline{\hat{a}}) = 0$ so that

$$0 = \mathfrak{L}^{*\mu}_{\hat{\mathsf{C}}_0} D\mathcal{F}(\mathbf{i}\hat{f}, \underline{\hat{\psi}}, \mathbf{i}\underline{\hat{a}}) = \mathfrak{L}^{*\mu}_{\hat{\mathsf{C}}_0}(\mathfrak{L}_{\hat{\mathsf{C}}_0}(\mathbf{i}\hat{f}) + \underline{\hat{\psi}} \oplus \mathbf{i}\underline{\hat{a}}) = \mathfrak{L}^{*\mu}_{\hat{\mathsf{C}}_0}\mathfrak{L}_{\hat{\mathsf{C}}_0}(\mathbf{i}\hat{f}).$$

Thus,

$$0 = \int_{\hat{N}_T} \langle \mathfrak{L}^{*\mu}_{\hat{\mathsf{C}}_0}\mathfrak{L}_{\hat{\mathsf{C}}_0}(\mathbf{i}\hat{f}), \mathbf{i}\hat{f}\rangle \mathbf{m}_{2\mu} dv(\hat{g}) = \int_{\hat{N}_T} \langle \mathfrak{L}^{*}_{\hat{\mathsf{C}}_0}\mathbf{m}_{2\mu}\mathfrak{L}_{\hat{\mathsf{C}}_0}(\mathbf{i}\hat{f}), \mathbf{i}\hat{f}\rangle dv(\hat{g})$$

$$= \int_{\hat{N}_T} |\mathfrak{L}_{\hat{\mathsf{C}}_0}(\mathbf{i}\hat{f})|^2 \mathbf{m}_{2\mu} dv(\hat{g}) \pm \int_{\partial\hat{N}_T} (\hat{f}\frac{d}{dt}\hat{f})\mathbf{m}_{2\mu}(T) dv(g).$$

By letting $T \to \infty$ we obtain

$$0 = \int_{\hat{N}} |\mathfrak{L}_{\hat{\mathsf{C}}_0}(\mathbf{i}\hat{f})|^2 \mathbf{m}_{2\mu} dv(\hat{g})$$

so that $\mathbf{i}\hat{f} \in \ker \mathfrak{L}_{\hat{\mathsf{C}}_0} \cong T_1\hat{G}_0$. This equality forces $\underline{\hat{\psi}} = 0$ and $\underline{\hat{a}} = 0$.

• **Surjectivity** We need the following technical result. Its proof will be presented after we complete the proof of Lemma 4.3.8.

**Lemma 4.3.9.** *The range of the bounded linear operator*

$$\mathfrak{L}_{\hat{\mathsf{C}}_0} : \{\mathbf{i}\hat{f} \in L^{1,2}_{\mu,ex}(M, \mathbf{i}\mathbb{R});\ \partial_\infty \mathbf{i}\hat{f} \in T_1 G_\infty\} \to L^2_\mu(\mathbb{S}^+_{\hat{\sigma}} \oplus \mathbf{i}T^*\hat{N})$$

*is closed.*

If we assume the lemma then we deduce that any $\hat{\psi} \oplus \mathbf{i}\hat{a} \in L^2_\mu(\mathbb{S}^+_{\hat{\sigma}} \oplus \mathbf{i}T^*\hat{N})$ decomposes $L^2_\mu$-orthogonally as

$$\hat{\psi} \oplus \mathbf{i}\hat{a} = \mathfrak{L}_{\mathsf{C}_\infty}(\mathbf{i}\hat{f}) + \underline{\hat{\psi}} \oplus \mathbf{i}\underline{\hat{a}}$$

where $\mathfrak{L}^{*_\mu}_{\hat{\mathsf{C}}_0}(\underline{\hat{\psi}} \oplus \mathbf{i}\underline{\hat{a}}) = 0$ and $\mathbf{i}\hat{f}$ is unique up to an element of $\ker \mathfrak{L}_{\hat{\mathsf{C}}_0}$.

**Lemma 4.3.10.** *If*

$$\hat{\psi} \oplus \mathbf{i}\hat{a} \in L^{2,2}_\mu$$

*then*

$$\mathbf{i}\hat{f} \in L^{3,2}_{\mu,ex}.$$

Observe that if $\hat{\psi} \oplus \mathbf{i}\hat{a} \in L^{2,2}_\mu$ then Lemma 4.3.10 implies $\underline{\hat{\psi}} \oplus \mathbf{i}\underline{\hat{a}} \in L^{1,2}_\mu$, thus proving the surjectivity of $D\mathcal{F}$.

**Proof of Lemma 4.3.10** Observe that $f_\infty := \partial_\infty \hat{f}$ is a constant function on $N$ and thus extends in an obvious fashion to $\hat{N}$. Set

$$\hat{f}_0 := \hat{f} - f_\infty.$$

We use the equality

$$\mathfrak{L}^{*_\mu}_{\mathsf{C}_0}\mathfrak{L}_{\mathsf{C}_0}(\mathbf{i}\hat{f}_0) = u := \mathfrak{L}^{*_\mu}_{\mathsf{C}_0}(\hat{\psi} \oplus \mathbf{i}\hat{a}) - \mathfrak{L}^{*_\mu}_{\mathsf{C}_0}\mathfrak{L}_{\mathsf{C}_0} f_\infty \in L^{1,2}_\mu.$$

Along a cylinder $[T-2, T+2] \times N$, $T > 3$, we have

$$\mathfrak{L}^{*_\mu}_{\mathsf{C}_0}\mathfrak{L}_{\mathsf{C}_0}(\mathbf{i}\hat{f}_0) = \Big(\mathfrak{L}^*_{\hat{\mathsf{C}}_0}\mathfrak{L}_{\hat{\mathsf{C}}} + 2\mu\mathfrak{L}_{\hat{\mathsf{C}}_0}\Big)(\mathbf{i}\hat{f}_0) = u$$

so that using interior elliptic estimates we deduce

$$\|\hat{f}_0\|_{L^{3,2}((T-1,T+1)\times N)} \leq C\Big(\|\hat{f}_0\|_{L^2((T-2,T+2)\times N)} + \|u\|_{L^{1,2}((T-2,T+2)\times N)}\Big)$$

$$\leq Ce^{-\mu T}\Big(\|e^{\mu t}\hat{f}_0(t)\|_{L^2((T-2,T+2)\times N)} + \|e^{\mu t}u(t)\|_{L^{1,2}((T-2,T+2)\times N)}\Big).$$

Thus

$$\|e^{\mu t}\hat{f}_0\|_{L^{3,2}((T-1,T+1)\times N)} \leq Ce^{\mu T}\|\hat{f}_0\|_{L^{3,2}((T-1,T+1)\times N)}$$

$$\leq C\Big(\|e^{\mu t}\hat{f}_0(t)\|_{L^2((T-2,T+2)\times N)} + \|e^{\mu t}u(t)\|_{L^{1,2}((T-2,T+2)\times N)}\Big).$$

If we now square the above inequality and then sum over $T = 2, 3, \cdots$ we obtain an estimate of the $L^{3,2}_\mu$-norm of $\hat{f}_0$ in terms of the $L^{1,2}_\mu$-norm of $u$ and the weaker $L^2_\mu$-norm of $\hat{f}_0$. This completes the proof of the claim. ∎

We can now apply the implicit function theorem to conclude that there exists an open neighborhood $W$ of $(1, 0) \in \widehat{\mathfrak{G}}_{\mu,ex}(\mathsf{C}_\infty) \times \hat{\mathcal{S}}_{\hat{\mathsf{C}}_0}$ such that the restriction of $\mathcal{F}$ to $W$ is a submersion. Since $\ker D_{(1,0)}F \cong T_1\hat{G}_0$ we deduce that the fibers of $\mathcal{F} : W \to \mathcal{F}(W)$ are smooth manifolds of dimension $\dim \hat{G}_0$. In particular, if $\hat{G}_0 = \mathbf{1}$ then $\mathcal{F}$ is a local diffeomorphism.

Suppose $\hat{G}_0 = S^1$ so that $\hat{\psi}_0 = 0$. We have to prove that each fiber of $\mathcal{F} : W \to \mathcal{F}(W)$ consists of a single $\hat{G}_0$-orbit. Let $\mathcal{F}(\exp(\mathbf{i}\hat{f}_1); \hat{\psi}_1, \mathbf{i}\hat{a}_1) = \mathcal{F}(\exp(\mathbf{i}\hat{f}_2); \hat{\psi}_2, \mathbf{i}\hat{a}_2)$, i.e.

$$\exp(\mathbf{i}\hat{f})\hat{\psi}_1 = \hat{\psi}_2, \;\; \hat{a}_1 - \hat{a}_2 = 2\hat{d}\hat{f}$$

where $\hat{f} := \hat{f}_1 - \hat{f}_2$. Since $(\hat{\psi}_j, \mathbf{i}\hat{a}_j) \in \mathcal{S}_{\hat{\mathsf{C}}_0}$ we deduce

$$\mathfrak{L}^{*\mu}_{\hat{\mathsf{C}}_0}(\hat{\psi}_j, \mathbf{i}\hat{a}_j) = 0 \Longleftrightarrow \hat{d}^{*_\mu}\hat{a}_j = 0.$$

This implies

$$\hat{d}^{*_\mu}\hat{d}\hat{f} = 0.$$

Using again an integration by parts argument as before (over $\hat{N}_T$, $T \to \infty$) we conclude $\hat{d}\hat{f} = 0$, which leads to the desired conclusion. This concludes the proof of Lemma 4.3.8. ∎

**Proof of Lemma 4.3.9** Suppose we are given

$$\hat{f}_n \in L^{1,2}_{\mu,ex}, \partial_\infty \hat{f}_n \in T_1 G_\infty$$

such that

$$\mathfrak{L}_{\hat{\mathsf{C}}_0}(\mathbf{i}\hat{f}_n) \xrightarrow{L^2_\mu} (\underline{\hat{\psi}}, \mathbf{i}\underline{\hat{a}}), \;\; n \to \infty. \tag{4.3.2}$$

We have to show there exists

$$\mathbf{i}\hat{f} \in L^{1,2}_{\mu,ex}, \partial_\infty \hat{f} \in T_1 G_\infty$$

such that

$$\mathfrak{L}_{\hat{\mathsf{C}}_0}(\mathbf{i}\hat{f}) = (\underline{\hat{\psi}}, \mathbf{i}\underline{\hat{a}}).$$

First of all, observe that it suffices to consider only the case

$$\partial_\infty \hat{f}_n = 0.$$

Indeed, we can write

$$\hat{f}_n = \hat{f}^0_n + \partial_\infty \hat{f}_n$$

and

$$\mathfrak{L}_{\hat{\mathsf{C}}_0}(\mathbf{i}\hat{f}_n) = \mathfrak{L}_{\hat{\mathsf{C}}_0}(\mathbf{i}\hat{f}_n^0) + \mathfrak{L}_{\hat{\mathsf{C}}_0}(\mathbf{i}\partial_\infty \hat{f}_n) = \mathfrak{L}_{\hat{\mathsf{C}}_0}(\mathbf{i}\hat{f}_n^0) + (\exp(\mathbf{i}\partial_\infty \hat{f}_n)\hat{\psi}_0, 0).$$

A subsequence of $\partial_\infty \hat{f}_n$ converges modulo $2\pi\mathbb{Z}$ to a constant $\omega$ and clearly

$$(\exp(\mathbf{i}\omega)\hat{\psi}_0, 0) = \mathfrak{L}_{\hat{\mathsf{C}}_0}(\mathbf{i}\omega).$$

Thus, it suffices to consider only the situation $\hat{f}_n \in L^{1,2}_\mu$. The condition (4.3.2) implies

$$d\hat{f}_n \xrightarrow{L^2_\mu} -\underline{\hat{a}}.$$

Now observe that we have the following

**A priori estimate** *There exists $C > 0$ such that*

$$\|\hat{g}\|_{L^{1,2}_\mu(\mathbb{R}_+\times N)} \leq C\|\hat{d}\hat{g}\|_{L^2_\mu(\mathbb{R}_+\times N)}, \tag{4.3.3}$$

$\forall \hat{g} \in L^2_\mu(\mathbb{R}_+ \times N) \cap L^{1,2}_{loc}(\mathbb{R}_+ \times N)$.

To prove the above inequality we will use a trick[2] in [**151**, Prop. (2.39)]. Observe first that we only need to prove a $L^2_\mu$-bound for $\hat{g}$ since

$$\|\hat{g}\|^2_{L^{1,2}_\mu} = \|\hat{g}\|^2_{L^2_\mu} + \|\hat{d}\hat{g}\|^2_{L^2_\mu}.$$

Set $\hat{b} := \hat{d}\hat{g}$ and observe that

$$\frac{d}{dt}\hat{g} = \partial_t \lrcorner \hat{b}$$

which implies

$$|\hat{g}(t)| = |\hat{g}(t) - \hat{g}(\infty)| \leq \int_t^\infty |\hat{b}(s)| ds.$$

Thus

$$\int_0^\infty |\hat{g}(t)|^2 e^{2\mu t} dt \leq \int_0^\infty \Big(\int_t^\infty |\hat{b}(s)| ds\Big)^2 e^{2\mu t} dt$$

(use the Cauchy-Schwarz inequality for the interior integral, $0 < \nu < \mu$)

$$\leq \int_0^\infty \Big(\int_t^\infty |\hat{b}(s)|^2 e^{2\nu s} ds\Big)\Big(\int_t^\infty e^{-2\nu s} ds\Big) e^{2\mu t} dt$$

$$= \frac{1}{2\nu}\int_0^\infty \Big(\int_t^\infty |\hat{b}(s)|^2 e^{2\nu s} ds\Big) e^{2(\mu-\nu)t} dt$$

(switch the order of integration)

$$= \frac{1}{2\nu}\int_0^\infty \Big(\int_0^s e^{2(\mu-\nu)t} dt\Big) |\hat{b}(s)|^2 e^{2\nu s} ds$$

[2] I am indebted to Stephen Bulloch for drawing my attention to this trick.

$$= \frac{1}{2\nu}\int_0^\infty \frac{1}{2(\mu-\nu)}\bigl(e^{2(\mu-\nu)s}-1\bigr)|\hat{b}(s)|^2e^{2\nu s}ds$$

$$= \frac{1}{4\nu(\mu-\nu)}\Bigl(\int_0^\infty |\hat{b}(s)|^2e^{2\mu s}ds - \int_0^\infty |\hat{b}(s)|^2e^{2\nu s}ds\Bigr)$$

$$\leq \frac{1}{4\nu(\mu-\nu)}\Bigl(\int_0^\infty |\hat{b}(s)|^2e^{2\mu s}ds\Bigr).$$

To obtain the a priori estimate we only need to integrate the above inequality over $N$. ■

Using (4.3.3) we deduce

$$\|\hat{f}_n - \hat{f}_m\|_{L^{1,2}_\mu(\mathbb{R}\times N)} \leq C\|\underline{\hat{a}}_n - \underline{\hat{a}}_m\|_{L^2_\mu(\mathbb{R}_+\times N)}, \quad \forall n,m>0.$$

Since $(\underline{\hat{a}}_n)$ is $L^2_\mu$-Cauchy sequence we deduce that $(\hat{f}_n)$ converges in the $L^{1,2}_\mu$-norm to $\hat{f}$ satisfying (weakly) the differential equation

$$\hat{d}\hat{f} = -\underline{\hat{a}}.$$

This shows $\mathfrak{L}_{\hat{\mathsf{C}}_0}(\mathbf{i}\hat{f}) = (\underline{\hat{\psi}}, \mathbf{i}\underline{\hat{a}})$, which concludes the proof of Lemma 4.3.9. ■

**Remark 4.3.11.** (a) Observe that if $\hat{g}\in L^{1,2}_{loc}(\hat{N})$ is such that $\hat{d}\hat{g}\in L^2_\mu$ then the above proof shows that $\hat{g}\in L^2_{ex}(\hat{N})$ and

$$\|\hat{g}-\partial_\infty\hat{g}\|_{L^2_\mu} \leq C\|\hat{d}\hat{g}\|_{L^2_\mu}.$$

This is essentially the content of the key technical result [**132**, Lemma 5.2] proved there by entirely different means.

(b) Suppose $\hat{E}\to\hat{N}$ is a Hermitian vector bundle equipped with a cylindrical structure $(\hat{\vartheta},\hat{\nabla}^0)$. Fix $\mu>0$. The above proof shows that there exists a positive constant $C$ with the following property: for every $u\in L^2(\hat{E})$ such that $\hat{\nabla}^0u\in L^2_\mu(T^*\hat{N}\otimes\hat{E})$ we have $\|\hat{u}\|_{L^2_\mu}\leq C\|\hat{\nabla}^0\hat{u}\|_{L^2_\mu}$. Iterating the above procedure to the bundles $T^*\hat{N}^{\otimes k}\otimes\hat{E}$ we deduce

$$\|\hat{u}\|_{L^2_\mu} \leq C_k\|(\hat{\nabla}^0)^{\otimes k}\hat{u}\|_{L^2_\mu} \tag{4.3.4}$$

for all $\hat{u}\in L^2(\hat{E})$.

**Exercise 4.3.3.** Prove the claims in the above remark.

* * *

We can now complete the proof of Proposition 4.3.7. We need to prove that there exists a small $\hat{G}_0$-invariant neighborhood $\hat{V}$ of $0 \in \hat{\mathcal{S}}_{\hat{\mathsf{C}}_0}$ such that every $\widehat{\mathfrak{G}}_{\mu,ex}(\mathsf{C}_\infty)$-orbit intersects $\hat{\mathsf{C}}_0 + \hat{V}$ along at most one orbit. In other words, we need to prove that, for $\hat{V}$ as above, each fiber of the map

$$\mathcal{F} : \widehat{\mathfrak{G}}_{\mu,ex}(\mathsf{C}_\infty) \times \hat{V} \to \partial_\infty^{-1}(\mathsf{C}_\infty)$$

consists of a *single* $\hat{G}_0$-orbit. Observe that according to Lemma 4.3.8 this statement is true for the restriction of $\mathcal{F}$ to a $\hat{G}_0$-invariant neighborhood $\hat{U}_0 \times \hat{V}_0$ of $(1, 0) \in \widehat{\mathfrak{G}}_{\mu,ex}(\mathsf{C}_\infty) \times \hat{\mathcal{S}}_{\hat{\mathsf{C}}_0}$. We will argue by contradiction.

Suppose there exist sequences $(\hat{\psi}_n, \mathbf{i}\hat{a}_n), (\hat{\phi}_n, \mathbf{i}\hat{b}_n) \in \hat{V}_0$ and $\hat{\gamma}_n \in \widehat{\mathfrak{G}}_{\mu,ex}(\mathsf{C}_\infty)$ with the following properties.

$$(\hat{\psi}_n, \mathbf{i}\hat{a}_n), (\hat{\phi}_n, \mathbf{i}\hat{b}_n) \to 0 \text{ in } L^{2,2}_\mu. \tag{4.3.5}$$

$$\hat{\mathsf{C}}_0 + (\hat{\psi}_n, \mathbf{i}\hat{a}_n) = \hat{\gamma}_n \cdot (\hat{\mathsf{C}}_0 + (\hat{\phi}_n, \mathbf{i}\hat{b}_n)), \ \ \forall n. \tag{4.3.6}$$

We will show that $\hat{\gamma}_n \in \hat{G}_0$, $\forall n \gg 0$. We will rely on the following auxiliary result.

**Lemma 4.3.12.** *$\hat{\gamma}_n$ belongs to the identity component of $\widehat{\mathfrak{G}}_{\mu,ex}(\mathsf{C}_\infty)$ for all $n \gg 0$.*

Let us first show why this result implies $\hat{\gamma}_n \in \hat{G}_0$ for all $n \gg 0$. Using Lemma 4.3.12 we can write

$$\hat{\gamma}_n = \exp(\mathbf{i}\hat{f}_n), \ \ \hat{f}_n \in L^{3,2}_{\mu,ex}.$$

We can also assume that the constant function $f_n := \partial_\infty \hat{f}_n$ lies in the interval $[0, 2\pi]$. By extracting a subsequence we can assume

$$f_n \to f_\infty.$$

Using (4.3.6) we deduce

$$\hat{d}(\hat{f}_n - f_n) = \hat{b}_n - \hat{a}_n. \tag{4.3.7}$$

The a priori estimate (4.3.3) implies

$$\|\hat{f}_n - f_n\|_{L^{1,2}_\mu} \le C\|\hat{b}_n - \hat{a}_n\|_{L^2_\mu} \to 0.$$

The equality (4.3.7) also implies

$$\|\hat{d}(\hat{f}_n - f_n)\|_{L^{2,2}_\mu} \le \|\hat{b}_n - \hat{a}_n\|_{L^{2,2}_\mu}.$$

We conclude that $\hat{f}_n$ converges in $L^{3,2}_{\mu,ex}$ to the constant function $\hat{f}_\infty \equiv f_\infty$. Using (4.3.6) we deduce

$$\exp(\mathbf{i}\hat{f}_\infty) \cdot \hat{\mathsf{C}}_0 = \hat{\mathsf{C}}_0$$

so that $\exp(\mathbf{i}\hat{f}_\infty) \in \hat{G}_0$. This proves that, for large $n$, $\hat{\gamma}_n$ lies in the $\hat{G}_0$-invariant neighborhood $\hat{U}_0$ of $1 \in \widehat{\mathfrak{G}}_{\mu,ex}(\mathsf{C}_\infty)$. Thus, for all $n \gg 0$ $(1, (\hat{\psi}_n, \mathbf{i}\hat{a}_n))$ and $(\hat{\gamma}_n, (\hat{\psi}_n, \mathbf{i}\hat{b}_n))$ lie in the same fiber of the restriction of $\mathcal{F}$ to $\hat{U}_0 \times \hat{V}_0$. This shows $\hat{\gamma}_n \in \hat{G}_0$, thus completing the proof of Proposition 4.3.7. ■

**Proof of Lemma 4.3.12** The equality (4.3.6) shows that

$$\|(\hat{d}\hat{\gamma}_n)/\hat{\gamma}_n\|_{L^2_\mu} \to 0$$

so that it suffices to prove that there exists $c > 0$ such that

$$\left\|\frac{\hat{d}\hat{\gamma}}{\hat{\gamma}}\right\|_{L^2_\mu} \geq c \tag{4.3.8}$$

for all $\hat{\gamma} \in \widehat{\mathfrak{G}}_{\mu,ex}(\mathsf{C}_\infty)$ which do not lie in the identity component of 1.

Observe that $\Omega_{\hat{\gamma}} := (\hat{d}\hat{\gamma})/\hat{\gamma}$ is closed and $\hat{\gamma}$ lies in the identity component of $\widehat{\mathfrak{G}}_{\mu,ex}(\hat{\mathsf{C}}_\infty)$ if and only if there exists $\hat{f} \in L^{3,2}_\mu$ such that

$$\Omega_{\hat{\gamma}} := \mathbf{i}\hat{d}\hat{f}.$$

Set

$$I_{\hat{\gamma}} : L^{1,2}_\mu(\hat{N}) \to \mathbb{R}, \;\; \hat{f} \mapsto \|\Omega_{\hat{\gamma}} + \mathbf{i}\hat{d}f\|_{L^2_\mu}.$$

This functional is smooth, strictly convex, and coercive, i.e.

$$I_{\hat{\gamma}}(\hat{f}) \to \infty \;\text{ as }\; \|\hat{f}\|_{L^{1,2}_\mu} \to \infty.$$

(The coercivity is a consequence of (4.3.3).) The variational principle [**19**, III.20] (or [**105**, Prop. 9.3.16]) implies there exists a *unique* $\hat{f}_{\hat{\gamma}} \in L^{1,2}_\mu$ such that

$$\|\Omega_{\hat{\gamma}} + \mathbf{i}\hat{d}\hat{f}_{\hat{\gamma}}\|_{L^2_\mu} = \min I_{\hat{\gamma}}.$$

$\hat{f}_{\hat{\gamma}}$ is characterized by the variational equation

$$\hat{d}^{*_\mu}\hat{d}\hat{f}_{\hat{\gamma}} = \mathbf{i}\hat{d}^{*_\mu}\Omega_{\hat{\gamma}}.$$

Arguing exactly as in the proof of Lemma 4.3.10 we deduce $\hat{f}_{\hat{\gamma}} \in L^{3,2}_\mu$. Set

$$[\hat{\gamma}] := \exp(\mathbf{i}\hat{f}_{\hat{\gamma}})\hat{\gamma}, \;\; [\Omega_{\hat{\gamma}}] := \Omega_{[\hat{\gamma}]}.$$

Observe $[\hat{\gamma}]$ is in the same component as $\hat{\gamma}$ but

$$\|\,[\Omega_{\hat{\gamma}}]\,\|_{L^2_\mu} \leq \|\Omega_{\hat{\gamma}}\|_{L^2_\mu}.$$

Notice also that the assumption (4.3.1) implies that $[\Omega_{\hat{\gamma}}]$ lies in the *finite-dimensional* kernel of the Fredholm operator

$$(\hat{d} + \hat{d}^{*_\mu}) : L^{1,2}_\mu(\mathbf{i}\Lambda^*T^*\hat{N}) \to: L^2_\mu(\mathbf{i}\Lambda^*T^*\hat{N}).$$

The set

$$\{[\Omega_{\hat{\gamma}}];\ \ \hat{\gamma} \in \widehat{\mathfrak{G}}_{\mu,ex}(\mathsf{C}_\infty)\}$$

is an Abelian subgroup $S$ of $\ker(\hat{d} + \hat{d}^{*_\mu})$ isomorphic to the *discrete* group of components of $\widehat{\mathfrak{G}}_{\mu,ex}(\mathsf{C}_\infty)$. The constant $c$ in (4.3.8) is given by

$$\inf\{\|s\|_{L^2_\mu};\ \ s \in S \setminus \{0\}\ \} > 0. \ \blacksquare$$

It is now time to put together the results we proved so far to describe a topology on the set $\hat{\mathcal{C}}_{\mu,ex}/\widehat{\mathfrak{G}}_{\mu,ex}$. The results we proved so far amount essentially to a "straightening statement": each orbit has an open invariant neighborhood equivariantly diffeomorphic to an open invariant neighborhood of the zero section of a $\widehat{\mathfrak{G}}_{\mu,ex}$-equivariant vector bundle over $\widehat{\mathfrak{G}}_{\mu,ex}$. Let us provide the details.

Fix $\hat{\mathsf{C}}_0 \in \hat{\mathcal{C}}_{\mu,ex}$ and set $\mathsf{C}_\infty := \partial_\infty \hat{\mathsf{C}}_0$. To describe a neighborhood of $\widehat{\mathfrak{G}}_{\mu,ex} \cdot \hat{\mathsf{C}}_0$ we need to fix several objects.

• A small open neighborhood $U_\infty$ of $0 \in \mathcal{S}_{\mathsf{C}_\infty}$ such that every $\mathfrak{G}_\sigma$-orbit intersects $\mathsf{C}_\infty + U_\infty$ along at most one $G_\infty$-orbit.

• A small open neighborhood $\hat{V}$ of $0 \in \hat{\mathcal{S}}_{\hat{\mathsf{C}}_0}$ such that every $\widehat{\mathfrak{G}}_{\mu,ex}(\mathsf{C}_\infty)$-orbit on $\partial_\infty^{-1}(\mathsf{C}_\infty)$ intersects $\hat{\mathsf{C}}_0 + \hat{V}$ along at most one $\hat{G}_0$-orbit. Set

$$\hat{U}_0 := \hat{U}_0(\hat{V}, U_\infty) = \hat{V} + i_0 U_\infty$$

where $i_0 : \mathcal{C}_\sigma \to \hat{\mathcal{C}}_{\mu,ex}$ is the extension map defined as in §4.1.4.

**Lemma 4.3.13.** *The set $\hat{\mathcal{W}} := \widehat{\mathfrak{G}}_{\mu,ex} \cdot \hat{U}_0$ is an open neighborhood of $\widehat{\mathfrak{G}}_{\mu,ex} \cdot \hat{\mathsf{C}}_0$ in $\hat{\mathcal{C}}_{\mu,ex}$.*

**Sketch of proof** Since $\hat{\mathcal{W}}$ is $\widehat{\mathfrak{G}}_{\mu,ex}$-invariant it suffices to show that there exists an open neighborhood $\mathcal{V}$ of $\hat{\mathsf{C}}_0$ such that $\hat{\mathcal{W}} = \widehat{\mathfrak{G}}_{\mu,ex} \cdot \hat{\mathcal{V}}$. To construct the neighborhood $\hat{\mathcal{V}}$ we consider as in Lemma 4.3.8 a map

$$\mathcal{F} : \widehat{\mathfrak{G}}_{\mu,ex} \times \hat{U}_0 \to \hat{\mathcal{C}}_{\mu,ex},\ \ \mathcal{F}(\hat{\gamma}; \hat{\psi}, \mathbf{i}\hat{a}) = (\hat{\gamma}(\hat{\psi}_0 + \hat{\psi}), \hat{A}_0 + \mathbf{i}\hat{a} - 2\hat{d}\hat{\gamma}/\hat{\gamma}).$$

Using the implicit function theorem (whose applicability can be established using the same arguments as in the proof of Lemma 4.3.8) we can then show there exists a neighborhood $\hat{\mathcal{N}}$ of $\mathbf{1} \in \widehat{\mathfrak{G}}_{\mu,ex}$ such that the restriction of $\mathcal{F}$ to $\hat{\mathcal{N}} \times \hat{U}_0$ is a submersion. Then $\hat{\mathcal{V}} := \mathcal{F}(\hat{\mathcal{N}} \times \hat{U}_0)$ is an open neighborhood of $\hat{\mathsf{C}}_0$ in $\hat{\mathcal{C}}_{\hat{\sigma},\mu}$ and $\hat{\mathcal{W}} = \widehat{\mathfrak{G}}_{\mu,ex} \cdot \hat{\mathcal{V}}$. $\blacksquare$

There is a tautological left $\widehat{\mathfrak{G}}_{\mu,ex}$-action on $\widehat{\mathfrak{G}}_{\mu,ex} \times \hat{U}_0$ and the above map $\mathcal{F}$ is $\widehat{\mathfrak{G}}_{\mu,ex}$-equivariant. Observe that the group $\hat{G}_0$ acts *freely* on $\widehat{\mathfrak{G}}_{\mu,ex} \times \hat{U}_0$

by

$$\hat{\gamma}_0 \cdot (\hat{\gamma}, \hat{\underline{\mathsf{C}}}) := (\hat{\gamma} \cdot \hat{\gamma}_0^{-1}, \hat{\gamma}_0 \cdot \hat{\underline{\mathsf{C}}})$$

$\forall \hat{\gamma}_0 \in \hat{G}_0$, $\hat{\gamma} \in \widehat{\mathfrak{G}}_{\mu,ex}$, $\hat{\underline{\mathsf{C}}} \in \hat{U}_0$. This action commutes with the above $\widehat{\mathfrak{G}}_{\mu,ex}$ action and, moreover, $\mathcal{F}$ is $\hat{G}_0$-invariant. We let the reader check the following fact.

**Exercise 4.3.4.** Each fiber of $\mathcal{F}$ consists of a single $\hat{G}_0$-orbit.

We deduce the following local linearization statement.

**Proposition 4.3.14.** *The induced map*

$$\mathcal{F} : (\widehat{\mathfrak{G}}_{\mu,ex} \times \hat{U}_0)/\hat{G}_0 \to \hat{\mathcal{W}}$$

*is a $\widehat{\mathfrak{G}}_{\mu,ex}$-equivariant diffeomorphism.*

A neighborhood of $(1,0) \in \Big((\widehat{\mathfrak{G}}_{\mu,ex} \times \hat{U}_0)/\hat{G}_0\Big)/\widehat{\mathfrak{G}}_{\mu,ex}$ is homeomorphic to $\hat{U}_0/\hat{G}_0$. This has the following consequence.

**Corollary 4.3.15.** *A neighborhood of $\hat{\mathsf{C}}_0$ in $\hat{\mathcal{W}}/\widehat{\mathfrak{G}}_{\mu,ex}$ (equipped with the quotient topology) is homeomorphic to $\hat{U}_0/\hat{G}_0$.*

Sometimes it is convenient to have a based version of this result. Fix a base point $* \in N$ and form the groups

$$\mathfrak{G}_\sigma(*) := \{\gamma \in \mathfrak{G}_\sigma;\ \ \gamma(*) = 1\}$$

and

$$\widehat{\mathfrak{G}}_{\mu,ex}(*) := \partial_\infty^{-1}(\mathfrak{G}_\sigma(*)).$$

Using the short exact sequence

$$1 \hookrightarrow \widehat{\mathfrak{G}}_{\mu,ex}(*) \hookrightarrow \widehat{\mathfrak{G}}_{\mu,ex} \twoheadrightarrow S^1 \to 1$$

(where the second arrow is given by $\hat{\gamma} \mapsto \partial_\infty \hat{\gamma}(*)$) we obtain a fibration

$$\begin{array}{ccc} \widehat{\mathfrak{G}}_{\mu,ex}(*) & \hookrightarrow & \widehat{\mathfrak{G}}_{\mu,ex} \times \hat{U}_0 \\ & & \Big\downarrow \mathbf{p} \\ & & S^1 \times \hat{U}_0. \end{array}$$

The projection $\mathbf{p}$ is $\hat{G}_0$-equivariant and we get a fibration

$$\begin{array}{ccc} \widehat{\mathfrak{G}}_{\mu,ex}(*) & \hookrightarrow & (\widehat{\mathfrak{G}}_{\mu,ex} \times \hat{U}_0)/\hat{G}_0 \\ & & \Big\downarrow \mathbf{p} \\ & & (S^1 \times \hat{U}_0)/\hat{G}_0. \end{array}$$

The last diagram has the following consequence.

**Corollary 4.3.16.** *The based gauge group $\widehat{\mathfrak{G}}_{\mu,ex}(*)$ acts freely on $\hat{\mathcal{C}}_{\hat{\sigma},\mu}$ and the quotient is naturally a smooth Banach manifold equipped with a smooth $S^1$-action. A neighborhood of $\hat{\mathsf{C}}_0$ in this based quotient is $S^1$-equivariantly diffeomorphic to*

$$(S^1 \times \hat{U}_0)/\hat{G}_0.$$

*Moreover, we have a natural homeomorphism*

$$\hat{\mathcal{C}}_{\hat{\sigma},\mu}/\widehat{\mathfrak{G}}_{\mu,ex} \cong \Big(\hat{\mathcal{C}}_{\hat{\sigma},\mu}/\widehat{\mathfrak{G}}_{\mu,ex}(*)\Big)/S^1.$$

The asymptotic boundary map $\partial_\infty : \hat{\mathcal{C}}_{\hat{\sigma},\mu} \to \mathcal{C}_\sigma$ fits nicely in this picture. Observe first that

$$\partial_\infty(\hat{\gamma} \cdot \hat{\mathsf{C}}) = (\partial_\infty \hat{\gamma}) \cdot (\partial_\infty \hat{\mathsf{C}}), \quad \forall \hat{\gamma} \in \hat{\gamma}_\mu(*)$$

and thus we get a smooth map

$$\partial_\infty : \hat{\mathcal{C}}_{\hat{\sigma},\mu}/\widehat{\mathfrak{G}}_{\mu,ex}(*) \to \mathcal{C}_\sigma/\mathfrak{G}^\partial_\sigma(*). \tag{4.3.9}$$

This map is locally described by

$$\partial_\infty : \hat{U}_0 \to U_\infty$$

which is clearly a submersion. Observe also that the map (4.3.9) is onto.

### 4.3.2. The Kuranishi picture.

The ambient configuration space

$$\hat{\mathcal{C}}_{\mu,ex}/\widehat{\mathfrak{G}}_{\mu,ex}$$

has a rich local and *infinitesimal* structure. We now want to analyze whether the set of gauge equivalence classes of finite energy monopoles has a natural local structure compatible in a natural way with the local structure of the ambient space.

We first need to define the appropriate functional set-up for the Seiberg-Witten map (whose zeros will be our finite energy monopoles). To construct such a set-up we will rely on the nondegeneracy assumption (**N**). Denote by $\mathfrak{Z}_\sigma \subset \mathcal{C}_\sigma$ the set of 3-monopoles on $N$. The nondegeneracy assumption implies that $\mathfrak{Z}_\sigma$ is a Banach manifold.

Define

$$\hat{\mathcal{C}}_{\mu,sw} := \partial_\infty^{-1}(\mathfrak{Z}_\sigma)$$

and

$$\hat{\mathfrak{Y}}_\mu := L^{1,2}_\mu(\hat{\mathbb{S}}^-_{\hat{\sigma}} \oplus \mathbf{i}\Lambda^2_+ T^*\hat{N}).$$

Observe that $\hat{\mathcal{C}}_{\mu,sw}$ is a smooth $\widehat{\mathfrak{G}}_{\mu,ex}$-invariant submanifold of $\hat{\mathcal{C}}_{\mu,ex}$.

At this point we want to draw the attention to a very confusing fact having to do with the cylindrical structure of $\mathbf{i}\Lambda^2_+ T^*\hat{N}$ described in Example 4.1.24 of §4.1.6. Recall that along the neck $\mathbb{R}_+ \times N$ we have the bundle *isometry*

$$\mathcal{I} : \Lambda^2_+ T^*\hat{N} \stackrel{\cong}{\to} \pi^*\Lambda^1 T^*N, \;\; \omega \mapsto \sqrt{2}\, \lrcorner_t\, \omega$$

where $\pi$ is the natural projection $\mathbb{R} \times N \to N$.

The following fact indicates that, for essentially metric reasons, we have to be very careful how we interpret the term $q(\psi)$, as an endomorphism or as a differential form.

**Exercise 4.3.5.** (a) Show that if $e^1, e^2, e^3$ is a local oriented orthonormal frame of $T^*N$ then for every $\psi \in \Gamma(\mathbb{S}_\sigma)$ we have

$$\mathbf{c}^{-1}(q(\psi)) = \frac{1}{2}\sum_i \langle \psi, \mathbf{c}(e^i)\psi \rangle e^i.$$

(b) Show that for every $t > 0$ and every $\hat{\psi} \sim (\psi(t)) \in \Gamma(\hat{\mathbb{S}}^+_{\hat{\sigma}}) \cong \Gamma(\pi^*\mathbb{S}_\sigma)$

$$\sqrt{2}\mathcal{I}\Big(\hat{\mathbf{c}}^{-1}\big(q(\hat{\psi})\big)\Big)\,|_{t\times N} = \mathbf{c}^{-1}\big(q(\psi(t))\big).$$

**Hint for (b):** Use part (a) and the identity in Exercise 1.3.2.

The Seiberg-Witten equations define a natural map

$$\widehat{SW} : \mathcal{C}_{\mu,sw} \to \hat{\mathfrak{Y}}_\mu, \quad (\hat{\psi}, \hat{A}) \mapsto \hat{\mathfrak{D}}_{\hat{A}}\hat{\psi} \oplus \Big(\sqrt{2}(F^+_{\hat{A}} - \frac{1}{2}\hat{\mathbf{c}}^{-1}(q(\hat{\psi}))\Big).$$

Using Exercise 4.3.5 the reader can immediately check that indeed $\widehat{SW}(\hat{\mathsf{C}}) \in \hat{\mathfrak{Y}}_\mu$ for all $\hat{\mathsf{C}} \in \hat{\mathcal{C}}_{\mu,sw}$ and that $\widehat{SW}$ is twice continuously differentiable. Set

$$\widehat{\mathfrak{M}}_\mu := \widehat{SW}^{-1}(0)/\widehat{\mathfrak{G}}_{\mu,ex}, \;\; \widehat{\mathfrak{M}}_\mu(*) := \widehat{SW}^{-1}(0)/\widehat{\mathfrak{G}}_{\mu,ex}(*).$$

We want to analyze the local structure of $\widehat{\mathfrak{M}}_\mu$ and $\widehat{\mathfrak{M}}_\mu(*)$.

Suppose $\hat{\mathsf{C}}_0$ is a smooth finite energy monopole. The results in §4.2.4 show that, modulo a $L^{3,2}_{loc}$-gauge transformation, we can assume $\hat{\mathsf{C}}_0 \in \hat{\mathcal{C}}_{\mu,sw}$.

Denote by $\underline{\widehat{SW}}_{\hat{\mathsf{C}}_0}$ the linearization of $\widehat{SW}$ at $\hat{\mathsf{C}}_0$. We obtain a differential complex

$$(\widehat{\mathcal{K}}_{\hat{\mathsf{C}}_0}) \qquad 0 \to T_1\widehat{\mathfrak{G}}_{\mu,ex} \stackrel{\frac{1}{2}\mathfrak{L}_{\hat{\mathsf{C}}_0}}{\longrightarrow} T_{\hat{\mathsf{C}}_0}\hat{\mathcal{C}}_{\mu,sw} \stackrel{\underline{\widehat{SW}}_{\hat{\mathsf{C}}_0}}{\longrightarrow} T_0\mathfrak{Y}_\mu \to 0.$$

**Proposition 4.3.17.** *The complex $\widehat{\mathcal{K}}_{\hat{\mathsf{C}}_0}$ is Fredholm.*

**Proof** Let us first introduce a bit of terminology. A *Hilbert complex* is a differential complex

$$0 \to H_0 \stackrel{d}{\to} H_1 \stackrel{d}{\to} H_2 \to \cdots$$

in which the spaces of cochains $H_i$ are Hilbert spaces and the differentials are *bounded* linear maps. A *Fredholm complex* is a Hilbert complex with finite-dimensional cohomology. (For more on Hilbert complexes we refer to [**20**].) The following result is left to the reader as an exercise.

**Lemma 4.3.18.** *Suppose*

$$0 \to (C_0, d_0) \stackrel{f}{\to} (C_1, d_1) \stackrel{g}{\to} (C_2, d_2) \to 0$$

*is a short exact sequence of Fredholm complexes where the morphisms $f$ and $g$ are bounded linear maps with closed ranges. If two of the complexes are Fredholm then so is the third and, moreover,*

$$\chi(C_1, d_1) = \chi(C_0, d_0) + \chi(C_2, d_2)$$

*where $\chi$ denotes the Euler-Poincaré characteristic of the associated $\mathbb{Z}$-graded cohomology space.*

The complex $\widehat{(\mathcal{K}_{\hat{\mathsf{C}}_0})}$fits in a short exact sequence

$$(\mathbf{E}) \qquad 0 \to F \stackrel{i}{\hookrightarrow} \widehat{\mathcal{K}}_{\hat{\mathsf{C}}_0} \stackrel{\partial_\infty}{\longrightarrow} B \to 0$$

defined as follows.

- $F = F_{\hat{\mathsf{C}}_0}$:

$(F)$

$$0 \hookrightarrow L^{3,2}_\mu(\hat{N}, \mathbf{i}\mathbb{R}) = T_1\widehat{\mathfrak{G}}_\mu \stackrel{\mathfrak{L}_{\hat{\mathsf{C}}_0}}{\longrightarrow} L^{2,2}_\mu(\hat{\mathbb{S}}^+_{\hat{\sigma}} \oplus \mathbf{i}T^*\hat{N}) = T_{\hat{\mathsf{C}}_0}\partial^{-1}_\infty(\mathsf{C}_\infty) \stackrel{\underline{\widehat{SW}}_{\hat{\mathsf{C}}_0}}{\longrightarrow} \hat{\mathfrak{Y}}_\mu \to 0.$$

- $B = B(\hat{\mathsf{C}}_0)$:

$$(B) \qquad 0 \to T_1\mathfrak{G}_\sigma \stackrel{\frac{1}{2}\mathfrak{L}_{\mathsf{C}_\infty}}{\longrightarrow} T_{\mathsf{C}_\infty}\mathfrak{Z}_\sigma \to 0 \to 0.$$

Denote by $d(\mathsf{C}_\infty)$ the dimension of the component of $\mathfrak{M}_\sigma$ containing $\mathsf{C}_\infty$. We leave the reader to check the following elementary facts.

**Exercise 4.3.6.** (a) Prove that (**E**) is exact and the maps $i$ and $\partial_\infty$ have closed ranges.
(b) Prove that $B$ is Fredholm and $\chi(B) = \dim \mathbf{Stab}(\mathsf{C}_\infty) - d(\mathsf{C}_\infty)$.

We see that Proposition 4.3.17 is a consequence of the following result.

**Lemma 4.3.19.** *The differential complex $F$ is Fredholm if $0 < \mu < \mu_0(\sigma, g)$.*

**Proof** The arguments in the proof of Lemma 4.3.9 (especially the estimate (4.3.3)) show that the differential $\mathfrak{L}_{\hat{\mathsf{C}}_0}$ in $F$ has closed range if $0 < \mu < \mu_0(\sigma, g)$. Moreover $\ker \mathfrak{L}_{\hat{\mathsf{C}}_0} = T_1\mathbf{Stab}\,(\hat{\mathsf{C}}_0)$. Thus it suffices to show that

$$\underline{\widehat{SW}} : L^{2,2}_\mu(\hat{\mathbb{S}}^+_{\hat{\sigma}} \oplus \mathbf{i}T^*\hat{N}) \to \hat{\mathfrak{Y}}_\mu$$

has closed, finite codimensional range and $\dim\Big(\ker \underline{\widehat{SW}}_{\hat{\mathsf{C}}_0}/\mathrm{Range}\,(\mathfrak{L}_{\hat{\mathsf{C}}_0})\Big) < \infty$.

Using Lemma 4.3.10 we deduce that any $\underline{\hat{\mathsf{C}}} \in L^{2,2}_\mu(\hat{\mathbb{S}}^+_{\hat{\sigma}} \oplus \mathbf{i}T^*\hat{N})$ decomposes uniquely as

$$\underline{\hat{\mathsf{C}}} = \underline{\hat{\mathsf{C}}}_0 + \underline{\hat{\mathsf{C}}}^\perp$$

where

$$\underline{\hat{\mathsf{C}}}^\perp \in \mathrm{Range}\,(\mathfrak{L}_{\hat{\mathsf{C}}_0}) \subset \ker \underline{\widehat{SW}}_{\hat{\mathsf{C}}_0}$$

and

$$\underline{\hat{\mathsf{C}}}_0 \in L^{2,2}_\mu(\hat{\mathbb{S}}^+_{\hat{\sigma}} \oplus \mathbf{i}T^*\hat{N}), \;\; \mathfrak{L}^{*_\mu}_{\hat{\mathsf{C}}_0}\underline{\hat{\mathsf{C}}}_0 = 0.$$

Thus it suffices to show that the operator

$$\hat{\mathcal{T}}_{\hat{\mathsf{C}}_0,\mu} := \underline{\widehat{SW}}_{\hat{\mathsf{C}}_0} \oplus \frac{1}{2}\mathfrak{L}^{*_\mu}_{\hat{\mathsf{C}}_0} : L^{2,2}_\mu(\hat{\mathbb{S}}^+_{\hat{\sigma}} \oplus \mathbf{i}T^*\hat{N}) \to \hat{\mathfrak{Y}}_\mu \oplus L^{1,2}_\mu(N, \mathbf{i}\mathbb{R})$$

is Fredholm. To do so, we will rely on Lockhart-McOwen Theorem 4.1.16.

Let us first observe that $\hat{\mathcal{T}}_{\hat{\mathsf{C}}_0,\mu}$ is an a-$APS$ operator. Set

$$\hat{\mathsf{C}}_0 = (\hat{\psi}_0(t), \hat{A}_0), \;\; \mathsf{C}_\infty := (\psi_\infty, A_\infty) = \partial_\infty \hat{\mathsf{C}}_0$$

and

$$\underline{\hat{\mathsf{C}}} = \begin{bmatrix} \underline{\psi}(t) \\ \\ \mathbf{i}dt \wedge u(t) + \mathbf{i}\underline{a}(t) \end{bmatrix} \in \begin{matrix} L^{2,2}_\mu(\hat{\mathbb{S}}^+_{\hat{\sigma}}) \\ \oplus \\ L^{2,2}_\mu(\mathbf{i}T^*(\mathbb{R}_+ \times N)) \end{matrix}$$

Along the neck we can write $\hat{A}_0 = A_\infty + \mathbf{i}dt \wedge v(t) + \mathbf{i}a(t)$, $v, a \in L^{k,2}_\mu$ for all $k \in \mathbb{Z}_+$, $0 < \mu < \mu_0$. The operator $\underline{\widehat{SW}}_{\hat{\mathsf{C}}_0} \oplus \frac{1}{2}\mathfrak{L}^{*_\mu}_{\hat{\mathsf{C}}_0}$ has the form

$$\Big(\underline{\widehat{SW}}_{\hat{\mathsf{C}}_0} \oplus \frac{1}{2}e^{-2\mu t}\mathfrak{L}^*_{\hat{\mathsf{C}}_0}e^{2\mu t}\Big)\Big(\underline{\psi}(t) \oplus \mathbf{i}dt \wedge u(t) + \mathbf{i}a(t)\Big)$$

$$= \begin{bmatrix} \mathfrak{D}_{\hat{A}_0}\underline{\psi}(t) + \frac{1}{2}\hat{\mathbf{c}}\big(\,\mathbf{i}dt\wedge u(t) + \mathbf{i}a(t)\,\big)\hat{\psi}_0 \\ \\ \sqrt{2}\hat{d}^+(\,\mathbf{i}dt\wedge u(t) + \mathbf{i}\underline{a}(t)\,) - \frac{1}{\sqrt{2}}\hat{\mathbf{c}}^{-1}\big(\dot{q}(\,\hat{\psi}_0,\underline{\psi}\,)\big) \\ \\ -\mathbf{i}\hat{d}^*\Big(\,dt\wedge u(t) + \underline{a}(t)\,\Big) + 2\mathbf{i}\mu u(t) - \frac{\mathbf{i}}{2}\mathfrak{Im}\langle\,\hat{\psi}_0,\underline{\psi}\,\rangle \end{bmatrix}$$

(use Exercise 4.3.5 and the computations in Example 4.1.24)

$$= \begin{bmatrix} J & 0 & 0 \\ 0 & 1 & 0 \\ 0 & 0 & 1 \end{bmatrix} \begin{bmatrix} \hat{\nabla}^0_t\underline{\psi}(t) - \Big(\mathfrak{D}_{A_\infty}\underline{\psi} + \frac{\mathbf{i}}{2}(\mathbf{c}(\underline{a}(t)) - u(t))\hat{\psi}_0\Big) & +\frac{\mathbf{i}}{2}v(t)\underline{\psi}(t) \\ \\ \mathbf{i}(\,\partial_t\underline{a}(t) + *d\underline{a}(t) - du(t)\,) & -\frac{1}{2}\mathbf{c}^{-1}\dot{q}(\hat{\psi}_0,\underline{\psi}) \\ \\ \mathbf{i}\Big(\partial_t u(t) - d^*\underline{a}(t) + 2\mu u(t)\Big) & -\frac{\mathbf{i}}{2}\mathfrak{Im}\langle\,\hat{\psi}_0,\underline{\psi}\,\rangle \end{bmatrix}$$

$$= \begin{bmatrix} J & 0 & 0 \\ 0 & 1 & 0 \\ 0 & 0 & 1 \end{bmatrix} \left( \partial_t \begin{bmatrix} \underline{\psi}(t) \\ \mathbf{i}\underline{a}(t) \\ \mathbf{i}u(t) \end{bmatrix} - \begin{bmatrix} \mathfrak{D}_{A_\infty} & 0 & 0 \\ 0 & -*d & d \\ 0 & d^* & -2\mu \end{bmatrix} \begin{bmatrix} \underline{\psi}(t) \\ \mathbf{i}\underline{a}(t) \\ \mathbf{i}u(t) \end{bmatrix} \right.$$

$$\left. - \begin{bmatrix} \frac{\mathbf{i}}{2}(\,\mathbf{c}(\underline{a}(t)) - u(t)\,)\hat{\psi}_0(t) \\ \\ \frac{1}{2}\mathbf{c}^{-1}\dot{q}(\hat{\psi}_0(t),\underline{\psi}(t)) \\ \\ \frac{\mathbf{i}}{2}\mathfrak{Im}\langle\,\hat{\psi}_0(t),\underline{\psi}(t)\,\rangle \end{bmatrix} \right) + \begin{bmatrix} \frac{\mathbf{i}v(t)}{2}J & 0 & 0 \\ 0 & 0 & 0 \\ 0 & 0 & 0 \end{bmatrix} \begin{bmatrix} \underline{\psi}(t) \\ \mathbf{i}\underline{a}(t) \\ \mathbf{i}u(t) \end{bmatrix}$$

Proposition 4.2.35 shows that $\|\psi(t) - \psi_\infty\|_{L^{k,2}_\mu} + \|a\|_{L^{k,2}_\mu} + \|v(t)\|_{L^{k,2}_\mu} < \infty$, for all $k \in \mathbb{Z}_+$. The above computation now implies that $\hat{\mathcal{T}}_{\hat{\mathsf{C}}_0,\mu}$ is an a-*APS* operator and, using (4.2.2), we deduce

$$\vec{\partial}_\infty \hat{\mathcal{T}}_{\hat{\mathsf{C}}_0,\mu} = \mathcal{T}_{\mathsf{C}_\infty,\mu} := \begin{bmatrix} \underline{\mathbf{SW}}_{\mathsf{C}_\infty} & -\frac{1}{2}\mathfrak{L}_{\mathsf{C}_\infty} \\ \\ -\frac{1}{2}\mathfrak{L}^*_{\mathsf{C}_\infty} & -2\mu \end{bmatrix}.$$

We want to show that $\ker(\mu + \mathcal{T}_{\mathsf{C}_\infty,\mu}) = 0$ for all $0 < \mu < \mu_-(\sigma, g)$.

Suppose $\dot{\mathsf{C}} \oplus \mathbf{i}f \in \ker(\nu + \mathcal{T}_{\mathsf{C}_\infty,\mu})$, $\nu \in \mathbb{R}$. This means

$$\left\{ \begin{array}{c} \underline{\mathbf{SW}}_{\mathsf{C}_\infty}(\dot{\mathsf{C}}) - \frac{1}{2}\mathfrak{L}_{\mathsf{C}_\infty}(\mathbf{i}f) = -\nu\mathbf{i}\dot{\mathsf{C}} \\ \\ \mathfrak{L}^*_{\mathsf{C}_\infty}(\dot{\mathsf{C}}) + 4\mu\mathbf{i}f = 2\nu\mathbf{i}f \end{array} \right. . \tag{4.3.10}$$

Observe that $\mathfrak{L}^*_{\mathsf{C}_\infty}\underline{\mathbf{SW}}_{\mathsf{C}_\infty} = (\underline{\mathbf{SW}}_{\mathsf{C}_\infty}\mathfrak{L}_{\mathsf{C}_\infty})^* = 0$. If we apply $\mathfrak{L}^*_{\mathsf{C}_\infty}$ to the first equation in (4.3.10) we deduce

$$\mathfrak{L}^*_{\mathsf{C}_\infty}\mathfrak{L}_{\mathsf{C}_\infty}(\mathbf{i}f) = 2\nu\mathfrak{L}^*_{\mathsf{C}_\infty}\dot{\mathsf{C}} = 4\nu(\nu - 2\mu)(\mathbf{i}f).$$

Let us now require that $\nu$ is such that

$$4\nu(\nu-2\mu)<0.$$

This implies $f\equiv 0$ and forces

$$\mathfrak{L}^*_{\mathsf{C}_\infty}\dot{\mathsf{C}}=0,\ \ \underline{\mathbf{SW}}_{\mathsf{C}_\infty}(\dot{\mathsf{C}})=-\nu\dot{\mathsf{C}}.$$

Suppose additionally that

$$0<\nu<\mu_-(\sigma,g)\leq\mu_-(\mathsf{C}_\infty).$$

This implies $\dot{\mathsf{C}}=0$.

Now, if $\nu:=\mu<\mu_-(\sigma,g)$ then automatically both requirements are satisfied because

$$4\nu(\nu-2\mu)=-4\mu^2<0.$$

We deduce that $\ker(\mu+\mathcal{T}_{\mathsf{C}_\infty,\mu})=0$ as soon as $0<\mu<\mu_0(\sigma,g)$. The Lockhart-McOwen Theorem 4.1.16 now implies that $\hat{\mathcal{T}}_{\hat{\mathsf{C}}_0}$ is Fredholm if $0<\mu<\mu_0(\sigma,g)\leq\mu_-(\sigma,g)$. This completes the proof of Lemma 4.3.18 and of Proposition 4.3.17. ■

Set

$$H^i_{\hat{\mathsf{C}}_0}:=H^i(\widehat{\mathcal{K}}_{\hat{\mathsf{C}}_0}),\ \ i=0,1,2.$$

The finite-dimensional space $H^2_{\hat{\mathsf{C}}_0}$ is called the *obstruction space* at $\hat{\mathsf{C}}_0$. Observe also that

$$H^0_{\hat{\mathsf{C}}_0}\cong T_1\hat{G}_0.$$

The results in §4.3.1 show that the quotient $\hat{\mathcal{C}}_{\mu,sw}/\widehat{\mathfrak{G}}_{\mu,ex}$ equipped with the quotient topology has a nice local structure. Suppose $\hat{\mathsf{C}}_0\in\hat{\mathcal{C}}_{\mu,sw}$ is a finite energy monopole. Set

$$\mathcal{S}^{sw}_{\mathsf{C}_\infty}:=\mathcal{S}_{\mathsf{C}_\infty}\cap T_{\mathsf{C}_\infty}\mathfrak{Z}_\sigma.$$

Then there exist a small neighborhood $V_\infty$ of $0\in\mathcal{S}^{sw}_{\mathsf{C}_\infty}$ and a small neighborhood $\hat{V}_0$ of $0\in\hat{\mathcal{S}}_{\hat{\mathsf{C}}_0}$ such that if

$$\hat{U}_0:=\hat{\mathsf{C}}_0+\hat{V}+i_0V_\infty$$

then a neighborhood of $[\hat{\mathsf{C}}_0]$ in $\hat{\mathcal{C}}_{\mu,sw}/\widehat{\mathfrak{G}}_{\mu,ex}$ is homeomorphic to the quotient $\hat{U}_0/\hat{G}_0$. The results in §4.3.1 show that additionally

$$T_{\hat{\mathsf{C}}_0}\hat{\mathcal{C}}_{\mu,sw}=\left(\hat{\mathcal{S}}_{\hat{\mathsf{C}}_0}+i_0(\mathcal{S}^{sw}_{\mathsf{C}_\infty})\right)+\mathrm{Range}\left(\mathfrak{L}_{\hat{\mathsf{C}}_0}:T_1\widehat{\mathfrak{G}}_{\mu,ex}\to T_{\hat{\mathsf{C}}_0}\hat{\mathcal{C}}_{\mu,sw}\right)$$

and

$$\left(\hat{\mathcal{S}}_{\hat{\mathsf{C}}_0}+i_0(\mathcal{S}^{sw}_{\mathsf{C}_\infty})\right)\cap\mathrm{Range}\left(\mathfrak{L}_{\hat{\mathsf{C}}_0}:T_1\widehat{\mathfrak{G}}_{\mu,ex}\to T_{\hat{\mathsf{C}}_0}\hat{\mathcal{C}}_{\mu,sw}\right)=0.$$

Thus, to understand the nature of a small neighborhood of $[\hat{\mathsf{C}}_0]$ in $\widehat{\mathfrak{M}}_\mu$ it suffices to understand the nature of the set of small solutions of the nonlinear equation

$$\mathcal{F}(\underline{\hat{\mathsf{C}}}) = 0 \tag{4.3.11}$$

where

$$\mathcal{F} : \left(\hat{\mathcal{S}}_{\hat{\mathsf{C}}_0} + i_0(\mathcal{S}^{sw}_{\mathsf{C}_\infty})\right) \to \hat{\mathfrak{Y}}_\mu, \;\; \mathcal{F}(\underline{\hat{\mathsf{C}}}) := \widehat{SW}(\hat{\mathsf{C}}_0 + \underline{\hat{\mathsf{C}}}).$$

Proposition 4.3.17 shows that the linearization of $\mathcal{F}$ at 0 is a Fredholm map and, moreover,

$$\ker D_0\mathcal{F} \cong H^1_{\hat{\mathsf{C}}_0}, \;\; \operatorname{coker} D_0\mathcal{F} \cong H^2_{\hat{\mathsf{C}}_0}.$$

Arguing exactly as in §2.2.2 we deduce that there exist a small $\hat{G}_0$-invariant open neighborhood $\hat{\mathcal{N}}$ of $0 \in H^1_{\hat{\mathsf{C}}_0}$ and a $\hat{G}_0$-invariant map

$$\hat{Q}_{\hat{\mathsf{C}}_0} : \hat{\mathcal{N}} \to H^2_{\hat{\mathsf{C}}_0}$$

such that $\hat{Q}_{\hat{\mathsf{C}}_0}(0) = 0$ and $\hat{Q}^{-1}_{\hat{\mathsf{C}}_0}(0)/\hat{G}_0$ is homeomorphic to an open neighborhood of $[\hat{\mathsf{C}}_0]$ in $\widehat{\mathfrak{M}}_\sigma$.

**Definition 4.3.20.** (a) The monopole $\hat{\mathsf{C}}_0$ is called *regular* if its obstruction space is trivial, $H^2_{\hat{\mathsf{C}}_0} = 0$. $\hat{\mathsf{C}}_0$ is called *strongly regular* if $H^2(F) = 0$.

(b) The integer

$$d(\hat{\mathsf{C}}_0) := -\chi(\widehat{\mathcal{K}}_{\hat{\mathsf{C}}_0}) = \dim_{\mathbb{R}} H^1_{\hat{\mathsf{C}}_0} - \dim_{\mathbb{R}} H^0_{\hat{\mathsf{C}}_0} - \dim_{\mathbb{R}} H^2_{\hat{\mathsf{C}}_0}$$

is called the *virtual dimension* at $\hat{\mathsf{C}}_0$ of the moduli space $\widehat{\mathfrak{M}}_\mu$.

**Remark 4.3.21.** The long exact sequence associated to (**E**) shows that there is a surjective map $H^2(F) \twoheadrightarrow H^2_{\hat{\mathsf{C}}_0}$ so that a strongly regular monopole is also regular.

The above discussion has the following consequence.

**Corollary 4.3.22.** *(a) If $\hat{\mathsf{C}}_0$ is a regular irreducible monopole then a small neighborhood of $[\hat{\mathsf{C}}_0] \in \widehat{\mathfrak{M}}_\mu$ is homeomorphic to $\mathbb{R}^{d(\hat{\mathsf{C}}_0)}$.*

*(b) If $\hat{\mathsf{C}}_0$ is a strongly regular irreducible then there exist a small neighborhood $\hat{U}_0$ of $[\hat{\mathsf{C}}_0] \in \widehat{\mathfrak{M}}_\mu$ and a small neighborhood $U_\infty$ of $\mathsf{C}_\infty \in \mathfrak{M}_\sigma$ such that $\hat{U}_0 \cong \mathbb{R}^{d(\hat{\mathsf{C}}_0)}$, $U_\infty = \partial_\infty(\hat{U}_0)$ and the induced map $\partial_\infty : \hat{U}_0 \to U_\infty$ is a submersion.*

**Example 4.3.23.** We want to point out some subtleties hidden in (**E**). Consider the special situation when $(N, g)$ is the sphere $S^3$ equipped with the round metric $g$ of radius 1. $Spin^c(N)$ consists of a single structure $\sigma$

and the pair $(\sigma, g)$ is good since $g$ has positive scalar curvature. $\mathfrak{M}_\sigma$ consists of single reducible monopole $\mathsf{C}_0 = (0, A_0)$. We deduce

$$H^0(B) \cong \mathbb{R}, \quad H^1(B) = 0.$$

Suppose $\hat{\mathsf{C}}_0 \in \hat{\mathcal{C}}_{\mu,sw}$ is a smooth irreducible monopole on $\hat{N}$. Then $\partial_\infty \hat{\mathsf{C}}_0 = \mathsf{C}_0$ and the sequence (**E**) leads to a short exact sequence

$$0 \to H^0(B) = \mathbb{R} \to H^1(F) \to H^1_{\hat{\mathsf{C}}_0} \to 0. \tag{4.3.12}$$

A superficial look at the complex $(F)$ might lead one to believe that $H^1(F)$ is intended to be the tangent space at $\hat{\mathsf{C}}_0$ to the fiber of

$$\partial_\infty : \widehat{\mathfrak{M}}_\mu \to \mathfrak{M}_\sigma.$$

Thus one would expect that $H^1(F)$ would inject into $H^1_{\hat{\mathsf{C}}_0}$, intended to be $T_{\hat{\mathsf{C}}_0}\widehat{\mathfrak{M}}_\mu \to \mathfrak{M}_\sigma$. However, the sequence (4.3.12) shows that the natural map $H^1(F) \to H^1_{\hat{\mathsf{C}}_0}$ is not injective since $\dim H^1(F) = \dim H^1_{\hat{\mathsf{C}}_0} + 1$. How can this be possible?

The explanation is simple. The fiber of the map $\partial_\infty : \widehat{\mathfrak{M}}_\mu \to \mathfrak{M}_\sigma$ over $\mathsf{C}_0$ should be understood as the set of monopoles on $\hat{N}$ modulo the group

$$\widehat{\mathfrak{G}}_{\mu,ex}(\mathsf{C}_0) = \partial_\infty^{-1}(\mathbf{Stab}(\mathsf{C}_0)).$$

A careful look at $(F)$ shows that it involves a *smaller* group $\widehat{\mathfrak{G}}_\mu$ which fits in a short exact sequence

$$\mathbf{1} \hookrightarrow \widehat{\mathfrak{G}}_\mu \hookrightarrow \widehat{\mathfrak{G}}_{\mu,ex}(\mathsf{C}_0) \to S^1 \to \mathbf{1}.$$

To correct our initial intuition of $H^1(F)$ we should think of it as intended to be the tangent space to the fibers of

$$\partial_\infty : \widehat{\mathfrak{M}}_\mu(*) \to \mathfrak{M}_\sigma(*).$$

In our case $\mathfrak{M}_\sigma(*) = \mathfrak{M}_\sigma$.

In the remaining part of this subsection we want to provide alternate descriptions of the cohomology spaces intervening in the long exact sequence associated with (**E**). These interpretations (more precisely Propositions 4.3.27 and 4.3.30) constitute the main difference between the approach to gluing we propose in this book and the traditional one pioneered by T. Mrowka, [**99**]. They are responsible for substantial simplifications to the whole gluing procedure. Our first result should be obvious.

**Lemma 4.3.24.** *We have natural isomorphisms*

$$H^1(F_{\hat{\mathsf{C}}_0}) \cong \ker_\mu(\hat{\mathcal{T}}_{\hat{\mathsf{C}}_0,\mu}),$$

$$H^2(F_{\hat{\mathsf{C}}_0}) \cong \ker_\mu(\hat{\mathcal{T}}^{*_\mu}_{\hat{\mathsf{C}}_0,\mu}).$$

**Lemma 4.3.25.** *There exists a natural exact sequence*

$$0 \to U_0 \to \ker_{ex}(\hat{\mathcal{T}}_{\hat{\mathsf{C}}_0,\mu}) \to H^1_{\hat{\mathsf{C}}_0} \to 0$$

*where $U_0$ is the kernel of the natural map $H^1(F_{\hat{\mathsf{C}}_0}) \to H^1_{\hat{\mathsf{C}}_0}$ or, equivalently, the cokernel of the map $\partial_\infty : T_1\hat{G}_0 \to T_1 G_\infty$.*

**Proof** The proof consists of two parts. We will first construct a natural map

$$\ker_{ex}(\hat{\mathcal{T}}_{\hat{\mathsf{C}}_0,\mu}) \to H^1_{\hat{\mathsf{C}}_0}$$

and then we will prove it leads to the above exact sequence. The details will be carried out in several steps.

**Step 1** If $\hat{\underline{\mathsf{C}}} \in \ker_{ex}(\hat{\mathcal{T}}_{\hat{\mathsf{C}}_0,\mu})$ then $\hat{\underline{\mathsf{C}}} \in T_{\hat{\mathsf{C}}_0}\hat{\mathcal{C}}_{\mu,sw}$ , i.e. $\hat{\underline{\mathsf{C}}}$ is *strongly* cylindrical.

Suppose that along the neck $\hat{\underline{\mathsf{C}}}$ has the form

$$\hat{\underline{\mathsf{C}}} = (\underline{\psi}(t), \mathbf{i}\underline{a}(t) + \mathbf{i}u(t)dt).$$

Since $\hat{\underline{\mathsf{C}}} \in \ker_{ex}(\hat{\mathcal{T}}_{\hat{\mathsf{C}}_0,\mu})$ we deduce

$$\partial_\infty \hat{\underline{\mathsf{C}}} = (\underline{\psi}(\infty), \mathbf{i}\underline{a}(\infty), \mathbf{i}u(\infty)) \in \ker \mathcal{T}_{\mathsf{C}_\infty,\mu}.$$

To prove that $u(\infty) = 0$ it suffices to show that if $(\psi, \mathbf{i}a, \mathbf{i}u) \in \ker \mathcal{T}_{\mathsf{C}_\infty}$ then $u = 0$. This follows easily by looking at (4.3.10) in which $\nu = 0$. The details can be safely left to the reader. Thus, we have a well defined map

$$\Upsilon : \ker_{ex}(\hat{\mathcal{T}}_{\hat{\mathsf{C}}_0,\mu}) \hookrightarrow \ker\Big(\underline{\widehat{SW}}_{\hat{\mathsf{C}}_0} : T_{\hat{\mathsf{C}}_0}\hat{\mathcal{C}}_{\mu,sw} \to \hat{\mathfrak{Y}}_\mu\Big) \twoheadrightarrow H^1_{\hat{\mathsf{C}}_0}.$$

**Step 2** $\Upsilon$ *is onto.* Observe first that the long exact sequence associated to (**E**) implies that we can represent each cohomology class $\tau \in H^1_{\hat{\mathsf{C}}_0}$ by an element $\hat{\underline{\mathsf{C}}} \in T_{\hat{\mathsf{C}}_0}\hat{\mathcal{C}}_{\mu,sw}$ such that

$$\underline{\mathbf{SW}}_{\mathsf{C}_\infty}\partial_\infty\hat{\underline{\mathsf{C}}} = 0, \;\; \mathfrak{L}^*_{\mathsf{C}_\infty}\partial_\infty\hat{\underline{\mathsf{C}}} = 0.$$

Next observe that since $\hat{\mathsf{C}}$ is strongly cylindrical we have

$$\partial_\infty\Big(\mathfrak{L}^{*_\mu}_{\hat{\mathsf{C}}_0}\hat{\underline{\mathsf{C}}}\Big) = \mathfrak{L}^*_{\mathsf{C}_\infty}\partial_\infty\hat{\underline{\mathsf{C}}} = 0$$

so that

$$\mathfrak{L}^{*_\mu}_{\hat{\mathsf{C}}_0}\hat{\underline{\mathsf{C}}} \in L^{1,2}_\mu(\hat{N}, \mathbf{i}\mathbb{R}).$$

Arguing as in the proof of Lemma 4.3.9 we deduce that the densely defined, selfadjoint operator

$$\Delta_{\hat{\mathsf{C}}_0,\mu} := \mathfrak{L}^{*_\mu}_{\hat{\mathsf{C}}_0}\mathfrak{L}_{\hat{\mathsf{C}}_0} : L^{2,2}_\mu(\hat{N}, \mathbf{i}\mathbb{R}) \subset L^2_\mu(\hat{N}, \mathbf{i}\mathbb{R}) \to L^2_\mu(\hat{N}, \mathbf{i}\mathbb{R})$$

has closed range. Clearly its kernel is trivial so that it is also surjective. Arguing as in the proof of Lemma 4.3.10 we deduce that

$$\Delta^{-1}_{\hat{\mathsf{C}}_0,\mu}\left(L^{1,2}_\mu\right) = L^{3,2}_\mu.$$

Thus we can find $\mathbf{i}\hat{f}_0 \in L^{3,2}_\mu(\hat{N}, \mathbf{i}\mathbb{R})$ such that

$$\Delta_{\hat{\mathsf{C}}_0,\mu}(\mathbf{i}f_0) = \mathfrak{L}^{*_\mu}_{\hat{\mathsf{C}}_0}\underline{\hat{\mathsf{C}}}.$$

If we set

$$\underline{\hat{\mathsf{C}}}' := \underline{\hat{\mathsf{C}}} - \mathfrak{L}_{\hat{\mathsf{C}}_0}(\mathbf{i}\hat{f}_0)$$

then $\underline{\widehat{SW}}_{\hat{\mathsf{C}}_0}\hat{\mathsf{C}}' = 0$ so that $\hat{\mathsf{C}}$ and $\hat{\mathsf{C}}'$ define the same element in $H^1_{\hat{\mathsf{C}}_0}$. Moreover

$$\mathfrak{L}^{*_\mu}_{\hat{\mathsf{C}}_0}\hat{\mathsf{C}}' = \mathfrak{L}^{*_\mu}_{\hat{\mathsf{C}}_0}\hat{\mathsf{C}} - \Delta_{\hat{\mathsf{C}}_0,\mu}(\mathbf{i}\hat{f}_0) = 0$$

so that $\hat{\mathsf{C}}' \in \ker_{ex}\hat{\mathcal{T}}_{\hat{\mathsf{C}}_0,\mu}$. This proves that $\Upsilon$ is onto.

**Step 3** $\ker\Upsilon = \ker(H^1(F_{\hat{\mathsf{C}}_0}) \to H^1_{\hat{\mathsf{C}}_0})$. From the natural inclusion

$$H^1(F_{\hat{\mathsf{C}}_0}) = \ker_\mu(\hat{\mathcal{T}}_{\hat{\mathsf{C}}_0,\mu}) \subset \ker_{ex}(\hat{\mathcal{T}}_{\hat{\mathsf{C}}_0,\mu})$$

we deduce that

$$\ker(H^1(F_{\hat{\mathsf{C}}_0}) \to H^1_{\hat{\mathsf{C}}_0}) \subset \ker\Upsilon.$$

Conversely, suppose $\Upsilon(\underline{\hat{\mathsf{C}}}) = 0 \in H^1_{\hat{\mathsf{C}}_0}$. In particular, this implies

$$\partial_\infty\underline{\hat{\mathsf{C}}} = 0,$$

i.e. $\underline{\hat{\mathsf{C}}} \in L^2_\mu \iff \underline{\hat{\mathsf{C}}} \in H^1(F_{\hat{\mathsf{C}}_0})$. ■

**Remark 4.3.26.** It is perhaps instructive to describe the image of $U_0$ in $\ker_{ex}\hat{\mathcal{T}}_{\hat{\mathsf{C}}_0,\mu}$. Suppose for simplicity that $N$ is connected, $\hat{\mathsf{C}}_0$ is irreducible but $\mathsf{C}_\infty$ is reducible. Then $U_0 \subset H^1(F_{\hat{\mathsf{C}}_0})$ is spanned by the the infinitesimal variation $\mathfrak{L}_{\hat{\mathsf{C}}_0}(\mathbf{i})$. To find its harmonic representative (i.e. describe the element in $\ker_\mu\hat{\mathcal{T}}_{\hat{\mathsf{C}}_0,\mu}$ defining the same class in $H^1(F_{\hat{\mathsf{C}}_0})$) it suffices to solve the equation

$$\Delta_{\hat{\mathsf{C}}_0,\mu}(\mathbf{i}\hat{f}) = \Delta_{\hat{\mathsf{C}}_0,\mu}(\mathbf{i})$$

with unique solution

$$L^{3,2}_\mu \ni \mathbf{i}\varphi_0 := \Delta^{-1}_{\hat{\mathsf{C}}_0,\mu}\left(\Delta_{\hat{\mathsf{C}}_0,\mu}(\mathbf{i})\right).$$

Then the harmonic representative of $\mathfrak{L}_{\hat{\mathsf{C}}_0}(\mathbf{i})$ is $\mathfrak{L}_{\hat{\mathsf{C}}_0}(\mathbf{i} - \mathbf{i}\varphi_0)$. Observe that $\hat{f}_0 := 1 - \varphi_0$ is the *unique* function $f \in L^{3,2}_{\mu,ex}$ satisfying the equations

$$\Delta_{\hat{\mathsf{C}}_0,\mu}(\mathbf{i}\hat{f}_0) = 0, \;\; \partial_\infty\hat{f}_0 = 1. \;\blacksquare$$

Lemma 4.3.25 has one small "defect". More precisely, it describes a geometric object, the virtual tangent space $H^1_{\hat{\mathsf{C}}_0}$, in terms of the quantity $\ker \hat{\mathcal{T}}_{\hat{\mathsf{C}}_0,\mu}$ which depends on the choice of $\mu$ dictated by functional analytic considerations. Our next result will remove this defect. Set

$$\hat{\mathcal{T}}_{\hat{\mathsf{C}}_0} := \underline{\widehat{SW}}_{\hat{\mathsf{C}}_0} \oplus \frac{1}{2}\mathfrak{L}^*_{\hat{\mathsf{C}}_0}.$$

Observe that the a-$APS$ operator $\hat{\mathcal{T}}_{\hat{\mathsf{C}}_0}$ can be formally obtained from $\hat{\mathcal{T}}_{\hat{\mathsf{C}}_0,\mu}$ by setting $\mu = 0$. Moreover, the decomposition

$$\ker \mathcal{T}_{\mathsf{C}_\infty} = T_{\mathsf{C}_\infty}\mathfrak{M}_\sigma \oplus T_1 G_\infty$$

produces a decomposition of the boundary map

$$\partial_\infty : \ker_{ex} \hat{\mathcal{T}}_{\hat{\mathsf{C}}_0} \to \ker \mathcal{T}_{\mathsf{C}_\infty}$$

into components

$$\partial^0_\infty : \ker_{ex} \hat{\mathcal{T}}_{\hat{\mathsf{C}}_0} \to T_1 G_\infty, \quad \partial^c_\infty : \ker_{ex} \hat{\mathcal{T}}_{\hat{\mathsf{C}}_0} \to T_{\mathsf{C}_\infty}\mathfrak{M}_\sigma.$$

**Remark 4.3.27.** Using (4.1.22) of §4.1.5 with $G = 1$ we deduce that we have the orthogonal decomposition

$$\partial^0_\infty(\ker_{ex} \hat{\mathcal{T}}_{\hat{\mathsf{C}}_0}) \oplus \partial^0_\infty(\ker_{ex} \hat{\mathcal{T}}^*_{\hat{\mathsf{C}}_0}) = T_1 G_\infty.$$

Now observe that if $(\Psi, \mathbf{i}f) \in L^{1,2}_{\mu,ex}(\mathbb{S}^-_{\hat{\sigma}} \oplus \mathbf{i}\Lambda^2_+ T^*\hat{N}) \oplus L^{1,2}_{\mu,ex}(\mathbf{i}\Lambda^0 T^*\hat{N})$ belongs to $\ker_{ex} \hat{\mathcal{T}}^*_{\hat{\mathsf{C}}_0}$ then $\mathbf{i}f \in T_1\hat{G}_0$ (see the the proof of Proposition 4.3.30). Thus

$$\partial^0_\infty(\ker_{ex} \hat{\mathcal{T}}_{\hat{\mathsf{C}}_0}) \cong T_1(G_\infty/\partial_\infty \hat{G}_0).$$

As an example, suppose $\hat{\mathsf{C}}_0$ is reducible, $\hat{\mathsf{C}}_0 = (0, \hat{A}_0)$. Then

$$\hat{\mathcal{T}}_{\hat{\mathsf{C}}_0} = \mathfrak{D}_{\hat{A}_0} \oplus \mathbf{ASD}.$$

The above observation implies that any 1-form $\omega \in \ker_{ex} \mathbf{ASD}$ is strongly cylindrical. This is in perfect agreement with the equality (4.1.28) proved in Example 4.1.24 of §4.1.6.

**Proposition 4.3.28.** *There exists a natural short exact sequence*

$$(H_1) \qquad 0 \to H^1_{\hat{\mathsf{C}}_0} \to \ker_{ex} \hat{\mathcal{T}}_{\hat{\mathsf{C}}_0} \to T_1(G_\infty/\partial_\infty \hat{G}_0) \to 0.$$

*In particular*

$$\ker_{ex} \hat{\mathcal{T}}_{\hat{\mathsf{C}}_0,\mu} \cong \ker_{ex} \hat{\mathcal{T}}_{\hat{\mathsf{C}}_0}.$$

**Proof** We discuss separately three cases.

**Case A.** $\hat{\mathsf{C}}_0$ is reducible. In view of Lemma 4.3.25 we only have to prove $\ker_{ex}\hat{\mathcal{T}}_{\hat{\mathsf{C}}_0,\mu}\cong H^1_{\hat{\mathsf{C}}_0}\cong\ker_{ex}\hat{\mathcal{T}}_{\hat{\mathsf{C}}_0}$. Set

$$\hat{V}:=\hat{\mathbb{S}}^+_{\hat{\sigma}}\oplus\mathbf{i}\Lambda^1T^*\hat{N}.$$

Along the neck it decomposes as

$$\hat{V}\cong\mathbb{S}_\sigma\oplus\mathbf{i}\pi^*\Lambda^1T^*N\oplus\mathbf{i}dt\pi^*\underline{\mathbb{R}}_N$$

where $\pi:\mathbb{R}_+\times N\to N$ is the natural projection. Over the neck, each section $\underline{\hat{\mathsf{C}}}$ of $\hat{V}$ splits as

$$\underline{\hat{\mathsf{C}}}=\underline{\psi}(t)\oplus(\mathbf{i}\underline{a}(t)+\mathbf{i}u(t)dt).$$

Denote by $T_\mu$ the automorphism of $\hat{V}$ which is the identity off the neck while along the neck it has the form

$$T_\mu\Big(\underline{\psi}(t)\oplus(\mathbf{i}\underline{a}(t)+\mathbf{i}u(t)dt)\Big)=\underline{\psi}(t)\oplus(\mathbf{i}\underline{a}(t)+\mathbf{i}\mathbf{m}_{2\mu}u(t)dt).$$

A simple computation shows that since $\hat{\mathsf{C}}_0$ is reducible we have

$$\mathfrak{L}^{*\mu}_{\hat{\mathsf{C}}_0}=\mathbf{m}_{-2\mu}\mathfrak{L}^*_{\hat{\mathsf{C}}_0}T_\mu$$

and

$$\widehat{\underline{SW}}_{\hat{\mathsf{C}}_0}T_\mu\underline{\hat{\mathsf{C}}}=T_\mu\widehat{\underline{SW}}_{\hat{\mathsf{C}}_0}\underline{\hat{\mathsf{C}}}.$$

We thus have a well defined *bijection*

$$\ker\hat{\mathcal{T}}_{\hat{\mathsf{C}}_0,\mu}\to\ker\hat{\mathcal{T}}_{\hat{\mathsf{C}}_0},\ \ \underline{\hat{\mathsf{C}}}\mapsto T_\mu\underline{\hat{\mathsf{C}}}$$

which maps $\ker_{-\varepsilon}\hat{\mathcal{T}}_{\hat{\mathsf{C}}_0,\mu}$ *injectively* into $\ker_{-\mu-\varepsilon}\hat{\mathcal{T}}_{\hat{\mathsf{C}}_0}$. Its inverse maps the space $\ker_{-\mu-\varepsilon}\hat{\mathcal{T}}_{\hat{\mathsf{C}}_0}$ *injectively* into $\ker_{-\mu-\varepsilon}\hat{\mathcal{T}}_{\hat{\mathsf{C}}_0,\mu}$. To conclude the proof of Case A we only need to recall Proposition 4.1.17 which states that if $\mu$ is sufficiently small then

$$\ker_{ex}\hat{\mathcal{T}}_{\hat{\mathsf{C}}_0,\mu}\cong\ker_{-\varepsilon}\hat{\mathcal{T}}_{\hat{\mathsf{C}}_0,\mu}\cong\ker_{-\mu-\varepsilon}\hat{\mathcal{T}}_{\hat{\mathsf{C}}_0,\mu},\ \ \ker_{ex}\hat{\mathcal{T}}_{\hat{\mathsf{C}}_0}\cong\ker_{-\mu-\varepsilon}\hat{\mathcal{T}}_{\hat{\mathsf{C}}_0}.$$

**Case B.** $\mathsf{C}_\infty$ is irreducible, and thus so is $\hat{\mathsf{C}}_0$. We have to show $\ker_{ex}\hat{\mathcal{T}}_{\hat{\mathsf{C}}_0}\cong H^1_{\hat{\mathsf{C}}_0}$. Note that any $\underline{\hat{\mathsf{C}}}\in\ker_{ex}\hat{\mathcal{T}}_{\hat{\mathsf{C}}_0}$ tautologically defines a cohomology class in $H^1_{\hat{\mathsf{C}}_0}$. We want to show that the induced map $\ker_{ex}\hat{\mathcal{T}}_{\hat{\mathsf{C}}_0}\to H^1_{\hat{\mathsf{C}}_0}$ is an isomorphism.

Observe first that this map is $1-1$. Indeed, if

$$\underline{\hat{\mathsf{C}}}\in\ker_{ex}\hat{\mathcal{T}}_{\hat{\mathsf{C}}_0}\ \text{ and }\ \underline{\hat{\mathsf{C}}}=\mathfrak{L}_{\hat{\mathsf{C}}_0}(\mathbf{i}f)$$

for some $f\in L^{3,2}{}_{\mu},ex$ then $\Delta_{\hat{\mathsf{C}}_0}(\mathbf{i}f):=\mathfrak{L}^*_{\hat{\mathsf{C}}_0}\mathfrak{L}_{\hat{\mathsf{C}}_0}(\mathbf{i}f)=0$. Multiplying the last equality by $\mathbf{i}f$ and integrating by parts on $\hat{N}_{T\to\infty}$ we deduce $\mathfrak{L}_{\hat{\mathsf{C}}_0}(\mathbf{i}f)=0$.

To show that this map is onto we construct a right inverse $\Gamma$. More precisely, if $\underline{\hat{\mathsf{C}}} \in T_{\hat{\mathsf{C}}_0}\mathcal{C}_{\mu,sw}$ satisfies $\underline{\widehat{SW}}_{\hat{\mathsf{C}}_0}$ then we set

$$\Gamma(\underline{\hat{\mathsf{C}}}) = \underline{\hat{\mathsf{C}}} - \mathfrak{L}_{\hat{\mathsf{C}}_0}\Delta_{\hat{\mathsf{C}}_0}^{-1}\mathfrak{L}_{\hat{\mathsf{C}}}^*\underline{\hat{\mathsf{C}}}$$

where we regard $\Delta_{\hat{\mathsf{C}}_0}$ as a bounded Fredholm operator

$$\Delta_{\hat{\mathsf{C}}_0} : L^{3,2}(\hat{N}) \to L^{1,2}(\hat{N}).$$

(It is Fredholm since $\Delta_{\hat{\mathsf{C}}_0} = \Delta + \frac{1}{4}|\hat{\psi}_0|^2$ and $\partial_\infty\hat{\psi}_0 \neq 0$.) As such it has trivial index and kernel and $\Delta_{\hat{\mathsf{C}}_0}(L_\mu^{1,2}) \subset L_\mu^{3,2}$.

**Case C.** $\hat{\mathsf{C}}_0$ is irreducible but $\mathsf{C}_\infty$ is reducible. In view of Remark 4.3.27 we only have to prove that

$$H^1_{\hat{\mathsf{C}}_0} \cong K_0 := \ker\Big(\partial_\infty^0 : \ker_{ex}\hat{\mathcal{T}}_{\hat{\mathsf{C}}_0} \to T_1 G_\infty\Big).$$

Clearly $K_0 \subset T_{\hat{\mathsf{C}}_0}\mathcal{C}_{\mu,sw}$, that is, every $\underline{\hat{\mathsf{C}}} \in K_0$ is asymptotically strongly cylindrical, and thus we get a tautological map

$$K_0 \to H^1_{\hat{\mathsf{C}}_0}.$$

Arguing as in Case B we deduce that this map is $1-1$. To prove that this map is onto we construct a right inverse $\Gamma$ formally identical to the one in Case B,

$$\Gamma(\underline{\hat{\mathsf{C}}}) = \underline{\hat{\mathsf{C}}} - \mathfrak{L}_{\hat{\mathsf{C}}_0}\Delta_{\hat{\mathsf{C}}_0}^{-1}\mathfrak{L}_{\hat{\mathsf{C}}}^*\underline{\hat{\mathsf{C}}},$$

where this time we regard $\Delta_{\hat{\mathsf{C}}_0}$ as a bounded Fredholm operator

$$\Delta_{\hat{\mathsf{C}}_0} : L_\mu^{3,2} \to L_\mu^{1,2}$$

of trivial index and kernel. (Note that since $\partial_\infty\hat{\psi}_0 = 0$ the operator $\Delta_{\hat{\mathsf{C}}_0}$ is no longer Fredholm in the functional framework $L^{3,2} \to L^{1,2}$.) ■

We conclude this section by presenting a similar description of $H^2(F_{\hat{\mathsf{C}}_0})$ in terms of $\ker_{ex}\hat{\mathcal{T}}^*_{\hat{\mathsf{C}}_0}$.

**Proposition 4.3.30.** *There exists a natural short exact sequence*

$$(H_2) \qquad 0 \to H^2(F_{\hat{\mathsf{C}}_0}) \to \ker_{ex}\hat{\mathcal{T}}^*_{\hat{\mathsf{C}}_0} \xrightarrow{\partial_\infty^0} \mathrm{Range}(T_1\hat{G}_0 \overset{\partial_\infty}{\to} T_1 G_\infty) \to 0$$

*where the upper $*$ denotes the formal adjoint.*

**Proof** Let us first observe that

$$H^2(F_{\hat{C}_0}) = \ker_\mu \hat{\mathcal{T}}^{*_\mu}_{\hat{C}_0,\mu} \cong \ker(\hat{\mathcal{T}}_{\hat{C}_0,\mu}\hat{\mathcal{T}}^{*_\mu}_{\hat{C}_0,\mu} : L^{2,2}_\mu \to L^2_\mu)$$

and

$$\hat{\mathcal{T}}_{\hat{C}_0,\mu}\hat{\mathcal{T}}^*_{\hat{C}_0,\mu} = \underline{\widehat{SW}}_{\hat{C}_0}\underline{\widehat{SW}}^{*_\mu}_{\hat{C}_0} \oplus \frac{1}{4}\Delta_{\hat{C}_0,\mu}$$

where we recall that

$$\Delta_{\hat{C}_0,\mu} := \mathfrak{L}^{*_\mu}_{\hat{C}_0}\mathfrak{L}_{\hat{C}_0}.$$

Since $\ker(\Delta_{\hat{C}_0,\mu} : L^{3,2}_\mu \to L^{1,2}_\mu) = 0$ we deduce

$$(\Psi, \mathbf{i}\hat{f}) \in \ker_\mu \hat{\mathcal{T}}^{*_\mu}_{\hat{C}_0,\mu} \Longleftrightarrow \hat{f} \equiv 0 \ \text{ and } \ \underline{\widehat{SW}}^*_{\hat{C}_0}(\mathbf{m}_{2\mu}\Psi) = 0.$$

We conclude that the correspondence

$$\ker_\mu \hat{\mathcal{T}}^{*_\mu}_{\hat{C}_0,\mu} \ni (\Psi, \mathbf{i}\hat{f}) \overset{\varphi}{\mapsto} (\mathbf{m}_{2\mu}\Psi, 0)$$

induces a map

$$\varphi : \ker_\mu \hat{\mathcal{T}}^{*_\mu}_{\hat{C}_0,\mu} \to \ker_{-\mu}(\hat{\mathcal{T}}^*_{\hat{C}_0} = \underline{\widehat{SW}}^*_{\hat{C}_0} + \frac{1}{2}\mathfrak{L}_{\hat{C}_0}) = \ker_{ex}(\hat{\mathcal{T}}^*_{\hat{C}_0}).$$

Clearly $\partial^0_\infty \circ \varphi = 0$.

Conversely, suppose

$$(\Psi, \mathbf{i}\hat{f}) \in \ker_{ex} \hat{\mathcal{T}}^*_{\hat{C}_0} \Longleftrightarrow \underline{\widehat{SW}}^*_{\hat{C}_0}(\underline{\hat{C}}) + \frac{1}{2}\mathfrak{L}_{\hat{C}_0}(\mathbf{i}\hat{f}) = 0$$

and $\hat{f} \in L^2_\mu$ (i.e. $\partial^0_\infty(\Psi \oplus \mathbf{i}\hat{f}) = 0$). Apply $\mathfrak{L}^*_{\hat{C}_0}$ to both sides of the above equation and use the identity $\mathfrak{L}^*_{\hat{C}_0}\underline{\widehat{SW}}^*_{\hat{C}_0} \equiv 0$ to deduce

$$\mathfrak{L}^*_{\hat{C}_0}\mathfrak{L}_{\hat{C}_0}(\mathbf{i}\hat{f}) = 0.$$

Since $\hat{f}, \mathfrak{L}_{\hat{C}_0}(\mathbf{i}\hat{f}) \in L^2_\mu$ we can integrate the last equality by parts over $\hat{N}_{T,\infty}$ and we deduce

$$\int_{\hat{N}} |\mathfrak{L}_{\hat{C}_0}(\mathbf{i}\hat{f})|^2 d\hat{v} = 0 \Longleftrightarrow \mathfrak{L}_{\hat{C}_0}(\mathbf{i}\hat{f}) = 0 \Longleftrightarrow \hat{f} \equiv 0 \ \text{ (since } \hat{f} \in L^2_\mu).$$

The fact that the map

$$\ker_{ex} \hat{\mathcal{T}}^*_{\hat{C}_0} \overset{\partial^0_\infty}{\longrightarrow} \operatorname{Range}(T_1\hat{G}_0 \overset{\partial_\infty}{\to} T_1 G_\infty) \tag{4.3.13}$$

is onto now follows from Remark 4.3.27. Proposition 4.3.30 is proved. ■

**Remark 4.3.31.** Proposition 4.3.30 shows that we have a natural map $H^2(F_{\hat{\mathsf{C}}_0}) \to T_{\mathsf{C}_\infty}\mathfrak{M}_\sigma$ which for simplicity we will denote by $\partial_\infty$. Observe also that if $\hat{\mathsf{C}}_0$ is reducible there exists $(0, \mathbf{i}\hat{f}) \in \ker_{ex} \hat{\mathcal{T}}^*_{\hat{\mathsf{C}}_0}$ such that

$$\partial_\infty \hat{f} = 1.$$

If $(\Psi_1, \mathbf{i}\hat{f}_1)$, $(\Psi_2, \mathbf{i}\hat{f}_2)$ are two such elements then

$$(\Psi_1 - \Psi_2, \mathbf{i}\hat{f}_1 - \mathbf{i}\hat{f}_2) \in \varphi\Big(H^2(F_{\hat{\mathsf{C}}_0})\Big) \subset \ker_{ex} \hat{\mathcal{T}}^*_{\hat{\mathsf{C}}_0}$$

so that $\hat{f}_1 = \hat{f}_2$. The function $\hat{f}_0 = \hat{f}_1 = \hat{f}_2$ is uniquely determined by the equations

$$\hat{f}_0 \in L^{3,2}_{\mu,ex}, \;\; \mathfrak{L}^*_{\hat{\mathsf{C}}_0}\mathfrak{L}_{\hat{\mathsf{C}}_0}(\mathbf{i}\hat{f}_0) = 0, \;\; \partial_\infty \hat{f}_0 = 1.$$

Notice also that we have a unitary isomorphism

$$\partial_\infty \ker_{ex} \hat{\mathcal{T}}^*_{\hat{\mathsf{C}}_0} \cong \partial_\infty H^2(F_{\hat{\mathsf{C}}_0}) \oplus T_1 G_\infty.$$

More precisely, if $(\Psi, \mathbf{i}\hat{f}) \in \ker_{ex} \hat{\mathcal{T}}^*_{\hat{\mathsf{C}}_0}$ is such that $\partial_\infty \hat{f} = 1$ then

$$\partial_\infty \ker_{ex} \hat{\mathcal{T}}^*_{\hat{\mathsf{C}}_0} = \mathrm{span}_{\mathbb{R}}\Big\{\partial_\infty(\Psi, \mathbf{i}\hat{f}),\; \partial_\infty H^2(F_{\hat{\mathsf{C}}_0}) \oplus 0\Big\}$$

$$= \mathrm{span}_{\mathbb{R}}\Big\{0 \oplus \mathbf{i},\; \partial_\infty H^2(F_{\hat{\mathsf{C}}_0}) \oplus 0\Big\}.$$

**4.3.3. Virtual dimensions.** Suppose $\hat{\mathsf{C}}_0 = (\hat{\psi}_0, \hat{A}_0) \in \hat{\mathcal{C}}_{\mu,sw}$ is a monopole. Set $\mathsf{C}_\infty = (\psi_\infty, A_\infty) = \partial_\infty \hat{\mathsf{C}}_0$ and $d(\mathsf{C}_\infty) = \dim T_{\mathsf{C}_\infty}\mathfrak{M}_\sigma$. We want to describe a general procedure for computing the virtual dimension $d(\hat{\mathsf{C}}_0)$.

Using Lemma 4.3.18 and Exercise 4.3.6 we deduce

$$d(\hat{\mathsf{C}}_0) = -\chi(F) + d(\mathsf{C}_\infty) - \dim G_\infty$$

$$= \mathrm{ind}\Big(\underline{\widehat{SW}}_{\hat{\mathsf{C}}_0} \oplus \frac{1}{2}\mathfrak{L}^{*_\mu}_{\hat{\mathsf{C}}_0} : L^{2,2}_\mu(\hat{\mathbb{S}}^+_{\hat{\sigma}} \oplus \mathbf{i}\Lambda^1 T^*\hat{N}) \to L^{1,2}_\mu(\hat{\mathbb{S}}^-_{\hat{\sigma}} \oplus \mathbf{i}\Lambda^2_+ T^*\hat{N} \oplus \mathbf{i}\underline{\mathbb{R}})\Big)$$

$$+ d(\mathsf{C}_\infty) - \dim G_\infty$$

(use Proposition 4.1.17)

$$= I_{APS}(\hat{\mathcal{T}}_{\hat{\mathsf{C}}_0,\mu}) + d(\mathsf{C}_\infty) - \dim G_\infty$$

(use the excision formula (4.1.19) of §4.1.4)

$$= I_{APS}(\hat{\mathcal{T}}_{\hat{\mathsf{C}}_0}) + d(\mathsf{C}_\infty) - \dim G_\infty - SF(\mathcal{T}_{\mathsf{C}_\infty} \to \mathcal{T}_{\mathsf{C}_\infty,\mu}). \tag{4.3.14}$$

To proceed further let us first notice the following result, whose proof will be presented a bit later.

**Lemma 4.3.32.**

$$SF\Big(\mathcal{T}_{\mathsf{C}_\infty} \to \mathcal{T}_{\mathsf{C}_\infty,\mu}\Big) = -\dim G_\infty.$$

Thus

$$d(\hat{\mathsf{C}}_0) = I_{APS}(\hat{\mathcal{T}}_{\hat{\mathsf{C}}_0}) + d(\mathsf{C}_\infty). \tag{4.3.15}$$

Now denote by $\hat{\mathcal{T}}^0_{\hat{\mathsf{C}}_0}$ the operator obtained by setting $\hat{\psi}_0 = 0$ in the description of $\hat{\mathcal{T}}_{\hat{\mathsf{C}}_0}$. Observe that along the neck $\hat{\mathcal{T}}^0_{\hat{\mathsf{C}}_0}$ has the form

$$\hat{\mathcal{T}}^0_{\hat{\mathsf{C}}_0}\begin{bmatrix} \underline{\psi}(t) \\ \mathbf{i}\underline{a}(t) \\ \mathbf{i}u(t) \end{bmatrix}$$

$$= \begin{bmatrix} J & 0 & 0 \\ 0 & 1 & 0 \\ 0 & 0 & 1 \end{bmatrix} \left( \partial_t \begin{bmatrix} \underline{\psi}(t) \\ \mathbf{i}\underline{a}(t) \\ \mathbf{i}u(t) \end{bmatrix} - \begin{bmatrix} \mathfrak{D}_{A_\infty} & 0 & 0 \\ 0 & -*d & d \\ 0 & d^* & 0 \end{bmatrix} \begin{bmatrix} \underline{\psi}(t) \\ \mathbf{i}\underline{a}(t) \\ \mathbf{i}u(t) \end{bmatrix} \right)$$

$$+ \begin{bmatrix} \frac{\mathbf{i}v(t)}{2}J & 0 & 0 \\ 0 & 0 & 0 \\ 0 & 0 & 0 \end{bmatrix} \begin{bmatrix} \underline{\psi}(t) \\ \mathbf{i}\underline{a}(t) \\ \mathbf{i}u(t) \end{bmatrix}.$$

This shows $\hat{\mathcal{T}}^0_{\hat{\mathsf{C}}_0}$ is an a-$APS$ operator and

$$\mathcal{T}^0_{\mathsf{C}_\infty} := \vec{\partial}_\infty \hat{\mathcal{T}}^0_{\hat{\mathsf{C}}_0} = \begin{bmatrix} \mathfrak{D}_{A_\infty} & 0 \\ 0 & -\mathbf{SIGN} \end{bmatrix}.$$

Set

$$\mathcal{P}_{\mathsf{C}_\infty} := \mathcal{T}_{\mathsf{C}_\infty} - \mathcal{T}^0_{\mathsf{C}_\infty}.$$

Observe that $\mathcal{P}_{\mathsf{C}_\infty}$ is a zeroth order symmetric operator described by

$$\mathcal{P}_{\mathsf{C}_\infty}\begin{bmatrix} \underline{\psi} \\ \mathbf{i}\underline{a} \\ \mathbf{i}u \end{bmatrix} = \begin{bmatrix} \frac{\mathbf{i}}{2}(\,\mathbf{c}(\underline{a}) - u\,)\psi_\infty \\ \frac{1}{2}\mathbf{c}^{-1}\dot{q}(\psi_\infty, \underline{\psi}) \\ \frac{\mathbf{i}}{2}\mathfrak{Im}\langle\,\psi_\infty, \underline{\psi}\,\rangle \end{bmatrix}.$$

Denote by $\varphi(\mathsf{C}_\infty)$ the spectral flow of the family $\mathcal{T}^0_{\mathsf{C}_\infty} + t\mathcal{P}_{\mathsf{C}_\infty}$, $t \in [0,1]$. Using the excision formula (4.1.19) we deduce

$$d(\hat{\mathsf{C}}_0) = I_{APS}(\hat{\mathcal{T}}^0_{\hat{\mathsf{C}}_0}) + d(\mathsf{C}_\infty) - \varphi(\mathsf{C}_\infty).$$

$\hat{\mathcal{T}}^0_{\hat{\mathsf{C}}_0}$ is the direct sum of the *complex* operator $\mathfrak{D}_{\hat{A}_0}$ and the *real* operator **ASD**. Since we are interested in *real indices* we have

$$I_{APS}(\hat{\mathcal{T}}^0_{\hat{\mathsf{C}}_0}) = 2I_{APS}(\mathfrak{D}_{\hat{A}_0}) + I_{APS}(\mathbf{ASD}).$$

Denote by $\eta_{sign}(g)$ the eta invariant of **SIGN** and by $\eta_{dir}(\mathsf{C}_\infty)$ the eta invariant of $\mathfrak{D}_{A_\infty}$. We set

$$\mathbf{F}(\mathsf{C}_\infty) := 4\eta_{dir}(\mathsf{C}_\infty) + \eta_{sign}(g).$$

Using (4.1.3) of §4.1.2, (4.1.30) of §4.1.6 we deduce

$$I_{APS}(\hat{\mathcal{T}}^0_{\hat{\mathsf{C}}_0}) = \frac{1}{4}\int_{\hat{N}} -\frac{1}{3}p_1(\hat{\nabla}^{\hat{g}}) + c_1(\hat{A}_0)^2 - \Big(\dim_{\mathbb{C}} \ker \mathfrak{D}_{A_\infty} + \eta_{dir}(\mathsf{C}_\infty)\Big)$$

$$-\frac{1}{2}\Big(\chi_{\hat{N}} + \tau_{\hat{N}} + b_0(N) + b_1(N)\Big)$$

Using the signature formula of Atiyah-Patodi-Singer (see [**6**] and also (4.1.34) of §4.1.6) we deduce

$$\int_{\hat{N}} \frac{1}{3}p_1(\hat{\nabla}^{\hat{g}}) = \eta_{sign}(g) + \tau_{\hat{N}}$$

and we conclude

$$I_{APS}(\hat{\mathcal{T}}^0_{\hat{\mathsf{C}}_0}) = \frac{1}{4}\Big(\int_{\hat{N}} c_1(\hat{A}_0)^2 - (2\chi_{\hat{N}} + 3\tau_{\hat{N}})\Big) - \frac{1}{4}\mathbf{F}(\mathsf{C}_\infty)$$

$$-\dim_{\mathbb{C}} \ker \mathfrak{D}_{A_\infty} - \frac{1}{2}(b_0(N) + b_1(N)).$$

Putting together all of the above we obtain the following formula:

$$(\mathbf{VDim}) \quad \begin{aligned} d(\hat{\mathsf{C}}_0) = \frac{1}{4}\Big(\int_{\hat{N}} c_1(\hat{A}_0)^2 - (2\chi_{\hat{N}} + 3\tau_{\hat{N}})\Big) - \frac{1}{2}\Big(b_0(N) + b_1(N)\Big) \\ + d(\mathsf{C}_\infty) - \varphi(\mathsf{C}_\infty) - \dim_{\mathbb{C}} \ker \mathfrak{D}_{A_\infty} - \frac{1}{4}\mathbf{F}(\mathsf{C}_\infty). \end{aligned}$$

The first line in (**VDim**) consists of the *soft* terms, those which do not involve functional analytic terms. The second line consists of the *hard* terms and their computation often requires nontrivial analytical work.

**Remark 4.3.33.** (a) Observe that the integral term in (**VDim**) would formally give the virtual dimension of the moduli space if $\hat{N}$ were compact. The remaining contribution depends only on the geometry of the asymptotic boundary $N$ and we will refer to it as the *boundary correction*. We will denote it by $\beta(\mathsf{C}_\infty)$. The boundary correction is additive with respect to disjoint unions which shows that formula (**VDim**) also includes the case when the asymptotic boundary is disconnected.

(b) Assume $N$ is connected so that $b_0(N) = 1$. If $\mathsf{C}_\infty$ is reducible then, using the nondegeneracy assumption (**N**), we can simplify somewhat the virtual dimension formula because $\ker \mathfrak{D}_{A_\infty} = 0$, $d(\mathsf{C}_\infty) = b_1(N)$ and $\varphi(\mathsf{C}_\infty) = 0$. We deduce

$$\textbf{(VDim}_\mathbf{r}\textbf{)} \qquad \begin{aligned} d(\hat{\mathsf{C}}_0) = &\frac{1}{4}\Big(\int_{\hat{N}} c_1(\hat{A}_0)^2 - (2\chi_{\hat{N}} + 3\tau_{\hat{N}})\Big) \\ &+\frac{1}{2}\Big(b_1(N) - 1\Big) - \frac{1}{4}\mathbf{F}(\mathsf{C}_\infty). \end{aligned}$$

(c) The exact value of the term $\mathbf{F}(\mathsf{C}_\infty)$ is very difficult to compute in general although it is known in many concrete situations; see [**107, 108, 115**]. Consider more generally the quantity

$$\mathbf{F} : \mathfrak{A}_\sigma \times \text{Metrics on } N \to \mathbb{R}, \quad (A, g) \mapsto 4\eta(\mathfrak{D}_A) + \eta_{sign}(g).$$

$\mathbf{F}(A, g)$ satisfies the variational formula

$$\mathbf{F}(A_1, g_1) - \mathbf{F}(A_0, g_0) = 8SF(\mathfrak{D}_{A_t}) - \frac{1}{8\pi^2}\int_N (A_1 - A_0) \wedge (F_{A_0} + F_{A_1})$$

where $A_t := (1-t)A_0 + tA_1$, $g(t)$ is a smooth path of metrics on $g$ such that $g(i) = g_i$, $i = 0, 1$, and $\mathfrak{D}_{A_t}$ is the Dirac operator determined by $A_t$ and the metric $g(t)$. In particular, we deduce that $\mathbf{F}(A, g) \mod 8\mathbb{Z}$ is independent of $g$. Moreover, if $A_0, A_1$ are flat connections then

$$\mathbf{F}(A_0, g) = \mathbf{F}(A_1, g).$$

When $\sigma$ is defined by a *spin* structure and $A$ is the trivial connection, then $\mathbf{F}(A, g)$ is a special case of the Kreck-Stolz invariant, [**68**]. The above variational formula coupled with the Weitzenböck formula shows that this invariant is constant on the path components of the space of metrics of positive scalar curvature. In the paper [**68**], M. Kreck and S. Stolz have shown that the higher dimensional counterpart of $\mathbf{F}$ actually distinguishes such path components.

(d) The above variational formula shows that the boundary correction term *may not be* $\mathfrak{G}_\sigma$ *invariant* because

$$\mathbf{F}(\gamma \cdot \mathsf{C}_\infty) - \mathbf{F}(\mathsf{C}_\infty) = \frac{1}{2\pi^2}\int_N \frac{d\gamma}{\gamma} \wedge F_{A_\infty} = 2\int_N \gamma^*(\frac{1}{2\pi}d\theta) \wedge c_1(A).$$

A simple application of Stokes' formula shows that $\beta(\mathsf{C}_\infty)$ is $\mathfrak{G}_\sigma^\partial$-invariant, where we recall that $\mathfrak{G}_\sigma^\partial$ consists of the gauge transformations on $N$ which extend to $\hat{N}$.

(e) The notation $\beta(\mathsf{C}_\infty)$ is a bit misleading since it does not take into account the dependence of $\beta(\mathsf{C}_\infty)$ on the *orientation* of $N$. When changing the orientation we have to replace $\mathbf{F}(\mathsf{C}_\infty)$ by $-\mathbf{F}(\mathsf{C}_\infty)$. $\varphi(\mathsf{C}_\infty)$ changes as well, but in a less obvious fashion (see Exercise 4.3.7).

**Proof of Lemma 4.3.32** Assume for simplicity that $N$ is connected so that $\dim G_\infty \in \{0,1\}$. We first need to understand the spectrum of $\mathcal{T}_{\mathsf{C}_\infty,t\mu}$, $t \in [0,1]$, $\mu$ positive and very small. Equivalently this means solving the equation

$$\mathcal{T}_{\mathsf{C}_\infty,t\mu}\begin{bmatrix}\dot{\mathsf{C}}\\ \mathbf{i}f\end{bmatrix} = \nu\begin{bmatrix}\dot{\mathsf{C}}\\ \mathbf{i}f\end{bmatrix} \Longleftrightarrow \begin{cases}\underline{\mathbf{SW}}_{\mathsf{C}_\infty}(\dot{\mathsf{C}}) - \frac{1}{2}\mathfrak{L}_{\mathsf{C}_\infty}(\mathbf{i}f) = \nu\mathbf{i}\dot{\mathsf{C}}\\ \mathfrak{L}^*_{\mathsf{C}_\infty}(\dot{\mathsf{C}}) + 4t\mu\mathbf{i}f = -2\nu\mathbf{i}f\end{cases} \tag{4.3.16}$$

As in 4.3.2 we deduce

$$\Delta_{\mathsf{C}_\infty}(\mathbf{i}f) := \mathfrak{L}^*_{\mathsf{C}_\infty}\mathfrak{L}_{\mathsf{C}_\infty}(\mathbf{i}f) = 4\nu(\nu+2t\mu)(\mathbf{i}f). \tag{4.3.17}$$

The spectrum of the symmetric second order elliptic operator $\Delta_{\mathsf{C}_\infty}$ is discrete and consists only of *nonnegative* eigenvalues of finite multiplicities. We will distinguish two cases.

**Case 1** $\mathsf{C}_\infty$ *is irreducible, so that* $\dim G_\infty = 0$. In this case we have

$$\ker \Delta_{\mathsf{C}_\infty} \cong T_1 G_\infty = 0.$$

If $\dot{\mathsf{C}} \oplus \mathbf{i}f \in \ker \mathcal{T}_{\mathsf{C}_\infty,t\mu}$ then using (4.3.17) we deduce $f \equiv 0$. Using this information back in (4.3.16) we deduce

$$\underline{\mathbf{SW}}_{\mathsf{C}_\infty}(\dot{\mathsf{C}}) = 0.$$

This shows that $\ker \mathcal{T}_{\mathsf{C}_\infty,t\mu} = \ker \mathcal{T}_{\mathsf{C}_\infty}$, for all $t \in [0,1]$ and thus the spectral flow of the family $\mathcal{T}_{\mathsf{C}_\infty,t\mu}$ is equal to $0 = -\dim G_\infty$.

**Case 2** $\mathsf{C}_\infty$ *is reducible*, so that $\dim\ker \Delta_{\mathsf{C}_\infty} = \dim T_1 G_\infty$. Moreover

$$\ker \mathcal{T}_{\mathsf{C}_\infty} = \Big\{\dot{\mathsf{C}} \oplus \mathbf{i}f;\ \underline{\mathbf{SW}}_{\mathsf{C}_\infty}(\dot{\mathsf{C}}) \oplus \mathfrak{L}^*_{\mathsf{C}_\infty}(\dot{\mathsf{C}}) = 0,\ \ \mathfrak{L}_{\mathsf{C}_\infty}(\mathbf{i}f) = 0\Big\}.$$

Fix $t \in (0,1]$. We claim that

$$\ker \mathcal{T}_{\mathsf{C}_\infty,t\mu} = \Big\{\dot{\mathsf{C}} \oplus \mathbf{i}f;\ f \equiv 0,\ \underline{\mathbf{SW}}_{\mathsf{C}_\infty}(\dot{\mathsf{C}}) \oplus \mathfrak{L}^*_{\mathsf{C}_\infty}(\dot{\mathsf{C}}) = 0\Big\}. \tag{4.3.18}$$

Using (4.3.17) with $\nu = 0$ we deduce

$$\Delta_{\mathsf{C}_\infty}(\mathbf{i}f) \Longleftrightarrow \mathfrak{L}_{\mathsf{C}_\infty}(\mathbf{i}f) = 0.$$

Using this information in the first equation of (4.3.16) we deduce $\underline{\mathbf{SW}}_{\mathsf{C}_\infty}(\dot{\mathsf{C}}) = 0$. Now apply $\mathfrak{L}_{\mathsf{C}_0}$ to both sides of the second equation in (4.3.16). Using the equality $\mathfrak{L}_{\mathsf{C}_\infty}(\mathbf{i}f) = 0$ we conclude

$$\mathfrak{L}_{\mathsf{C}_\infty}\mathfrak{L}^*_{\mathsf{C}_\infty}(\dot{\mathsf{C}}_0) = 0.$$

We now take the inner product of the above equality with $\dot{\mathsf{C}}$ and then we integrate by parts over $N$ to deduce that

$$\int_N |\mathfrak{L}^*_{\mathsf{C}_\infty}\dot{\mathsf{C}}|^2 dv(g) = 0 \Longleftrightarrow \mathfrak{L}^*_{\mathsf{C}_\infty}\dot{\mathsf{C}} = 0.$$

Using the last equality in the second equation of (4.3.16) we deduce

$$t\mu f = 0 \Longleftrightarrow f \equiv 0$$

which proves our claim.

The equality (4.3.18) shows that there is no contribution to the spectral flow of the family $\mathcal{T}_{\mathsf{C}_\infty, t\mu}$ for $t \in (0, 1]$. The only contribution to the spectral flow can occur at $t = 0$. Since

$$\dim \ker \mathcal{T}_{\mathsf{C}_\infty} - \dim \ker \mathcal{T}_{\mathsf{C}_\infty, t\mu} = 1$$

and since the spectral flow contributions at $t = 0$ are nonpositive we deduce that this contribution is either 0 or $-1$.

To decide which is the correct alternative we need to understand the eigenvalues $\nu_t$ of $\ker \mathcal{T}_{\mathsf{C}_\infty, t\mu}$ such that

$$\nu_t \nearrow 0 \ \text{ as } t \searrow 0.$$

If $\nu_t$ is such an eigenvalue then $4\nu_t(\nu_t + 2t\mu)$ must be a very small eigenvalue of $\Delta_{\mathsf{C}_\infty}$, so that

$$\nu_t(\nu_t + 2t\mu) = 0.$$

The requirement $\nu_t < 0$ forces $\nu_t = -2t\mu$ and $\mathfrak{L}_{\mathsf{C}_\infty}(\mathbf{i}f) = 0$. Applying $\mathfrak{L}_{\mathsf{C}_\infty}$ to both sides of the second equation in (4.3.16) we deduce as before that

$$\mathfrak{L}^*_{\mathsf{C}_\infty} \dot{\mathsf{C}} = 0 \Longleftrightarrow \dot{\mathsf{C}} \in \mathcal{S}_{\mathsf{C}_\infty}.$$

Using the first equation in (4.3.16) we deduce

$$\underline{\mathbf{SW}}_{\mathsf{C}_\infty}(\dot{\mathsf{C}}) = -2t\mu\dot{\mathsf{C}}, \ \ \dot{\mathsf{C}} \in \mathcal{S}_{\mathsf{C}_\infty},$$

so that $\dot{\mathsf{C}}$ is an eigenvector of $\underline{\mathbf{SW}}_{\mathsf{C}_\infty} : \mathcal{S}_{\mathsf{C}_\infty} \to \mathcal{S}_{\mathsf{C}_\infty}$ corresponding to $-2t\mu$. Since

$$2t\mu < 2\mu < \mu_-(g) \leq \mu_-(\mathsf{C}_\infty)$$

(where $-\mu_-(\mathsf{C}_\infty)$ is the negative eigenvalue of $\underline{\mathbf{SW}}_{\mathsf{C}_\infty} : \mathcal{S}_{\mathsf{C}_\infty} \to \mathcal{S}_{\mathsf{C}_\infty}$ closest to zero) we deduce that $\dot{\mathsf{C}} \equiv 0$. Thus $-2t\mu$ is a simple eigenvalue of $\mathcal{T}_{\mathsf{C}_\infty, t\mu}$ and the corresponding eigenspace is

$$\{\dot{\mathsf{C}} \oplus \mathbf{i}f; \ \ \dot{\mathsf{C}} \equiv = 0, \ \ f \equiv const.\}.$$

This shows that the spectral flow contribution at $t = 0$ is $-1$ and thus

$$SF(\mathcal{T}_{\mathsf{C}_\infty, t\mu}; t \in [0, 1]) = -1 = -\dim G_\infty. \ \blacksquare$$

**Example 4.3.34.** Suppose $(N, g)$ is the sphere $S^3$ equipped with the round metric. There exists a unique *spin*$^c$ structure $\sigma$ on $N$ and the pair $(\sigma, g)$ is good. Denote by $\mathsf{C}_0$ the unique (modulo $\mathfrak{G}_\sigma$) monopole on $N$. $\mathsf{C}_0$ is reducible, $\mathsf{C}_0 = (0, A_0)$. Observe also that (4.1.37) (with $\ell = -1$) implies that $\mathbf{F}(\mathsf{C}_0) = 0$. Alternatively, $S^3$ admits an orientation reversing isometry, so that the spectra of both $\mathfrak{D}_{A_0}$ and **SIGN** are symmetric with respect to the origin and thus their eta invariants vanish. Using ($\mathbf{VDim_r}$) we deduce that the boundary correction determined by $\mathsf{C}_0$ is

$$\beta(\mathsf{C}_0) = -\frac{1}{2}. \quad \blacksquare$$

**Example 4.3.35.** Suppose $(N, g)$ is the 3-manifold $S^1 \times S^2$ equipped with the product of the canonical metrics on $S^1$ and $S^2$. $g$ has positive scalar curvature so that $(\sigma, g)$ is good for every $\sigma \in Spin^c(N)$. Since $H^1(N, \mathbb{Z}_2) = \mathbb{Z}_2$ there exist exactly two isomorphism classes of *spin* structures on $N$ but the induced *spin*$^c$ structures are isomorphic since $H^2(N, \mathbb{Z})$ has no 2-torsion.

Any monopole on $N$ is reducible so that the only *spin*$^c$ structure $\sigma$ for which there exist monopoles is the class $\sigma_0$ induced by the *spin* structures. The moduli space $\mathfrak{M}_{\sigma_0}$ is diffeomorphic to a circle.

Remark 4.3.33 (c, d) shows that the boundary correction term is $\mathfrak{G}_{\sigma_0}$-invariant and, moreover, it is identical for all $\mathsf{C} \in \mathfrak{M}_{\sigma_0}$. One can show that $\eta_{dir}(\mathsf{C}) = 0$ (see [**107**, Appendix C]) and $\eta_{sign}(g) = 0$ (see [**67**]). Since $b_1(N) = 1$ we deduce from ($\mathbf{VDim_r}$) that

$$\beta(\mathsf{C}) = 0, \ \ \forall \mathsf{C} \in \mathfrak{M}_{\sigma_0}. \quad \blacksquare$$

**Example 4.3.36.** Suppose $\hat{N} = \mathbb{R} \times N$. A finite energy monopole $\hat{\mathsf{C}}_0$ over $\hat{N}$ is called a *tunneling*. Observe that $\partial_\infty \hat{N} = (-N) \cup N$. A *spin*$^c$ structure $(\sigma_-, \sigma_+)$ on $\partial_\infty \hat{N}$ extends to $\hat{N}$ if and only if $\sigma_- = \sigma_+ = \sigma$. Its asymptotic limit is a $\mathfrak{G}^\partial$-orbit of pairs of $\sigma$-monopoles $(\mathsf{C}_-, \mathsf{C}_+)$, where $\mathfrak{G}^\partial$ consists of pairs $(\gamma_-, \gamma_+) \in \mathfrak{G}_\sigma \times \mathfrak{G}_\sigma$ such that $\gamma_-$ and $\gamma_+$ belong to the same component of $\gamma_\sigma$. We want to emphasize that a priori it is possible that $\mathsf{C}_-$ and $\mathsf{C}_+$ may be $\mathfrak{G}_\sigma$-equivalent. Set

$$\partial_\infty^\pm \hat{\mathsf{C}}_0 := \mathsf{C}_\pm$$

and

$$G_\pm = \mathbf{Stab}\,(\mathsf{C}_\pm).$$

Modulo a gauge transformation we can assume $\hat{\mathsf{C}}_0$ is temporal:

$$\hat{\mathsf{C}}_0 = (\mathsf{C}(t))_{t\in\mathbb{R}}.$$

The operator $\hat{\mathcal{T}}_{\hat{\mathsf{C}}_0}$ has the *APS* form $G(\partial_t - \mathcal{T}_{\mathsf{C}(t)})$. Using (4.3.14) and Lemma 4.3.32 we deduce

$$d(\hat{\mathsf{C}}_0) = I_{APS}(\hat{\mathcal{T}}_{\hat{\mathsf{C}}_0}) + d(\mathsf{C}_-) + d(\mathsf{C}_+)$$

$$\stackrel{(4.1.16)}{=} -\dim\ker \mathcal{T}_{\mathsf{C}_-} - SF(\mathcal{T}_{\mathsf{C}(t)}) + d(\mathsf{C}_-) + d(\mathsf{C}_+)$$

$$(\dim\ker \mathcal{T}_{\mathsf{C}_-} = d(\mathsf{C}_-) + \dim G_-)$$
$$= -SF(\mathcal{T}_{\mathsf{C}(t)}) + d(\mathsf{C}_+) - \dim G_-.$$

In particular, if $d(\mathsf{C}_\pm) = 0$ then

$$d(\hat{\mathsf{C}}_0) = -SF(\mathcal{T}_{\mathsf{C}(t)}) - \dim G_-. \blacksquare$$

As indicated in Remark 4.3.33 (e), the term $\varphi(\mathsf{C}_\infty)$ behaves less trivially when changing the orientation of $N$. One can use the computations in the above example to describe this behavior.

**Exercise 4.3.7.** Suppose $N$ is a compact, connected, orientable 3-manifold and $\mathsf{C}_\infty$ is an irreducible monopole on $N$. Denote by $\varphi_\pm(\mathsf{C}_\infty)$ the contributions $\varphi$ in (**VDim**) corresponding to the two choices of orientation on $N$. Show that

$$\varphi_+(\mathsf{C}_\infty) + \varphi_-(\mathsf{C}_\infty) = \dim_{\mathbb{R}} \ker \mathcal{T}_{\mathsf{C}_\infty} - \dim_{\mathbb{R}} \ker \mathcal{T}^0_{\mathsf{C}_\infty}$$
$$= d(\mathsf{C}_\infty) - \dim_{\mathbb{R}} \ker \mathcal{T}^0_{\mathsf{C}_\infty}.$$

**4.3.4. Reducible finite energy monopoles.** Assume for simplicity that $N$ is connected and suppose $\hat{\mathsf{C}}_0 = (0, \hat{A}_0) \in \hat{\mathcal{C}}_{\mu,sw}$ is a reducible monopole. This is equivalent to requiring that $\hat{A}_0$ is strongly a-cylindrical and

$$F^+_{\hat{A}_0} = 0.$$

Then

$$\hat{\mathcal{T}}_{\hat{\mathsf{C}}_0} = \mathfrak{D}_{\hat{A}_0} \oplus \mathbf{ASD}.$$

Using Proposition 4.3.28 and the computations in Example 4.1.24 we deduce

$$H^1_{\hat{\mathsf{C}}_0} \cong \ker_{ex} \mathfrak{D}_{\hat{A}_0} \oplus \ker_{ex} \mathbf{ASD} \cong \ker_{ex} \mathfrak{D}_{\hat{A}_0} \oplus \ker_{ex}(\hat{d} + \hat{d}^*)|_{\Omega^1(\hat{N})}$$

(use (4.1.28))

$$\cong \ker_{ex} \mathfrak{D}_{\hat{A}_0} \oplus H^1(\hat{N}, N) \oplus L^1_{top}. \tag{4.3.19}$$

Denote by $H^2_+(\hat{N})$ the self-dual part of $\ker_{L^2}(\hat{d} + \hat{d}^*)|_{\Omega^2(\hat{N})}$. Using Proposition 4.3.30 and the computations in Example 4.1.24 we deduce

$$\begin{aligned} H^2(F(\hat{\mathsf{C}}_0)) &\cong \ker_{ex} \mathfrak{D}^*_{\hat{A}_0} \oplus \ker_{ex} \mathbf{ASD}^* \\ &\cong \ker_{ex} \mathfrak{D}^*_{\hat{A}_0} \oplus H^2_+(\hat{N}) \oplus L^2_{top}. \end{aligned} \tag{4.3.20}$$

We deduce the following consequence.

**Corollary 4.3.37.** *If* $\ker_{ex} \mathfrak{D}^*_{\hat{A}_0} = 0$, $L^2_{top} = 0$ *and* $\hat{b}_+ := \dim H^2_+(\hat{N}) = 0$ *then* $\hat{\mathsf{C}}_0$ *is strongly regular.*

We now want to investigate in greater detail the subset

$$\widehat{\mathfrak{M}}_{\mu}^{red} \subset \widehat{\mathfrak{M}}_{\mu}$$

consisting of reducible monopoles. Observe that

$$\widehat{\mathfrak{M}}_{\mu}^{red} = \Big\{(0, \hat{A}) \in \hat{\mathcal{C}}_{\mu,sw};\ \ F_{\hat{A}}^{+} = 0\Big\}/\widehat{\mathfrak{G}}_{\mu,ex}.$$

Observe first that it is a connected space since it is a quotient of the linear affine subspace

$$F_{\hat{A}}^{+} = 0.$$

Set

$$\hat{\mathfrak{A}}_{\mu,sw} := \Big\{\hat{A};\ \ (0, \hat{A}) \in \hat{\mathcal{C}}_{\mu,sw}\Big\}.$$

There exists a natural *affine* map

$$\mathcal{F} : \hat{\mathfrak{A}}_{\mu,sw} \to L_{\mu}^{1,2}(\mathbf{i}\Lambda_{+}^{2}T^{*}\hat{N}),\ \ \hat{A} \mapsto F_{\hat{A}}^{+}$$

and $\widehat{\mathfrak{M}}_{\mu}^{red}$ can be identified with

$$\mathcal{F}^{-1}(0)/\widehat{\mathfrak{G}}_{\mu,ex}.$$

Given $\hat{A} \in \mathcal{F}^{-1}(0)$ we get as in §4.3.2 a Fredholm complex

$$\text{(K)} \qquad 1 \hookrightarrow T_1\widehat{\mathfrak{G}}_{\mu,ex} \to T_{\hat{A}}\hat{\mathfrak{A}}_{\mu,sw} \to L_{\mu}^{1,2}(\mathbf{i}\Lambda_{+}^{2}T^{*}\hat{N}) \to 0.$$

We denote its cohomology by $H_{\hat{A}}^{k}$ and we set

$$\chi(K) := H_{\hat{A}}^{0} - H_{\hat{A}}^{1} + H_{\hat{A}}^{2}.$$

Observe that $H_{\hat{A}}^{0}$ is the tangent space to the stabilizer of $\hat{A}$, which is $S^1$. Thus

$$\dim H_{\hat{A}}^{0} = 1.$$

Since $\mathcal{F}$ is *affine* we deduce that the Kuranishi map associated to this deformation picture is trivial. On the other hand, the stabilizer of $\hat{A}$ acts trivially on $H_{\hat{A}}^{1}$ and thus, if nonempty, $\widehat{\mathfrak{M}}_{\mu}^{red}$ is a connected, smooth manifold of dimension

$$\dim \widehat{\mathfrak{M}}_{\mu}^{red} = \dim H_{\hat{A}}^{1} - \dim H_{\hat{A}}^{2} = -\chi(K) + 1.$$

As in §4.3.2 we can embed (K) in an exact sequence of Fredholm complexes similar to (**E**). Denote by $\mathfrak{M}_{\sigma}^{red}$ the similar space of reducible $\sigma$-monopoles on $\partial_{\infty}\hat{N}$. Arguing *exactly* as in the proof of (4.3.15) of §4.3.3 we deduce that

$$-\chi(K) = I_{APS}(\mathbf{ASD}) + \dim \mathfrak{M}_{\sigma}^{red}$$

$$\stackrel{(4.1.30)}{=} -\frac{1}{2}\Big(\chi_{\hat{N}} + \tau_{\hat{N}} + b_0(N) + b_1(N)\Big) + b_1(N)$$

$$= -\frac{1}{2}\Big(\chi_{\hat{N}} + \tau_{\hat{N}} + 1 - b_1(N)\Big).$$

We have thus proved the following result.

**Proposition 4.3.38.** *If $\widehat{\mathfrak{M}}_\mu^{red}$ is nonempty then it is a smooth, connected manifold of dimension*

$$\widehat{\mathfrak{M}}_\mu^{red} = \frac{1}{2}\Big(b_1(N) + 1 - \chi_{\hat{N}} - \tau_{\hat{N}}\Big).$$

In the next section we will have more to say about the existence of reducibles.

**Example 4.3.39.** Consider again the manifold $\hat{N}_\ell$, $\ell = -1$, discussed in Example 4.1.27. Recall that $\hat{N}_{-1}$ is obtained from a disk bundle $D_{-1}$ of degree $-1$ over $S^2$ by attaching an infinite cylinder

$$\mathbb{R}_+ \times \partial D_{-1} \cong \mathbb{R}_+ \times S^3.$$

Since $H^1(S^3) = H^2(S^3) = 0$ we deduce $L^1_{top} = L^2_{top} = 0$ and since the intersection form of $\hat{N}_{-1}$ is negative definite we deduce $\hat{b}_+ = 0$. Moreover, $H^1(\hat{N}_{-1}, N_1) = 0$

Fix a *spin*$^c$ structure $\hat{\sigma}$ on $\hat{N}_{-1}$. In Example 4.1.27 we have equipped $\hat{N}_{-1}$ with a positive scalar curvature cylindrical metric and we have shown that for every reducible finite energy $\hat{\sigma}$-monopole $\hat{\mathsf{C}}_0 = (0, \hat{A}_0)$ on $\hat{N}_{-1}$ we have

$$\ker_{ex} \mathfrak{D}_{\hat{A}_0} = 0.$$

Set $\mathsf{C}_0 := \partial_\infty \hat{\mathsf{C}}_0$. Arguing exactly as in the proof of (4.1.36) we obtain

$$8 \dim \ker_{ex} \mathfrak{D}^*_{\hat{A}_0} = \mathbf{F}(\mathsf{C}_0) + \tau_{\hat{N}_{-1}} - \int_{\hat{N}_{-1}} c_1(\hat{A}_0)^2 = -1 - \int_{\hat{N}_{-1}} c_1(\hat{A}_0)^2.$$

Thus $H^1_{\hat{\mathsf{C}}_0} = 0$ and $\hat{\mathsf{C}}_0$ is strongly regular if and only if

$$c_1(\hat{\sigma}) \cdot c_1(\hat{\sigma}) = \int_{\hat{N}_{-1}} c_1(\hat{A}_0)^2 = -1.$$

If we identify $H^2(D_\ell, \mathbb{Z}) \cong H^2(D_{-1}, \partial D_{-1}; \mathbb{Z}) \cong \mathbb{Z}$ with generator $\mathbf{u}_0$, the Poincaré dual of the zero section of $D_\ell$, we see that the above equality is possible if and only if

$$c_1(\hat{\sigma}) = \pm \mathbf{u}_0.$$

We now want to prove that for any *spin*$^c$ structure $\hat{\sigma}$ over $\hat{N}_{-1}$ there exists a unique (modulo $\widehat{\mathfrak{G}}_{\mu,ex}$) finite energy $\hat{\sigma}$-monopole, which necessarily is reducible.

Observe first that according to Proposition 4.3.38 the space of reducibles is either empty or a smooth, connected manifold of dimension

$$\frac{1}{2}(1 + b_1(S^3) - \chi_{\hat{N}_{-1}} - \tau_{\hat{N}_{-1}}) = 0$$

so that it consists of at most one point.

Denote by $\sigma$ the unique *spin*$^c$-structure on $N_1 = \partial_\infty \hat{N}_{-1} \cong S^3$ and denote by $A_0$ the trivial connection on the trivial line bundle $\det(\sigma)$. We can form the energy functional defined in (2.4.8)

$$\mathcal{E}_\sigma(\psi, A) = \frac{1}{2}\int_{S^3}(A - A_0) \wedge F_A + \frac{1}{2}\int_{S^3} \mathfrak{Re}\,\langle \mathfrak{D}_A\psi, \psi\rangle dv.$$

The energy of the unique $\sigma$-monopole $\mathsf{C}_0 = (0, A_0)$ is 0. Now extend $A_0$ to a strongly cylindrical connection $\hat{A}_0$ on $\det(\hat{\sigma})$. If $\hat{\mathsf{C}} = (\hat{\psi}, \hat{A})$ is a finite energy $\hat{\sigma}$-monopole then according to Proposition 4.3.2 we have

$$\int_{\hat{N}_{-1}} \Big(|\hat{\nabla}^{\hat{A}}\hat{\psi}|^2 + \frac{1}{8}|q(\hat{\psi})|^2 + |F_{\hat{A}}|^2 + \frac{\hat{s}}{4}|\hat{\psi}|^2\Big) d\hat{v}$$

$$= E(\hat{\mathsf{C}}) = \int_{\hat{N}_{-1}} F_{\hat{A}} \wedge F_{\hat{A}} \leq \int_{\hat{N}_{-1}} |F_{\hat{A}}|^2 d\hat{v}.$$

Since $\hat{s} > 0$ we conclude that $\hat{\psi} \equiv 0$.

To establish the existence part it suffices to show there exists $\mathbf{i}\hat{a} \in L^{2,2}_\mu(\mathbf{i}\Lambda^1 T^*\hat{N}_{-1})$ such that if $\hat{A} := \hat{A}_0 + \mathbf{i}\hat{a}$ then

$$F^+_{\hat{A}} = 0 \Longleftrightarrow \mathbf{i}\hat{d}^+\hat{a} = -F^+_{\hat{A}_0}.$$

Look at the operator

$$\mathbf{ASD} : L^{2,2}_\mu(\Lambda^1 T^*\hat{N}_{-1}) \to L^{1,2}_\mu(\mathbf{i}(\Lambda^2_+ \oplus \Lambda^0)T^*\hat{N}_{-1}).$$

According to Proposition 4.1.17 its cokernel is isomorphic to $\ker_{ex}\mathbf{ASD}^* = 0$, which shows that the above operator is onto. Since $F^+_{\hat{A}_0} \in L^{1,2}_\mu$ (it has compact support) we can find $\hat{a} \in L^{2,2}_\mu(\Lambda^1 T^*\hat{N}_{-1})$ such that

$$\mathbf{i}\hat{d}^+\hat{a} = -F^+_{\hat{A}_0} \text{ and } \hat{d}^*\hat{a} = 0 \Longleftrightarrow \mathbf{ASD}(\mathbf{i}\hat{a}) = (-\sqrt{2}F^+_{\hat{A}_0}) \oplus 0.$$

This proves that reducible monopoles do exist.

Suppose $\hat{\mathsf{C}}_0$ is the unique finite energy $\hat{\sigma}$-monopole. Thus the reducibles are isolated points in $\widehat{\mathfrak{M}}_\mu$. Using the virtual dimension formula ($\mathbf{VDim_r}$) we deduce that

$$d(\hat{\mathsf{C}}_0) = \frac{1}{4}\int_{\hat{N}_{-1}} c_1(\hat{A}_0)^2 - \frac{1}{4}\Big(2\chi_{\hat{N}_{-1}} + 3\tau_{\hat{N}_{-1}}\Big) - \frac{1}{2} = \frac{1}{4}\int_{\hat{N}_{-1}} c_1(\hat{A}_0)^2 - \frac{3}{4} < 0.$$

If we denote by $\hat{\sigma}_n$ the *spin*$^c$ structure such that $c_1(\hat{\sigma}_n) = (2n+1)\mathbf{u}_0$ then the above formula becomes

$$d(\hat{\mathsf{C}}_0, \hat{\sigma}_n) = -(n^2 + n + 1). \tag{4.3.21}$$

This formula covers all *spin*$^c$ classes on $\hat{N}_{-1}$ since the intersection form of $\hat{N}_{-1}$ is odd.

**Example 4.3.40.** Consider the cylindrical manifold $\hat{N}$ diffeomorphic to the unit open ball $B^4 \subset \mathbb{R}^4$ equipped with a positive scalar curvature metric $\hat{g}$ such that $\partial_\infty \hat{g}$ is the round metric on $\partial_\infty \hat{N} \cong S^3$. $Spin^c(\hat{N})$ consists of a single structure $\hat{\sigma}_0$ and, exactly as in the previous example we deduce that modulo gauge there exists a unique finite energy monopole $\hat{\mathsf{C}}_0$ which is reducible, $\hat{\mathsf{C}}_0 = (0, \hat{A}_0)$. Set $\mathsf{C}_0 = \partial_\infty \hat{\mathsf{C}}_0$.

Since $\hat{g}$ has positive scalar curvature we deduce as before that $\ker_{ex} \mathfrak{D}_{\hat{A}_0} = 0$. Moreover, as in the previous example we have

$$8 \dim \ker_{ex} \mathfrak{D}^*_{\hat{A}_0} = \mathbf{F}(\mathsf{C}_0) + \tau_{\hat{N}} - \int_{\hat{N}} c_1(\hat{A}_0)^2 = 0.$$

Using Corollary 4.3.37 we deduce that $\hat{\mathsf{C}}_0$ is a strongly regular, reducible monopole.

**Example 4.3.41.** Consider the disk bundle $D^2 \times S^2 \to S^2$. It is a 4-manifold with boundary $N := S^1 \times S^2$ which we equip with the product metric $g$ as in Example 4.3.35. We form $\hat{N}$ by attaching the cylinder $\mathbb{R}_+ \times N$ to the boundary of $D^2 \times S^2$. As in Example 4.1.27 we can equip $\hat{N}$ with a cylindrical metric $\hat{g}$ of positive scalar curvature which along the neck has the form $dt^2 + g$.

The only *spin*$^c$ structure on $N$ which admits monopoles is the structure $\sigma_0$ induced by the spin structures on $N$. In this case all monopoles are reducible and

$$\mathfrak{M}_\sigma \cong S^1.$$

The structure $\sigma_0$ on $N$ is induced by pullback from $S^2$ and thus it can be extended to $\hat{N}$. On the other hand, since the map

$$H^2(\hat{N}, \mathbb{Z}) \to H^2(N, \mathbb{Z})$$

is one-to-one there exists exactly one extension $\hat{\sigma}_0$ of $\sigma_0$ to $\hat{N}$ satisfying

$$c_1(\hat{\sigma}_0) = 0.$$

Arguing as in Example 4.3.39 we deduce that all finite energy $\hat{\sigma}$-monopoles are reducible. According to Proposition 4.3.38, the expected dimension of $\widehat{\mathfrak{M}}^{red}_\mu$ is

$$\frac{1}{2}(b_1(N) + 1 - 2) = 0$$

so that there exists at most one finite energy $\hat{\sigma}$-monopole which must be reducible. Reducibles do exist because $\det(\hat{\sigma}_0)$ admits flat connections.

Suppose $\hat{\mathsf{C}}_0 = (0, \hat{A}_0)$ is a reducible monopole so that $\hat{A}_0$ is flat. From the long exact cohomology sequence of $(\hat{N}, N)$ we deduce that $H^1(\hat{N}, N) = 0$ and the morphism $H^2(D^2 \times S^2) \to H^2(S^1 \times N)$ is onto, i.e. $L^2_{top} \cong \mathbb{R}$. Thus $\hat{\mathsf{C}}_0$ is not strongly regular.

If $\mathsf{C}_0 := \partial_\infty \hat{\mathsf{C}}_0$ then exactly as in the previous example we deduce

$$8 \dim \ker_{ex} \mathfrak{D}^*_{\hat{A}_0} = \mathbf{F}(\mathsf{C}_0) + \tau_{\hat{N}} - \int_{\hat{N}} c_1(\hat{A}_0)^2.$$

In Example 4.3.35 we have shown that $\mathbf{F}(\mathsf{C}_0) = 0$ and since $\tau_{hN} = 0$ we deduce

$$8 \dim \ker_{ex} \mathfrak{D}^*_{\hat{A}_0} = -\int_{\hat{N}_{-1}} c_1(\hat{A}_0)^2 = 0.$$

According to $(\mathbf{VDim_r})$ we have

$$d(\hat{\mathsf{C}}_0) = -\frac{1}{4}(2\chi_{\hat{N}} + 3\tau_{\hat{N}}) + \frac{1}{2}(b_1(N) - 1) = -1.$$

## 4.4. Moduli spaces of finite energy monopoles: Global aspects

We now have quite a detailed understanding of the local structure of the moduli space of finite energy monopoles. For applications to topology we need to know some facts about the global structure of this space.

In this section we will discuss some global problems. As always we will work under the nondegeneracy assumption $(\mathbf{N})$.

**4.4.1. Genericity results.** In 4.3.2 we developed criteria to recognize when the moduli space of finite energy monopoles is smooth. As in the compact case, there are two sources of singularities. The main problem is due to the obstruction spaces $H^2_{\hat{\mathsf{C}}_0}$ and a second, less serious, problem is due to the presence of reducibles. We will deal first with the reducibles issue.

In the compact case we found a cheap way to avoid the reducibles by perturbing the Seiberg-Witten equations. We follow a similar strategy in the noncompact case.

Fix a cylindrical *spin*$^c$-structure $\hat{\sigma}$ on $\hat{N}$ with $\sigma := \partial_\infty \hat{\sigma}$ such that there exists at least one reducible finite energy monopole $\hat{\mathsf{C}}_0 = (0, \hat{A}_0)$. For every sufficiently regular, compactly supported 2-form $\eta$ on $\hat{N}$ we form the perturbed Seiberg-Witten equations

$$\widehat{SW}_\eta(\hat{\psi}, \hat{A}) = 0 \Longleftrightarrow \begin{cases} \hat{\mathfrak{D}}_{\hat{A}}\hat{\psi} = 0 \\ \hat{\mathbf{c}}(F^+_{\hat{A}} + \mathbf{i}\eta^+) = \frac{1}{2}q(\hat{\psi}) \end{cases}$$

We will refer to the solutions of these equations as $\eta$-monopoles. Since $\eta$ is supported away from the neck the finite energy $\eta$-monopoles can be organized in the same fashion as the unperturbed ones and we obtain a moduli space $\widehat{\mathfrak{M}}_\mu(\eta)$.

The reducible $\eta$-monopoles are described by the zeros of the map

$$\mathcal{F}_\eta : \hat{\mathfrak{A}}_{\mu,sw} \to L^{1,2}_\delta(\mathbf{i}\Lambda^2 T^*\hat{N}), \quad \hat{A} \mapsto F^+_{\hat{A}} + \mathbf{i}\eta^+.$$

If $\mathcal{F}_\eta(\hat{A}_0 + \mathbf{i}\hat{a}) = 0$ then

$$\hat{d}^+\hat{a} = -\eta^+.$$

To decide whether the above equation admits a solution $\mathbf{i}\hat{a} \in T_{\hat{A}_0}\hat{\mathfrak{A}}_{\mu,sw}$ we need to understand the cokernel of the map

$$\hat{d}^+ : T_{\hat{A}_0}\hat{\mathfrak{A}}_{\mu,sw} \to L^{1,2}_\delta(\mathbf{i}\Lambda^2 T^*\hat{N}). \tag{4.4.1}$$

This map is part of the complex (K) and thus it has closed range and its cokernel is isomorphic to $H^2_{\hat{A}_0}$.

To compute its dimension observe that

$$\dim H^1_{\hat{A}_0} - \dim H^2_{\hat{A}_0} = \frac{1}{2}\Big(b_1(N) + 1 - \chi_{\hat{N}} - \tau_{\hat{N}}\Big)$$

and, exactly as in Proposition 4.3.28, we have

$$\dim H^1_{\hat{A}_0} = \dim \ker_{ex} \mathbf{ASD}.$$

The computations in Example 4.1.24 imply that

$$\dim \ker_{ex} \mathbf{ASD} = \dim \ker_{L^2} \mathbf{ASD} + \dim \partial_\infty \ker_{ex} \mathbf{ASD} = \hat{b}^3 + l^1.$$

Referring to the notations in Example 4.1.24 we can further write

$$\dim H^2_{\hat{A}_0} = \hat{b}^3 + l^1 - \frac{b_1 + 1 - \chi_{\hat{N}} - \tau_{\hat{N}}}{2}$$

$$= \hat{b}^3 + l^1 - \frac{b_1 - \hat{b}^2 - \hat{b}_+ + \hat{b}_- + \hat{b}^1 + \hat{b}^3}{2}$$

$$= l^1 - \frac{b_1 - \hat{b}^2 - \hat{b}_+ + \hat{b}_- + \hat{b}^1 - \hat{b}^3}{2}$$

$$= l^1 - \frac{b_1 - 2\hat{b}_+ - r + l^1 - l^3}{2}$$

$(r = l^2,\ l^1 + l^2 = b_1,\ l^3 = 0)$

$$= \frac{2\hat{b}_+ + l^1 + l^2 + l^3 - b_1}{2} = \hat{b}_+.$$

Thus if $\hat{b}^+ = 0$ then $H^2_{\hat{A}_0} = 0$ and, exactly as in the compact case, the reducible cannot be perturbed away because $\mathcal{F}_\eta$ is surjective.

Suppose now $\hat{b}_+ > 0$. We can identify $H^2_{\hat{A}_0}$ with the $L^2_\mu$-orthogonal complement of the range of the map (4.4.1). This is a finite-dimensional space

$$V \subset L^2_\mu(\mathbf{i}\Lambda^2_+ T^*\hat{N}).$$

Now, fix a sufficiently large positive integer $k_0$ and define

$$\mathcal{N} := \Big\{\eta \in L^{k_0,2}_\mu(\mathbf{i}\Lambda^2 T^*\hat{N});\ \langle \eta^+, v\rangle_{L^2_\mu} = 0,\ \ \forall v \in V,\ \ \operatorname{supp}(\eta) \cap \text{neck} = \emptyset\Big\}.$$

We see that $\mathcal{N}$ is a nontrivial closed subspace of $L^{k_0,2}_\mu(\mathbf{i}\Lambda^2 T^*\hat{N})$ and for any $\eta \in \mathcal{N}$ there are no reducible $\eta$-monopoles.

Using the Sard-Smale transversality theorem as in §2.2.3 we can prove the following genericity result.

**Proposition 4.4.1.** *Suppose $\hat{b}_+ > 0$. There exists a generic subset $\tilde{\mathcal{N}} \subset \mathcal{N}$ such that if $\eta \in \tilde{\mathcal{N}}$ all $\eta$-monopoles are irreducible and* **strongly regular**. *In particular, for $\eta \in \tilde{\mathcal{N}}$ the moduli space $\widehat{\mathfrak{M}}_\mu(\eta)$ is a smooth manifold.*

**Idea of proof** Denote by $\Delta$ the diagonal of $\mathfrak{M}_\sigma \times \mathfrak{M}_\sigma$ and consider

$$\mathcal{F} : \mathcal{N} \times \hat{\mathcal{C}}^{irr}_{\mu,sw}/\widehat{\mathfrak{G}}_{\mu,ex} \times \mathfrak{M}_\sigma \to \hat{\mathfrak{Y}}_\mu \times \mathfrak{M}_\sigma \times \mathfrak{M}_\sigma,$$

$$\mathcal{F}(\eta, \hat{\mathsf{C}}, \mathsf{C}) = (\widehat{SW}_\eta(\hat{\mathsf{C}}), \partial_\infty \hat{\mathsf{C}}, \mathsf{C}).$$

One has to show that $\mathcal{F}$ is transversal to $0 \times \Delta \subset \hat{\mathfrak{Y}}_\mu \times \mathfrak{M}_\sigma \times \mathfrak{M}_\sigma$ and then apply Sard-Smale to the natural projection

$$\pi : \mathcal{N} \times \hat{\mathcal{C}}^{irr}_{\mu,sw}/\widehat{\mathfrak{G}}_{\mu,ex} \times \mathfrak{M}_\sigma \to \mathcal{N}$$

restricted to the smooth submanifold $\mathcal{F}^{-1}(0 \times \Delta)$. The details are very similar to the proof in §2.2.3 with a slight complication arising from the noncompact background. It should be a good exercise for the reader to practice the techniques developed in this chapter. ■

**Remark 4.4.2.** The strong regularity implies more than the smoothness of the moduli spaces of finite energy monopoles. Assume $\hat{b}_+ > 0$ and suppose for simplicity that $0 \in \tilde{\mathcal{N}}$ so that each finite energy monopole $\hat{\mathsf{C}}_0 \in \widehat{\mathfrak{M}}_\mu$ is strongly regular. Set $\mathsf{C}_\infty = \partial_\infty \hat{\mathsf{C}}_0$. The sequence (**E**) leads to a long exact sequence

$$0 \hookrightarrow T_1 G_\infty \to H^1(F(\hat{\mathsf{C}}_0)) \to H^1_{\hat{\mathsf{C}}_0} \to T_{\mathsf{C}_\infty} \to 0. \tag{4.4.2}$$

Now set $\mathfrak{M}^\partial_\sigma := \mathfrak{Z}_\sigma / \mathfrak{G}^\partial_\sigma$. $\mathfrak{M}_\sigma$ is a quotient of $\mathfrak{M}^\partial_\sigma$ modulo the action of the discrete group

$$H^1(N, \mathbb{Z})/H^1(\hat{N}, \mathbb{Z})$$

and we have a natural map

$$\partial_\infty : \widehat{\mathfrak{M}}_\mu \to \mathfrak{M}^\partial_\sigma.$$

The sequence (4.4.2) shows that the strong regularity forces the above map to be a submersion.

**4.4.2. Compactness properties.** Because the background space $\hat{N}$ is noncompact it is a priori (and a posteriori) possible that the moduli space $\widehat{\mathfrak{M}}_\mu$ is noncompact. In the present subsection we will try to understand in some detail the main sources of noncompactness.

Fix a cylindrical *spin*$^c$ structure $\hat{\sigma}$ on $\hat{N}$ with $\sigma := \partial_\infty \hat{\sigma}$. For $0 < \mu < \mu_0(\sigma, g)$ we denote by $\widehat{\mathfrak{M}}_\mu$ the moduli space of $\widehat{\mathfrak{G}}_{\mu,ex}$-orbits of finite energy $\hat{\sigma}$-monopoles topologized with the $L^{2,2}_{\mu,ex}$-topology.

Recall that in 4.2.3 we have introduced the quotient $\tilde{\mathfrak{M}}_\sigma := \mathfrak{Z}_\sigma / \mathfrak{G}^1_\sigma$, where $\mathfrak{G}^1_\sigma$ denotes the identity component of $\mathfrak{G}_\sigma$. $\tilde{\mathfrak{M}}_\sigma$ is a covering space of $\mathfrak{M}^\partial_\sigma$ and we denote by

$$\pi : \tilde{\mathfrak{M}}_\sigma \to \mathfrak{M}^\partial_\sigma$$

the natural projection. The group $H^1(N, \mathbb{Z})$ of components of $\mathfrak{G}_\sigma$ acts on $\tilde{\mathfrak{M}}_\sigma$ with quotient $\mathfrak{M}_\sigma$. Similarly, $\mathfrak{M}^\partial_\sigma$ is a quotient of $\tilde{\mathfrak{M}}_\sigma$ modulo a discrete group: the image of $H^1(\hat{N}, \mathbb{Z})$ in $H^1(N, \mathbb{Z})$. The map $\partial_\infty$ induces a continuous map

$$\partial_\infty : \widehat{\mathfrak{M}}_\mu \to \mathfrak{M}^\partial_\sigma.$$

We already see one (mild) source of noncompactness: the moduli space $\mathfrak{M}^\partial_\sigma$.

The three-dimensional energy functional $\mathcal{E}$ defines a continuous function on $\tilde{\mathfrak{M}}_\sigma$ with discrete range

$$\cdots \mathcal{E}_{-1} < \mathcal{E}_0 < \mathcal{E}_1 < \cdots .$$

Denote by $\tilde{\mathfrak{M}}_{\sigma,k}$ the subset of $\tilde{\mathfrak{M}}_\sigma$ where $\mathcal{E} \equiv \mathcal{E}_k$. Set

$$\mathfrak{M}^\partial_{\sigma,k} := \pi\left(\tilde{\mathfrak{M}}_{\sigma,k}\right).$$

Since $\mathcal{E}$ is invariant under the gauge transformations on $N$ which extend to $\hat{N}$ it descends to a continuous function on $\mathfrak{M}^\partial_\sigma$ and the sets $\mathfrak{M}^\partial_{\sigma,k}$ are precisely its fibers.

The energy functional defines a continuous function

$$E : \widehat{\mathfrak{M}}_{\hat{\sigma}} \to \mathbb{R}, \ \ \hat{\mathsf{C}} \mapsto E(\hat{\mathsf{C}}).$$

Proposition 4.3.2 shows that $E(\hat{\mathsf{C}})$ depends only on the component of $\mathfrak{M}^{\partial}_{\sigma,k}$ containing $\partial_\infty \hat{\mathsf{C}}$. We conclude that the range of $E$ is discrete since it injects into the set of critical values of the three-dimensional energy functional $\mathcal{E}_\sigma$. We will refer to the range of $E$ as the $(\hat{\sigma})$-*energy spectrum*. The energy spectrum is

$$\Big\{C + \mathcal{E}_k; \ \ k \in \mathbb{Z}\Big\}$$

where $C$ is a constant independent of $k$. Now denote by $\widehat{\mathfrak{M}}^k_\mu$ the subspace

$$\widehat{\mathfrak{M}}^k_\mu := \partial_\infty^{-1} \mathfrak{M}^{\partial}_{\sigma,k}.$$

Clearly, if the energy spectrum is infinite then the moduli space $\widehat{\mathfrak{M}}_\mu$ cannot be compact for obvious reasons. We would like to investigate the compactness properties of the energy level sets.

As in §4.2.3 define the energy density

$$\rho : \widehat{\mathfrak{M}}_\mu \to C^\infty(\hat{N}, \mathbb{R}),$$

$$\hat{\mathsf{C}} = (\hat{\psi}, \hat{A}) \mapsto \rho_{\hat{\mathsf{C}}} := |\hat{\nabla}^{\hat{A}} \hat{\psi}|^2 + \frac{1}{8}|q(\hat{\psi})|^2 + |F_{\hat{A}}|^2 + \frac{\hat{s}}{4}|\hat{\psi}|^2.$$

The Main Energy Identity in Lemma 2.4.4 shows that for every $\hat{\mathsf{C}} \in \widehat{\mathfrak{M}}_\mu$ the density $\rho_{\hat{\mathsf{C}}}$ is positive on the cylindrical neck. Remarkably, the **Key Estimate** in Lemma 2.2.3 continues to hold in the noncompact situation as well. More precisely, we have

$$\sup_{x \in \hat{N}} |\hat{\psi}(x)|^2 \leq 2 \sup_{x \in \hat{N}} |\hat{s}(x)|. \tag{4.4.3}$$

To prove (4.4.3) we set $u(x) := |\hat{\psi}(x)|^2$. As in Lemma 2.2.3 we observe that $u$ satisfies the differential inequality

$$\Delta_{\hat{N}} + \frac{1}{4}u^2 + \frac{\hat{s}}{2}u \leq 0.$$

If we compactify $\hat{N}$ to $\overline{\hat{N}}$ by adding $\{\infty\} \times N$ then $u$ extends to a continuous function on $\overline{\hat{N}}$ and thus it achieves a maximum at a point $x_0 \in \overline{\hat{N}}$. If $x_0 \in \hat{N}$ then we conclude exactly as in the proof of Lemma 2.2.3. If $x_0 \in \{\infty\} \times N$ then since $\hat{\psi}\,|_{\infty \times N}$ is a 3-monopole we deduce from Remark 4.2.4 in §4.2.2 that

$$u(x_0) \leq 2 \sup_{x \in N} |s(x)| \leq 2S_0, \ \ S_0 := \sup_{x \in \hat{N}} |\hat{s}(x)|.$$

Set $\hat{N}_T := \hat{N} \setminus (T, \infty) \times N$ and fix $E_0 > 0$. If $E(\hat{\mathsf{C}}) \leq E_0$ then since $\rho_{\hat{\mathsf{C}}}$ is positive on the neck we deduce

$$-S_0^2 \mathrm{vol}\,(\hat{N}_T) \leq \int_{\hat{N}_T} \Big( |\hat{\nabla}^{\hat{A}} \hat{\psi}|^2 + \frac{1}{8} |q(\hat{\psi})|^2 + |F_{\hat{A}}|^2 \Big) d\hat{v} - S_0^2 \mathrm{vol}\,(\hat{N}_T)$$

$$\stackrel{(4.4.3)}{\leq} \int_{\hat{N}_0} \rho_{\hat{\mathsf{C}}} d\hat{v} \leq E_0.$$

Thus, there exists a constant $C_0$ which depends only on the geometry of $\hat{N}$, $E_0$ and $T$ such that

$$\int_{\mathbb{R}_+ \times N} \rho_{\hat{\mathsf{C}}} d\hat{v} + \int_{\hat{N}_T} \Big( |\hat{\nabla}^{\hat{A}} \hat{\psi}|^2 + \frac{1}{8} |q(\hat{\psi})|^2 + |F_{\hat{A}}|^2 \Big) d\hat{v} \leq C_0, \tag{4.4.4}$$

$\forall \hat{\mathsf{C}} \in \widehat{\mathfrak{M}}_\mu$ s.t. $E(\hat{\mathsf{C}}) \leq E_0$. To proceed further we need the following technical result. Fix a smooth, strongly cylindrical, reference connection $\hat{A}_0$ on $\det(\hat{\sigma})$.

**Lemma 4.4.3.** *Fix the constants $E_0, T > 0$. Then there exists a positive constant $C > 0$ which depends only on $E_0$, $T$, $\hat{A}_0$ and the geometry of $\hat{N}$ with the following property.*

*For every $\hat{\mathsf{C}} = (\hat{\psi}, \hat{A}) \in \widehat{\mathfrak{M}}_\mu$ satisfying $E(\hat{\mathsf{C}}) \leq E_0$ there exists $\hat{\gamma} \in \widehat{\mathfrak{G}}_{\mu,ex}$ such that if $(\hat{\phi}, \hat{B}) = \hat{\gamma} \cdot \hat{\mathsf{C}}$ then*

$$\|\hat{B} - \hat{A}_0\|_{L^{3,2}(\hat{N}_T)} \leq C.$$

Roughly speaking, the above lemma states that if the energy of $(\hat{\psi}, \hat{A})$ on $\hat{N}_T$ is not too large then the gauge orbit of $\hat{A}$ cannot be too far from the gauge orbit of the reference connection $\hat{A}_0$. Thus, high (but) finite energy monopoles are far away from the reference configuration.

**Proof** Assume for simplicity that $T = 0$. The proof relies on elements of the Hodge theory for manifolds with boundary as presented, e.g., in [**98**, Chap. 7]. Set $\mathbf{i}\hat{a} := \hat{A} - \hat{A}_0$. The 1-form $\hat{a}$ decomposes uniquely as a sum of mutually $L^2$-orthogonal terms

$$\hat{a} = 2\hat{d}u + 2\hat{d}^* \hat{b} + 2\Omega$$

where $u \in L^{1,2}(\hat{N}_1)$, $\hat{b} \in L^{1,2}(\Lambda^2 T^* \hat{N}_1)$, $\Omega \in L^{1,2}(\Lambda^1 T^* \hat{N}_1)$ are constrained by the conditions

$$u\,|_{\partial \hat{N}_1} = 0, \quad \lrcorner_t \hat{b}|_{\partial \hat{N}_1} = 0, \quad \hat{d}\Omega = \hat{d}^* \Omega = 0.$$

$\Omega$ defines an element in the group $H^1(\hat{N}_1, \mathbb{R})$, which can be identified with the vector space spanned by the harmonic 1-forms in $L^{1,2}(\Lambda T^* \hat{N}_1)$. Denote by $[\Omega]$ a harmonic 1-form representing an element in $H^1(\hat{N}_1, 2\pi\mathbb{Z})$ closest to $\Omega$. We can find a map $\hat{\gamma} : \hat{N}_1 \to S^1$ (smooth up to the boundary) such that

$$\frac{2\hat{d}\hat{\gamma}}{\hat{\gamma}} = 2\mathbf{i}\hat{d}\hat{v} + 2\mathbf{i}[\Omega]$$

where $\hat{v} \in L^{3,2}(\hat{N}_1, \mathbb{R})$, $\hat{v}\,|_{\partial \hat{N}_1} = 0$. Consider the gauge transformation

$$\hat{\beta} := e^{\mathbf{i}(\hat{u}-\hat{v})}\hat{\gamma}.$$

Observe that

$$\hat{A} - \frac{2\hat{d}\hat{\beta}}{\hat{\beta}} = \hat{A}_0 + 2\mathbf{i}\hat{d}^*\hat{b} + 2\mathbf{i}(\Omega - [\Omega]).$$

Using [**98**, Thm. 7.7.9] we deduce that there exists a positive constant $\nu$ depending only on the geometry of $\hat{N}_1$ such that

$$\|\hat{d}^*\hat{b}\|_{L^2(\hat{N}_1)} \leq \nu \|\hat{d}\hat{d}^*\hat{b}\|_{L^2(\hat{N}_1)} = \nu \|F_{\hat{A}} - F_{\hat{A}_0}\|_{L^2(\hat{N}_1)}.$$

Using (4.4.4) we deduce

$$\|F_{\hat{A}}\|_{L^2(\hat{N}_1)} \leq C$$

so that

$$\|\hat{A} - \frac{2\hat{d}\hat{\beta}}{\hat{\beta}} - \hat{A}_0\|_{L^2(\hat{N}_1)} \leq C(1 + \|\Omega - [\Omega]\|_{L^2(\hat{N}_1)}) \leq C'$$

where $C'$ is a positive constant depending only on the geometry of $\hat{N}_1$ and $E_0$. We can now find a gauge transformation $\hat{\gamma}_1 \in \widehat{\mathfrak{G}}_{\mu,ex}$ such that

$$\hat{\gamma}_1 \equiv \hat{\beta} \ \text{ on } \ \hat{N}_{1/2} = \hat{N} \setminus (1/2, \infty) \times N.$$

Set $(\hat{\phi}, \hat{B}) := \hat{\gamma} \cdot \hat{\mathsf{C}}$ and $\mathbf{i}\hat{\delta} := \hat{B} - \hat{A}_0$. Observe that on $\hat{N}_{1/2}$ we have

$$\hat{d}^*\hat{\delta} = 0, \ \ \hat{d}\hat{\delta} = -\mathbf{i}(F_{\hat{B}} - F_{\hat{A}_0}), \ \ \|\hat{\delta}\|_{L^2(\hat{N}_{1/2})} \leq C'.$$

Using interior elliptic estimates for the operator $\hat{d} + \hat{d}^*$ we deduce

$$\|\hat{\delta}\|_{L^{1,2}(\hat{N}_1/4)} \leq C\big(\|\hat{d}\hat{\delta} \oplus \hat{d}^*\hat{\delta}\|_{L^2(N_{1/2})} + \|\hat{\delta}\|_{L^2(\hat{N}_{1/2})}\big) \leq C''.$$

We can now bootstrap the a priori $L^{1,2}$-bound to a $L^{3,2}$-bound using the Seiberg-Witten equations, as we have done many times in this chapter. ■

**Remark 4.4.4.** We only want to mention that one can use the techniques in [**141**] to give a different (albeit related) proof of Lemma 4.4.3. The results in [**141**] require $L^p$-bounds on curvature where $p > 2$. However, since our gauge group is Abelian the arguments in [**141**] extend without difficulty to $L^2$-bounds as well.

Using Lemma 4.4.3 and the estimate (4.4.3) we can obtain after a standard bootstrap the following result.

**Lemma 4.4.5.** *Fix $E_0, T > 0$. Then there exists $C$ which depends only on $E_0$, $T$ and the geometry of $\hat{N}$ such that, for every $\hat{\mathsf{C}} = (\hat{\psi}, \hat{A}) \in \widehat{\mathfrak{M}}_\mu$ satisfying $E(\hat{\mathsf{C}}) \leq E_0$, there exists $\hat{\gamma} \in \widehat{\mathfrak{G}}_{\mu,ex}$ such that $\frac{d}{dt}\hat{\gamma} = 0$ for $t \geq T+2$ and if we set $(\hat{\phi}, \hat{B}) := \hat{\gamma} \cdot \hat{\mathsf{C}}$ then*

$$\hat{d}^*(\hat{B} - \hat{A}_0) = 0 \ \ on \ \ \hat{N}_{T+1}$$

*and*

$$\|\hat{B} - \hat{A}_0\|_{L^{3,2}(\hat{N}_T)} + \|\hat{\phi}\|_{L^{3,2}(\hat{N}_T)} \leq C.$$

Along the neck any $\hat{\mathsf{C}} \in \hat{\mathcal{C}}_{\mu,sw}$ has the form

$$(\psi(t), A_0 + \mathbf{i}a(t) + \mathbf{i}f(t)dt)$$

where $(\psi_\infty, A_0 + \mathbf{i}a(\infty)) \in \mathfrak{Z}_\sigma$. For $T > 0$ we set

$$S_T(\hat{\mathsf{C}}) := \|\psi(t) - \psi(\infty)\|_{L^{3,2}_\mu([T,\infty)\times N)} + \|a(t) - a(\infty)\|_{L^{3,2}_\mu([T,\infty)\times N)}$$

$$+\|f(t)\|_{L^{3,2}_\mu([T,\infty)\times N)}.$$

It induces a function

$$[S_T] : \widehat{\mathfrak{M}}_\mu \to \mathbb{R}_+, \ \ [S_T]([\hat{\mathsf{C}}]) := \inf\Big\{ S_T(\hat{\gamma} \cdot \hat{\mathsf{C}}); \ \ \hat{\gamma} \in \widehat{\mathfrak{G}}_{\mu,ex} \Big\}.$$

According to Theorem 4.2.33 $[S_T]([\hat{\mathsf{C}}]) < \infty$ for all $\hat{\mathsf{C}} \in \widehat{\mathfrak{M}}_\mu$.

**Lemma 4.4.6.** *Fix $T > 0$. For any constants $E_0, S_0 > 0$ the set*

$$\Big\{ [\hat{\mathsf{C}}] \in \widehat{\mathfrak{M}}_\mu; \ \ E(\hat{\mathsf{C}}) \leq E_0, \ \ [S_T]([\hat{\mathsf{C}}]) \leq S_0 \Big\}$$

*is precompact.*

**Proof** Consider a sequence of smooth monopoles

$$\hat{\mathsf{C}}_n = (\hat{\psi}_n, \hat{A}_n) \in \hat{\mathcal{C}}_{\mu,sw}$$

such that

$$E(\hat{\mathsf{C}}_0) \leq E_0, \ \ S_T(\hat{\mathsf{C}}_n) \leq S_1 := S_0 + 1.$$

Set $\mathbf{i}\hat{a}_n := \hat{A}_n - \hat{A}_0$. According to Lemma 4.4.5 we can assume there exists a constant depending only on $E_0$ and the geometry of $\hat{N}_0$ such that

$$\|\hat{a}_n\|_{L^{3,2}(\hat{N}_T)} + \|\hat{\psi}_n\|_{L^{3,2}(\hat{N}_T)} \leq C, \ \ \forall n. \tag{4.4.5}$$

Along the neck we write $\hat{a}_n = a_n(t) + f_n(t)dt$ and set

$$\mathsf{C}_n := (\psi_n(\infty), A_0 + \mathbf{i}a_n(\infty)).$$

We can also assume $d^*a_n(\infty) = 0$, for otherwise we can replace $\mathsf{C}_n$ by $e^{\mathbf{i}f}\mathsf{C}_n$ for a suitable function $f : N \to \mathbb{R}$. (For any $\varepsilon > 0$ we can extend $f$ to $\hat{f}$ on $\hat{N}$ such that, for all $n$, $|S_T(e^{\mathbf{i}\hat{f}}\hat{\mathsf{C}}_n) - S_T(\hat{\mathsf{C}}_n)| < \varepsilon$.) We then deduce that $\forall n$

$$\|a_n(\infty)\|_{L^2(N)} \leq \|a_n(T)\|_{L^2(N)} + \|a_n(T) - a_n(\infty)\|_{L^2(N)}$$

$$\leq \|a_n(T)\|_{L^2(N)} + const \cdot S_T(\hat{\mathsf{C}}_n)$$

and

$$\|\psi_n(\infty)\|_{L^2(N)} \leq \|\psi_n(T)\|_{L^2(N)} + \|\psi_n(T) - \psi_n(\infty)\|_{L^2(N)}$$

$$\leq \|\psi_n(0)\|_{L^2(N)} + const \cdot S_T(\hat{\mathsf{C}}_n).$$

On the other hand, the estimate (4.4.5) implies that

$$\|a_n(T)\|_{L^2(N)} + \|\psi_n(T)\|_{L^2(N)} \leq C, \quad \forall n.$$

Thus

$$\|a_n(\infty)\|_{L^2(N)} + \|\psi_n(\infty)\|_{L^2(N)} \leq C, \quad \forall n.$$

Since $(\psi_n(\infty), A_0 + \mathbf{i}a_n(\infty))$ is a 3-monopole and $d^*a_n(\infty) = 0$ we deduce

$$\|a_n(\infty)\|_{L^{3,2}(N)} + \|\psi_n(\infty)\|_{L^{3,2}(N)} \leq C, \quad \forall n.$$

We can now conclude using the compact embeddings

$$L^{3,2}_\mu(\hat{N}) \to L^{2,2}_\mu(\hat{N}), \quad L^{3,2}(N) \to L^{2,2}(N). \quad \blacksquare$$

In Theorem 4.2.37 we have introduced the capture level $\hbar > 0$ and a constant $\mathfrak{t} > 0$ such that if $\hat{\mathsf{C}} \in \hat{\mathcal{C}}_{\mu,sw}$ is a smooth monopole satisfying

$$\int_{[T,\infty)\times N} \rho_{\hat{\mathsf{C}}} < \hbar$$

then

$$[S_T]([\hat{\mathsf{C}}]) \leq \mathfrak{t}.$$

For every $\hat{\mathsf{C}} \in \widehat{\mathfrak{M}}_\mu$ define $T(\hat{\mathsf{C}}) > 0$ as the smallest nonnegative number $T$ such that

$$\int_{[T,\infty)\times N} \rho_{\hat{\mathsf{C}}} d\hat{v} \leq \hbar.$$

We will refer to $T(\hat{\mathsf{C}})$ as the *capture moment* of $\hat{\mathsf{C}}$. Lemma 4.4.6 has the following consequence.

**Lemma 4.4.7.** *The set*

$$\left\{\hat{\mathsf{C}} \in \widehat{\mathfrak{M}}_\mu; \;\; E(\hat{\mathsf{C}}) \leq E_0, \;\; T(\hat{\mathsf{C}}) \leq T_0\right\}$$

*is precompact.*

The last results indicate that in order to proceed further we need a detailed study of the finite energy monopoles on cylinders of longer and longer lengths. This study will also be relevant when we discuss the gluing problem.

For each positive integer $n$ consider a tube

$$C_n := (a_n, b_n) \times N, \quad -\infty \leq a_n < b_n \leq \infty,$$

such that $\ell_n := (b_n - a_n) \to \infty$ as $n \to \infty$. Continue to denote by $\sigma$ the *spin*$^c$ structure induced by $\sigma$ on $C_n$. Consider now for each $n$ a $\sigma$-monopole $\hat{\mathsf{C}}_n$ on $C_n$ such that

$$-\infty < E_n := E(\hat{\mathsf{C}}_n) < \infty$$

and $E_n \to E_\infty \in \mathbb{R}_+$ as $n \to \infty$. Define a density $\mu_n$ on $\mathbb{R}$ by

$$\mu_n(t) := \begin{cases} \frac{1}{2}\int_{t\times N} \rho_{\hat{\mathsf{C}}_n} dv_N, & t \in [a_n, b_n] \\ 0 & \text{otherwise} \end{cases}.$$

Observe that $\mu_n$ are nonnegative $L^1$-functions on $\mathbb{R}$ and

$$\int_{\mathbb{R}} \mu_n(t)dt = \frac{1}{2}E_n.$$

Observe also that if $t \in (a_n, b_n)$ then

$$\mu_n(t) := \|SW(\hat{\mathsf{C}}_n(t))\|^2_{L^2(N)}.$$

According to the concentration-compactness principle of P.L. Lions [**80**, **81**], we have the following alternatives as $n \to \infty$.

*There exists a subsequence of $\mu_n$ (which we continue to denote by $\mu_n$) satisfying one and only one of the following possibilities (see Figure 4.7).*

- **Tight-compactness** *There exists a sequence $t_n \in \mathbb{R}$ such that*

$$\forall \varepsilon > 0, \;\; \exists T > 0 : \int_{[t_n - R, t_n + T]} \mu_n(t)dt \geq E_\infty - \varepsilon, \;\; \forall n \geq n(\varepsilon).$$

- **Vanishing**

$$\lim_{n\to\infty} \sup_{\tau\in\mathbb{R}} \int_{[\tau - T, \tau + T]} \mu_n(t)dt = 0, \;\; \forall T > 0.$$

- **Dichotomy** *There exists $0 < \lambda < E_\infty$ such that for all $\varepsilon > 0$ there exists $n_\varepsilon > 0$, $R_\varepsilon$, $t_n = t_{n,\varepsilon} \in \mathbb{R}$ and $d_n := d_{n,\varepsilon}$ satisfying for $n \geq n_\varepsilon$*

$$\begin{cases} \left|\int_{t_n - R_\varepsilon}^{t_n + R_\varepsilon} \mu_n dt - \lambda\right| \leq \varepsilon, \;\; \left|\int_{t_n - R_\varepsilon - d_n}^{t_n + R_\varepsilon + d_n} \mu_n dt - \lambda\right| \leq \varepsilon, \\ d_{n,\varepsilon} \to \infty. \end{cases} \tag{4.4.6}$$

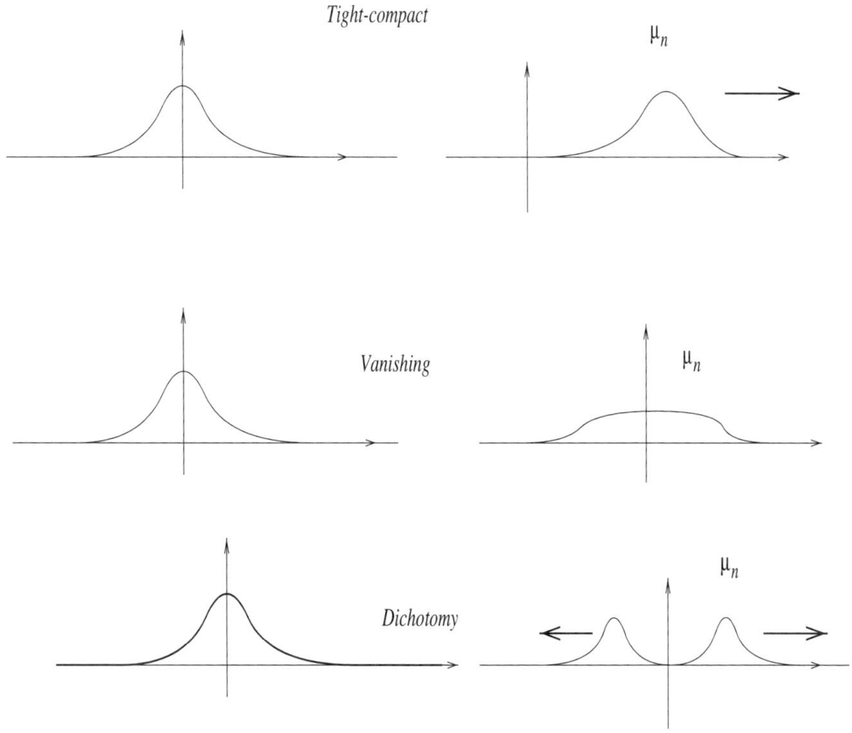


**Figure 4.7.** *Concentration compactness alternatives*

We call $\lambda$ above the *splitting level* of the dichotomy.

**Remark 4.4.8.** In [**103**] it is proved that the sequences $t_{n,\varepsilon}$ can be chosen independent of $\varepsilon$, which is what we will assume in the sequel.

**Lemma 4.4.9.** *The* **Vanishing** *alternative cannot occur if* $E_\infty > 0$.

**Proof** Suppose vanishing occurs. Then for every $\varepsilon > 0$ we can find $n(\varepsilon) > 0$ such that for all $n > n(\varepsilon)$ the integral of $\mu_n$ over any interval of length 4 is $< \varepsilon$. Using Corollary 4.2.15 we deduce that if $\varepsilon$ is sufficiently small then

$$\mu_n(t) \le C\varepsilon, \quad \forall t \in [a_n + 1, b_n - 1].$$

This shows that the path $t \mapsto \hat{\mathsf{C}}\,|_{t\times N}$ stays in a small neighborhood of a connected component of $\tilde{\mathfrak{M}}_\sigma$ for $t \in [a_n + 1, b_n - 1]$. Thus

$$0 < \mathcal{E}(\hat{\mathsf{C}}(b_n - 1)) - \mathcal{E}(\hat{\mathsf{C}}(a_n + 1)) < C_\varepsilon$$

where $C_\varepsilon \to 0$ as $\varepsilon \to 0$. This leads to a contradiction since

$$\begin{aligned} E_n &= E_{\hat{\mathsf{C}}_n}([a_n, a_n + 1]) + E_{\hat{\mathsf{C}}_n}([a_n + 1, b_n - 1]) + E_{\hat{\mathsf{C}}_n}([b_n - 1, b_n]) \\ &\le 2\varepsilon + C_\varepsilon. \ \blacksquare \end{aligned}$$

**Lemma 4.4.10.** *If the sequence $\mu_n$ is tight then by extracting a subsequence we can find a sequence $t_n \in \mathbb{R}$ such that $a_n - t_n \to A_\infty \in [-\infty, \infty]$, $b_n - t_n \to B_\infty \in [-\infty, \infty]$ , a sequence of gauge transformations $\hat{\gamma}_n$ on $C_n$ and a monopole $\hat{\mathsf{C}}$ on $[A_\infty, B_\infty] \times N$ such that*

$$E(\hat{\mathsf{C}}) = E_\infty$$

*and*

$$(\hat{\gamma}_n \cdot \hat{\mathsf{C}}_n)(t + t_n) \to \hat{\mathsf{C}}$$

*in $L^{1,2}_{loc}([A_\infty, B_\infty] \times N)$.*

**Proof** The Seiberg-Witten equations on cylinders are translation invariant so that by suitable translations we can assume the sequence $t_n$ in the description of **Tight-compactness** is identically zero. Also, assume for simplicity that $A_\infty = -\infty$ and $B_\infty = \infty$.

Fix $\varepsilon > 0$ smaller than the *capture level* $\hbar$. We deduce that there exists $T > 1$ such that for all $n \gg 0$

$$\int_{-\infty}^{-T+1} \mu_n(t)dt + \int_{T-1}^{\infty} \mu_n(t)dt \le \varepsilon, \quad \int_{-T-2}^{T+2} \mu_n(t)dt \ge E_n - \varepsilon$$

Arguing as in the proof of Lemma 4.4.6 we deduce that there exists $\hat{\gamma}_n \in \widehat{\mathfrak{G}}_{\mu,ex}(\mathbb{R} \times N)$ such that $\hat{\gamma}_n \cdot \hat{\mathsf{C}}_n$ is bounded in $L^{3,2}([-T-1, T+1] \times N)$. Relabel $\hat{\mathsf{C}}_n := \hat{\gamma}_n \cdot \hat{\mathsf{C}}_n$ so that, in the new notation, $\hat{\mathsf{C}}_n$ is bounded in $L^{3,2}([-T, T] \times N)$.

The arguments in §4.2.4 and in the proof of Lemma 4.4.6 show that there exist smooth 3-monopoles $\mathsf{C}_n^\pm$ and a smooth function

$$\hat{f}_n : \mathbb{R} \times N \to \mathbb{R}$$

such that $\hat{f}_n \equiv 0$ on $[-T, T] \times N$ and $e^{\mathbf{i}\hat{f}_n} \cdot \hat{\mathsf{C}}_n(t)$ stays in a tiny $L^{2,2}$-neighborhood of $\mathsf{C}_n^-$ for all $t \in [a_n, -T+1]$ and $e^{\mathbf{i}\hat{f}_n} \cdot \hat{\mathsf{C}}_n(t)$ stays in a tiny neighborhood of $\mathsf{C}_n^+$ for all $t \in [T-1, b_n]$.

Lemma 4.2.24 (or rather (4.2.44) in §4.2.5) shows that there exists a constant $C > 0$ independent of $n$ such that for every interval $I \subset \mathbb{R}$ of length $\le 1$ the $L^{2,2}(I \times N)$-norm of $\hat{\mathsf{C}}_n$ is bounded from above by $C$. It is now clear that a subsequence of $e^{\mathbf{i}\hat{f}_n} \cdot \hat{\mathsf{C}}_n$ converges strongly in $L^{1,2}_{loc}$ to a monopole $\hat{\mathsf{C}}$ on $\mathbb{R} \times N$. The tightness condition implies $E(\hat{\mathsf{C}}) = E_\infty$. ■

**Exercise 4.4.1.** Prove that the convergence in the above result can be improved to a strong $L^{2,2}_{loc}$-convergence.

**Remark 4.4.11.** The above $L^{2,2}_{loc}$-convergence has a built-in uniformity. More precisely, the rate of convergence on cylindrical pieces of length 1

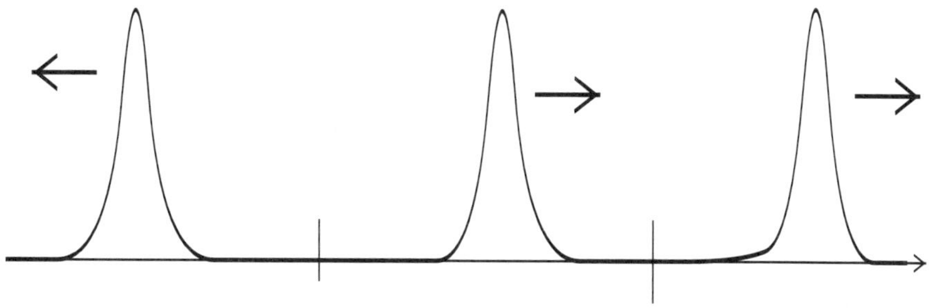

**Figure 4.8.** *Multiple splittings*

is bounded from above, meaning that for any $\varepsilon > 0$ there exists $n_\varepsilon > 0$ such that

$$\|\hat{\gamma}_n \hat{\mathsf{C}}_n(\bullet + t_n) - \hat{\mathsf{C}}(\bullet)\|_{L^{2,2}([T,T+1]\times N)} < \varepsilon$$

for all $n > n_\varepsilon$ and any admissible $T$.

We now have to deal with the dichotomy alternative. The "di-" prefix may be misleading. It is possible that the energy splits in several "bumps" each carrying a nontrivial amount of energy as in Figure 4.8. We want to first show that there are nontrivial constraints on how the dichotomy can occur.

If the energy spectrum consists of at least two values we define the *energy gap*

$$\delta := \min\Big\{\mathcal{E}_m - \mathcal{E}_k;\ \ m > k\Big\}.$$

Observe that the compactness of $\mathfrak{M}_\sigma$ coupled with the gauge change law (2.4.9) implies that $\delta > 0$. For every sufficiently small $\chi$ surround the closed sets $\tilde{\mathfrak{M}}_{\sigma,k}$ by tiny, mutually disjoint open neighborhoods $\mathcal{O}_k(\chi)$ such that if $\mathsf{C} \in \mathcal{O}_k$ then

$$|\mathcal{E}(\mathsf{C}) - \mathcal{E}_k| < \delta/8$$

and

$$\mathrm{dist}_{L^{2,2}}([\mathsf{C}], \tilde{\mathfrak{M}}_{\sigma,k}) \le \chi,\ \ \forall [\mathsf{C}] \in \mathcal{O}_k(\chi).$$

According to Proposition 4.2.16 we can find $\hbar(\chi) > 0$ such that if $\|SW(\mathsf{C})\|^2_{L^2} \le \hbar(\chi)$ then $\mathsf{C}$ modulo $\mathfrak{G}^1_\sigma$ belongs to one of the open sets $\mathcal{O}_k(\chi)$.

Suppose now that the dichotomy occurs. Fix a very small $\chi > 0$ and $\varepsilon > 0$ such that $0 < \varepsilon \ll \hbar(\chi)$. Set

$$\lambda_n := \min(\ell_n, d_n).$$

By suitable $t$-translations we can arrange that the sequence $t_n$ in the definition of dichotomy is identically zero. For each $n \gg 0$ we have

$$\int_{R_\varepsilon}^{R_\varepsilon+d_n} \mu_n(t)dt + \int_{-R_\varepsilon-d_n}^{-R_\varepsilon} \mu_n(t)dt \leq \varepsilon \tag{4.4.7}$$

and

$$\lambda - \varepsilon \leq \int_{-R_\varepsilon}^{R_\varepsilon} \mu_n(t)dt \leq \lambda + \varepsilon. \tag{4.4.8}$$

We can now split the interval $I_n = [a_n, b_n]$ into several parts:

$$I_n' := [a_n, b_n] \cap [-R_\varepsilon - \lambda_n/2, R_\varepsilon + \lambda_n/2], \quad J_n := I_n \setminus I_n'.$$

The set $J_n$ has at most two components and the dichotomy assumption guarantees that as $n \to \infty$ the measure of $J_n$ increases indefinitely. We cannot exclude the possibility that one of the components of $J_n$ has bounded size as $n \to \infty$. Define $J_n^0$ as the union of $I_n'$ with the (possibly empty) asymptotically bounded component of $J_n$. We set

$$[c_n, d_n] := J_n^0.$$

Observe that

$$\lambda - \varepsilon < \int_{J_n^0} \mu_n(t) \leq \lambda + 2\varepsilon.$$

$I_n \setminus J_n^0$ has at most two components and each of them increase indefinitely as $n \to \infty$.Three situations can occur (see Figure 4.9).

**A.** $I_n \setminus J_n^0$ has two components $J_n^{\pm 1}$ and their sizes increases indefinitely as $n \to \infty$.

**B.** The complement of $[-R_\varepsilon, R_\varepsilon]$ in $I_n$ consists of two intervals of indefinitely increasing sizes but $I_n \setminus J_n^0$ is an interval $J_n^1$ whose size increases indefinitely as $n \to \infty$.

**C.** Exactly one of the components of the complement of $[-R_\varepsilon, R_\varepsilon]$ in $I_n$ increases indefinitely as $n \to \infty$.

We will discuss the three cases separately.

**A.** Using (4.4.8) and Corollary 4.2.15 we deduce that $\hat{\mathsf{C}}_n|_{\partial J_n \times N}$ is very close to a pair of critical points of $\mathcal{E}$. Since the energy of $\hat{\mathsf{C}}_n$ over $J_n^0 \times N$ (which is $\approx \lambda$) can be expressed as

$$E_{\hat{\mathsf{C}}_n}([c_n, d_n] \times N) = \mathcal{E}(\hat{\mathsf{C}}_n(d_n)) - \mathcal{E}(\hat{\mathsf{C}}_n(c_n))$$

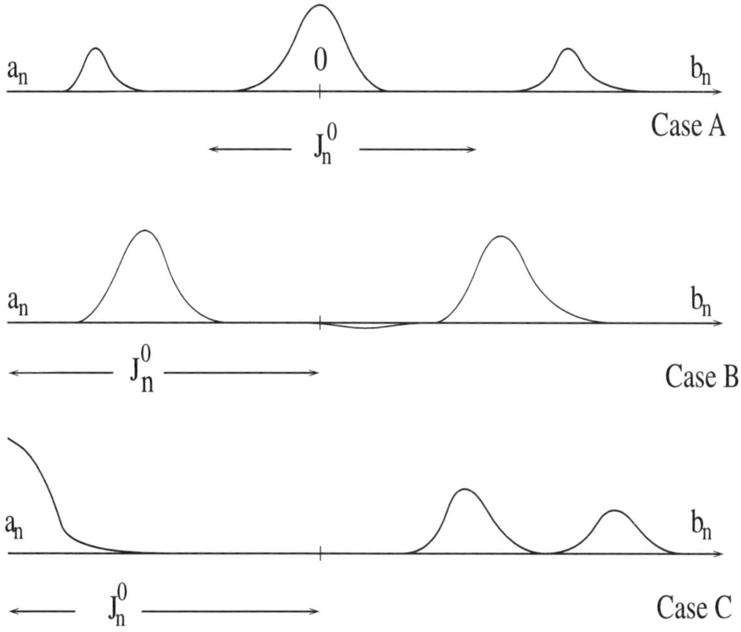


**Figure 4.9.** *Dichotomy alternatives*

we deduce that it is very close to the difference of two critical values of $\mathcal{E}$. Since $\lambda > 0$ these two critical values have to be distinct. We reach the conclusion that

$$\lambda > \delta/2.$$

Thus the splitting energy $\lambda$ is bounded from below by a strictly positive constant which depends only on the geometry of $N$.

**B.** We argue as before to conclude that for large $n$ the energy on the two intervals $J_n^0$ and $J_n^1$ is bounded from below by $\delta/2$.

**C.** The restriction $\hat{\mathsf{C}}_n^0$ of $\hat{\mathsf{C}}_n$ to $J_n^0 \times N$ defines a new sequence of monopoles on larger and larger domains. This sequence is tightly-compact and thus it converges to a nontrivial finite energy monopole on a semi-infinite interval.

**Definition 4.4.12.** A *right semi-tunneling* is a finite energy monopole on a cylinder $[a, \infty) \times N$. A *left semi-tunneling* is a finite energy monopole on a cylinder $(-\infty, b) \times N$.

In Figure 4.9 $\hat{C}_n^0$ converges to a right semi-tunneling. If we time reverse the situations depicted in this figure we see that left semi-tunnelings are also possible.

The next result summarizes the previous discussion.

**Lemma 4.4.13.** *If* **Dichotomy** *occurs then we can partition* $[a_n, b_n]$ *into* $k \leq 3$ *intervals* $J_n^i$, $1 \leq i \leq k$, *with the following properties.*

*(a)*

$$\lim_{n\to\infty} \text{length}(J_n^i) = \infty.$$

*(b) If we set* $\hat{C}_n^i := \hat{C}_n \mid_{J_n^i \times N}$ *then either* $(\hat{C}_n^i)$ *is tight and converges to a nontrivial (semi)-tunneling or* $(\hat{C}_n^i)$ *is not tight and* $E(\hat{C}_n^i) \geq \delta/2$.

If we iterate this discussion we deduce that there exist a positive integer $k$ constrained by

$$k < \frac{2E_\infty}{\delta} + 2$$

and a partition $I_n = [a_n, b_n]$ into $k$ intervals

$$I_n := I_n^1 \cup I_n^2 \cup \cdots \cup I_n^k$$

such that

$$\lim_{n\to\infty} \text{length}(I_n^j) = \infty, \ \ \forall 1 \leq j \leq k$$

and $\hat{C}_n^j := (\hat{C}_n \mid_{I_n^j \times N})$ is tight. Modulo gauge transformations and time translations the sequences $(\hat{C}_n^j)$ converge $L^{2,2}_{loc}$ to *nontrivial* (semi-)tunnelings $\hat{C}_\infty^j$ with the following properties.

- $\lim_{n\to\infty} E(\hat{C}_n^j) = E(\hat{C}_\infty^j), \ \forall j$.

- $\hat{C}_\infty^j$ is a *tunneling* for every $1 < j < k$.

- $\hat{C}_\infty^1$ is either a tunneling or a right semi-tunneling while $\hat{C}_\infty^k$ is either a tunneling or a left semi-tunneling.

- $\partial_\infty^+ \hat{C}_\infty^j = \partial_\infty^- \hat{C}_\infty^{j+1}$, for all $1 \leq j < k$.

- If $a_n = -\infty$ (resp. $b_n = \infty$) for all $n$ then $\hat{C}_\infty^1$ (resp. $\hat{C}_\infty^k$) must be a tunneling.

The above discussion has the following important consequence

**Proposition 4.4.14.** *If* $\widehat{\mathfrak{M}}_\mu^k$ *is noncompact then there exists a* nontrivial *tunneling* $\hat{C}_0$ *such that*

$$\partial_\infty^+ \hat{C}_0 \in \tilde{\mathfrak{M}}_{\sigma,k}.$$

**Proof** Suppose $\widehat{\mathfrak{M}}^k_\mu$ is not compact. Pick a sequence $\hat{\mathsf{C}}_n \in \widehat{\mathfrak{M}}^k_\mu$ with no convergent subsequence. Lemma 4.4.7 shows that the sequence $\hat{\mathsf{C}}_n \mid_{\mathbb{R}_+\times N}$ cannot be tight and vanishing cannot occur. Dichotomy is the only alternative and the previous discussion implies the existence of tunnelings with the required properties. ■

We want to present a few applications of the above result. Suppose $\sigma$ is such that $c_1(\det\sigma)$ is a torsion class. Then $\mathcal{E}_\sigma$ is $\mathfrak{G}_\sigma$-invariant and since $\mathfrak{M}_\sigma$ is compact we deduce that $\mathcal{E}$ has only finitely many critical values

$$\mathcal{E}_1 < \mathcal{E}_2 < \cdots < \mathcal{E}_m.$$

**Corollary 4.4.15.** *The space*

$$\Big\{[\hat{\mathsf{C}}] \in \widehat{\mathfrak{M}}_\mu;\ \ \mathcal{E}(\partial_\infty[\hat{\mathsf{C}}]) = \mathcal{E}_1\Big\}$$

*is compact.*

**Proof** If $\hat{\mathsf{C}}$ is a nontrivial $\sigma$-tunneling then $[\partial^\pm_\infty\hat{\mathsf{C}}] \in \mathfrak{M}_\sigma$ and

$$\mathcal{E}(\partial^+_\infty\hat{\mathsf{C}}) - \mathcal{E}(\partial^-_\infty\widehat{\mathfrak{M}}) > 0.$$

In particular, there cannot exist tunnelings towards $\sigma$-monopoles of smallest energy. The corollary now follows from Proposition 4.4.14. ■

**Corollary 4.4.16.** *Suppose the metric $g$ on $N$ has positive scalar curvature. Then for every $\sigma \in Spin^c(N)$, $\hat{\sigma} \in Spin^c(\hat{N})$ such that $\partial_\infty\hat{\sigma} = \sigma$ the space $\widehat{\mathfrak{M}}_\mu(\hat{\sigma})$ is either compact or empty.*

**Proof** If $\widehat{\mathfrak{M}}_\mu(\hat{\sigma}) \neq \emptyset$ then $\mathfrak{M}_\sigma \neq \emptyset$. Since $g$ has positive scalar curvature all the $\sigma$-monopoles are reducibles and thus $c_1(\det\sigma)$ is a torsion class. Moreover, according to Proposition 4.2.10 $\mathfrak{M}_\sigma$ is a $b_1(N)$-dimensional torus. The energy functional $\mathcal{E}_\sigma$ has only one critical value. The compactness now follows from the previous corollary. ■

**4.4.3. Orientability issues.** When the background manifold $\hat{N}$ is compact, we established the orientability of the moduli space of monopoles relying on two facts.

- The moduli space of monopoles is compact.
- The family of linearizations $\Big\{\mathcal{T}_{\hat{\mathsf{C}}};\ \ \hat{\mathsf{C}} \in \widehat{\mathfrak{M}}_\sigma\Big\}$ of the Seiberg-Witten equation can be deformed *through Fredholm operators* to an orientable family of Fredholm operators.

When $\hat{N}$ is a cylindrical manifold none of the above facts is true in general and thus a general approach to orientability requires new techniques. The possible noncompactness is not a very serious obstacle since one can naturally embed the moduli spaces of finite energy monopoles into some compact metric spaces. The deformation issue is a more serious problem and requires delicate analysis. The references we are aware of at this time (July 1999) are rather sketchy on the orientability issue which is discussed in special cases by ad-hoc methods.

We will not attempt to provide a comprehensive treatment of this problem since it is beyond the scope of these notes. Instead, we will discuss in detail only the situations arising in the topological applications we will present later on.

Suppose $(\hat{N}, \hat{g})$ is a cylindrical manifold such that $\hat{b}_+(\hat{N}) > 0$ and $(N, g) := \partial_\infty(\hat{N}, \hat{g})$ has *positive scalar curvature*. (The concrete examples we have in mind are $N = S^3, S^1 \times S^2$ with their natural metrics.) Assume $\hat{\sigma}$ is a *spin*$^c$ structure on $\hat{N}$ such that $\sigma := \partial_\infty \hat{\sigma}$ supports reducible monopoles (i.e. $c_1(\det \sigma)$ is a torsion class). The moduli space $\mathfrak{M}_\sigma$ consists only of reducible monopoles and is diffeomorphic to a $b_1(N)$-dimensional torus. We assume that we have generically perturbed the Seiberg-Witten equations on $\hat{N}$ as in §4.3.1 such that the resulting moduli space $\widehat{\mathfrak{M}}_\mu(\hat{\sigma})$ consists only of strongly regular irreducible monopoles. This implies that $\widehat{\mathfrak{M}}_\mu(\hat{\sigma})$ is a smooth manifold, the asymptotic boundary map

$$\partial_\infty : \widehat{\mathfrak{M}}_\mu(\hat{\sigma}) \to \mathfrak{M}^\partial_\sigma$$

is a submersion and the dimension of each component of $\widehat{\mathfrak{M}}_\mu$ is given by the virtual dimension formula. We want to warn the reader that, contrary to the compact case, the moduli space $\widehat{\mathfrak{M}}_\mu$ may consist of several components of different virtual (and in this case actual) dimensions. We assume for simplicity that $\eta \equiv 0$ is such a generic perturbation.

Before we proceed with our orientability discussion let us first point out an interesting result. We will present some of its topological implications in §4.6.2.

**Corollary 4.4.17.** *If $N$ is equipped with a metric of positive scalar curvature and the image of $H^1(\hat{N}, \mathbb{Z})$ in $H^1(N, \mathbb{Z})$ has infinite index then*

$$\widehat{\mathfrak{M}}_\mu(\hat{\sigma}) = \emptyset.$$

**Proof** Set $G := H^1(N, \mathbb{Z})/H^1(\hat{N}, \mathbb{Z})$. The universal cover of $\mathfrak{M}^\partial_\sigma$ is $\tilde{\mathfrak{M}}_\sigma \cong \mathbf{H}^1(N, g)$ (= monopoles modulo gauge transformations homotopic to the identity). We deduce that $\mathfrak{M}_\sigma^\partial$ is connected and

$$\mathfrak{M}_\sigma := \mathfrak{M}_\sigma^\partial / G.$$

In particular, we deduce that $\mathfrak{M}_\sigma^\partial$ is *noncompact* and *connected.* Thus, there cannot exist submersions from a *compact* smooth manifold $M$ to $\mathfrak{M}_\sigma^\partial$ so that $\widehat{\mathfrak{M}}_\mu(\hat{\sigma})$ must be empty. ■

For each $\hat{\mathsf{C}} \in \widehat{\mathfrak{M}}$ with $\mathsf{C}_\infty := \partial_\infty \hat{\mathsf{C}}$, the tangent space $T_{\hat{\mathsf{C}}}\widehat{\mathfrak{M}}_\mu$ fits in a long exact sequence derived from (**E**),

$$0 \to H^1(F(\hat{\mathsf{C}})) \to T_{\hat{\mathsf{C}}}\widehat{\mathfrak{M}}_\mu \to T_{\mathsf{C}_\infty}\mathfrak{M}_\sigma \to 0.$$

To describe orientations on $T_{\hat{\mathsf{C}}}\widehat{\mathfrak{M}}_\mu$ we need to describe orientations on $H^1(F(\hat{\mathsf{C}}))$ and $T\mathfrak{M}_\sigma$. It is clear that $\mathfrak{M}_\sigma$ can be oriented by specifying an orientation on $H^1(N, \mathbb{R})$.

To orient $H^1(F(\hat{\mathsf{C}}))$ observe that

$$\det H^1(F(\hat{\mathsf{C}})) \cong \det \hat{\mathcal{T}}_{\hat{\mathsf{C}},\mu}$$

where we regard $\hat{\mathcal{T}}_{\hat{\mathsf{C}},\mu}$ as an unbounded Fredholm operator $L^2_\mu \to L^2_\mu$. Thus, we need to study the orientability of the family of Fredholm operators

$$\widehat{\mathfrak{M}}_\mu \ni \hat{\mathsf{C}} \mapsto \hat{\mathcal{T}}_{\hat{\mathsf{C}},\mu}.$$

The computations in §4.3.2 show that if $\hat{\mathsf{C}} = (\hat{\psi}, \hat{A})$ and $\mathsf{C}_\infty = (\psi_\infty, A_\infty)$ ($\psi_\infty \equiv 0$ since all monopoles on $N$ are reducible) then we can write

$$\hat{\mathcal{T}}_{\hat{\mathsf{C}},\mu} = \begin{bmatrix} \mathfrak{D}_{\hat{A}} & 0 \\ 0 & \mathbf{ASD}_\mu \end{bmatrix} + \hat{\mathcal{P}}_{\hat{\mathsf{C}}}$$

where $\mathbf{ASD}_\mu := \sqrt{2}\hat{d}^+ \oplus (-d^{*_\mu})$ and $\hat{\mathcal{P}}_{\hat{\mathsf{C}}}$ is a zeroth order operator. Set $\hat{\mathcal{T}}^s_{\hat{\mathsf{C}},\mu} := \hat{\mathcal{T}}_{\hat{\mathsf{C}},\mu} - (1-s)\hat{\mathcal{P}}_{\hat{\mathsf{C}}}$. We let the reader check that the family of operators

$$[0,1] \times \widehat{\mathfrak{M}}_\mu \ni (s, \hat{\mathsf{C}}) \mapsto \hat{\mathcal{T}}^s_{\hat{\mathsf{C}},\mu} \in \text{Bounded Operators } L^{1,2}_\mu \to L^2_\mu$$

is continuous. Since

$$\vec{\partial}_\infty \hat{\mathcal{T}}^s_{\hat{\mathsf{C}},\mu} = \begin{bmatrix} \mathfrak{D}_{A_\infty} & 0 & 0 \\ 0 & -*d & d \\ 0 & d^* & -2\mu \end{bmatrix}$$

is independent of $s$ we deduce that all the operators $\hat{\mathcal{T}}^s_{\hat{\mathsf{C}},\mu}$ are Fredholm.[3] The orientability of $\hat{\mathsf{C}} \mapsto \hat{\mathcal{T}}_{\hat{\mathsf{C}},\mu}$ is thus equivalent to the orientability of

$$\hat{\mathcal{T}}^0_{\hat{\mathsf{C}},\mu} := \begin{bmatrix} \mathfrak{D}_{\hat{A}} & 0 \\ 0 & \mathbf{ASD}_\mu \end{bmatrix}.$$

The first component of the above operator acts on complex spaces and thus defines a naturally oriented family. The second component is independent of $\hat{\mathsf{C}}$ and thus is orientable. To fix an orientation we need to specify orientations on $\ker_\mu \mathbf{ASD}_\mu$ and $\ker_\mu \mathbf{ASD}^{*_\mu}_\mu$. Arguing as in the proofs of Propositions 4.3.28 and 4.3.30 we deduce

$$\ker_\mu \mathbf{ASD}_\mu \cong \ker_{ex} \mathbf{ASD}, \quad \ker_\mu \mathbf{ASD}^{*_\mu}_\mu \cong \ker_{ex} \mathbf{ASD}/H^0(\hat{N}, \mathbb{R}).$$

The computations in Example 4.1.24 show that $\ker_{ex} \mathbf{ASD}/H^0(\hat{N}, \mathbb{R})$ fits in a short exact sequence

$$0 \to H^2_+(\hat{N}) \to \ker_{ex} \mathbf{ASD}/H^0(\hat{N}, \mathbb{R}) \to L^2_{top} \to 0$$

where $L^2_{top}$ denotes the image of $H^2(\hat{N}, \mathbb{R})$ in $H^2(N, \mathbb{R})$ while $H^2_+(\hat{N})$ denotes a maximal positive subspace of the intersection form on $H^2(\hat{N}, N; \mathbb{R})$.

Similarly $\ker_{ex} \mathbf{ASD}$ can be included in a short exact sequence

$$0 \to H^1_{L^2}(\hat{N}) \to \ker_{ex} \mathbf{ASD} \to L^1_{top} \to 0$$

where $H^1_{L^2}(\hat{N})$ denotes the image of $H^1(\hat{N}, N; \mathbb{R}) \to H^1(\hat{N}, \mathbb{R})$ while $L^1_{top}$ denotes the image of $H^1(\hat{N}, \mathbb{R}) \to H^1(N, \mathbb{R})$.

**Proposition 4.4.18.** *Suppose $(N, g)$ has positive scalar curvature. Then $\widehat{\mathfrak{M}}_\mu$ is orientable. We can fix an orientation on it by choosing orientations on*

$$H^1(N, \mathbb{R}),\ L^1_{top},\ L^2_{top}, H^1_{L^2}(\hat{N}),\ H^2_+(\hat{N}). \tag{4.4.9}$$

**Remark 4.4.19.** Using the long exact sequence of the pair $(\hat{N}, N)$ we see that the spaces in the above proposition are naturally related. We let the reader to verify that a choice of orientations on $H^1(\hat{N}, \mathbb{R})$, $H^1(N, \mathbb{R})$ and $H^2_+(\hat{N}, \mathbb{R})$ naturally induces orientations on the spaces (4.4.9).

## 4.5. Cutting and pasting of monopoles

We have traveled a long road and we have gathered a lot of information about the finite energy monopoles. This section is the culmination of all this work. We will describe how to glue two finite energy monopoles into a monopole on a closed compact manifold (pasting) and then we will explain

[3] **Warning:** If $\mathsf{C}_\infty$ were irreducible then the operator $\hat{\mathcal{T}}^s_{\hat{\mathsf{C}},\mu}$ may *not* be Fredholm for all $s$.

why all monopoles on a closed manifold partitioned by a hypersurface split into finite energy monopoles (cutting).

**4.5.1. Some basic gluing constructions.** Consider again the situation in §4.1.5. Suppose $(\hat{N}, \hat{g})$ is a cylindrical manifold, $(N, g) := \partial_\infty(\hat{N}, \hat{g})$. We want to emphasize one aspect relating to the notion of cylindrical structure which was muted in our original definition. More precisely, a cylindrical structure presupposes the existence of an isometry $\varphi$ between the complement of a precompact open set $D \subset \hat{N}$ and the cylinder $\mathbb{R}_+ \times N$. The complete notation of a cylindrical structure ought to be

$$(\hat{N}, D, N, \hat{g}, g, \varphi)$$

but that would push the pedantry to dangerous levels. This notation (which will certainly not be used in the sequel) has one conceptual advantage. It shows that there is a "quasi"-action by pullback of the group of diffeomorphisms of $N$ on the space of cylindrical structures. We use the term "quasi" since a diffeomorphism $f$ of $N$ may not extend to a diffeomorphism of $\hat{N}_1$. However, there will always exist a metric $\hat{g}_f$ on $\hat{N}_1$ such that

$$\hat{g}_f \,|_{(0,1)\times N} \cong dt^2 + f^* g.$$

This "quasi"-action induces a genuine action on the space of equivalence classes of cylindrical manifolds where we declare two cylindrical manifolds $\hat{N}_1$ and $\hat{N}_2$ to be equivalent when there exists an orientation preserving diffeomorphism $\Psi : \hat{N}_1 \to \hat{N}_2$ which restricts to an isometry along the necks.

Similarly, if $(\hat{E}, \vartheta, E)$ is a cylindrical vector bundle on $\hat{N}$ with $E := \partial_\infty \hat{E}$ there exists a natural action of $\mathrm{Aut}\,(E)$ on the space of isomorphism classes of cylindrical structures on $\hat{E}$.

As in §4.1.5, consider two cylindrical manifolds

$$(\hat{N}_i, D_i, N_i, \hat{g}_i, g_i, \varphi_i), \;\; (N_i, g_i) = \partial_\infty(\hat{N}_i, \hat{g}_i), \;\; i = 1, 2.$$

Recall that $(\hat{N}_i, \hat{g}_i)$ are compatible if $N_1 \cong -N_2$ (as oriented manifolds) and $g_1 = g_2$. More precisely, this means there exists an orientation reversing isometry

$$\phi : \partial_\infty(\hat{N}_1, \hat{g}_1) \to \partial_\infty(\hat{N}_2, \hat{g}_2).$$

We set $N := N_1 (\cong -N_2)$. Observe that the above "quasi"-action is hidden inside the above definition of compatibility.

For every $r \gg 0$ we chop the half-cylinders $(r+2, \infty) \times N_i$ and glue the resulting manifolds $N_i(r+2)$ over a cylinder $(r, r+2) \times N$ to form a closed manifold $\hat{N}(r)$ with a long cylinder. The diffeomorphism class of $\hat{N}(r)$ depends on $\phi$ but in order to simplify the notation we will not indicate this in writing.

A simple rescaling argument shows that there exists a constant $C > 0$ which depends only on the geometry of $\hat{N}_i$ such that for all $r > 100$ we have

$$(4.5.1)\quad \|u\|_{L^p(\hat{N}(r))} \leq Cr^{\frac{1}{2}+\frac{1}{p}}\|u\|_{L^{1,2}(\hat{N}(r))}, \quad \forall u \in L^{1,2}(\hat{N}(r)),\ \ 1 < p \leq 6.$$

Suppose $(\hat{E}_i, \vartheta_i, E_i) \to \hat{N}_i$ are compatible cylindrical manifolds as defined in §4.1.5. They can be glued in an obvious fashion to form a bundle $\hat{E}(r) \to \hat{N}(r)$. For every $p \in (1, \infty)$, $k \in \mathbb{Z}_+$ and $\mu > 0$ there exists a natural linear map

$$\Delta = \Delta(\hat{E}_1, \hat{E}_2) : L^p_{\mu,ex}(\hat{E}_1) \times L^{k,p}_{\mu,ex}(\hat{E}_2) \to L^{k,p}(E),$$

$$\Delta(\hat{u}_1, \hat{u}_2) = \partial_\infty \hat{u}_1 - \partial_\infty \hat{u}_2.$$

The pairs of sections $(\hat{u}_1, \hat{u}_2) \in \ker \Delta(E_1, E_2)$ are called *compatible pairs.* In 4.1.5 we have constructed a gluing map

$$\#_r : \ker \Delta(\hat{E}_1, \hat{E}_2) \to L^{k,p}_\mu(\hat{E}(r)), \quad (\hat{u}_1, \hat{u}_2) \mapsto \hat{u}_1 \#_r \hat{u}_2$$

defined by the cut off construction (4.1.20) (see Figure 4.10)

$$\hat{u}_1 \#_r \hat{u}_2 := \hat{u}_i(r) \#_r \hat{u}_2(r).$$

The gluing construction extends to compatible asymptotically cylindrical first order p.d.o. $\hat{L}_i$ to produce a first order p.d.o. $\hat{L}_1 \#_r \hat{L}_2$ on $\hat{E}(r)$.

**Lemma 4.5.1.** *Suppose $\hat{L}_i$ are compatible asymptotically cylindrical operators. For any $k \in \mathbb{Z}_+$ and any $p \in (1, \infty)$, $\mu > 0$ there exists a constant which depends only on $k, p, \mu$ and the coefficients of $\hat{L}_i$ such that if $\hat{u}_i \in L^{k+1,p}_{\mu,ex}(\hat{E}_i)$ satisfy*

$$\partial_\infty \hat{u}_1 = \partial_\infty \hat{u}_2, \quad \hat{L}_i \hat{u}_i = 0, \quad i = 1, 2,$$

*then*

$$\|\hat{L}_1 \#_r \hat{L}_2(\hat{u}_1 \#_r \hat{u}_2)\|_{L^{k,p}(\hat{E}(r))} \leq Ce^{-\mu r}\Big(\|\hat{u}_1\|_{L^{k+1,p}_{\mu,ex}} + \|\hat{u}_2\|_{L^{k+1,p}_{\mu,ex}}\Big).$$

**Proof** For simplicity we will consider only the case $k = 0$. Fix $p \in (1, \infty)$ and $\mu > 0$. We can write

$$\hat{L}_1 := \hat{L}^0_1 + \hat{A}_1$$

where $\hat{L}^0_1$ is a cylindrical operator and $\hat{A}_1$ is a bundle morphism which belongs to $\bigcap_{m \in \mathbb{Z}_+} L^{m,p}_\mu$.

The manifold $\hat{N}(r)$ consists of three parts (see Figure 4.11):

$$\hat{N}(r)^- \cong \hat{N}_1 \setminus (r, \infty) \times N, \quad \hat{N}(r)^+ \cong \hat{N}_2 \setminus (r, \infty) \times N$$

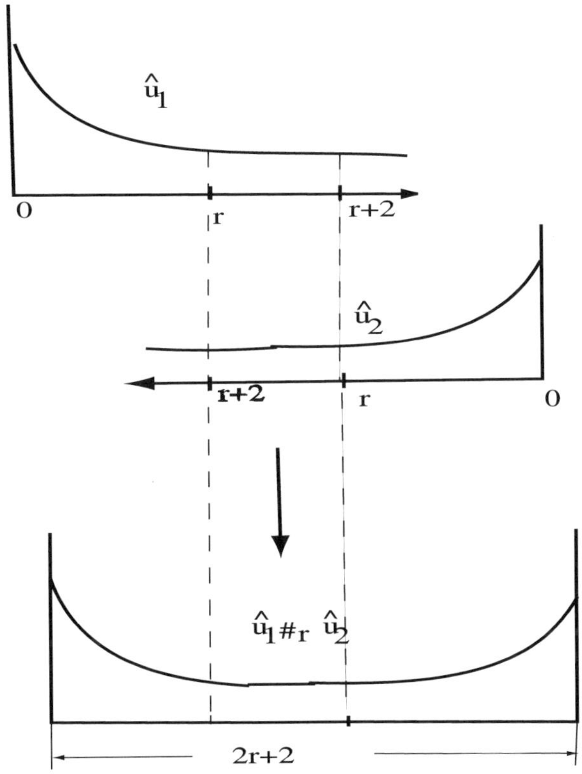


**Figure 4.10.** *Gluing compatible sections*

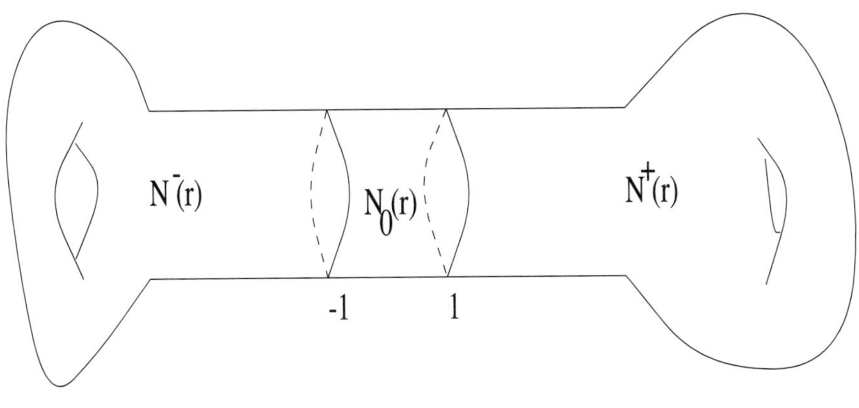


**Figure 4.11.** *The three regions of $\hat{N}(r)$*

and the overlap region

$$\hat{N}_0(r) \cong (-1, 1) \times N.$$

Over $\hat{N}(r)^-$ we have

$$\hat{L}_1\#_r\hat{L}_2 \equiv \hat{L}_1, \quad \hat{u}_1\#_r\hat{u}_2 \equiv \hat{u}_1.$$

A similar thing happens over $\hat{N}(r)^+$. Thus, the section $\hat{L}_1\#_r\hat{L}_2(\hat{u}_1\#_r\hat{u}_2)$ is supported on $\hat{N}_0(r)$. To ease the presentation identify the region

$$\hat{N}_0^-(r) := (-1,0)\times N \subset \hat{N}_0(r)$$

with the region $(r, r+1)\times N \subset \hat{N}_1$. Over $\hat{N}_0^-(r)$ we have

$$\hat{u}_1\#_r\hat{u}_2 = \alpha(t-r)\Big(\hat{u}_1 - \partial_\infty\hat{u}_1\Big) + \partial_\infty\hat{u}_1$$

and

$$\hat{L}_1\#_r\hat{L}_2 = \hat{L}_1 - \beta(t-r)\hat{A}_1 = \hat{L}_1^0 + \alpha(t-r)\hat{A}_1$$

where $\alpha(t)$ and $\beta(t)$ are depicted in Figure 4.4 of §4.1.4. A symmetric statement is true over $\hat{N}_0^+(r) := (0,1)\times N \subset \hat{N}(r)^0$.

To simplify the presentation we will use the symbol $q_1 \sim q_2$ to denote two quantities $q_1$, $q_2$ over $\hat{N}(r)^0$ such that

$$\|q_1 - q_2\|_{L^p(\hat{N}(r)^0)} \leq Ce^{-\mu r}\Big(\|\hat{u}_1\|_{L^{1,p}_{\mu,ex}} + \|\hat{u}_2\|_{L^{1,p}_{\mu,ex}}\Big)$$

where $C > 0$ is a constant depending only on $p$, $\mu > 0$ and the coefficients of $\hat{L}_i$.

We deduce that over $\hat{N}_0^-(r)$ we have

$$\hat{L}_1\#_r\hat{L}_2(\hat{u}_1\#_r\hat{u}_2) = \Big(\hat{L}_1 - \beta\hat{A}_1\Big)\big(\,\alpha(\hat{u}_1 - \partial_\infty\hat{u}_1) + \partial_\infty\hat{u}_1\,\big)$$

$$= \hat{L}_1(\alpha(\hat{u}_1 - \partial_\infty\hat{u}_1)) + \hat{L}_1\partial_\infty\hat{u}_1 - \beta\hat{A}_1(\,\alpha(\hat{u}_1 - \partial_\infty\hat{u}_1) + \partial_\infty\hat{u}_1\,)$$

$$\sim \hat{L}_1(\alpha(\hat{u}_1 - \partial_\infty\hat{u}_1)) + \hat{L}_1\partial_\infty\hat{u}_1 = \hat{L}_1(\alpha\hat{u}_1 + \beta\partial_\infty\hat{u}_1)$$

$$(\alpha+\beta = 1,\ \|\partial_\infty\hat{u}_1 - \hat{u}_1\|_{L^{1,p}(\hat{N}_0^-(r))} \leq Ce^{-\mu r}\|\hat{u}_1\|_{L^{1,p}_{\mu,ex}})$$

$$= \hat{L}_1\hat{u}_1 + \hat{L}_1(\beta(\partial_\infty\hat{u}_1 - \hat{u}_1)) = \hat{L}_1(\beta(\partial_\infty\hat{u}_1 - \hat{u}_1)) \sim 0. \ \blacksquare$$

**Remark 4.5.2.** Completely similar arguments can be used to prove the more general estimate

$$\Big\|(\hat{L}_1\#_r\hat{L}_2)(\hat{u}_1\#_r\hat{u}_2) - (\hat{L}_1\hat{u}_1)\#_r(\hat{L}_2\hat{u}_2)\Big\|_{L^{k,p}(\hat{N}(r))} \leq Ce^{-\mu r}\Big(\|\hat{u}_1\|_{L^{k+1,p}_{\mu,ex}} + \|\hat{u}_2\|_{L^{k+1,p}_{\mu,ex}}\Big). \tag{4.5.2}$$

**Exercise 4.5.1.** Prove the estimate in the above remark.

**Exercise 4.5.2.** Suppose $\omega_1, \omega_2$ are two compatible, asymptotically strongly cylindrical differential forms on $\hat{N}_1$ and $\hat{N}_2$ respectively. Show that

$$\hat{d}(\omega_1 \#_r \omega_1) = (\hat{d}\omega_1)\#_r(\hat{d}\omega_2).$$

Finally, we would like to explain how to glue cylindrical $spin^c$-structures. We refer back to §4.1.1 for the detailed description of the notion of cylindrical $spin^c$ structure. To figure out what to expect we begin with a simple argument.

Suppose we have two compatible cylindrical manifolds $\hat{N}_1$, $\hat{N}_2$. As before, form $\hat{N}(r)$ for $r \gg 0$. Let us (noncanonically) identify $Spin^c(\hat{N}(r))$ with $H^2(\hat{N}(r), \mathbb{Z})$ or, equivalently, with the group $\mathrm{Pic}^\infty(\hat{N}(r))$ of isomorphism classes of smooth complex line bundles over $\hat{N}(r)$. This group can be recovered from the two pieces of the decomposition using the Mayer-Vietoris sequence

$$H^1(\hat{N}(r), \mathbb{Z}) \to H^1(\hat{N}_1, \mathbb{Z}) \oplus H^1(\hat{N}_2, \mathbb{Z}) \xrightarrow{\Delta_1} H_1(N, \mathbb{Z})$$

$$\xrightarrow{\delta_1} H^2(\hat{N}(r), \mathbb{Z}) \xrightarrow{r_2} H^2(\hat{N}_1, \mathbb{Z}) \oplus H^2(\hat{N}_2, \mathbb{Z}) \xrightarrow{\Delta_2} H^2(N, \mathbb{Z}).$$

The arrow $r_2$ indicates that a line bundle $\hat{\sigma}$ on $\hat{N}(r)$ induces by restriction line bundles $\hat{\sigma}_i$ on $\hat{N}_i$ while the arrow $\Delta_2$ shows that these line bundles induce isomorphic line bundles on the dividing hypersurface $N$. Denote by $\sigma$ this isomorphism class. The arrow $\delta_1$ shows that in order to recover $\hat{\sigma}$ we need to glue $\hat{\sigma}_i$ using an automorphism $\varphi$ of $\sigma$

$$\hat{\sigma} = \hat{\sigma}_1 \#_\varphi \hat{\sigma}_2.$$

On the space of automorphisms of $\sigma$ we can now define an equivalence relation $\sim$ generated by

$$\varphi \sim \tau \Longleftrightarrow \varphi \circ \tau^{-1} \text{is homotopic to an automorphism of } \sigma$$

$$\text{which extends to } \hat{N}_1 \text{ or } \hat{N}_2.$$

The arrow $\delta_1$ shows that the isomorphism class of $\hat{\sigma}_1 \#_\varphi \hat{\sigma}_2$ depends only on the equivalence class of $\varphi$. (*Can you see this directly?*) If we set

$$G := H^1(N, \mathbb{Z}) \text{ and } G_i := \mathrm{Range}(H^1(\hat{N}_i, \mathbb{Z}) \to G)$$

then we deduce that the space of $\sim$-equivalence classes is isomorphic to $G/(G_1 + G_2)$. More intuitively, define

$$\Delta : \mathrm{Pic}^\infty_{cyl}(\hat{N}_1) \times \mathrm{Pic}^\infty_{cyl}(\hat{N}_2) \to \mathrm{Pic}^\infty(N), \ (\hat{L}_1, \hat{L}_2) \mapsto \hat{L}_1 |_N - \hat{L}_2 |_N .$$

Then the restriction map $r_2$ defines a fibration

$$\mathrm{Pic}^\infty(\hat{N}(r)) \to \ker \Delta$$

with fiber the space of gluing parameters $H^1(\hat{N},\mathbb{Z})/(G_1+G_2)$,

$$G/(G_1+G_2) \hookrightarrow \mathrm{Pic}^\infty(\hat{N}(r)) \to \ker \Delta.$$

Let us now refine this construction. We define a cylindrical structure on a line bundle $\hat{E} \to \hat{N}(r)$ as a pair $(E,\vartheta)$ where $E$ is a line bundle over $N$ while $\vartheta$ is an isomorphism

$$\vartheta : \hat{E}\,|_{(-r,r)\times N} \to \pi^* E$$

where $\pi : (-r,r)\times N \to N$ denotes the natural projection. Denote by $\mathrm{Pic}^\infty_{cyl}(\hat{N}(r))$ the Abelian group of isomorphism classes of cylindrical line bundles over $\hat{N}(r)$. Again, we have a surjective forgetful morphism

$$\Phi : \mathrm{Pic}^\infty_{cyl}(\hat{N}(r)) \to \mathrm{Pic}^\infty(\hat{N}(r))$$

whose kernel consists of the isomorphism classes of cylindrical structures on the trivial bundle over $\hat{N}(r)$. It is isomorphic to the quotient of $H^1(N,\mathbb{Z})$ modulo the subgroup generated by the classes which extend to $\hat{N}(r)$

$$\ker \Phi \cong G/(G_1 \cap G_2).$$

Denote by $C$ the cylinder $(-1,1)\times N$. A cylindrical structure on line bundle $\hat{L}$ over $C$ is by definition a pair $(L,\vartheta)$ where $L$ is a line bundle over $N$ and $\vartheta$ is an ismorphism

$$\vartheta : \hat{L} \to \pi^* L.$$

Observe that the forgetful morphism $\mathrm{Pic}^\infty_{cyl}(C) \to \mathrm{Pic}^\infty(C)$ is an isomorphism. Now set

$$\Delta_{cyl} : \mathrm{Pic}^\infty_{cyl}(\hat{N}_1) \times \mathrm{Pic}^\infty_{cyl}(\hat{N}_2) \to \mathrm{Pic}^\infty_{cyl}(C),\ \ (\hat{L}_1,\hat{L}_2) \mapsto \hat{L}_1\,|_C - \hat{L}_2\,|_C\,.$$

Two cylindrical line bundles $\hat{L}_i$ on $\hat{N}_i$ are called *compatible* if $(\hat{L}_1,\hat{L}_2) \in \ker \Delta_{cyl}$. The restriction maps $\mathrm{Pic}^\infty_{cyl}(\hat{N}(r)) \to \mathrm{Pic}^\infty_{cyl}(\hat{N}_i)$ produce a map

$$r : \mathrm{Pic}^\infty_{cyl}(\hat{N}(r)) \to \ker \Delta_{cyl}.$$

This map is clearly onto. Its kernel consists of cylindrical line bundles over $\hat{N}(r)$ which restrict to *trivial cylindrical* line bundles on $\hat{N}_i$. All these bundles on $\hat{N}(r)$ are obtained by gluing two trivial cylindrical line bundles $\hat{L}_i$ on $\hat{N}_i$ using a clutching map $\gamma : N \to S^1$ over $N$. We obtain in this fashion a *cylindrical* line bundle $\hat{L}_1\#_\gamma\hat{L}_2$ over $\hat{N}(r)$. This line bundle is trivial *as a cylindrical bundle* over $\hat{N}(r)$ if and only if $\gamma$ extends to $\hat{N}(r)$. This proves that the kernel of $r$ is isomorphic to $G/(G_1 \cap G_2)$. We have the following diagram:

$$\begin{array}{ccccccccc}
 & & & & 0 & & & & \\
 & & & & \downarrow & & & & \\
 & & & & G/(G_1 \cap G_2) & & & & \\
 & & & & \downarrow & & & & \\
0 & \longrightarrow & G/(G_1 \cap G_2) & \longrightarrow & \mathrm{Pic}^{\infty}_{cyl}(\hat{N}(r)) & \xrightarrow{r} & \ker \Delta_{cyl} & \longrightarrow & 0 \\
 & & \downarrow & & \downarrow \Phi & & \downarrow \Phi & & \\
0 & \longrightarrow & G/(G_1 + G_2) & \longrightarrow & \mathrm{Pic}^{\infty}(\hat{N}(r)) & \xrightarrow{r} & \ker \Delta & \longrightarrow & 0 \\
 & & & & \downarrow & & & & \\
 & & & & 0 & & & &
\end{array}$$

This diagram is commutative and its rows and columns are exact. It shows that $\ker \Delta_{cyl}$ is isomorphic to $\mathrm{Pic}^{\infty}(\hat{N}(r))$ and, in fact, there is a natural isomorphism

$$\#_r : \ker \Delta_{cyl} \to \mathrm{Pic}^{\infty}(\hat{N}(r))$$

called the *gluing map*. We have thus proved the following result.

**Proposition 4.5.3.** *Any complex line bundle decomposes uniquely as a connect sum of* **cylindrical** *line bundles on* $\hat{N}_i$.

We can now define the notion of cylindrical *spin*$^c$-structure on $\hat{N}(r)$ in an obvious fashion. The space of isomorphism classes of such structures is a $\mathrm{Pic}^{\infty}_{cyl}(\hat{N}(r))$-torsor. By fixing one such structure we can now reduce the decomposition problem for *spin*$^c$-structures to the analogous problem for line bundles. We have the following result.

**Proposition 4.5.4.** *Any spin*$^c$ *structure on* $\hat{N}(r)$ *uniquely decomposes as a connected sum of compatible* **cylindrical** *spin*$^c$ *structures on* $\hat{N}_i$.

**4.5.2. Gluing monopoles: Local theory.** Consider two compatible cylindrical 4-manifolds $\hat{N}_1$ and $\hat{N}_2$. Suppose $(N, g)$ satisfies the nondegeneracy assumption (**N**). Fix $\mu > 0$ sufficiently small. Form the closed manifold $\hat{N}(r)$, $r \gg 0$, and fix $\hat{\sigma} \in Spin^c(\hat{N}(r))$ so that

$$\hat{\sigma} = \hat{\sigma}_1 \# \hat{\sigma}_2$$

where $\hat{\sigma}_1$ and $\hat{\sigma}_2$ are compatible cylindrical *spin*$^c$-structures on $\hat{N}_1$ and $\hat{N}_2$ respectively. Now choose strongly cylindrical connections $\hat{A}_{0,i}$ on $\det(\hat{\sigma}_i)$ and set

$$\hat{A}_0 = \hat{A}_0(r) := \hat{A}_{0,1} \#_r \hat{A}_{0,2}.$$

If $\hat{\mathsf{C}}_i \in \hat{\mathcal{C}}_{\mu,ex}(\hat{N}_i)$ we set

$$\|\hat{\mathsf{C}}_i\|_{k,p} := \|\hat{\mathsf{C}}_i - (0, \hat{A}_{0,i})\|_{L^{k,p}_{\mu,ex}}.$$

Suppose $\hat{\mathsf{C}}_i \in \hat{\mathcal{C}}_{\mu,sw}(\hat{N}_i, \hat{\sigma}_i)$ are two smooth monopoles such that

$$\partial_\infty \hat{\mathsf{C}}_1 = \partial_\infty \hat{\mathsf{C}}_2.$$

As in the previous subsection we can form

$$\hat{\mathsf{C}}_r = (\hat{\psi}_r, \hat{A}_r) := \hat{\mathsf{C}}_1 \#_r \hat{\mathsf{C}}_2 = (\hat{\psi}_1 \#_r \hat{\psi}_2, \hat{A}_1 \#_r \hat{A}_2).$$

The configuration $\hat{\mathsf{C}}_r \in \hat{\mathcal{C}}_{\hat{\sigma}}(\hat{N}(r))$ may not be a monopole but it almost satisfies the Seiberg-Witten equations. Arguing as in the proof of Lemma 4.5.1 we deduce the following result.

**Lemma 4.5.5.** *There exist constants $C > 0$ and $r_0 > 0$ which depend only on the geometry of $\hat{N}_i$ such that*

$$\|\mathfrak{D}_{\hat{A}_r} \hat{\psi}_r\|_{L^{1,2}(\hat{N}(r))} + \|F^+_{\hat{A}_r} - \frac{1}{2} q(\hat{\psi}_r)\|_{L^{1,2}(\hat{N}(r))} \leq C e^{-\mu r}\Big(\|\hat{\mathsf{C}}_1\|_{2,2} + \|\hat{\mathsf{C}}_2\|_{2,2}\Big),$$

$\forall r > r_0$.

**Exercise 4.5.3.** Prove Lemma 4.5.5.

Naturally, we would like to know whether there exist genuine monopoles near $\hat{\mathsf{C}}_r$. In other words, we would like to investigate the $L^{2,2}$-small solutions $\underline{\hat{\mathsf{C}}}$ of the nonlinear equation

$$\widehat{SW}(\hat{\mathsf{C}}_r + \underline{\hat{\mathsf{C}}}) = 0, \;\; \mathfrak{L}^*_{\hat{\mathsf{C}}_r}(\underline{\hat{\mathsf{C}}}) = 0.$$

Form the nonlinear map

$$\mathcal{N} : L^{2,2}\Big(\hat{\mathbb{S}}^+_{\hat{\sigma}} \oplus \mathbf{i}T^*\hat{N}(r)\Big) \to L^{1,2}\Big(\hat{\mathbb{S}}^-_{\hat{\sigma}} \oplus \mathbf{i}\Lambda^2_+ T^*\hat{N}(r)\Big)$$

given by

$$\mathcal{N}(\underline{\hat{\mathsf{C}}}) := \widehat{SW}(\hat{\mathsf{C}}_r + \underline{\hat{\mathsf{C}}}) \oplus \mathfrak{L}^*_{\hat{\mathsf{C}}_r}(\underline{\hat{\mathsf{C}}}).$$

Denote by $\hat{\mathcal{T}}_r = \hat{\mathcal{T}}_{\hat{\mathsf{C}}_r}$ the linearization of $\mathcal{N}$ at 0

$$\hat{\mathcal{T}}_r(\underline{\hat{\mathsf{C}}}) = \underline{\widehat{SW}}_{\hat{\mathsf{C}}_r}(\underline{\hat{\mathsf{C}}}) \oplus \mathfrak{L}^*_{\hat{\mathsf{C}}_r}(\underline{\hat{\mathsf{C}}}).$$

Observe that

$$\hat{\mathcal{T}}_r := \hat{\mathcal{T}}_{\hat{\mathsf{C}}_1} \#_r \hat{\mathcal{T}}_{\hat{\mathsf{C}}_2}. \tag{4.5.3}$$

Now set

$$R(\hat{\mathsf{C}}) := \mathcal{N}(\hat{\mathsf{C}}) - \mathcal{N}(0) - \hat{\mathcal{T}}_r \underline{\hat{\mathsf{C}}}.$$

Using (4.5.1) with $p = 4$ we deduce the following result.

**Lemma 4.5.6.** *There exists a constant $C > 0$ which depends only on the geometry of $\hat{N}_i$ such that*

$$\|R(\hat{\underline{\mathsf{C}}})\|_{L^{1,2}(\hat{N}(r))} \leq Cr^{3/2}\|\hat{\underline{\mathsf{C}}}\|^2_{L^{2,2}(\hat{N}(r))}, \quad \forall \hat{\underline{\mathsf{C}}} \in L^{2,2}\Big(\hat{\mathbb{S}}^+_{\hat{\sigma}} \oplus \mathbf{i}T^*\hat{N}(r)\Big)$$

$$\|R(\hat{\underline{\mathsf{C}}}_1) - R(\hat{\underline{\mathsf{C}}}_2)\|_{L^{1,2}(\hat{N}(r))}$$

$$\leq Cr^{3/2}\Big(\|\hat{\underline{\mathsf{C}}}_1\|_{L^{2,2}(\hat{N}(r))} + \|\hat{\underline{\mathsf{C}}}_2\|_{L^{2,2}(\hat{N}(r))}\Big)\|\hat{\underline{\mathsf{C}}}_1 - \hat{\underline{\mathsf{C}}}_2\|_{L^{2,2}(\hat{N}(r))},$$

$\forall \hat{\underline{\mathsf{C}}}_1, \hat{\underline{\mathsf{C}}}_2 \in L^{2,2}\Big(\hat{\mathbb{S}}^+_{\hat{\sigma}} \oplus \mathbf{i}T^*\hat{N}(r)\Big)$.

To shorten the presentation we set

$$\mathfrak{X}^k_+ := L^{k,2}\Big(\hat{\mathbb{S}}^+_{\hat{\sigma}} \oplus \mathbf{i}T^*\hat{N}(r)\Big), \quad \mathfrak{X}^k_- := L^{k,2}\Big(\hat{\mathbb{S}}^-_{\hat{\sigma}} \oplus \mathbf{i}\Lambda^2_+T^*\hat{N}(r)\Big),$$

$$\mathfrak{X}^k := \mathfrak{X}^k_+ \oplus \mathfrak{X}^k_-.$$

According to Lemma 4.5.6, $\mathcal{N}$ is a continuous map $\mathfrak{X}^2_+ \to \mathfrak{X}^1_-$ differentiable at 0.

We can now form the closed, densely defined operator

$$\hat{L}_r : \mathfrak{X}^0 \to \mathfrak{X}^0$$

with block decomposition

$$\hat{L}_r := \begin{bmatrix} 0 & \hat{\mathcal{T}}^*_r \\ \hat{\mathcal{T}}_r & 0 \end{bmatrix}.$$

$\hat{L}_r$ is the analytical realization of a Dirac type operator. It is selfadjoint and induces bounded Fredholm operators

$$\mathfrak{X}^{k+1} \to \mathfrak{X}^k.$$

Denote by $\mathcal{H}_r$ the subspace of $\mathfrak{X}^0$ spanned by the eigenvectors of $\hat{L}_r$ corresponding to eigenvalues in the interval $(-r^{-2}, r^{-2})$. $\mathcal{H}_r$ consists entirely of smooth sections. The decomposition $\mathfrak{X}^0 = \mathfrak{X}^0_+ \oplus \mathfrak{X}^0_-$ induces a decomposition

$$\mathcal{H}_r = \mathcal{H}^+_r \oplus \mathcal{H}^-_r.$$

We denote by $\mathfrak{Y}(r)$ the orthogonal complement of $\mathcal{H}_r$ in $\mathfrak{X}^0$. $\mathfrak{Y}(r)$ is also equipped with a Sobolev filtration

$$\mathfrak{Y}^k(r) := \mathfrak{Y}^0 \cap \mathfrak{X}^k.$$

Again we have a decomposition

$$\mathfrak{Y}^k(r) := \mathfrak{Y}^k_+(r) \oplus \mathfrak{Y}^k_-(r).$$

Denote by $P_\pm$ the orthogonal projection $\mathfrak{X}_\pm \to \mathcal{H}^\pm_r$ and set $Q_\pm := 1 - P_\pm$. Observe that

$$Q_\pm(\mathfrak{X}^k) = \mathfrak{Y}^k_\pm(r).$$

For each $\hat{\underline{\mathsf{C}}} \in \mathfrak{X}_+^0$ we set

$$\hat{\underline{\mathsf{C}}}_0 := P_+\hat{\underline{\mathsf{C}}},\ \ \hat{\underline{\mathsf{C}}}^\perp := Q_+\hat{\underline{\mathsf{C}}}.$$

Observe that

$$P_-\hat{\mathcal{T}}_r(\hat{\underline{\mathsf{C}}}) = \hat{\mathcal{T}}_r(\hat{\underline{\mathsf{C}}}_0),\ \ Q_-(\hat{\mathcal{T}}_r\hat{\underline{\mathsf{C}}}) = \hat{\mathcal{T}}_r\hat{\underline{\mathsf{C}}}^\perp. \tag{4.5.4}$$

Moreover, for every $k \in \mathbb{Z}_+$, $\hat{\mathcal{T}}_r$ induces a bounded operator

$$\mathfrak{Y}_+^{k+1} \to \mathfrak{Y}_-^k$$

with bounded inverse $S$ and there exists $C = C_k > 0$ such that

$$\|Su\|_{L^{k+2,1}(\hat{N}(r))} \le C_k r^2 \|u\|_{L^{k,2}(\hat{N}(r))},\ \ \forall u \in \mathfrak{Y}_-^k. \tag{4.5.5}$$

The equation $\mathcal{N}(\hat{\mathsf{C}}) = 0$ is equivalent to the pair of equations

$$P_-\mathcal{N}(\hat{\mathsf{C}}) = 0 \ \text{ and } \ Q_-\mathcal{N}(\hat{\mathsf{C}}) = 0.$$

Using the identities (4.5.4) we can rewrite the above equations as

$$Q_-\mathcal{N}(0) + \hat{\mathcal{T}}_r\hat{\underline{\mathsf{C}}}^\perp + Q_-R(\hat{\underline{\mathsf{C}}}^\perp + \hat{\underline{\mathsf{C}}}_0) = 0 \tag{4.5.6a}$$

$$P_-\mathcal{N}(0) + \hat{\mathcal{T}}_r\hat{\underline{\mathsf{C}}}_0 + P_-R(\hat{\underline{\mathsf{C}}}^\perp + \hat{\underline{\mathsf{C}}}_0) \tag{4.5.6b}$$

Set $U^\perp := -SQ_-\mathcal{N}(0)$. Fix $\hat{\underline{\mathsf{C}}}_0$. We can rewrite (4.5.6a) as an equation for $\underline{\mathsf{C}}^\perp$

$$\hat{\underline{\mathsf{C}}}^\perp = \mathcal{F}(\hat{\underline{\mathsf{C}}}^\perp) := U^\perp - SQ_-R(\hat{\underline{\mathsf{C}}}^\perp + \hat{\underline{\mathsf{C}}}^0). \tag{4.5.7}$$

One should think of $\mathcal{F}$ as a family of functions $\mathcal{F}_{\underline{\mathsf{C}}_0}(\hat{\underline{\mathsf{C}}}^\perp)$ parameterized by $\hat{\underline{\mathsf{C}}}_0$. Using Lemma 4.5.6 and (4.5.5) we deduce

$$\begin{aligned} &\|\mathcal{F}(\hat{\underline{\mathsf{C}}}_1^\perp) - \mathcal{F}(\hat{\underline{\mathsf{C}}}_2^\perp)\|_{2,2} \\ &\le Cr^{5/2}\Big(\|\hat{\underline{\mathsf{C}}}_1^\perp + \hat{\underline{\mathsf{C}}}_0\|_{2,2} + \|\hat{\underline{\mathsf{C}}}_2^\perp + \hat{\underline{\mathsf{C}}}_0\|_{2,2}\Big)\|\hat{\underline{\mathsf{C}}}_1 - \hat{\underline{\mathsf{C}}}_2\|_{2,2}. \end{aligned} \tag{4.5.8}$$

Lemma 4.5.1 coupled with (4.5.5) shows that

$$\|\mathcal{F}(0)\|_{2,2} \le Cr^2e^{-\mu r}.$$

Thus

$$\|\mathcal{F}(\hat{\mathsf{C}}^\perp)\|_{2,2} \le \|\mathcal{F}(0)\|_{2,2} + \|\mathcal{F}(\hat{\mathsf{C}}^\perp) - \mathcal{F}(0)\|_{2,2}$$

$$\le Cr^2e^{-\mu r} + Cr^{5/2}\|\hat{\underline{\mathsf{C}}}^\perp + \hat{\underline{\mathsf{C}}}_0\|_{2,2}\|\hat{\underline{\mathsf{C}}}^\perp\|_{2,2}.$$

Observe that there exists $r = r(\mu) > 0$ such that for all $r > r(\mu)$ we have

$$\mathcal{F}_{\hat{\underline{\mathsf{C}}}_0}\Big\{\|\hat{\underline{\mathsf{C}}}^\perp\|_{2,2} \le r^{-3}\Big\} \subset \Big\{\|\hat{\underline{\mathsf{C}}}^\perp\|_{2,2} \le r^{-3}\Big\},\ \ \forall\, \|\hat{\underline{\mathsf{C}}}_0\|_{2,2} \le r^{-3}.$$

Moreover, according to (4.5.8) the induced map

$$\mathcal{F}_{\hat{\underline{\mathsf{C}}}_0} : \left\{ \|\hat{\underline{\mathsf{C}}}^{\perp}\|_{2,2} \leq r^{-3} \right\} \to \left\{ \|\hat{\underline{\mathsf{C}}}^{\perp}\|_{2,2} \leq r^{-3} \right\}$$

is a contraction. Set

$$B_{\perp}(r^{-3}) := \left\{ \|\hat{\underline{\mathsf{C}}}^{\perp}\|_{2,2} \leq r^{-3} \right\} \subset \mathfrak{Y}^2_+(r),$$

$$B_0(r^{-3}) := \left\{ \|\hat{\underline{\mathsf{C}}}_0\|_{2,2} \leq r^{-3} \right\} \subset \mathcal{H}_r^+.$$

For each $\hat{\mathsf{C}}_0 \in B_0(r^{-3})$ the fixed point equation (4.5.7) has an unique solution

$$\hat{\mathsf{C}}^{\perp} = \Phi(\hat{\mathsf{C}}_0) \in B_{\perp}(r^{-3}).$$

We let the reader verify that $\Phi$ depends differentiably upon $\hat{\mathsf{C}}_0$.

Now define the *Kuranishi map* $\hat{\mathsf{C}}_0 \mapsto \kappa_r(\hat{\mathsf{C}}_0)$ by making the substitution

$$\hat{\mathsf{C}}^{\perp} \to \Phi(\hat{\mathsf{C}}_0)$$

in (4.5.6b), that is

$$\kappa_r : B_0(r^{-3}) \to \mathcal{H}_r^-,$$

$$\hat{\underline{\mathsf{C}}}_0 \mapsto P_-\mathcal{N}(0) + \hat{\mathcal{T}}_r\hat{\underline{\mathsf{C}}}_0 + P_-R(\Phi(\hat{\underline{\mathsf{C}}}_0) + \hat{\underline{\mathsf{C}}}_0)$$

$$= P_-SW\Big(\hat{\mathsf{C}}_r + \hat{\underline{\mathsf{C}}}_0 + \Phi(\hat{\underline{\mathsf{C}}}_0)\Big).$$

The space $\mathcal{H}_r^-$ is called the *obstruction space*. The Kuranishi map $\kappa_r$ has the following significance. The part of the graph of $\Phi$ sitting above the zero set $\kappa_r^{-1}(0)$ consists of *all* the monopoles on $\hat{N}_r$ located in the local slice at $\hat{\mathsf{C}}_r$ at a $L^{2,2}$-distance $\leq r^{-3}$ from $\hat{\mathsf{C}}_r$. If $k_r \equiv 0$ (in which case we say that the gluing is unobstructed ) then the set of monopoles near $\hat{\mathsf{C}}_r$ is described by the graph of $\Phi$.

The results in §4.1.5 give more accurate information on the size and location of the Hilbert subspaces $\mathcal{H}_r^{\pm}$. More precisely, we have the short asymptotically exact sequence

$$0 \to \mathcal{H}_r^+ \longrightarrow^a \ker_{ex} \hat{\mathcal{T}}_{\hat{\mathsf{C}}_1} \oplus \ker_{ex} \hat{\mathcal{T}}_{\hat{\mathsf{C}}_1} \xrightarrow{\Delta} \hat{L}_1^+ + \hat{L}_2^+ \to 0$$

where $\hat{L}_i^+$ is the range of the asymptotic boundary map $\partial_\infty : \ker_{ex} \hat{\mathcal{T}}_{\hat{\mathsf{C}}_i} \to \ker \mathcal{T}_{\mathsf{C}_\infty}$. Similarly, we have a short asymptotically exact sequence

$$0 \to \mathcal{H}_r^- \longrightarrow^a \ker_{ex} \hat{\mathcal{T}}^*_{\hat{\mathsf{C}}_1} \oplus \ker_{ex} \hat{\mathcal{T}}^*_{\hat{\mathsf{C}}_2} \overset{\Delta}{\longrightarrow^a\!\!\to} \hat{L}_1^- + \hat{L}_2^- \to 0$$

where $\hat{L}_i^-$ is the range of $\partial_\infty : \ker_{ex} \hat{\mathcal{T}}^*_{\hat{\mathsf{C}}_i} \to \ker \mathcal{T}_{\mathsf{C}_\infty}$. Using the notation and results in §4.3.2 we set

$$L_i^+ := \partial^c_\infty \ker_{ex} \hat{\mathcal{T}}_{\hat{\mathsf{C}}_i} \to T_{\mathsf{C}_\infty}\mathfrak{M}_\sigma,$$

$$\mathfrak{C}_i^+ := \partial_\infty^0 \ker_{ex} \hat{\mathcal{T}}_{\hat{\mathsf{C}}_i} \to T_1 G_\infty \cong \operatorname{coker}(T_1 \hat{G}_i \stackrel{\partial_\infty}{\to} T_1 G_\infty),$$

$$L_i^- = \partial_\infty^c \ker_{ex} \hat{\mathcal{T}}^*_{\hat{\mathsf{C}}_i} \to T_{\mathsf{C}_\infty} \mathfrak{M}_\sigma,$$

$$\mathfrak{C}_i^- = \partial_\infty^0 \ker_{ex} \hat{\mathcal{T}}_{\hat{\mathsf{C}}_i} \to T_1 G_\infty \cong \operatorname{Range}(T_1 \hat{G}_i \stackrel{\partial_\infty}{\to} T_1 G_\infty).$$

The results in Propositions 4.3.28 and 4.3.30 imply that we can identify $H^1_{\hat{\mathsf{C}}_i}$ with the subspace $\ker(\partial_\infty^0 : \ker_{ex} \hat{\mathcal{T}}_{\hat{\mathsf{C}}_i} \to T_1 G_\infty)$ and

$$L_i^+ \cong \partial_\infty^c H^1_{\hat{\mathsf{C}}_i}, \quad L_i^- = \partial_\infty^c H^2(F_{\hat{\mathsf{C}}_i}).$$

To put the above facts in some geometric perspective we need to recall the results in Propositions 4.3.28 and 4.3.30. Denote by $\hat{G}_i$ the stabilizer of $\hat{\mathsf{C}}_i$ and by $G_\infty$ the stabilizer of $\mathsf{C}_\infty$. We then have the following commutative diagrams in which both the rows and the columns are exact. $S_r$ denotes the splitting map defined in §4.1.5 while $\Delta$ denotes the difference between the asymptotic limits.

- *Virtual tangent space diagram*

$$(\mathbf{T})\qquad \begin{array}{ccccc}
 & 0 & & 0 & & 0 \\
 & \downarrow & & \downarrow & & \downarrow \\
0 \to & \ker \Delta_+^c & \xrightarrow{S_r} & H^1_{\hat{\mathsf{C}}_1} \oplus H^1_{\hat{\mathsf{C}}_2} & \xrightarrow{\Delta_+^c} & L_1^+ + L_2^+ \to 0 \\
 & \downarrow & & \downarrow & & \downarrow \\
0 \to & \mathcal{H}_r^+ & \xrightarrow{S_r} & \ker_{ex} \hat{\mathcal{T}}_{\hat{\mathsf{C}}_1} \oplus \ker_{ex} \hat{\mathcal{T}}_{\hat{\mathsf{C}}_2} & \xrightarrow{\Delta} & \hat{L}_1^+ + \hat{L}_2^+ \to 0 \\
 & \downarrow & & \downarrow{\scriptstyle \Upsilon_1 \oplus \Upsilon_2} & & \downarrow \\
0 \to & \ker \Delta_+^0 & \xrightarrow{S_r} & \mathfrak{C}_1^+ \oplus \mathfrak{C}_2^+ & \xrightarrow{\Delta_+^0} & \mathfrak{C}_1^+ + \mathfrak{C}_2^+ \to 0 \\
 & \downarrow & & \downarrow & & \downarrow \\
 & 0 & & 0 & & 0
\end{array}$$

• *Obstruction diagram*

$$
\text{(O)}\qquad
\begin{array}{ccccccc}
 & & 0 & & 0 & & 0 \\
 & & \downarrow & & \downarrow & & \downarrow \\
0 \to & \ker \Delta^c_- & \xrightarrow{S_r} & H^2(F_{\hat{\mathsf{C}}_1}) \oplus H^2(F_{\hat{\mathsf{C}}_2}) & \xrightarrow{\Delta^c_-} & L_1^- + L_2^- & \to 0 \\
 & \downarrow & & \downarrow & & \downarrow & \\
0 \to & \mathcal{H}_r^- & \xrightarrow{S_r} & \ker_{ex} \hat{\mathcal{T}}^*_{\hat{\mathsf{C}}_1} \oplus \ker_{ex} \hat{\mathcal{T}}^*_{\hat{\mathsf{C}}_2} & \xrightarrow{\Delta} & \hat{L}_1^- + \hat{L}_2^- & \to 0 \\
 & \downarrow & & \downarrow{\scriptstyle \partial^0_\infty \oplus \partial^0_\infty} & & \downarrow & \\
0 \to & \ker \Delta^0_- & \xrightarrow{S_r} & \mathfrak{C}_1^- \oplus \mathfrak{C}_2^- & \xrightarrow{\Delta^0_-} & \mathfrak{C}_1^- + \mathfrak{C}_2^- & \to 0 \\
 & \downarrow & & \downarrow & & \downarrow & \\
 & 0 & & 0 & & 0 &
\end{array}
$$

The Lagrangian condition (4.1.22) establishes certain relationships between the above two sequences.

• *Complementarity equations*

$$
\text{(L)}\qquad L_i^+ \oplus L_i^- = T_{\hat{\mathsf{C}}_\infty} \mathfrak{M}_\sigma, \;\; \mathfrak{C}_i^+ \oplus \mathfrak{C}_i^- = T_1 G_\infty,
$$

$$
\mathfrak{C}_i^+ \cong \operatorname{coker}(T_1 \hat{G}_i \xrightarrow{\partial_\infty} T_1 G_\infty), \;\; \mathfrak{C}_i^- = \operatorname{Range}(T_1 \hat{G}_i \xrightarrow{\partial_\infty} T_1 G_\infty), \;\; i = 1, 2,
$$

$$
L_i^- = \left(L_i^+\right)^\perp, \;\; \mathfrak{C}_i^- = \left(\mathfrak{C}_i^+\right)^\perp.
$$

Suppose that at least one of the monopoles $\hat{\mathsf{C}}_i$ is irreducible, say $\hat{\mathsf{C}}_i$. Then $\mathfrak{C}_1^\perp = 0$ and $\ker \Delta^0_- = 0$. The diagram (**O**) implies

$$
\mathcal{H}_r^- \cong \ker \Delta^c \subset H^2(F_{\hat{\mathsf{C}}_0}) \oplus H^2(F_{\hat{\mathsf{C}}_0}).
$$

Our next result summarizes the facts we have established so far. A local gluing result of this nature was proved for the first time by Tom Mrowka in his dissertation [**99**], in a slightly different form and in the Yang-Mills context, relying on conceptually different methods.

**Theorem 4.5.7. (Local gluing theorem)** *Suppose $\hat{\mathsf{C}}_i \in \hat{\mathcal{C}}_{\mu,sw}(\hat{N}_i, \hat{\sigma}_i)$, $i = 1, 2$, are two finite energy monopoles with compatible asymptotic limits such that at least one of them is irreducible. Then the following hold.*

*(a)*

$$
\mathcal{H}_r^- \cong \ker \Delta^c \subset H^2(F_{\hat{\mathsf{C}}_0}) \oplus H^2(F_{\hat{\mathsf{C}}_0}).
$$

*(b) There exists $r_0 > 0$ (depending only on the geometry of $\hat{N}_i$ and $\|\hat{\mathsf{C}}_i\|_{2,2}$) with the following property. For every $r > r_0$ there exist smooth maps*

$$\kappa_r : B_0(r^{-3}) \subset \mathcal{H}_r^+ \to \mathcal{H}_r^-, \quad \Phi : B_0(r^{-3}) \subset \mathcal{H}_r^+ \to \mathfrak{Y}(r)^+$$

*such that the variety*

$$\Big\{ \underline{\hat{\mathsf{C}}} = \hat{\mathsf{C}}_r + \underline{\hat{\mathsf{C}}}^0 \oplus \underline{\hat{\mathsf{C}}}^\perp; \quad \|\underline{\hat{\mathsf{C}}}^0\|_{2,2} \leq r^{-3}, \ \kappa_r(\underline{\hat{\mathsf{C}}}^0) = 0, \ \underline{\hat{\mathsf{C}}}^\perp = \Phi(\underline{\hat{\mathsf{C}}}^0) \Big\}$$

*coincides with the set of monopoles $\hat{\mathsf{C}}$ on $\hat{N}(r)$ satisfying*

$$\mathfrak{L}^*_{\hat{\mathsf{C}}_r}(\hat{\mathsf{C}} - \hat{\mathsf{C}}_r) = 0, \quad \|\hat{\mathsf{C}} - \hat{\mathsf{C}}_r\|_{2,2} \leq r^{-3}$$

*where $\hat{\mathsf{C}}_r := \hat{\mathsf{C}}_1 \#_r \hat{\mathsf{C}}_2$ and $\mathcal{H}_r^\pm$ are determined from the diagram (**T**).*

**Remark 4.5.8.** The obstruction space $\mathcal{H}_r^-$ can also be described as the space spanned by the eigenvectors of $\hat{\mathcal{T}}_r\hat{\mathcal{T}}_r^*$ corresponding to *very small eigenvalues*, i.e. eigenvalues in $[0, r^{-4})$. (As pointed out in §4.1.5 the eigenvalues determining $\mathcal{H}_r^-$ are in reality a lot smaller than $r^{-4}$, in fact smaller than any $r^{-n}$ as $r \to \infty$.) Notice that

$$\hat{\mathcal{T}}_r\hat{\mathcal{T}}_r^* : L^{2,2}\begin{pmatrix} \mathbb{S}^-_{\hat{\sigma}} \oplus \mathbf{i}\Lambda^2_+ T^*\hat{N}(r) \\ \bigoplus \\ \mathbf{i}\Lambda^0 T^*\hat{N}(r) \end{pmatrix} \to L^2 \begin{pmatrix} \mathbb{S}^-_{\hat{\sigma}} \oplus \mathbf{i}\Lambda^2_+ T^*\hat{N}(r) \\ \bigoplus \\ \mathbf{i}\Lambda^0 T^*\hat{N}(r) \end{pmatrix}$$

has the block decomposition

$$\hat{\mathcal{T}}_r\hat{\mathcal{T}}_r^* = \begin{bmatrix} \underline{\widehat{SW}}_r \underline{\widehat{SW}}_r^* & \underline{\widehat{SW}}_r \circ \mathfrak{L}_{\hat{\mathsf{C}}_r} \\ & \\ \mathfrak{L}^*_{\hat{\mathsf{C}}_r} \circ \underline{\widehat{SW}}_r^* & \mathfrak{L}^*_{\hat{\mathsf{C}}_r}\mathfrak{L}_{\hat{\mathsf{C}}_r} \end{bmatrix}$$

where $\underline{\widehat{SW}}_r$ denotes the linearization of the Seiberg-Witten equations at $\hat{\mathsf{C}}_r$. Now witness a small miracle.

$$\underline{\widehat{SW}}_r \circ \mathfrak{L}_{\hat{\mathsf{C}}_r}(\mathbf{i}\hat{f}) = \frac{d}{dt}|_{t=0}\, \widehat{SW}(e^{\mathbf{i}t\hat{f}} \cdot \hat{\mathsf{C}}_r)$$

$$= \frac{d}{dt}|_{t=0} \Big( e^{\mathbf{i}t\hat{f}} \cdot \hat{\mathfrak{D}}_{\hat{A}_r}\hat{\psi}_r, \sqrt{2}F^+_{\hat{A}_r} - \frac{1}{\sqrt{2}}q(\hat{\psi}_r)\Big) - \Big(\mathbf{i}\hat{f}\hat{\mathfrak{D}}_{\hat{A}_r}\hat{\psi}_r, 0\Big).$$

This shows that the off-diagonal terms in the above description of $\hat{\mathcal{T}}_r\hat{\mathcal{T}}_r^*$ are *zeroth order* operators !!! Since

$$\|\hat{\mathfrak{D}}_{\hat{A}_r}\hat{\psi}_r\|_{L^{2,2}(\hat{N}(r))} \leq Ce^{-\mu r}$$

we deduce that their norm is exponentially small. We can now write

$$\hat{\mathcal{T}}_r\hat{\mathcal{T}}_r^* = \begin{bmatrix} \underline{\widehat{SW}}_r \underline{\widehat{SW}}_r^* & 0 \\ 0 & \mathfrak{L}^*_{\hat{\mathsf{C}}_r}\mathfrak{L}_{\hat{\mathsf{C}}_r} \end{bmatrix} + W_r =: V_r + W_r$$

where $W_r$ is bounded, symmetric and $\|W_r\| = O(e^{-\mu r})$. Denote (temporarily) by $\tilde{\mathcal{H}}_r^-$ the space spanned by the eigenvectors of $V_r$ corresponding to eigenvalues in $[0, r^{-4})$. We can now use the perturbation results in [**60**] to deduce that the gap distance between $\mathcal{H}_r^-$ and $\tilde{\mathcal{H}}_r^-$ converges to zero as $r \to \infty$. In applications it thus suffices to work with $\tilde{\mathcal{H}}_r^-$ rather than $\mathcal{H}_r^-$. The space $\tilde{\mathcal{H}}_r^-$ has an additional structure deriving from the diagonal structure of $V_r$. More precisely, $\tilde{\mathcal{H}}_r^-$ splits into a direct sum

$$\textit{very small eigenvalues of } \underline{\widehat{SW}}_r \underline{\widehat{SW}}_r^* \oplus \textit{very small eigenvalues of } \mathfrak{L}^*_{\hat{\mathsf{C}}_r}\mathfrak{L}_{\hat{\mathsf{C}}_r}.$$

We deduce from this picture that the operator $\mathfrak{L}^*_{\hat{\mathsf{C}}_r}\mathfrak{L}_{\hat{\mathsf{C}}_r}$ does not have very small eigenvalues if at least one of $\hat{\mathsf{C}}_i$ is irreducible. The reason is simple: any eigenvector corresponding to such an eigenvalue will contribute nontrivially to the kernel of $\Delta^0_-$ in the diagram (**O**). We conclude that for any $\varepsilon > 0$ there exists $R = R_\varepsilon > 0$ such that for all $r > R_\varepsilon$ we have

$$\|\mathfrak{L}_{\hat{\mathsf{C}}_r}(\mathbf{i}\hat{f})\|^2_{L^2(\hat{N}(r))} = \langle \mathfrak{L}^*_{\hat{\mathsf{C}}_r}\mathfrak{L}_{\hat{\mathsf{C}}_r}(\mathbf{i}\hat{f}), (\mathbf{i}\hat{f})\rangle \geq r^{-2-\varepsilon}\|\hat{f}\|^2_{L^2(\hat{N}(r))},$$

$\forall \hat{f} \in L^{1,2}(\hat{N}(r))$. ■

We left out one technical issue in the above discussion. More precisely, we cannot a priori eliminate the possibility that some of the monopoles constructed in Theorem 4.5.7 are gauge equivalent. It is true that they lie in the slice $\ker \mathfrak{L}^*_{\hat{\mathsf{C}}_r}$ but it is possible that the neighborhood in which they are situated is so large that one gauge orbit intersects it several times. We will now show that this is not the case by providing an *explicit, $r$-dependent estimate* of the diameter of the local slice at $\hat{\mathsf{C}}_r$.

**Lemma 4.5.9.** *There exists $r_0 > 0$ such that for all $r > r_0$ the configurations*

$$\hat{\mathsf{C}}_r + \Xi, \;\; \|\Xi\|_{L^{2,2}(\hat{N}(r))} \leq r^{-3}, \;\; \mathfrak{L}^*_{\hat{\mathsf{C}}_r}\Xi = 0$$

*are pairwise gauge inequivalent.*

**Proof** We argue by contradiction. We assume that for all $r > 0$ there exist

$$\hat{\gamma}_r \in \widehat{\mathfrak{G}}^{3,2}_{\hat{\sigma}} \neq 1$$

and $\Xi_{1,r} \neq \Xi_{2,r}$ such that

$$\text{(4.5.9)} \quad \hat{\gamma}_r \cdot \left(\hat{\mathsf{C}}_r + \Xi_{1,r}\right) = \hat{\mathsf{C}}_r + \Xi_{2,r}, \;\; \mathfrak{L}^*_{\hat{\mathsf{C}}_r}\Xi_{i,r} = 0, \;\; \|\Xi_{i,r}\|_{L^{2,2}(\hat{N}(r))} \leq r^{-3}.$$

Set $\hat{\mathsf{C}}_r =: (\hat{\psi}_r, \hat{A}_r)$, $\Xi_{i,r} =: (\underline{\hat{\psi}}_{i,r}, \mathbf{i}\hat{a}_{i,r})$ and $\Xi_r := \Xi_{2,r} - \Xi_{1,r}$. Observe that

$$\text{(4.5.10)} \qquad \|\Xi_{i,r}\|_{2,2} = O(r^{-3}) \text{ as } r \to \infty.$$

Denote by $c_r$ the average value of $\hat{\gamma}_r : \hat{N}(r) \to \mathbb{C}$. We can regard $c_r$ as the orthogonal projection of $\hat{\gamma}_r$ onto the kernel of $\hat{d} + \hat{d}^*$. Using the estimate in Exercise 4.1.6 of §4.1.6 we deduce

$$\|\hat{\gamma}_r - c_r\|_{L^2}^2 = O\Big(r^{1+\varepsilon}\|\hat{d}\hat{\gamma}_r\|_{L^2}^2\Big).$$

The equality (4.5.9) implies

$$2\hat{d}(\hat{\gamma}_r - c_r) = 2\hat{d}\hat{\gamma}_r = \mathbf{i}\hat{\gamma}_r(\hat{a}_{2,r} - \hat{a}_{1,r}) \tag{4.5.11}$$

so that

$$\|\hat{d}\hat{\gamma}_r\|_2 = O(r^{-3}).$$

Hence

$$\|\hat{\gamma}_r - c_r\|_2^2 = O(r^{-5+\varepsilon}). \tag{4.5.12}$$

Now use (4.5.10), (4.5.12) and interior elliptic estimates for the elliptic equation (4.5.11) to deduce that there exists $C > 0$ such that for any open set $U \subset \hat{N}(r)$ of diameter $< 1$ we have

$$\|\hat{\gamma}_r - c_r\|_{L^{3,2}(U)} \leq Cr^{-5/2+\varepsilon}.$$

Using the Sobolev embedding $L^{3,2}(U) \to L^\infty(U)$ (where the embedding constant can be chosen **independent** of $U$ and $r$) we deduce

$$\|\hat{\gamma}_r - c_r\|_{L^\infty(\hat{N}(r))} = O(r^{-5/2+\varepsilon}).$$

The last estimate shows that $\hat{\gamma}_r$ is very close (in the sup-norm) to being constant and thus it can be represented as

$$\hat{\gamma}_r = \exp(\mathbf{i}\hat{f}_r).$$

Denote by $c_r'$ the point on the unit circle $S^1 \subset \mathbb{C}$ and pick $\varphi_r \in [0, 2\pi]$ such that $\exp(\mathbf{i}\varphi_r) = c_r'$. Observe that we can choose $\hat{f}_r$ so that

$$\|\hat{f}_r - \varphi_r\|_{L^\infty(\hat{N}(r))} = O(r^{-5/2+\varepsilon}).$$

We can now rewrite (4.5.9) as

$$\mathbf{i}\hat{a}_{1,r} - 2\mathbf{i}\hat{d}\hat{f}_r = \mathbf{i}\hat{a}_{2,r}, \;\; \exp(\mathbf{i}\hat{f}_r)(\psi_r + \underline{\hat{\psi}}_{1,r}) = \hat{\psi}_r + \underline{\hat{\psi}}_{2,r}.$$

These two equalities have to be supplemented by the slice conditions

$$0 = \mathfrak{L}^*_{\hat{\mathsf{C}}_r}(\Xi_{i,r}) = 2\hat{d}^*\hat{a}_{i,r} + \mathfrak{Im}\langle\hat{\psi}_r, \underline{\hat{\psi}}_{i,r}\rangle.$$

A simple computation leads to the equality

$$-4\hat{d}^*\hat{d}\hat{f}_r + \mathfrak{Im}\langle\hat{\psi}_r, (e^{\mathbf{i}\hat{f}_r} - 1)(\hat{\psi}_r + \underline{\hat{\psi}}_{2,r})\rangle.$$

We can further rewrite the above as

$$4\hat{d}^*\hat{d}\hat{f}_r = -\sin(\hat{f}_r)|\hat{\psi}_r|^2 + \mathfrak{Im}\langle\hat{\psi}_r, (e^{\mathbf{i}\hat{f}_r} - 1)\underline{\hat{\psi}}_{2,r}\rangle. \tag{4.5.13}$$

Set $\xi_r := \hat{f}_r - \varphi_r$. We have

$$4\hat{d}^*\hat{d}\xi_r = -\sin(\varphi_r)|\hat{\psi}_r|^2 - \big(\sin(\hat{f}_r) - \sin(\varphi_r)\big)|\hat{\psi}_r|^2 + \mathfrak{Im}\langle\hat{\psi}_r, (e^{\mathbf{i}\hat{f}_r} - 1)\hat{\underline{\psi}}_{2,r}\rangle.$$

Multiply the last equality by 1 and integrate by parts over $\hat{N}(r)$. Since $\|\hat{\underline{\psi}}_{2,r}\|_{L^2} = O(r^{-3})$ and $\|\sin(\hat{f}_r) - \sin(\varphi_r)\|_{L^\infty} = O(r^{-5/2+\varepsilon})$ we deduce

$$|\sin(\varphi_r)| \int_{\hat{N}(r)} |\hat{\psi}_r|^2 d\,\mathrm{vol} = O\Big(r^{-5/2+\varepsilon}\Big) \int_{\hat{N}(r)} |\hat{\psi}_r|^2 d\,\mathrm{vol}\Big).$$

Thus

$$|\sin(\varphi_r)| = O(r^{-5/2+\varepsilon}).$$

Thus either $|\varphi_r| = O(r^{-5/2+\varepsilon})$ or $|\varphi_r - \pi| = O(r^{-5/2+\varepsilon})$. We can exclude the second possibility by using the equality

$$e^{\mathbf{i}\hat{f}_r}(\hat{\psi}_r + \hat{\underline{\psi}}_{1,r}) = \hat{\psi}_r + \hat{\underline{\psi}}_{2,r}$$

and the fact that $\hat{\psi}_r$ does not vanish identically; better yet, $\|\hat{\psi}_r\|_\infty$ is bounded away from zero independent of $r$. (Recall that $\hat{\mathsf{C}}_r$ is an almost monopole obtained by gluing two finite energy monopoles at least one of which was irreducible.) Hence

$$\|\hat{f}_r\|_{L^\infty} = O(r^{-5/2+\varepsilon}). \tag{4.5.14}$$

We can rewrite the equality (4.5.13) as

$$\begin{aligned} \mathfrak{L}^*_{\hat{\mathsf{C}}_r}\mathfrak{L}_{\hat{\mathsf{C}}}(\hat{f}_r) &= 4\hat{d}^*\hat{d}\hat{f}_r + |\hat{\psi}_r|\hat{f}_r \\ &= (\hat{f}_r - \sin(\hat{f}_r))|\hat{\psi}_r|^2 + \mathfrak{Im}\langle\hat{\psi}_r, (e^{\mathbf{i}\hat{f}_r} - 1)\hat{\underline{\psi}}_{2,r}\rangle. \end{aligned} \tag{4.5.15}$$

Using interior elliptic estimates for the above equation we deduce that there exists $C > 0$ such that if $U \subset \hat{N}(r)$ is an open subset of diameter $< 1$ then

$$\|\hat{f}_r\|_{L^{3,2}(U)} \leq Cr^{-5/2+\varepsilon}. \tag{4.5.16}$$

Multiplying the equality (4.5.15) by $\hat{f}_r$ we deduce

$$\langle\Delta_{\hat{\mathsf{C}}_r}\hat{f}_r, \hat{f}_r\rangle \leq Cr^{-5/2+\varepsilon}\|\hat{f}_r\|^2_{L^2}.$$

Using the eigenvalue estimate in Remark 4.5.8 we deduce

$$\|\hat{f}_r\|^2_{L^2} \leq Cr^{2+\varepsilon}\langle\Delta_{\hat{\mathsf{C}}_r}\hat{f}_r, \hat{f}_r\rangle_{L^2}.$$

The last two estimates contradict each other for $r \gg 0$. This concludes the proof of Lemma 4.5.9. ■

We have thus proved the following result.

**Corollary 4.5.10.** *There exist $r_1 > 0$ and for every $r > r_1$ an open neighborhood $U_r$ of $0 \in \mathcal{V}_r$ such that the set*

$$\Big\{\hat{\mathsf{C}}_r + \underline{\hat{\mathsf{C}}};\ \mathfrak{L}^*_{\hat{\mathsf{C}}_r}\underline{\hat{\mathsf{C}}} = 0,\ \ \underline{\hat{\mathsf{C}}} = \underline{\hat{\mathsf{C}}}_0 + \Phi(\underline{\hat{\mathsf{C}}}_0),\ \ \underline{\hat{\mathsf{C}}}_0 \in U_r,\ \ \kappa_r(\underline{\hat{\mathsf{C}}}_0) = 0\Big\}$$

*is homeomorphic to an open set in the moduli space $\widehat{\mathfrak{M}}_{\hat{\sigma}_1\#\hat{\sigma}_2}$.* ■

We will refer to the open subsets of $\widehat{\mathfrak{M}}_{\hat{\sigma}_1\#\hat{\sigma}_2}$ described in the above corollary as *splitting neighborhoods.*

**Remark 4.5.11.** The choice of size $r^{-3}$ in the definition of $\Phi_r$ and $\kappa_r$ is by no means unique or natural. Our proof shows that if we replace $r^{-3}$ by $r^{-n}$, $n \geq 3$, everywhere in the statement of Theorem 4.5.7 we will still get a valid result.

To give the reader an idea of the strength of the gluing theorem we consider several special cases.

**Example 4.5.12.** *Both $\hat{\mathsf{C}}_1$ and $\hat{\mathsf{C}}_2$ are irreducible, strongly regular and $\hat{\mathsf{C}}_\infty$ is irreducible.* In this case, the middle column in (**O**) is identically zero and we deduce that the obstruction space $\mathcal{H}_r^-$ is trivial. Thus, $\kappa_r \equiv 0$ and the set of monopoles close to $\hat{\mathsf{C}}_r := \hat{\mathsf{C}}_1\#_r\hat{\mathsf{C}}_2$ can be represented as the graph of a smooth map

$$\Phi : B_0(r^{-3}) \subset \mathcal{H}_r^+ \to \mathfrak{Y}^2_+(r)$$

where $\underline{\hat{\mathsf{C}}}$ is implicitly defined by the fixed point equation (4.5.7). Moreover, the dimension and location of $\mathcal{H}_r^+$ can be determined from the diagram (**T**), which in this case simplifies to

$$0 \to \mathcal{H}_r^+ \longrightarrow^a H^1_{\hat{\mathsf{C}}_1} \oplus H^1_{\hat{\mathsf{C}}_2} \overset{\Delta}{\longrightarrow}{}^a\, T_{\mathsf{C}_\infty}\mathfrak{M}_\sigma \to 0.$$

To see why $L_1^+ + L_2^+ = T_{\mathsf{C}_\infty}\mathfrak{M}_\sigma$ observe that in our special case we have $L_i^- = 0$ and thus, using (**L**), we conclude $L_i^+ = T_{\hat{\mathsf{C}}_\infty}\mathfrak{M}_\sigma$. The smooth manifold filled by the monopoles close to $\hat{\mathsf{C}}_r$ has dimension

$$d(\hat{\mathsf{C}}_1) + d(\hat{\mathsf{C}}_2) - d(\mathsf{C}_\infty).$$

Observe that all the monopoles on $\hat{N}(r)$ constructed in this way are *regular.*

**Example 4.5.13.** *Both $\hat{\mathsf{C}}_1$ and $\hat{\mathsf{C}}_2$ are irreducible, strongly regular but $\hat{\mathsf{C}}_\infty$ is reducible.* The obstruction space $\mathcal{H}_r^-$ is trivial and the monopoles near $\hat{\mathsf{C}}_r$ form a manifold of the same dimension as $\mathcal{H}_r^+$, which is

$$d(\hat{\mathsf{C}}_1) + d(\hat{\mathsf{C}}_2) - d(\mathsf{C}_\infty) + \dim G_\infty.$$

Again, all the monopoles near $\hat{\mathsf{C}}_r$ are irreducible and regular.

**Example 4.5.14.** Suppose *both* $\hat{\mathsf{C}}_i$ *are strongly regular,* $\hat{\mathsf{C}}_1$ *is irreducible but* $\hat{\mathsf{C}}_2$ *is reducible.* Again we deduce that the obstruction space $\mathcal{H}_r$ vanishes. The monopoles near $\hat{\mathsf{C}}_r$ form a manifold of dimension

$$\dim \mathcal{H}_r^+ = d(\hat{\mathsf{C}}_1) + d(\hat{\mathsf{C}}_2) - d(\mathsf{C}_\infty) + \dim G_\infty.$$

Set

$$d(\hat{\mathsf{C}}_1)\#d(\hat{\mathsf{C}}_2) := d(\hat{\mathsf{C}}_1) + d(\hat{\mathsf{C}}_2) - d(\mathsf{C}_\infty) + \dim G_\infty.$$

The above three examples show that if both $\hat{\mathsf{C}}_i$ are strongly regular and at least one is irreducible then the set of monopoles near $\hat{\mathsf{C}}_r$ is a smooth manifold of dimension $d(\hat{\mathsf{C}}_1)\#d(\hat{\mathsf{C}}_2)$. All these monopoles are both irreducible and regular. We can formally write

$$d(\hat{\mathsf{C}}_1\#_r\hat{\mathsf{C}}_2) = d(\hat{\mathsf{C}}_1)\#d(\hat{\mathsf{C}}_2).$$

**4.5.3. The local surjectivity of the gluing construction.** The gluing process described in the previous subsection constructed certain open subsets (splitting neighborhoods) of the moduli spaces of monopoles on a 4-manifold with a very long neck. This splitting process we are about to present will show that if the 4-manifold is sufficiently stretched then these splitting neighborhoods cover the *entire* moduli space.

Consider again the Riemannian manifold $\hat{N}(r)$ introduced in the previous subsection. If $\hat{\mathsf{C}} = (\hat{\psi}, \hat{A})$ is a monopole on $\hat{N}(r)$ then, according to Proposition 2.1.4, its energy

$$E(\hat{\mathsf{C}}) := \int_{\hat{N}(r)} \Big( |\hat{\nabla}^{\hat{A}}\hat{\psi}|^2 + \frac{1}{8}|q(\hat{\psi})|^2 + |F_{\hat{A}}|^2 + \frac{\hat{s}(\hat{g}_r)}{4}|\hat{\psi}|^2 \Big) dv(\hat{g}_r)$$

is a topological invariant, depending only on the *spin*$^c$ structure and not on the metric. On the other hand, $\|\hat{s}(\hat{g}_r)\|_{L^\infty}$ is independent of $r$ and because $\|\hat{\psi}\|_{L^\infty} \leq 2\|\hat{s}(\hat{g}_r)\|_{L^\infty}$ we deduce that the energy of $\hat{\mathsf{C}}$ on any open set of $\hat{N}(r)$ of volume $O(1)$ as $r \to \infty$ is $O(1)$ as $r \to \infty$. If we take this open set to be the complement of the long neck we conclude that the energy of $\hat{\mathsf{C}}$ on the long neck is bounded from above by a constant independent of $r$.

The discussion in §4.4.2 shows that any sequence $(\hat{\mathsf{C}}_n)$ of monopoles on $\hat{N}(r_n \to \infty)$ splits as $n \to \infty$ into a chain

$$\hat{\mathsf{C}}_0, \hat{\mathsf{C}}_1, \hat{\mathsf{C}}_2, \cdots, \hat{\mathsf{C}}_k, \hat{\mathsf{C}}_{k+1}$$

where $\hat{\mathsf{C}}_0$ is a finite energy monopole on $\hat{N}_1$, $\hat{\mathsf{C}}_{k+1}$ is a finite energy monopole on $\hat{N}_2$ and $\hat{\mathsf{C}}_1, \cdots, \hat{\mathsf{C}}_k$ are tunnelings on $\mathbb{R} \times N$ such that

$$\partial_\infty^+ \hat{\mathsf{C}}_i = \partial_\infty^- \hat{\mathsf{C}}_{i+1}.$$

Assume for simplicity that **tunnelings do not exist**. We deduce that the moduli spaces of finite energy monopoles on $\hat{N}_i$ are compact and, moreover, as $r \to \infty$ the monopoles on $\hat{N}(r)$ will split into a pair of finite energy monopoles $\hat{\mathsf{C}}_1$ and $\hat{\mathsf{C}}_2$ with matching asymptotic limits, $\partial_\infty \hat{\mathsf{C}}_1 = \partial_\infty \hat{\mathsf{C}}_1 \in \mathfrak{M}_\sigma$. Denote by $\mathcal{P}$ the set of such pairs.

Given such a pair $(\hat{\mathsf{C}}_1, \hat{\mathsf{C}}_2)$, the local gluing theorem postulates the existence of $r_0 = r_0(\hat{\mathsf{C}}_1, \hat{\mathsf{C}}_2) > 0$ and for each $r > r_0$ the existence of an open set $\mathcal{U}_{\hat{\mathsf{C}}_1, \hat{\mathsf{C}}_2, r} \subset \widehat{\mathfrak{M}}_{\hat{\sigma}}(\hat{g}_r)$ with the property

$$\mathcal{U}_{\hat{\mathsf{C}}_1, \hat{\mathsf{C}}_2, r} = \Big\{ \hat{\mathsf{C}} \in \widehat{\mathfrak{M}}_{\hat{\sigma}}(\hat{g}_r);\ \ \mathrm{dist}_{L^{2,2}}([\hat{\mathsf{C}}], [\hat{\mathsf{C}}_1 \#_r \hat{\mathsf{C}}_2]) < r^{-3} \Big\}.$$

Since $\mathcal{P}$ is compact we deduce that there exists $R_0 > 0$ such that

$$r_0(\hat{\mathsf{C}}_1, \hat{\mathsf{C}}_2) < R_0, \ \ \forall (\hat{\mathsf{C}}_1, \hat{\mathsf{C}}_2) \in \mathcal{P}.$$

For each $r > R_0$ we set

$$\mathcal{U}_r := \bigcup_{(\hat{\mathsf{C}}_1, \hat{\mathsf{C}}_2) \in \mathcal{P}} \mathcal{U}_{\hat{\mathsf{C}}_1, \hat{\mathsf{C}}_2, r} \subset \widehat{\mathfrak{M}}_{\hat{\sigma}}(\hat{g}_r).$$

We can now state the main result of this subsection.

**Theorem 4.5.15.** *Assume $\hat{N}_1$ and $\hat{N}_2$ are equipped with* **real analytic** *structures. Then there exists $R_1 > 0$ such that*

$$\mathcal{U}_r = \widehat{\mathfrak{M}}_{\hat{\sigma}}(\hat{g}_r), \ \ \forall r > R_1.$$

**Sketch of proof** The method we will employ in the proof is a substantially sharper variation of the strategy used in [**26**, Sec. 2.2] to establish a similar fact.

Consider a sequence $\hat{\mathsf{C}}_{r \to \infty}$ of monopoles on $\hat{N}(r)$ which splits as $r \to \infty$ to a pair $(\hat{\mathsf{C}}_1, \hat{\mathsf{C}}_2) \in \mathcal{P}$. Let us explain in some detail the meaning of this statement.

Identify the long neck of $\hat{N}(r)$ with the long cylinder $[-r, r] \times N$. The splitting implies that there exists $\tau > 0$ *independent of* $r$ with the following property: if we denote by $\hat{\mathsf{C}}_1^r$ (resp. $\hat{\mathsf{C}}_2^r$) the restriction of $\hat{\mathsf{C}}_r$ to the portion of $\hat{N}(r)$ containing $[-r, \tau] \times N$ (resp. $[\tau, r] \times N$) then $\hat{\mathsf{C}}_r^i$ converges in $L^{2,2}_{loc}$ to $\hat{\mathsf{C}}_i$ (with the additional uniformity explained in Remark 4.4.11). Denote by $\hat{G}_i$ the stabilizer of $\hat{\mathsf{C}}_i$.

We want to prove that for all $r \gg 0$ there exists $(\hat{\mathsf{C}}_1(r), \hat{\mathsf{C}}_2(r)) \in \mathcal{P}$ such that

$$\partial_\infty \hat{\mathsf{C}}_i = \partial_\infty \hat{\mathsf{C}}_i(r) =: \mathsf{C}_\infty$$

and

$$\mathrm{dist}_{L^{2,2}}([\hat{\mathsf{C}}_r], [\hat{\mathsf{C}}_1(r)\#_r\hat{\mathsf{C}}_2(r)]) < r^{-3}.$$

Assume for simplicity that $\tau = 0$. It will be convenient to regard $\hat{\mathsf{C}}_i$ as monopoles on the truncated manifold $\hat{N}_i(r) = \hat{N}_i \setminus (r, \infty) \times N$.

Define the configurations $\hat{\mathsf{C}}_{i,r} \in \hat{\mathcal{C}}_{\mu,sw}(\hat{N}_i)$ by

$$\hat{\mathsf{C}}_{i,r} = \alpha_r \hat{\mathsf{C}}_i^r + (1 - \alpha_r)\mathsf{C}_\infty$$

where $\alpha_r = \alpha(t - r + 1)$ and $\alpha$ is depicted in Figure 4.4 of 4.1.4.

Using the estimate (4.2.35) in Remark 4.2.29 of §4.2.4 coupled with the uniform $L^{2,2}_{loc}$-convergence of $\hat{\mathsf{C}}_i^r$ we deduce after some elementary manipulations that

$$\|\widehat{SW}(\hat{\mathsf{C}}_{i,r})\|_{L^{1,2}_\mu} = O(e^{-\mu r}), \quad \mathrm{dist}_{L^{2,2}_\mu}(\hat{\mathsf{C}}_i, \hat{\mathsf{C}}_{i,r}) = o(1) \quad \text{as } r \to \infty. \tag{4.5.17}$$

**Exercise 4.5.4.** Prove the above estimates.

**Hint:** Consult [**26**, Sec. 2.2] for inspiration.

To proceed further we need to use some of the constructions (and notation) in §4.3.1 and §4.3.2. Denote by $\hat{\mathcal{S}}_i$ the global "slice"

$$\hat{\mathcal{S}}_i = \ker \mathfrak{L}^{*_\mu}_{\hat{\mathsf{C}}_i} \cap L^{2,2}_\mu.$$

Using Proposition 4.3.7 we deduce that there exists a $L^{2,2}_\mu$-small neighborhood $\hat{V}_i$ of $0 \in \hat{\mathcal{S}}_i$ such that every orbit of $\widehat{\mathfrak{G}}_\mu$ on $\hat{\mathcal{C}}_{\mu,sw}(\hat{N}_i)$ intersects $\hat{\mathsf{C}}_i + \hat{V}_i$ along at most one point. Modulo $\widehat{\mathfrak{G}}_\mu$ we can assume that $\hat{\mathsf{C}}_{i,r} \in \hat{\mathsf{C}}_i + \hat{V}_i$. Set $\Xi_{i,r} := \hat{\mathsf{C}}_{i,r} - \hat{\mathsf{C}}_i \in \hat{\mathcal{S}}_i$.

Now denote by $\mathfrak{Y}_i^+ \subset \hat{\mathcal{S}}_i$ the $L^2_\mu$-orthogonal complement of $H^1(F_{\hat{\mathsf{C}}_i})$ in $\hat{\mathcal{S}}_i$, by $\mathfrak{Y}_i^-$ the $L^2_\mu$-orthogonal complement of $H^2(F_{\hat{\mathsf{C}}_0})$ in its natural ambient space and by $\widehat{\mathfrak{M}}_i(\mathsf{C}_\infty)$ the moduli space of $\widehat{\mathfrak{G}}_\mu$-equivalence classes of finite energy monopoles $\hat{\mathsf{C}}$ on $\hat{N}_i$ such that $\partial_\infty \hat{\mathsf{C}} = \mathsf{C}_\infty$. We have the usual Kuranishi local description of a neighborhood of $\hat{\mathsf{C}}_i$ in $\widehat{\mathfrak{M}}_i(\mathsf{C}_\infty)$. More precisely, there exist a small neighborhood $U_i$ of $0 \in H^1(F_{\hat{\mathsf{C}}_i})$, a smooth map

$$\Phi_i : U_i \to \mathfrak{Y}_i, \quad \Phi_i(0) = 0$$

and a **real analytic map** $\kappa_i : U_i \to H^2(F_{\hat{\mathsf{C}}_2})$ such that the set

$$\Big\{\hat{\mathsf{C}}_i + u + \Phi_i(u); \quad u \in U_i, \ \kappa_i(u) = 0\Big\}$$

is homeomorphic to an open neighborhood of $\hat{\mathsf{C}}_i \in \widehat{\mathfrak{M}}_i(\mathsf{C}_\infty)$. Moreover, there exists $C > 0$ such that

$$\|\kappa_i(u)\|_{L^{1,2}_\mu} \leq C\|\widehat{SW}(\hat{\mathsf{C}}_i + u)\|_{L^{1,2}_\mu}, \quad \forall u \in U_i. \tag{4.5.18}$$

**Exercise 4.5.5.** Use the fixed-point strategy in the proof of Theorem 4.5.7 to establish (4.5.18).

Decompose $\Xi_{i,r} = \Xi^0_{i,r} + \Xi^\perp_{i,r} \in H^1(F_{\hat{\mathsf{C}}_i}) \oplus \mathfrak{Y}_i^+$. Since $\widehat{SW}(\hat{\mathsf{C}}_i + \Xi_{i,r}) = O(e^{-\mu r})$ and $\hat{\mathcal{T}}_{\hat{\mathsf{C}}_i,\mu} = \underline{\widehat{SW}}_{\hat{\mathsf{C}}_i} \oplus \mathfrak{L}^{*_\mu}_{\hat{\mathsf{C}}_i}$ has closed range we deduce

$$\|\Xi^\perp_{i,r}\|_{L^{2,2}_\mu} = O(e^{-\mu r}).$$

Thus

$$\|\widehat{SW}(\hat{\mathsf{C}}_i + \Xi^0_{i,r})\|_{L^{1,2}_\mu} = O(e^{-\mu r}). \tag{4.5.19}$$

The iterative construction of $\Phi_i$ via the Banach fixed point theorem shows that for every $u \in U_i$ and every sufficiently small $\Xi^\perp \in \mathfrak{Y}_i^+$ we have

$$\|\Phi_i(u) - \Xi^\perp\|_{L^{2,2}_\mu} \leq C\|Q_-\widehat{SW}(\hat{\mathsf{C}}_i + u + \Xi^\perp)\|_{L^{1,2}}$$

where $Q_-$ denotes the orthogonal projection onto $\mathfrak{Y}_i^-$. In particular, we deduce that

$$\|\Phi_i(\Xi^0_{i,r})\|_{L^{2,2}_\mu} = O(e^{-\mu r}). \tag{4.5.20}$$

The estimates (4.5.18) and (4.5.19) imply that

$$\|\kappa_i(\Xi_{i,r})\| = O(e^{-\mu r}).$$

Since $\kappa_r$ is real analytic we can use Łojasewicz' inequality (see [**15, 86**]) to deduce that there exists $p > 0$ such that

$$\operatorname{dist}(\Xi^0_{i,r}, k_i^{-1}(0)) = O(\|\kappa_i(\Xi^0_{i,r})\|^p) = O(e^{-p\mu r}) \quad \text{as} \quad r \to \infty.$$

Using (4.5.20) we can now conclude that

$$\operatorname{dist}_{L^{2,2}_\mu}(\hat{\mathsf{C}}_{i,r}, \widehat{\mathfrak{M}}_i(\mathsf{C}_\infty)) = O(e^{-cr})$$

for some $c > 0$. Thus, we can find $\hat{\mathsf{C}}_i(r) \in \widehat{\mathfrak{M}}_i(\mathsf{C}_\infty)$ such that

$$\operatorname{dist}_{L^{2,2}_\mu}(\hat{\mathsf{C}}_{i,r}, \hat{\mathsf{C}}_i(r)) = O(e^{-cr}).$$

This implies immediately that there exists $R_1 > 0$ which depends only on the geometry of $\hat{N}_i$ such that for all $r > R_1$ we have

$$\operatorname{dist}_{L^{2,2}}(\hat{\mathsf{C}}_r, \hat{\mathsf{C}}_1(r)\#_r\hat{\mathsf{C}}_2(r)) \leq Ce^{-cr} < r^{-3}.$$

This completes the proof of Theorem 4.5.15. ■

**4.5.4. Gluing monopoles: Global theory.** It is now the time to put together the facts established in the previous two subsections. There is a wide range of situations possible and we will not attempt to formulate the most general result. In this subsection we will deal only with two generic situations which display most of the relevant features of the general gluing problem.

Again we consider the cylindrical manifolds $(\hat{N}_1, \hat{g}_1)$ and $(\hat{N}_2, \hat{g}_2)$ with $N = \partial_\infty \hat{N}_i$, $g := \partial_\infty \hat{g}_i$ together with compatible cylindrical *spin*$^c$ structures $\hat{\sigma}_i$, $\partial_\infty \hat{\sigma}_1 = \partial_\infty \hat{\sigma}_2 = \sigma$. We form as before the Riemannian manifold $(\hat{N}(r), \hat{g}_r)$ with a long cylindrical neck.

**CASE 1.** We will first consider the situation characterized by the following conditions.

$\mathbf{A}_1$ *$(g, \sigma)$ is good.*

$\mathbf{A}_2$ *There exist no $(g, \sigma)$-tunnelings on $\mathbb{R} \times N$.*

$\mathbf{A}_3$ $b_+(\hat{N}_i) > 0$.

$\mathbf{A}_4$ *All finite energy monopoles on $\hat{N}_i$ are irreducible and strongly regular.*

Observe that $\mathbf{A}_1$ and $\mathbf{A}_2$ are automatically satisfied if $g$ has positive scalar curvature. The genericity discussion in §4.4.1 shows that we can arrange so that $\mathbf{A}_4$ is fulfilled using generic compactly supported perturbations of the Seiberg-Witten equations. We need to introduce some notation.

- $\hat{\mathfrak{Z}}_i \subset \hat{\mathcal{C}}_{\mu,ex}(\hat{N}_i, \hat{\sigma}_i)$ – the set of finite energy $\hat{\sigma}_i$-monopoles on $\hat{N}_i$.
- $\mathfrak{Z} \subset \mathcal{C}_\sigma(N)$ – monopoles on $N$.
- $\widehat{\mathfrak{G}}_i^{\hat{N}}$ – the subgroup of $\widehat{\mathfrak{G}}_{\mu,ex}(\hat{N}_i)$ consisting of gauge transformations which extend (in an intuitively clear sense) to $\hat{N}(r)$, $r \gg 0$.
- $\mathfrak{G}^{\hat{N}}$ – the subgroup of $\mathfrak{G}_\sigma(N)$ consisting of gauge transformations which extend to $\hat{N}(r)$.
- $\widehat{\mathfrak{M}}_i^{\hat{N}} = \mathfrak{Z}_i / \widehat{\mathfrak{G}}_i^{\hat{N}}$, $\mathfrak{M}_\sigma^{\hat{N}} = \mathfrak{Z} / \mathfrak{G}^{\hat{N}}$.

We get a map

$$\partial_\infty \times \partial_\infty : \hat{\mathfrak{Z}}_1 \times \hat{\mathfrak{Z}}_2 \to \mathfrak{Z} \times \mathfrak{Z}$$

and we set

$$\hat{\mathfrak{Z}}_\Delta := (\partial_\infty \times \partial_\infty)^{-1}(\Delta_{\mathfrak{Z}}) \subset \hat{\mathfrak{Z}}_1 \times \hat{\mathfrak{Z}}_2.$$

Clearly, $\hat{\mathfrak{Z}}_\Delta$ consists of pairs of compatible finite energy monopoles. The diagonal $\Delta_{\mathfrak{Z}}$ decomposes into a reducible and irreducible part

$$\Delta_{\mathfrak{Z}} = \Delta_{\mathfrak{Z}}^{red} \cup \Delta_{\mathfrak{Z}}^{irr}.$$

We get a corresponding decomposition

$$\hat{\mathfrak{Z}}_\Delta = \hat{\mathfrak{Z}}_{\Delta_{red}} \cup \hat{\mathfrak{Z}}_{\Delta_{irr}}$$

where $\Delta_{\mathfrak{Z}}$ denotes the diagonal of $\mathfrak{Z} \times \mathfrak{Z}$.

Define

$$\widehat{\mathfrak{G}}_\Delta := \{(\hat{\gamma}_1, \hat{\gamma}_2) \in \widehat{\mathfrak{G}}_{\mu,ex}(\hat{N}_1) \times \widehat{\mathfrak{G}}_{\mu,ex}(\hat{N}_2);\ \ \partial_\infty \hat{\gamma}_1 \cdot (\partial_\infty \hat{\gamma}_2)^{-1} = 1\}.$$

The group $\widehat{\mathfrak{G}}_\Delta$ acts freely on $\hat{\mathfrak{Z}}_\Delta$. Correspondingly, we get a quotient

$$\widehat{\mathfrak{M}}_\Delta := \hat{\mathfrak{Z}}_\Delta / \widehat{\mathfrak{G}}_\Delta.$$

The gluing map produces a family of group morphisms

$$\#_r : \widehat{\mathfrak{G}}_\Delta \to \widehat{\mathfrak{G}}(\hat{N}_r)$$

and a family of smooth maps

$$\#_r : \mathfrak{Z}_\Delta \to \hat{\mathcal{C}}(\hat{N}_r).$$

Since $\#_r$ commutes with the exterior derivative (see Exercise 4.5.2 in §4.5.1) we deduce that these maps are almost-equivariant in the sense that

$$\#_r(\hat{\gamma}_1 \cdot \hat{\mathsf{C}}_1, \hat{\gamma}_2 \cdot \hat{\mathsf{C}}_2) = (\hat{\gamma}_1 \#_r \hat{\gamma}_2) \cdot (\hat{\mathsf{C}}_1 \#_r \hat{\mathsf{C}}_2),$$

$\forall (\hat{\gamma}_1, \hat{\gamma}_2) \in \widehat{\mathfrak{G}}_\Delta$, $(\hat{\mathsf{C}}_1, \hat{\mathsf{C}}_2) \in \mathfrak{Z}_\Delta$. We deduce that the gluing map descends to a well defined map

$$\#_r : \widehat{\mathfrak{M}}_\Delta \to \hat{\mathcal{C}}(\hat{N}_r)/\widehat{\mathfrak{G}}(\hat{N}_r).$$

The quotient $\widehat{\mathfrak{M}}_\Delta$ decomposes as an union

$$\widehat{\mathfrak{M}}_\Delta = \widehat{\mathfrak{M}}_{\Delta_{red}} \cup \widehat{\mathfrak{M}}_{\Delta_{irr}}.$$

This construction has a based counterpart. We can regard a based configuration on $\hat{N}_i$ as a pair $(\hat{\mathsf{C}}_i, \xi_i)$ where $\hat{\mathsf{C}}_i \in \hat{\mathcal{C}}_{\mu,sw}(\hat{N}_i)$ and $\xi_i \in S^1$. The group $\widehat{\mathfrak{G}}_{\mu,ex}(\hat{N}_i)$ acts on such a pair according to the rule

$$\hat{\gamma} \cdot (\hat{\mathsf{C}}_i, \xi_i) = (\hat{\gamma} \cdot \hat{\mathsf{C}}_i, \hat{\gamma}(*)^{-1}\xi_i)$$

where $*$ is a fixed point in $\partial_\infty \hat{N}_i$. On $\hat{N}(r)$ we choose the base point $*$ to lie on the mid-section $\{0\} \times N$ of the long neck $[-r, r] \times N$.

Denote by $\mathfrak{Z}_i^\flat$ the set of based, finite energy monopoles on $\hat{N}_i$ and by $\mathfrak{Z}^\flat$ the set of based monopoles on the separating hypersurface $N$. We get a map

$$\partial_\infty^\flat \times \partial_\infty^\flat : \mathfrak{Z}_1^\flat \times \mathfrak{Z}_2^\flat \to \mathfrak{Z}^\flat \times \mathfrak{Z}^\flat$$

and we denote by

$$\mathfrak{Z}^\flat_\Delta := (\partial^\flat_\infty \times \partial^\flat_\infty)^{-1}(\Delta^\flat)$$

the set of compatible, finite energy based monopoles.

The gluing operation $\#_r$ between compatible finite energy monopoles has an obvious $S^1$-equivariant, based counterpart. Given $(\hat{\mathsf{C}}_1, \hat{\mathsf{C}}_2, \xi_\infty) \in \mathfrak{Z}^\flat_\Delta$ we can form the based almost monopole

$$\hat{\mathsf{C}}_r := (\hat{\mathsf{C}}_1 \#_r \hat{\mathsf{C}}_2, \xi_\infty)$$

on $\hat{N}(r)$. In this fashion we obtain a set $\#_r \mathfrak{Z}^\flat_\Delta \subset \hat{\mathcal{B}}_{\hat{\sigma},irr}(*)$ of *based* gauge orbits on $\hat{N}(r)$ and a surjective map

$$\#^\flat_r : \mathfrak{Z}^\flat_\Delta / \widehat{\mathfrak{G}}_\Delta(*) =: \widehat{\mathfrak{M}}_\Delta(*) \to \Big(\widehat{\mathfrak{G}}(\hat{N}_r, *) \cdot \#_r \mathfrak{Z}^\flat_\Delta\Big) / \widehat{\mathfrak{G}}(\hat{N}_r, *).$$

Clearly

$$\#_r \mathfrak{Z}_\Delta := \#_r \mathfrak{Z}^\flat_\Delta / S^1.$$

The space $\widehat{\mathfrak{M}}^{\hat{N}}_i$ is a cover of $\widehat{\mathfrak{M}}_\mu(\hat{N}_i)$ while $\mathfrak{M}^{\hat{N}}_\sigma$ is a cover of $\mathfrak{M}_\sigma$. We have natural maps

$$\partial_\infty : \widehat{\mathfrak{M}}^{\hat{N}}_i \to \mathfrak{M}^{\hat{N}}_\sigma, \quad \partial^\flat_\infty : \widehat{\mathfrak{M}}^{\hat{N}}_i(*) \to \mathfrak{M}^{\hat{N}}_\sigma(*).$$

The map $\partial^\flat_\infty$ is $S^1$-equivariant. Denote by $\Delta_\sigma$ the diagonal in $\mathfrak{M}^{\hat{N}}_\sigma \times \mathfrak{M}^{\hat{N}}_\sigma$, by $\Delta^\flat_\sigma$ the diagonal in $\mathfrak{M}^{\hat{N}}_\sigma(*) \times \mathfrak{M}^{\hat{N}}_\sigma(*)$ and by $\pi$ the Cartesian product

$$\pi = \partial_\infty \times \partial_\infty : \widehat{\mathfrak{M}}^{\hat{N}}_1 \times \widehat{\mathfrak{M}}^{\hat{N}}_2 \to \mathfrak{M}^{\hat{N}}_\sigma \times \mathfrak{M}^{\hat{N}}_\sigma.$$

Similarly we have

$$\pi_\flat : \partial^\flat_\infty \times \partial^\flat_\infty : \widehat{\mathfrak{M}}^{\hat{N}}_1(*) \times \widehat{\mathfrak{M}}^{\hat{N}}_2(*) \to \mathfrak{M}^{\hat{N}}_\sigma(*) \times \mathfrak{M}^{\hat{N}}_\sigma(*).$$

Assumption $\mathbf{A}_4$ implies that $\pi$ is transversal to $\Delta_\sigma$ so that the fiber product $\pi^{-1}(\Delta_\sigma)$ is a smooth manifold. $\Delta_\sigma$ decomposes into a reducible and irreducible part

$$\Delta_\sigma = \Delta_{\sigma,red} \cup \Delta_{\sigma,irr}.$$

The next lemma, which is a consequence of the irreducibility of the monopoles on $\hat{N}_i$, relates the various objects defined above.

**Lemma 4.5.16.** *There exist natural isomorphisms*

$$\widehat{\mathfrak{M}}_{\Delta^{irr}} \cong \pi^{-1}(\Delta_{\sigma,irr}) \subset \widehat{\mathfrak{M}}^{\hat{N}}_1 \times \widehat{\mathfrak{M}}^{\hat{N}}_2,$$

$$\widehat{\mathfrak{M}}_{\Delta^{red}} \cong \pi_\flat^{-1}(\Delta_{\sigma,red})/S^1 \subset (\widehat{\mathfrak{M}}^{\hat{N}}_1(*) \times \widehat{\mathfrak{M}}^{\hat{N}}_2(*))/S^1$$

*where $S^1$ acts diagonally on the product.*

**Exercise 4.5.6.** Prove the above lemma.

The isomorphisms in the above lemma can be combined in a single one

$$\widehat{\mathfrak{M}}_{\Delta}(*) \cong \pi_{\flat}^{-1}(\Delta_{\sigma}^{\flat}).$$

We have an $S^1$-equivariant gluing map

$$\pi_{\flat}^{-1}(\Delta_{\sigma}) \cong \widehat{\mathfrak{M}}_{\Delta}(*) \xrightarrow{\#_r} \mathcal{B}_{\hat{\sigma},irr}(\hat{N}_r, *) := \hat{\mathcal{C}}_{\hat{\sigma},irr}/\widehat{\mathfrak{G}}(\hat{N}_r, *).$$

Denote by $\widehat{\mathfrak{M}}_r$ the moduli space of $(\hat{\sigma}_1\#\hat{\sigma}_2, \hat{g}_r)$-monopoles on $\hat{N}(r)$.

**Theorem 4.5.17. (Global gluing theorem)** *There exists $r_0 > 0$ with the following properties.*

*(i) For all $r > r_0$ all the monopoles in $\widehat{\mathfrak{M}}_r$ are irreducible and regular.*

*(ii) For all $r > r_0$ the smooth manifolds $\widehat{\mathfrak{M}}_r(*)$ and $\#_r\widehat{\mathfrak{M}}_{\Delta}(*)$ are $S^1$-equivariantly isotopic inside $\hat{\mathcal{B}}_{\hat{\sigma},irr}(\hat{N}_r, *)$.*

**Proof** We can associate to each $\hat{\mathsf{C}}_r \in \#_r\mathfrak{Z}_{\Delta}$ a virtual tangent space $\mathcal{H}^+_{\hat{\mathsf{C}}_r}$, described by a diagram of the type (**T**), and an obstruction space $\mathcal{H}^-_{\hat{\mathsf{C}}_r}$, described by a diagram of the type (**O**).

Since the moduli spaces $\widehat{\mathfrak{M}}_{\hat{\sigma}_i,\mu}$ are compact these diagrams are asymptotically exact (uniformly in $\hat{\mathsf{C}}_r(\hat{\mathsf{C}}_1, \hat{\mathsf{C}}_2)$) as $r \to \infty$. In particular, we deduce that there exists $R_0 > 0$ such that $\mathcal{H}^-_{\hat{\mathsf{C}}_r} = 0$, $\forall r > R_0$ and all $\hat{\mathsf{C}}_r \in \#_r\mathfrak{Z}_{\Delta}$. Moreover, the diagram (**T**) shows that the map

$$\#_r\mathfrak{Z}_{\Delta} \ni \hat{\mathsf{C}}_r \mapsto \dim_{\mathbb{R}} \mathcal{H}^+_{\hat{\mathsf{C}}_r}$$

is continuous and the family $\left\{\mathcal{H}^+_{\hat{\mathsf{C}}_r};\ \ \hat{\mathsf{C}}_r \in \#_r\mathfrak{Z}_{\Delta}\right\}$ forms a smooth vector bundle $\mathcal{H}^+_r \to \#_r\mathfrak{Z}_{\Delta}$.

A quick inspection of the diagram (**T**) shows that $\mathcal{H}^+_r \cong T\#_r\mathfrak{Z}_{\Delta}$ in a very strong sense: there exists $\theta : [1, \infty) \to \mathbb{R}_+$ such that $\theta(r) \to 0$ as $r \to \infty$ and

$$\sup_{\hat{\mathsf{C}}_r \in \#_r\mathfrak{Z}_{\Delta}} \text{gap dist}_{L^2}(T_{\hat{\mathsf{C}}_r}\#_r\mathfrak{Z}_{\Delta}, \mathcal{H}^+_{\hat{\mathsf{C}}_r}) \leq \theta(r), \ \ \forall r \gg 0.$$

Thus, for all intents and purposes we can identify $T\#_r\mathfrak{Z}_{\Delta}$ with $\mathcal{H}^+_r$.

The space $\mathfrak{Y}^2_+(\hat{\mathsf{C}}_r)$ introduced in the proof of the local gluing theorem is orthogonal (or uniformly almost orthogonal) to $T_{\hat{\mathsf{C}}_r}\#_r\mathfrak{Z}_{\Delta}$ and thus the collection

$$\mathfrak{Y}^+_r = \left\{\mathfrak{Y}^2_+(\hat{\mathsf{C}}_r);\ \ \hat{\mathsf{C}}_r \in \#_r\mathfrak{Z}_{\Delta}\right\}$$

defines an infinite-dimensional vector bundle over $\#_r\mathfrak{Z}_{\Delta}$: the normal bundle corresponding to the embedding

$$\#_r\mathfrak{Z}_{\Delta} \hookrightarrow \hat{\mathcal{B}}_{\hat{\sigma},irr}.$$

We leave the reader to verify that Lemma 4.5.9 implies that the exponential map $\mathfrak{Y}_r^+ \to \hat{\mathcal{B}}_{\hat{\sigma},irr}$ defined by the embedding

$$\#_r \mathfrak{Z}_\Delta \hookrightarrow \hat{\mathcal{B}}_{\hat{\sigma},irr}$$

induces a diffeomorphism from the bundle of disks of radii $r^{-3}$ of $\mathfrak{Y}_r^+$ to a tubular neighborhood of $\#_r \mathfrak{Z}_\Delta \hookrightarrow \hat{\mathcal{B}}_{\hat{\sigma},irr}$.

The local gluing theorem produces for each $\hat{\mathsf{C}}_r \in \#_r \mathfrak{Z}_\Delta$ a local section $\Phi_{\hat{\mathsf{C}}_r}$ of $\mathfrak{Y}_r^+$ defined on a neighborhood of radius $r^{-3}$ centered at $\hat{\mathsf{C}}_r$. We can view $\Phi_{\hat{\mathsf{C}}_r}$ as a normal pushforward of a $r^{-3}$-sized neighborhood of $\hat{\mathsf{C}}_r$ into a small tubular neighborhood of $\#_r \mathfrak{Z}_\Delta \hookrightarrow \hat{\mathcal{B}}_{\hat{\sigma},irr}$. Set

$$\Psi_r(\hat{\mathsf{C}}_r) := \Phi_{\hat{\mathsf{C}}_r}(\hat{\mathsf{C}}_r).$$

Since this is an unobstructed gluing problem we deduce that $\Psi_r(\hat{\mathsf{C}}_r)$ is a genuine $\hat{\sigma}$-monopole. Moreover, according to Remark 4.5.11,

$$\mathrm{dist}_{L^{2,2}}(\hat{\mathsf{C}}_r, \Psi_r(\hat{\mathsf{C}}_r)) \leq r^{-n}, \;\; \forall r \gg 0, \; \forall \hat{\mathsf{C}}_r.$$

We can now invoke Corollary 4.5.10 and Theorem 4.5.15 to conclude that for large $r$ the space $\widehat{\mathfrak{M}}_{\hat{\sigma}}(\hat{g}_r)$ consists only of irreducible, regular monopoles and the map $\Psi_r$ is a diffeomorphism

$$\Psi_r : \#_r \mathfrak{Z}_\Delta \to \widehat{\mathfrak{M}}_{\hat{\sigma}}(\hat{g}_r).$$

Clearly $\Gamma_r := \Psi_r \circ \#_r : \#_r \mathfrak{Z}_\Delta \to \widehat{\mathfrak{M}}_{\hat{\sigma}}(\hat{g}_r)$ is a diffeomorphism. Since this diffeomorphism is defined by a small pushforward in the normal bundle it is clear that it can be completed to an isotopy. This construction lifts to an $S^1$-equivariant diffeorphism

$$\Gamma_r^\flat : \#_r \mathfrak{Z}_\Delta^\flat \to \widehat{\mathfrak{M}}_{\hat{\sigma}}(\hat{g}_r, *). \;\; \blacksquare$$

**Remark 4.5.18.** The covering spaces $\widehat{\mathfrak{M}}_i^{\hat{N}}$ (or $\mathfrak{M}_\sigma^{\hat{N}}$) may have infinite fibers if the index of

$$H^1(\hat{N}, \mathbb{Z}) \to H^1(\hat{N}_i, \mathbb{Z}) \;\text{ or }\; H^1(\hat{N}, \mathbb{Z}) \to H^1(N, \mathbb{Z})$$

is infinite. This would indicate that $\widehat{\mathfrak{M}}_\Delta$ and thus $\widehat{\mathfrak{M}}_{\hat{\sigma}}(\hat{g}_r)$ may be noncompact, which we know is not the case. How can we resolve this conflict?

First of all, if these coverings are infinite then it is possible that the moduli spaces $\widehat{\mathfrak{M}}_i$ are empty (see Corollary 4.4.17 for such an example). On the other hand, the maps $\partial_\infty^\flat : \widehat{\mathfrak{M}}_i^{\hat{N}}(*) \to \mathfrak{M}_\sigma^{\hat{N}}(*)$ have compact fibers and may not be onto. The intersection $\partial_\infty^\flat(\widehat{\mathfrak{M}}_1^{\hat{N}}) \cap \partial_\infty^\flat(\widehat{\mathfrak{M}}_2^{\hat{N}})$ can then be compact or even empty.

**CASE 2.** We now analyze one special case of degenerate gluing. More precisely, assume the following.

$\mathbf{B_1}$ *$(N, g)$ is the sphere $S^3$ equipped with the round metric.*

$\mathbf{B}_2$ $b_+(\hat{N}_1) > 0$, $b_+(\hat{N}_2) = 0$.

$\mathbf{B}_3$ *All the finite energy monopoles on $\hat{N}_1$ are irreducible and strongly regular.*

$\mathbf{B_4}$ *Up to gauge equivalence, there exists a unique finite energy $\hat{\sigma}_2$-monopole $\hat{\mathsf{C}}_2 = (0, \hat{A}_2)$ on $\hat{N}_2$ which is reducible and satisfies $H^1_{\hat{\mathsf{C}}_2} = 0$. We denote by $d_0$ its virtual dimension. (Observe that $d_0 \leq 0$.)*

Observe that, according to (4.3.19), the condition $\mathbf{B_4}$ implies

$$H^1(\hat{N}_2, N; \mathbb{R}) = 0 = H^1(\hat{N}_2, \mathbb{R}).$$

The identity (4.3.20) implies

$$H^2(F_{\hat{\mathsf{C}}_2}) \cong \ker_{ex} \mathfrak{D}^*_{\hat{A}_2}.$$

$H^2(F_{\hat{\mathsf{C}}_2})$ is a complex vector space, and thus equipped with a natural $S^1$-action. Set

$$h_2 := \dim_{\mathbb{C}} H^2(F_{\hat{\mathsf{C}}_2}) = -\frac{d_0 + 1}{2}.$$

Denote by

$$\mathcal{L} \to \widehat{\mathfrak{M}}_\mu(\hat{N}_1, \hat{\sigma}_1)$$

the complex line bundle associated to the principal $S^1$-bundle

$$\widehat{\mathfrak{M}}_\mu(\hat{N}_1, \hat{\sigma}_1 *) \to \widehat{\mathfrak{M}}_\mu(\hat{N}_1, \hat{\sigma}_1).$$

**Theorem 4.5.19. (Degenerate gluing)** *There exists $r_0 > 0$ with the following property. For every $r > r_0$ the moduli space $\widehat{\mathfrak{M}}_{\hat{\sigma}_1 \# \hat{\sigma}_2}(\hat{N}, \hat{g}_r)$ consists only of regular irreducible monopoles and there exists a $S^1$-equivariant map*

$$S_r : \widehat{\mathfrak{M}}_\mu(\hat{N}_1, \hat{\sigma}_1, *) \to \mathbb{C}^{h_2}$$

*such that its zero set is a smooth $S^1$-invariant submanifold of $\widehat{\mathfrak{M}}_\mu(\hat{N}_1, \hat{\sigma}_1, *)$ $S^1$-equivariantly diffeomorphic to $\widehat{\mathfrak{M}}_{\hat{\sigma}_1 \# \hat{\sigma}_2}(\hat{N}, \hat{g}_r, *)$. In particular, this means there exists a section $s_r$ of the vector bundle $\mathcal{L}^{h_2} \to \widehat{\mathfrak{M}}_\mu(\hat{N}_1, \hat{\sigma}_1)$ whose zero locus is a smooth submanifold diffeomorphic to $\widehat{\mathfrak{M}}_{\hat{\sigma}_1 \# \hat{\sigma}_2}(\hat{N}, \hat{g}_r)$.*

**Sketch of proof** We use the same notation as in **CASE 1**. Observe first that assumption $\mathbf{B_1}$ implies that there exist a unique *spin$^c$* structure $\sigma_0$ on $N$ and an unique $\sigma_0$-monopole $\mathsf{C}_\infty$ which is reducible and regular. In particular

$$T_{\mathsf{C}_\infty} \mathfrak{M}_{\sigma_0} = 0, \;\; T_1 G_\infty \cong \mathbb{R}.$$

Moreover, since $H^1(N,\mathbb{Z}) = 0$ we deduce that any gauge transformation on $\hat{N}_1$ extends to $\hat{N}$.

Suppose $(\hat{\mathsf{C}}_1, \hat{\mathsf{C}}_2) \in \mathfrak{Z}_\Delta$. Then we can form $\hat{\mathsf{C}}_r := \hat{\mathsf{C}}_1 \#_r \hat{\mathsf{C}}_2$. There are many cancellations in the diagrams (**L**) and (**O**) associated to $\hat{\mathsf{C}}_r$. More precisely, we have

$$H^1_{\hat{\mathsf{C}}_2} = 0, \;\; L_i^{\pm} = 0, \;\; \mathfrak{C}_2^+ = 0, \;\; \mathfrak{C}_1^- = 0, \;\; H^2(F_{\hat{\mathsf{C}}_1}) = 0.$$

We deduce that $\ker \Delta_+^0 = 0$, $\ker \Delta_+^c \cong H^1_{\hat{\mathsf{C}}_1}$ such that

$$\mathcal{H}_r^+(\hat{\mathsf{C}}_r) \cong H^1_{\hat{\mathsf{C}}_1}.$$

Observe that Lemma 4.5.9 implies that the subspace $\mathcal{H}_r^+(\hat{\mathsf{C}}_r) \subset T_{\hat{\mathsf{C}}_r}\hat{\mathcal{C}}(\hat{N}_r)$ and the tangent space to the $\widehat{\mathfrak{G}}(\hat{N}_r)$-orbit through $\hat{\mathsf{C}}_r$ are transversal. Moreover,

$$T_{\hat{\mathsf{C}}_r}\Big(\#_r(\mathfrak{Z}_\Delta)\Big) = T_{\hat{\mathsf{C}}_r}\Big(\widehat{\mathfrak{G}}(\hat{N}_r)\cdot \hat{\mathsf{C}}_r\Big) + \mathcal{H}_r^+(\hat{\mathsf{C}}_r)$$

and the assignment

$$\mathfrak{Z}_\Delta \ni (\hat{\mathsf{C}}_1, \hat{\mathsf{C}}_2) \mapsto \mathcal{H}_r^+(\hat{\mathsf{C}}_1 \#_r \hat{\mathsf{C}}_2) \subset T_{\hat{\mathsf{C}}_1 \#_r \hat{\mathsf{C}}_2}\hat{\mathcal{C}}(\hat{N}_r)$$

is a $\widehat{\mathfrak{G}}_\Delta$-equivariant subbundle of $T\hat{\mathcal{C}}(\hat{N}_r)\,|_{\#_r \mathfrak{Z}_\Delta}$. This shows that the family $(\mathcal{H}_r^+(\hat{\mathsf{C}}_1 \#_r \hat{\mathsf{C}}_2))$ descends to a smooth vector bundle

$$\mathcal{H}_r^+ \to \#_r\widehat{\mathfrak{M}}_\Delta := \big(\widehat{\mathfrak{G}}(\hat{N}_r)\cdot \#_r \mathfrak{Z}_\Delta\big)/\widehat{\mathfrak{G}}(\hat{N}_r).$$

For large $r$ we have diffeomorphisms

$$\#_r\widehat{\mathfrak{M}}_\Delta \cong \widehat{\mathfrak{M}}_\Delta \cong \widehat{\mathfrak{M}}_1 \cong \widehat{\mathfrak{M}}_1^{\hat{N}}.$$

Moreover, the bundle $\mathcal{H}_r^+ \to \#_r\widehat{\mathfrak{M}}_\Delta$ is isomorphic to the tangent bundle of $\widehat{\mathfrak{M}}_1$.

To see this observe first that $T_{\hat{\mathsf{C}}_1}\widehat{\mathfrak{M}}_1 \cong H^1_{\hat{\mathsf{C}}_1}$. Next, the *compactness* of $\widehat{\mathfrak{M}}_1$ implies that we have uniformity with respect to $\hat{\mathsf{C}}_1$ as $r \to \infty$ in the Cappell-Lee-Miller gluing theorem. Thus, the family

$$\mathfrak{Z}_\Delta \ni (\hat{\mathsf{C}}_1, \hat{\mathsf{C}}_2) \mapsto \mathcal{H}_r^+(\hat{\mathsf{C}}_1 \#_r \hat{\mathsf{C}}_2)$$

is homotopic as $r \to \infty$ to the family

$$\mathfrak{Z}_\Delta \ni (\hat{\mathsf{C}}_1, \hat{\mathsf{C}}_2) \mapsto H^1_{\hat{\mathsf{C}}_1} = T_{\hat{\mathsf{C}}_1}\widehat{\mathfrak{M}}_1.$$

Using the obstruction diagram (**O**) we conclude similarly that

$$\mathcal{H}_r^-(\hat{\mathsf{C}}_r) \cong H^2(F_{\hat{\mathsf{C}}_2}).$$

Fix $(\hat{\mathsf{C}}_1^0, \hat{\mathsf{C}}_2^0) \in \mathfrak{Z}_\Delta$ and set

$$V_r^0 := \mathcal{H}_r(\hat{\mathsf{C}}_1^0 \#_r \hat{\mathsf{C}}_2^0) \subset \mathfrak{Y}_r := L^{1,2}(\mathbb{S}^-_{\hat{\sigma}_1 \#_r \hat{\sigma}_2} \oplus \mathbf{i}\Lambda_+^2 T^*\hat{N}_r).$$

According to the Cappell-Lee-Miller gluing theorem, Theorem 4.1.22, there exists $r_0 = r_0(\hat{\mathsf{C}}_1, \hat{\mathsf{C}}_2) > 0$ so that for $r > r_0(\hat{\mathsf{C}}_1, \hat{\mathsf{C}}_2)$ the last isomorphism is described by an *explicit* map

$$I^-_{\hat{\mathsf{C}}_r, r} : \mathcal{H}^-_r(\hat{\mathsf{C}}_r) \to V^0_r \cong H^2(F_{\hat{\mathsf{C}}^0_2}).$$

In fact, since $\widehat{\mathfrak{M}}_\mu(\hat{N}_1, \hat{\sigma}_1)$ is compact, we have

$$R_0 := \sup_{(\hat{\mathsf{C}}_1, \hat{\mathsf{C}}_2) \in \mathfrak{Z}_\Delta} r_0(\hat{\mathsf{C}}_1, \hat{\mathsf{C}}_2) < \infty$$

so that for all $r > R_0$ there exists an isomorphism $I_{\hat{\mathsf{C}}_1, r} : \mathcal{H}^-_r(\hat{\mathsf{C}}_r) \to V^0_r$ depending continuously on $\hat{\mathsf{C}}_1$. This means that for $r > R_0$ the collection

$$\mathfrak{Z}_\Delta \ni (\hat{\mathsf{C}}_1, \hat{\mathsf{C}}_2) \mapsto \mathcal{H}^-_r(\hat{\mathsf{C}}_1 \#_r \hat{\mathsf{C}}_2)$$

forms a *trivial* complex vector bundle $\mathcal{H}^-_r$ of rank $h_2$ over $\mathfrak{Z}_\Delta$. Using the diffeomorphism

$$\#_r : \mathfrak{Z}_\Delta \to \#_r(\mathfrak{Z}_\Delta) \subset \hat{\mathcal{C}}(\hat{N}_r)$$

we can think of $\mathcal{H}^-_r$ as a vector bundle over $\#_r(\mathfrak{Z}_\Delta)$.

If $(\hat{\gamma}_1, \hat{\gamma}_2) \in \widehat{\mathfrak{G}}_\Delta$ then

$$\mathcal{H}^-_r((\hat{\gamma}_1 \#_r \hat{\gamma}_2) \cdot \hat{\mathsf{C}}_r) = (\hat{\gamma}_1 \#_r \hat{\gamma}_2) \cdot \mathcal{H}^-_r(\hat{\mathsf{C}}_r).$$

Two configurations in $\#_r \mathfrak{Z}_\Delta$ belong to the same $\widehat{\mathfrak{G}}(\hat{N}_r)$-orbit if and only if they belong to the same $\#_r \widehat{\mathfrak{G}}_\Delta$-orbit. Since $\#_r \mathfrak{Z}_\Delta$ consists only of irreducible configurations we can thus think of $\mathcal{H}^-_r$ as a $\widehat{\mathfrak{G}}(\hat{N}_r)$-equivariant subbundle of the infinite-dimensional vector bundle $\mathcal{W}_r$ over $\widehat{\mathfrak{G}}(\hat{N}_r) \cdot \#_r \mathfrak{Z}_\Delta$ with standard fiber $\mathfrak{Y}_r$. Although the bundle $\mathcal{H}^-_r$ is trivial, *it is not equivariantly trivial.* To see this, we present an alternate description of the bundle $\mathcal{H}^-_r$.

Denote by **Grass** the Grassmannian of complex $h_2$-dimensional subspaces of $L^{1,2}(\mathbb{S}^-_{\hat{\sigma}_1 \#_r \hat{\sigma}_2}) \subset \mathfrak{Y}_r$. The action of $\widehat{\mathfrak{G}}(\hat{N}_r)$ on $\mathfrak{Y}_r$ induces an action on **Grass**. The isomorphisms $I_{\hat{\mathsf{C}}_r, r}$ can be regarded as a $\#_r \widehat{\mathfrak{G}}_\Delta$-equivariant map

$$\phi : \#_r \mathfrak{Z}_\Delta \to \mathbf{Grass}$$

whose image lies in the $\widehat{\mathfrak{G}}(\hat{N}_r)$-orbit of $V^0_r$. The bundle $\mathcal{H}^-_r$ is defined by the $\widehat{\mathfrak{G}}(\hat{N}_r)$-equivariant extension of $\phi$

$$\Phi : \widehat{\mathfrak{G}}(\hat{N}_r) \cdot \#_r \mathfrak{Z}_\delta \to \mathbf{Grass}.$$

The stabilizer of $V^0_r \in$ **Grass** with respect to the action of $\widehat{\mathfrak{G}}(\hat{N}_r)$ is the subgroup $S^1$ of constant gauge transformations. It is convenient to think of $S^1$ as given by the obvious inclusion $S^1 \to \widehat{\mathfrak{G}}(\hat{N}_r)$ which splits the short exact sequence

$$1 \hookrightarrow \widehat{\mathfrak{G}}(\hat{N}_r, *) \hookrightarrow \widehat{\mathfrak{G}}(\hat{N}_r) \xrightarrow{\mathbf{ev}_*} S^1 \twoheadrightarrow 1.$$

The quotient $\left(\widehat{\mathfrak{G}}(\hat{N}_r)\cdot \#_r\mathfrak{Z}_\Delta\right)/\widehat{\mathfrak{G}}(\hat{N}_r,*)$ is the space of gauge equivalence classes of based almost monopoles on $\hat{N}_r$,

$$\left(\widehat{\mathfrak{G}}(\hat{N}_r)\cdot \#_r\mathfrak{Z}_\delta\right)/\widehat{\mathfrak{G}}(\hat{N}_r,*)\cong \#_r\widehat{\mathfrak{M}}_\Delta(*).$$

The bundle $\mathcal{H}_r^-$ descends to a bundle $[\mathcal{H}_r^-]\to \#_r\widehat{\mathfrak{M}}_\Delta$ which is the bundle associated to the $S^1$-fibration

$$\#\widehat{\mathfrak{M}}_\Delta(*)\to \#_r\widehat{\mathfrak{M}}_\Delta$$

via the natural action of $S^1$ on $V_r^0$,

$$[\mathcal{H}_r^-]\cong \#_r\widehat{\mathfrak{M}}_\Delta\times_{S^1}V_0^r\cong \#_r\widehat{\mathfrak{M}}_\Delta\times_{S^1}\mathbb{C}^{h_2}\cong \mathcal{L}^{h_2}\to \#\widehat{\mathfrak{M}}_\Delta.$$

Denote by $\Xi_r$ the orthogonal complement of $\mathcal{H}_r^+$ in $T\mathcal{B}_{\hat{\sigma}_1\#_r\hat{\sigma}_2,irr}$. We can regard $\Xi_r$ as the normal bundle of the embedding

$$\#_r\widehat{\mathfrak{M}}_\Delta\hookrightarrow \mathcal{B}_{\hat{\sigma}_1\#_r\hat{\sigma}_2,irr}.$$

Using the exponential map we can identify a tubular neighborhood $\mathcal{U}_r$ (of diameter $\approx r^{-3}$) of

$$\#_r\widehat{\mathfrak{M}}_\Delta\subset \mathcal{B}_{\hat{\sigma}_1\#\hat{\sigma}_2,irr}$$

with a neighborhood $\mathcal{V}_r$ of the zero section of $\Xi_r$. Observe that we have a natural projection $\pi:\mathcal{U}_r\to \#_r\widehat{\mathfrak{M}}_\Delta$ which we can use to pull back $\mathcal{H}_r^-$ to a vector bundle $\pi^*[\mathcal{H}_r^-]\to\mathcal{U}_r$.

The Seiberg-Witten equations over $\hat{N}$ define a section $\widehat{SW}$ of an infinite dimensional vector bundle $\mathcal{W}_r$ over $\mathcal{B}_{\hat{\sigma}_1\#\hat{\sigma}_2,irr}$ with standard fiber $\mathfrak{Y}_r$. According to Remark 4.5.8 we can regard $[\mathcal{H}_r^-]$ as a subbundle of $\mathcal{W}_r$. We denote by $P_-$ the $L^2$-orthogonal projection

$$P_-:\mathcal{W}_r\to[\mathcal{H}_r^-].$$

Arguing as in the proof of Theorem 4.5.17 we deduce from the local gluing theorem that there exists a smooth section $\Psi_r:\mathcal{U}_r\cong\mathcal{V}_r\subset\Xi_r\to[\mathcal{H}_r^-]$ such that, for all $[\hat{\mathsf{C}}_r]\in \#_r\widehat{\mathfrak{M}}_\Delta$, we have

$$\widehat{SW}\left(\hat{\mathsf{C}}_r+\Psi_r(\hat{\mathsf{C}}_r)\right)\in \pi^*\mathcal{H}_r^-\,|_{\hat{\mathsf{C}}_r+\Psi_r(\hat{\mathsf{C}}_r)},\ \ \forall\hat{\mathsf{C}}_r\in \#_r\mathfrak{Z}_\Delta. \tag{4.5.21}$$

Set

$$\#_r\widehat{\mathfrak{M}}'_\Delta:=\left\{\hat{\mathsf{C}}_r+\Psi_r(\hat{\mathsf{C}}_r);\ \ \hat{\mathsf{C}}_r\in \#_r\widehat{\mathfrak{M}}_\Delta\right\}\subset\mathcal{U}_r.$$

Observe that

$$\#_r\widehat{\mathfrak{M}}'_\Delta\cong \#_r\widehat{\mathfrak{M}}_\Delta\cong \widehat{\mathfrak{M}}_\mu(\hat{N}_1,\hat{\sigma}_1)$$

and moreover, according to (4.5.21), the restriction of the Seiberg-Witten section $\widehat{SW}$ to $\#_r\widehat{\mathfrak{M}}'_\Delta$ defines a smooth section of the vector bundle $\pi^*[\mathcal{H}_r^-]$. This is a smooth section $s_r$ of

$$\mathcal{L}^{h_2} \to \#_r\widehat{\mathfrak{M}}'_\Delta \cong \#_r\widehat{\mathfrak{M}}_\Delta.$$

Its zero set is precisely $\widehat{\mathfrak{M}}_{\hat{\sigma}_1\#\hat{\sigma}_2}(\hat{N}, \hat{g}_r)$, which is generically a smooth manifold. ■

The above theorem has an immediate corollary which will be needed in the next section. Suppose $\hat{N}$ is a compact, smooth, oriented 4-manifold and $\hat{N}_1$ is the cylindrical 4-manifold obtained from $\hat{N}$ by deleting a small ball and attaching the infinite cylinder $\mathbb{R}_+ \times S^3$. Denote by $\hat{N}_2$ the cylindrical 4-manifold with positive scalar curvature obtained by attaching the infinite cylinder $\mathbb{R}_+ \times S^3$ to a small ball. Observe that $\hat{N}_1\#_r\hat{N}_2$ is diffeomeorphic to $\hat{N}$. Moreover, if $\hat{\sigma}_2$ denotes the unique cylindrical *spin*$^c$-structure on $\hat{N}_2$ then the correspondence

$$Spin^c_{cyl}(\hat{N}_1) \ni \hat{\sigma}_1 \mapsto \hat{\sigma}_1\#\hat{\sigma}_2 \in Spin^c(\hat{N})$$

is a bijection. We will denote its inverse, $Spin^c(\hat{N}) \to Spin^c(\hat{N}_1)$, by

$$\hat{\sigma} \mapsto \hat{\sigma}\,|_{\hat{N}_1}\,.$$

**Corollary 4.5.20.** *Suppose $b_+(\hat{N}) > 0$. Then the $S^1$ bundles*

$$S^1 \hookrightarrow \widehat{\mathfrak{M}}_{\hat{\sigma}}(\hat{N}, \hat{g}_r, *) \to \widehat{\mathfrak{M}}_{\hat{\sigma}}(\hat{N}, \hat{g}_r)$$

*and*

$$S^1 \hookrightarrow \widehat{\mathfrak{M}}_\mu(\hat{N}_1, \hat{\sigma}\,|_{\hat{N}_1}, *) \to \widehat{\mathfrak{M}}_\mu(\hat{N}_1, \hat{\sigma}\,|_{\hat{N}_1})$$

*are naturally isomorphic.*

**Proof** The conditions $\mathbf{B}_1$ and $\mathbf{B}_2$ are clearly satisfied. $\mathbf{B}_3$ is generically satisfied. Finally, according to Example 4.3.40 in §4.3.4, condition $\mathbf{B}_4$ is also satisfied, with $h_2 = 0$. The corollary now follows immediately from Theorem 4.5.19. ■

## 4.6. Applications

We have some good news for the reader who has survived the avalanche of technicalities in this chapter. It's payoff time!

We will illustrate the power of the results we have established so far by proving some beautiful topological results. All the gluing problems in Seiberg-Witten theory follow the same pattern.

A major limitation of the cutting and pasting technique has its origin in the difficulties involved in describing the various terms arising in the diagrams (**T**), (**O**), (**L**). A good understanding of both the geometric and topological background is always a make or break factor.

**4.6.1. Vanishing results.** The simplest topological operation one can perform on smooth manifolds is the connected sum. It is natural then to ask how this operation affects the Seiberg-Witten invariants. The first result of this section provides the surprisingly simple answer.

**Theorem 4.6.1. (Connected sum theorem)** *Suppose $M_1$ and $M_2$ are two compact, oriented smooth manifolds such that $b_+(M_i) > 0$. Then*

$$\mathbf{sw}_{M_1\#M_2}(\sigma) = 0, \;\; \forall \sigma \in Spin^c(M_1\#M_2).$$

Before we present the proof of this result let us mention a surprising consequence.

**Corollary 4.6.2.** *No compact symplectic 4-manifold $M$ can be decomposed as a connected sum $M_1\#M_2$ with $b_+(M_i) > 0$.*

**Proof** The result is clear if $b_+(M) = 1$ since $b_+(M_1\#M_2) = b_+(M_1) + b_+(M_2)$. If $b_+(M) > 1$ then, according to Taubes' Theorem 3.3.29 not all the Seiberg-Witten invariants of $M$ are trivial. ■

**Remark 4.6.3.** (a) The smooth 4-manifolds which cannot be decomposed as $M_1\#M_2$ with $b_+(M_i) > 0$ are called irreducible. We can rephrase the above corollary by saying that all the symplectic 4-manifolds are irreducible. It was believed, or rather hoped, that the symplectic manifolds exhaust the list of irreducible 4-manifolds and all other can be obtained from them by some basic topological operations, much as in the two-dimensional case where all compact oriented surfaces are connected sums of tori.

This belief was shattered by Z. Szabó in [**131**], who constructed the first example of a simply connected, irreducible, symplectic 4-manifold. Immediately after that, R. Fintushel and R. Stern showed in [**36**] that the phenomenon discovered by Szabó was not singular and developed a very elegant machinery to produce irreducible manifolds, most of which are not symplectic.

(b) Up to this point we knew only one vanishing theorem: positive scalar curvature $\Rightarrow$ trivial Seiberg-Witten invariants. The connected sum theorem, however, has a different flavor since the vanishing is a consequence of a *topological* condition rather than of a *geometric* one.

**Proof of Theorem 4.6.1** Set $\hat{N} := M_1\#M_2$. Observe that $b_+(\hat{N}) > 1$ so that the Seiberg-Witten invariants of $\hat{N}$ are metric independent.

Denote by $\hat{N}_i$ the manifold obtained from $M_i$ by deleting a small ball and then attaching the infinite cylinder $\mathbb{R}_+ \times S^3$. Observe that

$$\hat{N} \cong_{diffeo} \hat{N}_1 \#_{S^3,r} \hat{N}_2.$$

On $S^3$ there exists a single $spin^c$ structure and any two cylindrical structures $\hat{\sigma}_i \in Spin^c_{cyl}(\hat{N}_i)$ are compatible. Thus

$$Spin^c(\hat{N}) \cong Spin^c_{cyl}(\hat{N}_1) \times Spin^c(\hat{N}_2).$$

The manifolds $\hat{N}_1$ and $\hat{N}_2$ (generically) satisfy all the assumptions of the Global Gluing Theorem 4.5.17 and thus

$$\widehat{\mathfrak{M}}_{\hat{\sigma}_1\#\hat{\sigma}_2}(\hat{N}, \hat{g}_r) \cong \widehat{\mathfrak{M}}_\mu(\hat{N}_1, \hat{\sigma}_1, *) \times \widehat{\mathfrak{M}}_\mu(\hat{N}_2, \hat{\sigma}_2 *)/S^1.$$

Moreover, according to the computation in Example 4.5.13 we have (componentwise)

$$\dim \widehat{\mathfrak{M}}_{\hat{\sigma}_1\#\hat{\sigma}_2}(\hat{N}, g_r) = \dim \widehat{\mathfrak{M}}_\mu(\hat{N}_1, \hat{\sigma}_1) + \dim \widehat{\mathfrak{M}}_\mu(\hat{N}_2, \hat{\sigma}_2) + 1.$$

The left-hand side of the above equality can be zero if and only if one of the two dimensions on the right-hand side is negative, forcing the corresponding moduli space to be (generically) empty. Thus, if $\hat{\sigma} \in Spin^c(\hat{N})$ is such that the expected dimension $d(\hat{\sigma}) = 0$ then the corresponding moduli space is generically empty so that $\mathbf{sw}_{\hat{N}}(\hat{\sigma}) = 0$.

To deal with the case $d(\hat{\sigma}_1\#\hat{\sigma}_2) > 0$ we follow an approach we learned from Frank Connolly. Suppose $\hat{\sigma}_0 = \hat{\sigma}_1\#\hat{\sigma}_2 \in Spin^c(\hat{N})$ is such that $d(\hat{\sigma}_0) = 2n > 0$. Then

$$\mathbf{sw}_{\hat{N}}(\hat{\sigma}_0) = \int_{\widehat{\mathfrak{M}}_{\hat{\sigma}_0}} \Omega_0^n$$

where $\Omega_0 \in H^2(\widehat{\mathfrak{M}}_{\hat{\sigma}_0}, \mathbb{Z})$ is the first Chern class of the base point fibration

$$S^1 \hookrightarrow X_0 := \widehat{\mathfrak{M}}_{\hat{\sigma}_0}(*) \xrightarrow{p_0} \widehat{\mathfrak{M}}_{\hat{\sigma}_0}.$$

Denote by $\Omega_i$, $i = 1, 2$, the first Chern class of the base point fibration

$$S^1 \hookrightarrow X_i := \widehat{\mathfrak{M}}_\mu(\hat{N}_i, \hat{\sigma}_i, *) \xrightarrow{p_i} \widehat{\mathfrak{M}}_\mu(\hat{N}_i, \hat{\sigma}_i).$$

It is convenient to think of $\Omega_j$, $j = 0, 1, 2$, as differential forms. The pullbacks $p^*\Omega_j$ are exact and there exist 1-forms $\theta_j$ such that

$$d\theta_j = p^*\Omega_j$$

and

$$\int_{\widehat{\mathfrak{M}}_{\hat{\sigma}_j}} \Omega^m = \int_{\widehat{\mathfrak{M}}_{\hat{\sigma}_j}(*)} \theta \wedge (d\theta_j)^m, \quad \forall m \in \mathbb{Z}_+, \quad j = 0, 1, 2.$$

(Above, we have tacitly used the fact that the manifolds $\widehat{\mathfrak{M}}_{\hat{\sigma}_j}$ are orientable.) The 1-forms $\theta_j$ have a simple geometric interpretation: they are global angular forms of the corresponding $S^1$-fibrations. In topology these forms also go by the name of transgression forms.

On the other hand, we can regard $\theta_0$ as a global angular form for the diagonal $S^1$-action on

$$X := \widehat{\mathfrak{M}}_\mu(\hat{N}_1, \hat{\sigma}_1, *) \times \widehat{\mathfrak{M}}_\mu(\hat{N}_2, \hat{\sigma}_2, *)$$

so that we can choose

$$\theta_0 = \frac{1}{2}\Big(\theta_1 + \theta_2\Big) + \text{exact form}.$$

Thus

$$\mathbf{sw}_{\hat{N}}(\hat{\sigma}_0) = \frac{1}{2^{n+1}} \int_{X_0} (\theta_1 + \theta_2) \wedge (d\theta_1 + d\theta_2)^n$$

$$= \frac{1}{2^{n+1}} \int_{X_1 \times X_2} (\theta_1 + \theta_2) \wedge (d\theta_1 + d\theta_2)^n.$$

For $j = 0, 1, 2$ set $m_j := \dim X_j$ and $c_0 := 2^{-(n+1)}$. Observe that when $\widehat{\mathfrak{M}}_\mu(\hat{N}_i) \neq \emptyset$ its dimension must be nonnegative and we have

$$m_1, m_2 > 0, \quad m_0 = n + 1 = m_1 + m_2. \tag{4.6.1}$$

Using Newton's binomial formula we deduce

$$\mathbf{sw}_{\hat{N}}(\hat{\sigma}_0) = c_0 \sum_{k=0}^{m_0-1} \binom{m_0 - 1}{k} \Big(\int_{X_1} \theta_1 \wedge (d\theta_1)^k\Big)\Big(\int_{X_2} (d\theta_2)^{m_0-1-k}\Big)$$

$$+ c_0 \sum_{k=0}^{m_0-1} \binom{m_0 - 1}{k} \Big(\int_{X_1} (d\theta_1)^k\Big)\Big(\int_{X_2} \theta_2 \wedge (d\theta_2)^{m_0-1-k}\Big).$$

The integrals involving only powers of $(d\theta_j)$ vanish because these are exact forms. We deduce

$$\mathbf{sw}_{\hat{N}}(\hat{\sigma}_0) = c_0 \int_{X_1} \theta_1 \wedge (d\theta_1)^n + c_0 \int_{X_2} \theta_2 \wedge (d\theta_2)^n.$$

Using (4.6.1) we now deduce $n + 1 > \max(m_1, m_2)$ so that both integrals above vanish. ■

**Remark 4.6.4.** For a proof of the connected sum theorem not relying on gluing and pasting techniques we refer to [**120**].

We conclude this subsection with another vanishing result implied by a topological constraint. This result will be considerably strengthened in the next subsection.

Before we state the result let us mention that an element $x$ of an Abelian group $G$ is called *essential* if it generates an *infinite* cyclic group.

**Proposition 4.6.5.** *Suppose $\hat{N}$ is a compact, oriented, smooth 4-manifold satisfying the following conditions.*

*(a) $b_+(\hat{N}) > 1$.*

*(b) There exists a smoothly embedded $S^2 \hookrightarrow \hat{N}$ with trivial self-intersection and defining an essential element in $H^2(\hat{N}, \mathbb{Z})$.*

*Then all the Seiberg-Witten invariants of $\hat{N}$ are trivial.*

**Proof** Observe that because the self-intersection of $S^2 \hookrightarrow \hat{N}$ is trivial it admits a small tubular neighborhood $U$ diffeomorphic to the trivial disk bundle $D^2 \times S^2$. Set $\hat{N} := \partial U \cong S^1 \times S^2$ and equip it with the product metric $g$.

Denote by $(\hat{N}_1, \hat{g}_1)$ the manifold obtained from $\hat{N}$ by removing $U$ and attaching the infinite cylinder $\mathbb{R}_+ \times N$. Moreover, we choose $\hat{g}_1$ such that $\partial_\infty \hat{g}_1 = g$. Also, denote by $(\hat{N}_2, \hat{g}_2)$ the cylindrical manifold obtained by attaching the cylinder $\mathbb{R}_+ \times N$ to $U$ and such that $\partial_\infty \hat{g}_2 = g$.

Observe that $\hat{N}$ is diffeomorphic to $\hat{N}_1 \#_r \hat{N}_2$ for any $r > 0$. Suppose there exists a *spin*$^c$ structure $\hat{\sigma}$ on $\hat{N}$ such that

$$\mathbf{sw}_{\hat{N}}(\hat{\sigma}) = 0.$$

Since $b_+(\hat{N}) > 1$ this implies that $\widehat{\mathfrak{M}}_{\hat{\sigma}}(\hat{N}, \hat{g}_r) \neq \emptyset$, $\forall r > 0$. In particular, if we use the unique decomposition

$$\hat{\sigma} = \hat{\sigma}_1 \# \hat{\sigma}_2$$

we conclude that $\widehat{\mathfrak{M}}_\mu(\hat{N}_1, \hat{\sigma}_1) \neq \emptyset$. At this point we want to invoke the following topological result, whose proof we postpone.

**Lemma 4.6.6.** *The image of $H^1(\hat{N}_1, \mathbb{Z}) \to H^1(N, \mathbb{Z})$ has infinite index.*

The last result and the positivity of the scalar curvature of $N$ now place us in the setting of Corollary 4.4.17 of §4.4.3 which implies that $\widehat{\mathfrak{M}}_\mu(\hat{N}_1, \hat{\sigma}_1)$ is *empty*. This contradiction completes the proof of Proposition 4.6.5. ■

**Proof of Lemma 4.6.6** We will prove the dual homological statement, namely that the image of $H_3(\hat{N}_1, N, \mathbb{Z}) \to H_2(N, \mathbb{Z})$ has infinite index.

Observe that

$$H_2(N, \mathbb{Z}) = H_2(S^1 \times S^2, \mathbb{Z}) \cong \mathbb{Z}$$

with generator $S^2 \hookrightarrow N \cong S^1 \times S^2$. Next, notice that the inclusion

$$N \hookrightarrow \hat{N}$$

induces an injection

$$H_2(N, \mathbb{Z}) \hookrightarrow H_2(\hat{N}, \mathbb{Z})$$

whose image is generated by the cycle $S^2 \hookrightarrow \hat{N}$. Denote by $k[S^2]$ the generator of the image $H_3(\hat{N}_1, N, \mathbb{Z}) \to H_2(N, \mathbb{Z})$. Thus, there exists a cycle $c \in H_3(\hat{N}_1, N, \mathbb{Z})$ such that

$$\partial c = k[S^2] \in H_2(N, \mathbb{Z}).$$

This cycle determines a three-dimensional chain $c$ on $\hat{N}$ such that

$$\partial c = k[S^2 \hookrightarrow \hat{N}]$$

so that $k[S^2 \hookrightarrow \hat{N}] = 0 \in H_2(\hat{N}, \mathbb{Z})$. Since the homology class $[S^2 \hookrightarrow \hat{N}]$ is essential we deduce $k = 0$ so that the morphism $H_3(\hat{N}_1, N, \mathbb{Z}) \to H_2(N, \mathbb{Z})$ is trivial. ∎

**4.6.2. Blow-up formula.** In the previous subsection we have shown that the connected sum of two 4-manifolds with positive $b_+$'s has trivial Seiberg-Witten invariants. This raises the natural question of understanding what happens when one of the manifolds is negative definite. In this case we know that the intersection form is diagonal, exactly as the intersection form of a connected sum of $\overline{\mathbb{CP}}^2$'s.

In this final subsection we will investigate one special case of this new problem. More precisely, we will determine the Seiberg-Witten invariants of $M\#\overline{\mathbb{CP}}^2$ in terms of the Seiberg-Witten invariants of $M$. As explained in Chapter 2, the connected sum $M\#\overline{\mathbb{CP}}^2$ can be interpreted as the blow-up of $M$ at some point. It is thus natural to refer to the main result of this subsection as the *blow-up formula*.

Suppose $M$ is a compact, oriented, smooth 4-manifold such that $b_+(M) > 1$. Denote by $\hat{N}_1$ the manifold obtained from $M$ by removing a small ball and then attaching the infinite cylinder $\mathbb{R}_+ \times S^3$. Observe that

$$Spin^c(M) \cong Spin^c_{cyl}(\hat{N}_1).$$

Now denote by $\hat{N}_2$ the manifold obtained from $\overline{\mathbb{CP}}^2$ by removing a small disk and then attaching the cylinder $\mathbb{R}_+ \times S^3$. Again we have

$$Spin^c(\overline{\mathbb{CP}}^2) \cong Spin^c_{cyl}(\hat{N}_2).$$

Moreover, any two *spin*$^c$ structures $\hat{\sigma}_i \in Spin^c_{cyl}(\hat{N}_i)$ are compatible and the induced map

$$Spin^c_{cyl}(\hat{N}_1) \times Spin^c_{cyl}(\hat{N}_2) \to Spin^c(M\#\overline{\mathbb{CP}}^2),$$

$$(\hat{\sigma}_1, \hat{\sigma}_2) \mapsto \hat{\sigma}_1 \# \hat{\sigma}_2$$

is a bijection.

The manifold $\hat{N}_2$ can also be obtained as in Example 4.3.39 in §4.3.4 by attaching $\mathbb{R}_+ \times S^3$ to the boundary of the Hopf disk bundle over $S^2$. If we now regard $S^3$ as the total space of the degree $-1$ circle bundle over $S^2$ we can equip it with a metric $g$ of positive scalar curvature as in Example 4.1.27. (The round metric is included in the constructions of Example 4.1.27.) Fix cylindrical metrics $\hat{g}_i$ on $\hat{N}_i$ such that $\hat{g}_2$ has positive scalar curvature and

$$\partial_\infty \hat{g}_1 = g = \partial_\infty \hat{g}_2.$$

The manifold $\overline{\mathbb{CP}}^2$ is equipped with a canonical *spin*$^c$ structure $\sigma_{can}$ induced by the complex structure on $\mathbb{CP}^2$. The map

$$Spin^c(\hat{N}_2) \ni \hat{\sigma} \mapsto c_1(\det(\hat{\sigma})) \in H^2(\hat{N}_2, \mathbb{Z}) \cong \mathbb{Z}$$

is a bijection onto $2\mathbb{Z}+1 \subset \mathbb{Z}$ where the generator of $H^2(\hat{N}_2, \mathbb{Z})$ is chosen such that $c_1(\sigma_{can}) = 1$. For each $n \in \mathbb{Z}$ denote by $\hat{\sigma}_n$ the unique cylindrical *spin*$^c$ structure on $\hat{N}_2$ such that $c_1(\hat{\sigma}_n) = (2n+1)$. Observe that $c_1(\bar{\sigma}_{can}) = -1$ so that $\bar{\sigma}_{can} = \sigma_{-1}$.

**Theorem 4.6.7. (Blow-up Formula)** *For every $\hat{\sigma} \in Spin^c(M)$ we have*

$$|\mathbf{sw}_{M\#\overline{\mathbb{CP}}^2}(\hat{\sigma}\#\hat{\sigma}_n)| = \begin{cases} 0 & if \quad d(\hat{\sigma}) < \pm n(n+1) \\ |\mathbf{sw}_M(\hat{\sigma})| & if \quad d(\hat{\sigma}) \geq n(n+1) \end{cases}.$$

**Corollary 4.6.8.** *If $\mathbb{B}_M \subset Spin^c(M)$ denotes the set of basic classes of $M$ then*

$$\mathbb{B}_{M\#\overline{\mathbb{CP}}^2} = \Big\{\hat{\sigma}\#\hat{\sigma}_n;\ \ \hat{\sigma} \in \mathbb{B}_M,\ n \in \mathbb{Z}\ d(\hat{\sigma}) \geq n(n+1)\Big\}.$$

*In particular, $\mathbb{B}_M \neq \emptyset \Longleftrightarrow \mathbb{B}_{M\#\overline{\mathbb{CP}}^2} \neq \emptyset$.*

**Proof of the Blow-up Formula** The computations in Example 4.3.39 show that the moduli space $\widehat{\mathfrak{M}}(\hat{N}_2, \hat{\sigma}_n)$ consists of a single reducible monopole and the virtual dimension is $d_n = -(n^2+n+1)$. Moreover (see Example 4.5.14 in §4.5.2)

$$d(\hat{\sigma}\#\hat{\sigma}_n) = d(\hat{\sigma})\#d(\hat{\sigma}_n) := d(\hat{\sigma}) + d(\hat{\sigma}_n) + 1 = d(\hat{\sigma}) - n(n+1).$$

We prove first that

$$|\mathbf{sw}_{M\#\overline{\mathbb{CP}}^2}(\hat{\sigma}\#\hat{\sigma}_n)| = |\mathbf{sw}_M(\hat{\sigma})|$$

if $n = \pm 1$. We want to use Theorem 4.5.19. The computations in Example 4.3.39 show that the assumptions $\mathbf{B}_1, \mathbf{B}_2, \mathbf{B}_4$ are satisfied with $h_2 = 0$.

Moreover, $\mathbf{B}_3$ is generically satisfied. We deduce that we have an isomorphism between the $S^1$-bundles

$$P := \left\{ \widehat{\mathfrak{M}}_\mu(\hat{N}_1, \hat{\sigma}, *) \to \widehat{\mathfrak{M}}_\mu(\hat{N}_1, \hat{\sigma}) \right\}$$

and

$$P_n := \left\{ \widehat{\mathfrak{M}}_{\hat{\sigma}\#\hat{\sigma}_n}(M\#\overline{\mathbb{CP}}^2, \hat{g}_r, *) \to \widehat{\mathfrak{M}}_{\hat{\sigma}\#\hat{\sigma}_n}(M\#\overline{\mathbb{CP}}^2, \hat{g}_r) \right\}.$$

Using Corollary 4.5.20 we obtain an isomorphism of $S^1$-bundles

$$P = \left\{ \widehat{\mathfrak{M}}_\mu(\hat{N}_1, \hat{\sigma}, *) \to \widehat{\mathfrak{M}}_\mu(\hat{N}_1, \hat{\sigma}) \right\} \cong \left\{ \widehat{\mathfrak{M}}_{\hat{\sigma}}(M, \hat{g}_r, *) \to \widehat{\mathfrak{M}}_{\hat{\sigma}}(M, \hat{g}_r) \right\} = Q.$$

Thus we have $\Omega := c_1(Q) = c_1(P)$,

$$\mathbf{sw}_M(\hat{\sigma}) = \left\langle \Bigl(1 - \Omega\Bigr)^{-1}, [\widehat{\mathfrak{M}}_{\hat{\sigma}}(M)] \right\rangle$$

$$= \pm \left\langle \Bigl(1 - c_1(P_n)\Bigr)^{-1}, [\widehat{\mathfrak{M}}_{\hat{\sigma}\#\hat{\sigma}_n}(M\#\overline{\mathbb{CP}}^2)] \right\rangle = \mathbf{sw}_{M\#\overline{\mathbb{CP}}^2}(\hat{\sigma}\#\hat{\sigma}_n).$$

(The above integrations are well defined since all the manifolds involved are orientable.)

In general, set

$$X_n := \widehat{\mathfrak{M}}_{\hat{\sigma}\#\hat{\sigma}_n}(M\#\overline{\mathbb{CP}}^2, \hat{g}_r), \quad X := \widehat{\mathfrak{M}}_\mu(\hat{N}_1, \hat{\sigma}).$$

Example 4.3.39 shows that we can apply Theorem 4.5.19 for any *spin*$^c$ structure $\hat{\sigma}_n$ on $\hat{N}_2$ but if $n \neq \pm 1$ we will encounter obstructions to gluing. The manifold $X_n$ is thus the smooth zero set of a section $s_r$ of the vector bundle

$$\mathcal{O}_n := P \times_{S^1} \mathbb{C}^{h_2}, \quad h_2 := \frac{n(n+1)}{2}$$

over $X$. The cycle determined by $X_n$ in $X$ is therefore the Poincaré dual of the Euler class of this vector bundle. Observe that

$$e(\mathcal{O}_n) = c_1(P)^{h_2} = \Omega^{h_2}.$$

Consequently,

$$|\mathbf{sw}_{M\#\overline{\mathbb{CP}}^2}(\hat{\sigma}\#\hat{\sigma}_n)| = \left| \left\langle (1-\Omega)^{-1}, [s_r^{-1}(0)] \right\rangle \right|$$

$$= \left| \left\langle (1-\Omega)^{-1} e(\mathcal{O}_n), [X] \right\rangle \right| = \left| \left\langle \Omega^{h_2}(1-\Omega)^{-1}, [X] \right\rangle \right|$$

$$= \left| \left\langle (1-\Omega)^{-1}, [X] \right\rangle \right| = |\mathbf{sw}_M(\hat{\sigma})|. \quad \blacksquare$$

**Corollary 4.6.9. (Fintushel-Stern [34], Morgan-Szabó-Taubes [97])** *Suppose $M$ is a compact, oriented, smooth 4-manifold satisfying the following conditions.*

*(a)* $b_+(M) > 1$.

*(b) There exists an embedding $S^2 \hookrightarrow M$ which determines an essential element of $H_2(M, \mathbb{Z})$ with nonnegative self-intersection $d$.*

*Then all the Seiberg-Witten invariants of $M$ are trivial, i.e.* $\mathbb{B}_M = \emptyset$.

**Proof** Denote by $M_d$ the $d$-fold blow-up of $M$, $M_d := M \# d\overline{\mathbb{CP}}^2$. Each blow-up decreases self-intersections by 1 so that $M_d$ contains an essentially embedded 2-sphere with trivial self-intersection. According to Proposition 4.6.5 in the preceding subsection we have

$$\mathbb{B}_{M_d} = \emptyset.$$

We can now invoke Corollary 4.6.8 to conclude that $\mathbb{B}_M = \emptyset$. ■

**Remark 4.6.10.** The results of C.T.C Wall [**144**] imply that if $M$ is a simply connected manifold with indefinite intersection form and $c \in H_1(M, \mathbb{Z})$ is a *primitive class* (i.e. $H_2(M)/\mathbb{Z} \cdot c$ is torsion free) which is represented by an embedded 2-sphere and $c^2 = 0$ then

$$M \cong N \#(S^2 \times S^2) \text{ or } M \cong N \#(\mathbb{CP}^2 \# \overline{\mathbb{CP}}^2).$$

In particular, by the connected sum theorem the Seiberg-Witten invariants of $M$ must vanish. Corollary 4.6.9 shows that the Seiberg-Witten vanishing holds even without the *primitivity* assumption.

**Remark 4.6.11.** We have reduced the proof of Corollary 4.6.9 to the special case when the embedded sphere $S^2 \hookrightarrow M$ has self-intersection 0.

Stefano Vidussi has shown in [**143**] that such an essential sphere exists if and only if there exists a hypersurface $N \hookrightarrow M$ carrying a metric of positive scalar curvature such that $b_1(N) > 0$ and decomposing $M$ into two parts $M^{\pm}$ satisfying

$$b_1(M) + b_1(N) > b_1(M^+) + b_1(M^-).$$

We refer the reader to [**111, 143**] for details and generalizations of Corollary 4.6.9.

The above vanishing corollary has an intriguing topological consequence.

**Corollary 4.6.12.** *Let $M$ be a compact symplectic 4-manifold with*

$$b_+(M) > 1.$$

*If $\Sigma \hookrightarrow M$ is an embedded surface representing an essential element in $H_2(M, \mathbb{Z})$ with nonnegative self-intersection then its genus must be positive.*

**Proof** If the genus of $\Sigma$ were zero then, according to Corollary 4.6.9, the Seiberg-Witten invariants of $M$ would vanish. Taubes' theorem tells us this is not possible for a symplectic 4-manifold with $b_+ > 1$. ■

**Remark 4.6.13.** (a) The above genus estimate is optimal from different points of view. First of all, the genus bound is optimal since it is achieved by the fibers of an elliptic fibration. The condition on self-intersection being nonnegative cannot be relaxed without affecting the genus bound. For example, the exceptional divisor of the blow-up of a Kähler surface has self-intersection $-1$ and it is represented by an embedded sphere.

(b) The above minimal genus estimate has the following generalization known as the *adjunction inequality.*

*Suppose $M$ is a closed, oriented 4-manifold such that $b_+(M) > 1$. If $\Sigma \hookrightarrow M$ is an essentially embedded surface such that $\Sigma \cdot \Sigma \geq 0$ then for any basic class $\sigma \in \mathbb{B}_M$ we have*

$$2g(\Sigma) \geq 2 + \Sigma \cdot \Sigma - \langle c_1(\det \sigma), \Sigma \rangle.$$

(When $g(\Sigma) \geq 1$ we can drop the *essential* assumption.) One can imitate the proof of the Thom conjecture in §2.4.2 to obtain this result (see [**119**]). For a different proof, using the full strength of the cutting-and-pasting technique we refer to [**97**].

Observe that if $M$ is symplectic and the essential homology class $c \in H_2(M, \mathbb{Z})$ is represented by a symplectically embedded surface $\Sigma_0$ and $c \cdot c \geq 0$ then the adjunction equality implies

$$2g(\Sigma_0) = 2 + \Sigma_0 \cdot \Sigma_0 - \langle c_1(\det(\sigma)), \Sigma_0 \rangle.$$

In particular, if $\Sigma$ is any other embedded surface representing $c$ we deduce from the adjunction inequality that

$$g(\Sigma_0) \leq g(\Sigma).$$

This shows that if $\Sigma$ is a symplectically embedded surface such that $\Sigma \cdot \Sigma \geq 0$ then it is genus minimizing in its homology class.

In a remarkable work, [**114**], P. Ozsvath and Z. Szabó have shown that we can remove the nonnegativity assumption $\Sigma \cdot \Sigma \geq 0$ from the statement of the adjunction inequality provided we assume that $g(\Sigma) > 0$ and $X$ has *simple type*, i.e. if $\sigma \in \mathbb{B}_M$ is a basic class then $d(\sigma) = 0$. It is known that all symplectic manifolds have simple type; see[**97**].

**Exercise 4.6.1.** Use the blow-up formula and the techniques in §2.4.2 to prove the adjunction inequality in the case $\Sigma \cdot \Sigma \geq 0$.

The adjunction inequality implies the following generalization of Corollary 4.6.12.

**Corollary 4.6.14.** *Suppose $M$ is a symplectic manifold and $\Sigma \hookrightarrow M$ is an essentially embedded surface such that $\Sigma \cdot \Sigma \geq 0$. Then*

$$g(\Sigma) \geq 1 + \frac{1}{2}\Sigma \cdot \Sigma. \tag{4.6.2}$$

*In particular, for any $n \in \mathbb{Z}^*$ we have*

$$g(n\Sigma) \geq 1 + \frac{n^2}{2}\Sigma \cdot \Sigma.$$

Assume $b_+(M) > 1$. For every $c_0 \in \mathrm{Hom}\,(H^2(M,\mathbb{Z}),\mathbb{R})$ and every $a, b \in \mathbb{R}$ the set

$$S_{c_0}(a) := \{x \in H^2(M,\mathbb{Z});\ \ |\langle x, c_0\rangle| \leq a\}$$

represents a strip in the lattice $H^2(M,\mathbb{Z})$. The adjunction inequality shows that we have restrictions on the location of the set of basic classes. More precisely, for every essentially embedded surface $\Sigma \hookrightarrow M$ $(g(\Sigma) > 0$ if $\Sigma{\cdot}\Sigma < 0)$ we have

$$c_1(\mathbb{B}_M) \subset S_{[\Sigma]}(\mu(\Sigma)),\ \ \mu(\Sigma) := -\chi(\Sigma) - \Sigma \cdot \Sigma.$$

If $M$ also happens to be symplectic, then Taubes' Theorem 3.3.29 also implies

$$c_1(\mathbb{B}_M) \subset -\frac{1}{2}c_1(K_M) + S_\omega(\deg_\omega K_M).$$

**Exercise 4.6.2.** Suppose $M$ is a closed, oriented 4-manifold with $b_+(M) > 1$.

(a) Show that if $c \in H_2(M,\mathbb{Z})$ is a nontrivial homology class such that $c \cdot c = 0$ which is represented by a smoothly embedded torus $T^2 \hookrightarrow M$ then

$$\mathbb{B}_M \subset c^\perp := \big\{\sigma \in Spin^c(M);\ \ \langle c_1(\det \sigma), c\rangle = 0\big\}.$$

(b) Show that if $c \in H_2(M,\mathbb{Z})$ is represented by an embedded 2-torus and $c \cdot c = -2$ then either

$$\langle c_1(\mathbb{B}_M), c\rangle \subset \{-2, 0, 2\}$$

or

$$\langle c_1(\mathbb{B}_M), c\rangle \subset \{-1, 1\}.$$

(c) Show that the same conclusion continues to hold if $c \cdot c = -2$ and $c$ is represented by an embedded 2-sphere.

(d) Suppose $c$ is a homology class represented by an essentially embedded surface $\Sigma$. If

$$g(\Sigma) = 1 + \frac{1}{2}c \cdot c > 0$$

then $\mathbb{B}_M \subset c^\perp$. If moreover $0 < g(\Sigma) < 1 + \frac{1}{2}c \cdot c$ then $\mathbb{B}_M = \emptyset$.

# Epilogue

*A whole is that which has a beginning, a middle and an end.*

Aristotle, *Poetics*

We can now take a step back and enjoy the view. Think of the places we've been and of the surprises we've uncovered! I hope this long and winding road we took has strengthened the idea that Mathematics is One Huge Question, albeit that it appears in different shapes, colours and flavors in the minds of the eccentric group of people we call mathematicians.

I think the sights you've seen are so breathtaking that even the clumsiest guide cannot ruin the pleasure of the mathematical tourist. I also have some good news for the thrill seeker. There is a lot more out there and, hereafter, you are on your own. Still, I cannot help but mention some of the trails that have been opened and are now advancing into the Unknown. (This is obviously a biased selection.)

We've learned that counting the monopoles on a 4-manifold can often be an extremely rewarding endeavour. The example of Kähler surfaces suggests that individual monopoles are carriers of interesting geometric information. As explained in [**70**], even the knowledge that monopoles exist can lead to nontrivial conclusions. What is then the true nature of a monopole? The experience with the Seiberg-Witten invariants strongly suggests that the answers to this vaguely stated question will have a strong geometric flavour.

In dimension four, the remarkable efforts of C.H. Taubes [**136, 137, 138, 139**], have produced incredibly detailed answers and raised more refined questions.

One subject we have not mentioned in this book but which naturally arises when dealing with more sophisticated gluing problems is that of the gauge theory of 3-manifolds. There is a large body of work on this subject (see [**25, 43, 44, 70, 77, 78, 83, 88, 89, 91, 109, 111**] and the references therein) which has led to unexpected conclusions. The nature of 3-monopoles is a very intriguing subject and there have been some advances [**70, 72, 100, 108**], suggesting that these monopoles reflect many shades of the underlying geometry. These studies also seem to indicate that three-dimensional contact topology ought to have an important role in elucidating the nature of monopoles.

One important event unfolding as we are writing these lines is the incredible *tour de force* of Paul Feehan and Thomas Leness, who in a long sequence of very difficult papers ([**33**]) are establishing the original prediction of Seiberg and Witten that the "old" Yang-Mills theory is topologically equivalent to the new Seiberg-Witten theory. While on this subject we have to mention the equally impressive work in progress of Andrei Teleman [**140**] directed towards the same goal but adopting a different tactic. Both these efforts are loosely based on an idea of Pidstrigach and Tyurin. A new promising approach to this conjecture has been recently proposed by Adrian Vâjiac [**142**], based on an entirely different principle.

Gauge theory has told us that the low-dimensional world can be quite exotic and unruly. At this point there is no one generally accepted suggestion about how one could classify the smooth 4-manifolds but there is a growing body of counterexamples to most common sense guesses. Certain trends have developed and there is a growing acceptance of the fact that geometry ought to play a role in any classification scheme. In any case, the world is ready for the next Big Idea.

# Bibliography

[1] H. Amann: *Fixed point equations and nonlinear eigenvalue problems in ordered Banach spaces*, SIAM Rev., **18**(1976), 620-709.

[2] V.I. Arnold: *Mathematical Methods of Classical Mechanics*, Springer Texts, Springer Verlag, 1989.

[3] M.F. Atiyah: *K-Theory*, W.A. Benjamin Inc., New York, 1967.

[4] M.F. Atiyah, R. Bott: *Yang-Mills equations over Riemann surfaces*, Phil. Trans. Roy. Soc. London, A **308**(1982), 523-615.

[5] M.F. Atiyah, N.J. Hitchin, I.M. Singer: *Selfduality in four dimensional Riemannian geometry*, Proc. Roy., Soc. London, A**362**(1978), 425-461.

[6] M.F. Atiyah, V.K. Patodi, I.M. Singer: *Spectral asymmetry and Riemannian geometry I*, Math. Proc. Cambridge Philos. Soc. **77**(1975), 43-69.

[7] M.F. Atiyah, V.K. Patodi, I.M. Singer: *Spectral asymmetry and Riemannian geometry II*, Math. Proc. Cambridge Philos. Soc. **78**(1975), 405-432.

[8] M.F. Atiyah, V.K. Patodi, I.M. Singer: *Spectral asymmetry and Riemannian geometry III*, Math. Proc. Cambridge Philos. Soc. **79**(1976), 71-99.

[9] W. Barth, C. Peters, A. Van de Ven: *Compact Complex Surfaces*, Springer Verlag, 1884.

[10] A.Beauville: *Surfaces Algébriques Complexes*, Asterisque 54, Société Mathématique de Frances, 1978.

[11] P.H. Bérard: *From vanishing theorems to estimating theorems: the Bochner technique revisited*, Bull. A.M.S., **19**(1988), 371-406.

[12] N.Berline, E.Getzler, M. Vergne: *Heat Kernels and Dirac Operators*, Springer Verlag, 1992.

[13] O. Biquard: *Les équations de Seiberg-Witten sur une surface complexe non Kählérienne*, Comm. Anal. and Geom., **6**(1998), 173-196.

[14] J.M. Bismut, D. S. Freed: *The analysis of elliptic families. II Dirac operators, eta invariants, and the holonomy theorem*, Comm. Math. Phys. **107**(1986), 103-163.

[15] J. Bochnak, M. Coste, M-F. Roy: *Géométrie Algébrique Réelle*, Ergeb. der Math., Band. 12, Springer Verlag, 1987.

[16] B. Booss, K. Wojciechowski: *Elliptic Boundary Problems for Dirac Operators*, Birkhauser, 1993.

[17] R.Bott, L.Tu: *Differential Forms in Algebraic Topology*, Springer-Verlag, 1982.

[18] J.-P. Bourguignon, P. Gauduchon: *Spineurs, opérateurs de Dirac et variations de métriques*, Comm. Math. Phys.. **144**(1992), 581-599

[19] H. Brezis: *Analyse Fonctionelle. Théorie et Applications*, Masson, 1983.

[20] J. Brüning, M. Lesch: *Hilbert complexes*, J. Funct. Anal., **108**(1992), 88-132.

[21] R. Brussee: *The canonical class and $C^\infty$-properties of Kähler surfaces*, New York J. of Mathematics, **2**(1996), 103-146; alg-geom/9503004.

[22] J. Bryan: *Seiberg-Witten theory and $\mathbb{Z}/2^p$ actions on spin 4-manifolds*, Math. Res. Letters, **5**(1998), 165-183. dg-ga/9704010.

[23] H.-D. Cao, J. Zhou: *Equivariant cohomology and the wall crossing formulas in Seiberg-Witten theory*, Math. Res. Letters, **5**(1998), 711-721; math.DG/9804134.

[24] S. Cappell, R. Lee, E. Miller: *Self-adjoint elliptic operators and manifold decompositions. Part I: Low eigenmodes and stretching*, Comm. Pure Appl. Math., **49**(1996), 825-866.

[25] W.Chen: *Casson invariant and Seiberg-Witten gauge theory*, Turkish J. Math., **21**(1997), 61-81.

[26] W. Chen: *The Seiberg-Witten theory of homology 3-spheres*, dg-ga/9703009.

[27] F. Connolly, Lê H. V. , K. Ono: *Almost complex structures which are compatible with Kähler or symplectic structures*, Ann. of Glob. Anal. and Geom., **15**(1997), 325-334.

[28] S. K. Donaldson: *An application of gauge theory to four dimensional topology*, J. Diff. Geom., **18**(1983), 279-315.

[29] S.K. Donaldson, P.B. Kronheimer: *The Geometry of Four-Manifolds*, Clarendon Press, Oxford, 1990.

[30] B. A. Dubrovin, A.T. Fomenko, S.P Novikov: *Modern Geometry - Methods and Applications*, Vol.1-3, Springer Verlag, 1984, 1985, 1990.

[31] D. Eisenbud, J. Harris: *The Geometry of Schemes*, Graduate Texts in Math., vol. 197, Springer Verlag, 2000.

[32] N. Elkies: *A characterization of the $\mathbb{Z}^n$ lattice*, Math. Res. Lett., **2**(1995), 321-326.

[33] P.M.N. Feehan, T.G. Leness: *PU(2) monopoles etc.* papers available at `http://www.math.ohio-state.edu/~feehan/preprints.html`

[34] R. Fintushel, R.J. Stern: *Immersed spheres in 4-manifolds and the immersed Thom conjecture*, Turkish J. of Math., **19**(1995), 145-157.

[35] R. Fintushel, R. Stern: *Rational blowdowns of smooth 4-manifolds*, J. Diff. Geom., **46**(1997), 181-235.

[36] R. Fintushel, R. Stern: *Knots, links, and 4-manifolds*, Invent. Math., **134**(1998), 363-400.

[37] D.S. Freed, K. Uhlenbeck: *Instantons and 4-Manifolds*, Springer-Verlag, 1985.

[38] M.H. Freedman: *The topology of smooth four-dimensional manifolds*, J. Diff. Geom., **17**(1982), 357-453.

[39] R. Friedman: *Algebraic Surfaces and Holomorphic Vector Bundles*, Universitext, Springer Verlag, 1998.

[40] R. Friedman, J.W. Morgan: *Smooth Four-Manifolds and Complex Surfaces*, Ergeb. Math, Band 27, Springer Verlag, 1994.

[41] R. Friedman, J.W. Morgan: *Algebraic surfaces and Seiberg-Witten invariants*, J. Alg. Geom., **6**(1997), 445-479; alg-geom/9502026

[42] R. Friedman, J.W. Morgan: *Obstruction bundles, semiregularity and Seiberg Witten invariants*, Comm. Anal. and Geom., **7**(1999), 451-496; alg-geom/9509007

[43] K.A. Frøyshov: *The Seiberg-Witten equations and four manifolds with boundary*, Math. Res. Lett., **3** (1996), 373-390.

[44] Y. Fukumoto, M. Furuta: *Homology 3-spheres bounding acyclic 4-manifolds*, Max Planck Institut Preprint, 1997-110.
`http://www.mpim-bonn.mpg.de/html/preprints/preprints.html`

[45] M. Furuta: *Monopole equation and the 11/8 conjecture*, preprint, 1995.

[46] P. Gauduchon: *Hermitian connections and Dirac operators*, Boll. U.M.I. **11-B**(1997), Suppl.fasc.2, 257-288.

[47] D. Gilbarg, N.S. Trudinger: *Elliptic Partial Differential Equations Of Second Order*, Springer Verlag, 1983.

[48] P.B. Gilkey: *Invariance Theory, the Heat Equation and The Atiyah-Singer Index Theorem*, 2nd Edition, CRC Press, Boca Raton, 1995.

[49] P. Griffiths, J. Harris: *Principles of Algebraic Geometry*, John Wiley& Sons, 1978.

[50] M. Gromov, H.B. Lawson: *The classification of simply connected manifolds of positive scalar curvature*, Ann. of Math., **111**(1980), 423-434.

[51] R.E. Gompf, A.I. Stipsicz: *An Introduction to 4-Manifolds and Kirby Calculus*, Graduate Studies in Mathematics, vol.20, Amer. Math. Soc., 1999.

[52] G.H. Hardy, E.M Wright: *An Introduction to the Theory of Numbers*, Oxford University Press, 1962.

[53] R. Hartshorne: *Algebraic Geometry*, Graduate Texts in Mathematics **49**, Springer-Verlag, 1977.

[54] F. Hirzebruch: *Topological Methods in Algebraic Geometry*, Springer Verlag, New York, 1966.

[55] F. Hirzebruch, H. Hopf: *Felder von Flachenelementen in 4-dimensionalen Manigfaltigkeiten* Math. Ann. **136**(1958).

[56] F. Hirzebruch, W.D. Neumann, S.S. Koh: *Differentiable Manifolds and Quadratic Forms*, Lect. Notes in Pure and Appl. Math., No. 4, Marcel Dekker, 1971.

[57] M. Hutchings, C.H. Taubes: *An introduction to the Seiberg-Witten equations on symplectic manifolds*, in the volume "*Symplectic Geometry and Topology*", Y. Eliashberg, L. Traynor Eds, Amer. Math. Soc. 1999.

[58] G. Ionesei: *A gauge theoretic proof of the Abel-Jacobi theorem*, Canad. Math. Bull., **43**(2000), 183-192.

[59] V.A. Iskovskikh, I.R. Shafarevich: *Algebraic surfaces*, in the volume *Algebraic Geometry II*, Encyclopedia of Mathematical Sciences, vol. **35**, I.R Shafarevich (Ed.), Springer-Verlag, 1996.

[60] T. Kato: *Perturbation Theory for Linear Operators*, Springer Verlag, 1984.

[61] J.L. Kazdan, F.W. Warner: *Curvature functions for compact 2-manifolds*, Ann. Math., **99**(1974), 14-47.

[62] M. Kervaire, J. Milnor: *On 2-spheres in 4-manifolds*, Proc. Nat. Acad. Science USA, **47**(1961), 1651-1757.

[63] S. Kobayashi: *Differential Geometry of Complex Vector Bundles*, Princeton University Press, 1987.

[64] S. Kobayashi, K. Nomizu: *Foundations Of Differential Geometry*, Interscience Publishers, New York, 1963.

[65] K. Kodaira: *Collected Works. Vol. 3*, Iwanami Shoten Publishers and Princeton University Press, 1979.

[66] K. Kodaira, D.C. Spencer: *Groups of complex line bundles over compact Kähler varieties. Divisor class group on algebraic varieties*, Proc. Nat. Acad. Sci. USA **39**(1953), p. 868-877.

[67] M. Komuro: *On Atiyah-Patodi-Singer $\eta$-invariant for $S^1$-bundles over Riemann surfaces*, J. Fac. Sci. Univ. Tokyo Sect. IA Math., **30**(1984), 525-548.

[68] M. Kreck, S. Stolz: *Non connected moduli spaces of positive sectional curvature metrics*, J. Amer. Math. Soc., **6**(1993), 825-850.

[69] D. Kotschick: *The Seiberg-Witten invariants of symplectic four-manifolds*, Séminaire Bourbaki.

[70] P. Kronheimer: *Embedded surfaces and gauge theory in three and four dimensions, "Surveys in Differential geometry". vol III*, 243-298, International Press, 1998.
`http://www.math.harvard.edu/~kronheim/`

[71] P. Kronheimer, T. Mrowka: *The genus of embedded surfaces in the projective plane*, Math. Res. Letters **1**(1994), 797-808.

[72] P. Kronheimer, T. Mrowka: *Monopoles and contact structures*, Invent. Math., **130**(1997), 209-255.
`http://www.math.harvard.edu/~kronheim/`

[73] K. Lamotke: *The topology of complex projective varieties after S. Lefschetz*, Topology, **20**(1981), 15-51.

[74] H. B. Lawson, M.-L. Michelson: *Spin Geometry*, Princeton University Press, 1989.

[75] P. Lelong: *Intégration sur une ensemble analytique complexe*, Bull. Soc. Math. France, **85**(1957), 239-262.

[76] T.J. Li, A. Liu: *General wall crossing formula*, Math. Res. Letters, **2**(1995), 797-810.

[77] Y. Lim: *Seiberg-Witten invariants for 3-manifolds in the case $b_1 = 0$ or 1*, Pacific J. of Math. to appear

[78] Y. Lim: *The equivalence of Seiberg-Witten and Casson invariants for homology 3-spheres*, Math. Res. Letters, **6**(1999), 631-644.

[79] J.L. Lions, E. Magenes: *Problèmes aux Limites Nonhomogènes et Applications*, **vol.1**, Dunod, Paris, 1968.

[80] P.L. Lions: *The concentration-compactness principle in the calculus of variations. The locally compact case I*, Ann. Inst. H. Poincaré Anal Non Linéaire, **1**(1984), 109-145.

[81] P.L. Lions: *The concentration-compactness principle in the calculus of variations. The locally compact case II*, Ann. Inst. H. Poincaré Anal Non Linéaire, **1**(1984), 223-283.

[82] P. Lisca: *On the Donaldson polynomials of elliptic surfaces*, Math. Ann. **299**(1994), 629-639.

[83] P. Lisca: *Symplectic fillings and positive scalar curvature*, Geometry and Topology, **2**(1998), 103-116.
`http://www.maths.warwick.ac.uk/gt/`

[84] R. Lockhart: *Fredholm, Hodge and Liouville theorems on noncompact manifolds*, Trans. Amer. Math. Soc., **301**(1987), 1-35.

[85] R.B. Lockhart, R.C. McOwen: *Elliptic differential equations on noncompact manifolds*, Annali di Scuola Norm. Sup. di Pisa, **12**(1985), 409-448.

[86] S. Łojasiewicz: *Sur la problème de la division*, Studia Math., **18**(1959), 87-136.

[87] R. Mandelbaum: *4-dimensional topology: an introduction*, Bull. Amer. Math. Soc., **2**(1980), 1-159.

[88] M. Marcolli: *Seiberg-Witten Gauge Theory*, monograph, Texts and Readings in Mathematics, Hindustan Book Agency, 1999.

[89] M. Marcolli: *Equivariant Seiberg-Witten-Floer homology*, dg-ga 9606003.

[90] C.T. McMullen, C.H. Taubes: *4-manifolds with inequivalent symplectic forms and 3-manifolds with inequivalent fibrations*, Math. Res. Letters, **6**(1999), 681-695.
`http://abel.math.harvard.edu/HTML/Individuals/Curtis_T_McMullen.html`

[91] G. Meng, C.H. Taubes: *$\underline{SW}$ = Milnor torsion*, Math. Res. Letters, **3**1996, 661-674.

[92] J.W. Milnor: *Spin structures on manifolds*, L'Enseignement Math., **9**(1963), 198-203.

[93] J. Milnor, J.D. Stasheff: *Characteristic Classes*, Ann. Math. Studies 74, Princeton University Press, Princeton, 1974.

[94] J.W. Morgan: *The Seiberg-Witten Equations and Applications to the Topology of Smooth Manifolds*, Mathematical Notes, Princeton University Press, 1996.

[95] J. Morgan, T. Mrowka: *On the diffeomorphism classification of regular elliptic surfaces*, Intern. Math. Res. Notices, bound within Duke Math. J. **70**(1993), 119-124.

[96] J.W. Morgan, T. Mrowka, D. Ruberman: *The $L^2$-Moduli Space and a Vanishing Theorem for Donaldson Polynomial Invariants*, International Press, 1994.

[97] J.W.Morgan, Z. Szabó, C.H. Taubes: *A product formula for the Seiberg-Witen invariants and the generalized Thom conjecture*, J. Diff. Geom., **44**(1997), 706-788.

[98] C. Morrey: *Multiple Integrals in the Calculus of Variations*, Springer Verlag, 1966.

[99] T. Mrowka: *A local Mayer-Vietoris principle for Yang-Mills moduli spaces*, PhD Thesis, 1988.

[100] T. Mrowka, P. Ozsvath, B. Yu: *Seiberg-Witten monopoles on Seifert fibered spaces*, Comm. Anal. and Geom., **4**1997, 685-791.

[101] V. Muñoz: *Constraints for Seiberg-Witten basic classes of glued manifolds*, dg-ga/951102012.

[102] A. Newlander, L. Nirenberg: *Complex analytic coordinates on almost complex manifolds*, Ann. Math., **54**(1954), p.391-404.

[103] L. I. Nicolaescu : *An extension of the concentration-compactness lemma of P.L. Lions*, An. Şti. Univ. "Al. I. Cuza" Iaşi, Math., **36**(1990), 111-117.

[104] L.I. Nicolaescu: *The Maslov index, the spectral flow and decompositions of manifolds*, Duke Math. J., **80**(1995), 485-533.

[105] L.I. Nicolaescu: *Lectures on the Geometry of Manifolds*, World Sci. Pub. Co. 1996.

[106] L.I. Nicolaescu: *Adiabatic limits of the Seiberg-Witten equations on Seifert manifolds*, Comm. in Anal. and Geom., **6**(1998), 331-392. math.DG/9601107

[107] L.I. Nicolaescu: *Eta invariants of Dirac operators on circle bundles over Riemann surfaces and virtual dimensions of finite energy Seiberg-Witten moduli spaces* , Israel. J. Math. **114**(1999), 61-123; math.DG/9805046.

[108] L. I. Nicolaescu: *Finite energy Seiberg-Witten moduli spaces on 4-manifolds bounding Seifert fibrations*, dg/ga 9711006 (to appear in Comm. Anal. Geom.)

[109] L.I. Nicolaescu: *Seiberg-Witten theoretic invariants of lens spaces*, math.DG/9901071.

[110] L.I. Nicolaescu: *On the Cappell-Lee-Miller gluing theorem*, math.DG/9803154.

[111] H. Ohta, K. Ono: *Simple singularities and the topology of symplectically filling 4-manifold*, Comment. Math. Helv. **74**(1999), 575-590.

[112] C. Okonek, A. Teleman: *Seiberg-Witten invariants for manifolds with* $b_+ = 1$ *and the universal wall crossing formula*, Intern. J. Math., **7**(1996), 811-832.

[113] C. Okonek, A. Teleman: *Seiberg-Witten invariants and the rationality of complex surfaces*, Math. Z., **225**(1997), 139-149.

[114] P. Ozsvath, Z. Szabó: *The symplectic Thom conjecture*, Ann. of Math., **151**(2000), 93-124; math.DG/9811097.

[115] J. Park: *Seiberg-Witten invariants of rational blow-downs and geography problems of irreducible 4-manifolds*, PhD Dissertation, Michigan State University, 1996.

[116] T.H. Parker: *Gauge theories on four-dimensional Riemannian manifolds*, Comm. in Math. Physics, **85**(1982), 563-602.

[117] J. Roe: *Elliptic Operators, Topology and Asymptotic Methods*, Pitman Res. Notes in Math Series **179**, Longman Scientific and Technical, Harlow, 1988.

[118] V.A. Rohlin: *New results in the theory of four dimensional manifolds*, Dok. Akad. Nauk USSR, **84**(1952), 221-224.

[119] D. Salamon: *Spin Geometry and Seiberg-Witten Invariants*, monograph, to appear, Birkhäuser, 1999.

[120] D. Salamon: *Removable singularities and a vanishing result for Seiberg-Witten invariants*, Turkish J. Math., **20**(1996), 61-73.

[121] J.P. Serre: *A Course in Arithmetic*, Graduate Texts in Mathematics, vol. 7, Springer Verlag, 1996.

[122] I.R. Shafarevitch: *Algebra I. Basic Notions*, Encyclopedia of Mathematical Sciences, vol. **11**, Springer Verlag, 1990.

[123] L. Simon: *Asymptotics for a class of nonlinear evolution equations, with applications to geometric problems*, Ann. of Math., **118**(1983), 525-571.

[124] S. Smale: *An infinite dimensional version of Sard's theorem*, Amer. J. Math., **87**(1965), 861-866.

[125] S. Smale: *Generalized Poincaré conjecture in dimensions greater than four*, Ann. of Math., **76**(1961), 391-406.

[126] E. H. Spanier: *Algebraic Topology*, McGraw Hill, 1966.

[127] M. Spivak: *A Comprehensive Introduction to Differential Geometry*, vol. 1-5, Publish or Perish, 1979.

[128] N.E. Steenrod: *The Topology of Fibre Bundles*, Princeton University Press, 1951.

[129] A. Stipsicz, Z. Szabó: *The smooth classification of elliptic surfaces with* $b^+ > 1$, Duke Math. J. **75**(1994), 1-50.

[130] S. Stolz: *Simply connected manifolds of positive scalar curvature*, Ann. of Math., **136**(1992), 511-540.

[131] Z. Szabó: *Simply connected irreducible* 4*-manifolds with no symplectic structures*, Invent. Math., **132**(1998), 457-466.

[132] C.H. Taubes: *Gauge theory on asymptotically periodic* 4*-manifolds*, J. Diff. Geom., **25**(1987), 363-430.

[133] C.H. Taubes: $L^2$*-moduli spaces on* 4*-manifolds with cylindrical ends*, Monographs in Geometry and Topology, Vol. I, International Press, 1993.

[134] C.H. Taubes: *The Seiberg-Witten invarianst and symplectic forms*, Math. Res. Letters, **1**(1994), 809-822.

[135] C.H. Taubes: *More constraints on symplectic forms from Seiberg-Witten invariants*, Math. Res. Letters, **2**(1995), 9-13.

[136] C.H. Taubes: SW $\Rightarrow$ Gr*: from the Seiberg-Witten equations to pseudo-holomorphic curves.*, J. Amer. Math.Soc., **9**(1996), 845-918.

[137] C.H. Taubes: *Counting pseudo-holomorphic submanifolds in dimension* 4, J. Diff. Geom., **44**(1996), 819-893.

[138] C.H. Taubes: Gr $\Rightarrow$ SW*: from pseudo-holomorphic curves to Seiberg-Witten solutions*, preprint.

[139] C.H. Taubes: *Seiberg-Witten invariants and pseudoholomorphic subvarieties for self-dual harmonic* 2*-forms*, Geometry and Topology, **3**(1999), 167-210.
`http://www.maths.warwick.ac.uk/gt/`

[140] A. Teleman: *Moduli spaces of* $PU(2)$*-monopoles*, math.DG/9906163. More can be found at
`http://front.math.ucdavis.edu/search/author:Teleman`

[141] K.K. Uhlenbeck: *Connections with* $L^p$*-bounds on curvature*, Comm. Math. Phys., **83**(1982), 31-42.

[142] A. Vâjiac: *A derivation of Witten's conjecture relating Donaldson and Seiberg-Witten invariants*, preprint, hep-th/0003214.
`http://arXiv.org/abs/hep-th/0003214`

[143] S. Vidussi: *Seiberg-Witten theory for* 4*-manifolds decomposing along* 3*-manifolds of positive scalar curvature*, preprint, École Polytechnique, 99-5.
`http://math.polytechnique.fr/cmat/vidussi/`

[144] C.T.C. Wall: *Diffeomorphisms of* 4*-manifolds*, J. London Math. Math. Soc., **39**(1964), 131-140.

[145] C.T.C. Wall: *On simply connected* 4*-manifolds*, J. London Math. Math. Soc., **39**(1964), 141-149.

[146] A. Weil: *Introduction à L'Étude des Variétés Kählériennes*, Hermann, Paris, 1971.

[147] J.H.C. Whitehead: *On simply connected* 4*-dimensional polyhedra*, Comment. Math. Helv., **22**(1949), 48-92.

[148] E.T. Whittaker, G.N. Watson: *A course of Modern Analysis*, Cambridge Univ. Press, 1996.

[149] E. Witten: *Monopoles and four-manifolds*, Math.Res. Letters **1** (1994), p.769-796.

[150] K. Yosida: *Functional Analysis*, Springer Verlag, 1974.

[151] S. Zucker: $\ell^2$ *cohomology of warped products and arithmetic quotients*, Invent. Math., **70**(1982), 169-218.

# Index